Mathematics for the Trades
A Guided Approach

Eleventh Edition

Hal M. Saunders

Robert A. Carman

Director, Portfolio Management: Michael Hirsch
Courseware Portfolio Manager: Matt Summers
Content Producer: Lauren Morse
Managing Producer: Scott Disanno
Producer: Erin Carreiro
Manager, Courseware QA: Mary Durnwald
Manager, Content Development: Eric Gregg
Product Marketing Manager: Fiona Murray
Senior Author Support/Technology Specialist: Joe Vetere
Manager, Rights and Permissions: Gina Cheselka
Manufacturing Buyer: Carol Melville, LSC Communications
Art Director: Barbara Atkinson
Text Design, Production Coordination, Composition, and Illustrations: iEnergizer/Aptara®, Ltd.
Cover Image: Huntstock/Brand X/Getty Images (Carpenters measuring window frame alignment); Hero Images/Getty Images (Female helicopter mechanic repairing propeller); Gagliardi Images/Shutterstock (Patient undergoing X-ray test)

Attributions of third-party content appear on page 895, which constitutes an extension of this copyright page.

Library of Congress Cataloging-in-Publication Data

Names: Saunders, Hal M., author. | Carman, Robert A., author.
Title: Mathematics for the trades : a guided approach / Hal M. Saunders,
    Robert A. Carman.
Description: Eleventh edition. | New York, NY : Pearson, 2018. | Includes
    indexes.
Identifiers: LCCN 2017049542 | ISBN 9780134756967 (pbk.) | ISBN 0134756967
    (pbk.)
Subjects: LCSH: Mathematics—Textbooks.
Classification: LCC QA39.3 .C37 2018 | DDC 513/.14—dc23 LC record available at
https://lccn.loc.gov/2017049542

ISBN 13: 978-0-13-475696-7
ISBN 10:   0-13-475696-7

21 2023

*Dedicated to my mother Bernice who, at 95,
is still providing me with love and encouragement.
I could not have done it without you!*

Contents

vi   **Contents**

Preface

Mathematics for the Trades: A Guided Approach provides the practical mathematics skills needed in a wide variety of trade, technical, and other occupational areas, including plumbing, automotive, electrical and construction trades, machine technology, landscaping, HVAC, allied health, and many more. It is especially intended for students who find math challenging and for adults who have been out of school for a time. This text assists students by providing a direct, practical approach that emphasizes careful, complete explanations and actual on-the-job applications. It is intended to provide practical help with real math, beginning at each student's individual level of ability. Careful attention has been given to readability, and reading specialists have helped plan both the written text and the visual organization.

Several special features are designed with the math-challenged student in mind. Each chapter begins with a preview quiz keyed to the textbook, which shows concepts that will be covered in that chapter. A summary is provided for each chapter, and each chapter ends with a set of problems to check the student's progress. The book can be used for a traditional course, a course of self-study, or a tutor-assisted study program.

The format is clear and easy to follow. It respects the individual needs of each reader, providing immediate feedback at each step to ensure understanding and continued attention. The emphasis is on *explaining* concepts rather than simply *presenting* them. This is a practical presentation rather than a theoretical one.

A calculator is a necessary tool for workers in trade and technical areas. We have integrated calculators extensively into the text—in finding numerical solutions to problems (including specific keystroke sequences) and in determining the values of transcendental functions. We have taken care to first explain all concepts and problem solving without the use of the calculator and describe how to estimate and check answers. Many realistic problems in the exercise sets involve large numbers, repeated calculations, and large quantities of information and are thus ideally suited to calculator use. They are representative of actual job situations in which a calculator is needed. Detailed instruction on the use of calculators is included in special sections at the end of Chapters 1, 2, and 3 as well as being integrated into the text.

Special attention has been given to on-the-job math skills by using a wide variety of real problems and situations. Many problems parallel those that appear on professional and apprenticeship exams. The answers to the odd-numbered exercises are given in the back of the book.

New to This Edition

- We have added a new "Case Study" feature, which appears at the end of every chapter. Each Case Study is a multi-part problem that delves deeply into a specific real-world application and uses many of the mathematical skills covered in the chapter. Each of the chapter opening photos has been chosen to illustrate the subject matter of the Case Study, and the chapter introduction includes a brief preview of the Case Study.

- At the beginning of each section in a chapter there are two "Learning Catalytics" problems in the margin. These are designed to be quick and simple problems reviewing skills that are needed to learn the mathematics in the upcoming section. Instructors can use this feature to get immediate feedback on the readiness of their class to take on the new material.

- In Chapters 2 and 3, special attention has now been given to the order of operations with fractions and decimals.

- In Chapter 3, more examples and problems have been added that contain fractions and decimals mixed together. Also in this chapter, the table of decimal-fraction equivalents on page 168 is now used to find the closest fractional equivalent to a given decimal number.

- The old Section 5-3 has been split into two smaller sections. The new Section 5-3 focuses strictly on the metric system itself, while the new Section 5-4 covers conversions between the metric and the U.S. customary systems. The concept of "greatest possible error" has also been added to Chapter 5.

- At the request of many instructors, the use of significant digits has been de-emphasized in rounding instructions. However, this topic is still covered in Chapter 5 and it is still used in Chapter 10 for rounding with trig calculations.

- We have added a new section to Chapter 12 covering measures of dispersion, which includes the concepts of range and standard deviation, as well as a basic introduction to normal distribution. This section had appeared in some earlier editions of the textbook, and it has been extensively revised and updated for this edition.

- As with every new edition, hundreds of applied problems have either been added or revised in both the exercise sets and the worked examples so that they continue to be authentic, topical, up to date, and representative of a wide variety of occupational areas.

- Many of the generic sections of the exercise sets have been reorganized to achieve a better progression from easier to more difficult problems as well as to create a better odd–even balance.

- In addition to all of these specific changes and added features, thousands of smaller changes have been made to improve the clarity of the explanations and the consistency of the presentations.

Resources for Success

Get the Most Out of MyLab Math
for *Mathematics for the Trades*, Eleventh Edition by Hal Saunders and Robert Carman

Mathematics for Trades: A Guided Approach, Eleventh edition offers market-leading content written by author-educators, tightly integrated with the #1 choice in digital learning—MyLab Math. MyLab Math courses can be tailored to the needs of instructors and students, while weaving the author team's voice and unique approach into all elements of the course. Teaching students the mathematical skills needed in realistic trade applications in both the text and MyLab Math course enables students to succeed not just in this course, but in their future trade.

Take advantage of the following resources, many of which are new or improved, to get the most out of your MyLab Math course.

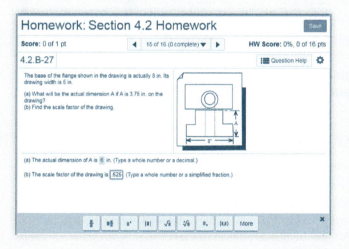

Motivate Students through Math in Context

NEW! A trade application question library provides a wide range of exercises available to assign to work for any instructor's class dynamics. Available in the MyLab Math Assignment Manager, this library of exercises now allows instructors to pull in additional application questions from particular trades or industries.

NEW! Case Study videos complement the Case Studies that have been added to the text in this revision. These videos demonstrate a real-world trade application using math concepts to bring the math to life for students. Videos can be assigned in MyLab as homework, or could be used in the classroom to kick off a lecture.

pearson.com/mylab/math

Resources for Success

Motivate and Support Students with Updated Resources

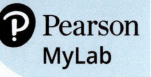

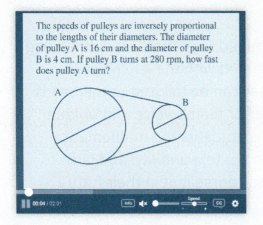

The speeds of pulleys are inversely proportional to the lengths of their diameters. The diameter of pulley A is 16 cm and the diameter of pulley B is 4 cm. If pulley B turns at 280 rpm, how fast does pulley A turn?

Updated video program provides section lectures for each section of the textbook. Lectures highlight key examples and exercises for every section of the textbook within a mobile-ready player that allows videos to be played on any device. Support students no matter where they are.

NEW! Learning Catalytics is an interactive student response tool that uses students' smartphones, tablets, or laptops to engage them in more sophisticated tasks and thinking. Available through the MyLab Math course, pre-made questions specific to *Mathematics for the Trades* are available to use with students. Margin references in the text indicate where an existing Learning Catalytics question would be applicable to use. To find these questions in Learning Catalytics, search for **Saunders#**, where **#** is the chapter number.

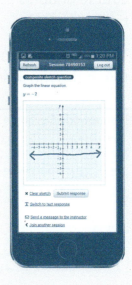

Resources for Instructors
The following resources can be downloaded from the Instructor's Resource Center on www.pearson.com/us/ higher-education, or in the MyLab Math course.

NEW! PowerPoints
For the first time, PowerPoints for every section of the text are available. Fully editable, these can be used as a starting point for instructors to use in lecture, or for students to use as a study tool.

Instructor's Solutions Manual
This manual contains detailed, worked-out solutions to all exercises in the text.

UPDATED! TestGen
TestGen® (www.pearsoned.com/testgen) enables instructors to build, edit, print, and administer tests using a computerized bank of questions developed to cover all the objectives of the text. Updated for the Eleventh edition, TestGen is now algorithmically based, allowing instructors to create multiple but equivalent versions of the same question or test with the click of a button. Instructors can also modify test bank questions or add new questions.

pearson.com/mylab/math

Acknowledgements

I would like to acknowledge the many people who have contributed to this Eleventh edition of *Mathematics for the Trades*. First and foremost, I am especially grateful to my editor, Matt Summers, who not only provided all the resources I needed, but who also gave me invaluable advice and encouragement from the first of the questionnaires to the last of the page proofs. Many other staff members at Pearson have played key roles in the revision process, including project manager Lauren Morse, editor-in-chief Michael Hirsch, producer Erin Carreiro, and product marketing manager Alicia Frankel, and I thank all of you for your contributions.

Special thanks go to Rachel Johnson from Northeast Wisconsin Technical College and to Cheryl Anderson from City College Montana State University for their creativity and expertise in providing most of the Case Studies. It was a pleasure working with both of you on this exciting new feature.

I would also like to express my appreciation to production supervisor Monica Moosang at Aptara, Inc. for efficiently managing the production process and providing such a great staff of compositors. And to copy editor Leanne Rancourt, you not only did an outstanding job of providing edits and suggestions, you were an absolute joy to work with.

A successful revision must be as error-free as possible, and there were two people who helped assure this outcome. My good friend Sally Pittman-Rabbin spent endless hours working all the new and revised problems so I could double-check my answers and correct any mistakes. And Deana Richmond not only triple-checked the mathematical accuracy, but proofread the entire revised text and provided many great suggestions for refining the wording of the explanations and improving consistency of the presentations. I offer my sincere thanks to both of you.

Last, but not least, I would like to thank the reviewers of this edition and of previous editions. Your thoughtful responses to the questionnaires strongly influenced the revisions that have been made.

Marylynne Abbott, Ozarks Technical Community College

Robert Ahntholz, Coordinator—Learning Center, Central City Occupational Center

Cheryl Anderson, City College of Montana State University at Billings

Michelle Askew, Lamar University

Peter Arvanites, Rockland Community College

Dean P. Athans, East Los Angeles College

Gerald Barkley, Glen Oaks Community College

Itzhak Ben-David, The Ohio State University

Joseph Bonee, Spencerian College

James W. Brennan, Boise State University

Frances L. Brewer, Vance-Granville Community College

Robert S. Clark, Spokane Community College

James W. Cox, Merced College

Harry Craft, Spencerian College

Elizabeth C. Cunningham, Bellingham Technical College

Jennifer Davis, Ivy Tech Community College

Greg Daubenmire, Las Positas College

Gordon A. DeSpain, San Juan College

B. H. Dwiggins, Systems Development Engineer, Technovate, Inc.

Kenneth R. Ebernard, Chabot College

Hal Ehrenreich, North Central Technical College

Donald Fama, Chairperson, Mathematics-Engineering Science Department, Cayuga Community College

Karmen Franklin, Ivy Tech Community College

James Graham, Industrial Engineering College of Chicago

Edward Graper, LeBow Co.

Mary Glenn Grimes, Bainbridge College

Ronald J. Gryglas, Tool and Die Institute

Valerie Gwinn, Central Piedmont Community College

Michele Hampton, Cuyahoga Community College

Ryan Harper, Spartanburg Community College

Vincent J. Hawkins, Chairperson, Department of Mathematics, Warwick Public Schools

Jason Hill, New Castle School of Trades

Bernard Jenkins, Lansing Community College

Chris Johnson, Spokane Community College

Rachel Johnson, Northeast Wisconsin Technical College

Judy Ann Jones, Madison Area Technical College

Steve Kessler, Cincinnati State Technical and Community College

Robert Kimball, Wake Technical College

Tami Kinkaid, Louisville Technical Institute

Karen Lambertus, Ivy Tech Community College

Shu Lin, University of Hawaii, Honolulu

Jonathan Loss, Catawba Valley Community College

J. Tad Martin, Technical College of Alamance

Rose Martinez, Bellingham Technical College

Gregory B. McDaniel, Texas State Technical College

Ashley McHale, Las Positas College

Krystal Miller, Pikes Peak Community College

Sharon K. Miller, North Central Technical College

David C. Mitchell, Seattle Central Community College

Voya S. Moon, Western Iowa Technical Community College

Robert E. Mullaney, Santa Barbara School District

Joe Mulvey, Salt Lake Community College

Jack D. Murphy, Penn College

R. O'Brien, State University of New York–Canton

Paul Oman, Ivy Tech Community College

Steven B. Ottmann, Southeast Community College

Emma M. Owens, Tri-County Technical College

Francois Nguyen, Saint Paul College

David A. Palkovich, Spokane Community College

William Poehler, Santa Barbara School District

Richard Powell, Fullerton College

Martin Prolo, San Jose City College

Ilona Ridgeway, Fox Valley Technical College

A. Gwinn Royal, Ivy Tech Community College

Kurt Schrampfer, Fox Valley Technical College

Richard C. Spangler, Developmental Instruction Coordinator, Tacoma Community College

Turi Suski, Fox Valley Technical College

Arthur Theobald, Bergen County Vocational School

Curt Vander Vere, Pennsylvania College of Technology

Pat Velicky, Florence-Darlington Technical College

Cathy Vollstedt, North Central Technical College

Joseph Weaver, Associate Professor, State University of New York

Kay White, Walla Walla Community College

Terrance Wickman, Eastfield College

Raymond E. Wilhite, Solano Community College

This book has benefited greatly from their excellence as teachers.

Hal M. Saunders

Santa Barbara, California

How to Use This Book

In this book you will find many questions—not only at the end of each chapter or section, but on every page. This textbook is designed for those who need to learn the practical math used in the trades, and who want it explained carefully and completely at each step. The questions and explanations are designed so that you can:

- Start either at the beginning or where you need to start.

- Work on only what you need to know.

- Move as fast or as slowly as you wish.

- Receive the guidance and explanation you need.

- Skip material that you already understand.

- Do as many practice problems as you need.

- Test yourself often to measure your progress.

In other words, if you find mathematics difficult and you want to be guided carefully through it, this book is designed for you.

This is no ordinary book. You cannot browse through it; you don't read it. You *work* your way through it.

This textbook has been designed for students who will work through it to achieve understanding and practical skills. The alert student will look for and use the helpful features described below.

You will learn to do mathematics problems because you will follow our examples. The signal that a worked example is coming up looks like this:

EXAMPLE 1   You should respond by following each step and questioning each step. If necessary, seek specific help from your instructor or tutor.   ●

When you are confident that you understand the process, move on to the section labeled

Your Turn   This is your chance to show that you are ready to try a similar problem or two on your own. Use the step-by-step procedure you learned in the *Example* to solve the problems.

Solution   The *Your Turn* section is followed by the answers or worked solutions to the problems in the *Your Turn*.

Further drill problems are often provided in another set, which is labeled as follows:

More Practice   These problems may be followed by worked solutions or by a list of answers.

Answers

Keep on the lookout for the following helpers:

Note   Every experienced teacher knows that certain mathematical concepts and procedures will present special difficulties for students. To help you with these, special notes are included in the text. A pencil symbol and a warning word appear in the left margin to indicate the start of the comment, and a bullet ● shows when it is completed. The word **Note**, as used at the start of this paragraph, calls your attention to conclusions or consequences that might be overlooked, common mistakes students make, or alternative explanations. ●

Careful   A **Careful** comment points out a common mistake that you might make and shows you how to avoid it. ●

Learning Help   A **Learning Help** gives you an alternative explanation or slightly different way of thinking about and working with the concepts being presented. ●

A Closer Look   This phrase signals a follow-up to an answer or a worked solution. It may provide a more detailed or an alternative look at the whole process. ●

Examples often include step-by-step explanations that look like this:

Step 1   The solution of each worked example is usually organized in a step-by-step format, similar to this paragraph.

Step 2   In each worked example, explanations for each step are provided alongside the corresponding mathematical operations.

Step 3   Color, ⟵ boxed comments , and other graphical aids ⟶ are used to highlight the important or tricky aspects of a solution, if needed.

A **check icon** appears in a problem solution to remind you to check your work.

The calculator is an important tool for the modern trades worker or technician. We assume in this textbook that once you have learned the basic operations of arithmetic you will use a calculator. Problems in the exercise sets or examples in the text that involve the use of a scientific calculator are preceded by the calculator symbol shown here.

Solutions often include a display of the proper calculator key sequences. For example, the calculation

$$\frac{85.7 + (12.9)^2}{71.6}$$

would be shown as

85.7 ⊕ 12.9 x^2 ⊜ ÷ 71.6 ⊜ → 3.521089385

Not every student owns the same brand or model of calculator, and there are often variations in the way different calculators operate. The calculator key sequences shown in this textbook are based on the way that the newer generation of scientific calculators operate, and they will also be appropriate for graphing calculators. On these calculators, you enter calculations exactly as they would be correctly written or spoken aloud. Furthermore, their windows can display at least two lines—one to show the sequence of keys that you pressed (except for the equals sign), and the other to show the final answer.

If you are wondering whether or not you own one of the newer-generation calculators, here is a simple test: Try to take the square root of 9 by pressing the square-root key first, then the 9, and then = or ENTER ($\checkmark$ 9 =). If you are able to do this and obtain the correct answer of 3, then you own the type of calculator that will work for the sequences shown in the text. If, on the other hand, you need to enter 9 first and then the square-root key to obtain 3, you own one of the older-generation calculators, and you will need to consult your owner's manual to learn the proper key sequences (or buy a new calculator!). Another clue is that the older style machines normally have only a one-line display window.

Exercises   At the conclusion of each section of each chapter you will find a set of problems covering the work of that section. These will include a number of routine or drill problems as well as applications or word problems. Each applied problem begins with an indication of the occupational area from which it has been taken.

Chapter Summary   A chapter summary is included at the end of each chapter. It contains a list of objectives with corresponding reviews and worked examples.

Problem Set   Following each chapter summary is a set of problems reviewing all of the material covered in the chapter.

> Important rules, definitions, equations, or helpful hints are often placed in a box like this so they will be easy to find.

If your approach to learning mathematics is to skim the text lightly on the way to puzzling through a homework assignment, you will have difficulty with this or any other textbook. If you are motivated to study mathematics so that you understand it and can use it correctly, this textbook is designed for you.

According to an old Spanish proverb, the world is an ocean and one who cannot swim will sink to the bottom. If the modern world of work is an ocean, the skill needed to keep afloat or even swim to the top is clearly mathematics. It is the purpose of this book to help you learn these basic skills.

Now, turn to page 1 and let's begin.

H. M. S.

PREVIEW

1 Arithmetic of Whole Numbers

Objective	Sample problems		For help, go to page
When you finish this chapter you will be able to:			
1. (a) Write whole numbers in words.	(a) Write 250,374 in words	_____	5
(b) Write, in numerical form, whole numbers that are spoken or written in words.	(b) Write, in numerical form: "one million, sixty-five thousand, eight"	_____	6
2. Round whole numbers.	Round 214,659 (a) to the nearest ten thousand	_____	6
	(b) to the nearest hundred	_____	
3. Add whole numbers.	(a) $67 + 58$	_____	7
	(b) $7009 + 1598$	_____	
4. Subtract whole numbers.	(a) $82 - 45$	_____	21
	(b) $4035 - 1967$	_____	
	(c) $14 + 31 + 59 - 67 + 22 + 37 - 19$	_____	
5. Multiply whole numbers.	(a) 64×37	_____	30
	(b) 305×243	_____	
	(c) 908×705	_____	

Name _____

Date _____

Course/Section _____

1

Objective	Sample problems		For help, go to page
6. Divide whole numbers.	(a) $2006 \div 6$	_____	41
	(b) $7511 \div 37$	_____	
7. Determine factors and prime factors.	(a) List all the factors of 12.	_____	46
	(b) Write 12 as a product of its prime factors.	_____	47
8. Use the correct order of operations with addition, subtraction, multiplication, and division.	(a) $6 + 9 \times 3$	_____	53
	(b) $35 - 14 \div 7$	_____	
	(c) $56 \div 4 \times 2 + 9 - 4$	_____	
	(d) $(23 - 7) \times 24 \div (12 - 4)$	_____	
9. Solve practical applications involving whole numbers.	**Machine Trades** A metal casting weighs 680 lb. 235 lb of metal is removed during shaping. What is its finished weight?	_____	

(Answers to these preview problems are given in the Appendix. Also, worked solutions to many of these problems appear in the chapter Summary. Don't peek.)

If you are certain that you can work *all* these problems correctly, turn to page 62 for a set of practice problems. If you cannot work one or more of the preview problems, turn to the page indicated to the right of the problem. For those who wish to master this material with the greatest success, turn to Section 1-1 and begin to work there.

Arithmetic of Whole Numbers

A century ago, the average person used numbers to tell time, count, and keep track of money. Today, most people need to develop technical skills based on their ability to read, write, and work with numbers in order to earn a living. Although we live in an age of computers and calculators, much of the simple arithmetic used in industry, business, and the skilled trades is still done mentally or by hand. In fact, many trade and technical areas require you to *prove* that you can do the calculations by hand before you can get a job.

In this opening chapter, we take a practical, how-to-do-it look at the four basic operations of whole numbers and the order of operations. Once we get past the basics, we will show you how to use a calculator to do such calculations.

In addition to learning the basic mathematical skills, the other main purpose of this text is to teach you how to apply these skills in practical, on-the-job situations. Every section features worked examples of practical applications, and the exercise sets contain problems from a variety of trade, technical, and other occupations.

CASE STUDY: Robin's New Car

At the end of each chapter you will be guided through a **Case Study**, a multi-step application that uses some of the math skills you learned in the chapter. The Case Study for Chapter 1 (p. 67) teaches you the difference between financing and leasing a new car and how to use the mathematics of whole numbers to help you choose the best option.

1-1   Reading, Writing, Rounding, and Adding Whole Numbers

learning|catalytics™

1. Write in numerical form: two hundred six thousand, seven hundred thirty
2. Round the number in problem 1 to the nearest thousand.

The simplest numbers are the whole numbers—the numbers we use for counting the number of objects in a group. The whole numbers are 0, 1, 2, 3, . . . , and so on.

EXAMPLE 1

How many letters are in the collection shown in the margin?

We counted 23. Notice that we can count the letters by grouping them into sets of ten:

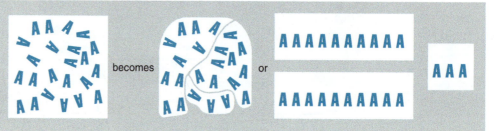

2 tens + 3 ones

20 + 3 or 23

Expanded Form

Mathematicians call this the **expanded form** of a number. For example,

$$46 = \quad 40 + 6 \qquad = 4 \text{ tens} + 6 \text{ ones}$$

$$874 = 800 + 70 + 4 \quad = 8 \text{ hundreds} + 7 \text{ tens} + 4 \text{ ones}$$

$$305 = 300 + 5 \qquad = 3 \text{ hundreds} + 0 \text{ tens} + 5 \text{ ones}$$

Only ten numerals or number symbols—0, 1, 2, 3, 4, 5, 6, 7, 8, and 9—are needed to write any number. These ten basic numerals are called the **digits** of the number. The digits 4 and 6 are used to write 46, the number 274 is a three-digit number, and so on.

Your Turn

Write out the following three-digit numbers in expanded form:

(a) 762 = ____ + ____ + ____ = ____ hundreds + ____ tens + ____ ones

(b) 425 = ____ + ____ + ____ = ____ hundreds + ____ tens + ____ ones

(c) 208 = ____ + ____ + ____ = ____ hundreds + ____ tens + ____ ones

Solutions

(a) *762 = 700 + 60 + 2 = 7 hundreds + 6 tens + 2 ones*

(b) *425 = 400 + 20 + 5 = 4 hundreds + 2 tens + 5 ones*

(c) *208 = 200 + 0 + 8 = 2 hundreds + 0 tens + 8 ones*

A Closer Look

Notice that the 2 in 762 means something different from the 2 in 425 or 208. In 762, the 2 signifies two ones. In 425, the 2 signifies two tens. In 208, the 2 signifies two hundreds. Ours is a *place-value* system of naming numbers: the value of any digit depends on the place where it is located.

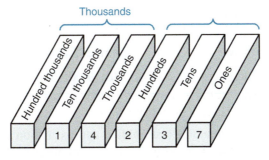

Being able to write a number in expanded form will help you to understand and remember the basic operations of arithmetic—even though you'll never find it on a blueprint or in a technical handbook.

Reading and Writing Whole Numbers

This expanded-form idea is useful especially in naming very large numbers. Any large number given in numerical form may be translated to words by using the following diagram:

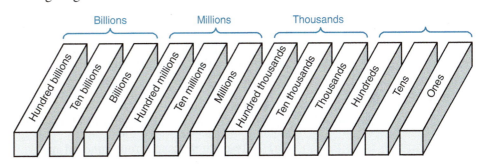

EXAMPLE 2 The number 14,237 can be placed in the diagram like this:

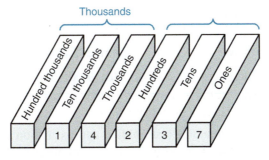

and read "fourteen thousand, two hundred thirty-seven."

EXAMPLE 3 The number 47,653,290,866 becomes

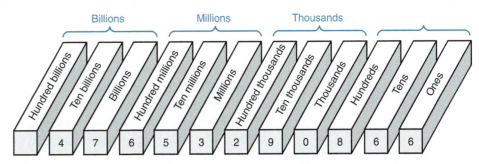

and is read "forty-seven billion, six hundred fifty-three million, two hundred ninety thousand, eight hundred sixty-six."

In each block of three digits read the digits in the normal way ("forty-seven," "six hundred fifty-three") and add the name of the block ("billion," "million"). Notice that the word "and" is not used in naming these numbers. ●

Your Turn   Use the diagram to name the following numbers.

(a) 4072   (b) 1,360,105

(c) 3,000,210   (d) 21,862,031,001

Answers   (a) Four thousand, seventy-two

(b) One million, three hundred sixty thousand, one hundred five

(c) Three million, two hundred ten

(d) Twenty-one billion, eight hundred sixty-two million, thirty-one thousand, one

---

Note   Notice in part (a), we did not insert a comma between the 4 and the 0. In a four-digit number such as this, the comma is optional, and we will usually omit it. ●

It is also important to be able to write numbers correctly when you hear them spoken or when they are written in words.

More Practice   Read each of the following aloud and then write them in correct numerical form.

(a) Fifty-eight thousand, four hundred six

(b) Two hundred seventy-three million, five hundred forty thousand

(c) Seven thousand, sixty

(d) Nine billion, six million, two hundred twenty-three thousand, fifty-eight

Answers   (a) 58,406   (b) 273,540,000

(c) 7060   (d) 9,006,223,058

Rounding Whole Numbers   In many situations a simplified approximation of a number is more useful than its exact value. For example, the accountant for a business may calculate its total monthly revenue as $247,563, but the owner of the business may find it easier to talk about the revenue as "about $250,000." The process of approximating a number is called *rounding*. Rounding numbers comes in handy when we need to make estimates or do "mental mathematics."

A number can be rounded to any desired place. For example, $247,563 is approximately

$247,560 rounded to the nearest ten,
$247,600 rounded to the nearest hundred,
$248,000 rounded to the nearest thousand,
$250,000 rounded to the nearest ten thousand, and
$200,000 rounded to the nearest hundred thousand.

To round a whole number, follow this step-by-step process:

			to the nearest hundred thousand	to the nearest ten thousand
EXAMPLE 4	Round 247,563			
	Step 1	Determine the place to which the number is to be rounded. Mark it on the right with a ∧.	$2∧47,563	$24∧7,563
	Step 2	If the digit to the right of the mark is less than 5, replace all digits to the right of the mark with zeros.	$200,000	
	Step 3	If the digit to the right of the mark is equal to or larger than 5, increase the digit to the left by 1 and replace all digits to the right with zeros.		$250,000

Your Turn   Try these for practice. Round

(a) 73,856 to the nearest thousand

(b) 64 to the nearest ten

(c) 4852 to the nearest hundred

(d) 350,000 to the nearest hundred thousand

(e) 726 to the nearest hundred

Solutions   (a) **Step 1**   Place a mark to the right of the thousands place. The digit 3 is in the thousands place.   73∧856

Step 2   Does not apply.

Step 3   The digit to the right of the mark, 8, is larger than 5. Increase the 3 to a 4 and replace all digits to the right with zeros.   74,000

(b) **Step 1**   Place a mark to the right of the tens place. The digit 6 is in the tens place.   6∧4

Step 2   The digit to the right of the mark, 4, is less than 5. Replace it with a zero.   60

(c) 4900   (d) 400,000   (e) 700

Addition of Whole Numbers   Adding whole numbers is fairly easy provided that you have stored in your memory a few simple addition facts. It is most important that you be able to add simple one-digit numbers mentally.

The following sets of problems in one-digit addition are designed to give you some practice. Work quickly. You should be able to answer all problems in a set in the time shown.

PROBLEMS   One-Digit Addition

A. Add.

7	5	2	5	8	2	3	8	9	7
3	6	9	7	8	5	6	7	3	6

6	8	9	3	7	2	9	9	7	4
4	5	6	5	7	7	4	9	2	7

9	2	5	8	4	9	6	4	8	8
7	6	5	9	5	5	6	3	2	3

5	6	7	7	5	6	2	3	6	9
8	7	5	9	4	5	8	7	8	8

7	5	9	4	3	8	4	8	5	7
4	9	2	6	8	6	9	4	8	8

Mastery time = 80 seconds

B. Add. Try to do all addition mentally.

2	7	3	4	2	6	3	5	9	5
5	3	6	5	7	7	4	7	6	2
4	2	5	8	9	8	4	8	3	8

6	5	4	8	6	9	7	4	8	1
2	4	2	1	8	3	1	9	4	8
7	5	9	9	8	5	6	1	6	7

1	9	3	1	7	2	9	9	8	5
9	9	1	6	9	9	8	5	3	4
2	1	4	3	6	1	2	1	3	7

Mastery time = 90 seconds

The answers are given in the Appendix.

✎ Note   The answer to an addition problem is called the **sum**. ●

Adding one-digit numbers is very important. It is the key to many mathematical computations—even if you do the work on a calculator. But you also must be able to add numbers with two or more digits. Suppose that, as a contractor, you need to find the total time spent on a job by two workers who have put in 31 hours and 48 hours, respectively. You will need to find the sum

31 hours + 48 hours = _____

Estimating   What is the first step? Start adding digits? Punch in some numbers on your trusty calculator? Rattle your abacus? None of these. The first step is to *estimate* your answer. The most important rule in any mathematical calculation is:

> Know the approximate answer to any calculation before you calculate it.

Never do an arithmetic calculation until you know roughly what the answer is going to be. Always know where you are going.

Rounding to the nearest 10 hours, the preceding sum can be estimated as 31 hours + 48 hours or approximately 30 hours + 50 hours or 80 hours, not 8 or 8000 or 800 hours. Having the estimate will keep you from making any major (and embarrassing) mistakes. Once you have a rough estimate of the answer, you are ready to do the arithmetic work.

Calculate 31 + 48 = _____

You don't really need an air-conditioned, solar-powered, talking calculator for that, do you?

You should set it up like this:

Tens column Ones column

$$\begin{array}{r} 3\ 1 \\ +\ 4\ 8 \\ \hline \end{array}$$

1. The numbers to be added are arranged vertically (up and down) in columns.

2. The right end or ones digits are placed in the ones column, the tens digits are placed in the tens column, and so on.

Avoid the confusion of $\begin{array}{r} 31 \\ +\ 48 \\ \hline \end{array}$ or $\begin{array}{r} 31 \\ +\ 48 \\ \hline \end{array}$

! Careful Most often the cause of errors in arithmetic is carelessness, especially in simple tasks such as lining up the digits correctly. ●

Once the digits are lined up the problem is easy.

$\begin{array}{r} 31 \\ +\ 48 \\ \hline 79 \end{array}$ **First,** find the sum of the ones column: $1 + 8 = 9$
Then, find the sum of the tens column: $3 + 4 = 7$

Does the answer agree with your original estimate? Yes. The estimate, 80, is roughly equal to the actual sum, 79.

What we have just shown you is called the *guess and check method* of doing mathematics.

1. **Estimate** the answer using rounded numbers.

2. **Work** the problem carefully.

 3. **Check** your answer against the estimate. If they disagree, repeat both Steps 1 and 2. The check icon reminds you to check your work.

Most students worry about estimating, either because they won't take the time to do it or because they are afraid they might do it incorrectly. Relax. You are the only one who will know your estimate. Do it in your head, do it quickly, and make it reasonably accurate. Step 3 helps you to find incorrect answers before you finish the problem. The guess *and* check method means that you never work in the dark; you always know where you are going.

✎ Note Estimating is especially important in practical math, where a wrong answer is not just a mark on a piece of paper. An error may mean time and money lost. ●

EXAMPLE 5 Here is a slightly more difficult problem:

27 lb + 58 lb = _____

First, estimate the answer. 27 + 58 is roughly

$$30 + 60 \text{ or about } 90.$$

The answer is about 90 lb.

Second, line up the digits in columns.
$$\begin{array}{r} 27 \\ + 58 \end{array}$$

(The numbers to be added, 27 and 58 in this case, are called *addends*.)

Third, add carefully.
$$\begin{array}{r} \overset{1}{2}7 \\ + 58 \\ \hline 85 \end{array}$$

✔️ **Finally,** check your answer by comparing it with the estimate. The estimate 90 lb is roughly equal to the answer 85 lb—at least you have an answer in the right ballpark. ●

🔍 **A Closer Look**   What does that little 1 above the tens digit mean? What really happens when you "carry" a digit? Let's look at it in detail. In expanded notation,

$$\begin{array}{l} 27 \rightarrow 2 \text{ tens } + 7 \text{ ones} \\ + 58 \rightarrow 5 \text{ tens } + 8 \text{ ones} \\ \hline \quad = 7 \text{ tens } + 15 \text{ ones} \end{array}$$

$$= 7 \text{ tens } + 1 \text{ ten } + 5 \text{ ones}$$

$$= 8 \text{ tens } \qquad + 5 \text{ ones}$$

$$= 85$$

The 1 that is carried over to the tens column is really a ten. ●

👉 **Learning Help**   Trades people often must calculate exact answers mentally. To do a problem such as 27 + 58, a trick called "balancing" works nicely. Simply add 2 to the 58 to get a "round" number, 60, and subtract 2 from 27 to balance, keeping the total the same. Therefore, 27 + 58 is the same as 25 + 60, which is easy to add mentally to get 85. ●

Your Turn   Try this one: 456 + 87 = _____

Solution   **First,** estimate the answer. 456 + 87 is roughly 460 + 90 or about 550.

Then, line up the digits in columns and add.
$$\begin{array}{r} \overset{1\ 1}{4}56 \\ + \ 87 \\ \hline 543 \end{array}$$

The answer is very close to the estimate.

We use the same procedure to add three or more numbers. Estimating and checking become even more important when the problem gets more complicated.

EXAMPLE 6   To add 536 + 1473 + 875 + 88

Estimate: Rounding each number to the nearest hundred,

$$500 + 1500 + 900 + 100 = 3000$$

Step 1   To help avoid careless errors, put the number with the most digits, 1473, on top. Put the number with the fewest digits, 88, on the bottom.

$$
\begin{array}{r}
1473 \\
536 \\
875 \\
+\quad 88 \\
\end{array}
$$

Step 2
$$
\begin{array}{r}
\overset{2}{1}473 \\
536 \\
875 \\
+\quad 88 \\
\hline
2
\end{array}
$$
$3 + 6 + 5 + 8 = 22$ Write 2; carry 2 tens

Step 3
$$
\begin{array}{r}
\overset{22}{1}473 \\
536 \\
875 \\
+\quad 88 \\
\hline
72
\end{array}
$$
$2 + 7 + 3 + 7 + 8 = 27$ Write 7; carry 2 hundreds

Step 4
$$
\begin{array}{r}
\overset{122}{1}473 \\
536 \\
875 \\
+\quad 88 \\
\hline
972
\end{array}
$$
$2 + 4 + 5 + 8 = 19$ Write 9; carry 1 thousand

Step 5
$$
\begin{array}{r}
\overset{122}{1}473 \\
536 \\
875 \\
+\quad 88 \\
\hline
2972
\end{array}
$$
$1 + 1 = 2$ Write 2.

☑ **Check:**   The estimate 3000 and the answer 2972 are very close.   ●

Your Turn   The following is a short set of problems. Add, and be sure to estimate your answers first. Check your answers with your estimates.

(a) $429 + 738 =$ _____   (b) $1446 + 867 =$ _____

(c) $82 + 2368 + 744 =$ _____   (d) $409 + 2572 + 3685 + 94 =$ _____

Solutions   (a) **Estimate:**   Rounding each number to the nearest hundred, $400 + 700 = 1100$
Line up the digits: 429
                                   $+\ 738$

Calculate:

Step 1
$$
\begin{array}{r}
\overset{1}{4}29 \\
+\ 738 \\
\hline
7
\end{array}
$$
$9 + 8 = 17$ Write 7; carry 1 ten.

Step 2
$$
\begin{array}{r}
\overset{1}{4}29 \\
+\ 738 \\
\hline
67
\end{array}
$$
$1 + 2 + 3 = 6$ Write 6.

Step 3
$$\overset{1}{429}$$
$$\underline{+\ 738}$$ $4 + 7 = 11$ Write 11.
$$1167$$

Check: The estimate 1100 and the answer 1167 are roughly equal.

(b) Estimate: $1400 + 900 = 2300$

Calculate:

Step 1
$$\overset{1}{1446}$$
$$\underline{+\ 867}$$ $6 + 7 = 13$ Write 3; carry 1 ten.
$$3$$

Step 2
$$\overset{11}{1446}$$
$$\underline{+\ 867}$$ $1 + 4 + 6 = 11$ Write 1; carry 1 hundred.
$$13$$

Step 3
$$\overset{111}{1446}$$
$$\underline{+\ 867}$$ $1 + 4 + 8 = 13$ Write 3; carry 1 thousand.
$$313$$

Step 4
$$\overset{111}{1446}$$
$$\underline{+\ 867}$$ $1 + 1 = 2$ Write 2.
$$2313$$

Check: The estimate 2300 and the answer 2313 are roughly equal.

(c) Estimate: Rounding each number to the nearest hundred,
$100 + 2400 + 700 = 3200$

Calculate:
$$\overset{111}{2368}$$
$$744$$
$$\underline{+\ \ \ 82}$$
$$3194$$

Check: The estimate 3200 and the answer 3194 are roughly equal.

(d) Estimate: $400 + 2600 + 3700 + 100 = 6800$

Calculate:
$$\overset{122}{2572}$$
$$3685$$
$$409$$
$$\underline{+\ \ \ 94}$$
$$6760$$

Check: The estimate 6800 and the answer 6760 are roughly equal.

Estimating answers is a very important part of any mathematics calculation, especially for the practical mathematics used in engineering, technology, and the trades. A successful builder, painter, or repairperson must make accurate estimates of job costs—business success depends on it. If you work in a technical trade, getting and keeping your job may depend on your ability to get the correct answer *every* time.

If you use a calculator to do the actual arithmetic, it is even more important to get a careful estimate of the answer first. If you plug a wrong number into the calculator, accidentally hit a wrong key, or unknowingly use a failing battery, the calculator may give you a wrong answer—lightning fast, but wrong. The estimate is your best insurance that a wrong answer will be caught immediately. Convinced?

Practical Applications   Once you learn a basic math skill, you must also learn how to apply that skill to a real-world situation in your job or in your daily life. To help you practice applying the math, practical applications, also known to students as word problems, will be provided throughout this text in the examples and in the exercises.

Word problems that call for addition often contain key questions such as "how many . . .?" or "what is the total number of . . .?" Look for these clues that indicate you must add to find the answer.

EXAMPLE 7   **Transportation**   In order to accommodate the weight of their equipment, the California Highway Patrol is now using an SUV instead of a sedan as their primary vehicle. The additional equipment they carry includes a push bumper (54 lb), a computer (12 lb), a gun tub with rifles (32 lb), a light bar (30 lb), a prisoner barrier (56 lb), radio and video equipment (270 lb), and other miscellaneous supplies such as flares, first aid supplies, and traffic cones (50 lb). What is the total weight of the additional equipment listed?

The phrase "What is the *total* weight . . ." indicates that addition is required to solve the problem.

Estimate: $50 + 10 + 30 + 30 + 60 + 270 + 50 = 500$

Calculate: To avoid confusion, place the only 3-digit number, 270, on top.

$$
\begin{array}{r}
{\scriptstyle 3\,1} \\
270 \\
54 \\
12 \\
32 \\
30 \\
56 \\
+\ \ 50 \\
\hline
504
\end{array}
$$

The new SUVs will need to carry 504 lb of additional equipment. Notice that the estimate is very close to the final answer.   ●

Your Turn   **Trades Management**   During the 3-year construction of the Wilshire Grand Center in Los Angeles, a large number of workers from 9 different trades were needed. First estimates were as follows: 2421 ironworkers, 1853 carpenters, 1466 concrete contractors, 1198 electricians, 908 mechanical contractors, 747 stone and tile contractors, 587 elevator contractors, 462 plumbers, and 121 glaziers. What was the total number of estimated trades workers needed for the project? (Source: Turner Construction.)

Solution   Again, the phrase "What was the total number . . ." indicates that addition is needed to solve the problem.

Estimate: $2400 + 1900 + 1500 + 1200 + 900 + 700 + 600 + 500 + 100 = 9800$

Calculate:
$$\begin{array}{r} \overset{4\ 4\ 4}{2421} \\ 1853 \\ 1466 \\ 1198 \\ 908 \\ 747 \\ 587 \\ 462 \\ +\ 121 \\ \hline 9763 \end{array}$$

A total of 9763 workers from these 9 trades were needed for the project. Notice that the original estimate is very close to the final answer.

Units of Measure

Units of measure are important in every trade. Just as you want to make sure that a customer is not thinking in centimeters while you are thinking yards, you will want to make sure that the measurements required for a project have consistent units.

An in-depth study of measurement units is presented in Chapter 5, and we require only a few basic conversions between units until then. However, because it is nearly impossible to discuss numbers without discussing their units, many of the practical word problems in Chapters 1 through 4 contain units of measure. Therefore, we are providing you with the following table listing the units of measure used in Chapters 1 through 4, along with their most common abbreviations.

Type of Measurement	U.S. Customary Units	Metric Units
Length or distance	inch (in.* or ″) foot (ft or ′) yard (yd) mile (mi)	millimeter (mm) centimeter (cm) meter (m) kilometer (km)
Weight	ounce (oz) pound (lb) ton (t)	microgram (µg) milligram (mg) gram (g) kilogram (kg)
Area	square inch (sq in.) square foot (sq ft) square yard (sq yd) acre (a)	square centimeter (sq cm) square meter (sq m) square kilometer (sq km) hectare (ha)
Capacity or volume	fluid ounces (fl oz) pint (pt) quart (qt) gallon (gal) bushel (bu) cubic inch (cu in.) cubic foot (cu ft) cubic yard (cu yd)	cubic centimeter (cu cm, cc) milliliter (mL) liter (L) cubic meter (cu m)

(continued)

Type of Measurement	U.S. Customary Units	Metric Units
Velocity or speed	miles per hour (mph or mi/hr) beats per minute (bpm) cycles per second (hertz) revolutions per minute (rpm or rev/min)	meters per second (m/sec) kilometers per hour (km/hr)
Temperature	degrees Fahrenheit (°F)	degrees Celsius (°C)
Power, energy, and heat	ohm (Ω) watt (W) volt (V) ampere (A) horsepower (hp) British thermal unit (Btu)	cubic foot per meter (cfm) kilohertz (kHz) picofarad (pF) kilowatt (kW)
Pressure	pounds per square inch (psi or lb/in.^{2})	pascal (Pa)
Amount of lumber	board feet (bf or fbm)	

*For abbreviations that might be mistaken for a word (e.g., "in" for inches), a period is included at the end of the abbreviation. For abbreviations that would not be mistaken for a word (e.g, "ft"), no period is added.

Now, try the following problems for practice in working with whole numbers.

Exercises 1-1   Reading, Writing, Rounding, and Adding Whole Numbers

A. Reading, Writing, and Rounding Whole Numbers

Write in words.

1. 357   2. 2304   3. 17,092   4. 207,630   5. 2,000,034

6. 10,007   7. 740,106   8. 5,055,550   9. 118,180,018

10. 6,709,210,046

Write as numbers.

11. Three thousand, six

12. Seventeen thousand, twenty-four

13. Eleven thousand, one hundred

14. Three million, two thousand, seventeen

15. Four million, forty thousand, six

16. Five billion, seven hundred twenty million, ten

Round as indicated.

17. 357 to the nearest ten

18. 4386 to the nearest hundred

19. 4386 to the nearest thousand

20. 15,472 to the nearest thousand

21. 225,799 to the nearest ten thousand

22. 32,408,792 to the nearest hundred thousand

B. Add.

1. 47 $\underline{\quad 22}$	2. 27 $\underline{\quad 31}$	3. 45 $\underline{\quad 35}$	4. 38 $\underline{\quad 55}$	5. 75 $\underline{\quad 48}$
6. 26 $\underline{\quad 98}$	7. 48 $\underline{\quad 84}$	8. 67 $\underline{\quad 69}$	9. 747 $\underline{\quad 59}$	10. 508 $\underline{\quad 95}$

11. 684	12. 432	13. 621	14. 189	15. 375
706	399	388	204	486

16. 4237	17. 5076	18. 7907	19. 3785	20. 6709
288	385	1395	7643	9006

21. 10674	22. 40026	23. 18745	24. 19876	25. 78044
397	7085	6972	4835	97684

26. 4728	27. 5818	28. 83754	29. 498321	30. 843592
683	244	66283	65466	710662
29	33	5984	95873	497381
			3604	25738

C. Arrange vertically and find the sum.

1. 487 + 29 + 526 = _____

2. 65 + 796 + 87 = _____

3. 322 + 46 + 5984 = _____

4. 7268 + 209 + 178 = _____

5. 5016 + 423 + 1075 = _____

6. 715 + 4293 + 184 + 19 = _____

7. 1706 + 387 + 42 + 307 = _____

8. 456 + 978 + 1423 + 3584 = _____

9. 6284 + 28 + 674 + 97 = _____

10. 6842 + 9008 + 57 + 368 = _____

11. 268 + 1593 + 88 + 2165 = _____

12. 8764 + 85 + 983 + 19 = _____

13. 4 + 6 + 11 + 7 + 14 + 3 + 9 + 6 + 4 = _____

14. 12 + 7 + 15 + 16 + 21 + 8 + 10 + 5 + 30 + 17 = _____

15. 1 + 2 + 3 + 4 + 5 + 6 + 7 + 8 + 9 + 10 = _____

16. 22 + 31 + 43 + 11 + 9 + 1 + 19 + 12 = _____

17. 75 + 4 + 81 + 12 + 14 + 65 + 47 + 22 + 37 = _____

18. 89,652 + 57,388 + 6506 = _____

19. 443,700 + 629,735 + 85,962 + 6643 = _____

20. 784,396 + 858,390 + 662,043 + 965,831 + 62,654 = _____

D. Practical Applications.

1. **Electrical Trades**   In setting up his latest wiring job, an electrician cut the following lengths of wire: 387, 913, 76, 2640, and 845 ft. Find the total length of wire used.

2. **Flooring and Carpeting**   The Acme Lumber Co. made four deliveries of 1-in. by 6-in. flooring: 3280, 2650, 2465, and 2970 fbm. What was the total number of board feet of flooring delivered?

3. **Machine Trades**   The stockroom has eight boxes of No. 10 hexhead cap screws. How many screws of this type are in stock if the boxes contain 346, 275, 84, 128, 325, 98, 260, and 120 screws, respectively?

4. **Trades Management**   In calculating her weekly expenses, a contractor found that she had spent the following amounts: materials, $53,860; labor, $7854; salaried help, $1542; overhead expense, $1832. What was her total expense for the week?

5. **Trades Management**   The head machinist at Tiger Tool Co. is responsible for totaling time cards to determine job costs. She found that five different jobs this week took 78, 428, 143, 96, and 384 minutes each. What was the total time in minutes for the five jobs?

6. **Roofing**   On a home construction job, a roofer laid 1480 wood shingles the first day, 1240 the second, 1560 the third, 1320 the fourth, and 1070 the fifth day. How many shingles did he lay in five days?

7. **Industrial Technology**   Eight individually powered machines in a small production shop have motors using 420, 260, 875, 340, 558, 564, 280, and 310 watts each. What is the total wattage used when (a) the total shop is in operation? (b) the three largest motors are running? (c) the three smallest motors are running?

8. **Automotive Trades**   A mechanic is taking inventory of oil in stock. He has 24 quarts of 10W-30, 8 quarts of 30W, 42 quarts of 20W-50, 16 quarts of 10W-40, and 21 quarts of 20W-40. How many total quarts of oil does he have in stock?

9. **Construction**   The Happy Helper building materials supplier has four piles of bricks containing 1250, 865, 742, and 257 bricks. What is the total number of bricks it has on hand?

10. **Machine Trades**   A machinist needs the following lengths of 1-in. diameter rod: 8 in., 14 in., 6 in., 27 in., and 42 in. How long a rod is required to supply all five pieces? (Ignore cutting waste.)

11. **Landscaping**   A new landscape maintenance business requires the following equipment:

    1 rototiller for $599
    1 gas trimmer for $309
    1 mower for $369
    1 hedge trimmer for $280

    What is the total cost of this equipment?

12. **Electrical Trades**   The Radius Electronics Company orders 325 resistors, 162 capacitors, 25 integrated circuit boards, and 68 transistors. Calculate the total number of parts ordered.

13. **Electronics**   When resistors are connected in series, the total resistance is the sum of the individual resistors. If resistances of 520, 1160, 49, and 1200 ohms are connected in series, calculate the total resistance.

14. **Electronics**   Kirchhoff's law states that the sum of the voltage drops around a closed circuit is equal to the source voltage; that is,

$$V_s = E_1 + E_2 + E_3 + E_4 + E_5.$$

Calculate the source voltage V_s for the circuit shown.

15. **Automotive Trades**   To balance an engine, a mechanic must know the total weight of a piston assembly, also known as the reciprocating weight of the assembly. A certain piston assembly consisted of a 485-gram piston, a 74-gram wrist-pin, 51 grams of compression and oil rings, and the small end of the connecting rod weighing 146 grams. What is the reciprocating weight of this assembly?

16. **Allied Health**   The following standard amounts of food were recorded in one patient's Intake/Output record: one carton of milk, 8 fl oz (fluid ounces); one carton of juice, 8 fl oz; one bowl of soup, 15 fl oz; one cup of coffee, 6 fl oz; one serving of sherbet, 12 fl oz; one glass of water, 8 fl oz. Determine the total fluid intake for this patient.

17. **Construction**   For working on remote construction sites, GVM Construction needs a generator that can supply a 1400-W (watt) circular saw, an 1800-W table saw, a 600-W hand drill, and a 100-W radio simultaneously. What total wattage does the generator need to supply?

18. **Automotive Trades**   The base price for a 2016 BMW 320i, including destination charge, is $34,145. The following is a list of optional packages and extras:

Item	Cost	Item	Cost
Track Handling Package	$2300	Driver Assistance Plus	$ 950
Premium Package	$3100	Navigation System	$1950
Sport Package	$1300	Lighting Package	$ 700
USB/Smartphone Integration	$ 350	Cold Weather Package	$ 800
Fold-Down Rear Seat	$ 475	Interior Color Upgrade	$1450

What would be the total pretax cost of the car if a customer added the following features:

(a) Track Handling Package, Navigation System, Cold Weather Package, and Fold-Down Rear Seat?

(b) Driver Assistance Plus, Sport Package, USB/Smartphone Integration, and Interior Color Upgrade?

(c) Premium Package, Lighting Package, Fold-Down Rear Seat, USB/Smartphone Integration, and Interior Color Upgrade?

19. **Sports and Leisure**   In the 1988 Olympics in Seoul, South Korea, Jackie Joyner-Kersee (USA) won the gold medal in the heptathlon with a world-record performance. The heptathlon consists of seven separate events performed over two days, and the points for each event are added to determine the total score. Jackie's points (and performances) for each event were as follows:

Day	Event	Performance	Points Earned
Day 1	100-meter hurdles	12.69 sec	1172
	High jump	1.86 m	1054
	Shot put	15.80 m	915
	200-meter dash	22.56 sec	1123
Day 2	Long jump	7.27 m	1264
	Javelin	45.66 m	776
	800-meter run	2 min 8.51 sec	987

(a) How many points did she earn on Day 1?

(b) How many points did she earn on Day 2?

(c) What was Jackie's world-record-setting point total?

E. Calculator Problems

You probably own a calculator and, of course, you are eager to put it to work doing practical math calculations. In this text we include problem sets for calculator users. These problems are taken from real-life situations and, unlike most textbook problems, involve big numbers and lots of calculations. If you think that having an electronic brain-in-a-box means that you do not need to know basic arithmetic, you will be disappointed. The calculator helps you to work faster, but it will not tell you *what* to do or *how* to do it.

Detailed instructions on using a calculator with whole numbers appears on page 68.

Here are a few helpful hints for calculator users:

1. Always *estimate* your answer before doing a calculation.

2. *Check* your answer by comparing it with the estimate or by the other methods shown in this text. Be certain that your answer makes sense.

3. If you doubt the calculator (they do break down, you know), put a problem in it whose answer you know, preferably a problem like the one you are solving.

1. **Electronics**   An electronics mixing circuit adds two given input frequencies to produce an output signal. If the input frequencies are 35,244 kHz and 61,757 kHz, calculate the frequency of the output signal.

2. **Manufacturing**   The following table lists the number of widget fasteners made by each of the five machines at the Ace Widget Co. during the last ten working days.

Day	Machine A	B	C	D	E	Daily Totals
1	347	402	406	527	237	
2	451	483	312	563	316	
3	406	511	171	581	289	
4	378	413	0	512	291	
5	399	395	452	604	342	
6	421	367	322	535	308	
7	467	409	256	578	264	
8	512	514	117	588	257	
9	302	478	37	581	269	
10	391	490	112	596	310	
Machine Totals						

(a) Complete the table by finding the number of fasteners produced each day. Enter these totals under the column "Daily Totals" on the right.

(b) Find the number of fasteners produced by each machine during the ten-day period and enter these totals along the bottom row marked "Machine Totals."

(c) Does the sum of the daily totals equal the sum of the machine totals?

3. Add the following as shown.

(a) $137427
   67429
   91006
    6070
    4894
     399

(b) $216847
   386738
    86492
    28104
     9757
     4875

(c) $693884
   675489
   560487
   276921
    44682
    47039

(d) $4299 + $137 + $20 + $177 + $63 + $781 + $1008 + $671 = ?

4. **Trades Management**   Joe's Air Conditioning Installation Co. has not been successful, and he is wondering if he should sell it and move to a better location. During the first three months of the year, his expenses were as follows:

Rent $4260
Supplies $2540
Part-time helper $2100
Transportation $948

Utilities $815
Advertising $750
Miscellaneous $187

His monthly income was:

January $1760
February $2650
March $3325

(a) What was his total expense for the three-month period?

(b) What was his total income for the three-month period?

(c) Now turn your calculator around to learn what Joe should do about this unhappy situation.

5. **Electrical Trades**   A mapper is a person employed by an electrical utility company who has the job of reading diagrams of utility installations and listing the materials to be installed or removed by engineers. Part of a typical job list might look like this:

INSTALLATION (in feet of conductor)

Location Code	No. 12 BHD (bare, hard-drawn copper wire)	#TX (triplex)	410 AAC (all-aluminum conductor)	110 ACSR (aluminum-core steel-reinforced conductor)	6B (No. 6, bare conductor)
A3	1740	40	1400		350
A4	1132		5090		2190
B1	500			3794	
B5		87	3995		1400
B6	4132	96	845		
C4		35		3258	2780
C5	3949		1385	1740	705

(a) How many total feet of each kind of conductor must the installer have to complete the job?

(b) How many feet of conductor are to be installed at each of the seven locations?

When you have completed these exercises, check your answers to the odd-numbered problems in the Appendix and then continue with Section 1-2.

1-2 Subtraction of Whole Numbers

Do the following without using a calculator:

1. Add: 2649 + 857
2. Subtract: 287 − 52

Subtraction is the reverse of addition.

Addition: $3 + 4 = \square$

Subtraction: $3 + \square = 7$

Written this way, a subtraction problem asks the question: How much must be added to a given number to produce a required amount?

Most often, however, the numbers in a subtraction problem are written using a minus sign (−).

To fill in the box in $3 + \square = 7$, we must perform the subtraction $7 - 3 = \square$. If we remember that $3 + 4 = 7$, then we will know that $7 - 3 = 4$.

✎ **Note**   The **difference** is the name given to the answer in a subtraction problem. ●

Solving simple subtraction problems depends on your knowledge of the addition of one-digit numbers.

For example, to solve the problem

$9 - 4 =$ _____

you probably go through a chain of thoughts something like this:

> Nine minus four. Four added to what number gives nine? Five? Try it: four plus five equals nine. Right.

Subtraction problems with small whole numbers will be easy for you if you know your addition tables. See how quickly and accurately you can mentally compute the following subtraction facts.

PROBLEMS   Subtraction Facts

5 −3	9 −1	15 −8	10 −5	7 −6	12 −9	4 −2	11 −6	13 −9	9 −2	12 −3
18 −9	10 −3	14 −7	11 −4	9 −7	16 −8	8 −6	6 −1	11 −7	10 −1	7 −4
11 −3	9 −4	12 −6	10 −2	15 −9	4 −1	8 −5	13 −6	11 −2	6 −3	10 −7
14 −8	11 −9	8 −7	17 −9	9 −6	12 −7	7 −1	6 −4	8 −1	5 −1	3 −2
2 −1	15 −7	8 −4	11 −5	6 −2	16 −7	10 −6	13 −4	13 −7	8 −3	17 −8
8 −2	12 −5	9 −3	15 −6	4 −3	14 −6	7 −2	13 −8	14 −9	5 −2	12 −5
14 −5	3 −1	13 −5	11 −8	12 −4	10 −8	9 −5	5 −4	6 −5	12 −8	7 −3

Mastery time: 3 minutes

The answers are given in the Appendix.

EXAMPLE 1   Here is a more difficult subtraction problem:

$47 - 23 =$ _____

The **first** step is to estimate the answer—remember?

$47 - 23$ is roughly $50 - 20$ or $30.$

The difference, your answer, will be about 30—not 3 or 10 or 300.

The **second** step is to write the numbers vertically, as you did with addition. Be careful to keep the ones digits in line in one column, the tens digits in a second column, and so on.

$$\begin{array}{cc} 4 & 7 \\ -2 & 3 \end{array}$$ Notice that the larger number is written above the smaller number.

Once the numbers have been arranged in this way, the difference may be written by performing the following two steps:

Step 1
$$\begin{array}{cc} 4 & 7 \\ -2 & 3 \\ \hline & 4 \end{array}$$ ones digits: $7 - 3 = 4$

Step 2
$$\begin{array}{cc} 4 & 7 \\ -2 & 3 \\ \hline 2 & 4 \end{array}$$ tens digits: $4 - 2 = 2$

✔ The difference is 24, which agrees roughly with our estimate.   ●

EXAMPLE 2   With some problems, it is necessary to rewrite the larger number before the problem can be solved. Let's try this one:

$64 - 37 =$ _____

First, estimate the answer. Rounding to the nearest ten, $64 - 37$ is roughly $60 - 40$ or 20.

Second, arrange the numbers vertically in columns.
$$\begin{array}{r} 64 \\ -\ 37 \\ \hline \end{array}$$

Because 7 is larger than 4, we must "borrow" one ten from the 6 tens in 64. We are actually rewriting 64 (6 tens + 4 ones) as 5 tens + 14 ones. In actual practice, our work would look like this:

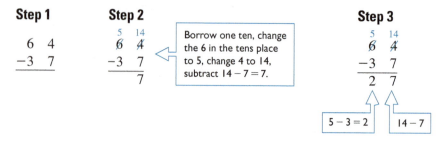

Step 1
$$\begin{array}{cc} 6 & 4 \\ -3 & 7 \\ \hline \end{array}$$

Step 2
$$\begin{array}{cc} {}^5 & {}^{14} \\ \not{6} & \not{4} \\ -3 & 7 \\ \hline & 7 \end{array}$$

Borrow one ten, change the 6 in the tens place to 5, change 4 to 14, subtract $14 - 7 = 7$.

Step 3
$$\begin{array}{cc} {}^5 & {}^{14} \\ \not{6} & \not{4} \\ -3 & 7 \\ \hline 2 & 7 \end{array}$$

$5 - 3 = 2$ $14 - 7$

✔ Double-check subtraction problems by adding the answer (the *difference*) and the smaller number (the *subtrahend*); their sum should equal the larger number.

Step 4   Check:
$$\begin{array}{r} 37 \\ +\ 27 \\ \hline 64 \end{array}$$

●

Learning Help If you need to get an exact answer to a problem such as $64 - 37$ mentally, add or subtract to make the smaller number, 37, a "round" number. In this case, add 3 to make it 40. Because we're subtracting, we want the *difference*, not the *total*, to be the same or balance. Therefore, we also add 3 to the 64 to get 67. The problem becomes $64 - 37 = (64 + 3) - (37 + 3) = 67 - 40$. Subtracting a round number is easy mentally: $67 - 40 = 27$. ●

When a subtraction problem involves three-digit numbers, the procedure is exactly the same. There is simply one additional step.

EXAMPLE 3 To subtract $426 - 128$, follow these steps:

Estimate: $400 - 100 = 300$

Step 1

$$\begin{array}{r} 4\ 2\ 6 \\ -\ 1\ 2\ 8 \\ \hline \end{array}$$

Step 2

$$\begin{array}{r} {}^{1}\ {}^{16} \\ 4\ \cancel{2}\ \cancel{6} \\ -\ 1\ 2\ 8 \\ \hline 8 \end{array}$$

Steps 3 and 4

$$\begin{array}{r} {}^{3}\ {}^{11}\ {}^{16} \\ \cancel{4}\ \cancel{2}\ \cancel{6} \\ -\ 1\ 2\ 8 \\ \hline 2\ 9\ 8 \end{array}$$

In this case we borrow twice. Borrow one ten from the 20 in 426 and make 16. Then borrow one hundred from the 400 in 426 to make 11 in the tens place.

$16 - 8 = 8$ Write 8

$11 - 2 = 9$ Write 9

$3 - 1 = 2$ Write 2

✓ The answer 298 is approximately equal to the estimate 300.

Double-check:

$$\begin{array}{r} {}^{1}\ {}^{1} \\ 1\ 2\ 8 \\ +\ 2\ 9\ 8 \\ \hline 4\ 2\ 6 \end{array}$$

●

Your Turn Subtract as indicated:

(a) $\begin{array}{r} 59 \\ -24 \\ \hline \end{array}$ (b) $\begin{array}{r} 71 \\ -39 \\ \hline \end{array}$ (c) $\begin{array}{r} 687 \\ -194 \\ \hline \end{array}$ (d) $\begin{array}{r} 902 \\ -\ 65 \\ \hline \end{array}$

Solutions (a) **Estimate:** $60 - 20 = 40$

Step 1

$$\begin{array}{r} 5\ 9 \\ -2\ 4 \\ \hline 5 \end{array}$$

Step 2

$$\begin{array}{r} 5\ 9 \\ -2\ 4 \\ \hline 3\ 5 \end{array}$$

✓ The difference is 35, which roughly agrees with our estimate.

(b) **Estimate:** $70 - 40 = 30$

Step 1

$$\begin{array}{r} {}^{6}\ {}^{11} \\ \cancel{7}\ \cancel{1} \\ -3\ 9 \\ \hline \end{array}$$

Borrow one ten from 70, change the 7 in the tens place to 6, change the 1 in the ones place to 11.

Steps 2 and 3

$$\begin{array}{r} {}^{6}\ {}^{11} \\ \cancel{7}\ \cancel{1} \\ -3\ 9 \\ \hline 3\ 2 \end{array}$$

$11 - 9 = 2$ Write 2

$6 - 3 = 3$ Write 3

✓ The answer 32 is approximately equal to the estimate 30. As a shortcut, mentally add 1 to each number,

$$71 - 39 = 72 - 40$$

then subtract.

$$72 - 40 = 32$$

(c) **Estimate:** $700 - 200 = 500$

Step 1	**Step 2**	**Step 3**
6 8 7	⁵6 ¹⁸8̶ 7	⁵6 ¹⁸8̶ 7
−1 9 4	−1 9 4	−1 9 4
3	9 3	4 9 3

✓ The answer is very close to the original estimate.

(d) **Estimate:** $900 - 50 = 850$

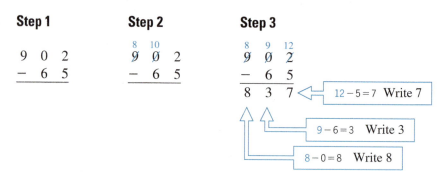

Step 1	**Step 2**	**Step 3**
9 0 2	⁸9̶ ¹⁰0̶ 2	⁸9̶ ⁹0̶ ¹²2̶
− 6 5	− 6 5	− 6 5
		8 3 7

12−5=7 Write 7
9−6=3 Write 3
8−0=8 Write 8

✓ The answer 837 is roughly equal to the estimate 850.

In problem (d) we first borrow one hundred from 900 to get a 10 in the tens place. Then we borrow one 10 from the tens place to get a 12 in the ones place.

Problems with Zero Digits   Let's work through a few examples of subtraction problems involving zero digits. These can be troublesome for some students.

EXAMPLE 4   (a) $400 - 167 = ?$

Step 1	**Step 2**	**Step 3**	**Check**
4 0 0	³4̶ ¹⁰0̶ 0	³4̶ ⁹0̶ ¹⁰0̶	1 6 7
−1 6 7	−1 6 7	−1 6 7	+2 3 3
		2 3 3	4 0 0

Do you see in Steps 2 and 3 that we have rewritten 400 as 300 + 90 + 10?

(b) $5006 - 2487 = ?$

Step 1	**Step 2**	**Step 3**	**Step 4**	**Check**
5 0 0 6	⁴5̶ ¹⁰0̶ 0 6	⁴5̶ ⁹0̶ ¹⁰0̶ 6	⁴5̶ ⁹0̶ ⁹0̶ ¹⁶6̶	2 4 8 7
−2 4 8 7	−2 4 8 7	−2 4 8 7	−2 4 8 7	+2 5 1 9
			2 5 1 9	5 0 0 6

More Practice   Subtract.

(a) 500   (b) 7008   (c) 30206
   234   3605   4738

Answers   (a)   266   (b) 3403   (c) 25,468

Practical Applications   When solving word problems, you must look for key words or phrases to help you decide what operation to use. Here are some typical questions that call for subtraction:

> "What is the *difference* between . . . ?"
> "How much *remains* . . . ?" or "How much *is left* . . . ?"
> "By how much is A *greater than* B?" or "What was *the increase* from A to B?"
> "How much do you *save*?"

As shown in the next example, finding a missing dimension in a drawing might also call for subtraction.

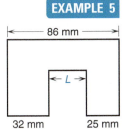

EXAMPLE 5   **Machine Trades**   Suppose we need to find the missing dimension L in the drawing in the margin. Along the top of the drawing, we have the total horizontal dimension of 86 mm. We can also see that the three lower horizontal dimensions must add up to this total. To find the missing dimension L, we must subtract the other two given dimensions from the total. Therefore, the missing dimension is

$$L = 86 \text{ mm} - 32 \text{ mm} - 25 \text{ mm}$$

Estimate: $90 - 30 - 30 = 30$

Step 1

  8 6
−3 2
‾‾‾‾‾
  5 4

Step 2

    4 14
  $\cancel{5}$ $\cancel{4}$
−2 5
‾‾‾‾‾
  2 9

The missing dimension L is 29 mm. This is very close to our estimate of 30 mm. ●

Your Turn   **Welding**   Three pieces measuring 14 in., 22 in., and 26 in. are cut from a steel bar 72 in. long. Allowing for a total of 1 in. for waste in cutting, what is the length of the piece remaining?

Solution   The key words that indicate subtraction are *cut*, *waste*, and *remaining*. To find how much remains, we must subtract the amounts cut and the amount wasted from the original length. The remaining length in inches will be

$$72 - 14 - 22 - 26 - 1$$

But instead of performing four subtraction problems, it might be easier to add all the pieces cut and wasted, and then subtract this total from the original length.

Step 1

  14
  22
  26
+ 1
‾‾‾‾‾‾
63 in. cut and wasted

Step 2

  72 ← Original length
−63
‾‾‾‾‾
   9 ← The remaining length is 9 inches.

Practical problems are not always simple and straightforward. They will often involve several steps and more than one operation, as shown in the next example. ●

EXAMPLE 6   **Life Skills**   When consumers purchase new automobiles these days, they are often faced with the decision of whether to buy the hybrid version of a particular car. The hybrid version usually costs more but saves money over time because of lower fuel costs.

Suppose that a potential car buyer is deciding between a Toyota Camry and a Camry Hybrid. He estimates that he will keep the car at least 5 years and then sell it. He wants to know if the hybrid model will save him money and, if so, how much. The dealer provides him with the following information:

	Toyota Camry LE	Camry Hybrid LE
Sales Price (MSRP)	$23,935	$27,655
5-Year Fuel Cost*	$ 5550	$ 3570
Insurance and Fees	$ 8200	$ 8350
Maintenance and Repairs	$ 4700	$ 4400
5-Year Resale Value	$ 8100	$ 8000

* *Note:* These costs are based on 12,000 annual miles of driving, mostly in the city, and gas costing $2.50 per gallon. (Source: KBB.com)

We can use these figures to get a rough estimate of the total **5-year cost** of each model.

First, find the sum of the expenditures: The sales price, fuel cost, insurance and fees, and maintenance and repairs.

Toyota Camry LE	Camry Hybrid LE
$23935	$27655
5550	3570
8200	8350
+ 4700	+ 4400
$42385	$43975

Then, subtract the 5-year resale values from these results.

Toyota Camry LE	Camry Hybrid LE
$42385	$43975
− 8100	− 8000
$34285	$35975

These figures represent rough estimates of the total 5-year cost of each vehicle. They show that if the buyer keeps the car he chooses for at least 5 years and then sells it, choosing the non-hybrid Camry LE will save him money.

Finally, find the difference between these two costs to find the total amount of savings.

Camry Hybrid LE: $35975
Toyota Camry LE: −   34285
                              $ 1690

Overall, the non-hybrid Camry LE will save the car buyer $1690 over 5 years based on initial price, fuel costs, insurance and fees, maintenance and repairs, and resale values. It should be noted that if the buyer decides to keep the car longer, the continued fuel savings of the hybrid might end up making that model a better buy.

Notice that this problem required two different operations and several steps to solve. Both addition and subtraction were used to calculate the 5-year cost of each model, and then subtraction was used to determine how much money was saved. ●

Now check your progress on subtraction in Exercises 1-2.

| Exercises 1-2 | **Subtraction of Whole Numbers** |

A. Subtract.

1. 42	2. 25	3. 34	4. 76	5. 64	6. 68
7	8	9	7	31	10

7. 75	8. 96	9. 40	10. 78	11. 51	12. 36
13	22	27	49	39	17

13. 42	14. 52	15. 65	16. 46	17. 84	18. 70
27	16	27	17	38	48

19. 34	20. 56	21. 546	22. 409	23. 476	24. 330
9	18	357	324	195	76

25. 504	26. 747	27. 400	28. 803	29. 632	30. 438
96	593	127	88	58	409

31. 6218	32. 6084	33. 13042	34. 57022
3409	386	524	980

35. 5007	36. 10000	37. 48093	38. 27004
266	386	500	4582

B. Practical Applications

1. **Painting** In planning for a particular job, a painter buys $486 worth of materials. When the job is completed, she returns some unused rollers and brushes for a credit of $27. What was the net amount of her bill?

2. **Construction** How many square feet (sq ft) of plywood remain from an original supply of 8000 sq ft after 5647 sq ft is used?

3. **Welding** A storage rack at the Tiger Tool Company contains 3540 ft of 1-in. stock. On a certain job 1782 ft is used. How much is left?

4. **Welding** Five pieces measuring 26, 47, 38, 27, and 32 cm are cut from a steel bar that was 200 cm long. Allowing for a total of 1 cm for waste in cutting, what is the length of the piece remaining?

5. **Trades Management** Taxes on a group of factory buildings owned by the Ace Manufacturing Company amounted to $875,977 eight years ago. Taxes on the same buildings last year amounted to $1,206,512. Find the increase in taxes.

6. **Trades Management** To pay their bills, the owners of Edwards Plumbing Company made the following withdrawals from their bank account: $72, $375, $84, $617, and $18. If the original balance was $5820, what was the amount of the new balance?

7. **Manufacturing** Which total volume is greater, four drums containing 72, 45, 39, and 86 liters, or three drums containing 97, 115, and 74 liters? By how much is it greater?

8. **Machine Trades** Determine the missing dimension (L) in the following drawings. In the figures, feet are abbreviated with the ($'$) symbol, and inches are abbreviated with the ($''$) symbol.

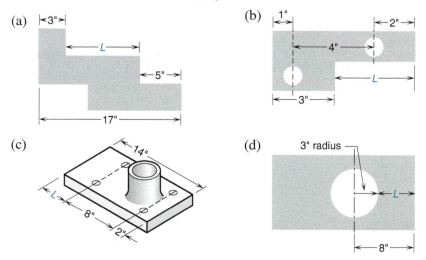

(a)   (b)   (c)   (d)

9. **Automotive Trades** A service department began the day with 238 gallons of coolant. During the day 64 gallons were used. How many gallons remained at the end of the day?

10. **Construction** A truck loaded with rocks weighs 14,260 lb. If the truck weighs 8420 lb, how much do the rocks weigh?

11. **Printing** A press operator has a total of 22,000 impressions to run for a job. If the operator runs 14,250 the first day, how many are left to run?

12. **Plumbing** In the following plumbing diagram, find pipe lengths A and B.

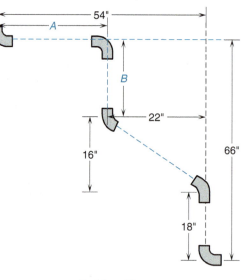

Problem 12

13. **Electronics** A *potentiometer* is a device that acts as a variable resistor, allowing a resistance to change from 0 ohms to some maximum value. If a 20,000-ohm potentiometer is set to 6500 ohms (Ω) as shown in the figure, calculate the resistance R. (*Hint: $R + 6500 = 20,000$.*)

14. **Allied Health** The white blood cell (WBC) count is an important indicator of health. Before surgery, the WBC count of a patient was 9472. After surgery, his WBC count had dropped to 5786. Calculate the difference in WBC count for this patient.

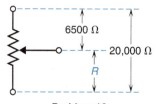

Problem 13

15. **Electronics** An electronic mixer can produce a signal with frequency equal to the difference between two input signals. If the input signals have frequencies of 1,350,000 and 850,000 hertz, calculate the frequency difference.

16. **Automotive Trades** A set of tires was rated at 40,000 miles. A car's odometer read 53,216 when the tires were installed and 91,625 when they needed replacing.

    (a) How long did they actually last?

    (b) By how many miles did they miss the advertised rating?

17. **Office Services** A gas meter read 8701 at the beginning of the month and 8823 at the end of the month. Find the difference between these readings to calculate the number of HCF (hundred cubic feet) used.

18. **General Interest** The tallest building in the United States is One World Trade Center in New York City at 1776 feet. The tallest completed building in the world is the Burj Khalifa in Dubai at 2722 feet. How much taller is the Burj Khalifa than One World Trade Center?

19. **Life Skills** A potential new-car buyer is trying to decide between a Hyundai Sonata and a Sonata Hybrid. Use the following figures for the Limited models to calculate her total 5-year cost for each model. Then determine which model will cost her less and by how much. (See Example 6 on page 26.)

	Sonata	Sonata Hybrid
Sales Price (MSRP)	$28,185	$30,935
5-Year Fuel Cost	$ 5444	$ 3953
Insurance and Fees	$ 8166	$ 9101
Maintenance and Repairs	$ 4716	$ 4168
5-Year Resale Value	$ 9419	$ 9107

(Source: KBB.com)

C. Calculator Problems

1. **Plumbing** The Karroll Plumbing Co. has 10 trucks and, for the month of April, the following mileage was recorded on each:

Truck No.	Mileage at Start	Mileage at End
1	58352	60027
2	42135	43302
3	76270	78007
4	40006	41322
5	08642	10002
6	35401	35700
7	79002	80101
8	39987	40122
9	10210	11671
10	71040	73121

Find the mileage traveled by each truck during the month of April and the total mileage of all vehicles.

2. Which sum is greater?

987654321		123456789
87654321		123456780
7654321		123456700
654321		123456000
54321	or	123450000
4321		123400000
321		123000000
21		120000000
1		100000000

3. **Trades Management**   If an electrician's helper earns $28,245 per year and she pays $3814 in withholding taxes, what is her take-home pay?

4. **Trades Management**   The revenue of the Smith Construction Company for the year is $3,837,672, and the total expenses are $3,420,867. Find the difference, Smith's profit, for that year.

5. **Life Skills**   Balance the following checking account record:

Date	Deposits	Withdrawals	Balance
7/1			$6375
7/3		$ 379	
7/4	$1683		
7/7	$ 474		
7/10	$ 487		
7/11		$2373	
7/15		$1990	
7/18		$ 308	
7/22		$1090	
7/26		$ 814	
8/1			A

(a) Find the new balance A.

(b) Keep a running balance by filling each blank in the balance column.

6. **Construction**   A water meter installed by the BetterBilt Construction Company at a work site read 9357 cubic feet on June 1 and 17,824 cubic feet on July 1. How much water was used at the site during the month of June?

When you have completed these exercises, check your answers to the odd-numbered problems in the Appendix, then turn to Section 1-3 to study the multiplication of whole numbers.

1-3   Multiplication of Whole Numbers

learning|**catalytics**™

1. Subtract: 572 − 286
2. Multiply: 6 × 9

In a certain football game, the West Newton Waterbugs scored five touchdowns at six points each. How many total points did they score through touchdowns? We can answer the question several ways:

1. Count points, ...

2. Add touchdowns, $6 + 6 + 6 + 6 + 6 = ?$

or

3. Multiply $5 \times 6 = ?$

We're not sure about the mathematical ability of the West Newton scorekeeper, but most people would multiply. Multiplication is a shortcut method of performing repeated addition.

How many points did they score?

Product

$$5 \times 6 = 30$$

Factors

 Note   In a multiplication problem the **product** is the name given to the result of the multiplication. The numbers being multiplied are the **factors** of the product. ●

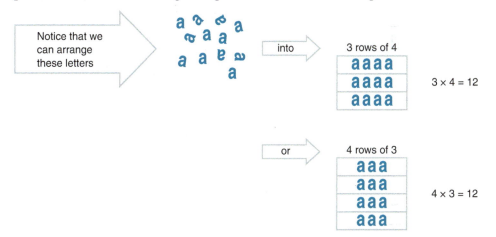

Notice that we can arrange these letters   into   3 rows of 4

aaaa
aaaa $3 \times 4 = 12$
aaaa

or   4 rows of 3

aaa
aaa $4 \times 3 = 12$
aaa
aaa

Changing the order of the factors does not change their product. This is called the **commutative property** of multiplication.

To become skillful at multiplication, you must know the basic one-digit multiplication facts from memory. Even if you use a calculator for your work, you need to know these products to make estimates and to check your work.

PROBLEMS   One-Digit Multiplication

Multiply as shown. Work quickly; you should be able to answer all problems in a set correctly in the time indicated.

A. Multiply.

6	4	9	6	3	9	7	8	2
2	8	7	6	4	2	8	3	7

6	8	5	5	2	3	9	7	3
8	2	9	6	5	3	8	5	6

7	5	4	7	4	8	6	9	8
4	3	9	7	2	5	7	6	8

5	3	5	9	9	2	7	4	4
4	2	5	3	9	2	3	6	4

Mastery time = 90 sec

B. Multiply.

2	6	3	5	6	4	8	2	7
8	5	3	7	3	7	6	6	9

8	4	2	3	5	6	9	5	8
4	5	9	8	5	4	5	2	9

3	7	5	6	9	2	7	8	2
5	7	8	9	4	4	6	8	2

5	3	2	8	6	4	9	7	3
5	9	3	7	6	3	9	2	7

Mastery time = 90 sec

Check your answers in the Appendix.

If you are not able to perform these one-digit multiplications quickly from memory, you should practice until you can do so. A multiplication table is given in the Appendix on page 883. Use it if you need it.

Multiplying by Zero and One   We omitted multiplication by zero and one from the basic facts drill because these products are very simple. The product of any number and zero is zero. For example,

$$0 \times 2 = 0$$

$$0 \times 7 = 0$$

$$395 \times 0 = 0$$

The product of any number and 1 is that same number. For example,

$$1 \times 2 = 2$$

$$6 \times 1 = 6$$

or even

$$1 \times 753 = 753$$

The Sexy Six

Here are the six most often missed one-digit multiplications:

Inside
digits

$9 \times 8 = 72$
$9 \times 7 = 63$ It may help you to notice that in these multiplications
$9 \times 6 = 54$ the "inside" digits, such as 8 and 7, are consecutive and
$8 \times 7 = 56$ the digits of the answer add to nine: $7 + 2 = 9$. This is
$8 \times 6 = 48$ true for *all* one-digit numbers multiplied by 9.
$7 \times 6 = 42$

Be certain that you have these memorized.

(There is nothing very sexy about them, but we did get your attention, didn't we?)

Multiplying by Larger Numbers   The multiplication of larger numbers is based on the one-digit number multiplication facts.

EXAMPLE 1   Consider the problem:

$34 \times 2 =$ _____

First, estimate the answer: $30 \times 2 = 60$. The actual product of the multiplication will be about 60.

Second, arrange the factors to be multiplied vertically, with ones digits in a single column, tens digits in a second column, and so on.

$$\begin{array}{cc} 3 & 4 \\ \times & 2 \end{array}$$

Finally, to make the process clear, let's write it in expanded form.

$$\begin{array}{c} 34 \\ \times\, 2 \end{array} \Rightarrow \begin{array}{c} 3\text{ tens } + 4\text{ ones} \\ \underline{\times\ 2} \\ 6\text{ tens } + 8\text{ ones } = 60 + 8 = 68 \end{array}$$

 The estimate 60 is roughly equal to the answer 68.

EXAMPLE 2   Here is a slightly more difficult problem. To find the product

$$\begin{array}{c} 28 \\ \times\, 3 \end{array}$$

First, estimate the answer: $30 \times 3 = 90$ The answer is about 90.

Then, proceed as follows:

$$\begin{array}{c} 28 \\ \times\, 3 \end{array} \Rightarrow \begin{array}{c} 2\text{ tens } + 8\text{ ones} \\ \underline{\times\ 3} \\ 6\text{ tens } + 24\text{ ones} \end{array}$$

$$\begin{aligned} &= 6\text{ tens } + 2\text{ tens } + 4\text{ ones} \\ &= 8\text{ tens } + 4\text{ ones} \\ &= 80 + 4 \\ &= 84 \end{aligned}$$

✔ 90 is roughly equal to 84.

Of course, we do not normally use the expanded form; instead we simplify the work like this:

$$\begin{array}{cc} \overset{2}{2} & 8 \\ \times & 3 \\ \hline 8 & 4 \end{array}$$
$3 \times 8 = 24$ Write 4 and carry 2 tens.
3×2 tens $= 6$ tens   6 tens $+ 2$ tens $= 8$ tens
Write 8.

Your Turn   Now try these problems to be certain you understand the process. Multiply as shown.

(a) $\begin{array}{c} 43 \\ \times\, 5 \end{array}$ (b) $\begin{array}{c} 29 \\ \times\, 6 \end{array}$ (c) $\begin{array}{c} 258 \\ \times\ \ 7 \end{array}$

Solutions   (a) **Estimate:** $40 \times 5 = 200$ The answer is roughly 200.

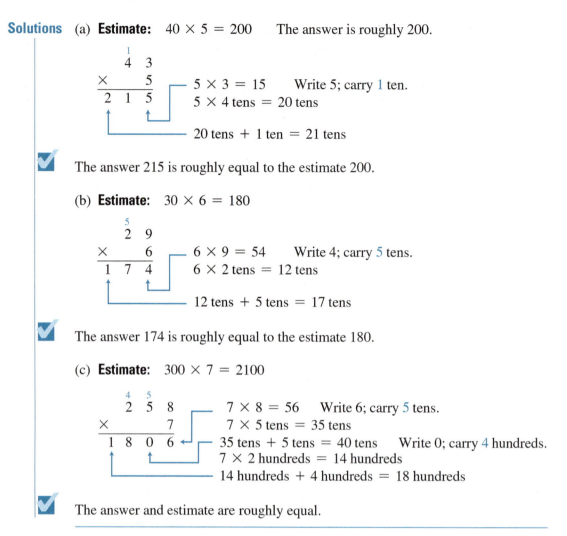

$$\begin{array}{r} \overset{1}{} \\ 4\ 3 \\ \times \quad 5 \\ \hline 2\ 1\ 5 \end{array}$$

$5 \times 3 = 15$ Write 5; carry 1 ten.
5×4 tens $= 20$ tens

20 tens $+ 1$ ten $= 21$ tens

✓ The answer 215 is roughly equal to the estimate 200.

(b) **Estimate:** $30 \times 6 = 180$

$$\begin{array}{r} \overset{5}{} \\ 2\ 9 \\ \times \quad 6 \\ \hline 1\ 7\ 4 \end{array}$$

$6 \times 9 = 54$ Write 4; carry 5 tens.
6×2 tens $= 12$ tens

12 tens $+ 5$ tens $= 17$ tens

✓ The answer 174 is roughly equal to the estimate 180.

(c) **Estimate:** $300 \times 7 = 2100$

$$\begin{array}{r} \overset{4}{}\ \overset{5}{} \\ 2\ 5\ 8 \\ \times \quad\ \ 7 \\ \hline 1\ 8\ 0\ 6 \end{array}$$

$7 \times 8 = 56$ Write 6; carry 5 tens.
7×5 tens $= 35$ tens
35 tens $+ 5$ tens $= 40$ tens   Write 0; carry 4 hundreds.
7×2 hundreds $= 14$ hundreds
14 hundreds $+ 4$ hundreds $= 18$ hundreds

✓ The answer and estimate are roughly equal.

Learning Help   Multiplying a two- or three-digit number by a one-digit number can be done mentally. For example, think of 43×5 as 40×5 plus 3×5, or $200 + 15 = 215$. For a problem such as 29×6, think 30×6 minus 1×6 or $180 - 6 = 174$. Tricks like this are very useful on the job when neither paper and pencil nor a calculator is at hand. ●

Two- and Three-Digit Multiplications   Calculations involving two- and three-digit multipliers are done in a similar way.

EXAMPLE 3   To multiply

$$\begin{array}{r} 89 \\ \times\ 24 \end{array}$$

First, estimate the answer: $90 \times 20 = 1800$

Second, multiply by the ones digit 4.

$$\begin{array}{r} \overset{3}{} \\ 8\ 9 \\ \times\ 2\ 4 \\ \hline 3\ 5\ 6 \end{array}$$

$4 \times 9 = 36$ Write 6; carry 3 tens.
4×8 tens $= 32$ tens
32 tens $+ 3$ tens $= 35$ tens

Third, multiply by the tens digit 2.

$$\begin{array}{r} {\scriptstyle 1} \\ {\scriptstyle 3} \\ 89 \\ \times\,24 \\ \hline 356 \\ 178 \\ \hline \end{array}$$

$2 \times 9 = 18$ Write 8; carry 1.
$2 \times 8 = 16$
$16 + 1 = 17$

— Leave a blank space here, because we are actually multiplying $89 \times 20 = 1780$.

Fourth, add the products obtained.

$$\begin{array}{r} 3\,5\,6 \\ 1\,7\,8 \\ \hline 2\,1\,3\,6 \end{array}$$

Finally, check it. The estimate and the answer are roughly the same, or at least in the same ballpark.

Notice that the product in the third step, 178, is written one digit space over from the product from the second step. When we multiplied 2×9 to get 18 in Step 3, we were actually multiplying $20 \times 9 = 180$, but the zero in 180 is usually omitted to save time. ●

Your Turn   Try these:

(a) $\begin{array}{r} 64 \\ \times\,37 \\ \hline \end{array}$ (b) $\begin{array}{r} 327 \\ \times\;45 \\ \hline \end{array}$

Solutions   (a) **Estimate:** $60 \times 40 = 2400$

$$\begin{array}{r} 64 \\ \times\,37 \\ \hline 448 \\ 192 \\ \hline 2368 \end{array}$$

$7 \times 4 = 28$ Write 8; carry 2.
$7 \times 6 = 42$ Add carry 2 to get 44; write 44.
$3 \times 4 = 12$ Write 2; carry 1.
$3 \times 6 = 18$ Add carry 1 to get 19; write 19.
— Add to obtain the answer.

(b) **Estimate:** $300 \times 50 = 15,000$

$$\begin{array}{r} 327 \\ \times\;45 \\ \hline 1635 \\ 1308 \\ \hline 14715 \end{array}$$

$5 \times 7 = 35$ Write 5; carry 3.
$5 \times 2 = 10$ Add carry 3 to get 13; write 3, carry 1.
$5 \times 3 = 15$ Add carry 1 to get 16; write 16.
$4 \times 7 = 28$ Write 8, carry 2.
$4 \times 2 = 8$ Add carry 2 to get 10; write 0, carry 1.
$4 \times 3 = 12$ Add carry 1 to get 13; write 13.

The product is 14,715.

Comparing Example 2 to Example 3, notice that adding a second digit to the bottom factor creates a second row in the solution. In the next example, we should see three rows in the solution in addition to the answer row.

EXAMPLE 4   To multiply

$$\begin{array}{r} 527 \\ \times\;231 \\ \hline \end{array}$$

First, estimate the answer. $500 \times 200 = 100,000$

Next, multiply 527 by each digit in 231, from right to left, creating a separate row for each product. Be sure to indent each row one space to the left of the row above it.

```
    527
  × 231
    527  ←——— 527 × 1
   1581  ←——— 527 × 3
  1054   ←——— 527 × 2
 121737
```

The product is 121,737. This agrees roughly with our estimate. ●

Your Turn   Multiply as indicated.

(a)   742   (b)   56   (c)   198   (d)   691   (e)   344
      × 8   × 17   × 45   × 382   × 207

Answers   (a) 5936   (b) 952   (c) 8910   (d) 263,962   (e) 71,208

A Closer Look   Did you have trouble with the zero in problem (e)? The zero simply creates a row of zeros in your solution. Your work should look like this:

```
    344
  × 207
   2408
    000  ←——— 0 × 4 = 0, 0 × 4 = 0, and 0 × 3 = 0
   688
  71208
```
●

Multiplication Shortcuts

There are hundreds of quick ways to multiply various numbers. Most of them are quick only if you are already a math whiz. If you are not, the shortcuts will confuse more than help you. Here are a few that are easy to do and easy to remember.

1. To multiply by 10, attach a zero to the right end of the first factor. For example,

    $34 \times 10 = 340$

    $256 \times 10 = 2560$

    Multiplying by 100 or 1000 is similar.

    $34 \times 100 = 3400$

    $256 \times 1000 = 256000$

2. To multiply by a number ending in zeros, carry the zeros forward to the answer. For example,

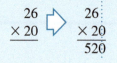

```
    26          26
  × 20        × 20
               520
```
Multiply 26 × 2 and attach the zero on the right. The product is 520.

(continued)

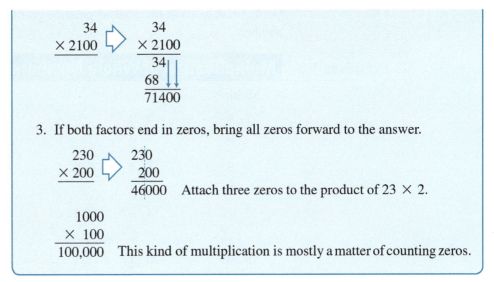

3. If both factors end in zeros, bring all zeros forward to the answer.

$$
\begin{array}{r}
230 \\
\times\ 200 \\
\end{array}
\quad\Longrightarrow\quad
\begin{array}{r}
230 \\
200 \\
\hline
46000 \\
\end{array}
\quad\text{Attach three zeros to the product of } 23 \times 2.
$$

$$
\begin{array}{r}
1000 \\
\times\ 100 \\
\hline
100{,}000 \\
\end{array}
\quad\text{This kind of multiplication is mostly a matter of counting zeros.}
$$

Practical Applications   When we are given a number of items that all have the same numerical value, we use multiplication to find a total.

EXAMPLE 5   (a) **Masonry**   If a mason works for 40 hours and earns $18 per hour, we multiply the number of hours by the hourly wage to find the total amount earned:

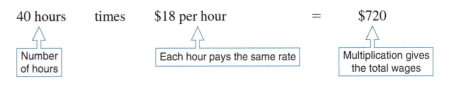

40 hours   times $18 per hour   = $720

Number of hours — Each hour pays the same rate — Multiplication gives the total wages

(b) **Electrical Trades**   If an electrician needs 12 pieces of wire that are each 18 inches long, we multiply to find the total length of wire required:

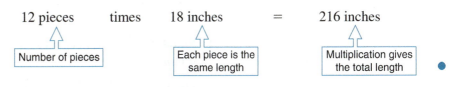

12 pieces   times   18 inches   =   216 inches

Number of pieces — Each piece is the same length — Multiplication gives the total length   ●

Your Turn   **Allied Health**   To maintain his weight, a man who exercises moderately needs to consume about 16 calories per day for every pound that he weighs. Suppose someone who exercises moderately weighs 175 pounds. How many calories per day should he consume in order to maintain his weight?

Solution   We know the number of pounds and the number of calories needed per pound, so we must multiply to find the total number of calories he should consume per day:

$$
\begin{array}{r}
175 \\
\times\ 16 \\
\hline
1050 \\
175 \\
\hline
2800 \\
\end{array}
\quad\text{calories per day}
$$

Go to Exercises 1-3 for a set of practice problems on the multiplication of whole numbers.

Exercises 1-3	**Multiplication of Whole Numbers**

A. Multiply.

1. 7
    6
	2. 7
	    8
		3. 6
		    8
			4. 8
			    9
				5. 9
				    7
					6. 7
					    7

7. 9
    6
	8. 4
	    7
		9. 5
		    9
			10. 12
			     7
				11. 37
				     8
					12. 24
					     7

13. 72
     8
	14. 47
	     9
		15. 64
		     5
			16. 39
			     4
				17. 58
				     5
					18. 94
					     6

19. 47
     6
	20. 77
	     4
		21. 32
		    13
			22. 46
			    14
				23. 72
				    11
					24. 68
					    16

25. 54
    26
	26. 17
	     19
		27. 48
		    15
			28. 64
			    27
				29. 90
				    56
					30. 86
					    83

31. 34
    57
	32. 66
	    25
		33. 59
		    76
			34. 29
			    32
				35. 78
				    49
					36. 94
					    95

B. Multiply.

1. 809
     9
	2. 609
	     7
		3. 8043
		      37
			4. 809
			    47
				5. 500
				    50

6. 407
    22
	7. 316
	    32
		8. 514
		    62
			9. 684
			    45
				10. 708
				     58

11. 305
    123
	12. 2006
	     125
		13. 807
		    111
			14. 542
			    600
				15. 7009
				     504

16. 563
    107
	17. 3706
	     102
		18. 560
		    203
			19. 2008
			     198
				20. 2043
				     670

C. Practical Applications

1. **Plumbing**   A plumber receives $75 per hour. How much is she paid for 40 hours of work?

2. **Electrical Trades**   What is the total length of wire on 14 spools if each spool contains 150 ft?

3. **Construction**   How many total linear feet of redwood are there in 65 2-in. by 4-in. boards each 20 ft long?

4. **Automotive Trades**   An auto body shop does 17 paint jobs at a special price of $859 each. How much money does the shop receive from these jobs?

5. **Office Services**   Three different-sized boxes of envelopes contain 50, 100, and 500 envelopes, respectively. How many envelopes total are there in 18 boxes of the first size, 16 of the second size, and 11 of the third size?

6. **Machine Trades**   A machinist needs 25 lengths of steel each 9 in. long. What is the total length of steel that he needs? No allowance is required for cutting.

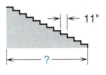

Problem 8

7. **Machine Trades** The Ace Machine Company advertises that one of its machinists can produce 2 parts per hour. How many such parts can 27 machinists produce if they work 45 hours each?

8. **Construction** What is the horizontal distance in inches covered by 12 stair steps if each is 11 in. wide? (See the figure.)

9. **Automotive Trades** Five bolts are needed to mount a tire. How many bolts are needed to mount all four tires on 60 cars?

10. **Electrical Trades** A voltmeter has a 4000 ohms per volt rating. This means that, for example, on the 3-volt scale, the meter reads 3 × 4000, or 12,000 ohms resistance. Find the resistance on the 50-volt range.

11. **Printing** A printer finds that nine 6-in. by 8-in. cards can be cut from a 19-in. by 25-in. sheet. How many such cards can be cut from 850 of the sheets?

12. **Industrial Technology** The Hold Tite fastener plant produces 738,000 cotter pins per shift. If it operates two shifts per day, five days each week, how many cotter pins will the plant produce in six weeks?

13. **Machine Trades** If a machine produces 16 screws per minute, how many screws will it produce in 24 hours?

14. **Electronics** For a transistor amplifier circuit, the collector current is calculated by multiplying the gain by the base current. For a circuit with a gain of 72, and base current 210 microamps, calculate the collector current.

15. **Roofing** In the Happy Homes development, cedar shingles are laid to expose 5 in. per course. If 23 courses of shingles are needed to cover a surface, how long is the surface?

16. **Electronics** A 78-ohm resistor R draws 3 amps of current i in a given circuit. Calculate the voltage drop E across the resistor by multiplying resistance by current.

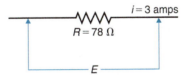

17. **Automotive Trades** The maximum resistance for a new spark plug wire is 850 ohms per inch. A 25-in. wire had a resistance of 21,500 ohms. Does this fall within the acceptable limit?

18. **Automotive Trades** According to the EPA rating, the average gas mileage for a certain car is expected to be 24 miles per gallon. How many total miles can be expected from a 16-gallon tank?

19. **Agriculture** The owners of KneeHi Farms expect a yield of 170 bushels of corn per acre. How much corn should they plan for if they plant 220 acres?

20. **HVAC** Home ventilation standards require a controlled ventilation rate of 15 cubic feet per minute per person. If a house is occupied by four people, what total ventilation rate is required?

21. **Electrical Trades** If a solar energy system consists of 96 solar panels, and each panel puts out 5 A (amperes) of current, how many total amperes does the system generate?

22. **Allied Health**   Pulse rate is generally expressed as beats per minute (bpm). To save time, health practitioners often make a quick check of pulse rate by counting the number of beats in 15 seconds, and then multiplying this number by 4 to determine beats per minute. If a nurse counts 23 pulse beats in 15 seconds, what is the patient's pulse rate in bpm?

23. **Allied Health**   During a blood donation drive in Goleta, California, 176 people donated blood. If each person donated the maximum recommended amount of 500 milliliters (mL), how much blood was collected?

24. **Allied Health**   To maintain her weight, a woman who exercises vigorously needs to consume about 18 calories per day for every pound that she weighs. If a woman weighs 155 pounds and exercises vigorously, how many calories per day should she consume in order to maintain her weight?

25. **Life Skills**   If a certain job pays $16 per hour, how much would it pay annually at 40 hours per week for 52 weeks per year?

26. **Retail Merchandising**   In retail management, large quantities are often ordered by the gross. A **gross** is 12 dozen items, and a dozen is 12 items. If a buyer ordered 8 gross pairs of socks, how many pairs of socks did she order?

27. **Water/Wastewater Treatment**   A rural water district is required to set aside enough water for two hours of fire protection at a rate of 250 gallons per minute. How many gallons must it set aside? (*Hint:* Be careful with your time units.)

28. **Allied Health**   At the height of allergy season, there can be at least 50 grains of pollen in 1 cubic foot of air. At this rate, if the average person inhales 16 cubic feet of air per hour, how many grains of pollen would he inhale in 24 hours?

29. **Construction**   Granite stones weigh approximately 170 pounds per cubic foot. How many pounds are needed for a job requiring 800 cubic feet of granite?

D. Calculator Problems

1. **Business and Finance**   If an investment company started in business with $1,000,000 of capital on January 1 and, because of mismanagement, lost an average of $873 every day for a year, what would be its financial situation on the last day of the year?

2. **Life Skills**   Which of the following pay schemes gives you the most money over a one-year period?

    (a) $100 per day

    (b) $700 per week

    (c) $400 for the first month and a $400 raise each month

    (d) 1 cent for the first two-week pay period, 2 cents for the second two-week period, 4 cents for the third two-week period, and so on, the pay doubling each two weeks.

3. Multiply.

    (a) $12,345,679 \times 9 =$
       $12,345,679 \times 18 =$
       $12,345,679 \times 27 =$

    (b) $15,873 \times 7 =$
       $15,873 \times 14 =$
       $15,873 \times 21 =$

(c) $1 \times 1 =$
$11 \times 11 =$
$111 \times 111 =$
$1111 \times 1111 =$
$11111 \times 11111 =$

(d) $6 \times 7 =$
$66 \times 67 =$
$666 \times 667 =$
$6666 \times 6667 =$
$66666 \times 66667 =$

Can you see the pattern in each of these?

4. **Machine Trades**   The Aztec Machine Shop has nine lathes each weighing 2285 lb, five milling machines each weighing 2570 lb, and three drill presses each weighing 395 lb. What is the total weight of these machines?

5. **Manufacturing**   The Omega Calculator Company makes five models of electronic calculators. The following table gives the hourly production output. Find the weekly (five eight-hour days) production costs for each model.

Model	Alpha	Beta	Gamma	Delta	Tau
Cost of production of each model	$6	$17	$32	$49	$78
Number produced during typical hour	117	67	29	37	18

6. **Construction**   The Passive House Standard is a strict building code used as a guideline for building energy-efficient homes. One Passive House guideline states that no more than 4755 Btu (British thermal units) per square foot should be used for heating and cooling annually. What would this annual maximum be for a 2800-square-foot house?

Check your answers to the odd-numbered problems in the Appendix.

1-4   Division of Whole Numbers

learning | catalytics™

1. Multiply: 38
              $\times\ 45$

2. Divide: $21 \div 3$

Division is the reverse of multiplication. It enables us to separate a given quantity into equal parts. The mathematical phrase $12 \div 3$ is read "twelve divided by three," and it asks us to separate a collection of 12 objects into 3 equal parts. The mathematical phrases

$$12 \div 3 \qquad 3\overline{)12} \qquad \frac{12}{3} \qquad \text{and} \qquad 12/3$$

all represent division, and they are all read "twelve divided by three."

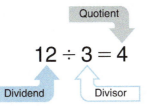

Quotient

$$12 \div 3 = 4$$

Dividend   Divisor

📝 **Note**   In this division problem, 12, the number being divided, is called the **dividend;** 3, the number used to divide, is called the **divisor;** and 4, the result of the division, is called the **quotient,** from a Latin word meaning "how many times." ●

One way to perform division is to reverse the multiplication process.

$24 \div 4 = \square$ means that $4 \times \square = 24$

If the one-digit multiplication facts are firmly in your memory, you will recognize immediately that □ = 6. Test your ability to reverse the one-digit multiplication facts by mentally computing the following basic division problems.

PROBLEMS   Basic Division Facts

Mentally calculate the following quotients. Work as quickly as possible. You should be able to complete the problems in the time indicated at the end of the set.

$12 \div 2 =$	$48 \div 8 =$	$16 \div 4 =$	$42 \div 7 =$	$54 \div 9 =$	$15 \div 5 =$	$24 \div 6 =$	$63 \div 9 =$
$16 \div 8 =$	$42 \div 6 =$	$9 \div 3 =$	$36 \div 6 =$	$35 \div 7 =$	$81 \div 9 =$	$28 \div 7 =$	$24 \div 3 =$
$48 \div 6 =$	$4 \div 2 =$	$8 \div 4 =$	$20 \div 5 =$	$54 \div 6 =$	$16 \div 2 =$	$30 \div 6 =$	$45 \div 9 =$
$8 \div 2 =$	$25 \div 5 =$	$24 \div 4 =$	$32 \div 8 =$	$6 \div 3 =$	$18 \div 9 =$	$6 \div 2 =$	$27 \div 3 =$
$30 \div 5 =$	$28 \div 4 =$	$18 \div 3 =$	$10 \div 2 =$	$56 \div 8 =$	$36 \div 9 =$	$21 \div 7 =$	$40 \div 5 =$
$14 \div 7 =$	$12 \div 3 =$	$18 \div 2 =$	$72 \div 8 =$	$63 \div 7 =$	$12 \div 4 =$	$45 \div 5 =$	$15 \div 3 =$
$36 \div 4 =$	$21 \div 3 =$	$24 \div 8 =$	$10 \div 5 =$	$27 \div 9 =$	$40 \div 8 =$	$49 \div 7 =$	$18 \div 6 =$
$20 \div 4 =$	$12 \div 6 =$	$35 \div 5 =$	$72 \div 9 =$	$14 \div 2 =$	$32 \div 4 =$	$56 \div 7 =$	$64 \div 8 =$

Mastery Time: 3 minutes 20 seconds

Check your answers in the Appendix.

Division by One and Zero   Recall that the product of any number and 1 is that same number. The same holds true for division. If you divide any number by 1, the quotient is the original number. For example, $8 \div 1 = 8$, $34 \div 1 = 34$, and so on.

However, division by zero has no meaning in arithmetic. (Any number) $\div$ 0 or $\dfrac{\text{(any number)}}{0}$ is not defined. But $0 \div$ (any number) $= \dfrac{0}{\text{(any number)}} = 0$.

Division with Larger Numbers   How do we divide numbers that are larger than 9×9 and therefore not in the multiplication table? Obviously, we need a better procedure.

EXAMPLE 1   Here is a step-by-step explanation of the division of whole numbers:

Divide $96 \div 8 =$ _____.

First, *estimate* the answer: $100 \div 10 = 10$. The quotient or answer will be about 10.

Second, arrange the numbers in this way:

$8\overline{)96}$

Notice that the order in which the numbers are written is reversed. With the $\div$ symbol, the divisor (8) is on the right. With the $\overline{)}$ symbol, the divisor is on the left.

Third, divide using the following step-by-step procedure:

Step 1 $8\overline{)96}^{\,1}$ 8 into 9? Once. Write 1 in the answer space above the 9.

Step 2 $8\overline{)96}^{\,1}$ Multiply $8 \times 1 = 8$ and write the product 8 under the 9.
             $\underline{8}$

Step 3 $\dfrac{1}{8) \ 96}$
$\dfrac{-8\downarrow}{16}$

Subtract $9 - 8 = 1$ and write 1. Bring down the next digit, 6.

Step 4 $\dfrac{12}{8) \ 96}$
$\dfrac{-8}{16}$

8 into 16? Twice. Write 2 in the answer space above the 6.

Step 5 $\dfrac{12}{8) \ 96}$
$\begin{array}{r}-8\\\hline 16\\-16\end{array}$

Multiply $8 \times 2 = 16$ and write the product 16 under the 16.

Step 6 $\dfrac{12}{8) \ 96}$
$\begin{array}{r}-8\\\hline 16\\-16\\\hline 0\end{array}$ ← The remainder is zero. 8 divides into 96 exactly 12 times.

✔ **Finally,** *check* your answer. The answer 12 is roughly equal to the original estimate 10. As a second check, multiply the divisor 8 and the quotient 12. Their product should be the original dividend number, $8 \times 12 = 96$, which is correct. ●

Your Turn   Practice this step-by-step division process by finding $112 \div 7$.

Solution   **Estimate:** $112 \div 7$ is roughly $100 \div 5$ or 20. The answer will be roughly 20.

$\begin{array}{r}016\\7) \ 112\\-7\downarrow\\\hline 42\\-42\\\hline 0\end{array}$

Step 1   7 into 1. Won't go, so put a zero in the answer place.

Step 2   7 into 11. Once. Write 1 in the answer space as shown.

Step 3 $7 \times 1 = 7$. Write 7 below 11.

Step 4 $11 - 7 = 4$. Write 4. Bring down the 2.

Step 5   7 into 42. Six. Write 6 in the answer space.

Step 6 $7 \times 6 = 42$. Subtract 42. The remainder is zero.

The leading zero in 016 can be ignored, so our quotient is 16.

✔ The answer 16 is roughly equal to our original estimate of 20. Check again by multiplying: $7 \times 16 = 112$. Finding a quick estimate of the answer will help you avoid making mistakes.

❗ **Careful**   Notice that once the first digit of the answer has been obtained, there will be an answer digit for every digit of the dividend.

$\begin{array}{c}16\\\uparrow\uparrow\\7)\overline{112}\end{array}$ ●

The remainder is not always zero, of course. This is shown in the next example.

EXAMPLE 2   To calculate $153 \div 4$,

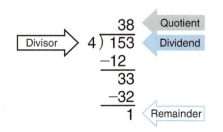

Write the answer as $38r1$ to indicate that the quotient is 38 and the remainder is 1.

☑ To check an answer with a remainder, first multiply the quotient by the divisor, then add the remainder. In this example,

$38 \times 4 = 152$

$152 + 1 = 153$, which checks.   ●

Your Turn   Ready for some guided practice? Try these problems.

(a) $59 \div 8$ (b) $206 \div 6$ (c) $3174 \div 6$

Solutions   (a) $59 \div 8 = $ _____

Estimate: $7 \times 8 = 56$.   The answer will be about 7.

$$\begin{array}{r} 7 \\ 8\overline{)\,59} \\ -56 \\ \hline 3 \end{array}$$

Step 1   8 into 5? No. There is no need to write the zero. 8 into 59? 7 times. Write 7 above the 9.

Step 2 $8 \times 7 = 56$. Subtract $59 - 56 = 3$

The quotient is 7 with a remainder of 3.

☑ Check it by multiplying and adding the remainder.

(b) $206 \div 6 = $ _____

Estimate: $6 \times 30 = 180$ The answer is about 30.

$$\begin{array}{r} 34 \\ 6\overline{)\,206} \\ -18 \\ \hline 26 \\ -24 \\ \hline 2 \end{array}$$

Step 1   6 into 2? No. 6 into 20? Three times. Write 3 above the zero in the answer.

Step 2 $6 \times 3 = 18$. Subtract $20 - 18 = 2$.

Step 3   Bring down 6. 6 into 26? 4 times. Write 4 in the answer.

Step 4 $6 \times 4 = 24$. Subtract $26 - 24 = 2$. The remainder is 2.

The quotient is 34 with a remainder of 2.

☑ The estimate 30 and the answer are roughly equal.

Double-check: $6 \times 34 = 204$ $204 + 2 = 206$.

(c) $3174 \div 6 = $ _____

Estimate: $6 \times 500 = 3000$ The answer will be roughly 500.

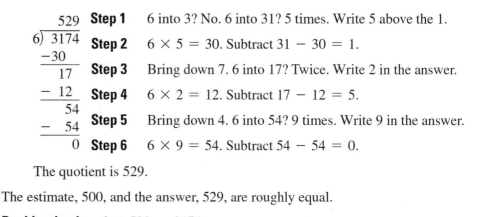

$$\begin{array}{r} 529 \\ 6\overline{)\ 3174} \\ -30 \\ \hline 17 \\ -\ 12 \\ \hline 54 \\ -\ 54 \\ \hline 0 \end{array}$$

Step 1   6 into 3? No. 6 into 31? 5 times. Write 5 above the 1.

Step 2 $6 \times 5 = 30$. Subtract $31 - 30 = 1$.

Step 3   Bring down 7. 6 into 17? Twice. Write 2 in the answer.

Step 4 $6 \times 2 = 12$. Subtract $17 - 12 = 5$.

Step 5   Bring down 4. 6 into 54? 9 times. Write 9 in the answer.

Step 6 $6 \times 9 = 54$. Subtract $54 - 54 = 0$.

The quotient is 529.

The estimate, 500, and the answer, 529, are roughly equal.

Double-check: $6 \times 529 = 3174$.

Now let's try dividing by a two-digit number.

EXAMPLE 3 $5084 \div 31 = \underline{\ \ ?\ \ }$

Estimate: This is roughly about the same as $5000 \div 30$ or $500 \div 3$ or about 200. The quotient will be about 200.

$$\begin{array}{r} 164 \\ 31\overline{)\ 5084} \\ -31 \\ \hline 198 \\ -186 \\ \hline 124 \\ -\ 124 \\ \hline 0 \end{array}$$

Step 1   31 into 5? No. 31 into 50? Yes, once. Write 1 above the zero.

Step 2 $31 \times 1 = 31$. Subtract $50 - 31 = 19$.

Step 3   Bring down 8. 31 into 198? (That is about the same as 3 into 19.) Yes, 6 times; write 6 in the answer.

Step 4 $31 \times 6 = 186$. Subtract $198 - 186 = 12$.

Step 5   Bring down 4. 31 into 124? (That is about the same as 3 into 12.) Yes, 4 times. Write 4 in the answer.

Step 6 $31 \times 4 = 124$. Subtract $124 - 124 = 0$.

The quotient is 164.

The estimate is reasonably close to the answer.

Double-check: $31 \times 164 = 5084$.

Notice that in Step 3, it is not at all obvious how many times 31 will go into 198. Again, you must make an educated guess and check your guess as you go along.   ●

Your Turn   Now try these yourself:

(a) $9465 \div 42$ (b) $56{,}313 \div 217$

Solutions   (a) **Estimate:**   This is roughly $9000 \div 45$, which is 200.

$$\begin{array}{r} 225 \\ 42\overline{)9465} \\ 84 \\ \hline 106 \\ 84 \\ \hline 225 \\ 210 \\ \hline 15 \end{array}$$

Step 1   42 into 9? No. 42 into 94? Yes, twice. Write 2 above the 4.

Step 2 $42 \times 2 = 84$.

Step 3 $94 - 84 = 10$. Bring down the 6. 42 goes into 106 twice. Write 2 above the 6. $2 \times 42 = 84$.

Step 4 $106 - 84 = 22$. Bring down the 5. 42 goes into 225 5 times. Write 5 above the 5 in the quotient. $5 \times 42 = 210$.

Step 5 $225 - 210 = 15$. There are no more numbers to bring down, so this is the remainder.

The quotient is 225 with a remainder of 15, or $225r15$. This is very close to our estimate.

(b) With larger numbers, making estimates becomes more difficult, but it is a good mental math exercise. This is roughly $56{,}000 \div 200$, which equals 280.

$$
\begin{array}{r}
259 \\
217\overline{)56313} \\
434 \\
\hline
1291 \\
1085 \\
\hline
2063 \\
1953 \\
\hline
110
\end{array}
$$

Step 1   217 goes into 563 twice. Write 2 above the 3.

Step 2 $2 \times 217 = 434$.

Step 3 $563 - 434 = 129$. Bring down the 1.

Step 4   217 goes into 1291 5 times. Write 5 above the 1.

Step 5 $5 \times 217 = 1085$. $1291 - 1085 = 206$. Bring down the 3.

Step 6   217 goes into 2063 9 times. Write 9 above the 3. $9 \times 217 = 1953$.

Step 7 $2063 - 1953 = 110$. This is the remainder.

The quotient is 259 with a remainder of 110, or $259r110$. This roughly matches our estimate.

Factors   It is sometimes useful in mathematics to be able to write any whole number as a product of other numbers. If we write

$6 = 2 \times 3$ 2 and 3 are called **factors** of 6.

Of course, we could also write

$6 = 1 \times 6$ and see that 1 and 6 are also factors of 6.

The factors of 6 are 1, 2, 3, and 6.

The factors of 12 are 1, 2, 3, 4, 6, and 12.

The factors of 30 are 1, 2, 3, 5, 6, 10, 15, and 30.

Any number is exactly divisible by its factors; that is, every factor divides the number with zero remainder.

Your Turn   List all the factors of each of the following whole numbers.

1. 4   2. 10   3. 24   4. 18   5. 14

6. 32   7. 20   8. 26   9. 44   10. 90

Answers   1.   1, 2, 4   2.   1, 2, 5, 10   3.   1, 2, 3, 4, 6, 8, 12, 24

4.   1, 2, 3, 6, 9, 18   5.   1, 2, 7, 14   6.   1, 2, 4, 8, 16, 32

7.   1, 2, 4, 5, 10, 20   8.   1, 2, 13, 26   9.   1, 2, 4, 11, 22, 44

10.   1, 2, 3, 5, 6, 9, 10, 15, 18, 30, 45, 90

Primes   For some numbers the only factors are 1 and the number itself. For example, the factors of 7 are 1 and 7 because $7 = 1 \times 7$. There are no other numbers that divide 7 with remainder zero. Such numbers are known as prime numbers. A **prime number** is one for which there are no factors other than 1 and the prime itself.

Here is a list of the first 15 prime numbers:

2, 3, 5, 7, 11, 13, 17, 19, 23, 29, 31, 37, 41, 43, 47

Notice that 1 is not listed. All prime numbers have two different factors: 1 and the number itself. The number 1 has only 1 factor—itself.

Prime Factors

The **prime factors** of a number are those factors that are prime numbers. For example, the prime factors of 6 are 2 and 3. The prime factors of 30 are 2, 3, and 5. You will see later that the concept of prime factors is useful when working with fractions.

To find the prime factors of a number a **factor tree** is often helpful.

EXAMPLE 4

Find the prime factors of 132.

First, write the number 132 and draw two branches below it.

Second, beginning with the smallest prime in our list, 2, test for divisibility. $132 = 2 \times 66$. Write the factors below the branches.

Next, repeat this procedure on the nonprime factor, 66, so that $66 = 2 \times 33$.

Finally, continue dividing until all branches end with a prime.

Therefore, $132 = 2 \times 2 \times 3 \times 11$.

Check by multiplying.

Your Turn

Write each whole number as a product of its prime factors. If the given number is a prime, label it with a letter P.

1. 9	2. 11	3. 15	4. 3	5. 17
6. 21	7. 23	8. 29	9. 52	10. 350

Answers

1. 3×3	2. P	3. 3×5	4. P	5. P
6. 3×7	7. P	8. P	9. $2 \times 2 \times 13$	10. $2 \times 5 \times 5 \times 7$

Practical Applications

Whenever we need to separate some numerical quantity into equal parts, we use division. The quantity that we are separating is the dividend. If the divisor is the numerical value of its equal parts, then the quotient will tell us how many equal parts there are. If the divisor is the number of equal parts, then the quotient will tell us the value of each equal part.

EXAMPLE 5

Carpentry Suppose a carpenter needs to cut pieces of wood, each 20 inches long, from a 144-inch-long board. To determine how many pieces he can cut, use division:

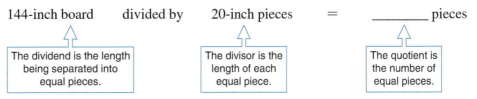

144-inch board   divided by   20-inch pieces   =   _____ pieces

| The dividend is the length being separated into equal pieces. | The divisor is the length of each equal piece. | The quotient is the number of equal pieces. |

Divide:
$$20)\overline{144} \quad \begin{array}{r} 7 \\ \hline \end{array}$$

$$\begin{array}{r} 7 \\ 20\overline{)144} \\ \underline{140} \\ 4 \end{array}$$

The quotient tells us that he can cut 7 of the 20-inch pieces from the board, and the remainder tells us that he will have 4 inches left over. ●

Learning Help 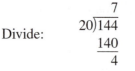 Another clue that indicates division is the word *per*. Whenever you are asked to find the number of "A per B," you will always need to divide the number A by the number B. ●

Your Turn **Electrical Trades**   An apprentice electrician was paid $1400 for 40 hours of work. How much was she paid per hour?

Solution The word *per* in the question "How much was she paid *per* hour" tells us that we must divide the total amount paid ($1400) by the number of hours (40).

$$\begin{array}{r} 35 \\ 40\overline{)1400} \\ \underline{120} \\ 200 \\ \underline{200} \\ 0 \end{array}$$

The apprentice was paid $35 per hour.

When new technologies are developed and improved, people naturally want to use them in their homes and businesses. Usually, they must make an initial investment to purchase the technology. However, these new technologies often provide a savings of ongoing costs over time. Eventually, this savings will "pay back" the initial investment. The term **payback time** is used to describe the amount of time it takes for the savings in ongoing costs to match the initial investment made to purchase the technology. After that, the continued savings will make the new technology a more cost-effective alternative.

EXAMPLE 6 **Electrical Trades**   A family made an initial investment of $14,850 to install solar panels on the roof of their house. Before using solar power, their monthly electricity bills averaged $120. After installing the solar panels, their monthly bills averaged $10. To calculate the payback time for the solar panels,

First, calculate the monthly savings.

$120 per month − $10 per month = $110 per month

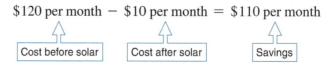

| Cost before solar | Cost after solar | Savings |

Then, divide the monthly savings into the initial investment to determine the payback time in months.

$$\begin{array}{r} 135 \\ 110\overline{)14850} \\ \underline{110} \\ 385 \\ \underline{330} \\ 550 \\ \underline{550} \\ 0 \end{array}$$

The payback time is 135 months. Notice that we used division to separate a total ($14,850) into equal monthly parts whose value ($110) was known. The quotient told us the number of equal parts (135 months).

Finally, we can convert this to years. Simply divide this answer by 12—the number of months in a year.

$$
\begin{array}{r}
11 \\
12\overline{)135} \\
\underline{12} \\
15 \\
\underline{12} \\
3
\end{array}
$$

The quotient, 11, is the number of whole years, and the remainder 3 is the number of additional months. The payback time for the solar panels can thus be expressed as 11 years and 3 months. After that, the family will be saving money on electricity. ●

Your Turn   **Automotive Trades**   Suppose that you have an older car that costs you about $3200 annually in gas alone. You have your eye on a new and more fuel-efficient car that is projected to cost you only $1800 per year in gas. If the new car costs $18,200, what is the payback time based solely on fuel cost?

Solution   **Step 1**   Calculate the yearly savings in gas: $3200 − $1800 = $1400

Step 2   Divide this result into the cost of the new car.

$$
\begin{array}{r}
13 \\
1400\overline{)18200} \\
\underline{1400} \\
4200 \\
\underline{4200} \\
0
\end{array}
$$

The payback time based solely on the cost of fuel is 13 years. This may seem like too long a time to make the purchase worthwhile, but there are other factors to consider. For example, the new car should save you a considerable amount of money in repairs and maintenance, and it will have a higher resale value over time. On the other hand, the new car will cost you more in insurance and annual registration fees. Finally, there are certain things that are priceless, such as the safety improvements of the newer vehicle and the excitement of having a new car!

Exercises 1-4 provide some practice in the division of whole numbers.

Some Division "Tricks of the Trade"

1. If a number is exactly divisible by 2, that is, divisible with zero remainder, it will end in an even digit, 0, 2, 4, 6, or 8.

   Example: We know that 374 is exactly divisible by 2 because it ends in the even digit 4.

   $$
   \begin{array}{r}
   187 \\
   2\overline{)374}
   \end{array}
   $$

(continued)

2. If a number is exactly divisible by 3, the sum of its digits is exactly divisible by 3.

   Example: The number 2784 has the sum of digits $2 + 7 + 8 + 4 = 21$. Because 21 is exactly divisible by 3, we know that 2784 is also exactly divisible by 3.

   $$\begin{array}{r} 928 \\ 3\overline{)2784} \end{array}$$

3. A number is exactly divisible by 4 if its last two digits are exactly divisible by 4 or are both zero.

   Example: We know that the number 3716 is exactly divisible by 4 because 16 is exactly divisible by 4.

   $$\begin{array}{r} 929 \\ 4\overline{)3716} \end{array}$$

4. A number is exactly divisible by 5 if its last digit is either 5 or 0.

   Example: The numbers 875, 310 and 33,195 are all exactly divisible by 5.

5. A number is exactly divisible by 8 if its last three digits are divisible by 8.

   Example: The number 35,120 is exactly divisible by 8 because the number 120 is exactly divisible by 8.

6. A number is exactly divisible by 9 if the sum of its digits is exactly divisible by 9.

   Example: The number 434,673 has the sum of digits 27. Because 27 is exactly divisible by 9, we know that 434,673 is also exactly divisible by 9.

   $$\begin{array}{r} 48297 \\ 9\overline{)434673} \end{array}$$

7. A number is exactly divisible by 10 if it ends with a zero.

   Example: The numbers 230, 7380, and 100,200 are all exactly divisible by 10.

Can you combine the tests for divisibility by 2 and 3 to get a rule for divisibility by 6?

| **Exercises 1-4** | **Division of Whole Numbers** |

A. Divide.

1. $63 \div 7$	2. $6 \div 6$	3. $32 \div 4$
4. $28 \div 7$	5. $54 \div 9$	6. $92 \div 8$
7. $\dfrac{72}{5}$	8. $\dfrac{37}{5}$	9. $\dfrac{71}{7}$
10. $245 \div 7$	11. $167 \div 7$	12. $228 \div 4$
13. $310 \div 6$	14. $3310 \div 3$	15. $\dfrac{1476}{7}$

16. $7\overline{)364}$ 17. $6\overline{)222}$ 18. $4\overline{)2018}$

19. $9\overline{)2000}$ 20. $7\overline{)1000}$ 21. $7\overline{)3507}$

22. $3\overline{)9003}$ 23. $6\overline{)3624}$ 24. $6\overline{)48009}$

B. Divide.

1. $322 \div 14$ 2. $382 \div 19$ 3. $936 \div 24$

4. $700 \div 28$ 5. $730 \div 81$ 6. $\dfrac{901}{17}$

7. $31\overline{)682}$ 8. $27\overline{)1724}$ 9. $42\overline{)371}$

10. $33\overline{)303}$ 11. $61\overline{)7320}$ 12. $2001 \div 21$

13. $16\overline{)904}$ 14. $2400 \div 75$ 15. $14\overline{)4275}$

16. $71\overline{)6005}$ 17. $53\overline{)6307}$ 18. $67\overline{)3354}$

19. $2016 \div 21$ 20. $47\overline{)94425}$ 21. $15\overline{)3000}$

22. $231\overline{)14091}$ 23. $24\overline{)2596}$ 24. $603\overline{)48843}$

25. $38\overline{)22800}$ 26. $102\overline{)2004}$ 27. $411\overline{)42020}$

28. $111\overline{)11111}$ 29. $405\overline{)7008}$

C. For each of the following whole numbers, (a) list all its factors, and (b) write it as a product of its prime factors.

1. 6   2. 16   3. 19   4. 27   5. 40   6. 48

D. Applied Problems

1. **Machine Trades**   A machinist has a piece of bar stock 243 in. long. If she must cut nine equal pieces, what is the length per piece? (Assume no waste to get a first approximation.)

2. **Carpentry**   How many braces 32 in. long can be cut from a piece of lumber 192 in. long?

3. **Construction**   If subflooring is laid at the rate of 85 sq ft per hour, how many hours will be needed to lay 1105 sq ft?

4. **Construction**   The illustration shows a stringer for a short flight of stairs. Determine the missing dimensions H and W.

5. **Construction**   How many joists spaced 16 in. o.c. (on center) are required for a floor 432 in. long? (Add one joist for a starter.)

6. **Roofing**   If rafters are placed 24 in. o.c., how many are required for both sides of a gable roof that is 576 in. long? Add one rafter on each side for a starter.

7. **Carpentry**   A stairway in a house is to be built with 18 risers. If the distance from the top of the first floor to the top of the second floor is 126 in., what is the height per step?

8. **Manufacturing**   What is the average horsepower of six engines having the following horsepower ratings: 385, 426, 278, 434, 323, and 392? (*Hint:* To find the average, add the six numbers and divide their sum by 6.)

9. **Trades Management**   Mr. Martinez, owner of the Maya Restaurant, hired a plumber who worked for 18 hours installing some new equipment. The total bill for labor and materials used was $4696. If the cost for materials was $3400, how much was the plumber paid per hour?

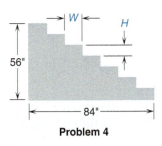

Problem 4

10. **Transportation**   A delivery truck traveling at an average rate of 54 miles per hour must cover a distance of 486 miles in one run. Allowing 2 hours for refueling and meals, how long will it take to complete the trip?

11. **Automotive Trades**   A mechanic needs to order 480 spark plugs. If there are ten plugs in a box, how many boxes does he need?

12. **Metalworking**   One cubic foot of a certain alloy weighs 375 lb. How many cubic feet does a 4875-lb block occupy?

13. **Printing**   A printer needs 13,500 sheets of paper for a job. If there are 500 sheets in a ream, how many reams does the printer need to get from the stockroom?

14. **Water/Wastewater Treatment**   A water tank contains 18,000 gallons. If 500 gallons per day are used from this tank, how long will it last?

15. **HVAC**   A geothermal heat pump uses loops of piping under a lawn to heat or cool a house. If a house requires 18 kW (kilowatts) of heating capacity and each loop has a capacity of 3 kW, how many loops of piping must be installed?

16. **Allied Health**   A home health care worker distributes one packet of multivitamins each week to each of his elderly patients, for a total of 52 packets per patient over an entire year. Last year he distributed a total of 1404 multivitamin packets. How many patients did he have?

17. **HVAC**   A family wants to replace their old oil-fired boiler with an air-source heat pump (ASHP) system. The initial cost of the ASHP system is $6900, but it will save the family $30 per month in heating and cooling costs. Determine the payback time both in months and in years and months.

18. **Agriculture**   Barley weighs 48 lb per bushel. How many bushels are in 816 lb of barley?

19. **Water/Wastewater Treatment**   A pump can deliver water at the rate of 200 gallons per minute. How long, in hours and minutes, will it take the pump to fill a 50,000-gallon tank?

20. **Electrical Trades**   A residential solar company charged a customer $15,600 to install rooftop solar panels. The customer estimated his monthly savings to be approximately $50. At this rate, what is the payback time in years?

21. **Carpentry**   Each 8-foot length of pine molding costs $22. How much will it cost to purchase 168 feet of molding?

22. **Automotive Trades**   The following table shows the driving range of five different electric cars. Assuming you begin with a fully charged battery, how many times will you have to recharge each car during a 760-mile trip? (*Hint:* Your answer must always be rounded to a whole number.)

Electric Car	Driving Range in Miles
Tesla Model S	265
Toyota RAV 4 EV SUV	103
Kia Soul EV	92
Nissan Leaf	84
Ford Focus Electric	76

E. Calculator Problems

1. Calculate the following.

(a) $\dfrac{4464}{48}$

(b) $\dfrac{169,722}{378}$

2. **Masonry** If 6-in. common red brick is on sale at $41 per hundred, calculate the cost of the 14,000 bricks needed on a construction project.

3. **Manufacturing** The specified shear strength of a $\frac{1}{8}$-in. 2117-T3 (AD) rivet is 344 lb. If it is necessary to ensure a strength of 6587 lb to a riveted joint in an aircraft structure, how many rivets must be used?

4. **Transportation** A train of 74 railway cars weighed a total of 9,108,734 lb. What was the average weight per car?

5. **Manufacturing** A high-speed stamping machine can produce small flat parts at the rate of 96 per minute. If an order calls for 297,600 parts to be stamped, how many hours will it take to complete the job? (60 minutes = 1 hour.)

6. **Painting** Following are four problems from the national apprentice examination for painting and decorating contractors:

(a) $11,877,372 \div 738$ (b) $87,445,005 \div 435$

(c) $1,735,080 \div 760$ (d) $206,703 \div 579$

7. **Masonry** Max, the master mason, is constructing a wall. From similar jobs he estimates that he can lay an average of 115 bricks an hour. How long will it take him to lay 4830 bricks?

Check your answers to the odd-numbered problems in the Appendix, then turn to Section 1-5 to study another important topic in the arithmetic of whole numbers.

1-5 Order of Operations

learning|**catalytics**™

1. Divide $24\overline{)768}$
2. Calculate: $8 \times 4 \div 2$

What is the value of the arithmetic expression

$3 \times 4 + 2 = ?$

If we multiply first, $(3 \times 4) + 2 = 12 + 2 = 14$

but if we add first, $3 \times (4 + 2) = 3 \times 6 = 18$

To avoid any possible confusion in situations like this, mathematicians have adopted a standard order of operations for arithmetic calculations. When two or more of the four basic arithmetic operations (addition, subtraction, multiplication, and division) are combined in the same calculation, follow these rules.

Rule 1 Perform any calculations shown inside parentheses first.

Rule 2 Perform all multiplications and divisions next, working from left to right.

Rule 3 Perform additions and subtractions last, working from left to right.

EXAMPLE 1   To calculate $3 \times 4 + 2$,

multiply first (Rule 2): $3 \times 4 = 12$. $3 \times 4 + 2 = 12 + 2$

Add last (Rule 3). $= 14$ ●

Your Turn   Perform the following calculations, using the order of operation rules.

(a) $5 + 8 \times 7$ (b) $12 \div 6 - 2$

(c) $(26 - 14) \times 2$ (d) $12 \times (6 + 2) \div 6 - 7$

Solutions   (a) Do the multiplication first (Rule 2): $8 \times 7 = 56$ $5 + 8 \times 7 = 5 + 56$
Then, add (Rule 3): $= 61$

(b) First, divide (Rule 2): $12 \div 6 = 2$ $12 \div 6 - 2 = 2 - 2$
Then, subtract (Rule 3). $= 0$

(c) Perform the operation in parentheses first (Rule 1):
$26 - 14 = 12$ $(26 - 14) \times 2 = 12 \times 2$
Then, multiply (Rule 2). $= 24$

(d) Work inside parentheses first (Rule 1): $12 \times (6 + 2) \div 6 - 7$
$6 + 2 = 8$ $= 12 \times 8 \div 6 - 7$
Then, multiply (Rule 2). $= 96 \div 6 - 7$
Next, divide (Rule 2). $= 16 - 7$
Finally, subtract (Rule 3). $= 9$

When division is written with a fraction bar or **vinculum,** all calculations above the bar or below the bar should be done before the division.

EXAMPLE 2 $\dfrac{24 + 12}{6 - 2}$ should be thought of as $\dfrac{(24 + 12)}{(6 - 2)}$

First, simplify the top and bottom to get $\dfrac{36}{4}$.

Then, divide to obtain the answer: $\dfrac{36}{4} = 9$.   ●

Note   This is very different from the calculation

$$\frac{24}{6} + \frac{12}{2} = 4 + 6 = 10$$

where we use Rules 2 and 3, doing the divisions first and the addition last. ●

More Practice   Perform the following calculations, using the three rules of order of operations.

(a) $240 \div (18 + 6 \times 2) - 2$ (b) $6 \times 5 + 14 \div (6 + 8) - 3$

(c) $\dfrac{28}{4} - \dfrac{45}{9}$ (d) $\dfrac{39 - 5 \times 3}{11 - 5}$

Solutions   (a) Perform the multiplication inside parentheses $240 \div (18 + 6 \times 2) - 2$
first (Rules 1 and 2): $6 \times 2 = 12$ $= 240 \div (18 + 12) - 2$
Add inside parentheses next (Rule 1): $18 + 12 = 30$ $= 240 \div 30 - 2$
Then, divide (Rule 2): $240 \div 30 = 8$ $= 8 - 2$
Finally, subtract (Rule 1). $= 6$

(b) First, work inside parentheses:

$6 + 8 = 14$

Next, multiply and divide from left to right:

$6 \times 5 = 30 \qquad 14 \div 14 = 1$

Finally, add and subtract from left to right.

$$6 \times 5 + 14 \div (6 + 8) - 3$$
$$= 6 \times 5 + 14 \div 14 - 3$$
$$= 30 + 14 \div 14 - 3$$
$$= 30 + 1 - 3$$
$$= 31 - 3 = 28$$

(c) Perform the first division (Rule 2):

Then do the second division (Rule 2):

Finally, subtract:

$$\frac{28}{4} - \frac{45}{9} = 7 - \frac{45}{9}$$
$$= 7 - 5$$
$$= 2$$

(d) Simplify the top first by multiplying (Rule 2)

and then subtracting (Rule 3).

Then, simplify the bottom.

Finally, divide.

$$\frac{39 - 5 \times 3}{11 - 5} = \frac{39 - 15}{11 - 5}$$
$$= \frac{24}{11 - 5}$$
$$= \frac{24}{6}$$
$$= 4$$

Practical Applications   The order of operations is involved in a great many practical applications.

EXAMPLE 3   **Office Services**   To encourage water conservation, a water district began charging its customers according to the following monthly block rate system:

Block	Quantity Used	Rate per HCF*
1	The first 20 HCF	$3
2	The next 30 HCF (from 21 through 50 HCF)	$4
3	The next 50 HCF (from 51 through 100 HCF)	$5
4	Usage in excess of 100 HCF	$6

*Hundred cubic feet

Suppose that a customer used 72 HCF in a month. The first 20 HCF will be charged at a rate of $3 per HCF. The next 30 HCF will be charged at a rate of $4 per HCF, and the remaining amount (72 HCF − 50 HCF) will be charged at a rate of $5 per HCF. We can calculate the customer's bill as follows:

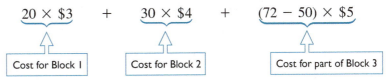

$$20 \times \$3 \quad + \quad 30 \times \$4 \quad + \quad (72 - 50) \times \$5$$

Cost for Block 1   Cost for Block 2   Cost for part of Block 3

Using the order of operations,

First, simplify within parentheses. $= 20 \times \$3 + 30 \times \$4 + 22 \times \$5$

Next, multiply from left to right. $= \$60 + \$120 + \$110$

Finally, add from left to right. $= \$290$

Your Turn   (a) **Construction**   A remodeling job requires 128 sq ft of countertops. Two options are being considered. The less expensive option is to use all Corian at $64 per sq ft. The more expensive option is to use 66 sq ft of granite at $150 per sq ft and 62 sq ft of laminate at $42 per sq ft. Write out a mathematical statement to calculate the difference in cost between these two options. Then calculate this difference.

(b) **Office Services**   Use the table in Example 3 to calculate the charge for using 116 HCF of water in a month.

Solutions   (a) The less expensive option is: $128 \times \$64 = \$8192$
        The more expensive option is: $66 \times \$150 + 62 \times \$42$
                                             $= \$9900 + \$2604$
                                             $= \$12{,}504$
        The difference is: $\$12{,}504 - \$8192 = \$4312$

(b) From the table, we see that 116 HCF includes all of Blocks 1, 2, and 3. The remainder $(116 - 100)$ HCF is charged to Block 4. The total monthly bill is:
$$20 \times \$3 + 30 \times \$4 + 50 \times \$5 + (116 - 100) \times \$6$$
To calculate, simplify
parentheses first. $= 20 \times \$3 + 30 \times \$4 + 50 \times \$5 + 16 \times \$6$
Then multiply. $= \$60 + \$120 + \$250 + \$96$
Finally add. $= \$526$

Now go to Exercises 1-5 for more practice on the order of operations.

Exercises 1-5   Order of Operations

A. Perform all operations in the correct order.

1. $2 + 8 \times 6$

2. $20 - 3 \times 2$

3. $40 - 20 \div 5$

4. $16 + 32 \div 4$

5. $16 \times 3 + 9$

6. $2 \times 9 - 4$

7. $48 \div 8 - 2$

8. $64 \div 16 + 8$

9. $(5 + 9) \times 3$

10. $(18 - 12) \div 6$

11. $24 \div (6 - 2)$

12. $9 \times (8 + 3)$

13. $16 + 5 \times (3 + 6)$

14. $8 + 3 \times (9 - 4)$

15. $(23 + 5) \times (12 - 8)$

16. $(17 - 9) \div (6 - 2)$

17. $6 + 4 \times 7 - 3$

18. $24 - 8 \div 2 + 6$

19. $5 \times 8 + 6 \div 6 - 12 \times 2$

20. $24 \div 8 - 14 \div 7 + 8 \times 6$

21. $2 \times (6 + 4 \times 9)$

22. $54 \div (8 - 3 \times 2)$

23. $(4 \times 3 + 8) \div 5$

24. $(26 \div 2 - 5) \times 4$

25. $8 - 4 + 2$

26. $24 \div 6 \times 2$

27. $18 \times 10 \div 5$

28. $22 + 11 - 7$

29. $12 - 7 - 3$

30. $48 \div 6 \div 2$

31. $12 - (7 - 3)$

32. $18 \div (3 \times 2)$

33. $\dfrac{36}{9} + \dfrac{27}{3}$

34. $\dfrac{36 - 27}{9 - 6}$

35. $\dfrac{44 + 12}{11 - 3}$

36. $\dfrac{44}{11} + \dfrac{12}{3}$

37. $\dfrac{6 + 12 \times 4}{15 - 3 \times 2}$

38. $\dfrac{36 - (7 - 4)}{5 + 3 \times 2}$

39. $\dfrac{12 + 6}{3 + 6} + \dfrac{24}{6} - 6 \div 6$

40. $8 \times 5 - \dfrac{2 + 4 \times 12}{18 - 4 \times 2} + 72 \div 9$

B. Practical Applications

1. **Painting**   A painter ordered 3 gallons of acrylic vinyl paint for $34 a gallon and 5 gallons of acrylic eggshell enamel for $39 a gallon. Write out a mathematical statement giving the total cost of the paint. Calculate the cost.

2. **Landscaping**   On a certain landscaping job, Steve charged a customer $468 for labor and $90 each for eight flats of plants. Write a single mathematical statement giving the total cost of this job, then calculate the cost.

3. **Electrical Trades**   An electrician purchased 12 dimmer switches at $30 each and received a $6 credit for each of the three duplex receptacles she returned. Write out a mathematical statement that gives the amount of money she spent, then calculate this total.

4. **Automotive Trades**   At the beginning of the day on Monday, the parts department has on hand 520 spark plugs. Mechanics in the service department estimate they will need about 48 plugs per day. A new shipment of 300 will arrive on Thursday. Write out a mathematical statement that gives the number of spark plugs on hand at the end of the day on Friday. Calculate this total.

5. **Trades Management**   A masonry contractor is preparing an estimate for building a stone wall and gate. He estimates that the job will take a 40-hour work week. He plans to have two laborers at $12 per hour and three masons at $20 per hour. He'll need $3240 worth of materials and wishes to make a profit of $500. Write a mathematical statement that will give the estimated cost of the job; then calculate this total.

6. **Allied Health**   A pregnant woman gained 4 pounds per month during her first trimester (3 months) of pregnancy, 3 pounds per month during her second trimester, and 2 pounds per month during her last trimester. What was her total weight gain during the pregnancy?

7. **Graphic Design**   Girilla Graphics charges $80 per hour for the work of its most senior graphic artists, $40 per hour for the work of a production designer, and $18 per hour for the work of a trainee. A recent project required 33 hours of work from a senior artist, 12 hours from a production designer, and 45 hours from a trainee. Write out a mathematical statement for calculating the total labor cost; then compute this cost.

8. **Culinary Arts**   A caterer must estimate the amount and cost of strawberries required for an anniversary dinner involving 200 guests. She estimates that she will need about 6 strawberries per person for decorations and cocktails, and about 10 strawberries per person for the dessert. A local farmer sells strawberries in flats containing 80 strawberries each for $7.00 per flat. How many flats will she need, and how much should she expect to pay for the strawberries?

9. **Sports and Leisure**   In the 2008 Olympics, the United States ended up with the most total medals, but China won the most gold medals. This created a controversy over which country had the best overall performance. One way to settle this issue objectively is to score the medals much like a high school or college track meet—that is, award 5 points for first place (gold), 3 points for second place (silver), and 1 point for third place (bronze). Use the following final medal counts to determine a score for each country based on this point system. According to these scores, which country "won" the 2008 Olympics?

	Gold	Silver	Bronze
China	51	21	28
United States	36	38	36

10. **Office Services**   Use the table in Example 3 to calculate the monthly water bill for the following amounts of water usage: (a) 32 HCF (b) 94 HCF (c) 121 HCF

11. **Automotive Trades**   A particular model of automobile is rated by the Environmental Protection Agency (EPA) at 22 mi/gal for city driving and 30 mi/gal for highway driving. How many gallons of gas would this car consume during 176 miles of city driving and 180 miles of highway driving?

C. Calculator Problems

1. $462 + 83 \times 95$

2. $425 \div 25 + 386$

3. $7482 - 1152 \div 12$

4. $1496 - 18 \times 13$

5. $(268 + 527) \div 159$

6. $2472 \times (1169 - 763)$

7. $612 + 86 \times 9 - 1026 \div 38$

8. $12 \times 38 + 46 \times 19 - 1560 \div 24$

9. $3579 - 16 \times (72 + 46)$

10. $273 + 25 \times (362 + 147)$

11. $864 \div 16 \times 27$

12. $973 - (481 + 327)$

13. $(296 + 18 \times 48) \times 12$

14. $(27 \times 18 - 66) \div 14$

15. $\dfrac{3297 + 1858 - 493}{48 \times 16 - 694}$

16. $\dfrac{391}{17} + \dfrac{4984}{89} - \dfrac{1645}{47}$

Check your answers to the odd-numbered problems in the Appendix, then turn to Problem Set 1 on page 62 for practice on the arithmetic of whole numbers, with many practical applications of this mathematics. If you need a quick review of the topics in this chapter, visit the chapter Summary first.

CHAPTER 1
SUMMARY

Arithmetic of Whole Numbers

Objective	Review
Write whole numbers in words then translate words to numbers. (p. 4)	Use place value and expanded form to help translate in both directions. **Example:** The number 250,374 can be placed in a diagram; 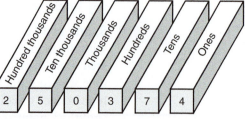 and read "two hundred fifty thousand, three hundred seventy-four." The number "one million, sixty-five thousand, eight" can be written as 1,065,008.

Objective	Review
Round whole numbers. (p. 6)	Place a $\wedge$ mark to the right of the place to which the number must be rounded. If the digit to the right of the mark is less than 5, replace all digits to the right of the mark with zeros. If the digit to the right of the mark is equal to or larger than 5, increase the digit to the left of the mark by 1 and replace all digits to the right with zeros.

Example: Round 214,659

(a) to the nearest ten thousand $21 \wedge 4659 = 210{,}000$

(b) to the nearest hundred $2146 \wedge 59 = 214{,}700$

Add whole numbers. (p. 7)	First, estimate your answer. Then line up the digits and add.

Example: Add: $7009 + 1598$

Estimate: $7000 + 1600 = 8600$

Line up and add:

$$\begin{array}{r} {}^{1\ 1}\\ 7009 \\ +1598 \\ \hline 8607 \end{array}$$

The answer is approximately equal to the estimate.

Subtract whole numbers. (p. 21)	First, estimate your answer. Then, line up the digits and subtract column by column from right to left. "Borrow" when necessary.

Example: Subtract: $4035 - 1967$

Estimate: $4000 - 2000 = 2000$

Line up and subtract:

$$\begin{array}{r} {}^{3\ 10}\\ 40\cancel{3}5 \\ -1967 \end{array} \qquad \begin{array}{r} {}^{3\ 9\ 13}\\ \cancel{4}0\cancel{3}5 \\ -1967 \end{array} \qquad \begin{array}{r} {}^{3\ 9\ 12\ 15}\\ \cancel{4}0\cancel{3}\cancel{5} \\ -\ 1967 \\ \hline 2068 \end{array}$$

Multiply whole numbers. (p. 30)	First, estimate your answer. Then multiply the top number by each digit in the bottom number. Leave a blank space on the right end of each row of the calculation. Finally, add the rows of products.

Example: Multiply: 305×243

Estimate: $300 \times 250 = 75{,}000$

$$\begin{array}{r} 305 \\ \times 243 \\ \hline 915 \\ 1220\ \ \\ 610\ \ \ \ \\ \hline 74115 \end{array}$$

$915 \leftarrow \boxed{305 \times 3}$

$1220 \leftarrow \boxed{305 \times 4}$

$610 \leftarrow \boxed{305 \times 2}$

The product is 74,115, and this agrees with the estimate.

Objective	**Review**
Divide whole numbers. (p. 41)	First, estimate your answer. Then divide, step-by-step, as shown in the following example.

Example: Divide $7511 \div 37$

Estimate: $8000 \div 40 = 200$

Step 1   37 into 7? No. 37 into 75? Yes, twice. Write 2 above the 5.

$$\begin{array}{r} 203 \\ 37\overline{)7511} \\ \underline{74} \\ 11 \\ \underline{0} \\ 111 \\ \underline{111} \\ 0 \end{array}$$

Step 2 $37 \times 2 = 74$. Subtract $75 - 74 = 1$

Step 3   Bring down 1. 37 into 11? No. Write 0 in the answer.

Step 4 $37 \times 0 = 0$. Subtract $11 - 0 = 11$

Step 5   Bring down 1. 37 into 111? Yes, three times. Write 3 in the answer.

Step 6 $37 \times 3 = 111$. Subtract $111 - 111 = 0$

The quotient is 203, which is very close to the estimate.

Determine factors and prime factors. (p. 46)

The factors of a whole number are all the numbers that divide it with zero remainder. The prime factors of a whole number are those factors that are prime; that is, they are evenly divisible only by themselves and 1.

Example: The factors of 12 are 1, 2, 3, 4, 6, and 12.

To write 12 as a product of its prime factors, use the factor tree shown.

$$12 = 2 \times 2 \times 3$$

(factor tree: 12 branches to 2 and 6; 6 branches to 2 and 3)

Use the correct order of operations with addition, subtraction, multiplication, and division. (p. 53)

Follow these steps in this order:

Step 1   Perform any calculations shown inside parentheses.

Step 2   Perform all multiplications and divisions, working from left to right.

Step 3   Perform all additions and subtractions, working from left to right.

Example:
$$\begin{aligned} & 8 + 56 \div 4 \times 2 + 8 - 3 \\ &= 8 + 14 \times 2 + 8 - 3 \\ &= 8 + 28 + 8 - 3 \\ &= 44 - 3 \\ &= 41 \end{aligned}$$

Divide: $56 \div 4 = 14$
Multiply next: $14 \times 2 = 28$
Add next: $8 + 28 + 8 = 44$
Finally, subtract: $44 - 3 = 41$

Solve practical applications involving whole numbers. (pp. 13, 25, 37, 47, 55)

Read the problem carefully, looking for key words or phrases that indicate which operation to use.

Example: A metal casting weighs 680 lb. What is the finished weight after 235 lb of metal is removed during shaping?

Solution: The word *removed* indicates that subtraction is involved.

$680 - 235 = 445$ lb

Example: A family wants to replace their old oil-fired boiler with an air-source heat pump (ASHP) system. The initial cost of the ASHP system is $6900, but it will save the family $30 per month in heating and cooling costs. Determine the payback time both in months and in years and months.

Solution: The payback time is the time it takes for the total monthly savings to match the initial cost of $6900. To solve, we must divide $6900 by $30.

Objective	Review

$$\begin{array}{r} 230 \\ 30\overline{)6900} \\ \underline{60} \\ 90 \\ \underline{90} \\ 0 \\ \underline{0} \\ 0 \end{array}$$

The payback time is 230 months. To convert this to years and months, we divide by 12.

$$\begin{array}{r} 19 \\ 12\overline{)230} \\ \underline{12} \\ 110 \\ \underline{108} \\ 2 \end{array}$$

The quotient 19 and the remainder 2 tells us that, in years and months, the payback time is 19 years and 2 months.

PROBLEM SET 1   Arithmetic of Whole Numbers

Answers to odd-numbered problems are given in the Appendix.

A.   Reading, Writing, and Rounding Whole Numbers

Write in words.

1. 593
2. 6710
3. 45,206
4. 137,589
5. 2,403,560
6. 970,001
7. 10,020
8. 1,528,643
9. 12,604,700,250

Write as numbers.

10. Six thousand, three hundred twenty-seven
11. Two hundred thirty thousand, fifty-six
12. Five million, ninety-eight thousand, one hundred seven
13. Sixty-four thousand, seven hundred
14. Eight hundred fifty-two million
15. Six billion, forty-seven million, nine hundred twenty thousand

Round as indicated.

16. 692 to the nearest ten
17. 5476 to the nearest hundred
18. 17,528 to the nearest thousand
19. 94,746 to the nearest hundred
20. 652,738 to the nearest hundred thousand
21. 705,618 to the nearest ten thousand

B.   Perform the arithmetic as shown.

1. $87 + 9$
2. $58 + 7$
3. $63 - 8$
4. $44 - 6$
5. $24 + 69$
6. $38 + 45$
7. $456 + 72$
8. $43 + 817$
9. $396 + 538$
10. $2074 + 906$
11. $43 - 28$
12. $93 - 67$
13. $734 - 85$
14. $315 - 119$
15. $543 - 348$
16. $3401 - 786$
17. 376×4
18. 489×7
19. 67×21
20. 45×82
21. 207×63
22. 314×926
23. 5236×44
24. 4018×392
25. $259 \div 7$
26. $1704 \div 8$
27. $42\overline{)2394}$
28. $34\overline{)2108}$
29. $1440 \div 160$
30. $11,309 \div 263$
31. $\dfrac{1314}{73}$
32. $\dfrac{23 \times 51}{17}$
33. $\dfrac{36 \times 91}{13 \times 42}$
34. $(18 + 5 \times 9) \div 7$
35. $120 - 40 \div 8$
36. $32 \div 4 + 16 \div 2 \times 4$
37. $3 \times 4 - 15 \div 3$
38. $\begin{array}{r} 139 \\ 407 \\ + \ 81 \end{array}$
39. $\begin{array}{r} 308 \\ 793 \\ + \ 144 \end{array}$

Name

_____

Date

_____

Course/Section

_____

40. 194
     271
   + 368

41. 256 + 89 + 965 + 42

42. 5127 + 386 + 2842 + 78 + 687

C. **For each of the following whole numbers, (a) list all its factors, and (b) write each as a product of its prime factors.**

  1. 8   2. 28   3. 31

  4. 35   5. 36   6. 42

D. **Practical Applications**

  1. **Electrical Trades**   From a roll of No. 12 wire, an electrician cut the following lengths: 6, 8, 20, and 9 ft. How many feet of wire did he use?

  2. **Machine Trades**   A machine shop bought 14 steel rods of $\frac{7}{8}$-in.-diameter steel, 23 rods of $\frac{1}{2}$-in. diameter, 8 rods of $\frac{1}{4}$-in. diameter, and 19 rods of 1-in. diameter. How many rods were purchased?

  3. **Construction**   Find the missing dimensions x and y in each drawing below. Assume that all corners are square.

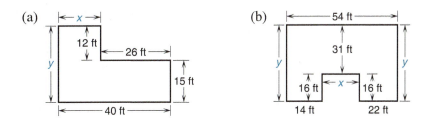

  4. **Allied Health**   A patient was instructed to take one tablet of blood pressure medication twice daily (one tablet in the morning and one in the evening). Before the patient leaves for a 12-week summer holiday, the medical technician must determine how many tablets the patient will need over her entire holiday. How many tablets will the patient require?

  5. **Electrical Trades**   Roberto has 210 ft of No. 14 wire left on a roll. If he cuts it into pieces 35 ft long, how many pieces will he get?

  6. **Flooring and Carpeting**   How many hours will be required to install the flooring of a 2160-sq-ft house if 90 sq ft can be installed in 1 hour?

  7. **Construction**   The weights of seven cement platforms are 210, 215, 245, 217, 220, 227, and 115 lb. What is the average weight per platform? (Add up and divide by 7.)

  8. **Flooring and Carpeting**   A room has 234 square feet of floor space. If carpeting costs $5 per sq ft, what will it cost to carpet the room?

  9. **Retail Merchandising**   A 4K Ultra HD TV set can be bought for $400 down and 12 payments of $110 each. What is its total cost?

  10. **Electronics**   The output voltage of an amplifier is calculated by multiplying the input voltage by the voltage gain of the amplifier. Calculate the output voltage, in millivolts (mV), for a circuit if the input voltage is 45 mV and the gain is 30.

11. **Automotive Trades** During a compression check on an engine, the highest compression in any cylinder was 136 pounds per square inch (psi), and the lowest was 107 psi. The maximum allowable difference between highest and lowest is 30 psi. Did the engine fall within the maximum allowed?

12. **Office Services** During the first three months of the year, the Print Rite Company reported the following sales of printers:

    January $55,724
    February $47,162
    March $62,473

    What is their sales total for this quarter of the year?

13. **Fire Protection** A fire control pumping truck can move 156,000 gallons of water in 4 hours of continuous pumping. What is the flow rate (gallons per minute) for this truck? (Be careful with your time units.)

14. **Construction** A pile of lumber contains 170 boards 10 ft long, 118 boards 12 ft long, 206 boards 8 ft long, and 19 boards 16 ft long.
    (a) How many boards are in the pile?
    (b) Calculate the total length, or linear feet, of lumber in the pile.

15. **Marine Technology** If the liquid pressure on a surface is 17 psi (pounds per square inch), what is the total force on a surface area of 167 square inches?

16. **Automotive Trades** A rule of thumb useful to auto mechanics is that a car with a pressure radiator cap will boil at a temperature of $(3 \times \text{cap rating}) + 212$ degrees. What temperature will a Dino V6 reach with a cap rating of 17?

17. **Machine Trades** If it takes 45 minutes to cut the teeth on a gear blank, how many hours will be needed for a job that requires cutting 32 such gear blanks?

18. **Electrical Trades** A 4-ft lighting track equipped with three fixtures costs $45. Fluorescent bulbs for the fixtures cost $15 each. Write out a mathematical statement that gives the total cost for three of these tracks, with bulbs, plus $9 shipping. Then calculate this total.

19. **Manufacturing** A gallon is a volume of 231 cubic inches. If a storage tank holds 380 gallons of oil, what is the volume of the tank in cubic inches?

20. **Office Services** The Ace Machine Tool Co. received an order for 15,500 flanges. If two dozen flanges are packed in a box, how many boxes are needed to ship the order?

21. **Manufacturing** At 8 A.M. the revolution counter on a diesel engine reads 460089. At noon it reads 506409. What is the average rate, in revolutions per minute, for the machine?

22. **Fire Protection** The pressure reading coming out of a pump is 164 lb. It is estimated that every 50-ft section of hose reduces the pressure by 7 lb. What will the estimated pressure be at the nozzle end of nine 50-ft sections?

23. **Carpentry** A fireplace mantel 6 ft long is to be centered along a wall 18 ft long. How far from each end of the wall should the ends of the mantel be?

24. **Electronics** When capacitors are connected in parallel in a circuit as shown, the total capacitance can be calculated by adding the individual

capacitor values C_1, C_2, C_3, C_4, and C_5. Find the total capacitance of this circuit, in picofarads (pF).

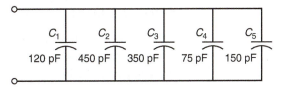

25. **Trades Management**   RAD Electric submitted a bid of $3300 for a certain job. Using billing for T&M (time and materials), RAD normally charges $85 per hour. If materials for the job come to $350, and the electrician actually worked for 36 hours, by how much money did RAD underbid the job?

26. **Life Skills**   An electrician is deciding between two job offers. One is with Company A, paying him $24 per hour plus health benefits. The other is with Company B, paying him $28 per hour with no health benefits. Health insurance would cost him $500 per month. Assume he works an 8-hour day and an average of 22 work days per month, then answer the following questions:
    (a) How much would he earn per month with Company A?
    (b) How much would he earn per month with Company B?
    (c) What are his net earnings with Company B after paying for his health benefits?
    (d) Which company provides the best overall compensation?

27. **Office Services**   Use the table in Example 3 on page 55 to determine the monthly cost for 87 HCF of water usage.

28. **Life Skills**   Gorilla glue comes in two sizes: an 8-ounce container for $11 and an 18-ounce container for $19.
    (a) How many of the smaller containers would you need in order to get 72 ounces of glue? How much would this cost?
    (b) How many of the larger containers would you need in order to get 72 ounces? How much would this cost?
    (c) How much do you save by purchasing 72 ounces in the larger-sized container instead of in the smaller-sized container?

29. **Automotive Trades**   A 2017 Subaru Outback 2.5i Touring model is EPA-rated at 25 mi/gal for city driving and 32 mi/gal for highway driving.
    (a) How many gallons of gas would be used on a trip that included 125 miles of city driving and 480 miles of highway driving?
    (b) At approximately $2 per gallon, what would be the total cost of gas for this trip?

30. **Landscaping**   A landscape company charges $55 per hour for the work of a designer, $40 per hour for the work of a foreman, and $25 per hour for the work of a laborer. A particular installation required 6 hours of design work, 11 hours from the foreman, and 33 hours of labor. Write out a mathematical statement for calculating the total labor cost and then compute this cost.

31. **Allied Health**   To maintain his weight, a sedentary person needs to consume about 13 calories per day for every pound that he weighs. Suppose a sedentary person weighs 220 pounds. How many calories per day should he consume to maintain his present weight?

32. **Construction**   A contractor estimates that it will require $16,400 of upgrades to make a home more energy efficient. She further estimates that the upgrades will save the residents $80 per month in energy bills. What is the payback time for these upgrades in months and in years and months?

33. **Life Skills**   A potential new-car buyer is debating between a Lexus ES 350 and its hybrid equivalent, the ES 300h. He plans to pay cash for the car, drive approximately 20,000 miles per year, and sell it in 5 years. He wants to know which car will be less expensive to own over that period of time. Use the following information to determine the total 5-year cost of each model. Which model has the lower 5-year cost and by how much? (See Example 6 on page 26.)

	ES 350	ES 300h
Sales Price (MSRP)	$38,950	$41,870
5-Year Fuel Cost	$ 7859	$ 6509
Insurance and Fees	$ 9768	$ 9599
Maintenance and Repairs	$ 5775	$ 5836
5-Year Resale Value	$13,839	$13,726

(*Source:* KBB.com)

34. **Carpentry**   A set of eight Stanley chisels was recently selling for $220. Purchased individually, the chisels were $33 each. How much would you save by purchasing the set of eight rather than buying them individually?

35. **Agriculture**   A bushel of oats weighs 32 lb. How many bushels are there in 786 lb of oats?

36. **Life Skills**   A 2016 Mazda 6 Touring model is priced at $26,850. If you decide to pay cash, you would keep the car for 3 years and then sell it at its depreciated value. The predicted depreciated value of the car 3 years from now is $10,885. Another option is to lease the car for 3 years. This would require a down payment of $1500, monthly payments of $330, a lease acquisition fee of $300, and a lease disposition fee of $500. Which option results in the lowest net cost to you and by how much?

37. **HVAC**   In designing a heating system for a house, an HVAC technician must first calculate the total heat loss of the house. To find total heat loss, she first multiplies the area of each type of interior surface by a factor that depends on two things: the type of surface and the difference between the coldest outdoor temperature and the desired indoor temperature. The sum of these products is the total heat loss of the house in Btu per hour. The following table shows the surface areas for a house and the factors for a given situation:

Surface	Area	Factor	Product
Exposed Walls	1344 sq ft	6	
Glass	220 sq ft	40	
Cold Ceiling	1664 sq ft	5	
Cold Floor	1664 sq ft	7	

Sum:

Calculate the total heat loss of the house.

CASE STUDY: Robin's New Car

Robin decides to give her old car to her daughter and replace it with a brand new Honda Accord Touring model with standard options. She cannot afford to pay cash, so she has two options: leasing the car or financing its purchase. She decides to compare the total net costs of the two options over a 3-year period.

1. Robin discovers that the MSRP (manufacturer's suggested retail price) of the car is $35,805 but, like most people, she wants to negotiate a lower price. According to KBB.com, a fair purchase price for her model is $32,195. She talks to the dealer, and he is willing to accept this price. How much does the KBB price save her compared to the MSRP?

Leasing. When you lease a vehicle, you are essentially renting it. You normally make a down payment in cash and then regular monthly payments. Your monthly payments include interest and a portion of the depreciation of the vehicle. In the case of a lease, the depreciation is the difference between the *capitalized cost* (the negotiated purchase price), minus the down payment, and what is known as the residual value. The *residual value* is the projected depreciated value of the vehicle when the lease ends. After the lease is up, you can either return the vehicle to the dealer or leasing company, or purchase it for its residual value. You will be told what the residual value will be when you sign the leasing contract. Although 3 years is the most common lease term, some leases last only 2 years, while others may last slightly longer than 3 years. Every lease also comes with a mileage limit, beyond which you must pay a certain amount per mile over this limit. There may be other fees associated with a lease, such as a lease acquisition fee and a lease disposition fee.

2. To lease the car, Robin must make a $3000 down payment and pay a $75 acquisition fee at the start of the lease and a $500 disposition fee at the end of the lease. She assumes she will drive about 10,000 miles per year, which is within the mileage limit of the lease. Her lease will last for 3 years (36 months) and her monthly payments will be $344. What is the total amount of money that Robin will pay for her 36-month lease?

Financing. When you finance the purchase of a new car, you also make monthly payments. These include interest and a portion of the total cost of the car, minus the down payment. You may choose a financing term that normally varies from 3 to 6 years. The interest rate depends on your credit rating as well as the length of the term.

3. In order to fairly compare the leasing option to the financing option, Robin asks the dealer to calculate her payments for a 3-year term. With Robin's credit rating, she must make a $3000 down payment and then 36 monthly payments of $849 each. What is the total amount of money Robin will pay to finance the purchase of the car?

Comparing Net Costs. When your lease ends, you have nothing. When your finance term ends, you own the car. Again, to make a fair comparison between our two options, we need to take this into account. Steps 4, 5, and 6 will help you calculate the net cost of financing. Then, in Step 7, you will be able to compare the net cost of the two options.

4. Robin returns to KBB.com and clicks on "5-Year Cost to Own." She resets the parameters to reflect her interest rate, down payment, 3-year term, and estimated 10,000 miles per year of driving. Under "Depreciation" she sees that the car will depreciate by $8340 the first year, $3554 the second year, and $2981 the third year. What is the total depreciation of the car over the 3-year period?

5. To see what the car will be worth after 3 years, subtract your answer to Step 4 from the MSRP of $35,805.

6. Finally, to determine the net cost of financing the purchase of the car, subtract your answer to Step 5 from your answer to Step 3.

7. Step 6 gave you the net cost of financing, and Step 2 gave you the net cost of leasing. Which appears to be the better deal? By how much?

8. Although Step 5 gives Robin an estimate of what the car will be worth after 3 years, there is no guarantee that she could actually sell the car for that price. If she did end up selling it after 3 years, what is the lowest sale price that would still make financing a better deal?

Using a Calculator, I: Whole-Number Calculations

Since its introduction in the early 1970s, the handheld electronic calculator has quickly become an indispensable tool in modern society. From clerks, carpenters, and shoppers to technicians, engineers, and scientists, people in almost every occupation now use calculators to perform mathematical tasks. With this in mind, we have included in this textbook special instructions, examples, and exercises demonstrating the use of a calculator wherever it is appropriate.

To use a calculator intelligently and effectively in your work, you should remember the following:

1. Whenever possible, make an estimate of your answer before entering a calculation. Then check this estimate against your calculator result. You may get an incorrect answer by accidentally pressing the wrong keys or by entering numbers in an incorrect sequence.

2. Always try to check your answers to equations and word problems by substituting them back into the original problem statement.

3. Organize your work on paper before using a calculator, and record any intermediate results that you may need later.

4. Round your answer whenever necessary.

This is the first of three sections designed to teach you how to use a calculator with the mathematics taught in this course. In addition to these special sections, the solutions to many worked examples include the appropriate calculator key sequences and displays. Most exercise sets contain problems that are designed to be solved with the aid of a calculator.

Selecting a Calculator

As noted in the "How to Use This Book" section (see page xvi), the calculator key sequences shown in this text are based on the newer generation of scientific calculators, as well as graphing calculators. A reasonably priced scientific calculator with at least two lines of display is ideal for this textbook, although for students planning to continue with more advanced mathematics, a graphing calculator may eventually be needed.

Becoming Familiar with Your Calculator

First check out the machine. Look for the "On" switch, usually located in the upper-right- or lower-left-hand corner of the keyboard. There may be a separate "Off" switch, or it may be a second function of the "On" switch. When the calculator is first turned on or cleared, it will either display a zero followed by a decimal point ($0.$), or a blinking cursor on the left end of the display window, or both. The lower portion of the keyboard will have both numerical keys (0, 1, 2, 3, . . . , 9) and basic function keys ($+$, $-$, $\times$, $\div$), as well as a decimal point key ($\cdot$), and either an equals ($=$) or ENTER key.

There is no standard layout for the remaining function keys, but some numerical keys and almost all non-numerical keys have a second function written above them. To activate these second functions, you must first press the 2nd key and then press the appropriate function key. The second function key is located in the upper-left-hand corner of the keyboard and may also be labeled 2ndF or SHIFT. Some calculators even have keys containing a third function, and a 3rd key is used to activate them. Unfortunately, there is no standard agreement among different models as to which operations are main functions and which are second functions. Therefore, we will not generally include the 2nd key in our key sequences, and we will leave it to the student to determine when to use this key.

If you have a calculator with two or more lines of display, you will find four arrow keys located in the upper-right portion of the keyboard arranged like this:

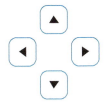

These calculators will have a CLEAR key, also labeled AC or simply C, and it may be combined with the "On" key. They will also have a delete (DEL) key with insert (INS) printed above it. The CLEAR key clears the entire calculation and the answer, but it is not generally necessary to clear the previous calculation in order to start a new one. To clear only part of a calculation, use the arrow keys to place the cursor at the appropriate spot, and then use DEL or INS as needed. The up-arrow key (▲) can also be used to retrieve a prior entry, even after the display has been cleared.

All calculators have two basic memory keys, one for storing information, labeled STO or Min, and one for retrieving information, labeled RCL or MR. These are often part of the same key, one being a second function of the other. Some calculators also have M+ and M− keys for adding and subtracting numbers to an existing memory total. A final pair of keys useful for basic whole number calculations are the parentheses keys. There is one for opening parentheses, labeled (, and one for closing parentheses, labeled).

Basic Operations   Every number is entered into the calculator digit by digit, left to right. For example, to enter the number 438, simply press the 4, 3, and 8 keys followed by the = key, and the display will read   438. . If you make an error in entering the number, press either the clear C or clear entry CE key and begin again. Note that the calculator does not know if the number you are entering is 4, 43, 438, or something even larger until you stop entering numerical digits and press a function key such as +, −, ×, ÷, or =.

EXAMPLE 1   To add 438 + 266

first, estimate the answer: 400 + 300 = 700

then, enter this sequence of keys,

4 3 8 + **2 6 6** = →   704.

On your calculator, the top line of the display will show all of your entries except for the equals sign. Therefore, in this case, the top line will show **438 + 266**. The next line will show the final answer, or in this case, **704**.

All four basic operations work the same way.

Try the following problems for practice.

Your Turn   (a) 789 − 375   (b) 246 × 97

(c) 2314 ÷ 26   (d) 46,129 + 8596

Answers   (a) **7 8 9** − **3 7 5** = →   414.

(b) **2 4 6** × **9 7** = →   23862.

(c) **2 3 1 4** ÷ **2 6** = →   89.

(d) **4 6 1 2 9** + **8 5 9 6** = →   54725.

Don't forget to estimate the answer before using the calculator.

Combined Operations

Scientific calculators are programmed to follow the correct order of operations. This means that you can enter a long string of calculations without having to find an intermediate result.

EXAMPLE 2

To calculate $2 + 3 \times 4$, we know from Section 1-5 that the multiplication ($3 \times 4 = 12$) must be performed first, and the addition ($2 + 12 = 14$) must be performed last. With a scientific calculator, simply enter the calculation exactly as it is written and then press the ⊟ key:

2 ⊞ 3 ⊠ 4 ⊟ → ☐☐☐☐☐☐ *14.*   ●

Your Turn

Enter each of the following calculations as written.

(a) $480 - 1431 \div 53$ (b) $72 \times 38 + 86{,}526 \div 69$

(c) $2478 - 726 + 598 \times 12$ (d) $271{,}440 \div 48 \times 65$

Answers

(a) **480** ⊟ **1431** ⊞ **53** ⊟ → ☐☐☐☐☐ *453.*

(b) **72** ⊠ **38** ⊞ **86526** ⊟ **69** ⊟ → ☐☐☐☐ *3990.*

(c) **2478** ⊟ **726** ⊞ **598** ⊠ **12** ⊟ → ☐☐☐☐ *8928.*

(d) **271440** ⊟ **48** ⊠ **65** ⊟ → ☐☐☐ *367575.*

🔍 A Closer Look

When the top line of the display of a two-line scientific calculator fills up, every additional entry causes characters to disappear from the left end of the display. You can make these entries reappear by pressing the left-arrow key (◄). Once you press ⊟, the top line shifts back to the start of the calculation, and overflow on the right end will disappear. If you then press the up-arrow key (▲), you can edit the calculation using the right- and left-arrow keys in combination with delete (DEL) and insert (INS). You can then press ⊟ at any time to get the new answer. ●

Parentheses and Memory

As you have already learned, parentheses, brackets, and other grouping symbols are used in a written calculation to signal a departure from the standard order of operations. Fortunately, every scientific calculator has parentheses keys. To perform a calculation with grouping symbols, either enter the calculation inside the parentheses first and press ⊟, or enter the calculation as it is written but use the parentheses keys.

EXAMPLE 3

To enter $12 \times (28 + 15)$ use either the sequence

28 ⊞ **15** ⊟ ⊠ **12** ⊟ → ☐☐☐☐ *516.*

or

12 ⊠ ⦅ **28** ⊞ **15** ⦆ ⊟ → ☐☐☐☐ *516.*

On most calculators, the ⊠ key is optional when multiplying an expression in parentheses. Try the following sequence to see if your calculator works this way:

12 ⦅ **28** ⊞ **15** ⦆ ⊟ → ☐☐☐☐ *516.*   ●

When division problems are given in terms of fractions, parentheses are implied but not shown.

EXAMPLE 4

The problem

$\dfrac{48 + 704}{117 - 23}$ is equal to $(48 + 704) \div (117 - 23)$

The fraction bar acts as a grouping symbol. The addition and subtraction in the top and bottom of the fraction must be performed before the division. Again you may use either parentheses or the $=$ key to simplify the top half of the fraction, but after pressing $=$ you must use parentheses to enter the bottom half. Here are the two possible sequences:

(48 + 704) ÷ (117 − 23) = → 8.

or

48 + 704 = ÷ (117 − 23) = → 8.

⎿────── Press $=$, or the calculator will divide
into 704 rather than into 48 + 704.

The memory keys provide a third option for this problem. Memory is used to store a result for later use. In this case, we can calculate the value of the bottom first, store it in memory, then calculate the top and divide by the contents of memory. The entire sequence looks like this:

117 − 23 = STO A   48 + 704 =   ÷ RCL A = → 8.

Calculate   Store it   Calculate   Divide by
the bottom.   in memory.   the top.   the bottom
                                            in memory.   ●

Although all memory storing begins with the STO key, and all memory retrieval begins with the RCL key, most calculators require you to then select a memory location, and the procedures for selecting these vary greatly from model to model. You may need to use your right-arrow key to select a location, such as A, B, C, and so on, from the display and then press $=$. Or you may need to press a number key or a key that has a letter of the alphabet written above it to select a location and then press ENTER. As shown in the previous key sequence, we shall always use the generic sequences "STO A" and "RCL A" to indicate storing and recalling results using a particular memory location. Be sure to consult your calculator's instruction manual to learn precisely how your calculator's memory works.

🔍 A Closer Look   Most calculators have a *last answer*, or ANS key, usually the second function of the negative (−) key. This key provides still another way to do the previous example without using memory. The sequence would look like this:

117 − 23 = (48 + 704) ÷ ANS = → 8.

After you press the first $=$, the calculator automatically stores the result of 117 − 23 as the "last answer," which acts as another memory location. It will stay there until you complete the next calculation, and it can be retrieved with the ANS function. Notice that we must use parentheses around 48 + 704 to avoid the need for $=$ after 704. Pressing $=$ at that point would replace the original "last answer." ●

Your Turn   Here are a few practice problems involving grouping symbols. Use the method indicated.

(a) $(4961 - 437) \div 52$ Do not use the parentheses keys.

(b) $56 \times (38 + 12 \times 17)$ Use the parentheses keys.

(c) $2873 - (56 + 83) \times 16$ Use the parentheses keys.

(d) $\dfrac{263 \times 18 - 41 \times 12}{18 \times 16 - 17 \times 11}$ Use memory.

(e) $12 \times 16 - \dfrac{7 + 9 \times 17}{81 - 49} + 12 \times 19$ Use parentheses.

(f) Repeat problem (e) using the memory to store the value of the fraction, then doing the calculation left to right.

(g) Repeat problem (d) using the last answer key.

Answers

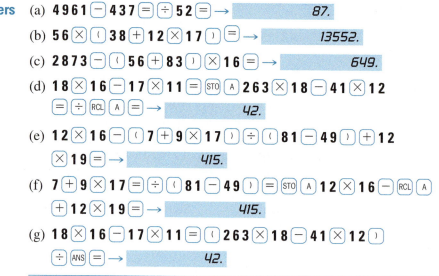

(a) **4961** ⊟ **437** ⊟ ⨌ **52** ⊟ → 87.

(b) **56** ⨯ ⦅ **38** ⊞ **12** ⨯ **17** ⦆ ⊟ → 13552.

(c) **2873** ⊟ ⦅ **56** ⊞ **83** ⦆ ⨯ **16** ⊟ → 649.

(d) **18** ⨯ **16** ⊟ **17** ⨯ **11** ⊟ STO A **263** ⨯ **18** ⊟ **41** ⨯ **12**
⊟ ⨌ RCL A ⊟ → 42.

(e) **12** ⨯ **16** ⊟ ⦅ **7** ⊞ **9** ⨯ **17** ⦆ ⨌ ⦅ **81** ⊟ **49** ⦆ ⊞ **12**
⨯ **19** ⊟ → 415.

(f) **7** ⊞ **9** ⨯ **17** ⊟ ⨌ ⦅ **81** ⊟ **49** ⦆ ⊟ STO A **12** ⨯ **16** ⊟ RCL A
⊞ **12** ⨯ **19** ⊟ → 415.

(g) **18** ⨯ **16** ⊟ **17** ⨯ **11** ⊟ ⦅ **263** ⨯ **18** ⊟ **41** ⨯ **12** ⦆
⨌ ANS ⊟ → 42.

2 Fractions

Objective	Sample problems	For help, go to page
When you finish this chapter you will be able to:		
1. Work with fractions.	(a) Write as a mixed number $\dfrac{31}{4}$. _____	78
	(b) Write as an improper fraction $3\dfrac{7}{8}$. _____	79
	Write as an equivalent fraction.	
	(c) $\dfrac{5}{16} = \dfrac{?}{64}$ _____	79
	(d) $1\dfrac{3}{4} = \dfrac{?}{32}$ _____	
	(e) Write in lowest terms $\dfrac{10}{64}$. _____	80
	(f) Which is larger, $1\dfrac{7}{8}$ or $\dfrac{5}{3}$? _____	83
2. Multiply and divide fractions.	(a) $\dfrac{7}{8} \times \dfrac{5}{32}$ _____	86
	(b) $4\dfrac{1}{2} \times \dfrac{2}{3}$ _____	
	(c) $\dfrac{3}{5}$ of $1\dfrac{1}{2}$ _____	
	(d) $\dfrac{3}{4} \div \dfrac{1}{2}$ _____	92

Name _____

Date _____

Course/Section _____

Objective	**Sample problems**	**For help, go to page**
	(e) $2\frac{7}{8} \div 1\frac{1}{4}$ _____	
	(f) $4 \div \frac{1}{2}$ _____	
3. Add and subtract fractions.	(a) $\frac{7}{16} + \frac{3}{16}$ _____	99
	(b) $1\frac{3}{16} + \frac{3}{4}$ _____	
	(c) $\frac{3}{4} - \frac{1}{5}$ _____	107
	(d) $4 - 1\frac{5}{16}$ _____	
4. Solve practical problems involving fractions.	**Carpentry** A tabletop is constructed using $\frac{3}{4}$-in. plywood with $\frac{5}{16}$-in. wood veneer on both sides. What is the total thickness of the tabletop? _____	

(Answers to these preview problems are given in the Appendix. Also, worked solutions to many of these problems appear in the chapter Summary.)

If you are certain that you can work *all* these problems correctly, turn to page 118 for a set of practice problems. If you cannot work one or more of the preview problems, turn to the page indicated to the right of the problem. For those who wish to master this material with the greatest success, turn to Section 2-1 and begin work there.

Fractions

The word *fraction* comes from a Latin word meaning "to break," and fraction numbers are used when we need to break down standard measuring units into smaller parts. In this chapter, you will learn how to perform all four basic operations with fractions as well as how to express fractions in various equivalent forms. You will also learn how fractions are applied in many different trade and technical occupations.

CASE STUDY: Restaurant Ownership

Fractions can be especially useful for those involved in the **culinary arts**. Recipes often contain amounts of ingredients expressed in fraction form, and fractions can be used to expand or downsize recipes. You will learn more about the role of fractions in the restaurant industry when you work through the Case Study at the end of this chapter.

2-1 # Working with Fractions

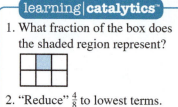
learning catalytics™

1. What fraction of the box does the shaded region represent?

2. "Reduce" $\frac{4}{8}$ to lowest terms.

Look at the following piece of lumber. What happens when we break it into equal parts?

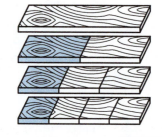

2 parts, each one-half of the whole

3 parts, each one-third of the whole

4 parts, each one-fourth of the whole

Suppose we divide this area [ ] into fifths by drawing vertical lines.

Notice that the five parts or "fifths" are equal in area. [ ]

A fraction is normally written as the division of two whole numbers: $\frac{2}{3}$, $\frac{3}{4}$, or $\frac{9}{16}$. One of the five equal areas below would be "one-fifth" or $\frac{1}{5}$ of the entire area.

[ ]

EXAMPLE 1   How would you label this portion of the area? [ ] = ?

$$\frac{3}{5} = \frac{3 \text{ shaded parts}}{5 \text{ total parts}}$$

The fraction $\frac{3}{5}$ implies an area equal to three of the five total parts.

$$\frac{3}{5} = 3 \times \left(\frac{1}{5}\right)$$

There are three equal parts, and the name of each part is $\frac{1}{5}$ or one-fifth.   ●

Your Turn   In this collection of letters, HHHHPPT, what fraction are Hs?

Solution   Fraction of Hs $= \dfrac{\text{number of Hs}}{\text{total number of letters}} = \dfrac{4}{7}$ (read it "four sevenths")

The fraction of Ps is $\frac{2}{7}$, and the fraction of Ts is $\frac{1}{7}$.

Numerator and Denominator   The two numbers that form a fraction are given special names to simplify talking about them. In the fraction $\frac{3}{5}$, the upper number 3 is called the **numerator** from the Latin *numero* meaning "number." It is a count of the number of parts.

The lower number 5 is called the **denominator** from the Latin *nomen* or "name." It tells us the name of the part being counted. The numerator and denominator are called the **terms** of the fraction.

$$\frac{3}{5}$$

◁ Numerator, the number of parts

◁ Denominator, the name of part, "fifths"

Learning Help   A handy memory aid is to remember that the denominator is the "down part"—D for down. ●

Your Turn   A paperback book costs $6 and I have $5. What fraction of its cost do I have?

numerator = _____, denominator = _____

Answer $ $ $ $ $ $5 is $\frac{5}{6}$ of the total cost.

$\underbrace{\hphantom{\$\ \$\ \$\ \$\ \$}}_{5}$ └numerator = 5, denominator = 6

More Practice   Complete the following sentences by writing in the correct fraction.

(a) If we divide a length into eight equal parts, each part will be _____ of the total length.

(b) Then three of these parts will represent _____ of the total length.

(c) Eight of these parts will be _____ of the total length.

(d) Ten of these parts will be _____ of the total length.

Answers   (a) $\dfrac{1}{8}$ (b) $\dfrac{3}{8}$ (c) $\dfrac{8}{8}$ (d) $\dfrac{10}{8}$

Proper and Improper Fractions   The original length is used as a standard for comparison, and any other length—smaller or larger—can be expressed as a fraction of it. A **proper fraction** is a number less than 1, as you would suppose a fraction should be. It represents a quantity less than the standard. In the preceding practice set, answer (a), $\frac{1}{8}$, and answer (b), $\frac{3}{8}$, are both proper fractions. Notice that for a proper fraction, the numerator is less than the denominator—the top number is less than the bottom number in the fraction.

An **improper fraction** is a number greater than or equal to 1 and represents a quantity greater than or equal to the standard. In the preceding practice set, answer (c), $\frac{8}{8}$, is an improper fraction because it is equal to 1. Answer (d), $\frac{10}{8}$, is an improper fraction because it is greater than 1. Notice that for an improper fraction, the numerator (top number) is greater than or equal to the denominator (bottom number).

Your Turn   Circle the proper fractions in the following list.

$$\frac{3}{2} \qquad \frac{3}{4} \qquad \frac{7}{8} \qquad \frac{5}{4} \qquad \frac{15}{12} \qquad \frac{1}{16} \qquad \frac{32}{32} \qquad \frac{7}{50} \qquad \frac{65}{64} \qquad \frac{105}{100}$$

Answer   You should have circled the following proper fractions: $\frac{3}{4}, \frac{7}{8}, \frac{1}{16}, \frac{7}{50}$. In each circled fraction the numerator (top number) is less than the denominator (bottom number). Each of these fractions represents a number less than 1.

Mixed Numbers   The improper fraction $\frac{7}{3}$ can be shown graphically as follows:

Unit standard = ⬜⬜⬜ is equal to 1

$\frac{1}{3}$ = ⬜

then, $\frac{7}{3}$ = ⬜⬜⬜⬜⬜⬜⬜ (seven; count 'em!)

We can rename this number by regrouping.

| $\frac{1}{3}$ | $\frac{1}{3}$ | $\frac{1}{3}$ | equals 1 |

| $\frac{1}{3}$ | $\frac{1}{3}$ | $\frac{1}{3}$ | equals 1 |

| $\frac{1}{3}$ | $\frac{7}{3} = 2 + \frac{1}{3}$ or $2\frac{1}{3}$

A **mixed number** is an improper fraction written as the sum of a whole number and a proper fraction.

$$\frac{7}{3} = 2 + \frac{1}{3} \text{ or } 2\frac{1}{3}$$

We usually omit the $+$ sign and write $2 + \frac{1}{3}$ as $2\frac{1}{3}$, and read it as "two and one-third." The numbers $1\frac{1}{2}$, $2\frac{2}{5}$, and $16\frac{2}{3}$ are all written as mixed numbers.

EXAMPLE 2   To write the improper fraction $\frac{13}{5}$ as a mixed number, divide the numerator by the denominator and form a new fraction as shown:

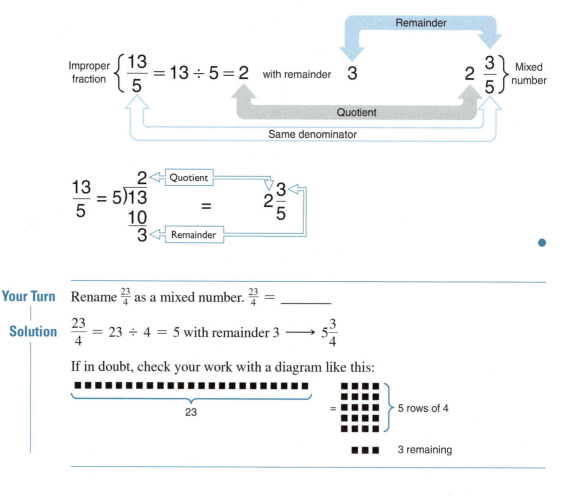

Improper fraction $\left\{ \dfrac{13}{5} = 13 \div 5 = 2 \right.$ with remainder 3 $2\,\dfrac{3}{5} \left. \right\}$ Mixed number

$$\frac{13}{5} = 5\overline{)13} \quad \begin{array}{c} 2 \\ \underline{10} \\ 3 \end{array} = 2\frac{3}{5}$$

Your Turn   Rename $\frac{23}{4}$ as a mixed number. $\frac{23}{4} = $ _____

Solution $\dfrac{23}{4} = 23 \div 4 = 5$ with remainder $3 \longrightarrow 5\dfrac{3}{4}$

If in doubt, check your work with a diagram like this:

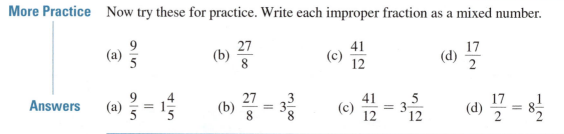

More Practice   Now try these for practice. Write each improper fraction as a mixed number.

(a) $\dfrac{9}{5}$ (b) $\dfrac{27}{8}$ (c) $\dfrac{41}{12}$ (d) $\dfrac{17}{2}$

Answers   (a) $\dfrac{9}{5} = 1\dfrac{4}{5}$ (b) $\dfrac{27}{8} = 3\dfrac{3}{8}$ (c) $\dfrac{41}{12} = 3\dfrac{5}{12}$ (d) $\dfrac{17}{2} = 8\dfrac{1}{2}$

The reverse process, rewriting a mixed number as an improper fraction, is equally simple.

EXAMPLE 3 We rewrite the mixed number $2\frac{3}{5}$ in improper fraction form as follows:

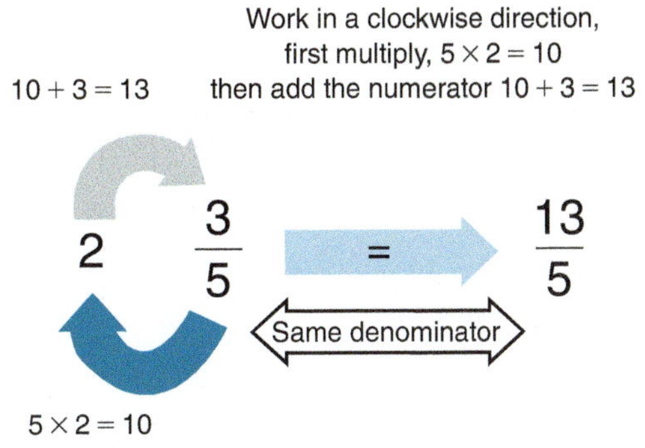
Work in a clockwise direction,
first multiply, 5 × 2 = 10
then add the numerator 10 + 3 = 13
10 + 3 = 13
2 3/5
=
13/5
Same denominator
5 × 2 = 10

Graphically, $2\frac{3}{5} = 1 + 1 + \frac{3}{5}$

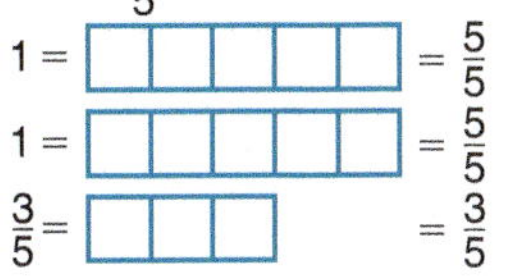
1 =
= 5/5
1 =
= 5/5
3/5 =
= 3/5

or $\frac{13}{5}$ (count them)

Your Turn Rewrite these mixed numbers as improper fractions.

(a) $3\frac{1}{5}$ (b) $4\frac{3}{8}$ (c) $1\frac{1}{16}$ (d) $5\frac{1}{2}$ (e) $15\frac{2}{3}$ (f) $9\frac{3}{4}$

Answers (a) **Step 1** $3 \times 5 = 15$

Step 2 $15 + 1 = 16$ ⟵ The new numerator

Step 3 $3\frac{1}{5} = \frac{16}{5}$ ⟵ The original denominator

(b) $4\frac{3}{8} = \frac{35}{8}$ (c) $1\frac{1}{16} = \frac{17}{16}$ (d) $5\frac{1}{2} = \frac{11}{2}$ (e) $15\frac{2}{3} = \frac{47}{3}$ (f) $9\frac{3}{4} = \frac{39}{4}$

Equivalent Fractions

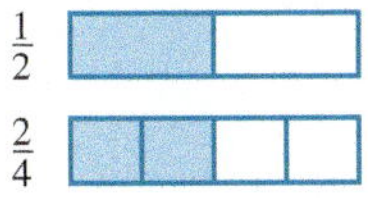
1/2
2/4

Two fractions are said to be **equivalent fractions** if they represent the same number. For example, $\frac{1}{2} = \frac{2}{4}$ because both fractions represent the same portion of some standard amount.

There is a very large set of fractions equivalent to $\frac{1}{2}$.

$$\frac{1}{2} = \frac{2}{4} = \frac{3}{6} = \frac{4}{8} = \frac{5}{10} = \cdots = \frac{46}{92} = \frac{61}{122} = \frac{1437}{2874} \quad \text{and so on}$$

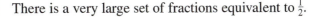

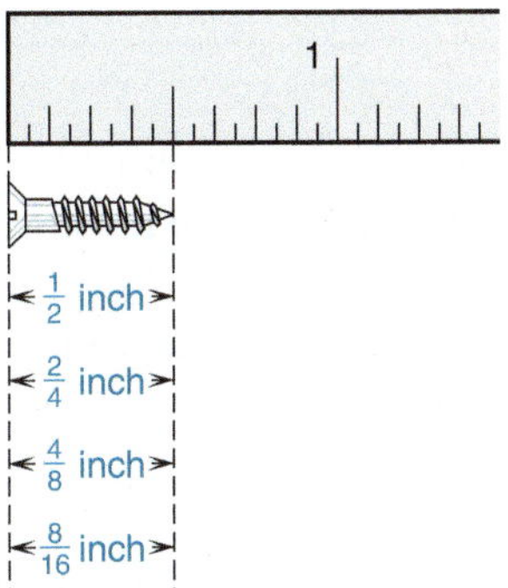
1
1/2 inch
2/4 inch
4/8 inch
8/16 inch

Each fraction is the same number, and we can use these fractions interchangeably in any mathematics problem.

To obtain a fraction equivalent to any given fraction, multiply the original numerator and denominator by the same nonzero number.

EXAMPLE 4   Rewrite the fraction $\frac{2}{3}$ as an equivalent fraction with denominator 18.

$$\frac{2}{3} = \frac{2 \times 6}{3 \times 6} = \frac{12}{18}$$

Multiply top and bottom by 6

Your Turn   Rewrite the fraction $\frac{3}{4}$ as an equivalent fraction with denominator equal to 20.

$$\frac{3}{4} = \frac{?}{20}$$

Solution $\frac{3}{4} = \frac{3 \times \square}{4 \times \square} = \frac{?}{20}$ $4 \times \square = 20$, so $\square$ must be 5.

$$= \frac{3 \times 5}{4 \times 5} = \frac{15}{20}$$

The number value of the fraction has not changed; we have simply renamed it.

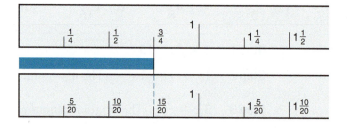

More Practice   Practice with these.

(a) $\frac{5}{8} = \frac{?}{16}$ (b) $\frac{7}{16} = \frac{?}{48}$ (c) $1\frac{2}{3} = \frac{?}{12}$ (d) $5\frac{1}{2} = \frac{?}{10}$

Solutions   (a) $\frac{5}{8} = \frac{5 \times 2}{8 \times 2} = \frac{10}{16}$ (b) $\frac{7}{16} = \frac{7 \times 3}{16 \times 3} = \frac{21}{48}$

(c) $1\frac{2}{3} = \frac{5}{3} = \frac{5 \times 4}{3 \times 4} = \frac{20}{12}$ (d) $5\frac{1}{2} = \frac{11}{2} = \frac{11 \times 5}{2 \times 5} = \frac{55}{10}$

Writing in Lowest Terms   Very often in working with fractions you will be asked to write a fraction in **lowest terms**. This means to replace it with the most simple fraction in its set of equivalent fractions. This is sometimes called "reducing" the fraction to lowest terms. To write $\frac{16}{32}$ in its lowest terms means to replace it with $\frac{1}{2}$. They are equivalent.

$$\frac{16}{32} = \frac{16 \div 16}{32 \div 16} = \frac{1}{2}$$

We have divided both the numerator and denominator of the fraction by 16.

In general, you would write a fraction in lowest terms as follows:

EXAMPLE 5   Write $\dfrac{30}{48}$ in lowest terms.

First, find the largest number that divides both the numerator and the denominator of the fraction exactly.

$$30 = 5 \times 6$$

$$48 = 8 \times 6$$

In this case, the factor 6 is the largest number that divides both parts of the fraction exactly.

Second, eliminate this common factor by dividing.

$$\frac{30}{48} = \frac{30 \div 6}{48 \div 6} = \frac{5}{8}$$

The fraction $\frac{5}{8}$ is the simplest fraction equivalent to $\frac{30}{48}$. No whole number greater than 1 divides both 5 and 8 exactly.   •

The process is exactly the same for improper fractions.

EXAMPLE 6   Write $\dfrac{105}{90}$ in lowest terms.

$$\frac{105}{90} = \frac{105 \div 15}{90 \div 15} = \frac{7}{6}$$

This process of eliminating a common factor is usually called *canceling*. When you cancel a factor, you divide both the top and the bottom of the fraction by that number. We would write $\dfrac{105}{90}$ as $\dfrac{7 \times \cancel{15}}{6 \times \cancel{15}} = \dfrac{7}{6}$. When you "cancel," you must divide by a *pair* of common factors: one factor in the numerator and the same factor in the denominator.   •

A Closer Look   Sometimes, it may be difficult to find the *largest* number that divides both the numerator and denominator. You can still reduce the fraction to lowest terms by using two or more steps. In Example 6, for instance, you might have just divided by 5 at first:

$$\frac{105}{90} = \frac{105 \div 5}{90 \div 5} = \frac{21}{18}$$

Checking this result, you can see that 3 is also a common factor. Now divide the numerator and denominator by 3.

$$\frac{21}{18} = \frac{21 \div 3}{18 \div 3} = \frac{7}{6}$$

Even though it took two steps to do it, the fraction is now written in lowest terms.   •

EXAMPLE 7   This one is a little tricky. Write $\dfrac{6}{3}$ in lowest terms.

$$\frac{6}{3} = \frac{6 \div 3}{3 \div 3} = \frac{2}{1} \qquad \text{or simply 2.}$$ •

Any whole number may be written as a fraction by using a denominator equal to 1.

$$3 = \frac{3}{1} \qquad 4 = \frac{4}{1} \qquad \text{and so on.}$$

Writing whole numbers in this way will be helpful when you learn to do arithmetic with fractions.

Your Turn   Write the following fractions in lowest terms.

(a) $\dfrac{6}{8}$ (b) $\dfrac{12}{16}$ (c) $\dfrac{2}{4}$ (d) $\dfrac{4}{12}$

(e) $\dfrac{15}{84}$ (f) $\dfrac{35}{21}$ (g) $\dfrac{12}{32}$ (h) $\dfrac{42}{7}$

Solutions   (a) $\dfrac{6}{8} = \dfrac{6 \div 2}{8 \div 2} = \dfrac{3}{4}$ (b) $\dfrac{12}{16} = \dfrac{12 \div 4}{16 \div 4} = \dfrac{3}{4}$

(c) $\dfrac{2}{4} = \dfrac{2 \div 2}{4 \div 2} = \dfrac{1}{2}$ (d) $\dfrac{4}{12} = \dfrac{4 \div 4}{12 \div 4} = \dfrac{1}{3}$

(e) $\dfrac{15}{84} = \dfrac{15 \div 3}{84 \div 3} = \dfrac{5}{28}$ (f) $\dfrac{35}{21} = \dfrac{35 \div 7}{21 \div 7} = \dfrac{5}{3}$ or $1\dfrac{2}{3}$

(g) $\dfrac{12}{32} = \dfrac{12 \div 4}{32 \div 4} = \dfrac{3}{8}$ (h) $\dfrac{42}{7} = \dfrac{42 \div 7}{7 \div 7} = \dfrac{6}{1}$ or 6

Practical Applications

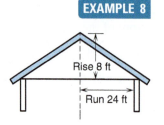

Rise 8 ft

Run 24 ft

EXAMPLE 8   **Roofing**   In roof construction, the **pitch** or steepness of a roof is defined as the fraction

$$\text{Pitch} = \frac{\text{rise*}}{\text{run}}$$

The **rise** is the increase in height and the **run** is the corresponding horizontal distance. Typically, roofers express the pitch as the amount of rise per foot, or per 12 inches, of run. Therefore, rather than write the pitch as a fraction in lowest terms, they express it as an equivalent fraction with a denominator of 12. For the roof shown,

$$\text{Pitch} = \frac{\text{rise}}{\text{run}} = \frac{8 \text{ ft}}{24 \text{ ft}} = \frac{8 \div 2}{24 \div 2} = \frac{4}{12}$$

Divide by 2 . . . to get a denominator of 12

Your Turn   (a) **Roofing**   Calculate the pitch of a storage shed roof having a rise of 6 ft over a run of 36 ft.

(b) **Allied Health**   A technician needs to dilute antiseptic for use in hand washing before and after surgery. The instructions say that 250 milliliters (mL) of antiseptic should be added to 750 mL water. What fraction of the final solution is water?

*This is the most commonly used definition of pitch today. However, in the past, pitch was often defined as $\dfrac{\text{rise}}{\text{span}}$, where the span is twice the run.

Solutions　(a)　Pitch $= \dfrac{6\text{ ft}}{36\text{ ft}} = \dfrac{6 \div 3}{36 \div 3} = \dfrac{2}{12}$

(b)　We wish to know what fraction of the final solution is water, so we write the amount of water, 750 mL, in the numerator and the *total* amount of solution, 750 mL + 250 mL, in the denominator.

$$\frac{750\text{ mL}}{750\text{ mL} + 250\text{ mL}} = \frac{750\text{ mL}}{1000\text{ mL}} = \frac{3}{4}$$

Comparing Fractions　If you were offered your choice between $\dfrac{2}{3}$ of a certain amount of money and $\dfrac{5}{8}$ of it, which would you choose?

EXAMPLE 9　Which is the larger fraction, $\dfrac{2}{3}$ or $\dfrac{5}{8}$?

Can you decide? Rewriting the fractions as equivalent fractions will help.

To compare two fractions, rename each by changing them to equivalent fractions with the same denominator.

$$\frac{2}{3} = \frac{2 \times 8}{3 \times 8} = \frac{16}{24} \qquad \text{and} \qquad \frac{5}{8} = \frac{5 \times 3}{8 \times 3} = \frac{15}{24}$$

Now compare the new fractions: $\dfrac{16}{24}$ is greater than $\dfrac{15}{24}$ and therefore $\dfrac{2}{3}$ is larger than $\dfrac{5}{8}$. ●

Learning Help　1.　The new denominator is the product of the original ones ($24 = 8 \times 3$).

2.　Once both fractions are written with the same denominator, the one with the larger numerator is the larger fraction (16 of the fractional parts is more than 15 of them). ●

Your Turn　Which of the following quantities is the larger?

(a)　$\dfrac{3}{4}$ in. or $\dfrac{5}{7}$ in.

(b)　$\dfrac{7}{8}$ or $\dfrac{19}{21}$

(c)　3 or $\dfrac{40}{13}$

(d)　$1\dfrac{7}{8}$ lb or $\dfrac{5}{3}$ lb

(e)　$2\dfrac{1}{4}$ ft or $\dfrac{11}{6}$ ft

Solutions　(a)　$\dfrac{3}{4} = \dfrac{21}{28}, \dfrac{5}{7} = \dfrac{20}{28}; \dfrac{21}{28}$ is larger than $\dfrac{20}{28}$, so $\dfrac{3}{4}$ in. is larger than $\dfrac{5}{7}$ in.

(b)　$\dfrac{7}{8} = \dfrac{147}{168}, \dfrac{19}{21} = \dfrac{152}{168}; \dfrac{152}{168}$ is larger than $\dfrac{147}{168}$, so $\dfrac{19}{21}$ is larger than $\dfrac{7}{8}$.

(c)　$3 = \dfrac{39}{13}; \dfrac{40}{13}$ is larger than $\dfrac{39}{13}$, so $\dfrac{40}{13}$ is larger than 3.

(d)　$1\dfrac{7}{8} = \dfrac{15}{8} = \dfrac{45}{24}, \dfrac{5}{3} = \dfrac{40}{24}; \dfrac{45}{24}$ is larger than $\dfrac{40}{24}$, so $1\dfrac{7}{8}$ lb is larger than $\dfrac{5}{3}$ lb.

(e)　$2\dfrac{1}{4} = \dfrac{9}{4} = \dfrac{54}{24}, \dfrac{11}{6} = \dfrac{44}{24}; \dfrac{54}{24}$ is larger than $\dfrac{44}{24}$, so $2\dfrac{1}{4}$ ft is larger than $\dfrac{11}{6}$ ft.

Practical Applications

EXAMPLE 10 **Construction** A $\frac{5}{8}$-in. drill bit is too small for a job, and a $\frac{3}{4}$-in. bit is too large. What size drill bit should be tried next?

To solve this problem, we must keep changing the given sizes to equivalent fractions until we can find a size between the two.

If we express them both in eighths, we have:

$$\frac{5}{8} \quad \text{and} \quad \frac{3}{4} = \frac{3 \times 2}{4 \times 2} = \frac{6}{8}$$

But there is no "in-between" size drill bit using these denominators. However, if we change them both to sixteenths, we have:

$$\frac{5}{8} = \frac{5 \times 2}{8 \times 2} = \frac{10}{16} \quad \text{and} \quad \frac{3}{4} = \frac{6}{8} = \frac{6 \times 2}{8 \times 2} = \frac{12}{16}$$

Now we can clearly see that an $\frac{11}{16}$-in. bit is larger than the $\frac{5}{8}$-in. bit and smaller than the $\frac{3}{4}$-in. bit. This "in-between" size bit should be tried next. ●

Your Turn **Construction** If a $\frac{3}{8}$-in. drill bit is too small for a job, and a $\frac{1}{2}$-in. bit is too large, what size bit should be tried next?

Solution To find a size between the two, we must change them both to sixteenths.

$$\frac{3}{8} = \frac{6}{16} \quad \text{and} \quad \frac{1}{2} = \frac{4}{8} = \frac{8}{16}$$

A $\frac{7}{16}$-in. bit is between the two in size and therefore should be tried next.

Now turn to Exercises 2-1 for some practice in working with fractions.

Exercises 2-1 **Working with Fractions**

A. Write as a mixed number.

1. $\frac{17}{4}$ 2. $\frac{8}{5}$ 3. $\frac{11}{8}$ 4. $\frac{40}{16}$ 5. $\frac{3}{2}$

6. $\frac{11}{3}$ 7. $\frac{100}{6}$ 8. $\frac{4}{3}$ 9. $\frac{80}{32}$ 10. $\frac{5}{2}$

B. Write as an improper fraction.

1. $2\frac{1}{3}$ 2. $7\frac{1}{2}$ 3. $8\frac{3}{8}$ 4. $1\frac{1}{16}$ 5. $2\frac{7}{8}$

6. 2 7. $2\frac{2}{3}$ 8. $4\frac{3}{64}$ 9. $4\frac{5}{6}$ 10. $1\frac{13}{16}$

C. Complete.

1. $\frac{7}{8} = \frac{?}{16}$ 2. $\frac{3}{4} = \frac{?}{16}$ 3. $\frac{1}{8} = \frac{?}{64}$ 4. $\frac{3}{8} = \frac{?}{64}$

5. $1\frac{1}{4} = \frac{?}{16}$ 6. $2\frac{7}{8} = \frac{?}{32}$ 7. $3\frac{3}{5} = \frac{?}{10}$ 8. $1\frac{1}{16} = \frac{?}{32}$

9. $1\frac{40}{60} = \frac{?}{3}$ 10. $4 = \frac{?}{6}$ 11. $2\frac{5}{8} = \frac{?}{16}$ 12. $2\frac{5}{6} = \frac{?}{12}$

D. Write in lowest terms.

1. $\frac{4}{16}$ 2. $\frac{4}{6}$ 3. $\frac{6}{16}$ 4. $\frac{18}{4}$ 5. $\frac{4}{10}$

6. $\frac{35}{30}$ 7. $\frac{24}{30}$ 8. $\frac{28}{7}$ 9. $4\frac{3}{12}$ 10. $\frac{34}{32}$

11. $\frac{42}{64}$ 12. $\frac{10}{35}$ 13. $\frac{15}{36}$ 14. $\frac{45}{18}$ 15. $\frac{38}{24}$

E. Which is larger?

1. $\frac{3}{5}$ or $\frac{4}{7}$ 2. $\frac{3}{2}$ or $\frac{13}{8}$ 3. $1\frac{1}{2}$ or $1\frac{3}{7}$ 4. $\frac{3}{4}$ or $\frac{13}{16}$

5. $\frac{7}{8}$ or $\frac{5}{6}$ 6. $2\frac{1}{2}$ or $1\frac{11}{8}$ 7. $1\frac{2}{5}$ or $\frac{6}{4}$ 8. $\frac{3}{16}$ or $\frac{25}{60}$

9. $\frac{13}{5}$ or $\frac{5}{2}$ 10. $3\frac{1}{2}$ or $2\frac{7}{4}$ 11. $\frac{3}{8}$ or $\frac{5}{12}$ 12. 4 or $\frac{37}{9}$

F.   Practical Applications

1. **Carpentry**   An apprentice carpenter measured the length of a 2-by-4 as $15\frac{6}{8}$ in. Express this measurement in lowest terms.

2. **Electrical Trades**   An electrical light circuit in John's welding shop had a load of 2800 watts. He changed the circuit to ten 150-watt bulbs and six 100-watt bulbs. What fraction represents a comparison of the new load with the old load?

3. The numbers $\frac{22}{7}$, $\frac{19}{6}$, $\frac{47}{15}$, $\frac{25}{8}$, and $\frac{41}{13}$ are all reasonable approximations to the number π. Which is the largest approximation? Which is the smallest approximation?

4. **Sheet Metal Trades**   Which is thicker, a $\frac{3}{16}$-in. sheet of metal or a $\frac{13}{64}$-in. fastener?

5. **Plumbing**   Is it possible to have a $\frac{7}{8}$-in. pipe with an inside diameter of $\frac{29}{32}$ in.? (See the figure.)

6. **Sheet Metal Trades**   Fasteners are equally spaced on a metal vent cover, with nine spaces between fasteners covering 24 in. Write the distance between spaces as a mixed number.

$$\frac{24}{9} = ?$$

7. **Printing**   A printer has 15 rolls of newsprint in the warehouse. What fraction of this total will remain if six rolls are used?

8. **Machine Trades**   A machinist who had been producing 40 parts per day increased the output to 60 parts per day by going to a faster machine. How many times faster is the new machine? Express your answer as a mixed number.

9. **Landscaping**   Before it can be used, a 12-ounce container of liquid fertilizer must be mixed with 48 ounces of water. What fraction of fertilizer is in the final mixture?

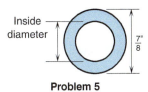

Inside
diameter

$\frac{7"}{8}$

Problem 5

10. **Roofing**   A ridge beam rises 18 in. over a horizontal run of 72 in. Calculate the pitch of the roof and express it as a fraction with a denominator of 12.

11. **Construction**   A $\frac{3}{4}$-in. drill bit is too large for a job, and an $\frac{11}{16}$-in. bit is too small. What size should be tried next?

12. **Allied Health**   During a visit to an elementary school, a health care worker determined that 15 of the 48 girls were overweight and 12 of 40 boys were overweight.

    (a) Express in lowest terms the fraction of girls who were overweight.

    (b) Express in lowest terms the fraction of boys who were overweight.

    (c) Is the fraction of overweight children higher among the girls or the boys? [*Hint:* Convert your answers to parts (a) and (b) to equivalent fractions with the same denominator.]

13. **General Trades**   The following ruler is graduated in sixteenths of an inch. For each lettered location, state the reading both in sixteenths and in lowest terms if possible.

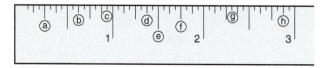

When you have had the practice you need, check your answers to the odd-numbered problems in the Appendix.

2-2   Multiplication of Fractions

learning|catalytics™

1. What is half of 18?

2. Write $\frac{24}{16}$ as a mixed number in lowest terms.

The simplest arithmetic operation with fractions is multiplication and, happily, it is easy to show graphically. The multiplication of a whole number and a fraction may be illustrated this way.

$$3 \times \frac{1}{4} = \frac{1}{4} + \frac{1}{4} + \frac{1}{4} = \frac{3}{4} \quad \text{three segments each } \frac{1}{4} \text{ unit long.}$$

Any fraction such as $\frac{3}{4}$ can be thought of as a product: $3 \times \frac{1}{4}$

The product of two fractions can also be shown graphically.

$$\frac{1}{2} \times \frac{1}{3} \quad \text{means} \quad \frac{1}{2} \text{ of } \frac{1}{3}$$

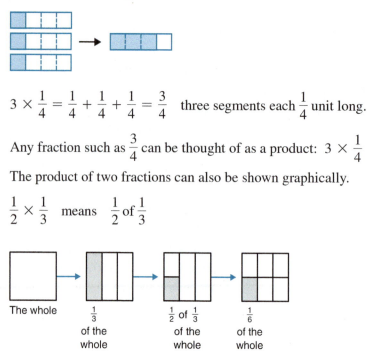

The product $\frac{1}{2} \times \frac{1}{3}$ is

$$\frac{1}{2} \times \frac{1}{3} = \frac{1}{6} = \frac{1 \text{ shaded area}}{6 \text{ equal areas in the square}}$$

In general, we calculate this product as

$$\frac{1}{2} \times \frac{1}{3} = \frac{1 \times 1}{2 \times 3}$$

> Multiply the numerators (top)
> Multiply the denominators (bottom)

$$\frac{1}{2} \times \frac{1}{3} = \frac{1}{6}$$

The product of two fractions is a fraction whose numerator is the product of their numerators and whose denominator is the product of their denominators.

EXAMPLE 1 **Fraction Times a Fraction**

Multiply $\frac{5}{6} \times \frac{2}{3}$.

$$\frac{5}{6} \times \frac{2}{3} = \frac{5 \times 2}{6 \times 3} = \frac{10}{18} = \frac{10 \div 2}{18 \div 2} = \frac{5}{9}$$

●

🖉 **Note** Always write your answer in lowest terms. In Example 1 you probably recognized that both 10 and 18 were evenly divisible by 2, so you canceled out that common factor. It will save you time and effort if you cancel common factors, such as the 2 above, *before* you multiply this way:

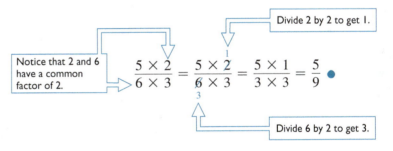

> Notice that 2 and 6 have a common factor of 2.
> Divide 2 by 2 to get 1.
> Divide 6 by 2 to get 3.

$$\frac{5 \times 2}{6 \times 3} = \frac{5 \times \overset{1}{2}}{\underset{3}{6} \times 3} = \frac{5 \times 1}{3 \times 3} = \frac{5}{9}$$ ●

❗ **Careful** You may cancel a pair of common factors only if one is in the numerator and the other is in the denominator. ●

EXAMPLE 2 **Whole Number Times a Fraction**

When multiplying a whole number by a fraction, first write the whole number as a fraction.

Multiply $6 \times \frac{1}{8}$.

Step 1 Rewrite 6 as a fraction with denominator 1. $\qquad 6 \times \frac{1}{8} = \frac{6}{1} \times \frac{1}{8}$

Step 2 Divide 6 and 8 by their common factor of 2. $\qquad = \frac{\overset{3}{6} \times 1}{1 \times \underset{4}{8}}$

Step 3 Multiply the remaining factors in each half of the fraction. $= \frac{3 \times 1}{1 \times 4} = \frac{3}{4}$ ●

EXAMPLE 3   **Mixed Number Times a Fraction**

Multiply $2\frac{1}{2} \times \frac{3}{4}$.

Step 1   Write the mixed number $2\frac{1}{2}$ as the improper fraction $\frac{5}{2}$.

$$2\frac{1}{2} \times \frac{3}{4} = \frac{5}{2} \times \frac{3}{4}$$

Step 2   Because no top and bottom pair have a common factor, we simply multiply straight across.

$$= \frac{5 \times 3}{2 \times 4} = \frac{15}{8}$$

Step 3   Write the product as a mixed number.

$$= 1\frac{7}{8}$$ ●

EXAMPLE 4   **Mixed Number Times a Mixed Number**

Multiply $3\frac{1}{3} \times 2\frac{1}{4}$.

Step 1   Write as improper fractions.

$$3\frac{1}{3} \times 2\frac{1}{4} = \frac{10}{3} \times \frac{9}{4}$$

Step 2   Divide out (cancel) two pairs of common factors: 10 and 4 have a common factor of 2; 9 and 3 have a common factor of 3.

$$= \frac{\overset{5}{\cancel{10}} \times \overset{3}{\cancel{9}}}{\underset{1}{\cancel{3}} \times \underset{2}{\cancel{4}}}$$

Step 3   Now multiply the remaining factors: $5 \times 3 = 15$ and $1 \times 2 = 2$.

$$= \frac{15}{2}$$

Step 4   Write the answer as a mixed number.

$$= 7\frac{1}{2}$$ ●

🔍 **A Closer Look**   In Step 2 of Example 4, it is not necessary to combine the two fractions into a single fraction before canceling. Your work would look like this:

✏️ **Note**   In many word problems the words *of* or *product of* appear as signals that you are to multiply. For example, the phrase "one-half of 16" means

$$\frac{1}{2} \times 16 = \frac{1}{2} \times \frac{16}{1} = \frac{1}{2} \times \frac{\overset{8}{\cancel{16}}}{\underset{1}{1}} = 8,$$ and the phrase "the product of $\frac{2}{3}$ and $\frac{1}{4}$"

should be translated as $\dfrac{2}{3} \times \dfrac{1}{4} = \dfrac{\overset{1}{\cancel{2}}}{3} \times \dfrac{1}{\underset{2}{\cancel{4}}} = \dfrac{1}{6}$. ●

Your Turn   Now test your understanding with these problems. Multiply as shown. Change any mixed numbers to improper fractions *before* you multiply.

(a) $\dfrac{3}{32}$ of $\dfrac{4}{15} =$ _____

(b) $\dfrac{15}{4} \times \dfrac{9}{10} =$ _____

(c) $\dfrac{3}{2}$ of $4 =$ _____

(d) $6 \times \dfrac{7}{8} =$ _____

(e) $3\dfrac{5}{6} \times \dfrac{3}{10} =$ _____

(f) $1\dfrac{4}{5} \times 1\dfrac{3}{4} =$ _____

(g) $\dfrac{5}{8} \times 4\dfrac{2}{3} \times \dfrac{9}{25} =$ _____

Solutions

(a) $\dfrac{\overset{1}{\cancel{3}}}{\underset{8}{\cancel{32}}} \times \dfrac{\overset{1}{\cancel{4}}}{\underset{5}{\cancel{15}}} = \dfrac{1}{40}$

(b) $\dfrac{\overset{3}{\cancel{15}}}{4} \times \dfrac{9}{\underset{2}{\cancel{10}}} = \dfrac{27}{8} = 3\dfrac{3}{8}$

(c) $\dfrac{3}{2} \times \dfrac{\overset{2}{\cancel{4}}}{1} = \dfrac{6}{1} = 6$

(d) $6 \times \dfrac{7}{8} = \dfrac{\overset{3}{\cancel{6}}}{1} \times \dfrac{7}{\underset{4}{\cancel{8}}} = \dfrac{21}{4} = 5\dfrac{1}{4}$

(e) $3\dfrac{5}{6} \times \dfrac{3}{10} = \dfrac{23}{\underset{2}{\cancel{6}}} \times \dfrac{\overset{1}{\cancel{3}}}{10} = \dfrac{23}{20} = 1\dfrac{3}{20}$

If you don't remember how to change a mixed number to an improper fraction see page 79.

(f) $1\dfrac{4}{5} \times 1\dfrac{3}{4} = \dfrac{9}{5} \times \dfrac{7}{4} = \dfrac{63}{20} = 3\dfrac{3}{20}$

(g) $\dfrac{5}{8} \times 4\dfrac{2}{3} \times \dfrac{9}{25} = \dfrac{\overset{1}{\cancel{5}}}{\underset{4}{\cancel{8}}} \times \dfrac{\overset{7}{\cancel{14}}}{\underset{1}{\cancel{3}}} \times \dfrac{\overset{3}{\cancel{9}}}{\underset{5}{\cancel{25}}} = \dfrac{21}{20} = 1\dfrac{1}{20}$

🔍 **A Closer Look**   Notice in solution (g) that we divided out a common factor of 5 from the *first* numerator and the *last* denominator. When more than two fractions are being multiplied, canceling may be done between non-adjacent fractions. ●

Practical Applications   As we mentioned in Section 1-3, we use multiplication to find a total when there are many items that share the same numerical characteristic. Remember also that the word *of* means multiplication.

EXAMPLE 5   **Retail Merchandising**   In retail management, goods are often bought and sold in quantities of dozens and fractions of a dozen. Suppose a buyer purchased $14\dfrac{2}{3}$ dozen shirts for a clothing store, and she needs to calculate the total number of shirts purchased.

Assume that the shirts are packed in boxes containing a dozen (12) shirts in each box. The buyer would have $14\dfrac{2}{3}$ boxes, so we can determine how many shirts were purchased by multiplying $14\dfrac{2}{3}$ by 12.

$$14\dfrac{2}{3} \times 12 = \dfrac{44}{\underset{1}{\cancel{3}}} \times \dfrac{\overset{4}{\cancel{12}}}{1} = 176$$

The buyer purchased 176 shirts.   ●

Your Turn   (a) **Graphic Design**   The copy for an advertisement must be reduced to $\dfrac{3}{4}$ of its original size in order to fit in a magazine. If the original dimensions of the ad are $3\dfrac{1}{2}$ in. by $5\dfrac{1}{4}$ in., what are the reduced dimensions?

(b) **Plumbing**   Polly the Plumber needs six lengths of PVC pipe each $26\dfrac{3}{4}$ in. long. What total length of pipe will she need?

Remember to write your answer in lowest terms.

Solutions   (a) In this problem, the word *of* is our clue that we need to multiply. We must multiply each original dimension by $\frac{3}{4}$.

$$\frac{3}{4} \times 3\frac{1}{2} = \frac{3}{4} \times \frac{7}{2} = \frac{21}{8} = 2\frac{5}{8} \text{ in.}$$

$$\frac{3}{4} \times 5\frac{1}{4} = \frac{3}{4} \times \frac{21}{4} = \frac{63}{16} = 3\frac{15}{16} \text{ in.}$$

The reduced dimensions of the ad are $2\frac{5}{8}$ in. by $3\frac{15}{16}$ in.

(b) Total length of pipe = number of pieces × length of each piece

$$= 6 \times 26\frac{3}{4} = 6 \times \frac{107}{4} = \frac{\overset{3}{\cancel{6}}}{1} \times \frac{107}{\underset{2}{\cancel{4}}} = \frac{321}{2} = 160\frac{1}{2} \text{ in.}$$

Exercises 2-2   **Multiplication of Fractions**

A. Multiply and write the answer in lowest terms.

1. $\frac{1}{2} \times \frac{1}{4}$ 2. $\frac{2}{5} \times \frac{2}{3}$ 3. $\frac{4}{5} \times \frac{1}{6}$ 4. $6 \times \frac{1}{2}$

5. $\frac{8}{9} \times 3$ 6. $\frac{11}{12} \times \frac{4}{15}$ 7. $\frac{8}{3} \times \frac{5}{12}$ 8. $\frac{7}{8} \times \frac{13}{14}$

9. $\frac{12}{8} \times \frac{15}{9}$ 10. $\frac{4}{7} \times \frac{49}{2}$ 11. $4\frac{1}{2} \times \frac{2}{3}$ 12. $6 \times 1\frac{1}{3}$

13. $2\frac{1}{6} \times 1\frac{1}{2}$ 14. $\frac{5}{7} \times 1\frac{7}{15}$ 15. $4\frac{3}{5} \times 15$ 16. $10\frac{5}{6} \times 3\frac{3}{10}$

17. $34 \times 2\frac{3}{17}$ 18. $7\frac{9}{10} \times 1\frac{1}{4}$ 19. $11\frac{6}{7} \times \frac{7}{8}$

20. $18 \times 1\frac{5}{27}$ 21. $\frac{1}{2} \times \frac{1}{2} \times \frac{1}{2}$ 22. $1\frac{4}{5} \times \frac{2}{3} \times \frac{1}{4}$

23. $\frac{1}{4} \times \frac{2}{3} \times \frac{2}{5}$ 24. $2\frac{1}{2} \times \frac{3}{5} \times \frac{8}{9}$ 25. $\frac{2}{3} \times \frac{3}{2} \times 2$

B. Find.

1. $\frac{1}{2}$ of $\frac{1}{3}$ 2. $\frac{1}{4}$ of $\frac{3}{8}$ 3. $\frac{2}{3}$ of $\frac{3}{4}$ 4. $\frac{7}{8}$ of $\frac{1}{2}$

5. $\frac{1}{2}$ of $1\frac{1}{2}$ 6. $\frac{3}{4}$ of $1\frac{1}{4}$ 7. $\frac{5}{8}$ of $2\frac{1}{10}$ 8. $\frac{5}{3}$ of $1\frac{2}{3}$

9. $\frac{4}{3}$ of $\frac{3}{4}$ 10. $\frac{3}{5}$ of $1\frac{1}{6}$ 11. $\frac{7}{8}$ of $1\frac{1}{5}$ 12. $\frac{3}{5}$ of 4

13. $\frac{7}{16}$ of 6 14. $\frac{5}{16}$ of $1\frac{1}{7}$ 15. $\frac{3}{8}$ of $2\frac{2}{3}$ 16. $\frac{15}{16}$ of $1\frac{3}{5}$

C. Practical Applications

1. **Flooring and Carpeting** Find the width of floor space covered by 38 boards with $3\frac{5}{8}$-in. exposed surface each.

2. **Construction** There are 14 risers in the stairs from the basement to the first floor of a house. Find the total height of the stairs if the risers are $7\frac{1}{8}$ in. high.

3. **Roofing** Shingles are laid so that 5 in. or $\frac{5}{12}$ ft is exposed in each layer. How many feet of roof will be covered by 28 courses?

4. **Carpentry** A board $5\frac{3}{4}$ in. wide is cut to three-fourths of its original width. Find the new width.

5. **Carpentry** What length of 2-in. by 4-in. material will be required to make six bench legs, each $28\frac{1}{4}$ in. long?

6. **Electrical Trades** Find the total length of 12 pieces of wire, each $9\frac{3}{16}$ in. long.

7. **Automotive Trades** If a car averages $22\frac{3}{4}$ miles to a gallon of gas, how many miles can it travel on 14 gallons of gas?

8. **Machine Trades** What is the shortest bar that can be used for making six chisels, each $6\frac{1}{8}$ in. in length?

9. **Manufacturing** How many pounds of grease are contained in a barrel if a barrel holds $46\frac{1}{2}$ gallons, and a gallon of grease weighs $7\frac{2}{3}$ lb?

10. **Plumbing** A plumber cut eight pieces of copper tubing from a coil. Each piece is $14\frac{3}{4}$ in. long. What is the total length used?

11. **Machine Trades** How far will a nut advance if it is given 18 turns on a $\frac{1}{4}$-in. 20-NF (National Fine thread) bolt? (*Hint:* The designation 20-NF means that the nut advances $\frac{1}{20}$ in. for each complete turn.)

12. **Landscaping** A landscaping project requires that you estimate the length of a drain line. You pace off the length as being 24 strides long. You know from prior experience that each of your strides is $\frac{3}{4}$ of a yard. What would be your estimate of the length of the drain line?

13. **Drafting** If $\frac{3}{8}$ in. on a drawing represents 1 ft, how many inches on the drawing will represent 26 ft?

14. **Manufacturing** What is the volume of a rectangular box with interior dimensions $12\frac{1}{2}$ in. long, $8\frac{1}{8}$ in. wide, and $4\frac{1}{4}$ in. deep? (*Hint:* Volume = length × width × height.)

15. **Machine Trades** How long will it take to machine 45 pins if each pin requires $6\frac{3}{4}$ minutes? Allow 1 minute per pin for placing stock in the lathe.

16. **Manufacturing** There are 231 cu in. in a gallon. How many cubic inches are needed to fill a container with a rated capacity of $4\frac{1}{3}$ gallons?

17. **Printing** In a print shop, 1 unit of labor is equal to $\frac{1}{6}$ hour. How many hours are involved in 64 units of work?

18. **Carpentry** Find the total width of 36 2-by-4s if the finished width of each board is actually $3\frac{1}{2}$ in.

19. **Photography** A photograph must be reduced to four-fifths of its original size to fit the space available in a newspaper. Find the length of the reduced photograph if the original was $6\frac{3}{4}$ in. long.

20. **Printing** A bound book weighs $1\frac{5}{8}$ lb. How many pounds will 20 cartons of 12 books each weigh?

21. **Printing** There are 6 picas in 1 in. If a line of type is $3\frac{3}{4}$ in. long, what is this length in picas?

22. **Water/Wastewater Treatment** The normal daily flow of raw sewage into a treatment plant is 32 million gallons per day (MGD). Because of technical problems one day, the plant had to cut back to three-fourths of its normal intake. What was this reduced flow?

23. **Masonry**   What is the height of 12 courses of $2\frac{1}{4}$-in. bricks with $\frac{3}{8}$-in. mortar joints? (12 rows of brick and 11 mortar joints)

24. **Plumbing**   To find the degree measure of the bend of a pipe fitting, multiply the fraction of bend by 360 degrees. What is the degree measure of a $\frac{1}{8}$ bend? A $\frac{1}{5}$ bend? A $\frac{1}{6}$ bend?

25. **Plumbing**   A drain must be installed with a grade of $\frac{1}{8}$ in. of vertical drop per foot of horizontal run. How much drop will there be for 26 ft of run?

26. **Machine Trades**   The center-to-center distance between consecutive holes in a strip of metal is $\frac{5}{16}$ in. What is the total distance x between the first and last centers as shown in the figure?

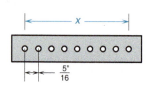

Problem 26

27. **Sheet Metal Trades**   The allowance for a wired edge on fabricated metal is $2\frac{1}{2}$ times the diameter of the wire. Calculate the allowance for a wired edge if the diameter of the wire is $\frac{3}{16}$ in.

28. **Allied Health**   One caplet of a certain cold medicine contains 240 mg of medication. How much medication is contained in $3\frac{1}{2}$ caplets?

29. **Allied Health**   From the age of 1 year to 3 years, a child's weight increases at an average rate of $4\frac{5}{8}$ pounds per year. How much weight would an average child be expected to gain in these two years?

30. **Graphic Design**   A graphic artist must produce 12 different illustrations for a report. Each illustration takes about $\frac{3}{4}$ of an hour to prepare, and she will need an additional 2 hours to insert the illustrations into the report. How many hours should she set aside for this project?

31. **Construction**   A patio $15\frac{1}{2}$ feet wide is being constructed next to a house. To drain water off the patio and away from the house, a slope of $\frac{1}{4}$ inch per foot is necessary. What total difference in height is required from one edge of the patio to the other?

32. **Carpentry**   A carpenter is drilling a pilot hole for a $\frac{1}{8}$-inch wood screw. Pilot holes must be slightly smaller than the actual size of the screw. Which of the following sized pilot holes should the carpenter drill: $\frac{9}{64}$ in., $\frac{5}{64}$ in., $\frac{5}{32}$ in., or $\frac{3}{16}$ in.?

33. **Retail Merchandising**   Taking inventory, the manager of a retail store found $6\frac{1}{6}$ dozen pairs of jeans in stock. How many pairs of jeans were there?

Check your answers to the odd-numbered problems in the Appendix, then continue in Section 2-3.

2-3   Division of Fractions

Addition and multiplication are both reversible arithmetic operations. For example,

4 + 5 and 5 + 4 both equal 9

2 × 3 and 3 × 2 both equal 6

The order in which you add or multiply is not important. This reversibility is called the **commutative property** of addition and multiplication.

In division this type of exchange is not allowed, and because it is not allowed many people find division very troublesome. In the division of fractions it is very important that you set up the problem correctly.

The phrase "8 divided by 4" can be written as

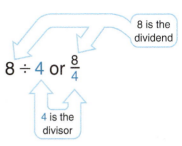

$$8 \div 4 \text{ or } \frac{8}{4}$$

In the previous problem you were being asked to divide a set of eight objects into sets of four objects.

EXAMPLE 1 In the division $5 \div \frac{1}{2}$, which number is the divisor?

The divisor is $\frac{1}{2}$.

The division $5 \div \frac{1}{2}$, read "5 divided by $\frac{1}{2}$," asks how many $\frac{1}{2}$-unit lengths are included in a length of 5 units.

5 units $\frac{1}{2}$ unit

Division answers the question: How many of the divisor are in the dividend?

$8 \div 4 = \square$ asks you to find how many 4s are in 8.
It is easy to see that $\square = 2$.

$5 \div \frac{1}{2} = \square$ asks you to find how many $\frac{1}{2}$s are in 5. Do you see that $\square = 10$?

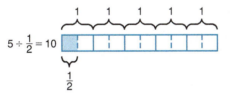

There are ten $\frac{1}{2}$-unit lengths contained in the 5-unit length.

Using a drawing like this to solve a division problem is difficult and clumsy. We need a simple rule.

Reciprocal We can simplify the division of fractions with the help of a new concept called the **reciprocal.** The reciprocal of a fraction is obtained by switching its numerator and denominator. (This is often called *inverting* the fraction.)

EXAMPLE 2 The reciprocal of $\frac{5}{6}$ is $\frac{6}{5}$.

The reciprocal of $\frac{1}{4}$ is $\frac{4}{1}$ or 4.

The reciprocal of 8 or $\frac{8}{1}$ is $\frac{1}{8}$.

To find the reciprocal of a mixed number, first convert it to an improper fraction.

The reciprocal of $2\frac{3}{5}$ or $\frac{13}{5}$ is $\frac{5}{13}$.

Your Turn   Find the reciprocal of each of the following numbers.

(a) $\frac{2}{5}$ (b) $\frac{7}{8}$ (c) $\frac{1}{6}$ (d) 7 (e) $2\frac{1}{5}$

Answers   (a) $\frac{5}{2}$ (b) $\frac{8}{7}$ (c) $\frac{6}{1}$ or 6   (d) $7 = \frac{7}{1}$ so its reciprocal is $\frac{1}{7}$.

(e) $2\frac{1}{5}$ is $\frac{11}{5}$ and its reciprocal is $\frac{5}{11}$.

We can now use the following simple rule to divide fractions:

> To divide by a fraction, multiply by its reciprocal.

EXAMPLE 3 $5 \div \frac{1}{2} = ?$

$$5 \div \frac{1}{2} \underset{\text{Reciprocal}}{\overset{\text{Multiply}}{=}} 5 \times \frac{2}{1} = \frac{10}{1} = 10 \text{ as shown graphically on the previous page}$$

EXAMPLE 4 $\frac{2}{5} \div \frac{4}{3} = ?$

$$\frac{2}{5} \div \frac{4}{3} \underset{\text{The reciprocal}}{\overset{\text{Multiply}}{=}} \frac{2}{5} \times \frac{3}{4} = \frac{\overset{1}{2} \times 3}{5 \times \underset{2}{4}} = \frac{3}{10}$$

⚠ Careful   When dividing fractions, never attempt to cancel common factors until *after* converting the division to multiplication by the reciprocal. ●

EXAMPLE 5 $\frac{5}{8} \div 3\frac{3}{4} = ?$

$$\frac{5}{8} \div 3\frac{3}{4} = \frac{5}{8} \div \frac{15}{4} = \frac{\overset{1}{5}}{\underset{2}{8}} \times \frac{\overset{1}{4}}{\underset{3}{15}} = \frac{1}{6}$$

We have converted division problems that are difficult to picture into simple multiplication.

✓ The final, and very important, step in every division is checking the answer. To check, multiply the divisor by the quotient and compare this answer with the original fraction or dividend.

If $\dfrac{5}{8} \div 3\dfrac{3}{4} = \dfrac{1}{6}$ then $3\dfrac{3}{4} \times \dfrac{1}{6}$ should equal $\dfrac{5}{8}$.

$$3\dfrac{3}{4} \times \dfrac{1}{6} = \dfrac{\overset{5}{\cancel{15}}}{4} \times \dfrac{1}{\underset{2}{\cancel{6}}} = \dfrac{5}{8}$$

●

Your Turn   Divide:

(a) $\dfrac{7}{8} \div \dfrac{3}{2} = $ _____   (b) $6 \div 4\dfrac{1}{2} = $ _____

Solutions

(a) $\dfrac{7}{8} \overset{\boxed{\text{Multiply}}}{\underset{\boxed{\text{Reciprocal}}}{\div}} \dfrac{3}{2} = \dfrac{7}{8} \times \dfrac{2}{3} = \dfrac{7 \times \overset{1}{\cancel{2}}}{\underset{4}{\cancel{8}} \times 3} = \dfrac{7}{12}$ ☑ $\dfrac{\cancel{3}}{2} \times \dfrac{7}{\underset{4}{\cancel{12}}} = \dfrac{7}{8}$

(b) $6 \div 4\dfrac{1}{2} = \dfrac{6}{1} \overset{\boxed{\text{Multiply}}}{\underset{\boxed{\text{Reciprocal}}}{\div}} \dfrac{9}{2} = \dfrac{6}{1} \times \dfrac{2}{9} = \dfrac{\overset{2}{\cancel{6}} \times 2}{1 \times \underset{3}{\cancel{9}}} = \dfrac{4}{3} = 1\dfrac{1}{3}$

☑ $4\dfrac{1}{2} \times \dfrac{4}{3} = \dfrac{\overset{3}{\cancel{9}}}{\underset{1}{\cancel{2}}} \times \dfrac{\overset{2}{\cancel{4}}}{\underset{1}{\cancel{3}}} = \dfrac{6}{1} = 6$

🖐 **Learning Help**   The chief source of confusion for many people in dividing fractions is deciding which fraction to invert. It will help if you:

1. Put every division problem in the form

   (dividend) ÷ (divisor)

   then find the reciprocal of the divisor, and finally, multiply to obtain the quotient.

2. Check your answer by multiplying. The product

   (divisor) × (quotient or answer)

   should equal the dividend. ●

More Practice   Here are a few more problems to test your understanding.

(a) $\dfrac{2}{5} \div \dfrac{3}{8}$ (b) $\dfrac{7}{40} \div \dfrac{21}{25}$ (c) $3\dfrac{3}{4} \div \dfrac{5}{2}$ (d) $4\dfrac{1}{5} \div 1\dfrac{4}{10}$

(e) $3\dfrac{2}{3} \div 3$ (f) Divide $\dfrac{3}{4}$ by $2\dfrac{5}{8}$.   (g) Divide $1\dfrac{1}{4}$ by $1\dfrac{7}{8}$.

Work carefully and check each answer.

Solutions   (a) $\dfrac{2}{5} \div \dfrac{3}{8} = \dfrac{2}{5} \times \dfrac{8}{3} = \dfrac{16}{15} = 1\dfrac{1}{15}$

The answer is $1\dfrac{1}{15}$.   ☑ $\dfrac{\overset{1}{\cancel{3}}}{\underset{1}{\cancel{8}}} \times \dfrac{\overset{2}{\cancel{16}}}{\underset{5}{\cancel{15}}} = \dfrac{2}{5}$

(b) $\dfrac{7}{40} \div \dfrac{21}{25} = \dfrac{\overset{1}{\cancel{7}}}{\cancel{40}} \times \dfrac{\overset{5}{\cancel{25}}}{\cancel{21}} = \dfrac{5}{24}$ ☑ $\dfrac{\overset{7}{\cancel{21}}}{\cancel{25}} \times \dfrac{\overset{1}{\cancel{5}}}{\cancel{24}} = \dfrac{7}{40}$

(c) $3\dfrac{3}{4} \div \dfrac{5}{2} = \dfrac{15}{4} \div \dfrac{5}{2} = \dfrac{\overset{3}{\cancel{15}}}{\underset{2}{\cancel{4}}} \times \dfrac{\overset{1}{\cancel{2}}}{\cancel{5}} = \dfrac{3}{2} = 1\dfrac{1}{2}$ ☑ $\dfrac{5}{2} \times \dfrac{3}{2} = \dfrac{15}{4} = 3\dfrac{3}{4}$

(d) $4\dfrac{1}{5} \div 1\dfrac{4}{10} = \dfrac{21}{5} \div \dfrac{14}{10} = \dfrac{\overset{3}{\cancel{21}}}{\cancel{5}} \times \dfrac{\overset{2}{\cancel{10}}}{\cancel{14}} = \dfrac{3}{1} \times \dfrac{\overset{1}{\cancel{2}}}{\cancel{2}} = 3$

☑ $\quad 1\dfrac{4}{10} \times 3 = \dfrac{\overset{7}{\cancel{14}}}{\underset{5}{\cancel{10}}} \times \dfrac{3}{1} = \dfrac{21}{5} = 4\dfrac{1}{5}$

(e) $3\dfrac{2}{3} \div 3 = \dfrac{11}{3} \div \dfrac{3}{1} = \dfrac{11}{3} \times \dfrac{1}{3} = \dfrac{11}{9} = 1\dfrac{2}{9}$

☑ $\quad 3 \times \dfrac{11}{9} = \dfrac{\overset{1}{\cancel{3}}}{1} \times \dfrac{11}{\underset{3}{\cancel{9}}} = \dfrac{11}{3} = 3\dfrac{2}{3}$

(f) $\dfrac{3}{4} \div 2\dfrac{5}{8} = \dfrac{3}{4} \div \dfrac{21}{8} = \dfrac{\overset{1}{\cancel{3}}}{\underset{1}{\cancel{4}}} \times \dfrac{\overset{2}{\cancel{8}}}{\underset{7}{\cancel{21}}} = \dfrac{2}{7}$ ☑ $2\dfrac{5}{8} \times \dfrac{2}{7} = \dfrac{\overset{3}{\cancel{21}}}{\underset{4}{\cancel{8}}} \times \dfrac{\overset{1}{\cancel{2}}}{\underset{1}{\cancel{7}}} = \dfrac{3}{4}$

(g) $1\dfrac{1}{4} \div 1\dfrac{7}{8} = \dfrac{5}{4} \div \dfrac{15}{8} = \dfrac{\overset{1}{\cancel{5}}}{\underset{1}{\cancel{4}}} \times \dfrac{\overset{2}{\cancel{8}}}{\underset{3}{\cancel{15}}} = \dfrac{2}{3}$

☑ $\quad 1\dfrac{7}{8} \times \dfrac{2}{3} = \dfrac{\overset{5}{\cancel{15}}}{\underset{4}{\cancel{8}}} \times \dfrac{\overset{1}{\cancel{2}}}{\cancel{3}} = \dfrac{5}{4} = 1\dfrac{1}{4}$

Why Do We Use the Reciprocal When We Divide Fractions?

Notice that $8 \div 4 = 2$ and $8 \times \dfrac{1}{4} = \dfrac{\overset{2}{\cancel{8}}}{1} \times \dfrac{1}{\underset{1}{\cancel{4}}} = 2$.

Dividing by a number gives the same result as multiplying by its reciprocal. The following shows why this is so.

The division $8 \div 4$ can be written $\dfrac{8}{4}$. Similarly, $\dfrac{1}{2} \div \dfrac{2}{3}$ can be written $\dfrac{\frac{1}{2}}{\frac{2}{3}}$.

To simplify this fraction, multiply by $\dfrac{\frac{3}{2}}{\frac{3}{2}}$ (which is equal to 1).

$$\dfrac{\frac{1}{2}}{\frac{2}{3}} = \dfrac{\frac{1}{2} \times \boxed{\frac{3}{2}}}{\frac{2}{3} \times \boxed{\frac{3}{2}}} = \dfrac{\frac{1}{2} \times \frac{3}{2}}{1} = \dfrac{1}{2} \times \dfrac{3}{2}$$

$$\dfrac{2}{3} \times \dfrac{3}{2} = \dfrac{2 \times 3}{3 \times 2} = \dfrac{6}{6} = 1$$

Therefore, $\dfrac{1}{2} \div \dfrac{2}{3} = \dfrac{1}{2} \times \dfrac{3}{2}$, where $\dfrac{3}{2}$ is the reciprocal of the divisor $\dfrac{2}{3}$.

Practical Applications Recall from Section 1-4 that division is used to separate a numerical quantity into equal parts (see page 47 for more detail). Questions involving division are often in the form "how many of X are in Y?"

EXAMPLE 6 **Electrical Trades** An electrician has five spools, each containing 10 ft of wire. He needs to cut pieces of wire that are $18\frac{1}{4}$ in. each. He wishes to know how many whole pieces of this length he can cut from the spools. Solve this problem as follows:

First, convert 10 feet to inches so that the units match. There are 12 inches in a foot, so

$$10\text{ ft} = 10\text{ ft} \times 12\text{ in. per foot} = 120\text{ in.}$$

Next, divide this by $18\frac{1}{4}$ in. to calculate the number of pieces of this length that can be cut from each spool.

$$120\text{ in.} \div 18\frac{1}{4}\text{ in. per piece}$$

Convert $18\frac{1}{4}$ to an improper fraction. $= \dfrac{120}{1} \div \dfrac{73}{4}$

Multiply by the reciprocal. $= \dfrac{120}{1} \times \dfrac{4}{73}$

$$= \dfrac{480}{73}$$

Convert back to a mixed number. $= 6\dfrac{42}{73}$ pieces

Then, because we can use only whole pieces, we ignore the fractional portion of the answer. There are six whole pieces in each spool.

Finally, multiply this answer by the number of spools.

$$6\text{ pieces per spool} \times 5\text{ spools} = 30\text{ whole pieces of }18\frac{1}{4}\text{-in. wire}$$

Your Turn **Carpentry** How many sheets of plywood, each $\frac{3}{4}$ in. thick, are in a stack 18 in. high?

Solution This question is in the form "how many of X are in Y?," so we divide Y by X to find the answer:

$$18 \div \frac{3}{4} = \frac{\overset{6}{\cancel{18}}}{1} \times \frac{4}{\underset{1}{\cancel{3}}} = 24 \quad \text{There are 24 sheets of plywood in the stack.}$$

Turn to Exercises 2-3 for a set of practice problems on dividing fractions.

Exercises 2-3 **Division of Fractions**

A. Divide and write the answer in lowest terms.

1. $\dfrac{5}{6} \div \dfrac{1}{2}$ 2. $6 \div \dfrac{2}{3}$ 3. $\dfrac{5}{12} \div \dfrac{4}{3}$ 4. $8 \div \dfrac{1}{4}$

5. $\dfrac{6}{16} \div \dfrac{3}{4}$ 6. $\dfrac{1}{2} \div \dfrac{1}{2}$ 7. $\dfrac{3}{16} \div \dfrac{6}{8}$ 8. $\dfrac{3}{4} \div \dfrac{5}{16}$

9. $1\frac{1}{2} \div \frac{1}{6}$ 10. $6 \div 1\frac{1}{2}$ 11. $3\frac{1}{7} \div 2\frac{5}{14}$ 12. $3\frac{1}{2} \div 2$

13. $6\frac{2}{5} \div 5\frac{1}{3}$ 14. $10 \div 1\frac{1}{5}$ 15. $8 \div \frac{1}{2}$ 16. $\frac{2}{3} \div 6$

17. $12 \div \frac{2}{3}$ 18. $\frac{3}{4} \div \frac{7}{8}$ 19. $5 \div \frac{2}{3}$ 20. $1\frac{1}{2} \div 2\frac{1}{2}$

21. $\frac{5}{16} \div \frac{3}{8}$ 22. $\frac{7}{12} \div \frac{2}{3}$ 23. $\frac{7}{32} \div 1\frac{3}{4}$ 24. $1\frac{2}{3} \div 1\frac{1}{4}$

B. Practical Applications

1. **Drafting**   How many feet are represented by a 4-in. line if it is drawn to a scale of $\frac{1}{2}$ in. = 1 ft?

2. **Drafting**   If $\frac{1}{4}$ in. on a drawing represents 1 ft 0 in., then $3\frac{1}{2}$ in. on the drawing will represent how many feet?

3. **Flooring and Carpeting**   How many boards $4\frac{5}{8}$ in. wide will it take to cover a floor 222 in. wide?

4. **Construction**   How many supporting columns $88\frac{1}{2}$ in. long can be cut from six pieces each 22 ft long? (*Hint:* Be careful of units.)

5. **Construction**   How many pieces of $\frac{1}{2}$-in. plywood are there in a stack 42 in. high?

6. **Masonry**   The front side of a standard brick measures $2\frac{1}{4}$ in. by 8 in. With a $\frac{1}{4}$-in. mortar joint, the visible dimensions of each brick increases to $2\frac{1}{2}$ in. by $8\frac{1}{4}$ in. To estimate the number of bricks needed per square foot (144 sq in.), you would need to perform the following calculation:

$$144 \div \left(2\frac{1}{2} \times 8\frac{1}{4} \right)$$

Determine this quantity rounded to the nearest whole number.

7. **Masonry**   If we allow $2\frac{5}{8}$ in. for the thickness of a course of brick, including mortar joints, how many courses of brick will there be in a wall $47\frac{1}{4}$ in. high?

8. **Plumbing**   How many lengths of pipe $2\frac{5}{8}$ ft long can be cut from a pipe 21 ft long?

9. **Machine Trades**   How many pieces $6\frac{1}{4}$ in. long can be cut from 35 metal rods each 40 in. long?

10. **Architecture**   The architectural drawing for a room measures $3\frac{5}{8}$ in. by $4\frac{1}{4}$ in. If $\frac{1}{8}$ in. is equal to 1 ft on the drawing, what are the actual dimensions of the room?

11. **Printing**   How many full $3\frac{1}{2}$-in. sheets can be cut from $24\frac{3}{4}$-in. stock?

12. **Machine Trades**   The feed on a boring mill is set for $\frac{1}{32}$ in. How many revolutions are needed to advance the tool $3\frac{3}{8}$ in.?

13. **Machine Trades**   If the pitch of a thread is $\frac{1}{18}$ in., how many threads are needed for the threaded section of a pipe to be $2\frac{1}{2}$ in. long?

14. **Construction**   The floor area of a room on a house plan measures $3\frac{1}{2}$ in. by $4\frac{5}{8}$ in. If the drawing scale is $\frac{1}{4}$ in. represents 1 ft, what is the actual size of the room?

15. **Construction**   The family room ceiling of a new Happy Home is to be taped and mudded. A $106\frac{1}{2}$ sq ft area has already been done, and this represents $\frac{3}{8}$ of the job. How large is the area of the finished ceiling?

16. **Plumbing**   A pipe fitter needs to divide a pipe $32\frac{5}{8}$ inches long into three pieces of equal length. Calculate the length of each piece.

17. **Automotive Trades**   Over a period of $3\frac{1}{2}$ years, $\frac{7}{16}$ in. of tread has worn off a tire. What was the average tread wear per year?

18. **Automotive Trades**   On a certain vehicle, each turn of the tie-rod sleeve changes the toe-in setting by $\frac{1}{8}$ in. How many turns of the sleeve are needed to change the toe-in setting by $\frac{5}{32}$ in.?

19. **Carpentry**   A carpenter must mount an electrical panel on an uneven surface. She must raise one corner of the panel $1\frac{1}{8}$ in. off the surface. If $\frac{3}{16}$-in. washers are used as spacers, how many washers are needed to mount the panel?

20. **Allied Health**   The average height of a 2-year old girl is $36\frac{1}{2}$ inches, and the average height of an 11-year old girl is 57 inches. What is the average annual growth in inches during this 9-year period? (*Hint:* $57 - 36\frac{1}{2} = 20\frac{1}{2}$)

21. **Landscaping**   An agricultural supply yard sells walk-on bark by the scoop, where one scoop is equivalent to $\frac{2}{3}$ of a cubic yard. How many scoops should be ordered if a landscaper needs (a) 4 cubic yards. (b) 7 cubic yards.

Check your answers to the odd-numbered problems in the Appendix, then continue in Section 2-4.

2-4   # Addition and Subtraction of Fractions

Addition   At heart, adding fractions is a matter of counting:

$$\frac{1}{5} + \frac{3}{5} = \frac{1+3}{5} = \frac{4}{5}$$

$\frac{1}{5}$ 1 fifth

$+$

$\frac{3}{5}$ + 3 fifths

$=$

$\frac{4}{5}$ = 4 fifths; count them.

EXAMPLE 1   Add $\dfrac{1}{8} + \dfrac{3}{8} =$ _____

This one is easy to see with measurements:

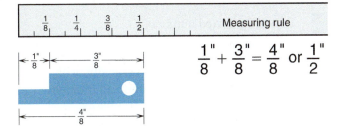

$$\frac{1"}{8} + \frac{3"}{8} = \frac{4"}{8} \text{ or } \frac{1"}{2}$$

Your Turn   Add $\dfrac{2}{7} + \dfrac{3}{7} = $ _____

Solution $\dfrac{2}{7} + \dfrac{3}{7} = \dfrac{2+3}{7} = \dfrac{5}{7}$

$\dfrac{2}{7}$ ▭▭▭▭▭▭▭   2 sevenths

$\dfrac{3}{7}$ ▭▭▭▭▭▭▭ $+$ 3 sevenths

$\dfrac{5}{7}$ ▭▭▭▭▭▭▭   5 sevenths or $\dfrac{5}{7}$

Adding Like Fractions   Fractions having the same denominator are called **like fractions**. In the preceding problem, $\frac{2}{7}$ and $\frac{3}{7}$ both have denominator 7 and are like fractions. Adding like fractions is easy: *first,* add the numerators to find the numerator of the sum and *second,* use the denominator the fractions have in common as the denominator of the sum.

$\dfrac{2}{9} + \dfrac{5}{9} = \dfrac{2+5}{9}$ ⟵ $\boxed{\text{Add numerators}}$

$\boxed{\text{Same denominator}}$

$= \dfrac{7}{9}$

We can use this same technique to add three or more like fractions.

EXAMPLE 2   Add $\dfrac{3}{12} + \dfrac{1}{12} + \dfrac{5}{12}$ like this:

$\dfrac{3}{12} + \dfrac{1}{12} + \dfrac{5}{12} = \dfrac{3+1+5}{12} = \dfrac{9}{12} = \dfrac{3}{4}$

Notice that we write the sum in lowest terms.   ●

Your Turn   Try these problems for exercise.

(a) $\dfrac{1}{16} + \dfrac{5}{16}$ (b) $\dfrac{7}{9} + \dfrac{5}{9} + \dfrac{4}{9} + \dfrac{8}{9}$

Solutions   (a) $\dfrac{1}{16} + \dfrac{5}{16} = \dfrac{1+5}{16} = \dfrac{6}{16} = \dfrac{3}{8}$

(b) $\dfrac{7}{9} + \dfrac{5}{9} + \dfrac{4}{9} + \dfrac{8}{9} = \dfrac{7+5+4+8}{9} = \dfrac{24}{9} = \dfrac{8}{3} = 2\dfrac{2}{3}$

When the addition problem involves mixed numbers or whole numbers, it may be easier to arrange the sum vertically, as shown in the next example.

EXAMPLE 3   To add the mixed numbers

$2\dfrac{1}{5} + 3\dfrac{3}{5}$

First, arrange the numbers vertically, with whole numbers in one column and fractions in another column.

$2\dfrac{1}{5}$

$+ \, 3\dfrac{3}{5}$

Then, add fractions and whole numbers separately.

$$2\frac{1}{5}$$
$$+\ 3\frac{3}{5}$$
$$\overline{5\frac{4}{5}} \longleftarrow \boxed{\text{Answer}} \bullet$$

If the sum of the fraction parts is greater than 1, an extra step will be required to simplify the answer. This is illustrated in Example 4.

EXAMPLE 4   Add $1\frac{5}{8} + 3\frac{7}{8}$.

If we arrange the mixed numbers vertically and find the sum of the whole numbers and fractions separately, we have:

$$1\frac{5}{8}$$
$$+\ 3\frac{7}{8}$$
$$\overline{4\frac{12}{8}}$$

Notice that the sum of the fraction parts is greater than 1. Therefore, we must simplify the answer as follows:

$$4\frac{12}{8} = 4 + \frac{12}{8} = 4 + \frac{3}{2} = 4 + 1\frac{1}{2} = 5\frac{1}{2} \qquad\qquad \bullet$$

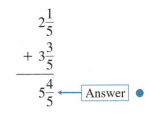 **Learning Help**   If you notice in advance that the sum of the fraction parts of the mixed numbers will be greater than or equal to 1, you may find it easier to first rewrite the mixed numbers as improper fractions and then add. For Example 4, this alternate solution would go as follows:

First, rewrite each mixed number as an improper fraction.
$$1\frac{5}{8} + 3\frac{7}{8} = \frac{13}{8} + \frac{31}{8}$$

Then, add.
$$= \frac{44}{8}$$

Finally, rewrite in lowest terms and convert back to a mixed number.
$$= \frac{11}{2} = 5\frac{1}{2} \bullet$$

Your Turn   Find each sum:

   (a) $6 + 3\frac{2}{3}$ (b) $1\frac{5}{8} + 2\frac{1}{8}$ (c) $4\frac{1}{6} + 2\frac{5}{6} + 3\frac{5}{6}$ (d) $3\frac{1}{4} + 2\frac{3}{4}$

Solutions   (a) 6 (b) $1\frac{5}{8}$

$$+\ 3\frac{2}{3}$$
$$\overline{9\frac{2}{3}}$$

$$+\ 2\frac{1}{8}$$

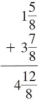   Always express the answer in lowest terms.

$$\overline{3\frac{6}{8}} = 3\frac{3}{4}$$

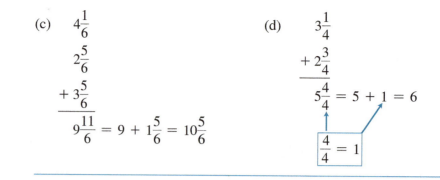

 A Closer Look   In problems (c) and (d), the sum of the fraction parts is greater than or equal to 1. If we use improper fractions to add, the solutions look like this:

(c) $4\frac{1}{6} + 2\frac{5}{6} + 3\frac{5}{6} = \frac{25}{6} + \frac{17}{6} + \frac{23}{6} = \frac{65}{6} = 10\frac{5}{6}$

(d) $3\frac{1}{4} + 2\frac{3}{4} = \frac{13}{4} + \frac{11}{4} = \frac{24}{4} = 6$ ●

Note   If the addition is done using improper fractions, large and unwieldy numerators may result. Be careful. ●

Adding Unlike Fractions   Fractions having different denominators are called **unlike fractions**. How do we add fractions whose denominators are not the same? For example, how do we add the unlike fractions $\frac{2}{3} + \frac{3}{4}$?

We must change these fractions to equivalent fractions with the same denominator.

$\frac{2}{3} = \frac{2 \times 4}{3 \times 4} = \frac{8}{12}$ We chose $\frac{4}{4}$ as a multiplier because it changes $\frac{2}{3}$ to an equivalent fraction with a denominator of 12.

$\frac{3}{4} = \frac{3 \times 3}{4 \times 3} = \frac{9}{12}$ We chose $\frac{3}{3}$ as a multiplier because it changes $\frac{3}{4}$ to an equivalent fraction with a denominator of 12.

The two fractions now have the same denominator.

Note   You should remember that we discussed equivalent fractions on page 79. Return for a quick review if you need it. ●

EXAMPLE 5   Add $\frac{2}{3} + \frac{3}{4}$ using the equivalent fractions.

$\frac{2}{3} + \frac{3}{4} = \frac{8}{12} + \frac{9}{12} = \frac{17}{12} = 1\frac{5}{12}$

We change the original fractions to equivalent fractions with the same denominator and then add as before.   ●

Least Common Denominator   How do you know what number to use as the new denominator? In general, you cannot simply guess at the best new denominator. We need a method for finding it from the denominators of the fractions to be added. The new denominator is called the **least common denominator**, abbreviated **LCD**. This number is also called the **least common multiple** of the original numbers.

EXAMPLE 6   Suppose that we want to add the fractions $\frac{1}{8} + \frac{5}{12}$.

The first step is to find the LCD of the denominators 8 and 12. To find the LCD, follow this procedure.

Step 1   Write each denominator as a product of its prime factors. If you need to review this concept, see page 47.

$$8 = 2 \cdot 2 \cdot 2 \qquad 12 = 2 \cdot 2 \cdot 3$$

Step 2   To form the LCD, write each factor that appears in either denominator, then repeat it for the most number of times it appears in any one denominator.

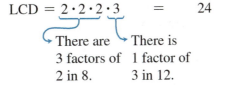

$$\text{LCD} = 2 \cdot 2 \cdot 2 \cdot 3 \qquad = \qquad 24$$

There are 3 factors of 2 in 8.   There is 1 factor of 3 in 12.

The LCD of 8 and 12 is 24. This means that 24 is the smallest number that is exactly divisible by both 8 and 12. ●

EXAMPLE 7   Find the LCD of the denominators 12 and 45.

Step 1   Find the prime factors of each number.

$$12 = 2 \cdot 2 \cdot 3 \qquad 45 = 3 \cdot 3 \cdot 5$$

Step 2   The LCD must contain the factors 2, 3, and 5. The factor 2 occurs twice in 12, the factor 3 occurs twice in 45, and the factor 5 occurs just once in 45.

$$\text{LCD} = 2 \cdot 2 \cdot 3 \cdot 3 \cdot 5 = 180$$

The number 180 is the smallest number that is exactly divisible by both 12 and 45. ●

Your Turn   Use the method described to find the LCD of the denominators 28 and 42.

Solution   **Step 1**   Find the prime factors of each number.

$$28 = 2 \cdot 2 \cdot 7 \qquad 42 = 2 \cdot 3 \cdot 7$$

Step 2   The factors 2, 3, and 7 all appear. The factor 2 appears at most twice (in 28) and both 3 and 7 appear at most just once in any single number. Therefore the LCD is

$$2 \cdot 2 \cdot 3 \cdot 7 = 84$$

A Calculator Method for Finding the LCD

Here is an alternative method for finding the LCD that you might find easier than the method just described. Follow these two steps.

Step 1   Choose the larger denominator and write down a few multiples of it.

Example: To find the LCD of 12 and 15, first write down a few of the multiples of the larger number, 15. The multiples are 15, 30, 45, 60, 75, and so on.

(continued)

Step 2 Test each multiple until you find one that is exactly divisible by the smaller denominator.

> **Example:** 15 is not exactly divisible by 12. 30 is not exactly divisible by 12. 45 is not exactly divisible by 12. 60 *is* exactly divisible by 12. The LCD is 60.

This method of finding the LCD has the advantage that you can use it with an electronic calculator. For this example the calculator steps would look like this:

$15 \div 12 =$	$\rightarrow$ 1.25	*Not* a whole number; therefore, *not* exactly divisible by 12.
$15 \times 2 \div 12 =$	$\rightarrow$ 2.5	Second multiple: answer is *not* a whole number.
$15 \times 3 \div 12 =$	$\rightarrow$ 3.75	Third multiple: answer is *not* a whole number.
$15 \times 4 \div 12 =$	$\rightarrow$ 5.	Fourth multiple: answer *is* a whole number; therefore, the LCD is 4×15 or 60.

With smaller numbers, this process may be done mentally.

Sometimes we must find the LCD for three or more different denominators.

EXAMPLE 8   Find the LCD of 4, 10, and 15.

Step 1   Write the prime factors of all three numbers.

$$4 = 2 \cdot 2 \qquad 10 = 2 \cdot 5 \qquad 15 = 3 \cdot 5$$

Step 2   The factors 2, 3, and 5 all appear and therefore must be included in the LCD. The factor 2 appears at most twice (in 4), and the factors 3 and 5 both appear at most just once in any single number. Therefore the LCD is

$$2 \cdot 2 \cdot 3 \cdot 5 = 60$$

This means that 60 is the smallest number that is exactly divisible by 4, 10, and 15. ●

Your Turn   Practice by finding the LCD for each of the following sets of denominators.

(a) 8 and 4   (b) 5 and 4   (c) 15 and 24

(d) 4, 5, and 6   (e) 4, 8, and 12   (f) 12, 15, and 21

Answers   (a) 8   (b) 20   (c) 120   (d) 60   (e) 24   (f) 420

To use the LCD to add fractions, rewrite the fractions with the LCD as the new denominator, then add the new equivalent fractions.

EXAMPLE 9   Add $\dfrac{1}{6} + \dfrac{5}{8}$.

First, find the LCD. The LCD of 6 and 8 is 24.

Next, rewrite the two fractions with denominator 24.

$$\frac{1}{6} = \frac{1 \times 4}{6 \times 4} = \frac{4}{24}$$

$$\frac{5}{8} = \frac{5 \times 3}{8 \times 3} = \frac{15}{24}$$

Finally, add the new equivalent fractions.

$$\frac{1}{6} + \frac{5}{8} = \frac{4}{24} + \frac{15}{24} = \frac{19}{24}$$

Your Turn   Add $\frac{3}{8} + \frac{1}{10}$.

Solution   The LCD of 8 and 10 is 40.

$$\frac{3}{8} = \frac{3 \times 5}{8 \times 5} = \frac{15}{40}$$

$$\frac{1}{10} = \frac{1 \times 4}{10 \times 4} = \frac{4}{40}$$

$$\frac{3}{8} + \frac{1}{10} = \frac{15}{40} + \frac{4}{40} = \frac{19}{40}$$

When mixed numbers contain unlike denominators, find the LCD of the fractions, rewrite the fractions with the LCD, and then add.

EXAMPLE 10   To add $2\frac{3}{8} + 3\frac{1}{6}$,

Step 1   Find the LCD of the fractions. The LCD of 8 and 6 is 24.

Step 2   Rewrite the two fractions with the LCD and arrange the sum vertically.

$$2\frac{9}{24} \quad \Leftarrow \boxed{\frac{3}{8} = \frac{3 \times 3}{8 \times 3} = \frac{9}{24}}$$

$$+3\frac{4}{24} \quad \Leftarrow \boxed{\frac{1}{6} = \frac{1 \times 4}{6 \times 4} = \frac{4}{24}}$$

Step 3   Add the whole numbers and fractions separately.

$$5\frac{13}{24} \quad \longleftarrow \boxed{\text{Answer}}$$

If the sum of the fraction parts of the mixed numbers is greater than or equal to 1, an additional step will be required to simplify the answer. This is illustrated in Example 11.

EXAMPLE 11   To add $5\frac{1}{2} + 3\frac{2}{3}$,

Step 1   The LCD of 2 and 3 is 6.

Step 2   Rewrite the two fractions with the LCD and arrange the sum vertically.

$$5\frac{3}{6} \quad \Leftarrow \boxed{\frac{1}{2} = \frac{1 \times 3}{2 \times 3} = \frac{3}{6}}$$

$$+3\frac{4}{6} \quad \Leftarrow \boxed{\frac{2}{3} = \frac{2 \times 2}{3 \times 2} = \frac{4}{6}}$$

Step 3   Add.

$$8\frac{7}{6}$$

Step 4   Simplify the answer. $\dfrac{8\,7}{6} = 8 + \dfrac{7}{6} = 8 + 1\dfrac{1}{6} = 9\dfrac{1}{6}$ ●

A Closer Look   Using improper fractions, the solution to Example 11 looks like this:

Step 1   Convert each mixed number to an improper fraction.

$$5\dfrac{1}{2} + 3\dfrac{2}{3} = \dfrac{11}{2} + \dfrac{11}{3}$$

Step 2   Rewrite each fraction with the LCD of 6.

$$= \dfrac{33}{6} + \dfrac{22}{6}$$

Step 3   Add.

$$= \dfrac{55}{6}$$

Step 4   Convert the result back to a mixed number.

$$= 9\dfrac{1}{6}$$ ●

Your Turn   Find the LCD, rewrite the fractions, and add.

(a) $\dfrac{1}{2} + \dfrac{1}{4}$ (b) $\dfrac{3}{8} + \dfrac{1}{4}$ (c) $\dfrac{1}{6} + \dfrac{2}{3} + \dfrac{5}{9}$

(d) $4\dfrac{3}{4} + \dfrac{1}{6}$ (e) $1\dfrac{4}{5} + 2\dfrac{3}{8}$

Solutions   (a) The LCD of 2 and 4 is 4.

$$\dfrac{1}{2} = \dfrac{?}{4} = \dfrac{1 \times 2}{2 \times 2} = \dfrac{2}{4}$$

$$\dfrac{1}{2} + \dfrac{1}{4} = \dfrac{2}{4} + \dfrac{1}{4} = \dfrac{3}{4}$$

(b) The LCD of 8 and 4 is 8.

$$\dfrac{1}{4} = \dfrac{?}{8} = \dfrac{1 \times 2}{4 \times 2} = \dfrac{2}{8}$$

$$\dfrac{3}{8} + \dfrac{1}{4} = \dfrac{3}{8} + \dfrac{2}{8} = \dfrac{5}{8}$$

(c) The LCD of 6, 3, and 9 is 18.

$$\dfrac{1}{6} = \dfrac{?}{18} = \dfrac{1 \times 3}{6 \times 3} = \dfrac{3}{18}$$

$$\dfrac{2}{3} = \dfrac{?}{18} = \dfrac{2 \times 6}{3 \times 6} = \dfrac{12}{18}$$

$$\dfrac{5}{9} = \dfrac{?}{18} = \dfrac{5 \times 2}{9 \times 2} = \dfrac{10}{18}$$

$$\dfrac{1}{6} + \dfrac{2}{3} + \dfrac{5}{9} = \dfrac{3}{18} + \dfrac{12}{18} + \dfrac{10}{18} = \dfrac{25}{18} = 1\dfrac{7}{18}$$

(d) The LCD of 4 and 6 is 12.

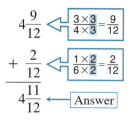

$$4\dfrac{9}{12} \quad \boxed{\dfrac{3 \times 3}{4 \times 3} = \dfrac{9}{12}}$$

$$+ \dfrac{2}{12} \quad \boxed{\dfrac{1 \times 2}{6 \times 2} = \dfrac{2}{12}}$$

$$4\dfrac{11}{12} \quad \longleftarrow \boxed{\text{Answer}}$$

(e) The LCD of 5 and 8 is 40.

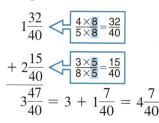

Using mixed numbers:

$$1\dfrac{32}{40} \quad \boxed{\dfrac{4 \times 8}{5 \times 8} = \dfrac{32}{40}}$$

$$+ 2\dfrac{15}{40} \quad \boxed{\dfrac{3 \times 5}{8 \times 5} = \dfrac{15}{40}}$$

$$3\dfrac{47}{40} = 3 + 1\dfrac{7}{40} = 4\dfrac{7}{40}$$

Using improper fractions:

$$1\dfrac{4}{5} + 2\dfrac{3}{8} = \dfrac{9}{5} + \dfrac{19}{8} = \dfrac{9 \times 8}{5 \times 8} + \dfrac{19 \times 5}{8 \times 5}$$

$$= \dfrac{72}{40} + \dfrac{95}{40} = \dfrac{167}{40} = 4\dfrac{7}{40}$$

Applications of Addition

EXAMPLE 12 **Construction** A construction plan calls for a stud $3\frac{1}{2}$-in. thick to be covered on both sides with $\frac{3}{8}$-in. Sheetrock. The contractor needs to know the total thickness of the resulting wall. The word *total* implies that addition must be used. To solve this problem, the contractor must add the stud thickness to two of the Sheetrock thicknesses:

$$3\frac{1}{2} + \frac{3}{8} + \frac{3}{8} = 3\frac{4}{8} + \frac{3}{8} + \frac{3}{8} \qquad \text{or} \qquad \begin{array}{r} 3\frac{4}{8} \\ \frac{3}{8} \\ + \ \frac{3}{8} \\ \hline 3\frac{10}{8} = 3 + \frac{5}{4} = 3 + 1\frac{1}{4} = 4\frac{1}{4} \end{array}$$

The total thickness of the wall is $4\frac{1}{4}$ in. ●

🔍 A Closer Look In the previous example, we could have multiplied $\frac{3}{8}$ by 2 and added this product to $3\frac{1}{2}$ instead of adding $\frac{3}{8}$ twice. ●

Your Turn **Manufacturing** The Monterey Canning Co. packs $8\frac{1}{2}$ ounces of tuna into each can. If the can itself weighs $1\frac{7}{8}$ ounces, what is the total weight of a can of Monterey tuna?

Solution To find the total weight, add the weight of the tuna to the weight of the can. The LCD of 2 and 8 is 8.

Using mixed numbers:

$$\begin{array}{r} 8\frac{4}{8} \ \boxed{\frac{1 \times 4}{2 \times 4} = \frac{4}{8}} \\ + 1\frac{7}{8} \\ \hline 9\frac{11}{8} = 9 + 1\frac{3}{8} = 10\frac{3}{8} \end{array}$$

Using improper fractions:

$$8\frac{1}{2} + 1\frac{7}{8} = \frac{17}{2} + \frac{15}{8}$$

$$= \frac{17 \times 4}{2 \times 4} + \frac{15}{8}$$

$$= \frac{68}{8} + \frac{15}{8}$$

$$= \frac{83}{8} = 10\frac{3}{8}$$

The total weight is $10\frac{3}{8}$ ounces.

Subtraction Once you have mastered the process of adding fractions, subtraction is very simple indeed. To find $\frac{3}{8} - \frac{1}{8}$, notice that the denominators are the same. We can subtract the numerators and write this difference over the common denominator.

$$\frac{3}{8} - \frac{1}{8} = \frac{3 - 1}{8} \qquad \overset{\longleftarrow}{\boxed{\text{Subtract numerators}}}$$

$$\overset{\longleftarrow}{\boxed{\text{Same denominator}}}$$

$$= \frac{2}{8} \quad \text{or} \quad \frac{1}{4}$$

To subtract fractions with unlike denominators, first find the LCD and then subtract the equivalent fractions with like denominators.

EXAMPLE 13   To find $\dfrac{3}{4} - \dfrac{1}{5}$,

first, determine that the LCD of 4 and 5 is 20.

Then, find equivalent fractions with a denominator of 20.

$$\frac{3}{4} = \frac{?}{20} = \frac{3 \times 5}{4 \times 5} = \frac{15}{20}$$

$$\frac{1}{5} = \frac{?}{20} = \frac{1 \times 4}{5 \times 4} = \frac{4}{20}$$

Finally, subtract the equivalent fractions:

$$\frac{3}{4} - \frac{1}{5} = \frac{15}{20} - \frac{4}{20} = \frac{15 - 4}{20} = \frac{11}{20}$$

Until the final step, the procedure is exactly the same as for addition.   ●

Your Turn   Subtract as indicated.

(a) $\dfrac{9}{20} - \dfrac{3}{20}$ (b) $\dfrac{5}{6} - \dfrac{3}{10}$

Solutions   (a) $\dfrac{9}{20} - \dfrac{3}{20} = \dfrac{9 - 3}{20} = \dfrac{6}{20} = \dfrac{3}{10}$ Be sure to write your final answer in lowest terms.

(b) The LCD of 6 and 10 is 30.

$$\frac{5}{6} = \frac{5 \times 5}{6 \times 5} = \frac{25}{30} \qquad \frac{3}{10} = \frac{3 \times 3}{10 \times 3} = \frac{9}{30}$$

$$\frac{5}{6} - \frac{3}{10} = \frac{25}{30} - \frac{9}{30} = \frac{25 - 9}{30} = \frac{16}{30} = \frac{8}{15}$$

If the fractions to be subtracted are given as mixed numbers, it is sometimes easier to work with them as improper fractions. The next two examples illustrate when to leave them as mixed numbers and when to convert them to improper fractions.

EXAMPLE 14   Consider the problem $5\dfrac{5}{6} - 3\dfrac{1}{6}$.

Because $\dfrac{5}{6}$ is larger than $\dfrac{1}{6}$, it will be easy to subtract using mixed numbers. Simply arrange the problem vertically and subtract the whole numbers and fractions separately.

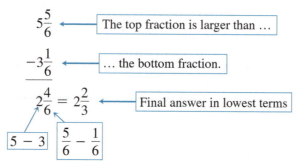

$$5\frac{5}{6} \quad \longleftarrow \boxed{\text{The top fraction is larger than \ldots}}$$

$$-3\frac{1}{6} \quad \longleftarrow \boxed{\ldots \text{the bottom fraction.}}$$

$$2\frac{4}{6} = 2\frac{2}{3} \quad \longleftarrow \boxed{\text{Final answer in lowest terms}}$$

$$\boxed{5 - 3} \quad \boxed{\frac{5}{6} - \frac{1}{6}}$$

●

EXAMPLE 15 Consider the problem $6\frac{1}{4} - 1\frac{3}{4}$.

Because $\frac{1}{4}$ is smaller than $\frac{3}{4}$, leaving them as mixed numbers will require the technique of borrowing. (See A Closer Look following this example.) To avoid this complication, convert the mixed numbers to improper fractions and then subtract.

$$6\frac{1}{4} - 1\frac{3}{4} = \frac{25}{4} - \frac{7}{4} = \frac{18}{4} = \frac{9}{2} = 4\frac{1}{2}$$

| Answer in lowest terms | | Answer converted to a mixed number | ●

A Closer Look In the previous example, if we try to subtract using mixed numbers, we must borrow before subtracting. Here is how the problem would be done:

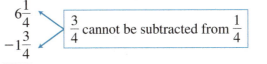

$$\begin{array}{r} 6\frac{1}{4} \\ -1\frac{3}{4} \\ \hline \end{array}$$

$\frac{3}{4}$ cannot be subtracted from $\frac{1}{4}$

| We must borrow 1 from the 6 so that it becomes 5. | → | $\begin{array}{r} 5\frac{5}{4} \\ -1\frac{3}{4} \\ \hline 4\frac{2}{4} = 4\frac{1}{2} \end{array}$ | ← | The 1 we borrow is added to the fraction part: $1 + \frac{1}{4} = \frac{4}{4} + \frac{1}{4} = \frac{5}{4}$ |

$4\frac{2}{4} = 4\frac{1}{2}$ ← | Answer in lowest terms |

If you wish to avoid having to borrow, always use improper fractions in this situation. ●

Your Turn Subtract as indicated.

(a) $9\frac{7}{8} - 3\frac{5}{8}$　　　　(b) $4\frac{1}{5} - 2\frac{4}{5}$

Solutions (a) Because $\frac{7}{8}$ is larger than $\frac{5}{8}$, we can proceed with the mixed numbers. Arranging the problem vertically, we have

$$\begin{array}{r} 9\frac{7}{8} \\ -3\frac{5}{8} \\ \hline 6\frac{2}{8} = 6\frac{1}{4} \end{array}$$

$6\frac{2}{8} = 6\frac{1}{4}$ ← | Answer in lowest terms |

(b) Because $\frac{1}{5}$ is smaller than $\frac{4}{5}$, we will use improper fractions to avoid having to borrow.

$$4\frac{1}{5} - 2\frac{4}{5} = \frac{21}{5} - \frac{14}{5} = \frac{7}{5} = 1\frac{2}{5}$$

When subtracting mixed numbers with unlike denominators, find the LCD of the fraction parts first. You can then decide which method is best.

EXAMPLE 16   (a) To subtract $3\frac{1}{2} - 1\frac{1}{3}$,

Step 1 Find the LCD. The LCD of 2 and 3 is 6.

Step 2 Find equivalent fractions with a denominator of 6.

$$\frac{1}{2} = \frac{1 \times 3}{2 \times 3} = \frac{3}{6} \qquad \frac{1}{3} = \frac{1 \times 2}{3 \times 2} = \frac{2}{6}$$

so that $3\frac{1}{2} - 1\frac{1}{3} = 3\frac{3}{6} - 1\frac{2}{6}$

Step 3 Because $\frac{3}{6}$ is larger than $\frac{2}{6}$, we can proceed with the mixed numbers.

$$\begin{array}{r} 3\frac{3}{6} \\ -1\frac{2}{6} \\ \hline 2\frac{1}{6} \end{array}$$ ⟵ Answer

(b) To subtract $5\frac{1}{4} - 2\frac{3}{8}$,

Step 1 Find the LCD. The LCD of 4 and 8 is 8.

Step 2 Only the first fraction must be rewritten.

$$\frac{1}{4} = \frac{1 \times 2}{4 \times 2} = \frac{2}{8}$$

so that $5\frac{1}{4} - 2\frac{3}{8} = 5\frac{2}{8} - 2\frac{3}{8}$

Step 3 Because $\frac{2}{8}$ is smaller than $\frac{3}{8}$, it will be easier to use improper fractions to perform the subtraction.

$$5\frac{2}{8} - 2\frac{3}{8} = \frac{42}{8} - \frac{19}{8} = \frac{23}{8} = 2\frac{7}{8}$$ ●

The following two-part example shows you what to do if one of the numbers is a whole number.

EXAMPLE 17   (a) If you are subtracting a whole number from a mixed number, simply subtract the numbers in their original form. To subtract $4\frac{5}{16} - 2$, arrange vertically and subtract as usual:

$$\begin{array}{r} 4\frac{5}{16} \\ -2 \\ \hline 2\frac{5}{16} \end{array}$$ ⟵ Answer

(b) If you are subtracting a mixed number from a whole number, use improper fractions to avoid borrowing. To subtract $8 - 3\frac{2}{3}$, recall that 8 in fraction form is $\frac{8}{1}$, so that

$$\boxed{\frac{8}{1} = \frac{8 \times 3}{1 \times 3} = \frac{24}{3}}$$

$$8 - 3\frac{2}{3} = \frac{8}{1} - \frac{11}{3} = \frac{24}{3} - \frac{11}{3} = \frac{13}{3} = 4\frac{1}{3}$$

More Practice Try these problems for practice in subtracting fractions.

(a) $\frac{7}{8} - \frac{5}{8}$ (b) $\frac{4}{5} - \frac{1}{6}$ (c) $9\frac{13}{16} - 3\frac{1}{4}$ (d) $6 - 2\frac{3}{4}$ (e) $7\frac{1}{6} - 2\frac{5}{8}$

Solutions (a) $\frac{7}{8} - \frac{5}{8} = \frac{7-5}{8} = \frac{2}{8} = \frac{1}{4}$

(b) The LCD of 5 and 6 is 30.

$$\frac{4}{5} = \frac{?}{30} = \frac{4 \times 6}{5 \times 6} = \frac{24}{30}$$

$$\frac{1}{6} = \frac{?}{30} = \frac{1 \times 5}{6 \times 5} = \frac{5}{30}$$

so $\dfrac{4}{5} - \dfrac{1}{6} = \dfrac{24}{30} - \dfrac{5}{30}$

$$= \frac{19}{30}$$

(c) The LCD of 4 and 16 is 16. Only the second fraction must be rewritten:

$$9\frac{13}{16} - 3\frac{1}{4} = 9\frac{13}{16} - 3\frac{4}{16} \quad \boxed{\frac{1}{4} = \frac{1 \times 4}{4 \times 4}}$$

$$9\frac{13}{16}$$
$$-3\frac{4}{16}$$
$$\overline{}$$
$$6\frac{9}{16} \quad \leftarrow \boxed{\text{Answer}}$$

$\boxed{(2 \times 4) + 3}$

(d) $6 - 2\frac{3}{4} = \frac{24}{4} - \frac{11}{4} = \frac{24 - 11}{4} = \frac{13}{4} = 3\frac{1}{4}$

$\boxed{6 = \frac{6}{1} = \frac{6 \times 4}{1 \times 4}}$

(e) The LCD of 6 and 8 is 24. $\dfrac{1}{6} = \dfrac{4}{24}$ $\dfrac{5}{8} = \dfrac{15}{24}$

Using improper fractions:

$$7\frac{1}{6} = 7\frac{4}{24} = \frac{172}{24} \quad \text{and} \quad 2\frac{5}{8} = 2\frac{15}{24} = \frac{63}{24}$$

so $7\dfrac{1}{6} - 2\dfrac{5}{8} = \dfrac{172}{24} - \dfrac{63}{24} = \dfrac{109}{24} = 4\dfrac{13}{24}$

Using the borrowing technique:

$$7\frac{4}{24} \rightarrow 6\frac{28}{24} \quad \leftarrow \boxed{\text{Borrow 1 from 7.} \quad 1 = \frac{24}{24} \quad \frac{24}{24} + \frac{4}{24} = \frac{28}{24}}$$
$$-2\frac{15}{24} \rightarrow -2\frac{15}{24}$$
$$\overline{}$$
$$4\frac{13}{24}$$

🔍 **A Closer Look**   In solution (e), choosing to use borrowing rather than improper fractions meant we did not have to deal with large numerators. ●

Order of Operations with Fractions   When a calculation with fractions involves two or more operations, be sure to follow the rules for the order of operations as explained in Section 1-5 (p. 53).

EXAMPLE 18   To calculate $\dfrac{5}{8} + \dfrac{11}{16} \div 4$, use Rules 2 and 3 from page 53.

First, divide. $\dfrac{5}{8} + \dfrac{11}{16} \div 4 = \dfrac{5}{8} + \dfrac{11}{16} \times \dfrac{1}{4}$

$$= \dfrac{5}{8} + \dfrac{11}{64}$$

Then, add. $= \dfrac{40}{64} + \dfrac{11}{64} = \dfrac{51}{64}$ ●

Your Turn   Calculate $\dfrac{2}{3}\left(1\dfrac{1}{2} - \dfrac{3}{4}\right)$ using the proper order of operations.

Solution   Rules 1 and 2 on page 53 tell us to simplify inside parentheses first and then multiply. The solution looks like this:

$$\dfrac{2}{3}\left(1\dfrac{1}{2} - \dfrac{3}{4}\right) = \dfrac{2}{3}\left(\dfrac{3}{4}\right) = \dfrac{1}{2}$$

Applications of Subtraction and the Order of Operations

EXAMPLE 19   **Plumbing**   The cross-section view of a pipe is shown in the figure. The pipe has a $2\dfrac{1}{2}$-in. O.D. (outside diameter) and a wall thickness of $\dfrac{1}{8}$ in. A plumber needs to know the I.D. (inside diameter).

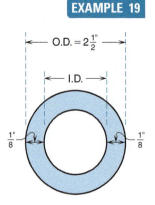

O.D. = $2\dfrac{1}{2}"$

I.D.

$\dfrac{1"}{8}$ $\dfrac{1"}{8}$

From the figure, notice that the I.D. plus *twice* the wall thickness equals the O.D. Therefore, to find the I.D., subtract twice the wall thickness from the O.D.

$$\text{I.D.} = 2\dfrac{1}{2} - 2\left(\dfrac{1}{8}\right) = 2\dfrac{1}{2} - \dfrac{2}{1} \times \dfrac{1}{8}$$

Step 1   According to the order of operations, we must first multiply. $= 2\dfrac{1}{2} - \dfrac{\overset{1}{\cancel{2}}}{1} \times \dfrac{1}{\underset{4}{\cancel{8}}}$

$$= 2\dfrac{1}{2} - \dfrac{1}{4}$$

Step 2   Rewrite with the LCD of 4. $= 2\dfrac{2}{4} - \dfrac{1}{4}$

Step 3   Subtract. $= 2\dfrac{1}{4}$

The inside diameter is $2\dfrac{1}{4}$ inches.   ●

Your Turn   **Machine Trades**   A bar $6\frac{3}{16}$ in. long is cut from a piece $24\frac{3}{8}$ in. long. If $\frac{5}{32}$ in. is wasted in cutting, what length remains?

Solution   *Cutting* and *wasting* both imply subtraction. So the problem becomes

$$24\frac{3}{8} - 6\frac{3}{16} - \frac{5}{32}$$

Rewritten with a common denominator of 32, we have

$$24\frac{12}{32} - 6\frac{6}{32} - \frac{5}{32}$$

According to the order of operations, we must subtract from left to right. Therefore, we subtract $24\frac{12}{32} - 6\frac{6}{32}$ first, and then subtract $\frac{5}{32}$ from the result.

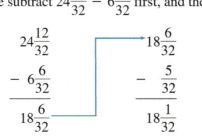

The remaining length is $18\frac{1}{32}$ in.

Now turn to Exercises 2-4 for a set of problems on adding and subtracting fractions.

Exercises 2-4   Addition and Subtraction of Fractions

A. Add or subtract as shown.

1. $\dfrac{1}{16} + \dfrac{3}{16}$ 2. $\dfrac{5}{12} + \dfrac{11}{12}$ 3. $\dfrac{5}{16} + \dfrac{7}{16}$

4. $\dfrac{2}{6} + \dfrac{3}{6}$ 5. $\dfrac{3}{4} - \dfrac{1}{4}$ 6. $\dfrac{13}{16} - \dfrac{3}{16}$

7. $\dfrac{3}{5} - \dfrac{1}{5}$ 8. $\dfrac{5}{12} - \dfrac{2}{12}$ 9. $\dfrac{5}{16} + \dfrac{3}{16} + \dfrac{7}{16}$

10. $\dfrac{1}{8} + \dfrac{3}{8} + \dfrac{7}{8}$ 11. $1\dfrac{7}{8} - \dfrac{3}{8}$ 12. $3\dfrac{9}{16} - 1\dfrac{5}{16}$

13. $\dfrac{1}{4} + \dfrac{1}{2}$ 14. $\dfrac{7}{16} + \dfrac{3}{8}$ 15. $\dfrac{5}{8} + \dfrac{1}{12}$

16. $\dfrac{5}{12} + \dfrac{3}{16}$ 17. $\dfrac{1}{2} - \dfrac{3}{8}$ 18. $\dfrac{5}{16} - \dfrac{3}{32}$

19. $\dfrac{15}{16} - \dfrac{1}{2}$ 20. $\dfrac{7}{16} - \dfrac{1}{32}$ 21. $\dfrac{3}{5} + \dfrac{1}{8}$

22. $\dfrac{2}{3} + \dfrac{4}{5}$ 23. $\dfrac{7}{8} - \dfrac{2}{5}$ 24. $\dfrac{4}{9} - \dfrac{1}{4}$

25. $\dfrac{1}{2} + \dfrac{1}{4} - \dfrac{1}{8}$ 26. $\dfrac{11}{16} - \dfrac{1}{8} - \dfrac{1}{3}$ 27. $1\dfrac{1}{2} + \dfrac{1}{4}$

28. $2\dfrac{7}{16} + \dfrac{3}{4}$ 29. $2\dfrac{1}{2} + 1\dfrac{5}{8}$ 30. $2\dfrac{8}{32} + 1\dfrac{1}{10}$

31. $2\dfrac{1}{3} + 1\dfrac{1}{5}$ 32. $1\dfrac{7}{8} + \dfrac{1}{4}$ 33. $4\dfrac{1}{8} - 1\dfrac{3}{4}$

34. $5\dfrac{3}{4} - 2\dfrac{1}{12}$ 35. $3\dfrac{1}{5} - 2\dfrac{1}{12}$ 36. $5\dfrac{1}{3} - 2\dfrac{2}{5}$

B. Add or subtract as shown.

1. $8 - 2\frac{7}{8}$

2. $3 - 1\frac{3}{16}$

3. $3\frac{5}{8} - \frac{13}{16}$

4. $\frac{1}{2} + \frac{1}{3} + \frac{1}{4} + \frac{1}{5}$

5. $\frac{1}{2} + \frac{1}{4} + \frac{1}{8}$

6. $6\frac{1}{2} + 5\frac{3}{4} + 8\frac{1}{8}$

7. $\frac{7}{8} + 2\frac{1}{2} - 1\frac{1}{4}$

8. $1\frac{3}{8}$ subtracted from $4\frac{3}{4}$

9. $2\frac{3}{16}$ less than $4\frac{7}{8}$

10. $6\frac{2}{3}$ reduced by $1\frac{1}{4}$

11. $2\frac{3}{5}$ less than $6\frac{1}{2}$

12. By how much is $1\frac{8}{7}$ larger than $1\frac{7}{8}$?

C. Practical Applications

1. **Construction**   The exterior wall of a small office building under construction consists of $\frac{1}{4}$-in. paneling, $\frac{5}{8}$-in. firecode Sheetrock, $5\frac{3}{4}$-in. studs, $\frac{1}{2}$-in. CDX plywood sheathing, $1\frac{1}{4}$-in. insulation board, and $\frac{5}{8}$-in. exterior surfacing. Calculate the total thickness of the wall.

2. **Carpentry**   A countertop is made of $\frac{5}{8}$-in. particleboard and is covered with $\frac{3}{16}$-in. laminated plastic. What width of metal edging is needed to finish off the edge?

3. **Welding**   A welder needs a piece of half-inch pipe $34\frac{3}{4}$ in. long. She has a piece that is $46\frac{3}{8}$ in. long. How much must she cut off from the longer piece?

4. **Plumbing**   If a piece of $\frac{3}{8}$-in.-I.D. (inside diameter) copper tubing measures $\frac{9}{16}$ in. O.D. (outside diameter), what is the wall thickness?

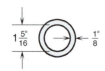

Problem 5

5. **Manufacturing**   What is the outside diameter of tubing whose inside diameter is $1\frac{5}{16}$ in. and whose wall thickness is $\frac{1}{8}$ in.? (See the figure.)

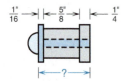

Problem 6

6. **Machine Trades**   How long a bolt is needed to go through a piece of tubing $\frac{5}{8}$ in. long, a washer $\frac{1}{16}$ in. thick, and a nut $\frac{1}{4}$ in. thick? (See the figure.)

7. **Office Services**   Newspaper ads are sold by the column inch (c.i.). What is the total number of column inches for a month in which a plumbing contractor has had ads of $6\frac{1}{2}$, $5\frac{3}{4}$, $3\frac{1}{4}$, $4\frac{3}{4}$, and 5 c.i.?

8. **Plumbing**   While installing water pipes, a plumber used pieces of pipe measuring $2\frac{3}{4}$, $4\frac{1}{8}$, $3\frac{1}{2}$, and $1\frac{1}{4}$ ft. How much pipe would remain if these pieces were cut from a 14-ft length of pipe? (Ignore waste in cutting.)

Problem 9

9. **Electrical Trades**   A piece of electrical pipe conduit has a diameter of $1\frac{1}{2}$ in. and a wall thickness of $\frac{3}{16}$ in. What is its inside diameter? (See the figure.)

10. **Machine Trades**   What is the total length of a certain machine part that is made by joining four pieces that measure $3\frac{1}{8}$, $1\frac{5}{32}$, $2\frac{7}{16}$, and $1\frac{1}{4}$ in.?

11. **Carpentry**   A blueprint requires four separate pieces of wood measuring $5\frac{3}{8}$, $8\frac{1}{4}$, $6\frac{9}{16}$, and $2\frac{5}{8}$ in. How long a piece of wood is needed to cut these pieces if we allow $\frac{1}{2}$ in. for waste?

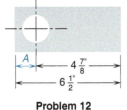

Problem 12

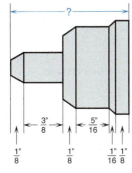

Problem 16

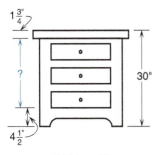

Problem 20

12. **Drafting**   Find the missing dimension *A* in the drawing shown.

13. **Machine Trades**   Two splice plates are cut from a piece of sheet steel that has an overall length of $18\frac{5}{8}$ in. The plates are $9\frac{1}{4}$ in. and $6\frac{7}{16}$ in. long. How much material remains from the original piece if each saw cut removes $\frac{1}{16}$ in.?

14. **Printing**   A printer has $2\frac{3}{4}$ rolls of a certain kind of paper in stock. He must do three jobs that require $\frac{5}{8}$, $1\frac{1}{2}$, and $\frac{3}{4}$ roll, respectively. Does he have enough?

15. **Automotive Trades**   The wheel stagger of an automobile is the difference between the axle-to-axle lengths on the right and left sides. If this length is $101\frac{1}{4}$ in. on the right side of a particular car and is $100\frac{7}{8}$ in. on the left side, find the wheel stagger of this automobile.

16. **Carpentry**   A cabinet 30 in. high must have a $4\frac{1}{2}$-in. base and a $1\frac{3}{4}$-in. top. How much space is left for the drawers? (See the figure.)

17. **Printing**   Before it was trimmed, a booklet measured $8\frac{1}{4}$ in. high by $6\frac{3}{4}$ in. wide. If each edge of the height and one edge of the width were trimmed $\frac{1}{4}$ in., what is the finished size?

18. **Carpentry**   A wall has $\frac{1}{2}$-in. paneling covering $\frac{3}{4}$-in. drywall attached to a $3\frac{3}{4}$-in. stud. What is the total thickness of the three components?

19. **Machine Trades**   The large end of a tapered pin is $2\frac{15}{16}$ in. in diameter, while the small end is $2\frac{3}{8}$ in. in diameter. Calculate the difference to get the amount of taper.

20. **Machine Trades**   Find the total length of the metal casting shown in the diagram.

21. **Carpentry**   A joiner is set to remove $\frac{7}{64}$ in. from the width of an oak board. If the board was $4\frac{5}{8}$ in. wide, find its width after joining once.

22. **Carpentry**   A rule of thumb used in constructing stairways is that the rise and the run should always add up to 17 inches. Applying this rule, what should be the run of a stairway if the rise is $7\frac{3}{4}$ in.?

23. **Automotive Trades**   During an oil change, $5\frac{1}{4}$ quarts of oil were needed to fill up an engine. During the next oil change, only $4\frac{1}{2}$ quarts of oil drained from the engine. How much oil was consumed between oil changes?

24. **Carpentry**   A 2-in. wood screw is used to join two pieces of a wooden workbench frame that are each $1\frac{1}{4}$ in. thick. How far into the second piece does the wood screw penetrate?

25. **Construction**   The concrete slab for a patio requires $4\frac{1}{3}$ cubic yards (cu yd) of concrete. If the truck delivering the concrete has a capacity of 9 cu yd and is full, how many cubic yards will remain in the concrete truck after delivery?

When you have finished these exercises, check your answers to the odd-numbered problems in the Appendix, and turn to Problem Set 2 on page 118 for practice working with fractions. If you need a quick review of the topics in this chapter, visit the chapter Summary first.

CHAPTER 2
SUMMARY

Fractions

Objective	Review
Write an improper fraction as a mixed number. (p. 78)	Divide the numerator by the denominator. Express the remainder in fraction form. **Example:** To write $\frac{31}{4}$ as a mixed number, divide as follows: $$\begin{array}{r} 7 \\ 4\overline{)31} \\ \underline{28} \\ 3 \end{array} \quad = \quad 7\frac{3}{4}$$ Quotient · Remainder
Write a mixed number as an improper fraction. (p. 79)	Multiply the whole number portion by the denominator and add the numerator. Place this total over the denominator. **Example:** $3\frac{7}{8}$ New Numerator $= 8 \times 3 + 7 \rightarrow \frac{31}{8}$
Write equivalent fractions with larger denominators. (p. 79)	Multiply both the numerator and the denominator by the same nonzero number. **Example:** $$\frac{5}{16} = \frac{?}{64} \qquad \frac{5 \times \square}{16 \times \square} = \frac{?}{64} \qquad \frac{5 \times 4}{16 \times 4} = \frac{20}{64}$$
Rewrite a fraction so that it is in lowest terms. (p. 80)	Divide the numerator and the denominator by their largest common factor. **Example:** $$\frac{10}{64} = \frac{10 \div 2}{64 \div 2} = \frac{5}{32}$$
Multiply fractions. (p. 86)	First, change whole or mixed numbers to improper fractions. Then, eliminate common factors. Finally, multiply numerator by numerator and denominator by denominator. **Example:** $$4\frac{1}{2} \times \frac{2}{3} = \frac{\overset{3}{\cancel{9}}}{\underset{1}{\cancel{2}}} \times \frac{\overset{1}{\cancel{2}}}{\underset{1}{\cancel{3}}} = \frac{3}{1} = 3$$
Divide fractions. (p. 92)	First, change whole or mixed numbers to improper fractions. Then, multiply by the reciprocal of the divisor. **Example:** $$2\frac{7}{8} \div 1\frac{1}{4} = \frac{23}{8} \div \frac{5}{4} = \frac{23}{\underset{2}{\cancel{8}}} \times \frac{\overset{1}{\cancel{4}}}{5} = \frac{23}{10} \quad \text{or} \quad 2\frac{3}{10}$$

Objective	Review
Add and subtract fractions. (p. 99)	If necessary, rewrite the fractions as equivalent fractions with the least common denominator (LCD). Then add or subtract the numerators and write the result as the numerator of the answer, with the LCD as denominator.

Example:

(a) Add $1\frac{3}{16} + \frac{3}{4}$ vertically.

The LCD of 16 and 4 is 16.

$$1\frac{3}{16} \longrightarrow 1\frac{3}{16}$$
$$+\frac{3}{4} \longrightarrow \frac{12}{16}$$
$$\overline{\qquad 1\frac{15}{16}} \longleftarrow \boxed{\text{Answer}}$$

$$\frac{3}{4} = \frac{3 \times 4}{4 \times 4} = \frac{12}{16}$$

(b) $\frac{3}{4} - \frac{1}{5} = \frac{15}{20} - \frac{4}{20} = \frac{11}{20}$

The LCD of 4 and 5 is 20.

$$\frac{1}{5} = \frac{1 \times 4}{5 \times 4} = \frac{4}{20}$$
$$\frac{3}{4} = \frac{3 \times 5}{4 \times 5} = \frac{15}{20}$$

| **Solve practical problems involving fractions.** (pp. 82, 84, 89, 97, 107, 112) | Read the problem carefully. Look for key words or phrases that indicate which operation to use, and then perform the calculation. |

Example: A tabletop is constructed using $\frac{3}{4}$-in. plywood with $\frac{5}{16}$-in. veneer on both sides. Calculate the total thickness of the tabletop.

The word *total* indicates addition. To solve, add two thicknesses of veneer to one thickness of plywood.

$$2\left(\frac{5}{16}\right) + \frac{3}{4} = \frac{5}{8} + \frac{3}{4}$$
$$= \frac{5}{8} + \frac{6}{8}$$
$$= \frac{11}{8} \text{ or } 1\frac{3}{8} \text{ in.}$$

PROBLEM SET 2   Fractions

Answers to odd-numbered problems are given in the Appendix.

A. Write as an improper fraction.

1. $1\dfrac{1}{8}$

2. $4\dfrac{1}{5}$

3. $1\dfrac{2}{3}$

4. $2\dfrac{3}{16}$

5. $3\dfrac{3}{32}$

6. $2\dfrac{1}{16}$

7. $1\dfrac{5}{8}$

8. $3\dfrac{7}{16}$

Write as a mixed number.

9. $\dfrac{10}{4}$

10. $\dfrac{19}{2}$

11. $\dfrac{25}{3}$

12. $\dfrac{9}{8}$

13. $\dfrac{25}{16}$

14. $\dfrac{21}{16}$

15. $\dfrac{35}{4}$

16. $\dfrac{7}{3}$

Write in lowest terms.

17. $\dfrac{6}{32}$

18. $\dfrac{8}{32}$

19. $\dfrac{12}{32}$

20. $\dfrac{18}{24}$

21. $\dfrac{5}{30}$

22. $1\dfrac{12}{21}$

23. $1\dfrac{16}{20}$

24. $3\dfrac{10}{25}$

Complete these.

25. $\dfrac{3}{4} = \dfrac{?}{12}$

26. $\dfrac{7}{16} = \dfrac{?}{64}$

27. $2\dfrac{3}{4} = \dfrac{?}{16}$

28. $1\dfrac{3}{8} = \dfrac{?}{32}$

29. $5\dfrac{2}{3} = \dfrac{?}{12}$

30. $1\dfrac{4}{5} = \dfrac{?}{10}$

31. $1\dfrac{1}{4} = \dfrac{?}{12}$

32. $2\dfrac{3}{5} = \dfrac{?}{10}$

Circle the larger number.

33. $\dfrac{7}{16}$ or $\dfrac{2}{15}$

34. $\dfrac{2}{3}$ or $\dfrac{4}{7}$

35. $\dfrac{13}{16}$ or $\dfrac{7}{8}$

36. $1\dfrac{1}{4}$ or $\dfrac{7}{6}$

37. $\dfrac{13}{32}$ or $\dfrac{3}{5}$

38. $\dfrac{2}{10}$ or $\dfrac{3}{16}$

39. $1\dfrac{7}{16}$ or $\dfrac{7}{4}$

40. $\dfrac{3}{32}$ or $\dfrac{1}{9}$

B. Multiply or divide as shown.

1. $\dfrac{1}{2} \times \dfrac{3}{16}$

2. $\dfrac{3}{4} \times \dfrac{2}{3}$

3. $\dfrac{7}{16} \times \dfrac{4}{3}$

4. $\dfrac{15}{64} \times \dfrac{1}{12}$

5. $1\dfrac{1}{2} \times \dfrac{5}{6}$

6. $3\dfrac{1}{16} \times \dfrac{1}{5}$

7. $\dfrac{3}{16} \times \dfrac{5}{12}$

8. $14 \times \dfrac{3}{8}$

9. $\dfrac{3}{4} \times 10$

10. $\dfrac{1}{2} \times 1\dfrac{1}{3}$

11. $18 \times 1\dfrac{1}{2}$

12. $16 \times 2\dfrac{1}{8}$

13. $2\dfrac{2}{3} \times 4\dfrac{3}{8}$

14. $3\dfrac{1}{8} \times 2\dfrac{2}{5}$

15. $\dfrac{1}{2} \div \dfrac{1}{4}$

16. $\dfrac{2}{5} \div \dfrac{1}{2}$

17. $4 \div \dfrac{1}{8}$

18. $8 \div \dfrac{3}{4}$

19. $\dfrac{2}{3} \div 4$

20. $1\dfrac{1}{2} \div 2$

21. $3\dfrac{1}{2} \div 5$

22. $1\dfrac{1}{4} \div 1\dfrac{1}{2}$

23. $2\dfrac{3}{4} \div 1\dfrac{1}{8}$

24. $3\dfrac{1}{5} \div 1\dfrac{5}{7}$

Name

_____

Date

_____

Course/Section

_____

C. Add or subtract as shown.

1. $\dfrac{3}{8} + \dfrac{7}{8}$ 2. $\dfrac{1}{2} + \dfrac{3}{4}$ 3. $\dfrac{3}{32} + \dfrac{1}{8}$ 4. $\dfrac{3}{8} + 1\dfrac{1}{4}$

5. $\dfrac{3}{5} + \dfrac{5}{6}$ 6. $\dfrac{5}{8} + \dfrac{1}{10}$ 7. $\dfrac{9}{16} - \dfrac{3}{16}$ 8. $\dfrac{7}{8} - \dfrac{1}{2}$

9. $\dfrac{11}{16} - \dfrac{1}{4}$ 10. $\dfrac{5}{6} - \dfrac{1}{5}$ 11. $\dfrac{7}{8} - \dfrac{3}{10}$ 12. $1\dfrac{1}{2} - \dfrac{3}{32}$

13. $2\dfrac{1}{8} + 1\dfrac{1}{4}$ 14. $1\dfrac{5}{8} + \dfrac{13}{16}$ 15. $6 - 1\dfrac{1}{2}$ 16. $3 - 1\dfrac{7}{8}$

17. $3\dfrac{2}{3} - 1\dfrac{7}{8}$ 18. $2\dfrac{1}{4} - \dfrac{5}{6}$ 19. $\dfrac{1}{2} + \dfrac{1}{3} + \dfrac{1}{5}$ 20. $1\dfrac{1}{2} + 1\dfrac{1}{4} + 1\dfrac{1}{5}$

21. $3\dfrac{1}{2} - 2\dfrac{1}{3}$ 22. $2\dfrac{3}{5} - 1\dfrac{4}{15}$ 23. $2 - 1\dfrac{3}{5}$ 24. $4\dfrac{5}{6} - 1\dfrac{1}{2}$

D. Practical Applications

1. **Welding** In a welding job, three pieces of 2-in. I-beam with lengths $5\frac{7}{8}$, $8\frac{1}{2}$, and $22\frac{3}{4}$ in. are needed. What is the total length of I-beam needed? (Do not worry about the waste in cutting.)

2. **Machine Trades** How many pieces of $10\frac{5}{16}$-in. bar can be cut from a stock 20-ft bar? The metal is torch cut and allowance of $\frac{3}{16}$ in. kerf (waste) should be made for each piece. (*Hint:* 20 ft = 240 in.)

3. **Welding** A piece of metal must be cut to a length of $22\frac{3}{8}$ in. $\pm \frac{1}{16}$ in. What are the longest and shortest acceptable lengths? (*Hint:* The symbol $\pm$ means to add $\frac{1}{16}$ in. to get the longest length and subtract $\frac{1}{16}$ in. to get the shortest length. Longest = $22\frac{3}{8}$ in. + $\frac{1}{16}$ in.; Shortest = $22\frac{3}{8}$ in. − $\frac{1}{16}$ in.)

4. **Automotive Trades** A damaged car is said to have "sway" when two corresponding diagonal measurements under the hood are different. If these diagonals are found to be $64\frac{1}{4}$ in. and $62\frac{7}{8}$ in., calculate the magnitude of the sway, the difference between these measurements.

5. **Machine Trades** A shaft $1\frac{7}{8}$ in. in diameter is turned down on a lathe to a diameter of $1\frac{3}{32}$ in. What is the difference in diameters?

6. **Machine Trades** A bar $14\frac{5}{16}$ in. long is cut from a piece $25\frac{1}{4}$ in. long. If $\frac{3}{32}$ in. is wasted in cutting, will there be enough left to make another bar $10\frac{3}{8}$ in. long?

7. **Manufacturing** A cubic foot contains roughly $7\frac{1}{2}$ gallons. How many cubic feet are there in a tank containing $34\frac{1}{2}$ gallons?

8. **Manufacturing** Find the total width of the three pieces of steel plate shown in the drawing.

9. **Machine Trades** What would be the total length of the bar formed by welding together the five pieces of bar stock shown in the drawing?

10. **Machine Trades** The Ace Machine Shop has the job of producing 32 zinger bars. Each zinger bar must be turned on a lathe from a piece of stock $4\frac{7}{8}$ in. long. How many feet of stock will the shop need?

11. **Carpentry** What is the thickness of a tabletop made of $\frac{3}{4}$-in. plywood and covered with a $\frac{3}{16}$-in. sheet of glass?

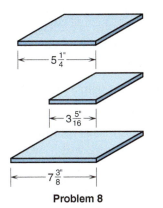

$5\frac{1}{4}"$

$3\frac{5}{16}"$

$7\frac{3}{8}"$

Problem 8

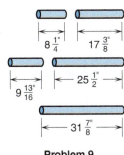

$8\frac{1}{4}"$ $17\frac{3}{8}"$

$9\frac{13}{16}"$ $25\frac{1}{2}"$

$31\frac{7}{8}"$

Problem 9

12. **Construction**   For the wooden form shown, find the lengths A, B, C, and D.

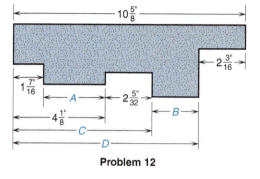

Problem 12

13. **Carpentry**   Find the spacing x between the holes in the figure shown.

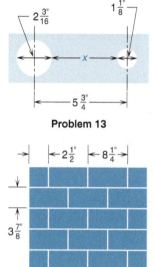

Problem 13

Problem 14

14. **Masonry**   Find the height of the five-course (five-bricks-high) brick wall shown in the drawing if each brick is $2\frac{1}{2}$ in. by $3\frac{7}{8}$ in. by $8\frac{1}{4}$ in. and all mortar joints are $\frac{1}{2}$ in.

15. **Masonry**   If the wall in problem 14 has 28 stretchers (bricks laid lengthwise), what is its length?

16. **Electrical Trades**   An electrical wiring job requires the following lengths of 14/2 BX cable: seven pieces each $6\frac{1}{2}$ ft long, four pieces each $34\frac{3}{4}$ in. long, and nine pieces each $19\frac{3}{8}$ in. long. What is the total length of cable needed?

17. **Printing**   An invitation must be printed on card stock measuring $4\frac{1}{4}$ in. wide by $5\frac{1}{2}$ in. long. The printed material covers a space measuring $2\frac{1}{8}$ in. wide by $4\frac{1}{8}$ in. long. If the printed material is centered in both directions, what are the margins?

18. **Printing**   As a rule of thumb, the top margin of a page of a book should be $\frac{2}{5}$ of the total margin, and the bottom margin should be $\frac{3}{5}$ of the total margin. If the print takes up $9\frac{1}{2}$ in. of an 11-in.-long page, what should the top and bottom margins be? (*Hint:* The total margin $= 11$ in. $- 9\frac{1}{2}$ in. $= 1\frac{1}{2}$ in.)

19. **Welding**   A 46-in. bar must have 9 equally spaced holes drilled through the centerline. If the centers of the two end holes are each $2\frac{1}{4}$ in. in from their respective ends, what should the center-to-center distance of the holes be? (*Hint:* There are 8 spaces between holes.)

20. **Construction**   If an I-beam is to be $24\frac{3}{8}$ in. long with a tolerance of $\pm\frac{1}{4}$ in., find the longest and shortest acceptable lengths.

21. **Machine Trades**   If a positioner shaft turns at 18 revolutions per minute, and the tool feed is $\frac{1}{16}$ in. per revolution, how long will it take to advance $7\frac{1}{2}$ in.?

22. **Sheet Metal Trades**   The total allowance for both edges of a grooved seam is three times the width of the seam. Half of this total is added to each edge of the seam. Find the allowance for each edge of a grooved seam if the width of the seam is $\frac{5}{16}$ in.

23. **Landscaping**   A landscaping project requires that you estimate the length and width of an area needed for a lawn. You pace off the length as being 18 strides long. You know from prior experience that each of your strides is $\frac{5}{6}$ of a yard. What would be your estimate of the length of the proposed lawn?

24. **Machine Trades**   A machine tech must bolt together two steel plates that are $\frac{7}{8}$ in. thick and $1\frac{1}{4}$ in. thick. If the bolt must be at least $\frac{1}{2}$ in. longer than the combined thickness of the plates, how long a bolt is needed?

25. **Automotive Trades**   On a certain vehicle, each turn of the tie-rod sleeve changes the toe-in setting by $\frac{3}{16}$ in. If the tie-rod sleeve makes $1\frac{1}{2}$ turns, how much will the toe-in setting change?

26. **Automotive Trades**   A certain engine has a stroke of $3\frac{1}{4}$ in. If the throw of the crankshaft is half the distance of the stroke, what is the throw of the crankshaft for this engine?

27. **Allied Health**   One bottle of NoGo pain medicine contains 60 tablets. How many tablets are contained in $3\frac{3}{4}$ bottles?

28. **Construction**   The standard slope for a drainage pipe carrying water is $\frac{1}{4}$ in. of vertical drop per foot of horizontal distance. How many inches of vertical drop would an 18-ft drainage pipe require?

29. **Carpentry**   Wood screws are used to attach a $\frac{3}{8}$-in. wood panel to a $\frac{3}{4}$-in. wood frame. Which of the following is the maximum screw length that will be less than the combined thickness of the panel and the frame: $\frac{7}{8}$ in., 1 in., $1\frac{1}{4}$ in., or $1\frac{1}{2}$ in.?

30. **Carpentry**   A carpenter making a window frame will use dual-pane glass for better insulation. If each pane of glass is $\frac{3}{32}$ in. thick and the space between panes is $\frac{5}{16}$ in., what overall thickness does the frame need to enclose?

31. **Flooring and Carpeting**   In the United States carpet comes in standard 12-ft widths. To fit the carpet to a room or a house, the installer may need to order several different lengths of these 12-ft widths and seam them together. One particular house required one length of $10\frac{1}{4}$ ft, one length of $8\frac{1}{2}$ ft, one length of $22\frac{2}{3}$ ft, and two lengths of $15\frac{1}{3}$ ft each.
    (a) What was the total length of carpet needed?
    (b) Multiply your answer to part (a) by 12 to find the total area of carpet needed in square feet.
    (c) Multiply your answer to part (b) by $6 per square foot to find the total cost of the carpet.

32. **Retail Merchandising**   In retail management, large quantities are often ordered by the gross (12 dozen). If a buyer ordered $7\frac{1}{3}$ gross scarves, how many scarves did the buyer order?

33. **Retail Merchandising**   The buyer for a retail store ordered $8\frac{3}{4}$ dozen pairs of shorts. How many pairs of shorts did she order?

34. **Carpentry**   A carpenter wishes to install 5 equally-wide panels of wainscot along the wall of a $207\frac{1}{2}$-inch-wide room. There are 6 vertical strips of wood, called *stiles*, that frame and separate the panels. The two end stiles are each $3\frac{1}{2}$-inches wide, and the four intermediate stiles are each $1\frac{7}{8}$ inches wide. How wide should each of the panels be between the stiles?

CASE STUDY: Restaurant Ownership

The restaurant industry is a large employer of both college students and college graduates. Over half of all adults have at one time worked in the restaurant industry, and there are currently about 14,400,000 restaurant industry employees in the United States. Surveys show that 46 out of every 100 current restaurant workers would like to own an eating establishment of their own someday (Haden, 2014), and 8 in 10 restaurant owners started their industry careers in entry-level positions (restaurant.org, 2016). Successful restaurant owners come from a variety of vocational backgrounds, bringing with them valuable problem-solving skills from their education and experience. Restaurant ownership not only requires a creative and talented chef, but also a knowledgeable and innovative business organizer.

Use the information in the preceding paragraph to estimate answers to questions 1 and 2.

1. Approximately how many current restaurant workers might want to own an eating establishment someday?

2. In Billings, Montana, there are 437 restaurants listed in the Yellow Pages. How many of these owners might have started in the restaurant industry in entry-level positions?

3. Creating a menu is one of the starting points of any restaurant. Once a recipe is selected for the menu, the proper amount of ingredients must be on hand. If a restaurant expects to serve 120 people the following meat loaf recipe, how much of each ingredient is required?

Carl's Meatloaf (Serves 8)

2 lb lean ground beef	$\frac{1}{2}$ tsp pepper
$1\frac{1}{4}$ cups chopped onion	1 egg
$\frac{3}{4}$ cup grated carrot	$\frac{2}{3}$ cup dry bread crumbs
1 cup chopped bell pepper	$\frac{5}{8}$ lb grated cheese
1 tsp salt	$\frac{1}{4}$ lb sliced mushrooms

4. In a restaurant, ingredient preparation is a large component of the work required in making a food dish. Key ingredients for several recipes can be prepared together. If each raw onion on average produces $\frac{3}{4}$ cup of chopped onion, how many onions must be chopped to produce 50 cups?

5. Quite often a restaurant can sell the spices or ingredients for a popular item pre-packaged for home preparation. Grannie's Restaurant is famous for their brownies, and they use the following recipe to make 96 brownies at a time:

Grannie's Brownies (Makes 96 brownies)

9 cups flour

$4\frac{1}{2}$ cups cocoa

2 cups sugar

$3\frac{3}{4}$ cups walnuts

4 cups chocolate chips

$\frac{3}{8}$ cup baking powder

continued...

Suppose they wish to sell a pre-packaged mix that makes 16 brownies. What size container would be required? (Express your answer to the nearest $\frac{1}{16}$ of a cup.)

6. The cost of ingredients for the brownie mix are as follows:
    50-pound bag of flour (1 pound equals $3\frac{1}{3}$ cups of flour): $13.50
    25-pound container of cocoa (1 pound equals $5\frac{1}{3}$ cups): $115.00
    50 pounds of sugar (1 pound equals $1\frac{7}{8}$ cups): $19.80
    25 pounds of walnuts (1 pound equals 4 cups): $87.75
    25 pounds of chocolate chips (1 pound equals $2\frac{2}{3}$ cups): $52.75
    5 pounds of baking powder (1 pound equals 2 cups): $7.00

    (a) What is the cost per cup for each ingredient?
    (b) Calculate the cost of ingredients for one complete package of the mix.
    (c) What would be the total cost of all the ingredients to prepare 100 packages of the mix?

7. Grannie's Restaurant plans to sell each package of brownie mix for $7.00. Industry guidelines recommend that the price of the ingredients be no more than 35% of the selling price of the food item. In fraction form, this means that the cost of the ingredients should be no more than $\frac{35}{100}$ of the selling cost, or $0.35 per $1.00.

    (a) What fraction of the planned selling price is the actual cost of the ingredients as calculated in 6(b)?
    (b) Is this fraction higher or lower than the recommended industry pricing?
    (c) What should the minimum selling price be in order to meet industry guidelines?

References:
Haden, J. (2014). Start a Killer Restaurant: 6 Tips. Retrieved from http://www.inc.com/jeff-haden/tyson-cole-how-to-start-successful-restaurant-6-tips.html.
Restaurant.org. (2016, December 12). http://www.restaurantorgtNews-ResearcftResearcWfacts-at-a-Glance.

Using a Calculator, II: Fractions

Fractions can be entered directly on most scientific calculators, and the results of arithmetic calculations with fractions can be displayed as fractions or decimals. If your calculator has an $\boxed{A\frac{b}{c}}$ key, you may enter fractions or mixed numbers directly into your machine without using the division key. Alternate names for this key include $\boxed{a\frac{b}{c}}$, $\boxed{\frac{n}{d}}$, and $\boxed{\frac{\blacksquare}{\square}}$.

Note   Most graphing calculators do not have a key such as $\boxed{A\frac{b}{c}}$ for entering fractions. Instead, you simply use the division key to enter them, and answers appear in decimal form. You can then convert them to fraction form using the function "▶ Frac" normally found on the "MATH" menu. ●

The calculator display will usually indicate a common or improper fraction with a ⌐, a ⌐, or a / symbol.

EXAMPLE 1   To enter the fraction $\dfrac{3}{4}$, we will show the following key sequence and answer display:

3 **4** $\boxed{=}$ → *3 / 4*

Depending on your model, you may also see one of the following displays:

$$3 \lrcorner 4 \qquad \text{or} \qquad 3 \ulcorner 4$$

If your calculator has a four-line display, you can enter the fraction in vertical form using the arrow keys, and the result will be displayed as $\frac{3}{4}$. See your instruction manual for further details. ●

If you enter an unsimplified fraction, pressing $=$ will automatically reduce it to lowest terms.

EXAMPLE 2   To enter the fraction $\frac{6}{16}$, we would have

6 $\boxed{A^b_c}$ **16** $\boxed{=}$ $\rightarrow$ $3/8$ ●

To enter a mixed number, use the $\boxed{A^b_c}$ key twice.

EXAMPLE 3   To enter $2\frac{7}{8}$, we will show the following key sequence and display:

2 $\boxed{A^b_c}$ **7** $\boxed{A^b_c}$ **8** $\boxed{=}$ $\rightarrow$ $2 \lrcorner 7/8$

On your calculator, you might also see this mixed number displayed in one of the following ways:

$$2\_7\lrcorner 8 \qquad \text{or} \qquad 2\lrcorner 7\lrcorner 8 \qquad \text{or} \qquad 2\ulcorner 7\ulcorner 8$$

Calculators with a four-line display are able to show the mixed number as

$$2\frac{7}{8}$$ ●

✎ **Note**   With a graphing calculator, you would enter $2\frac{7}{8}$ as **2** $\boxed{+}$ **7** $\boxed{\div}$ **8** $\boxed{=}$. If you convert this to a fraction, the graphing calculator will display it only as an improper fraction. ●

When you enter an improper fraction followed by $=$, the calculator will automatically convert it to a mixed number.

EXAMPLE 4   Entering the fraction $\frac{12}{7}$ followed by $=$ will result in a display of $1\frac{5}{7}$, like this:

12 $\boxed{A^b_c}$ **7** $\boxed{=}$ $\rightarrow$ $1 \lrcorner 5/7$

To convert this back to an improper fraction, look for a second function key labeled $\boxed{A^b_c \leftrightarrow \frac{d}{e}}$ or simply $\boxed{\frac{d}{c}}$. ●

EXAMPLE 5   The conversion of the previous answer (in Example 4) will look like one of the following:

$\boxed{2^{nd}}$ $\boxed{A^b_c \leftrightarrow \frac{d}{e}}$ $\boxed{=}$ $\rightarrow$ $12/7$ or $\boxed{2^{nd}}$ $\boxed{\frac{d}{c}}$ $\rightarrow$ $12\lrcorner 7$

Notice that in the second sequence, pressing the $=$ key is not required. Repeating either sequence will convert the answer back to a mixed number. We shall use the first sequence in the text. ●

To perform arithmetic operations with fractions, enter the calculation in the usual way.

EXAMPLE 6   Compute $\dfrac{26}{8} - 1\dfrac{2}{3}$ like this:

26 $\boxed{A^b_c}$ **8** $\boxed{-}$ **1** $\boxed{A^b_c}$ **2** $\boxed{A^b_c}$ **3** $\boxed{=}$ $\rightarrow$ $1 \lrcorner 7/12$ ●

✎ **Note**   Some models require you to press $\boxed{ENTER}$ or $\boxed{=}$ after keying in each fraction. Consult your instruction manual for details. ●

Your Turn   Work the following problems using your calculator. Where possible, express your answer both as a mixed number and as an improper fraction.

(a) $\dfrac{2}{3} + \dfrac{7}{8}$ (b) $1\dfrac{3}{4} - \dfrac{2}{5}$ (c) $\dfrac{25}{32} + 1\dfrac{7}{8}$ (d) $8\dfrac{1}{5} \div 2\dfrac{1}{6}$

(e) $\dfrac{17}{20} \times \dfrac{1}{3}$

Solutions

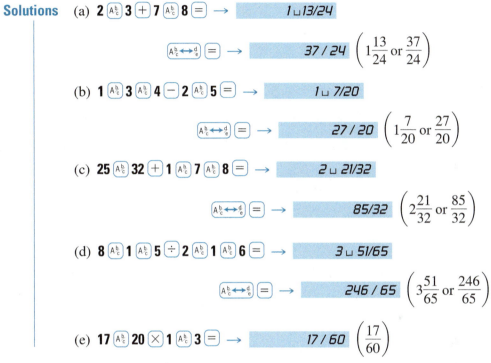

(a) **2** $\boxed{A_c^b}$ **3** $\boxed{+}$ **7** $\boxed{A_c^b}$ **8** $\boxed{=}$ $\rightarrow$ *1⎵13/24*

$\boxed{A_c^b \leftrightarrow \tfrac{d}{e}}$ $\boxed{=}$ $\rightarrow$ *37 / 24* $\left(1\dfrac{13}{24} \text{ or } \dfrac{37}{24}\right)$

(b) **1** $\boxed{A_c^b}$ **3** $\boxed{A_c^b}$ **4** $\boxed{-}$ **2** $\boxed{A_c^b}$ **5** $\boxed{=}$ $\rightarrow$ *1⎵7/20*

$\boxed{A_c^b \leftrightarrow \tfrac{d}{e}}$ $\boxed{=}$ $\rightarrow$ *27 / 20* $\left(1\dfrac{7}{20} \text{ or } \dfrac{27}{20}\right)$

(c) **25** $\boxed{A_c^b}$ **32** $\boxed{+}$ **1** $\boxed{A_c^b}$ **7** $\boxed{A_c^b}$ **8** $\boxed{=}$ $\rightarrow$ *2⎵21/32*

$\boxed{A_c^b \leftrightarrow \tfrac{d}{e}}$ $\boxed{=}$ $\rightarrow$ *85/32* $\left(2\dfrac{21}{32} \text{ or } \dfrac{85}{32}\right)$

(d) **8** $\boxed{A_c^b}$ **1** $\boxed{A_c^b}$ **5** $\boxed{\div}$ **2** $\boxed{A_c^b}$ **1** $\boxed{A_c^b}$ **6** $\boxed{=}$ $\rightarrow$ *3⎵51/65*

$\boxed{A_c^b \leftrightarrow \tfrac{d}{e}}$ $\boxed{=}$ $\rightarrow$ *246 / 65* $\left(3\dfrac{51}{65} \text{ or } \dfrac{246}{65}\right)$

(e) **17** $\boxed{A_c^b}$ **20** $\boxed{\times}$ **1** $\boxed{A_c^b}$ **3** $\boxed{=}$ $\rightarrow$ *17 / 60* $\left(\dfrac{17}{60}\right)$

3 Decimal Numbers

Objective	Sample problems	For help, go to page
When you finish this chapter you will be able to:		
1. (a) Write or pronounce a decimal number in words.	(a) Write in words 26.035	130
(b) Write a decimal in numerical form if it is given in words.	(b) Write as a decimal number: One hundred six and twenty-seven ten-thousandths	131
2. Add, subtract, multiply, and divide decimal numbers.	(a) $5.82 + 0.096$	133
	(b) $3.78 - 0.989$	136
	(c) $27 - 4.03$	
	(d) 7.25×0.301	142
	(e) $104.2 \div 0.032$	146
	(f) $0.09 \div 0.0004$	
	(g) $20.4 \div 6.7$ (round to three decimal places)	
3. Find averages.	Find the average of 4.2, 4.8, 5.7, 2.5, 3.6, 5.0	154

Name _____

Date _____

Course/Section _____

Objective	Sample problems	For help, go to page
4. Work with decimal fractions.	(a) Write as a decimal number $\frac{3}{16}$ _____	166
	(b) $1\frac{2}{3} + 1.785$ _____	
	(c) $4.1 \times 2\frac{1}{4}$ _____	
	(d) $1\frac{5}{16} \div 4.3$ (round to three decimal places) _____	
5. Solve practical problems involving decimal numbers.	(a) **Electrical Trades** Six recessed lights must be installed in a strip of ceiling. The housings cost $16.45 each, the trims cost $19.86 each, the lamps cost $11.99 each, miscellaneous hardware and wiring cost $33.45, and labor is estimated at 3.25 hours at $80 per hour. What will be the total cost of the job? _____	
	(b) **Machine Trades** A container of 175 bolts weighs 61.3 lb. If the container itself weighs 1.8 lb, how much does each bolt weigh? _____	

(Answers to these preview problems are given in the Appendix. Also, worked solutions to many of these problems appear in the chapter Summary.)

If you are certain that you can work *all* these problems correctly, turn to page 178 for a set of practice problems. If you cannot work one or more of the preview problems, turn to the page indicated to the right of the problem. Those who wish to master this material with the greatest success should turn to Section 3-1 and begin work there.

Decimal Numbers

Whether you are using mathematics as a consumer or as a trades worker, you will need to work with decimal numbers. Decimal numbers often provide us with a simpler way of dealing with fractional parts of quantities ranging from money to measurements. In this chapter, you will explore the relationship between fractions and decimals and learn how to perform all the basic operations with decimal numbers.

CASE STUDY: Fabricating a Workbench

In the Case Study at the end of this chapter, you will see how **welders** and others involved in fabrication and manufacturing often need to convert the fractional measurements given on blueprints to decimal form. They also need to use decimal operations to find missing dimensions and to determine the amounts of materials needed for a given product.

3-1 Addition and Subtraction of Decimal Numbers

Place Value of Decimal Numbers

1. Write as a decimal number: seven tenths
2. You have three dollars and seventy-five cents and you spend a dollar twenty. How much do you have left? (Write your answer as a decimal number without the dollar sign.)

By now you know that whole numbers are written in a form based on powers of ten. A number such as

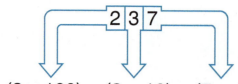

means $(2 \times 100) + (3 \times 10) + (7 \times 1)$ or $200 + 30 + 7$

This way of writing numbers can be extended to fractions. A *decimal* number is a fraction whose denominator is 10 or some power of 10, such as 10 or 100 or 1000, and so on.

A decimal number may have both a whole-number part and a fraction part. For example, the number 324.576 means

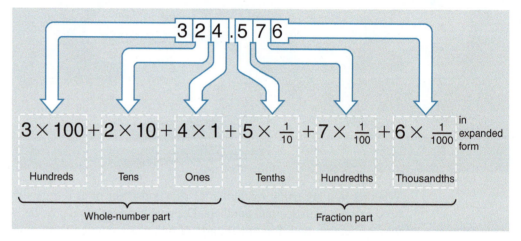

You are already familiar with this way of interpreting decimal numbers from working with money.

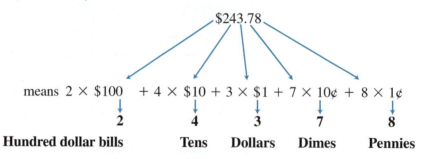

Writing a Decimal Number in Words

To write a decimal number in words, remember this diagram:

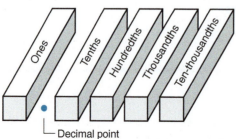

EXAMPLE 1   (a) $0.6 =$ Six tenths $= \dfrac{6}{10}$

(b) $7.05 =$ Seven and five hundredths $= 7\dfrac{5}{100}$

(c) $12.267 =$ Twelve and two hundred sixty-seven thousandths $= 12\dfrac{267}{1000}$

(d) $0.0065 =$ Sixty-five ten-thousandths $= \dfrac{65}{10{,}000}$

Decimal form

Fraction form

✎ **Note**   We usually write a decimal number less than 1 with a zero to the left of the decimal point. For Example 1(a), .6 is written 0.6, and for Example 1(d), .0065 is written 0.0065. It is easy to mistake .6 for 6, but the decimal point in 0.6 cannot be overlooked. ●

Your Turn   Write the following decimal numbers in words.

(a) 0.56   (b) 19.278   (c) 6.4   (d) 0.07   (e) 5.064

(f) **Automotive Trades**   The clearance between a piston and its bore is measured as 0.0018 in. Express this in words.

Answers   (a) Fifty-six hundredths

(b) Nineteen and two hundred seventy-eight thousandths

(c) Six and four tenths

(d) Seven hundredths

(e) Five and sixty-four thousandths

(f) Eighteen ten-thousandths of an inch

👈 **Learning Help**   The word *and* represents the decimal point. Everything preceding *and* is the whole-number part, and everything after *and* is the decimal part of the number. ●

Translating from Words to Numerical Form   It is also important to be able to perform the reverse process—that is, write a decimal in numerical form if it is given in words.

EXAMPLE 2   To write "six and twenty-four thousandths" as a numeral,

First, write the whole-number part (six). This is the part before the "and." Follow this by a decimal point—the "and."   6.

Next, draw as many blanks as the decimal part indicates. In this case, the decimal part is "thousandths," so we allow for three decimal places. Draw three blanks.   6.__ __ __

Finally, write a number giving the decimal (twenty-four). Write it so it *ends* on the far right blank. Fill in any blank decimal places with zeros.   6.__ 2 4

6.024   ●

Your Turn   Write each of the following as decimal numbers.

(a) Five thousandths

(b) One hundred and six tenths

(c) Two and twenty-eight hundredths

(d) Seventy-one and sixty-two thousandths

(e) Three and five hundred eighty-nine ten-thousandths

Answers   (a) 0.005   (b) 100.6   (c) 2.28   (d) 71.062   (e) 3.0589

Decimal Digits   In the decimal number 86.423 the digits 4, 2, and 3 are called *decimal digits*.

The number 43.6708 has four decimal digits: 6, 7, 0, and 8.

The number 5376.2 has one decimal digit: 2.

All digits to the right of the decimal point, those that name the fractional part of the number, are **decimal digits**.

Your Turn   How many decimal digits are included in each of these numbers?

(a) 1.4   (b) 315.7   (c) 0.425   (d) 324.0075

Answers   (a) one   (b) one   (c) three   (d) four

We will use the idea of decimal digits often in doing arithmetic with decimal numbers.

The decimal point is simply a way of separating the whole-number part from the fraction part. It is a place marker. In whole numbers, the decimal point usually is not written, but it is understood to be there.

The whole number 2 is written 2. as a decimal.

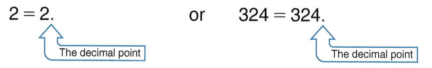

$$2 = 2. \qquad\qquad \text{or} \qquad 324 = 324.$$

The decimal point   The decimal point

This is very important. Many people make big mistakes in arithmetic because they do not know where that decimal point should go.

Very often, additional zeros are attached to the decimal number without changing its value. For example,

$8.5 = 8.50 = 8.5000$ and so on

$6 = 6. = 6.0 = 6.000$ and so on

The value of the number is not changed, but the additional zeros may be useful, as we shall see.

Expanded Form   Writing decimal numbers in expanded form will help us understand the addition process. The decimal number 0.267 can be written in expanded form as

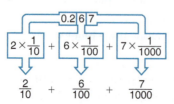

$$2 \times \frac{1}{10} + 6 \times \frac{1}{100} + 7 \times \frac{1}{1000}$$

$$\frac{2}{10} + \frac{6}{100} + \frac{7}{1000}$$

EXAMPLE 3 The decimal number 23.526 can be written in expanded form as

$$23.526 = 2 \times 10 + 3 \times 1 + 5 \times \frac{1}{10} + 2 \times \frac{1}{100} + 6 \times \frac{1}{1000}$$

$$= 20 + 3 + \frac{5}{10} + \frac{2}{100} + \frac{6}{1000}$$

●

Your Turn To help get these ideas clear in your mind, write the following in expanded form.

(a) 0.42 (b) 43.607 (c) 14.5060 (d) 235.22267

Answers (a) 0.42 $= 4 \times \frac{1}{10} + 2 \times \frac{1}{100}$

$= \frac{4}{10} + \frac{2}{100}$

(b) 43.607 $= 4 \times 10 + 3 \times 1 + 6 \times \frac{1}{10} + 0 \times \frac{1}{100} + 7 \times \frac{1}{1000}$

$= \quad 40 \quad + \quad 3 \quad + \quad \frac{6}{10} \quad + \quad \frac{0}{100} \quad + \quad \frac{7}{1000}$

(c) 14.5060 $= 10 + 4 + \frac{5}{10} + \frac{0}{100} + \frac{6}{1000} + \frac{0}{10,000}$

(d) 235.22267 $= 200 + 30 + 5 + \frac{2}{10} + \frac{2}{100} + \frac{2}{1000} + \frac{6}{10,000} + \frac{7}{100,000}$

Learning Help Notice that the denominators in the decimal fractions change by a factor of 10. For example,

3247 . 8956

3	×	1000	thousands	3000 . 0006	ten-thousandths	6	×	$\frac{1}{10,000}$
2	×	100	hundreds	200 . 005	thousandths	5	×	$\frac{1}{1000}$
4	×	10	tens	40 . 09	hundredths	9	×	$\frac{1}{100}$
7	×	1	ones	7 . 8	tenths	8	×	$\frac{1}{10}$

Each row changes by a factor of ten

●

Addition Because decimal numbers represent fractions with denominators equal to powers of ten, addition is very simple. Using expanded form, we can illustrate it this way:

$$2.34 = 2 + \frac{3}{10} + \frac{4}{100}$$
$$+5.23 = 5 + \frac{2}{10} + \frac{3}{100}$$
$$\overline{ = 7 + \frac{5}{10} + \frac{7}{100}} = 7.57$$

Adding like fractions

Of course, we do not need this clumsy business in order to add decimal numbers. As with whole numbers, we may arrange the digits in vertical columns and add directly.

EXAMPLE 4   Let's add 1.45 + 3.42.

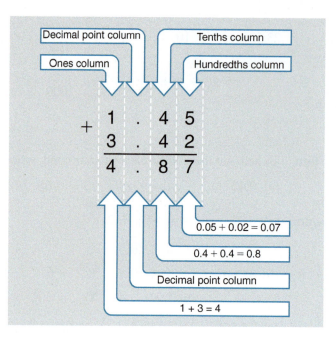

Digits of the same power of ten are placed in the same vertical column. Decimal points are always lined up vertically.

If one of the numbers is written with fewer decimal digits than the other, attach as many zeros as needed so that both have the same number of decimal digits.

$$\begin{array}{r} 2.345 \\ +1.5 \\ \hline \end{array} \quad \text{becomes} \quad \begin{array}{r} 2.345 \\ +1.500 \\ \hline \end{array} \quad \text{and } 6 + 3.08 \text{ becomes} \quad \begin{array}{r} 6.00 \\ +3.08 \\ \hline \end{array}$$

Except for the preliminary step of lining up decimal points, addition of decimal numbers is exactly the same process as addition of whole numbers.   ●

EXAMPLE 5   Here's how to add 13.2 + 1.57:

Decimal points are in line.

$$\begin{array}{r} 13.20 \\ 1.57 \\ \hline 14.77 \end{array}$$

Attach a zero to provide the same number of decimal digits as in the other addend.

Place answer decimal point in the same vertical line.   ●

✔   13 + 2 = 15, which agrees roughly with the answer.

Your Turn   Add the following decimal numbers.

(a) $4.02 + $3.67 = _____   (b) 23.007 + 1.12 = _____

(c) 14.6 + 1.2 + 3.15 = _____   (d) 5.7 + 3.4 = _____

(e) $9 + $0.72 + $6.09 = _____   (f) 0.07 + 6.79 + 0.3 + 3 = _____

Arrange each sum vertically, placing the decimal points in the same column, then add as with whole numbers.

Solutions

Decimal points are lined up vertically.

(a) $4.02
$3.67
$7.69

0.02 + 0.07 = 0.09   Add cents.
0.0 + 0.6 = 0.6   Add 10-cent units.
4 + 3 = 7   Add dollars.

As a check, notice that the sum is roughly $4 + $4 or $8, which agrees with the actual answer. Always check your answer by first estimating it, then comparing your estimate or rough guess with the final answer.

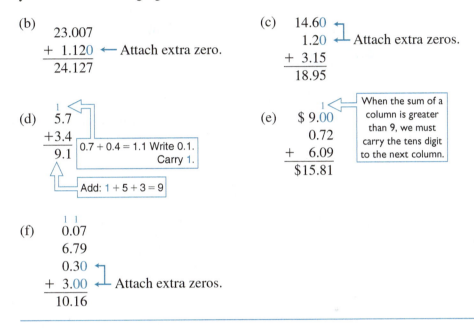

(b)
   23.007
+ 1.120 ← Attach extra zero.
   24.127

(c)
   14.60 ← Attach extra zeros.
    1.20
+ 3.15
   18.95

(d)
  ¹
  5.7
+3.4
  9.1   0.7 + 0.4 = 1.1 Write 0.1. Carry 1.

Add: 1 + 5 + 3 = 9

(e)
  ¹
$ 9.00   When the sum of a column is greater than 9, we must carry the tens digit to the next column.
  0.72
+ 6.09
$15.81

(f)
  ¹ ¹
  0.07
  6.79
  0.30 ← Attach extra zeros.
+ 3.00
  10.16

⊘ Careful   You must line up the decimal points vertically to be certain of getting a correct answer. ●

Applications of Decimal Addition

EXAMPLE 6   **Automotive Trades**   An automotive technician might be paid by the **flat-rate** method. This means that he is paid for the number of hours that each of his jobs *should* take, and not by the number of hours he actually works. In a particular week, a technician's daily flat rate hours were 8.75, 6.5, 9.4, 7.75, and 8. How many total hours of flat rate did the technician accumulate?

As we have seen with whole numbers and fractions, the word *total* indicates that we must *add* to find the answer.

  ² ¹
  8.75
  6.50 ←
  9.40 ← Attach extra zeros
  7.75
+ 8.00 ←
  40.40 hours

●

Your Turn   **Sheet Metal Trades**   A sheet metal worker uses 2.36, 7, 3.9, and 0.6 fluid ounces (fl oz) of cleaning concentrate on successive jobs. What total amount of concentrate has he used during this time?

Solution　Again, the question "What is the total amount . . . ?" indicates that we must add.

$$
\begin{array}{r}
\overset{1}{2}.36 \\
7.00 \\
3.90 \\
+\ 0.60 \\
\hline
13.86
\end{array}
$$

— Attach extra zeros.

13.86 fluid ounces of concentrate was used.

Subtraction　Subtraction is equally simple if you line up the decimal points carefully and attach any needed zeros before you begin work. As in the subtraction of whole numbers, you may need to "borrow" to complete the calculation.

EXAMPLE 7　(a) $437.56 - $41 = _____

$$
\begin{array}{r}
\$\,4\,\overset{3}{\cancel{3}}\,7.\overset{13}{\cancel{5}}\,6 \\
-\$\ \ \ 4\,1.0\,0 \\
\hline
\$\,3\,9\,6.5\,6
\end{array}
$$

— Decimal points are in a vertical line.
— Attach zeros (remember that $41 is $41. or $41.00).

Recall that we can check our subtraction by adding the answer ($396.56) to the number we subtracted ($41.00) to see if this sum is equal to the first number ($437.56).

(b) $19.452 - 7.3617 = $ _____

— Decimal points are in a vertical line.

$$
\begin{array}{r}
1\,9.\overset{3}{\cancel{4}}\,\overset{15}{\cancel{5}}\,\overset{1}{2}\,\overset{10}{\cancel{0}} \\
-\ \ 7.3\,6\,1\,7 \\
\hline
1\,2.0\,9\,0\,3
\end{array}
$$

← Attach zero.

— Answer decimal point is in same vertical line.

✓　$7.3617 + 12.0903 = 19.4520$

(c) $36 - 11.132 = $ _____

First, write 36 as 36.000.
Then, line up the decimal points and subtract.

$$
\begin{array}{r}
3\,\overset{5}{\cancel{6}}.\overset{9}{\cancel{0}}\,\overset{9}{\cancel{0}}\,\overset{10}{\cancel{0}} \\
-1\,1.1\,3\,2 \\
\hline
2\,4.8\,6\,8
\end{array}
$$

← Attach zeros.

✓　$11.132 + 24.868 = 36.000$

●

Your Turn　Try these problems to test yourself on the subtraction of decimal numbers.

(a) $37.66 - $14.57 = _____　(b) $248.3 - 135.921 = $ _____

(c) $7.304 - 2.59 = $ _____　(d) $20 - $7.74 = _____

Work carefully.

Solutions　(a)

— Line up decimal points.

$$
\begin{array}{r}
\$\,3\,7.\overset{5}{\cancel{6}}\,\overset{16}{\cancel{6}} \\
-\$\,1\,4.5\,7 \\
\hline
\$\,2\,3.0\,9
\end{array}
$$

✓　$14.57 + $23.09 = 37.66

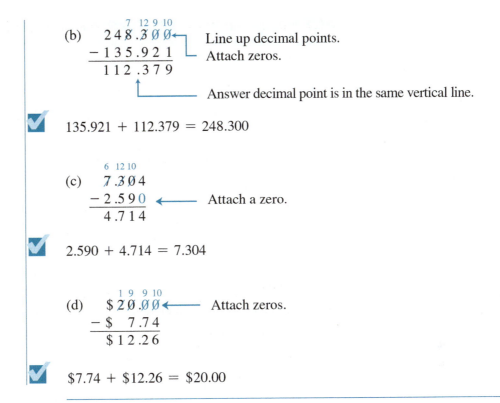

(b)
$$\begin{array}{r} \overset{7\ \ 12\ 9\ 10}{2\,4\,8\,.\,3\,\cancel{0}\,\cancel{0}} \\ -\ 1\,3\,5\,.\,9\,2\,1 \\ \hline 1\,1\,2\,.\,3\,7\,9 \end{array}$$

Line up decimal points.
Attach zeros.

Answer decimal point is in the same vertical line.

✓ 135.921 + 112.379 = 248.300

(c)
$$\begin{array}{r} \overset{6\ \ 12\ 10}{7\,.\,3\,\cancel{0}\,4} \\ -\ 2\,.\,5\,9\,0 \\ \hline 4\,.\,7\,1\,4 \end{array}$$

Attach a zero.

✓ 2.590 + 4.714 = 7.304

(d)
$$\begin{array}{r} \overset{1\ 9\ \ 9\ 10}{\$\,2\,\cancel{0}\,.\,\cancel{0}\,\cancel{0}} \\ -\ \$\ \ \,7\,.\,7\,4 \\ \hline \$\,1\,2\,.\,2\,6 \end{array}$$

Attach zeros.

✓ $7.74 + $12.26 = $20.00

Applications of Decimal Subtraction

EXAMPLE 8 **Sports and Leisure** In the 2012 Summer Olympics, Gabby Douglas of the United States was the all-around champion in women's gymnastics. Her routine on the uneven bars had a maximum value of 16.600 points, and the judges assessed her 0.867 points worth of deductions. What was her final score in the event?

The word *deductions* indicates that we must subtract to calculate her final score.

$$\begin{array}{r} \overset{5\ 15\ 9\ 10}{1\,6\,.\,6\,\cancel{0}\,\cancel{0}} \\ -\ \ \,0\,.\,8\,6\,7 \\ \hline 1\,5\,.\,7\,3\,3 \end{array}$$

Her final score was 15.733 points. ●

Your Turn **Machine Trades** If 0.037 in. of metal is machined from a rod exactly 10 in. long, what is the new length of the rod?

Solution The phrase *machined from* means "removed," which suggests subtraction.

$$\begin{array}{r} \overset{9\ \ 9\ \ 9\ 10}{\cancel{1}\,\cancel{0}\,.\,\cancel{0}\,\cancel{0}\,\cancel{0}} \\ -\ \ \,0\,.\,0\,3\,7 \\ \hline 9\,.\,9\,6\,3 \end{array}$$

Attach zeros.

✓ 0.037 + 9.963 = 10.000

The new length of the rod is 9.963 in.

Now, for a set of practice problems on addition and subtraction of decimal numbers, turn to Exercises 3-1.

Exercises 3-1   Addition and Subtraction of Decimal Numbers

A. Write in words.

1. 0.72	2. 8.7	3. 12.36	4. 0.05
5. 3.072	6. 14.091	7. 3.0024	8. 6.0083

Write as a decimal number.

9. Four thousandths

10. Three and four tenths

11. Six and seven tenths

12. Five hundredths

13. Twelve and eight tenths

14. Three and twenty-one thousandths

15. Ten and thirty-two hundredths

16. Forty and seven tenths

17. One hundred sixteen thousandths

18. Forty-seven ten-thousandths

19. Two and three hundred seventy-four ten-thousandths

20. Ten and two hundred twenty-two ten-thousandths

B. Add or subtract as shown.

1. $14.21 + 6.8$	2. $75.6 + 2.57$
3. $\$2.83 + \$12.19$	4. $\$52.37 + \$98.74$
5. $0.687 + 0.93$	6. $0.096 + 5.82$
7. $507.18 + 321.42$	8. $212.7 + 25.46$
9. $45.6725 + 18.058$	10. $390 + 72.04$
11. $19 - 12.03$	12. $7.83 - 6.79$
13. $\$33.40 - \$18.04$	14. $\$20.00 - \$13.48$
15. $75.08 - 32.75$	16. $40 - 3.82$
17. $\$30 - \$7.98$	18. $\$25 - \$0.61$
19. $130 - 16.04$	20. $19 - 5.78$
21. $37 + 0.09 + 3.5 + 4.605$	22. $183 + 3.91 + 45 + 13.2$
23. $\$14.75 + \$9 + \$3.76$	24. $148.002 + 3.4$
25. $68.708 + 27.18$	26. $35.36 + 4.347$
27. $47.04 - 31.88$	28. $180.76 - 94.69$
29. $26.45 - 17.832$	30. $92.302 - 73.6647$
31. $6.4 + 17.05 + 7.78$	32. $212.4 + 76 + 3.79$
33. $26.008 - 8.4$	34. $36.4 - 7.005$
35. $0.0046 + 0.073$	36. $0.038 + 0.00462$
37. $28.7 - 7.38 + 2.9$	38. $0.932 + 0.08 - 0.4$
39. $6.01 - 3.55 - 0.712$	40. $2.92 - 1.007 - 0.08$

C. Practical Applications

1. **Machine Trades** What is the combined thickness of these five shims: 0.008, 0.125, 0.150, 0.185, and 0.005 in.?

2. **Electrical Trades** The combined weight of a spool and the wire it carries is 13.6 lb. If the weight of the spool is 1.75 lb, what is the weight of the wire?

3. **Electrical Trades** The following are diameters of some common household wires: No. 10 is 0.102 in., No. 11 is 0.090 in., No. 12 is 0.081 in., No. 14 is 0.064 in., and No. 16 is 0.051 in.

   (a) The diameter of No. 16 wire is how much smaller than the diameter of No. 14 wire?

   (b) Is No. 12 wire larger or smaller than No. 10 wire? What is the difference in their diameters?

   (c) John measured the thickness of a wire with a micrometer as 0.059 in. Assuming that the manufacturer was slightly off, what wire size did John have?

4. **Trades Management** In estimating the cost of a job, a contractor included the following items:

Material	$ 877.85
Trucking	$ 62.80
Permits	$ 250.00
Labor	$1845.50
Profit	$ 450.00

   What was his total estimate for the job?

5. **Metalworking** Find A, B, and C.

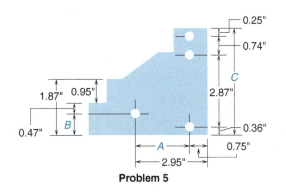

Problem 5

6. **Plumbing** A piece of pipe 8.4 in. long is cut from a piece 40.8 in. long. What is the length of the longer piece remaining if the width of the saw cut is 0.2 in.?

7. **Machine Trades** A certain machine part is 2.345 in. thick. What is its thickness after 0.078 in. is ground off?

8. **Masonry** Find the total cost of the materials for a certain masonry job if sand cost $16.63, mortar mix cost $99.80, and brick cost $1476.28.

9. **Masonry** The specifications for a reinforced masonry wall called for 1.5 sq in. of reinforcing steel per square foot of cross-sectional area. If the three pieces of steel being used had cross sections of 0.125, 0.200, and 1.017 sq in., did they meet the specifications for a 1-sq-ft area?

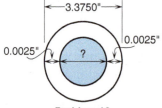

←——3.3750"——→

0.0025"   0.0025"

?

Problem 13

10. **Construction**   A plot plan of a building site showed that the east side of the house was 46.35 ft from the east lot line, and the west side of the house was 41.65 ft from the west lot line. If the lot was 156.00 ft wide along the front, how wide is the house?

11. **Automotive Trades**   A mechanic must estimate the total time for a particular servicing. She figures 0.3 hour for an oil change, 1.5 hours for a tune-up, 0.4 hour for a brake check, and 1.2 hours for air-conditioning service. What is the total number of hours of her estimate?

12. **Automotive Trades**   A heated piston measures 8.586 cm in diameter. When cold, it measures 8.573 cm in diameter. How much does it expand when heated?

13. **Automotive Trades**   A piston must fit in a bore with a diameter of 3.3750 in. What must be the diameter of the piston given the clearance shown in the diagram?

14. **Automotive Trades**   In squaring a damaged car frame, an autobody worker measured the diagonals between the two front cross-members. One diagonal was 196.1 cm and the other was 176.8 cm. What is the difference that must be adjusted?

15. **Machine Trades**   The diameter of a steel shaft is reduced 0.007 in. The original diameter of the shaft was 0.850 in. Calculate the reduced diameter of the shaft.

16. **Allied Health**   A bottle of injectable vertigo medication contains 30 mL. If 5.65 mL are removed, what volume of medicine remains?

17. **Construction**   A structural steel Lally column is mounted on a concrete footing as shown in the figure. Find the length of the Lally column.

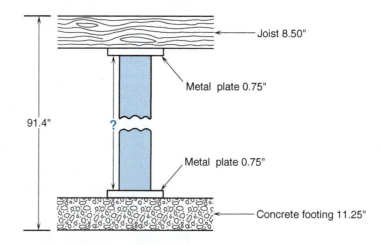

Joist 8.50"

Metal plate 0.75"

91.4"   ?

Metal plate 0.75"

Concrete footing 11.25"

18. **Allied Health**   A patient has body mass index (BMI) of 28.3. If the optimum range of BMI is between 18.5 and 25, how much above the upper limit is the patient's BMI?

19. **Automotive Trades**   A technician renovating a truck is replacing the spark plugs. The recommended spark plug gap is 0.044 in., but the new plugs have a gap of 0.035 in. By what amount must the electrodes be bent in order to achieve the recommended gap?

20. **Electronics**   A satellite dish must be angled at an elevation of 43.1 degrees above horizontal to capture the signal from a particular satellite. When installing the dish, the technician will first approximate the proper position and

then adjust the angle as needed. If the angle is initially set at 35.7 degrees, how many more degrees must the angle be increased so that the dish will capture the signal?

21. **Automotive Trades** An automotive technician is paid by the job, or flat rate, so that his hours will vary on a daily basis. During one particular day, he completed jobs that credited him with the following number of hours: 1.5, 0.6, 0.4, 1.0, 2.2, and 1.5. What was the total number of hours he accumulated that day?

22. **Sports and Leisure** In the 2012 Summer Olympics, American gymnast Aly Raisman was tied with Aliya Mustafina of Russia for the bronze medal at the conclusion of the women's all-around competition. They each had a total of 59.566 points, and their individual event scores were as follows:

	Vault	Uneven Bars	Balance Beam	Floor Exercise
Mustafina	15.233	16.100	13.633	14.600
Raisman	15.900	14.333	14.200	15.133

According to the tie-breaking procedure, each gymnast's lowest score was discarded, and the total score for their best three events determined the winner of the medal. Who won the bronze medal, and what was the margin of victory?

23. **Construction** A certain type of laminated glass consists of two outer layers of glass each 2.60 mm thick and an inner layer of adhesive that is 0.38 mm thick. What is the total thickness of this type of glass?

D. Calculator Problems (If you need help using a calculator with decimals, turn to page 187.)

1. **Life Skills** Balance this checkbook by finding the closing balance as of November 4.

Date	Balance	Withdrawals	Deposits
Oct. 1	$367.21		
Oct. 3		$167.05	
Oct. 4		$104.97	
Oct. 8			$357.41
Oct. 16		$ 87.50	
Oct. 18		$ 9.43	
Oct. 20		$ 30.09	
Oct. 22			$364.85
Oct. 27		$259.47	
Oct. 30		$100.84	
Nov. 2		$ 21.88	
Nov. 4	?		

2. **Automotive Trades** What is the actual cost of the following car?

Sticker price	$21,325.00
Destination and delivery	635.00
Leather interior	875.40
5-speed automatic transmission	900.00
CD player	575.95
Moonroof	735.50
Tax and license	1602.34
Less trade-in	$ 1780.00

3. Add as shown.

(a)	(b)	(c)
0.0067	1379.4	14.07
0.032	204.5	67.81
0.0012	16.75	132.99
0.0179	300.04	225.04
0.045	2070.08	38.02
0.5	167.99	4
0.05	43.255	16.899
0.0831	38.81	7.007
0.004	19.95	4.6

(d) $0.002 + 17.1 + 4.806 + 9.9981 - 3.1 + 0.701 - 1.001 - 14 - 8.09 + 1.0101$

Check your answers to the odd-numbered problems in the Appendix, then continue in Section 3-2.

3-2 Multiplication and Division of Decimal Numbers

learning|**catalytics**™

1. What is half of 0.8?
2. Calculate: $8 + 3 \times 6 - 4$
3. Round 1.683 to the nearest hundredth.

A decimal number is really a fraction with a power of 10 as denominator. For example,

$$0.5 = \frac{5}{10} \qquad 0.85 = \frac{85}{100} \qquad \text{and} \qquad 0.206 = \frac{206}{1000}$$

Multiplication of decimals is easy to understand if we think of it in this way:

$$0.5 \times 0.3 = \frac{5}{10} \times \frac{3}{10} = \frac{15}{100} = 0.15$$

Learning Help   To estimate the product of 0.5 and 0.3, remember that if two numbers are both less than 1, their product must be less than 1. ●

Multiplication   Of course it would be very, very clumsy and time-consuming to calculate every decimal multiplication this way. We need a simpler method. Here is the procedure most often used:

Step 1   Multiply the two decimal numbers as if they were whole numbers. Pay no attention to the decimal points.

Step 2   The sum of the decimal digits in the two numbers being multiplied will give you the number of decimal digits in the answer.

EXAMPLE 1   Multiply 3.2 by 0.41.

Step 1   Multiply, ignoring the decimal points.

$$\begin{array}{r} 32 \\ \times\ 41 \\ \hline 1312 \end{array}$$

Step 2   Count decimal digits in each number: 3.2 has *one* decimal digit (the 2), and 0.41 has *two* decimal digits (the 4 and the 1). The total number of decimal digits in the two factors is three. The answer will have *three* decimal digits. Count over *three* digits from right to left in the answer and insert the decimal point.

1.312   three decimal digits

✓ 3.2 × 0.41 is roughly 3 × ½ or about 1½. The answer 1.312 agrees with our rough guess. Remember, even if you use a calculator to do the actual work of arithmetic, you should *always* estimate your answer first and check it afterward.

For some products, there may be more decimal digits needed than there are digits in the result of Step 1. In these situations, we must attach zeros as placeholders to the left of this result. This technique is illustrated in Example 2.

EXAMPLE 2   Multiply: 0.3 × 0.2

Step 1   Multiply 3 × 2 = 6

Step 2   There is one decimal digit in each factor, so we need a total of *two* decimal digits in the product.

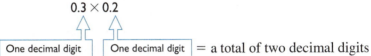

If we count over two places from the right of the 6, we must attach a zero as a place holder.

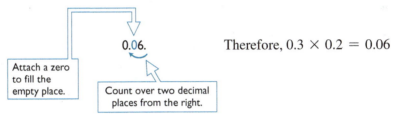

Therefore, 0.3 × 0.2 = 0.06

EXAMPLE 3   Multiply: 2.5 × 0.5

First, multiply 25 × 5 = 125.

Second, count decimal digits.

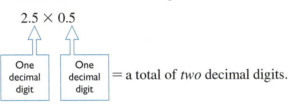

Count over *two* decimal digits from the right: 1.25.

The product is 1.25.

✓ 2 × ½ is about 1, so the answer seems reasonable.

Your Turn   Try these simple decimal multiplications.

(a) 0.1 × 0.1 = _____   (b) 10 × 0.6 = _____

(c) 2 × 0.003 = _____   (d) 0.04 × 0.005 = _____

Solutions   (a) 0.1 × 0.1   1 × 1 = 1

Count over *two* decimal digits from the right. Because there are not two decimal digits in the product, attach a zero on the left: 0.01

So 0.1 × 0.1 = 0.01.

✓ $\dfrac{1}{10} \times \dfrac{1}{10} = \dfrac{1}{100}$

(b) 10×0.6 $10 \times 6 = 60$

Count over *one* decimal digit from the right: 6.0

So $10 \times 0.6 = 6.0$.

↩ Learning Help   Notice that multiplication by 10 simply shifts the decimal place one digit to the right.

$10 \times 6.2 \quad = 62$
$10 \times 0.075 = 0.75$
$10 \times 8.123 = 81.23$ and so on ●

(c) 2×0.003 $2 \times 3 = 6$

Three decimal digits

Count over three decimal digits. Attach two zeros as placeholders.

0.006 $2 \times 0.003 = 0.006$

(d) 0.04×0.005 $4 \times 5 = 20$

Two decimal digits   Three decimal digits

Count over five decimal digits. Attach three zeros.

0.00020 $0.04 \times 0.005 = 0.00020 = 0.0002$

Total of five decimal digits   Five decimal digits

Multiplication of larger decimal numbers is performed in exactly the same manner.

EXAMPLE 4   Multiply 4.302×12.05

Estimate: $4 \times 12 = 48$ The answer will be about 48.

Multiply: 4302
              $\times 1205$
              5183910

(If you cannot do this multiplication correctly, turn to Section 1-3 for help with the multiplication of whole numbers.)

The two factors being multiplied have a total of five decimal digits (three in 4.302 and two in 12.05). Count over five decimal digits from the right in the answer.

51.83910

So $4.302 \times 12.05 = 51.83910 = 51.8391$

☑   The answer 51.8391 is approximately equal to the estimate of 48.   ●

Your Turn   Try these:

(a) $6.715 \times 2.002 =$ _____   (b) $3.144 \times 0.00125 =$ _____

Solutions   (a) **Estimate:** 6.7×2 is about 7×2 or 14.

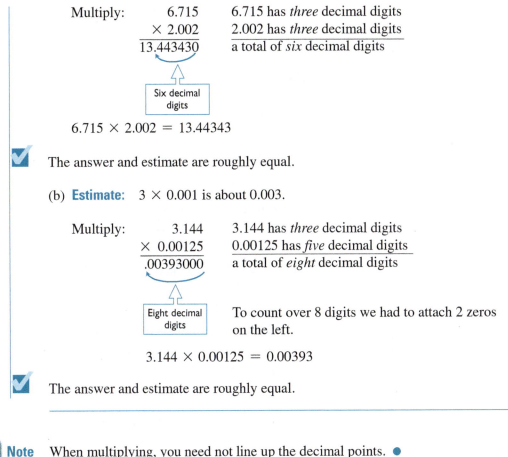

Multiply: 6.715   6.715 has *three* decimal digits
         × 2.002   2.002 has *three* decimal digits
        13.443430   a total of *six* decimal digits

Six decimal digits

6.715 × 2.002 = 13.44343

✓ The answer and estimate are roughly equal.

(b) **Estimate:** 3 × 0.001 is about 0.003.

Multiply: 3.144   3.144 has *three* decimal digits
       × 0.00125   0.00125 has *five* decimal digits
        .00393000   a total of *eight* decimal digits

Eight decimal digits

To count over 8 digits we had to attach 2 zeros on the left.

3.144 × 0.00125 = 0.00393

✓ The answer and estimate are roughly equal.

✎ **Note** When multiplying, you need not line up the decimal points. ●

One nice thing about a calculator is that it counts decimal digits and automatically adjusts the answer in any calculation. But to get an estimate of the answer, you must still understand the process.

Applications of Decimal Multiplication When you have a given number of items per unit and the number of units, multiply these two quantities to find the total number of items.

EXAMPLE 5 **Office Services** An automotive tech earns a base pay of $21.46 per hour plus time-and-a-half for overtime (weekly time exceeding 40 hours). If the tech worked 46.5 hours one week, what was her gross pay?

Here we must find two pay totals, one for regular time and one for overtime, and then add them together.

For regular time, we know the automotive tech's pay rate per hour ($21.46) and the number of non-overtime hours (40), so we must multiply to find her total pay.

Estimate: $20 × 40 = $800, so her non-overtime pay should be a little more than $800.

Calculation:

For her first 40 hours:

$21.46 ⟵— *two* decimal digits
  × 40 ⟵— *no* decimal digits
$858.40 ⟵— *two* decimal digits

Now we must calculate her overtime hourly wage. The phrase *time-and-a-half* means that her overtime hourly wage is 1.5 times her regular hourly wage. We estimate this to be about 1.5 × $20 or about $30 per hour.

$$\begin{array}{r} \$21.46 \\ \times\ 1.5 \\ \hline \$32.190 \end{array}$$

←——— *two decimal digits*
←——— *one decimal digit*
←——— *three decimal digits*

Next, multiply this overtime rate by the number of overtime hours:

46.5 total hours − 40 regular hours = 6.5 overtime hours

We estimate that $32.19 × 6.5 is approximately $30 × 7 or $210.

$$\begin{array}{r} \$32.19 \\ \times\ 6.5 \\ \hline \$209.235 \end{array}$$

←——— *two decimal digits*
←——— *one decimal digit*
←——— *three decimal digits*

For her gross pay, add the two amounts together:

$$\begin{array}{r} \$858.40 \\ +\ \ 209.235 \\ \hline \$1067.635 \end{array}$$ or approximately $1067.64   ●

You will solve a problem like Example 5 in Exercises 3-2, Problem F-21. Here is a simpler problem for you to try now.

Your Turn   Construction   A certain thickness of sheet glass weighs 1.53 lb/sq ft (pounds per square foot). How many pounds would 4 sheets weigh if they are each 7.5 sq ft in size?

Solution   This problem calls for us to multiply twice. When you multiply three or more numbers, the order in which you multiply does not matter. We will first perform the simpler multiplication to calculate the total number of square feet of glass. If 4 sheets each contain 7.5 sq ft, then 4 × 7.5 will give us the total number of square feet:

$$\begin{array}{r} 7.5 \\ \times\ \ 4 \\ \hline 30.0 \end{array}$$

7.5 ⟵ One decimal digit
30.0 ⟵ One decimal digit

There are 30 total square feet of glass. If each square foot weighs 1.53 lb, then we must again multiply to find the total weight:

$$\begin{array}{r} 1.53 \\ \times\ \ 30 \\ \hline 45.90 \end{array}$$

1.53 ⟵ Two decimal digits
45.90 ⟵ Two decimal digits

The weight of all 4 sheets is 45.9 lb.

Division   Division of decimal numbers is very similar to the division of whole numbers. For example,

$$6.8 \div 1.7 \quad \text{can be written} \quad \frac{6.8}{1.7}$$

and if we multiply both top and bottom of the fraction by 10 we have

$$\frac{6.8}{1.7} = \frac{6.8 \times 10}{1.7 \times 10} = \frac{68}{17}$$

But you should know how to divide these whole numbers.

$68 \div 17 = 4$

Therefore, $6.8 \div 1.7 = 4$.

To divide decimal numbers, use the following procedure.

Step 1	Write the divisor and dividend in standard long-division form.	$6.8 \div 1.7$ $1.7\overline{)6.8}$
Step 2	Shift the decimal point in the divisor to the right to make the divisor a whole number.	$1.7.\overline{)}$
Step 3	Shift the decimal point in the dividend the same number of places. (Attach zeros if necessary.)	$1.7.\overline{)6.8.}$
Step 4	Place the decimal point in the answer space directly above the new decimal position in the dividend.	$17.\overline{)68\uparrow}$
Step 5	Now divide exactly as you would with whole numbers. The decimal points in divisor and dividend may now be ignored. $6.8 \div 1.7 = 4$	$\begin{array}{r} 4. \\ 17.\overline{)68.} \\ -68 \\ \hline 0 \end{array}$

Notice in Steps 2 and 3 that we have simply multiplied both divisor and dividend by 10.

If there is a remainder in Step 5, we must add additional zeros to complete the division. This will be necessary in the next example.

EXAMPLE 6 Divide as shown: $1.38 \div 2.4$
Let's do it step-by-step.

Write the problem in standard long-division form.	$2.4\overline{)1.38}$
Shift both decimal points one digit to the right to make the divisor (2.4) a whole number (24).	$2.4.\overline{)1.3.8}$
Place the decimal point in the answer space.	$24.\overline{)13\overset{\cdot}{}8}$
Divide as usual. 24 goes into 138 five times.	$\begin{array}{r} .5 \\ 24.\overline{)13.8} \\ 12\,0 \\ \hline 1\,8 \end{array}$
There is a remainder (18), so we must keep going.	
Attach a zero to the dividend and bring it down.	$\begin{array}{r} .57 \\ 24.\overline{)13.80} \\ 12\,0 \\ \hline 1\,80 \\ 1\,68 \\ \hline \end{array}$
Divide 180 by 24. 24 goes into 180 seven times.	
Now the remainder is 12.	12
Attach another zero and bring it down	$\begin{array}{r} .575 \\ 24.\overline{)13.800} \\ 12\,0 \\ \hline 1\,80 \\ 1\,68 \\ \hline 120 \\ 120 \\ \hline 0 \end{array}$

24 goes into 120 exactly five times.

The remainder is zero. $1.38 \div 2.4 = 0.575$

  $1.38 \div 2.4$ is roughly $1 \div 2$ or 0.5.

Double-Check: $2.4 \times 0.575 = 1.38$. Always multiply the answer by the divisor to double-check your work. Do this even if you are using a calculator.   ●

 Note   In some problems, we may continue to attach zeros and divide yet never get a remainder of zero. We will examine problems like this later.   ●

EXAMPLE 7   How would you do this one?

$2.6 \div 0.052 = ?$

To make the divisor a whole number, shift the decimal point three digits to the right.

$$0.052.\overline{)2.6}$$

To shift the decimal place three digits in the dividend, we must attach two zeros to its right.

$$0.052.\overline{)2.600.}$$

Now place the decimal point in the answer space above that in the dividend and divide normally.

$$52.\overline{)2600.} \quad \begin{array}{r} 50. \\ \hline \end{array}$$
$$\underline{260} \quad \longleftarrow \quad 5 \times 52 = 260$$
$$0$$
$$0$$

$2.6 \div 0.052 = 50$ ☑ $0.052 \times 50 = 2.6$

Shifting the decimal point three digits and attaching zeros to the right of the decimal point in this way is equivalent to multiplying both divisor and dividend by 1000.   ●

 Note   In Example 7, inexperienced students think the problem is finished after the first division because they get a remainder of zero. You must keep dividing at least until all the places up to the decimal point in the answer are filled. The answer here is 50, not 5.   ●

Your Turn   Try these problems.

(a) $9 \div 0.02$ = _____   (b) $0.365 \div 18.25$ = _____

(c) $8.8 \div 3.2$ = _____   (d) $7.23 \div 6$ = _____

(e) $30.24 \div 0.42$ = _____   (f) $273.6 \div 0.057$ = _____

Solutions   (a) Shift the decimal point two places to the right. Divide 900 by 2.

$$0.02.\overline{)9.00.}$$

$$2.\overline{)900.} \quad \begin{array}{r} 450. \\ \hline \end{array} \qquad 9 \div 0.02 = 450$$

☑ $0.02 \times 450 = 9$

Dividing a whole number by a decimal is very troublesome for most people.

(b) $18.25.\overline{)0.36.5}$

$1825.\overline{)36.50} \quad \begin{array}{r} .02 \\ \hline \end{array}$
$\underline{36\ 50}$
0

1825 does not go into 365, so place a zero above the 5. Attach a zero after the 5.
1825 goes into 3650 twice. Place a 2 in the answer space above the zero.

$0.365 \div 18.25 = 0.02$

☑ $18.25 \times 0.02 = 0.365$

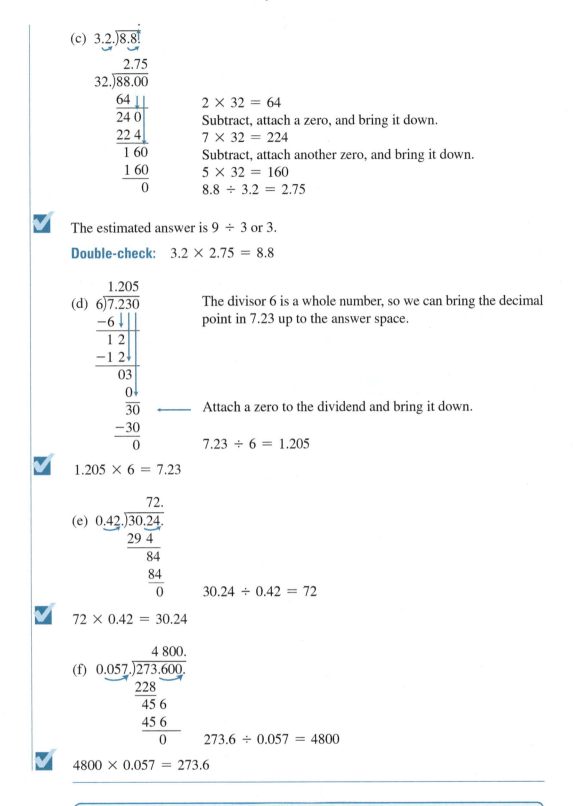

(c) 3.2.)8.8.

$$
\begin{array}{r}
2.75 \\
32.\overline{)88.00} \\
\end{array}
$$

64 $2 \times 32 = 64$
$\overline{24\ 0}$ Subtract, attach a zero, and bring it down.
$\underline{22\ 4}$ $7 \times 32 = 224$
$1\ 60$ Subtract, attach another zero, and bring it down.
$\underline{1\ 60}$ $5 \times 32 = 160$
0 $8.8 \div 3.2 = 2.75$

✓ The estimated answer is $9 \div 3$ or 3.

Double-check: $3.2 \times 2.75 = 8.8$

(d) $6\overline{)7.230}$ with quotient 1.205

The divisor 6 is a whole number, so we can bring the decimal point in 7.23 up to the answer space.

-6
$\overline{1\ 2}$
$-1\ 2$
$\overline{03}$
0
$\overline{30}$ ⟵ Attach a zero to the dividend and bring it down.
-30
$\overline{0}$ $7.23 \div 6 = 1.205$

✓ $1.205 \times 6 = 7.23$

(e) 0.42.)30.24. with quotient $72.$

$\begin{array}{r} 72. \\ 0.42.\overline{)30.24.} \\ 29\ 4 \\ \overline{84} \\ \underline{84} \\ 0 \end{array}$ $30.24 \div 0.42 = 72$

✓ $72 \times 0.42 = 30.24$

(f) 0.057.)273.600. with quotient $4\ 800.$

$\begin{array}{r} 4\ 800. \\ 0.057.\overline{)273.600.} \\ 228 \\ \overline{45\ 6} \\ \underline{45\ 6} \\ 0 \end{array}$ $273.6 \div 0.057 = 4800$

✓ $4800 \times 0.057 = 273.6$

Arithmetic "Tricks of the Trade"

1. To divide a number by 5, use the fact that 5 is one-half of 10. First, multiply the number by 2, then divide by 10 by shifting the decimal point one place to the left.

   Example: $64 \div 5$ $64 \times 2 \div 10 = 128 \div 10 = 12.8$

(continued)

2. To divide a number by 25, first multiply by 4, then divide by 100, shifting the decimal point two places left.

Example: $304 \div 25 = 304 \times 4 \div 100 = 1216 \div 100 = 12.16$

3. To divide a number by 20, first divide by 2, then divide by 10.

Example: $86 \div 20 = 86 \div 2 \div 10 = 43 \div 10 = 4.3$

4. To multiply a number by 20, first multiply by 2, then multiply by 10, shifting the decimal point one place to the right.

Example: $73 \times 20 = 73 \times 2 \times 10 = 146 \times 10 = 1460$

When doing "mental arithmetic" like this, it is important that you start with a rough estimate of the answer.

Applications Decimal Division   You may recall from previous chapters that division is used to answer the question, "How many of X are in Y?" The number X would be the divisor, and the number Y would be the dividend.

EXAMPLE 8   **Construction**   How many pieces of plywood 0.375 in. thick are in a stack 30 in. high?

This wording calls for division, where 0.375 is the divisor and 30 is the dividend.

$$\begin{array}{r} 80. \\ 0.375.\overline{\smash{)}30.000.} \\ \underline{30\ 00} \\ 0 \end{array}$$

✔ $80 \times 0.375 = 30$

There are 80 pieces of plywood in the stack.   ●

Division is also used when you need to calculate the number of "A *per* B." The number B is the divisor and the number A is the dividend.

Your Turn   **Retail Merchandising**   In retail sales, one measure of employee productivity is the average *sales per transaction*. If a salesperson had 42 transactions last week with gross sales of $3687.60, what were her average sales per transaction?

Solution   To calculate sales *per* transaction, we must *divide* gross sales by the number of transactions, or $3687.60 \div 42$.

$$\begin{array}{r} 87.80 \\ 42\overline{\smash{)}3687.60} \\ \underline{336} \\ 327 \\ \underline{294} \\ 33\ 6 \\ \underline{33\ 6} \\ 0 \\ \underline{0} \\ 0 \end{array}$$

The salesperson sold an average of $87.80 worth of merchandise per transaction.

Rounding Decimal Numbers   Calculations involving decimal numbers often result in answers having more decimal digits than are justified. When this occurs, we must round our answer. The process of rounding a decimal number is very similar to the procedure for rounding a whole number. The only difference is that, after rounding to a given decimal place, all digits to the right of that place are dropped. The following examples illustrate this difference.

EXAMPLE 9   Round 35,782.462 to the ⟶   **nearest thousand**   **nearest hundredth**

Step 1   Place a ∧ mark to the right of the place to which the number is to be rounded.   35∧782.462   35782.46∧2

Step 2   If the digit to the right of the mark is less than 5, replace all digits to the right of the mark with zeros.

35782.460

Drop this decimal digit zero.

If these zeros are decimal digits, discard them.

The rounded number is 35782.46

Step 3   If the digit to the right of the mark is equal to or larger than 5, increase the digit to the left by 1 and replace all digits to the right with zeros.

36 000.000

Drop these right-end decimal zeros

Drop all right-end decimal zeros, but keep zero placeholders.

36000

Keep these zeros as placeholders.

The rounded number is 36,000.

! **Careful**   Drop only the decimal zeros that are to the right of the ∧ mark. For example, to round 6.4086 to three decimal digits, we write 6.408∧6, which becomes 6.4090 or 6.409. We dropped the end zero because it was a decimal zero to the *right* of the mark. But we retained the other zero because it is needed as a placeholder.

Your Turn   Try rounding these numbers.

(a) Round   74.238   to two decimal places.

(b) Round   8.043   to two decimal places.

(c) Round   0.07354   to three decimal places.

(d) Round   7.98   to the nearest tenth.

Answers   (a) 74.238 is 74.24 to two decimal places.
(Write 74.23∧8 and note that 8 is larger than 5, so increase the 3 to 4 and drop the last digit because it is a decimal digit.)

(b) 8.043 is 8.04 to two decimal places.
(Write 8.04∧3 and note that 3 is less than 5, so drop it.)

(c) 0.07354 is 0.074 to three decimal places.
(Write 0.073 ∧ 54 and note that the digit to the right of the mark is 5; therefore, change the 3 to a 4 to get 0.074 ∧ 00. Finally, drop the digits on the right to get 0.074.)

(d) 7.98 is 8.0 to the nearest tenth.
(Write 7.9 ∧ 8 and note that 8 is greater than 5, so increase the 9 to 0 and the 7 to 8. Drop the digit in the hundredths place.)

There are a few very specialized situations where this rounding rule is not used:

1. Some engineers use a more complex rule when rounding a number that ends in 5.

2. In business, fractions of a cent are usually rounded up to determine selling price. Three items for 25 cents or $8\frac{1}{3}$ cents each is rounded to 9 cents each.

Our rule will be quite satisfactory for most of your work in arithmetic.

Rounding During Division   In some division problems, this process of dividing will never result in a zero remainder. At some point the answer must be rounded.

To round answers in a division problem, first continue the division so that your quotient has one place more than required, then round it.

EXAMPLE 10   In the division problem $4.7 \div 1.8 = ?$

to get an answer rounded to one decimal place, first divide to two decimal places:

$$1.8\overline{)4.7}$$

$$
\begin{array}{r}
2.61 \\
18\overline{)47.00} \quad \longleftarrow \text{Attach two zeros to the dividend} \\
\underline{36} \quad\quad \longleftarrow 2 \times 18 = 36 \\
11\,0 \\
\underline{10\,8} \quad \longleftarrow 6 \times 18 = 108 \\
20 \\
\underline{18} \quad \longleftarrow 1 \times 18 = 18
\end{array}
$$

Therefore, $4.7 \div 1.8 = 2.61 \ldots$. We can now round back to one decimal place to get 2.6.   ●

Note   When an answer has been rounded, we replace the $=$ sign with the $\approx$ (approximately equal) symbol. In Example 10, $4.7 \div 1.8 \approx 2.6$.   ●

Your Turn   For the following problem, divide and round your answer to two decimal places.

$6.84 \div 32.7 =$ _____

Careful now.

Solution $32.7\overline{)6.8.4}$

$$
\begin{array}{r}
.209 \quad \longleftarrow \text{Carry the answer to three decimal places.} \\
327\overline{)68.400} \quad \longleftarrow \text{Notice that two zeros must be attached to the dividend.} \\
\underline{65\,4} \quad\quad \longleftarrow 2 \times 327 = 654 \\
3\,00 \\
\underline{0} \quad\quad \longleftarrow 0 \times 327 = 0 \\
3\,000 \\
\underline{2\,943} \quad \longleftarrow 9 \times 327 = 2943
\end{array}
$$

0.209 rounded to two decimal places is 0.21.

$6.84 \div 32.7 \approx 0.21$

✓ $32.7 \times 0.21 = 6.867$, which is approximately equal to 6.84. (The check will not be exact because we have rounded.)

Applications of Rounding   The ability to round numbers is especially important for people who work in the practical, trade, or technical areas. You will need to round answers to practical problems if they are obtained "by hand" or with a calculator. Rounding is discussed in more detail in Chapter 5.

Here is a practical application involving rounding.

EXAMPLE 11   **HVAC**   A homeowner in the Midwest recently spent $9400 to convert his heating system from heating oil to natural gas. His heating costs fell from $2570 last winter using oil to $1120 this winter using gas. The homeowner wishes to know the payback time for his conversion cost rounded to the nearest tenth of a year.

First, we calculate his annual savings by subtracting his current winter's gas heating cost from his previous winter's oil heating cost:

$$\$2570 - \$1120 = \$1450$$

Then, to calculate the payback time for the cost of conversion ($9400), we divide this cost by the annual savings of $1450. We need to carry out the division to the hundredths place in order to round to the nearest tenth:

$$
\begin{array}{r}
6.48 \\
1450\overline{)9400.00} \\
8700 \\
\overline{700\,0} \\
580\,0 \\
\overline{120\,00} \\
116\,00 \\
\end{array}
$$

Finally, we round 6.48 to 6.5. The homeowner's payback time, to the nearest tenth of a year, is 6.5 years.   ●

Order of Operations   When a decimal calculation involves more than one operation, be sure to follow the order of operations as outlined in Section 1-5.

EXAMPLE 12   (a) To calculate $12.46 - 2.5 \times 1.7$

   First, multiply. $= 12.46 - 4.25$

   Then, subtract. $= 8.21$

   (b) To calculate $6.8 \div (3.5 - 2.7)$

   First, work inside parentheses. $= 6.8 \div 0.8$

   Then, divide. $= 8.5$ ●

Your Turn   Calculate:

(a) $16.5 + 8.4 \div 2.1$ (b) $\dfrac{(15.6 - 2.4) \times 2.5}{12}$

Solutions (a) **First,** divide. $16.5 + 8.4 \div 2.1 = 16.5 + 4$

Then, add. $= 20.5$

(b) **First,** work inside parentheses.

$$\frac{(15.6 - 2.4) \times 2.5}{12} = \frac{13.2 \times 2.5}{12}$$

Then, simplify the top. $= \dfrac{33}{12}$

Finally, divide. $= 2.75$

Averages   Suppose that you needed to know the diameter of a steel connecting pin. As a careful and conscientious worker, you would probably measure its diameter several times with a micrometer, and you might come up with a sequence of numbers like this (in inches):

$$1.3731, \ 1.3728, \ 1.3736, \ 1.3749, \ 1.3724, \ 1.3750$$

What is the actual diameter of the pin? The best answer is to find the **average** value or **arithmetic mean** of these measurements.

$$\text{Average} = \frac{\text{sum of the measurements}}{\text{number of measurements}}$$

For the preceding problem,

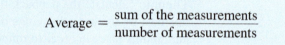

$$\text{Average} = \frac{1.3731 + 1.3728 + 1.3736 + 1.3749 + 1.3724 + 1.3750}{6}$$

$$= \frac{8.2418}{6} = 1.3736333\ldots$$

$$\approx 1.3736 \quad \text{rounded to four decimal places}$$

When you calculate the average of a set of numbers, the usual rule is to round the answer to the same number of decimal places as the least precise number in the set—that is, the one with the fewest decimal digits. If the numbers to be averaged are all whole numbers, the average will be a whole number.

EXAMPLE 13   The average of 4, 6, 4, and 5 is $\dfrac{19}{4} = 4.75$ or 5 when rounded.   ●

If one number of the set has fewer decimal places than the rest, round off to agree with the least precise number.

EXAMPLE 14   The average of 2.41, 3.32, 5.23, 3.51, 4.1, and 4.12 is

$\dfrac{22.69}{6} = 3.78166\ldots$ and this answer should be rounded to 3.8 to agree in precision with the least precise number, 4.1. (We will learn more about precision in Chapter 5.)

To avoid confusion, we will usually give directions for rounding.   ●

Your Turn   Try it. Find the average for each of these sets of numbers.

(a) 8, 9, 11, 7, 5   (b) 0.4, 0.5, 0.63, 0.2

(c) **Water/Wastewater Treatment**   California state-drinking-water-quality standards require that the running average of total trihalomethanes for any four consecutive quarters cannot exceed 80 micrograms per liter (μg/L). If the last four quarterly readings for this contaminant have been 101.2, 78.0, 70.2, and 74.7 μg/L, was the running average above or below the maximum level allowed?

Solutions   (a) Average $= \dfrac{8 + 9 + 11 + 7 + 5}{5} = \dfrac{40}{5} = 8$

(b) Average $= \dfrac{0.4 + 0.5 + 0.63 + 0.2}{4} = \dfrac{1.73}{4} = 0.4325 \approx 0.4$ rounded

(c) Average $= \dfrac{101.2 + 78.0 + 70.2 + 74.7}{4} = \dfrac{324.1}{4} = 81.025 \approx 81.0$ μg/L

This is greater than 80 μg/L and therefore violates the standards.

When averaging large sets of numbers in which some numbers occur multiple times, finding the sum of the numbers may involve both multiplication and addition. This is demonstrated in the next example.

EXAMPLE 15   **Manufacturing**   A manufacturing company recently made three purchases of an electronic component from a supplier. The company first purchased 60 of the components at a cost of $3.68 each. Its second purchase was for 48 of the components at $4.12 apiece. And its final purchase was for 24 of the components at $4.28 apiece. For pricing purposes, the manufacturer needed to know its average cost per component.

To determine the average cost, we cannot simply average the three prices because there were different quantities purchased at each price. Instead, we must calculate the total cost of all the components and divide this by the total number of components purchased.

Step 1   To calculate the total cost, multiply each price by the quantity purchased at that price and then find the sum of these products.

$$\text{Total cost} = 60 \times \$3.68 + 48 \times \$4.12 + 24 \times \$4.28$$

$$= \$220.80 + \$197.76 + \$102.72$$

$$= \$521.28$$

Step 2   Calculate the total number of components purchased.

$$\text{Total number of components} = 60 + 48 + 24 = 132$$

Step 3   Divide the total cost by the total number of components. This is the average cost per component.

$$\text{Average cost} = \frac{\$521.28}{132} = \$3.949 \ldots \approx \$3.95$$

Your Turn **Allied Health** A pharmacy purchased different quantities of a generic drug from four different suppliers. The quantities and costs are shown in the table.

Supplier	Number of Capsules	Cost per Capsule
A	200	$0.75
B	400	$0.60
C	150	$0.80
D	100	$1.10

Find the average cost per capsule paid by the pharmacy.

Solution **Step 1** Calculate the total cost of all four purchases.

$$\text{Total cost} = 200 \times \$0.75 + 400 \times \$0.60 + 150 \times \$0.80 + 100 \times \$1.10$$
$$= \$150 + \$240 + \$120 + \$110$$
$$= \$620$$

Step 2 Calculate the total number of capsules purchased.

$$\text{Total number of capsules} = 200 + 400 + 150 + 100 = 850$$

Step 3 Divide the total cost by the total number of capsules. This will be the average cost per capsule.

$$\text{Average cost} = \frac{\$620}{850} = \$0.7294\ldots \approx \$0.73$$

The pharmacy paid an average of 73 cents per capsule.

Now turn to Exercises 3-2 for a set of problems on the multiplication and division of decimal numbers.

Exercises 3-2 | ## Multiplication and Division of Decimals

A. Multiply or divide as shown.

1. 0.01×0.001
2. 10×2.15
3. 0.04×100

4. 0.3×0.3
5. 0.7×1.2
6. 0.005×0.012

7. 0.003×0.01
8. 7.25×0.301
9. 24.6×8.5

10. $0.2 \times 0.3 \times 0.5$
11. $0.6 \times 0.6 \times 6.0$
12. $2.3 \times 1.5 \times 1.05$

13. $3.618 \div 0.6$
14. $3.60 \div 0.03$
15. $4.40 \div 0.22$

16. $6.5 \div 0.05$
17. $0.0405 \div 0.9$
18. $0.378 \div 0.003$

19. $4 \div 0.01$
20. $2.59 \div 70$
21. $44.22 \div 6.7$

22. $104.2 \div 0.0320$
23. $484 \div 0.8$
24. 6.05×2.3

25. 0.0027×1.4
26. $0.0783 \div 0.27$
27. $0.00456 \div 0.095$

28. $800 \div 0.25$
29. $324 \div 0.0072$
30. $0.08322 \div 228$

31. $0.0092 \div 115$
32. 0.047×0.024
33. 2.375×12.1

34. $0.02 \times 0.06 \times 0.04$
35. $0.008 \times 0.4 \times 0.03$
36. 123.4×0.45

37. 0.062×27.5

B. Calculate using the proper order of operations.

1. $26.24 + 3.8 \times 2.6$

2. $15.8 - 4.4 \div 0.8$

3. $8.28 \div (6.3 - 4.8)$

4. $3.9 \times (1.25 + 2.72)$

5. $6.8 + 2.4 \times 1.5 - 3.6$

6. $\dfrac{(32.2 - 14.6) \times 3.5}{20}$

C. Round the following numbers as indicated.

1. 42.875 (two decimal places)

2. 0.5728 (nearest tenth)

3. 6.54 (one decimal place)

4. 117.6252 (three decimal places)

5. 79.135 (nearest hundredth)

6. 1462.87 (nearest whole number)

7. 3.64937 (four decimal places)

8. 19.4839 (nearest hundredth)

9. 0.2164 (nearest thousandth)

10. 56.826 (two decimal places)

D. Divide and round as indicated.

Round to two decimal digits.

1. $10 \div 3$

2. $5 \div 6$

3. $2.0 \div 0.19$

4. $3 \div 0.081$

5. $0.023 \div 0.19$

6. $12.3 \div 4.7$

7. $2.37 \div 0.07$

8. $6.5 \div 1.31$

Round to the nearest tenth.

9. $100 \div 3$

10. $21.23 \div 98.7$

11. $1 \div 4$

12. $100 \div 9$

13. $0.006 \div 0.04$

14. $1.008 \div 3$

Round to three decimal places.

15. $10 \div 70$

16. $0.09 \div 0.402$

17. $0.091 \div 0.0014$

18. $3.41 \div 0.257$

19. $6.001 \div 2.001$

20. $123.21 \div 0.1111$

E. Find the average for each of the following sets of numbers.

1. 76, 88, 64, 92, 85, 77, 82, 97

2. 6.58, 6.66, 6.64, 6.62, 6.59

3. 42.7, 55.8, 38.5, 46.9

4. 0.714, 0.716, 0.723, 0.720, 0.711, 0.719

5. 4.6, 5.6, 4.2, 5.1, 6.6, 5.3, 4.9, 5.7, 4.2, 4.3, 5.1, 5.4, 4.8, 4.5

6. 8.215, 8.213, 8.230, 8.223, 8.218, 8.222, 8.214, 8.216, 8.218

F. Practical Applications

1. **Painting** A gas-powered airless paint sprayer is advertised for $899. It can also be bought on an "installment plan" by making 24 payments of $39.75 each. How much extra do you pay by purchasing it on the installment plan?

2. **Allied Health** How much total medication will a patient receive if he is given four 0.075-mg tablets each day for 12 days?

3. **Metalworking** Find the average weight of five castings that weigh 17.0, 21.0, 12.0, 20.6, and 23.4 lb.

4. **Office Services** If you work 8.25 hours on Monday, 10.1 hours on Tuesday, 8.5 hours on Wednesday, 9.4 hours on Thursday, 6.5 hours on Friday, and 4.2 hours on Saturday, what is the average number of hours worked per day?

5. **Machine Trades**   For the following four machine parts, find W, the number of pounds per part; C, the cost of the metal per part; and T, the total cost.

Metal Parts	Number of Inches Needed	Number of Pounds per Inch	Cost per Pound	Pounds (W)	Cost per Part (C)
A	44.5	0.38	$0.98		
B	122.0	0.19	$0.89		
C	108.0	0.08	$1.05		
D	9.5	0.32	$2.15		
					$T =$

6. **Machine Trades**   How much does 15.7 sq ft of No. 16 gauge steel weigh if 1 sq ft weighs 2.55 lb?

7. **Construction**   A 4-ft by 8-ft sheet of $\frac{1}{4}$-in. plywood has an area of 32 sq ft. If the weight of $\frac{1}{4}$-in. plywood is 1.5 lb/sq ft (pounds per square foot), what is the weight of the sheet?

8. **Manufacturing**   Each inch of 1-in.-diameter cold-rolled steel weighs 0.22 lb. How much would a piece weigh that was 38 in. long?

9. **Transportation**   A truck can carry a load of 5000 lb. Assuming that it could be cut to fit, how many feet of steel beam weighing 32.6 lb/ft (pounds per foot) can the truck carry?

10. **Fire Protection**   One gallon of water weighs 8.34 lb. How much weight is added to a fire truck when its tank is filled with 750 gal of water?

11. **Electrical Trades**   Voltage values are often stated in millivolts [1 millivolt (mV) equals 0.001 volt] and must be converted to volts to be used in Ohm's law. If a voltage is given to be 75 mV, how many volts is this?

12. **Industrial Technology**   An industrial engineer must estimate the cost of building a storage tank. The tank requires 208 sq ft of material at $9.29 per square foot. It also requires 4.5 hours of labor at $26.40 per hour plus 1.6 hours of labor at $19.60 per hour. Find the total cost of the tank.

13. **Sheet Metal Trades**   How many sheets of metal are in a stack 5.00 in. high if each sheet is 0.0149 in. thick?

14. **Metalworking**   A casting weighs 3.68 lb. How many castings are contained in a load weighing 5888.00 lb?

15. **Agriculture**   A barrel partially filled with liquid fertilizer weighs 267.75 lb. The empty barrel weighs 18.00 lb, and the fertilizer weighs 9.25 lb/gal (pounds per gallon). How many gallons of fertilizer are in the barrel?

16. **Construction**   A 4-ft by 8-ft piece of $\frac{1}{2}$-in. gypsum board costs $9.75. Using this material, how much would it cost to purchase gypsum for a room with 52 ft of total wall width and an 8-ft ceiling?

17. **Flooring and Carpeting**   A room addition requires 320 sq ft of wood floor. Hardwood costs $4.59 per sq ft, while laminate costs $2.79 per sq ft. What is the total difference in cost between the two types of flooring?

18. **Office Services**   The Happy Hacker electronics warehouse purchases 460 8GB-USB flash drives, which are then packed in sets of five selling for $17.95 per set. Calculate the total revenue from the sale of these flash drives.

19. **Electronics** An oscillator is a device that generates an ac signal at some specified frequency. If the oscillator's output frequency is 7500 cycles per second (hertz), how many cycles does it generate in a 0.35-sec interval?

20. **Trades Management** A shop owner needs to move to a larger shop. He has narrowed his choice down to two possibilities: a 2400-sq-ft shop renting for $3360 per month and a 2800-sq-ft shop renting for $4060 per month. Which shop has the lower cost per square foot?

21. **Office Services** A shop tech earns a base pay of $19.88 per hour, plus time-and-a-half for overtime (time exceeding 40 hours). If he works 43.5 hours during a particular week, what is his gross pay?

22. **Automotive Trades** To calculate the torque at the rear wheels of an automobile, we multiply the torque of the engine by the drive ratio. If the engine torque is 305 foot-pounds and the drive ratio is 8.68 (to 1), find the torque at the rear wheels. (Round to the nearest ten foot-pounds.)

23. **Agriculture** Fertilizer is used at the Tofu Soy Farms at the rate of 3.5 lb per acre. How much fertilizer should be spread on a 760-acre field?

24. **Flooring and Carpeting** A carpet installer is paid $0.75 per square foot of carpet laid. How many square feet would he need to install in a week in order to earn $6000?

25. **HVAC** Regulations require a ventilation rate for houses of 0.35 air changes per hour—that is, 0.35 of the total volume of air must be changed. For a house with a volume of 24,000 cu ft, how many cubic feet of air must be changed per hour?

26. **Allied Health** A litter of seven puppies has been brought into a veterinary clinic to receive antibiotic for treatment of an infection. Each puppy should receive 0.3 milliliters (mL) of the antibiotic solution per kilogram (kg) of body weight. If the puppies weigh an average of 6.7 kg each, how much antibiotic solution should the veterinary technician prepare for the seven puppies? (Round to the nearest milliliter.)

27. **Flooring and Carpeting** Carpet installers often are paid by the square foot rather than by the hour. If a certain installer is paid $0.75 per square foot, how much would he earn on a day when he installed 864 sq ft of carpet?

28. **Sports and Leisure** To calculate the distance traveled by a bicycle with each pedal stroke, multiply the gear ratio by the circumference of the tires. If the gear ratio is 2.96 and the circumference of the tires is 81.5 in., how far does the bike roll during one pedal stroke? (Round to the nearest inch.)

29. **Graphic Design** Artists and architects often use the **golden rectangle** (also called the **golden ratio**) to determine the most visually pleasing proportions for their designs. The proportions of this rectangle specify that the longer side should be 1.618 times the shorter side. An artist wants to use this proportion for a rectangular design. If the shorter side of the design must be 3.5 in. long, what should the longer side measure? (Round to the nearest tenth.)

30. **Automotive Trades** A 2017 Hyundai Sonata Eco is EPA-rated at 28 mi/gal in the city and 36 mi/gal on the highway.

    (a) How many gallons of gas would be used on a trip consisting of 77 miles of city driving and 297 miles of highway driving?

    (b) At $2.50 per gallon, how much would the gas cost for this trip?

31. **Office Services**   Because of a drought emergency, a water district began charging its customers according to the following block rate system:

Block	Quantity Used	Rate per HCF*
1	The first 20 HCF	$3.90
2	The next 40 HCF (from 21 through 60 HCF)	$4.15
3	The next 60 HCF (from 61 through 120 HCF)	$4.90
4	Usage in excess of 120 HCF	$5.90

*HCF = Hundred cubic feet

What would a customer be charged for using:

(a) 38 HCF   (b) 96 HCF   (c) 132 HCF

(*Hint:* See Example 3 on page 55.)

32. **Construction**   A certain thickness of sheet glass weighs 1.53 lb per sq ft. How many pounds would 6 sheets weigh if they are each 2.5 sq ft in size?

33. **Construction**   The Passive House Standard is a strict building code used as a guideline for building energy-efficient homes. One Passive House guideline states that no more than 11.1 kwh (kilowatt-hours) per square foot should be used annually for the total energy use of the house. What would this annual maximum be for a 2460-square-foot house?

34. **Carpentry**   In a durability test, the Bosch RDN9V reciprocating-saw blade lasted for 14 cuts. If one of these blades costs $3.50, what does it cost per cut?

35. **Life Skills**   When you lease a car, your monthly payments include interest, just as if you were financing the purchase of the vehicle. However, the interest is determined by a number called a *money factor* rather than by an interest rate in percent form. To determine the approximate annual interest rate equivalent to the money factor, the general rule is to multiply the money factor by 2400. What is the annual interest rate equivalent to a money factor of 0.00114?

36. **Meteorology**   Marquette, Michigan, receives an average of 128.6 inches of snow per year. Assuming that it snows during six months of the year, what is the average monthly amount of snow that falls during these months? (Round to the nearest tenth.)

37. **Retail Merchandising**   As shown in **Your Turn** on page 150, one measure of salespeople's productivity is the average sales per transaction. If a salesperson had 36 transactions last week with gross sales of $2263.59, what were his average sales per transaction rounded to the nearest cent?

38. **HVAC**   A homeowner in the Northeast recently spent $8600 to convert his heating system from heating oil to natural gas. His heating costs fell from $2360 last winter using oil to $960 this winter using gas. At this rate, based solely on winter heating cost savings, what is his payback time in years? (Round up to the next year.)

39. **Life Skills**   An LCD television that is used for 6 hours per day consumes about 215 kilowatt-hours (kWh) of electricity per year. A high-definition cable box consumes an additional 170 kWh per year, and a high-definition digital video recorder adds another 275 kWh annually. If the average cost of a kilowatt-hour is $0.13 in this area, what is the total yearly electricity cost for these three items?

40. **Meteorology**   Three factors determine the amount of accumulated snow and ice that melts in a day. A process called sublimation, similar to evaporation, eliminates about 0.1 inch per day. On a clear day, sunshine accounts for up to 0.5 inches of melting per day if the underlying surface is dark pavement, but only 0.1 inch per day if the underlying surface is a lawn. Finally, the warmth of the air accounts for up to 0.15 inch per day for every degree above freezing (32 degrees Fahrenheit). What is the total maximum amount of snowmelt to be expected on dark pavement after four clear days in which the average air temperature is 40 degrees Fahrenheit?

41. **Retail Merchandising**   Supermarkets, discount stores, and drugstores use a measure called **sales per linear foot** in deciding how much shelf space to allot for different items. To calculate this measure, divide the gross sales of an item by the number of linear feet of shelf space that the item occupies. Consider the following figures for two brands of vitamins:

	Gross Sales	Linear Feet of Shelf Space
Brand A	$1830	24 ft
Brand B	$1180	18 ft

Calculate the sales per linear foot for each brand to the nearest cent. Which brand might be deserving of an increase in shelf space?

42. **Culinary Arts**   A restaurant manager made four purchases of salmon fillets last month. The number of pounds and the cost per pound for each purchase is shown in the table. Calculate the average cost per pound paid by the manager.

Purchase	Number of Pounds	Cost per Pound
1	20	$9.25
2	40	$11.50
3	28	$8.75
4	16	$8.50

43. **Agriculture**   When a citrus farmer sells tangerines to a packing house, the tangerines are graded according to size, and different prices are paid for the different sizes. The following table shows the number of pounds of tangerines sold in five different grades and the price per pound for each grade. Calculate the average price per pound paid for the tangerines.

Grade	Number of Pounds	Price per Pound
Colossal	4200	$1.00
Large	18,600	$1.50
Medium	21,200	$1.30
Small	3400	$0.60
Ponies	2600	$0.20

44. **Agriculture**   The table at the top of page 162 shows the number of gallons of water required to produce 1 oz of each type of food or drink. (Source: Mekkonen, M.M., and Hoekstra, A.Y., WaterStat, Water Footprint Network, Enschede, the Netherlands, UNESCO-IHE.)

Food or Drink	Gallons of Water to Produce 1 Ounce
Beef	12.75
Rice	1.95
Asparagus	2.44
Lettuce	0.10
Watermelon	0.21
Milk	0.65 (per fluid ounce)

How many gallons of water are required to produce a meal consisting of 8 oz of beef, 6 oz of rice, 7 oz of asparagus, 8 oz of lettuce, 10 oz of watermelon, and 8 fluid oz of milk?

45. **Electrical Trades**   For customers generating their own solar power, ZG&E charges them $3 per month per kilowatt (kW) of installed capacity and credits them $0.04 per kilowatt-hour (kWh) for excess electricity they export to the grid. Ready Edison charges customers a flat rate of $20 per month and credits them $0.06 per kWh for excess electricity they export to the grid. Determine the monthly bills for customers of both companies for each of the following situations:

   (a) Customer owns a 3-kW system and exports 120 kWh monthly to the grid.

   (b) Customer owns a 5-kW system and exports 300 kWh monthly to the grid.

G. Calculator Problems

   1. Divide. $9.87654321 \div 1.23456789$.

   Notice anything interesting? (Divide it to eight decimal digits.) You should be able to get the correct answer even if your calculator will not accept a nine-digit number.

   2. Divide. (a) $\dfrac{1}{81}$ (b) $\dfrac{1}{891}$ (c) $\dfrac{1}{8991}$

   Notice a pattern? (Divide them to about eight decimal places.)

   3. **Metalworking**   The outside diameter of a steel casting is measured six times at different positions with a digital caliper. The measurements are 4.2435, 4.2426, 4.2441, 4.2436, 4.2438, and 4.2432 in. Find the average diameter of the casting.

   4. **Printing**   To determine the thickness of a sheet of paper, five batches of 12 sheets each are measured with a digital micrometer. The thickness of each of the five batches of 12 is 0.7907, 0.7914, 0.7919, 0.7912, and 0.7917 mm. Find the average thickness of a single sheet of paper.

   5. **Construction**   The John Hancock Towers office building in Boston had a serious problem when it was built: When the wind blew hard, the pressure caused its windows to pop out! The only reasonable solution was to replace all 10,344 windows in the building at a cost of $6,000,000. What was the replacement cost per window? (Round to the nearest dollar.)

   6. **Culinary Arts**   In purchasing food for his chain of burger shops, José pays $1.89 per pound for cheese to be used in cheeseburgers. He calculates that an order of 5180 lb will be enough for 165,000 cheeseburgers.

   (a) How much cheese will he use on each cheeseburger? (Round to three decimal places.)

   (b) What is the cost, to the nearest cent, of the cheese used on a cheeseburger?

7. **Plumbing** The pressure P in psi (pounds per square inch) in a water system at a point 140.8 ft below the water level in a storage tank is given by the formula

$$P = \frac{62.4 \text{ lb per cu ft} \times 140.8 \text{ ft} \times 0.43}{144 \text{ sq in. per sq ft}}$$

Find P to the nearest tenth.

8. **Flooring and Carpeting** A carpet cleaner charges $50 for two rooms and $120 for five rooms. Suppose a two-room job contained 320 sq ft and a five-room job contained 875 sq ft. Calculate the difference in cost per square foot between these two jobs. (Round to the nearest tenth of a cent.)

9. **Electronics** Radar operates by broadcasting a high-frequency radio wave pulse that is reflected from a target object. The reflected signal is detected at the original source. The distance of the object from the source can be determined by multiplying the speed of the pulse by the elapsed time for the round trip of the pulse and dividing by 2. A radar signal traveling at 186,000 miles per second is reflected from an object. The signal has an elapsed round-trip time of 0.0001255 sec. Calculate the distance of the object from the radar antenna.

10. **Plumbing** Normal air pressure at sea level is about 14.7 psi. Additional pressure exerted by a depth of water is given by the formula

Pressure in psi $\approx 0.434 \times$ depth of water in ft

Calculate the total pressure on the bottom of a diving pool filled with water to a depth of 12 ft. (The sign $\approx$ means "approximately equal to.")

11. **Life Skills** A total of 12.4 gal was needed to fill a car's gas tank. The car had been driven 286.8 miles since the tank was last filled. Find the fuel economy in miles per gallon since this car was last filled. (Round to one decimal place.)

12. **Aviation** In November 2005, a Boeing 777-200LR set the world record for the longest nonstop flight by a commercial airplane. The jet flew east from Hong Kong to London, covering 13,422 miles in 22.7 hours. What was the jet's average speed in miles per hour? (Round to the nearest whole number.)

13. **Aviation** The jet referred to in problem 12 burned 52,670 gal of fuel during the 13,422-mile trip.

   (a) What was the fuel economy of the jet in miles per gallon? (Round to three decimal places.)

   (b) A more meaningful measure of fuel economy is **people-miles per gallon,** which is found by multiplying actual miles per gallon by the number of passengers on the plane. If there were 35 passengers on this historic flight, what was the fuel economy in people-miles per gallon? (Round to two decimal places.)

14. **Flooring and Carpeting** A certain type of prefinished oak flooring comes in 25-sq-ft cartons and is priced at $101.82 per carton. Assuming that you cannot purchase part of a carton, (a) how many cartons will you need to lay 860 sq ft of floor, and (b) what will the cost of the flooring be?

15. **Office Services** A gas company bills a customer according to a unit called *therms*. The company calculates the number of therms by multiplying the number of HCF (hundred cubic feet) used by a billing factor that varies according to geographical location. If a customer used 122 HCF during a particular month, and the customer's billing factor is 1.066, how many therms did the customer consume?

16. **Life Skills**   In the fall of 2016, the average cost of gas in California was $2.50 per gallon. At the same time in England, the average cost of gas (or "petrol," as they say) was 1.10 British pound per liter. One British pound was worth $1.28 at the time, and one gallon contains approximately 3.785 liters. Using this information, how much more, in dollars, did it cost to fill up a 16-gal tank in England than in California?

17. **Office Services**   Once the number of therms is established on a customer's gas bill (see problem 15), the monthly bill is calculated as follows:

Item	Charge
Customer charge	$0.16438 per day
Regulatory fee	$0.05012 per therm
Gas usage	$0.91071 per therm for the first 56 therms
	$1.09365 per therm for each therm over 56 therms

Suppose that during a 31-day month a customer used 130 therms. What would the customer's total bill be?

18. **Automotive Trades**   Most people measure the fuel efficiency of a vehicle using the number of miles per gallon. But to compare the effectiveness of different grades of gasoline—which, of course, have different prices—it may be more meaningful to calculate and compare the number of *miles per dollar* that different grades give you under similar driving conditions. Perform this calculation for parts (a), (b), and (c) and then answer part (d). (Round to the nearest hundredth.) (*Source:* Mike Allen, Senior Editor Automotive, *Popular Mechanics*)

(a) 278 miles on 12.4 gal of regular at $2.39 per gallon

(b) 304 miles on 13.1 gal of mid-grade at $2.49 per gallon

(c) 286 miles on 12.0 gal of premium at $2.59 per gallon

(d) Which grade gave the highest number of miles per dollar?

19. **Construction**   The **U-factor** of a window is a measure of how well the window resists the flow of heat. A lower U-factor indicates a better resistance to heat flow. European windows use different units of measure for their U-factor than do U.S. windows. To convert a European U-factor to a U.S. U-factor, simply divide by 5.678. If a European window has a U-factor of 1.363, what would the equivalent U.S. U-factor be? (Round to the nearest thousandth.)

20. **Retail Merchandising**   In retail sales, one measure of productivity is *sales per employee hour*. If a small clothing store had $68,992 of net sales last week when employees worked a total of 880 hours, what were the sales per employee hour for the week?

21. **Metalworking**   Rivet holes are being drilled in the wing panel of an aircraft. Although the diameter of the holes is specified at 0.250 in., they are acceptable if they are within the range of 0.245 in. to 0.255 in. To determine when a drill bit needs to be changed, the drill operator must take three measurements every 20 minutes and average them. If three consecutive averages show an upward trend, or if any one of the measurements falls outside the acceptable range, then the drill bit must be changed. The following table shows measurements for eight consecutive time periods. Find each average and determine if and when the drill bit should have been changed.

	Meas. 1	Meas. 2	Meas. 3
1	0.249	0.250	0.249
2	0.250	0.248	0.250
3	0.251	0.250	0.250
4	0.249	0.251	0.251
5	0.251	0.251	0.252
6	0.250	0.253	0.252
7	0.252	0.251	0.253
8	0.254	0.252	0.253

22. **Trades Management** A local water district purchased its annual supply of water from four different sources. The number of acre-feet purchased and the cost per acre-foot for each source are shown in the following table. Calculate the average cost of water per acre-foot. (Round to the nearest dollar.)

Source	Number of Acre-Feet	Cost per Acre-Foot
Reservoir A	3723	$520
Reservoir B	1531	$79
Well Water	248	$161
State Water	1320	$3288

23. **Transportation** The following table shows the cost of a ticket (converted to U.S. dollars) for several high-speed rail trips in various countries:

Trip	Length in Miles	Ticket Cost
Italy: Milan to Salerno	434	$108.50
Japan: Tokyo to Osaka	320	$118.40
Spain: Madrid to Barcelona	386	$154.40
France: Paris to Marseille	411	$131.50
U.S.: Boston to Washington	454	$227.00

(a) Calculate the cost per mile for each trip.

(b) What trip costs the most per mile?

(c) What trip costs the least per mile?

(d) For the proposed high-speed rail system in California, a ticket for the 438-mile trip from Los Angeles to San Francisco is projected to cost $86. How would this rank in cost per mile compared to the trips in the table?

24. **Carpentry** A tool-tester was testing the cost-effectiveness of fiber cement saw blades. Model A, costing $33, was able to effectively cut 742 linear feet; model B, costing $49, was able to effectively cut 965 linear feet; model C, costing $56 was able to effectively cut 1486 linear feet. Determine the cost per cut foot for each model and decide which was the most cost-effective blade. (Round to the nearest tenth of a cent.)

25. **General Interest** For measuring very large distances in outer space, astrophysicists have defined the **astronomical unit**, or AU. One AU is equal to approximately 93 million miles, or the distance from the earth to the sun. For example, Mars is 1.532 AU from Earth, while Jupiter is 4.278 AU from Earth. How much further away from Earth, in miles, is Jupiter than Mars?

When you have finished these exercises, check your answers to the odd-numbered problems in the Appendix, then continue with Section 3-3.

3-3 Decimal Fractions

learning|catalytics

1. Write $\frac{1}{4}$ as a decimal number.

2. What is one-third of 1.5?

We have learned that decimal numbers are simply fractions that are written in a different form. Because it is usually easier to work with decimals than with fractions, it is useful to be able to convert a fraction to its decimal equivalent.

Terminating Decimals To convert a fraction to a decimal number, simply divide the top by the bottom of the fraction (the numerator by the denominator). If the division has no remainder, the decimal number is called a **terminating decimal**.

EXAMPLE 1 To convert $\frac{5}{8}$ to decimal form, divide 5 by 8 as follows:

$$
\begin{array}{r}
.625 \\
8\overline{)5.000} \\
\underline{4\ 8} \\
20 \\
\underline{16} \\
40 \\
\underline{40} \\
0
\end{array}
$$

← Attach as many zeros as needed.

← Zero remainder; therefore, the decimal terminates or ends.

$\frac{5}{8} = 0.625$ in decimal form

If the denominator of the fraction is a power of ten (10 or 100 or 1000, etc.), then conversion to decimal form is even easier. Simply move the decimal point to the left as many places as there are zeros in the power of ten.

EXAMPLE 2 To convert $\frac{11}{1000}$ to decimal form, notice that the denominator is a power of ten with three zeros. We shift the decimal point three places to the left as follows:

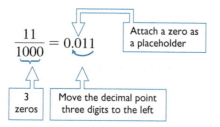

$$\frac{11}{1000} = 0.011$$

Attach a zero as a placeholder

3 zeros   Move the decimal point three digits to the left

Notice that we needed to attach one zero as a placeholder to the right of the decimal point. We can check to see that this is correct by using long division:

$$
\begin{array}{r}
.011 \\
1000\overline{)11.000} \\
\underline{0} \\
11\ 00 \\
\underline{10\ 00} \\
1\ 000 \\
\underline{1\ 000} \\
0
\end{array}
$$

Your Turn Convert the following fractions to decimal form.

(a) $\frac{4}{5}$ (b) $\frac{3}{100}$ (c) $\frac{29}{20}$ (d) $\frac{7}{16}$

Solutions (a) $\dfrac{4}{5} = \underline{\quad ? \quad}$ $5\overline{)4.0}$ = $.8$

$\dfrac{4}{5} = 0.8$

(b) The denominator is a power of ten with two zeros. Simply shift the decimal point two places to the left. You will need to attach one zero as a placeholder.

Placeholder

$\dfrac{3}{100} = .03 = 0.03$

Two zeros

Shift decimal point two places left

(c) $\dfrac{29}{20} = \underline{\quad ? \quad}$

$$\begin{array}{r} 1.45 \\ 20\overline{)29.00} \\ 20 \\ \hline 9\,0 \\ 8\,0 \\ \hline 1\,00 \\ 1\,00 \\ \hline 0 \end{array}$$

$\dfrac{29}{20} = 1.45$

(d) $\dfrac{7}{16} = \underline{\quad ? \quad}$

$$\begin{array}{r} .4375 \\ 16\overline{)7.0000} \\ 6\,4 \\ \hline 60 \\ 48 \\ \hline 120 \\ 112 \\ \hline 80 \\ 80 \\ \hline 0 \end{array}$$

$\dfrac{7}{16} = 0.4375$

Some fractions are used so often in practical work that it is important for you to know their decimal equivalents. The table on p. 168 shows the decimal equivalents for fractions that are used frequently in the trades—namely, those with denominators of 2, 4, 8, 16, 32, and 64. These are sometimes referred to as **shop fractions**. Although it would be difficult to memorize the entire table, it would be helpful for a trades worker to learn the decimal equivalents for the fractions that occur most often in his or her job.

Learning Help Notice that the number of decimal places in each decimal equivalent is consistent for each denominator. That is, all "halves" contain one decimal digit, all quarters contain two, all eighths contain three, all sixteenths contain four, all 32nds contain five, and all 64ths contain six decimal digits. ●

Study the ruler diagram at the top of the table and then try the following problems.

More Practice Quick now, without looking at the diagram, fill in the blanks in the problems shown.

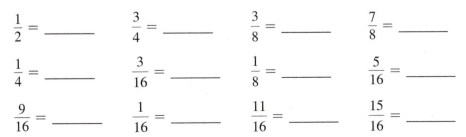

$\dfrac{1}{2} = \underline{\qquad}$ $\dfrac{3}{4} = \underline{\qquad}$ $\dfrac{3}{8} = \underline{\qquad}$ $\dfrac{7}{8} = \underline{\qquad}$

$\dfrac{1}{4} = \underline{\qquad}$ $\dfrac{3}{16} = \underline{\qquad}$ $\dfrac{1}{8} = \underline{\qquad}$ $\dfrac{5}{16} = \underline{\qquad}$

$\dfrac{9}{16} = \underline{\qquad}$ $\dfrac{1}{16} = \underline{\qquad}$ $\dfrac{11}{16} = \underline{\qquad}$ $\dfrac{15}{16} = \underline{\qquad}$

Check your work against the diagram when you are finished, then continue on p. 169.

Decimal-Fraction Equivalents

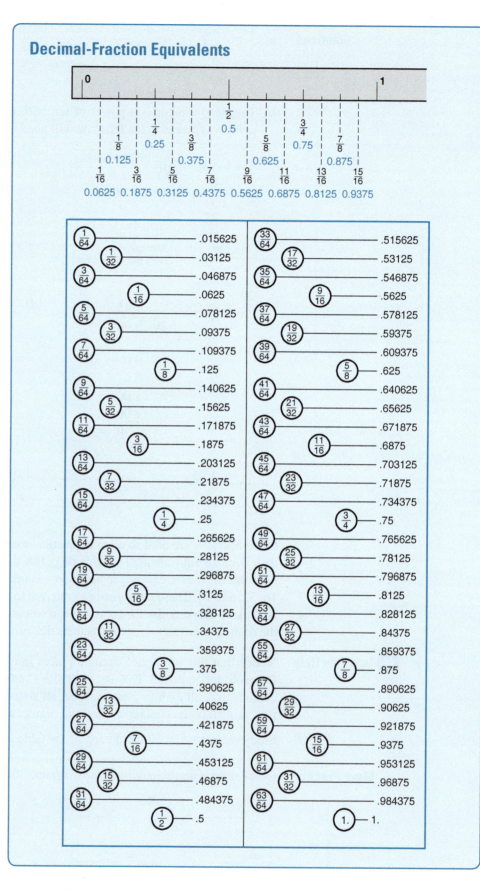

Another practical use of the table on page 168 is finding the fraction closest to a given decimal measurement. By using the table, we will only be concerned with the shop fractions—those with denominators of 2, 4, 8, 16, 32, and 64.

EXAMPLE 3   (a) To find the closest fractional equivalent to 0.741 in. from our table, we notice that 0.741 lies between 0.734375 $\left(\frac{47}{64}\right)$ and 0.75 $\left(\frac{3}{4}\right)$. To determine which of these is closer to 0.741, we will find the difference between each of the two options and 0.741. It is helpful to organize our work visually like this:

$$\text{Difference}$$

$$\frac{47}{64} = 0.734375$$

$$0.006625$$

$$\text{Given measurement: } 0.741$$

$$0.009$$

$$\frac{3}{4} = 0.75$$

Because 0.006625 is a smaller difference than 0.009, $\frac{47}{64}$ in. is the closest fractional equivalent to 0.741 in. from our table.

(b) Now suppose you need to approximate 0.741 in. with a ruler that measures only to the nearest sixteenth of an inch. To find the closest fractional equivalent on this ruler, we will consider only those values in the table with denominators of 2, 4, 8, and 16. Limiting ourselves to these denominators, we see that 0.741 lies between $\frac{11}{16}$ and $\frac{3}{4}$ $\left(\text{where } \frac{3}{4} = \frac{12}{16}\right)$. It is easy to see by inspection that 0.741 is much closer to 0.75 $\left(\text{or } \frac{3}{4}\right)$ than it is to 0.6875 $\left(\text{or } \frac{11}{16}\right)$, so $\frac{3}{4}$ in. will be the closest fractional equivalent to 0.741 in. if we can measure only to the nearest sixteenth of an inch.   ●

Note   In practical applications, such as finding the best size wrench for a certain size bolt, a minimal amount of clearance is required. In these cases, we would always choose the higher of the two options rather than the closer one.   ●

Your Turn   (a) Use the table on page 168 to find the closest fractional equivalent to 4.283 in.

(b) Find the closest fractional equivalent to 4.283 in. if we can measure only to the nearest sixteenth of an inch.

Solutions   (a) When the given measurement is greater than 1, simply ignore the whole number portion when using the table, and then reattach it in your final answer. In this case, we first use the table to find the closest fractional equivalent to 0.283. We see that 0.283 lies between 0.28125 $\left(\frac{9}{32}\right)$ and 0.296875 $\left(\frac{19}{64}\right)$. It is easy to tell by inspection that 0.283 is much closer to $\frac{9}{32}$ than to $\frac{19}{64}$. Reattaching the whole-number portion, we conclude that the closest fractional equivalent to 4.283 in. in the table is $4\frac{9}{32}$ in.

(b) To the nearest sixteenth of an inch, the decimal portion of 4.283 lies between $0.25 \left(\dfrac{1}{4} \text{ or } \dfrac{4}{16} \right)$ and $0.3125 \left(\dfrac{5}{16} \right)$. The closer of these two fractions is not so obvious, so we organize our work as in Example 3(a):

Difference

$$\dfrac{1}{4} = 0.25$$

0.033

Given measurement: 0.283

0.0295

$$\dfrac{5}{16} = 0.3125$$

The difference of 0.0295 is smaller than 0.033, so 0.283 is closer to $\dfrac{5}{16}$. Reattaching the whole-number portion, we conclude that $4\dfrac{5}{16}$ in. is the closest fractional equivalent to 4.283 in. if we can measure only to the nearest sixteenth of an inch.

Repeating Decimals   Decimal fractions that do not terminate always repeat in a sequence of digits. This type of decimal number is called a **repeating decimal**. For example,

$$\dfrac{1}{3} = 0.3333 \ldots$$

where the three dots are read "and so on," and they tell us that the digit 3 continues without end. The next example illustrates how we can tell when a decimal repeats.

EXAMPLE 4   To convert $\dfrac{3}{11}$ to decimal form, divide as follows:

```
     .2727
11)3.0000
   2 2
   ───
    80
    77
    ──
    30
    22
    ──
    80
    77      The remainder 3 is equal to the original dividend. This tells us
    ──      that the decimal quotient continues to repeat.
     3
```

$$\dfrac{3}{11} = 0.272727 \ldots$$

The three dots indicate that the digits "27" continue forever.   ●

Mathematicians often use a shorthand notation to show that a decimal repeats.

Write $\dfrac{1}{3} = 0.\overline{3}$

where the bar means that the digits under the bar repeat endlessly.

$$\frac{3}{11} = 0.\overline{27} \qquad \text{means} \qquad 0.272727\ldots$$

Here's an example in which only some of the decimal digits repeat.

EXAMPLE 5   Converting $\frac{11}{30}$ to decimal form, we have:

$$
\begin{array}{r}
.366 \\
30\overline{)11.000} \\
\underline{9\ 0} \\
2\ 00 \\
\underline{1\ 80} \\
200 \\
\underline{180} \\
20
\end{array}
$$

The remainders are the same, meaning that any further division will repeat the digit 6 in the quotient.

To write 0.366 . . . using bar notation, be careful to write the bar only over the 6:

$$\frac{11}{30} = 0.3\overline{6}$$

●

Your Turn   Express each of the following in decimal form using bar notation.

(a) $\frac{2}{3}$ (b) $\frac{5}{6}$ (c) $\frac{9}{11}$ (d) $\frac{7}{22}$ (e) $\frac{11}{15}$

Answers   (a) $0.\overline{6}$ (b) $0.8\overline{3}$ (c) $0.\overline{81}$ (d) $0.3\overline{18}$ (e) $0.7\overline{3}$

How to Write a Repeating Decimal as a Fraction

Suppose you need to reverse this process—that is, write a repeating decimal as a fraction. If the *entire* decimal portion repeats, there is a quick and easy way to do this.

First, use the digits that repeat as the numerator of the fraction.

Then, use the same number of consecutive 9s for the denominator.

For example,

$$0.\overline{7} = \frac{7}{9}$$

> One digit in the numerator

> One 9 in the denominator

$$0.\overline{27} = \frac{27}{99} = \frac{3}{11} \qquad \text{Two digits in 27; therefore, use 99 as the denominator.}$$

$$0.\overline{123} = \frac{123}{999} = \frac{41}{333} \qquad \text{Three digits in 123; therefore, use 999 as the denominator.}$$

Remember, this procedure works only when *all* of the decimal part repeats. Writing a repeating decimal such as $0.5\overline{3}$ as a fraction is more complicated and is beyond the scope of this textbook.

Rounding Decimal Fractions   In most practical work, we round decimal fractions to a specified number of decimal places. The rounding may occur before the decimal equivalent either terminates or repeats. For example, if you did the division on a calculator, you would see that $\frac{2}{13} = 0.1538461538\ldots$. Notice that the decimal digits begin to repeat after the sixth digit.

But unless we were doing highly technical work, we would round this decimal number before it begins to repeat.

- Rounded to one decimal place (the nearest tenth): $\frac{2}{13} = 0.2$

- Rounded to two decimal digits (the nearest hundredth): $\frac{2}{13} = 0.15$

- Rounded to three decimal places (the nearest thousandth): $\frac{2}{13} = 0.154$

Your Turn   Convert each of the following to decimal form rounded to three decimal digits.

(a) $\frac{1}{6}$ (b) $\frac{5}{11}$ (c) $\frac{9}{7}$ (d) $\frac{13}{32}$ (e) $\frac{5}{9}$ (f) $\frac{23}{80}$

Solutions   (a) $\frac{1}{6} = 0.1666\ldots$ or 0.167, rounded   (b) $\frac{5}{11} = 0.4545\ldots$ or 0.455, rounded

(c) $\frac{9}{7} = 1.2857\ldots$ or 1.286, rounded   (d) $\frac{13}{32} = 0.40625$ exactly or 0.406, rounded

(e) $\frac{5}{9} = 0.5555\ldots$ or 0.556, rounded   (f) $\frac{23}{80} = 0.2875$ exactly or 0.288, rounded

Calculations with Fractions and Decimals   When a calculation involves both fractions and decimals, it is usually simpler to convert the fractions to decimals and then perform the calculation. In practical work, we usually round the answer to agree with the number of decimal digits in the decimal number.

EXAMPLE 6   To add $2\frac{5}{8} + 3.28$

First, convert $2\frac{5}{8}$ to decimal form:

$$2\frac{5}{8} = 2.625$$

Then, add this to 3.28:

$$2.625 + 3.28 = 5.905$$

Finally, round the answer to agree with the number of decimal digits in the original decimal number:

$$5.905 \approx 5.91 \text{ rounded to two decimal places}$$ ●

Your Turn   (a) Multiply: $5\frac{1}{4} \times 7.3$ (b) Subtract: $9\frac{2}{5} - 4.73$

Solutions   (a) $5\frac{1}{4} \times 7.3 = 5.25 \times 7.3 = 38.325$, or 38.3 rounded to one decimal place.

(b) $9\frac{2}{5} - 4.73 = 9.4 - 4.73 = 4.67$

There is no need to round this answer because it has the same number of decimal digits as 4.73.

Practical Applications

EXAMPLE 7   **Water/Wastewater Treatment**   In water and wastewater treatment formulas, flow rates must be expressed in units of *millions* of gallons per day (MGD). However, in real-world situations, measurements of flow are frequently reported in *gallons* per day (gal/day). To convert gal/day to MGD, simply divide by 1,000,000. For example,

$$28,500 \text{ gal/day} = \frac{28,500}{1,000,000} \text{ MGD}$$

Because there are six zeros in 1,000,000, we divide by simply moving the decimal point six places to the left:

$$\frac{28,500}{1,000,000} = .028500. = 0.0285 \text{ MGD} \qquad \bullet$$

Your Turn   **Automotive Trades**   A certain cylinder is normally $3\frac{11}{32}$ in. in diameter. It is rebored to be 0.044 in. larger. What is the diameter of the rebored cylinder rounded to three decimal places?

Solution   **First,** convert the original diameter to decimal form. Round to three decimal places.

$$3\frac{11}{32} \text{ in.} = 3.34375 \text{ in.} = 3.344 \text{ in., rounded}$$

Then, add this to the increase in size.

3.344 in. + 0.044 in. = 3.388 in.

The rebored cylinder is approximately 3.388 in. in diameter.

Now turn to Exercises 3-3 for a set of problems on decimal fractions.

Exercises 3-3   Decimal Fractions

A. Write as decimal numbers. Use the bar notation for repeating decimals.

1. $\frac{1}{4}$	2. $\frac{2}{3}$	3. $\frac{3}{4}$	4. $\frac{2}{5}$
5. $\frac{3}{5}$	6. $\frac{5}{6}$	7. $\frac{2}{7}$	8. $\frac{4}{7}$
9. $\frac{6}{7}$	10. $\frac{3}{8}$	11. $\frac{6}{8}$	12. $\frac{1}{10}$
13. $\frac{3}{10}$	14. $\frac{2}{12}$	15. $\frac{5}{12}$	16. $\frac{3}{16}$
17. $\frac{6}{16}$	18. $\frac{9}{16}$	19. $\frac{13}{16}$	20. $\frac{3}{20}$
21. $\frac{7}{32}$	22. $\frac{13}{20}$	23. $\frac{11}{24}$	24. $\frac{39}{64}$
25. $\frac{3}{100}$	26. $\frac{213}{1000}$	27. $\frac{19}{1000}$	28. $\frac{17}{50}$

B. Calculate in decimal form. Round to agree with the number of decimal digits in the decimal number.

1. $2\frac{3}{5} + 1.785$

2. $\frac{1}{5} + 1.57$

3. $3\frac{7}{8} - 2.4$

4. $1\frac{3}{16} - 0.4194$

5. $2\frac{1}{2} \times 3.15$

6. $1\frac{3}{25} \times 2.08$

7. $3\frac{4}{5} \div 2.65$

8. $3.72 \div 1\frac{1}{4}$

9. $2.76 + \frac{7}{8}$

10. $16\frac{3}{4} - 5.842$

11. $3.14 \times 2\frac{7}{16}$

12. $1.17 \div 1\frac{3}{32}$

C. Practical Problems

1. **Allied Health**   If one tablet of calcium pantothenate contains 0.5 gram, how much is contained in $2\frac{3}{4}$ tablets? How many tablets are needed to make up 2.6 grams?

2. **Flooring and Carpeting**   Estimates of matched or tongue-and-groove (T&G) flooring—stock that has a tongue on one edge—must allow for the waste from milling. To allow for this waste, $\frac{1}{4}$ of the area to be covered must be added to the estimate when 1-in. by 4-in. flooring is used. If 1-in. by 6-in. flooring is used, $\frac{1}{6}$ of the area must be added to the estimate. (*Hint:* The floor is a rectangle. Area of a rectangle = length × width.)

    (a) A house has a floor size 28 ft long by 12 ft wide. How many square feet of 1-in. by 4-in. T&G flooring will be required to lay the floor?

    (b) A motel contains 12 units or rooms, and each room is 22 ft by 27 ft, or 594 sq ft. In five rooms the builder is going to use 4-in. stock and in the other seven rooms 6-in. stock—both being T&G. How many square feet of each size will be used?

3. **Roofing**   If a roofer lays $10\frac{1}{2}$ squares of shingles in $4\frac{1}{2}$ days, how many squares does he do in one day?

4. **Machine Trades**   A bearing journal measures 1.996 in. If the standard readings are in $\frac{1}{32}$-in. units, what is its probable standard size as a fraction?

5. **Interior Design**   Complete the following invoice for upholstery fabric:

    (a) $5\frac{1}{4}$ yd @ \$12.37 per yd = _____

    (b) $23\frac{3}{4}$ yd @ \$16.25 per yd = _____

    (c) $31\frac{5}{6}$ yd @ \$9.95 per yd = _____

    (d) $16\frac{2}{3}$ yd @ \$17.75 per yd = _____

    Total   _____

6. **Construction**   A land developer purchased a piece of land containing 437.49 acres. He plans to divide it into a 45-acre recreational area and lots of $\frac{3}{4}$ acre each. How many lots will he be able to form from this piece of land?

7. **Machine Trades**   Find the corner measurement A needed to make an octagonal end on a square bar as shown in the diagram.

8. **Automotive Trades**   A cylinder is normally $3\frac{7}{16}$ in. in diameter. It is rebored 0.040 in. larger. What is the size of the rebored cylinder? (Give your answer in decimal form.)

9. Plumbers deal with measurements in inches and common fractions of an inch, while surveyors use feet and decimal fractions of a foot. Often one trade needs to interpret the measurements of the other.

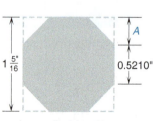

Problem 7

(a) **Construction** A drain has a run of 52 ft at a grade of $\frac{1}{8}$ in./in. The high end of the drain has an elevation of 126.70 ft. What is the elevation at the low end?

(b) **Plumbing** The elevation at one end of a lot is 84.25 ft, and the elevation at the other end is 71.70 ft. Express the difference in elevation in feet and inches and round to the nearest $\frac{1}{8}$ in.

10. **Plumbing** To fit a 45 degree connection, a plumber can approximate the diagonal length by multiplying the offset length by 1.414. Find the diagonal length of a 45 degree fitting if the offset length is $13\frac{11}{16}$ in. Express the answer as a fraction to the nearest $\frac{1}{16}$ in.

11. **Electronics** The frequency of a waveform or electronic device is the reciprocal of the period. If the period of a device is 40 milliseconds, calculate the frequency as a decimal numeral. The answer will be in units of kilohertz (kHz).

12. **Water/Wastewater Treatment** A reservoir is used at an average rate of 125,000 gallons per day. Convert this to units of MGD. (See Example 7 on p. 173.)

13. **Sheet Metal Trades** At one time the thickness of different gauges of stainless steel were given in fractions of an inch. Today, these thicknesses are listed as decimal numbers. Convert each of the following fractional thicknesses to decimal form. (Round to four decimal places.)

	Gauge	Fractional Thickness (in.)
(a)	6–0	$\frac{15}{32}$
(b)	2	$\frac{17}{64}$
(c)	7	$\frac{3}{16}$
(d)	11	$\frac{1}{8}$
(e)	20	$\frac{3}{80}$
(f)	32	$\frac{13}{1280}$

14. **Office Services** Machine helpers at the CALMAC Tool Co. earn $15.72 per hour. The company pays time-and-a-half for hours over 40 per week. Sunday hours are paid at double time. How much would you earn if you worked the following hours: Monday, 9; Tuesday, 8; Wednesday, $9\frac{3}{4}$; Thursday, 8; Friday, $10\frac{1}{2}$; Saturday, $5\frac{1}{4}$; Sunday, $4\frac{3}{4}$?

15. **Construction** A retaining wall requires 36.6 cu yd of concrete. The combined amounts of cement, sand, and gravel should be $1\frac{1}{2}$ times the amount of concrete needed. What total amount of cement, sand, and gravel will be required?

16. **Carpentry** For interior trim jobs, some carpenters prefer to use MDF (medium-density fiberboard) while others prefer finger-jointed trim. If MDF costs $1.12 per linear foot, and finger-jointed trim costs $1.68 per linear foot, what would be the total cost difference between the two materials for a job involving $38\frac{1}{4}$ feet of trim?

17. **General Trades** Use the table on page 168 to find the closest fractional equivalent to each of the following decimal measurements:

(a) 0.425 in. (nearest 8th of an inch)   (b) 3.618 in. (nearest 8th of an inch)

(c) 0.826 in. (nearest 16th of an inch)   (d) 0.315 in. (nearest 16th of an inch)

(e) 6.575 in. (nearest 32nd of an inch)   (f) 0.264 in. (nearest 32nd of an inch)

(g) 0.782 in. (nearest 64th of an inch)   (h) 7.184 in. (nearest 64th of an inch)

When you have finished these exercises, check your answers to the odd-numbered problems in the Appendix, and then turn to Problem Set 3 on page 178 for practice working with decimal numbers. If you need a quick review of the topics in this chapter, visit the chapter Summary first.

CHAPTER 3
SUMMARY

Decimal Numbers

Objective	Review
Write decimal numbers in words and translate words to numbers. (p. 130)	The word *and* represents the decimal point. In written form, the decimal portion of the number is followed by the place value of the far right digit. **Example:** (a) The number 26.035 is spoken or written in words as "twenty-six and thirty-five thousandths." (b) The number "one hundred six and twenty-seven ten-thousandths" is written in decimal form as 106.0027.
Add and subtract decimal numbers. (pp. 133–137)	Write the numbers in column format with the decimal points aligned vertically. Then add or subtract as with whole numbers. **Example:** (a) $5.82 + 0.096 \Rightarrow 5.820$ ← Attach a zero $ + 0.096$ $ \overline{5.916}$ (b) $27 - 4.03 \Rightarrow 27.00$ $ - 4.03$ Attach a decimal point and two zeros. $\Rightarrow \overset{6\ 9\,10}{27.\cancel{00}}$ $ - 4.03$ $ \overline{22.97}$
Multiply decimal numbers. (p. 142)	First, multiply, ignoring decimal points—they need not be aligned. Then, to find the total number of decimal digits in the product, count the total number of decimal digits in the factors. Attach additional zeros if needed. **Example:** $7.25 \times 0.301 = 7.25$ ← *two* decimal digits $ \times\ 0.301$ ← *three* decimal digits $ \overline{2.18225}$ ← *five* decimal digits

Objective	Review
Divide decimal numbers. (p. 146)	Shift the decimal point in the divisor to the right to make it a whole number. Shift the decimal point in the dividend the same number of places. Attach zeros if necessary. Divide as with whole numbers, placing the decimal point directly above the new position in the dividend. **Example:** $104.96 \div 0.032 = 0.032\overline{)104.96}$ $$0.032\overline{)104.960.} = 32\overline{)104960.}$$ $$\begin{array}{r} 3280. \\ 32\overline{)104960.} \\ \underline{96} \\ 89 \\ \underline{64} \\ 256 \\ \underline{256} \\ 0 \\ \underline{0} \\ 0 \end{array}$$
Calculate an average. (p. 154)	To find the average of a set of numbers, divide their sum by the number of items in the set. Round the answer to agree in precision with the least precise number in the set. **Example:** The average of the numbers 4.2, 4.8, 5.7, 2.5, 3.6, and 5.0 is $$\text{Average} = \frac{\text{sum}}{\text{number}} = \frac{25.8}{6} = 4.3$$
Round decimal numbers. (p. 151)	Follow the procedure for rounding whole numbers, except after rounding to a decimal place, drop all digits to the right of that place. **Example:** Round 16.2785 to the nearest hundredth. $$16.27 \wedge 85 = 16.28$$
Convert fractions to decimal numbers. (p. 166)	Divide the numerator by the denominator and round if necessary. Use bar notation for repeating decimals. **Example:** $\dfrac{3}{16} \Rightarrow$ $\begin{array}{r} .1875 \\ 16\overline{)3.0000} \\ \underline{1\ 6} \\ 1\ 40 \\ \underline{1\ 28} \\ 120 \\ \underline{112} \\ 80 \\ \underline{80} \end{array}$ $\qquad$ $\dfrac{5}{6} \Rightarrow$ $\begin{array}{r} .833\ldots = 0.8\overline{3} \\ 6\overline{)5.000} \\ \underline{4\ 8} \\ 20 \\ \underline{18} \\ 20 \\ \underline{18} \end{array}$
Solve practical problems involving decimal numbers. (pp. 135, 137, 145, 150, 153, 155, 173)	Read the problem statement carefully. Look for key words or phrases that indicate which operation(s) to use and then perform the calculation. **Example:** A box weighs 61.3 lb and contains 175 bolts. If the container itself weighs 1.8 lb, how much does each bolt weigh? **First,** find the weight of the bolts by subtracting the weight of the container: 61.3 lb − 1.8 lb = 59.5 lb **Then,** divide this amount by the total number of bolts to find the weight of each bolt: 59.5 lb ÷ 175 bolts = 0.34 lb per bolt

PROBLEM SET 3   Decimal Numbers

Answers to odd-numbered problems are given in the Appendix.

A. Write in words.

1. 0.9
2. 0.84
3. 23.164
4. 63.219
5. 9.3
6. 3.45
7. 10.0625
8. 15.037

Write as decimal numbers.

9. Seven hundredths

10. Eighteen thousandths

11. Two hundred and eight tenths

12. Sixteen and seventeen hundredths

13. Sixty-three and sixty-three thousandths

14. One hundred ten and twenty-one thousandths

15. Five and sixty-three ten-thousandths

16. Eleven and two hundred eighteen ten-thousandths

B. Add or subtract as shown.

1. $4.39 + 18.8$

2. $18.8 + 156.16$

3. $\$7.52 + \$11.77$

4. $26 + 0.06$

5. $3.68 - 1.74$

6. $\$12.46 - \$8.51$

7. $104.06 - 15.80$

8. $16 - 3.45$

9. $264.3 + 12.804$

10. $0.232 + 5.079$

11. $165.4 + 73.61$

12. $245.94 + 7.07$

13. $116.7 - 32.82$

14. $4.07 - 0.085$

15. $0.42 + 1.452 + 31.8$

16. $\frac{1}{2} + 4.21$

17. $3\frac{1}{5} + 1.08$

18. $1\frac{1}{4} - 0.91$

19. $3.045 - 1\frac{1}{8}$

20. $8.1 + 0.47 - 1\frac{4}{5}$

C. Calculate as shown.

1. 0.004×0.02

2. 0.06×0.05

3. 1.4×0.6

4. 3.14×12

5. $0.2 \times 0.6 \times 0.9$

6. $1.17 \div 4.5$

7. $12.8 \div 0.32$

8. $3.224 \div 2.6$

9. $187.568 \div 3.04$

10. $0.078 \div 0.3$

11. $0.6 \times 3.15 \times 2.04$

12. $3.78 + 4.1 \times 6.05$

13. $0.008 - 0.001 \div 0.5$

14. $3.1 \times 4.6 - 2.7 \div 0.3$

15. $\dfrac{4.2 + 4.6 \times 1.2}{2.73 \div 21 + 0.41}$

16. $\dfrac{7.2 - 3.25 \div 1.3}{3.5 + 5.7 \times 4.0 - 2.8}$

17. $1.2 + 0.7 \times 2.2 + 1.6$

18. $1.2 \times 0.7 + 2.2 \times 1.6$

Name

_____

Date

_____

Course/Section

_____

Round to two decimal digits.

19. 0.0371

20. 16.8449

21. 0.007 ÷ 0.03

22. 3.005 ÷ 2.01

23. 17.8 ÷ 6.4

24. 0.0041 ÷ 0.019

Round to three decimal places.

25. 27.0072

26. 1.1818

27. 0.04 ÷ 0.076

28. 234.1 ÷ 465.8

29. 17.6 ÷ 0.082

30. 0.051 ÷ 1.83

Round to the nearest tenth.

31. 47.233

32. 123.7666

33. 0.08 ÷ 0.053

34. 3.05 ÷ 0.13

35. 18.76 ÷ 4.05

36. 0.91 ÷ 0.97

D. **Write as a decimal number.**

1. $\frac{1}{16}$

2. $\frac{7}{8}$

3. $\frac{5}{32}$

4. $\frac{7}{16}$

5. $\frac{11}{8}$

6. $1\frac{3}{4}$

7. $2\frac{11}{20}$

8. $1\frac{1}{6}$

9. $2\frac{2}{3}$

10. $1\frac{1}{32}$

11. $2\frac{5}{16}$

Write in decimal form, calculate, round to two decimal digits.

12. $4.82 \div \frac{1}{4}$

13. $11.5 \div \frac{3}{8}$

14. $1\frac{3}{16} \div 0.62$

15. $2\frac{3}{4} \div 0.035$

16. $0.45 \times 2\frac{1}{8}$

17. $0.068 \times 1\frac{7}{8}$

E. **Find the average for each of the following sets of numbers.**

1. 92, 87, 76, 84, 94, 79, 88

2. 12.4, 15.8, 17.2, 19.6

3. 6.38, 6.72, 6.49, 6.26, 6.53

4. 0.387, 0.381, 0.388, 0.365, 0.375, 0.391, 0.366, 0.358, 0.388, 0.377

F. **Practical Applications**

1. **Machine Trades**   A $\frac{5}{16}$-in. bolt weighs 0.43 lb. How many bolts are there in a keg containing 125 lb of bolts?

2. **Metalworking**   Twelve equally spaced holes are to be drilled in a metal strip $34\frac{1}{4}$-in. long, with 2 in. remaining on each end. What is the distance, to the nearest hundredth of an inch, from center to center of two consecutive holes?

3. **Plumbing**   A plumber finds the length of pipe needed to complete a bend by performing the following multiplication:

     Length needed = 0.01745 × (radius of the bend) × (angle of bend)

     What length of pipe is needed to complete a 35° bend with a radius of 16 in.?

4. **Office Services**   During the first six months of the year the Busy Bee Printing Co. spent the following amounts on paper:

January $1070.16	February $790.12	March $576.77
April $600.09	May $1106.45	June $589.21

     What is the average monthly expenditure for paper?

5. **Machine Trades**   The *feed* of a drill is the distance it advances with each revolution. If a drill makes 310 rpm and drills a hole 2.125 in. deep in 3.25 minutes, what is the feed? (Round to four decimal places.)

$$\left(Hint: \text{Feed} = \frac{\text{total depth}}{(\text{rpm}) \times (\text{time of drilling in minutes})}. \right)$$

6. **Electrical Trades**   The resistance of an armature while it is cold is 0.208 ohm. After running for several minutes, the resistance increases to 1.340 ohms. Find the increase in resistance of the armature.

7. **Painting**   The Ace Place Paint Co. sells paint in steel drums each weighing 36.4 lb empty. If 1 gallon of paint weighs 9.06 lb, what is the total weight of a 50-gal drum of paint?

8. **Electrical Trades**   The diameter of No. 12 bare copper wire is 0.08081 in., and the diameter of No. 15 bare copper wire is 0.05707 in. How much larger is No. 12 wire than No. 15 wire?

9. **HVAC**   The R-value of blown fiberglass is 2.2 per inch of depth. The R-value of blown cellulose is 3.7 per inch. What is the total difference in the R-values of the two materials for 12 inches of blown-in insulation?

10. **General Trades**   What is the closest fractional equivalent of each of the following dimensions? (*Hint:* Use the table on page 168 and see Example 3 on page 169.)
    (a) 0.185 in. (nearest 8th of an inch)
    (b) 0.127 in. (nearest 16th of an inch)
    (c) 0.313 in. (nearest 32nd of an inch)
    (d) 0.805 in. (nearest 64th of an inch)

11. **Sheet Metal Trades**   The following table lists the thickness in inches of several gauges of standard sheet steel:

U.S. Gauge	Thickness (in.)
35	0.0075
30	0.0120
25	0.0209
20	0.0359
15	0.0673
10	0.1345
5	0.2092

    (a) What is the difference in thickness between 30 gauge and 25 gauge sheet?
    (b) What is the difference in thickness between seven sheets of 25 gauge and four sheets of 20 gauge?
    (c) What length of $\frac{3}{16}$-in.-diameter rivet is needed to join one thickness of 25 gauge sheet to a strip of $\frac{1}{4}$-in. stock? Add $1\frac{1}{2}$ times the diameter of the rivet to the length of the rivet to ensure that the rivet is long enough that a proper rivet head can be formed.
    (d) What length of $\frac{5}{32}$-in. rivet is needed to join two sheets of 20 gauge sheet steel? (Don't forget to add $1\frac{1}{2}$ times the rivet diameter.)

12. **Trades Management**   When the owners of the Better Builder Co. completed a small construction job, they found that the following expenses had been incurred: labor, $972.25; gravel, $86.77; sand, $39.41; cement, $280.96; and bricks, $2204.35. What total bill should they give the customer if they want to make a $225 profit on the job?

13. **Construction**   One linear foot of 12-in. I-beam weighs 25.4 lb. What is the length of a beam that weighs 444.5 lb?

14. **Sheet Metal Trades**   To determine the average thickness of a metal sheet, a sheet metal worker measures it at five different locations. Her measurements are 0.0401, 0.0417, 0.0462, 0.0407, and 0.0428 in. What is the average thickness of the sheet?

15. **Drafting**   A dimension in a technical drawing is given as $2\frac{1}{2}$ in. with a tolerance of ± 0.025 in. What are the maximum and minimum permissible dimensions written in decimal form?

16. **Trades Management**   The four employees of the Busted Body Shop earned the following amounts last week: $811.76, $796.21, $808.18, and $876.35. What is the average weekly pay for the employees of the shop?

17. **Electrical Trades**   If 14-gauge Romex cable sells for $0.19 per foot, what is the cost of 1210 ft of this cable?

18. **General Interest**   In 2015, many states, counties, and cities began raising the minimum wage from $9 per hour to $15 per hour. But was that enough? For a family of 2 adults and 2 children, the minimum "living wage" in 2015 was considered to be $66,645 per year. If both adults were working 40 hours per week for 50 weeks of the year, what hourly wage would they have needed to earn in order to match the living wage? (Round to the nearest cent.)

19. **Welding**   A welder finds that 2.1 cu ft of acetylene gas is needed to make one bracket. How much gas will be needed to make 27 brackets?

20. **Automotive Trades**   What is the outside diameter d of the tire shown in the figure?

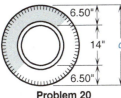

6.50"

14" d

6.50"

Problem 20

21. **Metalworking**   How many 2.34-in. spacer blocks can be cut from a 2-in. by 2-in. square bar 48 in. long? Allow $\frac{1}{4}$ in. waste for each saw cut.

22. **Automotive Trades**   At the Fixum Auto Shop, a mechanic is paid $22.50 per hour for a motor overhaul and is allotted 23 hours of labor time for the job. The mechanic completes that job in 18.5 hours, but he is still paid as if he had worked for 23 hours. Calculate his actual hourly compensation under these circumstances.

23. **Machine Trades**   A machinist estimates the following times for fabricating a certain part: 0.6 hour for setup, 2.4 hours of turning, 5.2 hours of milling, 1.4 hours of grinding, and 1.8 hours of drilling. What is the total time needed to make the part?

24. **Printing**   A 614-page book was printed on paper specified as 0.00175 in. thick and finished with a cover 0.165 in. thick. What was the total thickness of the bound book? (Be sure to count the cover twice and remember that there are two pages of the book for every one sheet of paper.)

25. **Allied Health**   The primary care physician has directed that his patient should receive a total weekly dose of 0.35 mg codeine sulfate. How many 0.035-mg tablets should she be given during a week?

26. **Fire Protection**   One gallon of water weighs approximately 8.34 lb. When sent to a particularly rough terrain, a Forestry Service truck is allowed to carry only 1.5 tons of weight. How many gallons of water can it carry? (1 ton = 2000 lb.)

27. **Plumbing**   To construct a certain pipe system, the following material is needed:

    6 1-in. elbows at $1.19 each
    5 tees at $0.99 each
    3 couplings at $0.49 each
    4 pieces of pipe each 6 ft long
    6 pieces of pipe each 2 ft long
    4 pieces of pipe each 1.5 ft long

    The pipe is schedule 40 1-in. PVC pipe, costing $0.39 per linear foot. Find the total cost of the material for the system.

28. **HVAC**   The thermal resistance per inch, or R-value, of the components of an exterior wall are as follows:

Brick	0.20	Fiberglass Batting	3.15
Plywood	1.25	Plaster	0.20
Sheetrock	0.90	Wood	1.25

    Calculate the total R-value of a wall consisting of 1-in. of wood, $2\frac{1}{2}$ in. of brick, $\frac{1}{2}$ in. of plywood, $\frac{1}{2}$ in. of Sheetrock, 6 in. of fiberglass batting, and $\frac{1}{4}$ in. of plaster. (Round to the nearest tenth.)

29. **Trades Management**   One way in which the owner of an auto shop evaluates the success of his business is to keep track of *revenue per square foot*. To calculate this indicator, he divides total revenue by the area (in square feet) of his shop. Last year, a 2260-sq-ft shop generated total revenue of $586,435. This year, the shop expanded to 2840 sq ft and generated total revenue of $734,579. During which year was the revenue per square foot the greater and by how much? (Round each calculation to the nearest cent.)

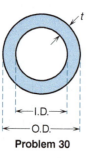

←I.D.→
←——O.D.——→
Problem 30

30. **Plumbing**   The inside diameter (I.D.) and outside diameter (O.D.) of a pipe are shown in the figure. The wall thickness of the pipe is the dimension labeled *t*. Calculate the wall thickness of schedule 120 pipe if its I.D. is 0.599 in. and its O.D. is 1.315 in.

31. **Electronics**   In the circuit shown in the figure, a solar cell is connected to a 750-ohm load. If the circuit current is 0.0045 ampere, calculate the voltage output of the solar cell by multiplying the load by the current.

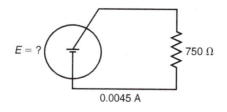

$E = ?$ 750 Ω

0.0045 A

32. **Office Services**   A shop technician earns $18.26 per hour plus time-and-a-half for overtime (time exceeding 40 hours). If he worked 42.5 hours during a particular week, what would be his gross pay?

33. **Life Skills**   A driver refueling her car fills the tank with 15.6 gal. If the odometer currently reads 63589.2, and it read 63186.7 when the tank was last filled, what was her fuel economy, in miles per gallon, between refuelings? (Round to one decimal place.)

34. **Automotive Trades**   An auto repair job consisted of a new timing belt and a brake job. The timing belt kit cost $379.95 and required 4.6 hours of labor. The brake pads cost $151.20 and required 1.8 hours of labor. If the labor rate is $95 per hour, find the total cost of all repairs before tax.

35. **Automotive Trades**   The *taper* of an automobile cylinder is caused by uneven wearing between the top and bottom of the cylinder. The size of this taper is calculated by subtracting the bore at the top of the cylinder from the bore at the bottom. If the bore at the top of a cylinder is 4.258 in. and the bore at the bottom is 4.276 in., what is the taper of this cylinder?

36. **Office Services**   A water district charges residential customers a monthly meter service charge of $51.60, plus an additional charge of $3.75 per HCF (hundred cubic feet) of water usage. What would be the total bill if a customer used 4600 cubic feet of water during a particular month?

37. **Automotive Trades**   An automotive technician, who is paid on a flat rate system, has logged the following daily hours during one particular week:

Monday	7.2	Thursday	8.8
Tuesday	8.4	Friday	7.4
Wednesday	6.6		

If he is paid $22.40 per hour, how much did he earn during the week?

38. **HVAC**   In March 2012, home heating oil sold for an average of $4.28 per gallon in New York State. In March 2016, it sold for an average of $2.41 per gallon. A typical heating oil tank holds 250 gal. How much less did it cost to fill this tank in 2016 than in 2012?

39. **Sports and Leisure**   In the 2008 Olympics, China's Chen Ruolin performed a nearly perfect dive in the final round of the women's 10-meter platform competition to overtake Canada's Emilie Heymans for the gold medal. The dive was a back $2\frac{1}{2}$ somersault with $1\frac{1}{2}$ twists, which was rated at a 3.4 degree of difficulty. Her scores from the seven judges were: 10.0, 10.0, 10.0, 9.5, 10.0, 9.5, and 9.5. To calculate the total score of a dive, the two best scores and the two worst scores are all thrown out, the remaining three are added, and that sum is multiplied by the degree of difficulty. Use this procedure to calculate Chen Ruolin's total score on this dive.

40. **Trades Management**   A business owner is looking for a new commercial space for his shop. An 8000-sq-ft space is advertised for a base price of $1.35 per square foot per month plus additional costs for property taxes and insurance. Suppose that property taxes and insurance are estimated to be $8856 per year.
    (a) What is the cost per month for property taxes and insurance?
    (b) Use your answer to part (a) to calculate the monthly cost per square foot for property taxes and insurance.
    (c) Add your answer to part (b) to the base price to determine the total cost per square foot per month.

41. **Office Services**   As shown in Example 3 on page 55, many public utilities charge customers according to a tier or block system in order to encourage conservation. For example, a certain electric utility divides

a customer's usage of kilowatt-hours (kWh) into five tiers according to the following table:

Tier	Usage	Cost per kWh
1	first 200 kWh	$0.02930
2	next 60 kWh	$0.05610
3	next 140 kWh	$0.16409
4	next 200 kWh	$0.21260
5	remaining kWh	$0.26111

Use this table to calculate the cost of using (a) 350 kWh and (b) 525 kWh. (Round to the nearest cent.)

42. **Sports and Leisure** In the 2012 Summer Olympics, Ally Raisman of the United States won three medals in women's gymnastics. During the all-around competition, her routine on the uneven bars had a maximum value of 15.900 points. The judges assessed her 1.567 points worth of deductions. What was her final score?

43. **Transportation** In 2014 in the South Coast Air Basin of the Los Angeles area, diesel trucks accounted for an average of 129.37 tons per day of nitrogen oxide pollution, while passenger cars accounted for an average of 34.77 tons per day. For the entire year (a non-leap year), how many more tons of this pollutant did diesel trucks create compared to passenger cars? (Source: South Coast Air Quality Management District.)

44. **Transportation** The following table lists both the seating capacities and the number of on-site parking spaces of four large concert venues in the Los Angeles area.

	Staples Center	Hollywood Bowl	The Forum	Nokia Live
Seating capacity	20,000	17,562	17,200	7100
On-site parking	3300	2800	3500	2500

Calculate the number of parking spaces per person for each venue when they are sold out. (*Hint:* This number will be much smaller than 1.) If you were attending a sold-out concert at each of these venues, at which one would you be least likely to find a parking space?

45. **Roofing** A roofer has calculated that 312.6 sq in. of net-free vent area (NFVA) must be provided by the ridge vents. If a certain ridge vent has a rating of 9.0 sq in. of NFVA per linear foot of vent, how many linear feet of venting should there be along the ridges? (Round to the nearest tenth of a foot.)

46. **General Interest** In January 2016, the state of Pennsylvania had the highest gasoline tax in the nation at 68.8 cents per gallon. Alaska had the lowest gasoline tax at 30.65 cents per gallon. If the cost of the fuel itself were the same, how much more would it cost you to fill up an 18-gallon tank in Pennsylvania compared to Alaska? (Express the answer in dollars and cents.)

47. **Life Skills** A certain electric company charges $0.04655 per kilowatt-hour (kWh) for baseline usage and $0.22123 per kWh for tier 5 consumption (usage of more than double the baseline amount). Find the difference between these two charges and round the answer to the nearest cent.

48. **Plumbing**   A plumber replaced a showerhead that used 3.75 gallons per minute (gpm) with one that uses only 1.5 gpm.
    (a) How many gallons will the new showerhead save during an 8-minute shower?
    (b) How many gallons will it save during a 30-day month if a family takes four 8-minute showers per day?
    (c) If 1 cubic foot ≈ 7.48 gallons, how many cubic feet of water will the new head save this family per month? (Round to the nearest hundredth.)
    (d) If water costs $0.05 per cubic foot, how much money will the new showerhead save the family in a year?

49. **Life Skills**   During the severe droughts in California during the 1980s, people were advised to take no more than 3 minutes for each shower.
    (a) Using the new showerhead from the previous problem, how many cubic feet of water per month will the same family save by taking four 3-minute showers per day?
    (b) If water costs $0.05 per cubic foot, how much money will the 3-minute showers save per month in comparison to the 8-minute showers?

50. **Automotive Trades**   Parts (a) and (b) of this problem help us compare the cost of operating a battery-powered electric car to a gas-powered car that averages 25 mi/gal.
    (a) Suppose that the battery-powered vehicle requires approximately 30 kilowatt-hours (kWh) of electricity for 100 miles of driving. If electricity costs $0.11 per kWh, what is the cost to charge the car for 100 miles?
    (b) The gas-powered car requires four gallons of gas for 100 miles of driving. If the average cost of a gallon of gas is $2.50, what is the cost to fuel this car for 100 miles?
    (c) Subtract your answer to part (a) from your answer to part (b) to determine how much you would save per 100 miles by driving the battery-powered car.
    (d) Suppose the battery-powered car was priced at $26,400 and the gas-powered car was priced at $22,800. Using your answer to part (c), determine how many miles of driving it would take for the fuel savings of the battery-powered car to make up for its higher initial cost.

51. **Agriculture**   The table below shows the number of gallons of water required to produce 1 ounce of each type of food or drink. (Source: Mekkonen, M. M., and Hoekstra, A. Y., WaterStat, Water Footprint Network, Enschede, the Netherlands, UNESCO-IHE.)

Food or Drink	Gallons of Water to Produce 1 Ounce
Pork	4.95
Pasta	1.99
Peas	5.34
Avocados	1.09
Mangos	3.42
Apple Juice	0.51 (per fluid ounce)

How many gallons of water are required to produce a meal consisting of 8 ounces of pork, 7 ounces of pasta, 4 ounces of peas, 6 ounces of avocado, 9 ounces of mangos, and 10 fluid ounces of apple juice?

52. **Construction**   Over the course of a building project, a contractor made four purchases of 2-by-4s. The number of board feet purchased and the cost per board foot are shown in the table below. Calculate the average cost per board foot.

Purchase	Number of Board Feet	Cost per Board Foot
1	50	$0.65
2	130	$0.70
3	120	$0.75
4	100	$0.72

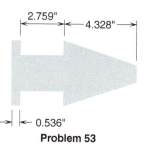

Problem 53

53. **Metalworking**   Find the total length of the piece of steel shown in the figure.

CASE STUDY: Fabricating a Workbench

Dwayne's Welding and Fabrication makes a variety of industrial and consumer products. They have decided to start a new line of workbenches to sell to home improvement retailers. Dwayne's designers hand him the blueprint detailing the frame of the workbench before the top and shelves are added. (See the blueprint.)

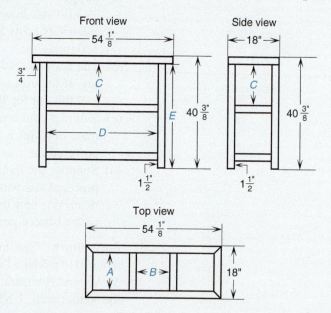

The specs call for $1\frac{1}{4}$-in. × $1\frac{1}{4}$-in. square steel tubing for the frame and $\frac{3}{32}$-in. thick sheet metal for the top and shelves. For pricing and shipping purposes, Dwayne needs to determine both the amount of materials needed and the weight of each workbench.

1. To simplify his calculations, Dwayne decides to convert all fractional dimensions to decimal form. List all fractional dimensions given on the blueprint in decimal form.

2. The designers omitted some crucial dimensions on the blueprint. Fill in the missing dimensions A, B, C, D, and E, assuming that there is equal spacing between shelves and between sections of the frame.

3. Dwayne next needs to determine the total amount of tubing needed to create this workbench. Assume that 18 cuts need to be made and that each cut wastes $\frac{1}{16}$ in. of tubing. What is the total number of inches of tubing needed for each workbench?

4. The fabrication team will be starting with 20-ft lengths of the tubing. How many 20-ft lengths are needed for each workbench? (*Hint:* To convert inches to feet, divide by 12.)

5. After the parts of the frame are welded together, the sheet metal will be added to the top and shelves. Determine the area of these rectangular surfaces by multiplying length by width. (For the shelves, disregard the corners that need to be cut out.) How many total square inches of sheet metal are needed?

6. Finally, for shipping purposes, Dwayne needs to determine the weight of the workbench. The square steel tubing for the frame weighs 0.15 pounds per inch. The sheet metal for the top and shelves weighs 0.026 pounds per square inch. What is the approximate total weight of one workbench? What have we not accounted for that may slightly increase the weight?

Using a Calculator, III: Decimals

To perform addition, subtraction, multiplication, and division of decimal numbers on a calculator, use the same procedures outlined at the end of Chapter 1, and press the $\cdot$ key to enter the decimal point. This key is normally located along the bottom row beneath the "2" key.

EXAMPLE 1

To calculate 243.78 + 196.1 × 2.75, use the keystroke sequence

243 · 78 + 196 · 1 × 2 · 75 = → 783.055

The decimal point key was shown as a special symbol here for emphasis, but in future calculator sequences, we will save space by not showing the decimal point as a separate key.

Notice that, in the preceding calculation, the numbers and operations were entered, and the calculator followed the standard order of operations. •

When you work with decimal numbers, remember that, for a calculator with a ten-digit display, if an answer is a nonterminating decimal or a terminating decimal with more than ten digits, the calculator will display a number rounded to ten digits. Knowledge of rounding is especially important in such cases because the result must often be rounded further.

Your Turn   Work the following problems for practice and round as indicated.

(a) $28.75 + $161.49 − $37.60

(b) 2.8 × 0.85 (Round to the nearest tenth.)

(c) 2347.68 ÷ 12.9 (Round to the nearest whole number.)

(d) 46.8 − 27.3 × 0.49 (Round to the nearest hundredth.)

(e) $\dfrac{16500 + 3700}{12 \times 68}$ (Round to the nearest tenth.)

(f) (3247.9 + 868.7) ÷ 0.816 (Round to the nearest ten.)

(g) 6 ÷ 0.07 × 0.8 + 900 (Round to the nearest hundred.)

Answers

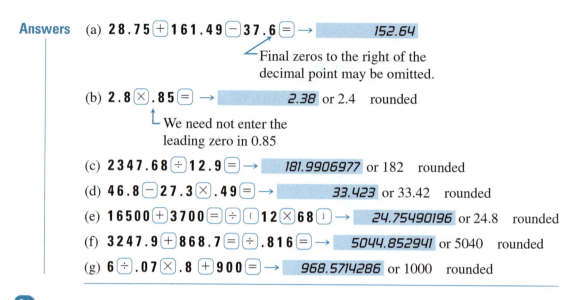

(a) $28.75\boxed{+}161.49\boxed{-}37.6\boxed{=}\rightarrow$ *152.64*

Final zeros to the right of the decimal point may be omitted.

(b) $2.8\boxed{\times}.85\boxed{=}\rightarrow$ *2.38* or 2.4   rounded

We need not enter the leading zero in 0.85

(c) $2347.68\boxed{\div}12.9\boxed{=}\rightarrow$ *181.9906977* or 182   rounded

(d) $46.8\boxed{-}27.3\boxed{\times}.49\boxed{=}\rightarrow$ *33.423* or 33.42   rounded

(e) $16500\boxed{+}3700\boxed{=}\boxed{\div}\boxed{(}12\boxed{\times}68\boxed{)}\rightarrow$ *24.75490196* or 24.8   rounded

(f) $3247.9\boxed{+}868.7\boxed{=}\boxed{\div}.816\boxed{=}\rightarrow$ *5044.852941* or 5040   rounded

(g) $6\boxed{\div}.07\boxed{\times}.8\boxed{+}900\boxed{=}\rightarrow$ *968.5714286* or 1000   rounded

✎ Note   Scientific calculators display decimal numbers between 0 and 1 as they are written in textbooks, with a leading 0 written before the decimal point. As mentioned earlier in the chapter, this helps prevent you from overlooking the decimal point. However, graphing calculators do not normally display this zero, so you must be extra careful when examining results on graphing calculators. ●

If you need to convert a decimal answer to fraction form, or vice versa, look for a second function key labeled $\boxed{F \leftrightarrow D}$. If your calculator does not have this key, simply press the $\boxed{A\frac{b}{c}}$ key to convert back and forth between fraction and decimal form.

EXAMPLE 2   (a) To calculate $\dfrac{2}{3} + \dfrac{4}{5}$ and convert the result to decimal form, enter whichever one of the following sequences works for your calculator:

$2\boxed{A\frac{b}{c}}3\boxed{+}4\boxed{A\frac{b}{c}}5\boxed{=}\boxed{F\leftrightarrow D}\boxed{=}\rightarrow$ *1.466666667*

(For some calculators, the $\boxed{=}$ is not required after $\boxed{F\leftrightarrow D}$.)

or $2\boxed{A\frac{b}{c}}3\boxed{+}4\boxed{A\frac{b}{c}}5\boxed{=}\boxed{A\frac{b}{c}}\rightarrow$ *1.466666667*

(b) To calculate $5.67 - 3.22$ and convert the result to fraction form, enter whichever one of the following sequences works for your calculator:

$5.67\boxed{-}3.22\boxed{=}\boxed{F\leftrightarrow D}\rightarrow$ *2⌴9/20*

or $5.67\boxed{-}3.22\boxed{=}\boxed{A\frac{b}{c}}\rightarrow$ *2⌴9/20*

The sequences shown in the text will use the $\boxed{F\leftrightarrow D}$ key.   ●

Your Turn   Calculate the following and then convert the result to the form indicated:

(a) $3\dfrac{7}{8} \div 1\dfrac{3}{16}$ (decimal, nearest hundredth)

(b) 7.2×4.3 (fraction, mixed number)

Solutions   (a) $3\boxed{A\frac{b}{c}}7\boxed{A\frac{b}{c}}8\boxed{\div}1\boxed{A\frac{b}{c}}3\boxed{A\frac{b}{c}}16\boxed{=}\boxed{F\leftrightarrow D}\boxed{=}\rightarrow$ *3.263157895* ≈ 3.26

(b) $7.2\boxed{\times}4.3\boxed{=}\boxed{F\leftrightarrow D}\boxed{=}\rightarrow$ *30⌴24/25* or $30\dfrac{24}{25}$

Ratio, Proportion, and Percent

Objective	Sample problems	For help, go to page
When you finish this chapter you will be able to:		
1. Calculate ratios.	(a) Find the ratio of the pulley diameters. 	_____ 194
	(b) **Automotive Trades** A small gasoline engine has a maximum cylinder volume of 520 cu cm and a compressed volume of 60 cu cm. Find the compression ratio.	_____ 195
2. Solve proportions.	(a) Solve for x. $$\frac{8}{x} = \frac{12}{15}$$	_____ 198
	(b) Solve for y. $$\frac{4.4}{2.8} = \frac{y}{9.1}$$	_____
3. Solve problems involving proportions.	(a) **Landscaping** A mixture of plant food must be prepared by combining three parts of a concentrate with every 16 parts water. How much water should be added to 12 ounces of concentrate?	_____ 201

Name _____

Date _____

Course/Section _____

Objective	Sample problems	For help, go to page
	(b) **Drafting** An architectural drawing of the living room of a house is $5\frac{1}{2}$ in. long and $4\frac{1}{2}$ in. wide. If the actual length of the living room is 22 ft, what is the actual width? _____	208
	(c) **Manufacturing** A large production job is completed by four machines in 6 hours. The same size job must be finished in 2 hours the next time it is ordered. How many machines should be used to accomplish this task? _____	213
4. Write fractions and decimal numbers as percents.	(a) Write $\frac{1}{4}$ as a percent. _____	223
	(b) Write 0.46 as a percent. _____	222
	(c) Write 5 as a percent. _____	
	(d) Write 0.075 as a percent. _____	
5. Convert percents to decimal numbers and fractions.	(a) Write 35% as a decimal number. _____	225
	(b) Write 0.25% as a decimal number. _____	
	(c) Write 112% as a fraction. _____	226
6. Solve problems involving percent.	(a) Find $37\frac{1}{2}\%$ of 600. _____	228
	(b) Find 120% of 45. _____	
	(c) What percent of 80 is 5? _____	231
	(d) 12 is 16% of what number? _____	232
	(e) If a measurement is given as 2.778 ± 0.025 in., state the tolerance as a percent. _____	252
	(f) **Machine Trades** If a certain kind of solder is 52% tin, how many pounds of tin are needed to make 20 lb of solder? _____	
	(g) **Painting** The paint needed for a redecorating job originally cost $144.75, but is discounted by 35%. What is its discounted price? _____	240

Objective	Sample problems	For help, go to page
	(h) **Machine Trades** A mower motor rated at 2.0 hp is found to deliver only 1.6 hp when connected to a transmission system. What is the efficiency of the transmission? _____	251
	(i) **Electrical Trades** If the voltage in a circuit is increased from 70 volts to 78 volts, what is the percent increase in voltage? _____	254

(Answers to these preview problems are given in the Appendix. Also, worked solutions to many of these problems appear in the chapter Summary.)

If you are certain that you can work *all* these problems correctly, turn to page 268 for a set of practice problems. If you cannot work one or more of the preview problems, turn to the page indicated to the right of the problem. Those who wish to master this material with the greatest success should turn to Section 4-1 and begin work there.

Ratio, Proportion, and Percent

In this chapter, we cover three important and interrelated topics: ratio, proportion, and percent. Examples of ratios in the trades include the compression ratio of an automobile, the gear ratio of a machine, scale drawings, the pitch of a roof, the mechanical advantage of a pulley system, and the voltage ratio in a transformer. Two equal ratios form a proportion, and these special equations are used to solve a wide variety of problems in which three quantities are known and a fourth one is unknown. Percents are especially important in business and consumer applications, but they are also useful in trades and technical work.

CASE STUDY: Hot Wheels

The Case Study for this chapter deals with Mattel's Hot Wheels toy racing cars. Because they are true-scale models of actual classic cars, the use of ratio and proportion is critical in the design of Hot Wheels. As you work through this Case Study, you will learn many interesting facts about Hot Wheels, and you will use many of the mathematical skills covered in this chapter.

4-1 Ratio and Proportion

Ratio A **ratio** is a comparison, using division, of two quantities of the same kind, both usually expressed in the same units. For example, the steepness of a hill can be written as the ratio of its height to its horizontal extent.

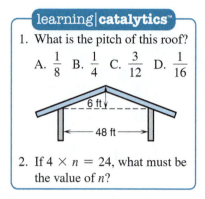

1. What is the pitch of this roof?

A. $\frac{1}{8}$ B. $\frac{1}{4}$ C. $\frac{3}{12}$ D. $\frac{1}{16}$

6 ft

48 ft

2. If $4 \times n = 24$, what must be the value of n?

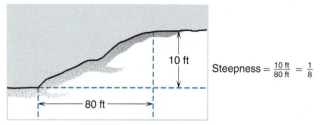

10 ft

80 ft

Steepness $= \frac{10 \text{ ft}}{80 \text{ ft}} = \frac{1}{8}$

The ratio is usually written as a fraction in lowest terms, and you would read this ratio as either "one-eighth" or "one to eight."

The **gear ratio** of a gear system is defined as

$$\text{Gear ratio} = \frac{\text{number of teeth on the driven gear}}{\text{number of teeth on the driving gear}}$$

EXAMPLE 1 **General Trades** Find the gear ratio of the system shown if A is the driven gear and B is the driving gear, and where A has 64 teeth and B has 16 teeth.

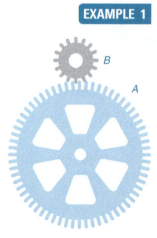

Gear ratio $= ?$

The ratio of the number of teeth on the driven gear to the number of teeth on the driving gear is

$$\text{Gear ratio} = \frac{64 \text{ teeth}}{16 \text{ teeth}} = \frac{4}{1}$$

Always write the fraction in lowest terms.

The gear ratio is 4 to 1. In technical work this is sometimes written as 4:1 and is read "four to one."   ●

Typical gear ratios on a passenger car are

First gear: 3.54:1

Second gear: 1.90:1

Third gear: 1.31:1

Reverse: 3.25:1

A gear ratio of 3.54:1 means that the engine turns 3.54 revolutions for each revolution of the drive shaft. If the drive or rear-axle ratio is 3.72, the engine needs to make 3.54 × 3.72 or approximately 13.2 revolutions in first gear to turn the wheels one full turn.

Here are a few important examples of the use of ratios in practical work.

Pulley Ratios A pulley is a device that can be used to transfer power from one system to another. A pulley system can be used to lift heavy objects in a shop or to connect a power source to a piece of machinery. The ratio of the pulley diameters determines the relative pulley speeds.

EXAMPLE 2 **General Trades** Find the ratio of the diameter of pulley A to the diameter of pulley B in the following drawing.

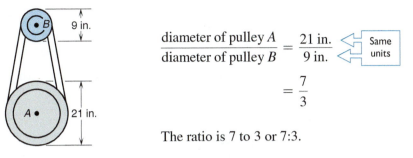

$$\frac{\text{diameter of pulley } A}{\text{diameter of pulley } B} = \frac{21 \text{ in.}}{9 \text{ in.}} \quad \text{Same units}$$

$$= \frac{7}{3}$$

The ratio is 7 to 3 or 7:3.

Notice that the units, inches, cancel from the ratio. A ratio is a fraction or decimal number, and it has no units. ●

Compression Ratio In an automobile engine there is a large difference between the volume of the cylinder space when a piston is at the bottom of its stroke and when it is at the top of its stroke. This difference in volumes is called the *engine displacement*. Automotive mechanics find it useful to talk about the compression ratio of an automobile engine. The **compression ratio** of an engine compares the volume of the cylinder at maximum expansion to the volume of the cylinder at maximum compression.

$$\text{Compression ratio} = \frac{\text{expanded volume}}{\text{compressed volume}}$$

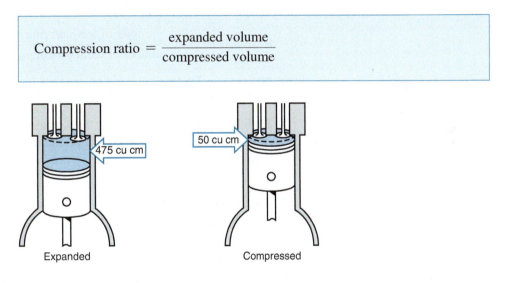

Expanded   Compressed

EXAMPLE 3 **Automotive Trades** Find the compression ratio of a gasoline engine if each cylinder has a maximum volume of 475 cu cm (cubic centimeters) and a minimum or compression volume of 50 cu cm.

$$\text{Compression ratio} = \frac{475 \text{ cu cm}}{50 \text{ cu cm}} = \frac{19}{2}$$

$$= 9.5$$

Compression ratios are always written so that the second number in the ratio is 1. This compression ratio would be written as $9\frac{1}{2}$ to 1. ●

EXAMPLE 4 **Roofing** As we learned in Chapter 2, the pitch of a roof is the ratio of rise to run, where the run is half the span. For the roof shown in the margin, the rise is 7 ft, the span is 48 ft, and therefore

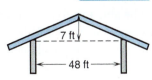

$$\text{Pitch} = \frac{\text{rise}}{\text{run}} = \frac{7 \text{ ft}}{24 \text{ ft}}$$

Roofers usually express pitch with a denominator of 12, so the pitch of this roof is $\frac{3.5}{12}$ or 3.5:12.   ●

Your Turn   Now, for some practice in calculating ratios, work the following problems.

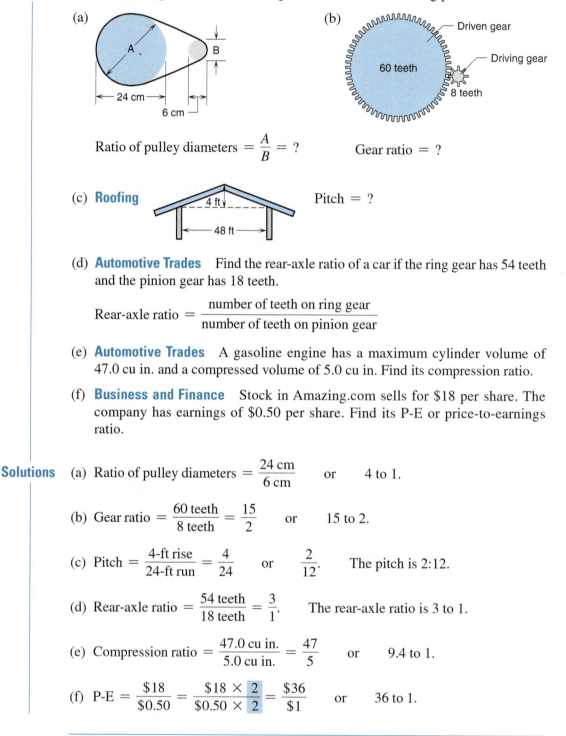

(a)

Ratio of pulley diameters $= \dfrac{A}{B} = ?$

(b)

Gear ratio $= ?$

(c) **Roofing**   Pitch $= ?$

(d) **Automotive Trades**   Find the rear-axle ratio of a car if the ring gear has 54 teeth and the pinion gear has 18 teeth.

$$\text{Rear-axle ratio} = \frac{\text{number of teeth on ring gear}}{\text{number of teeth on pinion gear}}$$

(e) **Automotive Trades**   A gasoline engine has a maximum cylinder volume of 47.0 cu in. and a compressed volume of 5.0 cu in. Find its compression ratio.

(f) **Business and Finance**   Stock in Amazing.com sells for $18 per share. The company has earnings of $0.50 per share. Find its P-E or price-to-earnings ratio.

Solutions   (a) Ratio of pulley diameters $= \dfrac{24\text{ cm}}{6\text{ cm}}$ or   4 to 1.

(b) Gear ratio $= \dfrac{60\text{ teeth}}{8\text{ teeth}} = \dfrac{15}{2}$ or   15 to 2.

(c) Pitch $= \dfrac{4\text{-ft rise}}{24\text{-ft run}} = \dfrac{4}{24}$ or $\dfrac{2}{12}$.   The pitch is 2:12.

(d) Rear-axle ratio $= \dfrac{54\text{ teeth}}{18\text{ teeth}} = \dfrac{3}{1}$.   The rear-axle ratio is 3 to 1.

(e) Compression ratio $= \dfrac{47.0\text{ cu in.}}{5.0\text{ cu in.}} = \dfrac{47}{5}$ or   9.4 to 1.

(f) P-E $= \dfrac{\$18}{\$0.50} = \dfrac{\$18 \times 2}{\$0.50 \times 2} = \dfrac{\$36}{\$1}$ or   36 to 1.

Simple Equations   To use ratios to solve a variety of problems, you must be able to solve a simple kind of algebraic equation. Consider this puzzle: "I'm thinking of a number. When I multiply my number by 3, I get 15. What is my number?" Solve the puzzle and check your solution with ours.

If you answered "five," you're correct. Think about how you worked it out. Most people take the answer 15 and do the "reverse" of multiplying by 3. That is, they divide by 3. In symbols the problem would look like this:

If $3 \times \square = 15$

then $\square = 15 \div 3 = 5$

EXAMPLE 5   Let's try another one: A number multiplied by 40 gives 2200. Find the number.

Using symbols again,

$40 \times \square = 2200$

so $\square = 2200 \div 40 = \dfrac{2200}{40} = 55$ ●

We have just solved a simple algebraic equation. But in algebra, instead of drawing boxes to represent unknown quantities, we use letters of the alphabet. For example, $40 \times \square$ could be written as $40 \times N$. Furthermore, we may signify multiplication in any one of these additional three ways:

- A raised dot   For example, $40 \cdot N$ or $8 \cdot 5$

- Writing a multiplier next to a letter   For example, $40N$

- Enclosing in parentheses one or both   For example, $8(5.6)$
  of the numbers being multiplied (This   or $\left(2\dfrac{1}{3}\right)\left(4\dfrac{5}{8}\right)$
  is especially useful when one or both
  of the numbers are fractions or decimals.)

EXAMPLE 6 $40 \times \square = 2200$

can be written $40 \cdot N = 2200$

or $40N = 2200$

To solve this type of equation, we divide 2200 by the multiplier of N, which is 40. ●

Your Turn   For practice, solve these equations.

(a) $25n = 275$ (b) $12y = 99$

(c) $88 = 8P$ (d) $15 = 30a$

Solutions   (a) $n = \dfrac{275}{25} = 11$ (b) $y = \dfrac{99}{12} = 8.25$

(c) Notice that P, the unknown quantity, is on the *right* side of the equation. You must divide the number by the multiplier of the unknown, so

$P = \dfrac{88}{8} = 11$

(d) Again, a, the unknown quantity, is on the right side, so

$a = \dfrac{15}{30} = 0.5$

! Careful   In problems like (d) some students mistakenly think that they must always divide the larger number by the smaller. That is not always correct. Remember, for an equation of this simple form, always divide by the number that multiplies the unknown. ●

Other types of equations will be solved in Chapter 7, but only these simple equations are needed for our work with proportions and percent.

Proportions   A **proportion** is an equation stating that two ratios are equal. For example,

$$\frac{1}{3} = \frac{4}{12} \quad \text{is a proportion.}$$

Notice that the equation is true because the fraction $\frac{4}{12}$ expressed in lowest terms is equal to $\frac{1}{3}$.

◆ Note   When you are working with proportions, it is helpful to have a way of stating them in words. For example, the proportion

$$\frac{2}{5} = \frac{6}{15} \quad \text{can be stated in words as}$$

"Two is to five as six is to fifteen." ●

When one of the four numbers in a proportion is unknown, it is possible to find the value of that number. In the proportion

$$\frac{1}{3} = \frac{4}{12}$$

notice what happens when we multiply diagonally:

$$\frac{1}{3} = \frac{4}{12} \rightarrow 1 \cdot 12 = 12 \qquad \frac{1}{3} = \frac{4}{12} \rightarrow 3 \cdot 4 = 12$$

These diagonal products are called the **cross-products** of the proportion. If the proportion is a true statement, the cross-products will always be equal. Here are more examples:

$$\frac{5}{8} = \frac{10}{16} \rightarrow 5 \cdot 16 = 80 \text{ and } 8 \cdot 10 = 80$$

$$\frac{9}{12} = \frac{3}{4} \rightarrow 9 \cdot 4 = 36 \text{ and } 12 \cdot 3 = 36$$

$$\frac{10}{6} = \frac{5}{3} \rightarrow 10 \cdot 3 = 30 \text{ and } 6 \cdot 5 = 30$$

This very important fact is called the cross-product rule.

The Cross-Product Rule

$$\text{If } \frac{a}{b} = \frac{c}{d} \quad \text{then } a \cdot d = b \cdot c.$$

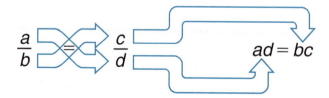

EXAMPLE 7   Find the cross-products for the proportion $\dfrac{3}{5} = \dfrac{12}{20}$.

The cross-products are $3 \cdot 20 = 60$ and $5 \cdot 12 = 60$.   ●

The cross-product rule can be used to solve proportions, that is, to find the value of an unknown number in the proportion.

EXAMPLE 8   Consider the proportion

$$\frac{x}{4} = \frac{12}{16}$$

Solving the proportion means finding the value of the unknown quantity x that makes the equation true. To do this:

First, use the cross-product rule.

If $\dfrac{x}{4} = \dfrac{12}{16}$ then $x \cdot 16 = 4 \cdot 12$ or $16x = 48$

Then, solve this equation using the division technique explained in Examples 5 and 6.

$$x = \frac{48}{16} = 3$$

    Finally, check your answer by replacing x with 3 in the original proportion equation.

$$\frac{3}{4} = \frac{12}{16}$$

Check 1:   Find the cross-products.

$3 \cdot 16 = 48$ and $4 \cdot 12 = 48$ The answer is correct.

Check 2:   Write $\frac{12}{16}$ in lowest terms.

$$\frac{12}{16} = \frac{12 \div 4}{16 \div 4} = \frac{3}{4}$$ ●

This procedure is the same no matter where the unknown is located.

EXAMPLE 9   Let's solve this proportion for N.

$$\frac{6}{N} = \frac{15}{10}$$

Step 1   Apply the cross-product rule.

If $\dfrac{6}{N} = \dfrac{15}{10}$ then $6 \cdot 10 = N \cdot 15$

or $15N = 60$

Step 2   Solve the equation.

$$N = \frac{60}{15} = 4$$

Step 3   Substitute 4 for N in the original proportion.

$$\frac{6}{4} = \frac{15}{10}$$

The cross-products are equal: $6 \cdot 10 = 60$ and $4 \cdot 15 = 60$.

Notice that, in lowest terms,

$$\frac{6}{4} = \frac{3}{2} \quad \text{and} \quad \frac{15}{10} = \frac{3}{2}$$

Note The unknown quantity can appear in any one of the four positions in a proportion. No matter where the unknown appears, solve the proportion the same way: write the cross-products and solve the resulting equation.

Your Turn Here are some problems for you to try. Solve each proportion.

(a) $\dfrac{28}{40} = \dfrac{x}{100}$ (b) $\dfrac{12}{y} = \dfrac{8}{50}$ (c) $\dfrac{12}{9} = \dfrac{32}{M}$

(d) $\dfrac{Y}{7} = \dfrac{3}{4}$ (e) $\dfrac{2\frac{1}{2}}{3\frac{1}{2}} = \dfrac{w}{2}$ (f) $\dfrac{12}{E} = \dfrac{0.4}{1.5}$

Solutions (a) $\dfrac{28}{40} = \dfrac{x}{100}$ $28 \cdot 100 = 40 \cdot x$ or $40x = 2800$

$x = 2800 \div 40$

$x = 70$

(b) $\dfrac{12}{y} = \dfrac{8}{50}$ $600 = 8y$ or $y = \dfrac{600}{8} = 75$

(c) $\dfrac{12}{9} = \dfrac{32}{M}$ $12M = 288$ $M = \dfrac{288}{12} = 24$

(d) $\dfrac{Y}{7} = \dfrac{3}{4}$ $4Y = 21$ $Y = \dfrac{21}{4} = 5\frac{1}{4}$ or 5.25

(e) $\dfrac{2\frac{1}{2}}{3\frac{1}{2}} = \dfrac{w}{2}$ $5 = 3.5w$ $w = \dfrac{5}{3.5} = 1\frac{3}{7}$

(f) $\dfrac{12}{E} = \dfrac{0.4}{1.5}$ $18 = 0.4E$ $E = \dfrac{18}{0.4} = 45$

A Closer Look You may have noticed that the two steps for solving a proportion can be simplified into one step. For example, in problem (a), to solve

$$\frac{28}{40} = \frac{x}{100} \qquad \text{we can write the answer directly as}$$

$$x = \frac{28 \cdot 100}{40}$$

Always divide by the number that is diagonally opposite the unknown, and multiply the two remaining numbers that are diagonally opposite each other.

This shortcut comes in handy when using a calculator. For example, with a calculator, problem (f) becomes

$1\,2\,\boxed{\times}\,1\,.\,5\,\boxed{\div}\,.\,4\,\boxed{=} \rightarrow$ **45.**

The numbers 12 and 1.5 are diagonally opposite each other in the proportion.

0.4 is diagonally opposite the unknown E.

Your Turn Practice this one-step process by solving the proportion

$$\frac{120}{25} = \frac{6}{Q}$$

Solution $Q = \dfrac{25 \cdot 6}{120} = 1.25$ Check it.

Practical Applications We can use proportions to solve a variety of problems involving ratios. If you are given the value of a ratio and one of its terms, it is possible to find the other term.

EXAMPLE 10 **Roofing** If the pitch of a roof is supposed to be 2:12 and the run is 20 ft, what must be the rise?

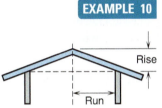

$$\text{Pitch} = \frac{\text{rise}}{\text{run}}$$

A ratio of 2:12 is equivalent to the fraction $\frac{2}{12}$; therefore, the equation becomes

$$\frac{2}{12} = \frac{\text{rise}}{20} \quad \text{or} \qquad \frac{2}{12} = \frac{R}{20}$$

Find the cross-products. $12R = 40$

Then solve for R. $R = \dfrac{40}{12}$

$$R = 3\frac{1}{3}\ \text{ft}$$

The rise is $3\frac{1}{3}$ ft or 3 ft 4 inches. $\left(\dfrac{1}{3}\ \text{ft} = 4\ \text{in.}\right)$ ●

The algebra you learned earlier in this chapter will enable you to solve any ratio problem of this kind. Here is another example.

EXAMPLE 11 **Allied Health** For medications provided in a liquid formulation, the amount of liquid given to a patient depends on the concentration of drug in the fluid. Medication X is available only in a liquid formulation with a concentration of 100 mg/mL (milligrams per milliliter). Suppose a physician orders 225 mg of medication X. To determine the amount of liquid required, first note that 100 mg/mL means that one milliliter of liquid contains 100 mg of the medication. Therefore, we can set up the following proportion:

$$\frac{1\ \text{mL}}{x\ \text{mL}} = \frac{100\ \text{mg}}{225\ \text{mg}}$$

Solving for x, we have $100x = 225$

$$x = 2.25\ \text{mL}$$

Therefore, the patient should take 2.25 mL of the liquid in order to receive 225 mg of medication X.   ●

EXAMPLE 12   **Manufacturing**   The pulley system of an assembly belt has a pulley diameter ratio of 4. If the larger pulley has a diameter of 15 in., what is the diameter of the smaller pulley?

If we set up the pulley ratio as
$$\text{Pulley ratio} = \frac{\text{diameter of larger pulley}}{\text{diameter of smaller pulley}}$$

Then we have
$$4 = \frac{15 \text{ in.}}{D} \qquad \text{or} \qquad \frac{4}{1} = \frac{15}{D}$$

Solving for D:
$$4D = 15$$

$$D = \frac{15}{4} \qquad \text{or} \qquad D = 3\frac{3}{4} \text{ in.} \qquad \bullet$$

Your Turn   Try these problems.

(a) **Manufacturing**   If the gear ratio on a mixing machine is 6:1 and the driving gear has 12 teeth, how many teeth are on the driven gear?

(b) **Automotive Trades**   The compression ratio of a classic Datsun 280Z is 8.3 to 1. If the compressed volume of the cylinder is 36 cu cm, what is the expanded volume of the cylinder?

(c) **Construction**   On a certain construction job, concrete is made using a volume ratio of 1 part cement to $2\frac{1}{2}$ parts sand and 4 parts gravel. How much sand should be mixed with 3 cu ft of cement?

(d) **Sports and Leisure**   Manager Sparky Spittoon of the Huntville Hackers wants his ace pitcher Lefty Groove to improve his strikeouts-to-walks ratio to at least 5:2. If he has 65 strikeouts so far this season, what is the maximum number of walks he can give up and still keep Sparky happy?

(e) **Roofing**   If the pitch of a roof must be 5:12, how much rise should there be over a run of 54 ft?

(f) **Allied Health**   Medication TTQ is available only in a liquid formulation with a concentration of 125 mg/5 mL (milligrams per 5 milliliters). How much of this liquid formulation should be given to a patient if 500 mg of TTQ is ordered by the physician?

Solutions   (a) Gear ratio $= \dfrac{\text{number of teeth on driven gear}}{\text{number of teeth on driving gear}}$

A ratio of 6:1 is equivalent to the fraction $\dfrac{6}{1}$.

Therefore,
$$\frac{6}{1} = \frac{x}{12}$$

or $72 = x$ The driven gear has 72 teeth.

(b) Compression ratio $= \dfrac{\text{expanded volume}}{\text{compressed volume}}$

$$8.3 = \frac{V}{36 \text{ cu cm}} \qquad \text{or} \qquad \frac{83}{10} = \frac{V}{36}$$

$$V = \frac{83 \cdot 36}{10} = 298.8 \text{ cu cm}$$

83 ⊠ 36 ⊟ 10 ⊜ → ⬚ *298.8.*

In any practical situation, this answer would be rounded to 300 cu cm.

(c) Ratio of cement to sand $= \dfrac{\text{volume of cement}}{\text{volume of sand}}$

$$\frac{1}{2\frac{1}{2}} = \frac{3 \text{ cu ft}}{S}$$

Therefore,

$$S = 3(2\tfrac{1}{2})$$

$$S = 7\tfrac{1}{2} \text{ cu ft}$$

(d) $\dfrac{\text{Strikeouts}}{\text{Walks}} = \dfrac{5}{2} = \dfrac{65}{w}$

$$5w = 130$$

$$w = 26 \qquad \text{Lefty should give up no more than 26 walks.}$$

(e) Pitch $= \dfrac{\text{rise}}{\text{run}}$

$$\frac{5}{12} = \frac{R}{54}$$

$$12R = 270$$

$$R = 22.5 \text{ ft}$$

(f) **First,** set up a proportion. $\dfrac{5 \text{ mL}}{x \text{ mL}} = \dfrac{125 \text{ mg}}{500 \text{ mg}}$

Then, solve. $125x = 2500$

$$x = 20 \text{ mL}$$

Therefore, 20 milliliters of liquid should be given to the patient.

Now turn to Exercises 4-1 for more practice on ratio and proportion.

Exercises 4-1   Ratio and Proportion

A. Complete the following tables.

1.

	Teeth on Driven Gear A	Teeth on Driving Gear B	Gear Ratio, $\dfrac{A}{B}$
(a)	35	5	
(b)	12	7	
(c)		3	2 to 1
(d)	21		$3\tfrac{1}{2}$ to 1
(e)	15		1 to 3
(f)		18	1 to 2
(g)		24	2:3
(h)	30		3:5
(i)	27	18	
(j)	12	30	

2.

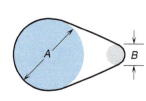

	Diameter of Pulley A	Diameter of Pulley B	Pulley Ratio, $\dfrac{A}{B}$
(a)	16 in.	6 in.	
(b)	15 in.	12 in.	
(c)		8 in.	2 to 1
(d)	27 cm		4.5 to 1
(e)		10 cm	4 to 1
(f)	$8\frac{1}{8}$ in.	$3\frac{1}{4}$ in.	
(g)	8.46 cm	11.28 cm	
(h)	20.41 cm		3.14 to 1
(i)		12.15 cm	1 to 2.25
(j)	4.45 cm		0.25 to 1

3.

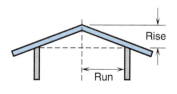

	Rise	Run	Pitch
(a)	8 ft	6 ft	
(b)		24 ft	4:12
(c)	7 ft		3:12
(d)	14 ft 4 in.	25 ft	
(e)	9 ft	15 ft	
(f)		20 ft	2.4:12
(g)	3 ft		1.8:12
(h)		30 ft 6 in.	2:12

B. Solve these proportion equations.

1. $\dfrac{3}{2} = \dfrac{x}{8}$

2. $\dfrac{6}{R} = \dfrac{5}{72}$

3. $\dfrac{y}{60} = \dfrac{5}{3}$

4. $\dfrac{2}{15} = \dfrac{8}{H}$

5. $\dfrac{5}{P} = \dfrac{30}{7}$

6. $\dfrac{1}{6} = \dfrac{17}{x}$

7. $\dfrac{A}{2.5} = \dfrac{13}{10}$

8. $\dfrac{27}{M} = \dfrac{3}{0.8}$

9. $\dfrac{2}{5} = \dfrac{T}{4.5}$

10. $\dfrac{0.12}{N} = \dfrac{2}{7}$

11. $\dfrac{138}{23} = \dfrac{18}{x}$

12. $\dfrac{3.25}{1.5} = \dfrac{A}{0.6}$

13. $\dfrac{x}{34.86} = \dfrac{1.2}{8.3}$

14. $\dfrac{2\frac{1}{2}}{R} = \dfrac{1\frac{1}{4}}{3\frac{1}{4}}$

15. $\dfrac{2\text{ ft }6\text{ in.}}{4\text{ ft }3\text{ in.}} = \dfrac{L}{8\text{ ft }6\text{ in.}}$

16. $\dfrac{6.2\text{ cm}}{x} = \dfrac{1.2\text{ in.}}{11.4\text{ in.}}$

17. $\dfrac{3\text{ ft }4\text{ in.}}{4\text{ ft }2\text{ in.}} = \dfrac{3.2\text{ cm}}{x}$

18. $\dfrac{3\frac{1}{2}\text{ in.}}{W} = \dfrac{1.4}{0.05}$

C. Solve.

1. **Automotive Trades**   The compression ratio in a certain engine is 9.6 to 1. If the expanded volume of a cylinder is 48 cu in., what is the compressed volume?

2. **Painting** A certain paint used for wood siding covers 420 sq ft per gallon when applying the first coat, 520 sq ft per gallon with the second coat, and 620 sq ft per gallon with the third coat. How many gallons are needed for 940 sq ft of siding if three coats are applied? (Assume that you cannot buy a fraction of a gallon.)

3. **Machine Trades** If 28 tapered pins can be machined from a steel rod 12 ft long, how many tapered pins can be made from a steel rod 9 ft long?

4. **Carpentry** If 6 lb of nails are needed for each thousand lath, how many pounds are required for 4250 lath?

5. **Masonry** For a certain kind of plaster work, 1.5 cu yd of sand are needed for every 100 sq yd of surface. How much sand will be needed for 350 sq yd of surface?

6. **Printing** The paper needed for a printing job weighs 12 lb per 500 sheets. How many pounds of paper are needed to run a job requiring 12,500 sheets?

7. **Machine Trades** A cylindrical oil tank 8 ft deep holds 420 gallons when filled to capacity. How many gallons remain in the tank when the depth of oil is $5\frac{1}{2}$ ft?

8. **Agriculture** A liquid fertilizer must be prepared by using one part of concentrate for every 32 parts of water. How much water should be added to 7 oz of concentrate?

9. **Welding** A 10-ft bar of I-beam weighs 208 lb. What is the weight of a 6-ft length?

10. **Photography** A photographer must mix a chemical in the ratio of 1 part chemical for every 7 parts of water. How many ounces of chemical should be used to make a 3-qt *total* mixture? (1 qt $=$ 32 oz)

11. **General Trades** If you earn $684.80 for a 32-hour work week, how much would you earn for a 40-hour work week at the same hourly rate?

12. **Machine Trades** A machinist can produce 12 parts in 40 minutes. How many parts can the machinist produce in 4 hours?

13. **Automotive Trades** The headlights on a car are set so the light beam drops 2 in. for each 25 ft measured horizontally. If the headlights are mounted 30 in. above the ground, how far ahead of the car will they hit the ground?

14. **Roofing** For a given depth, the breadth of a beam needed to support a roof is directly proportional to the width of the roof. If a 4-in. beam will support a 48-ft wide section of roof, what breadth is needed to support a 72-ft wide section of roof?

15. **Agriculture** To prepare a pesticide spray, 3.5 lb of BIOsid is added to 30 gal of water. How much BIOsid should be added to a spray tank holding 325 gal? (Round to the nearest 0.1 lb.)

16. **Painting** A painter must thin some paint for use in a sprayer. If the recommended rate is $\frac{1}{2}$ pint of water per gallon of paint, how many pints of water should be added to $5\frac{1}{2}$ gallons of paint?

17. **Allied Health** The label on a concentrated drug solution indicates that it contains 85 mg of medication in 5 mL. If the patient is to receive 220 mg of medication, how much of the solution should be given? (Round to the nearest tenth.)

18. **Automotive Trades**   In winter weather, fuel-line antifreeze must be added at a rate of one can per 8 gal of fuel. How many cans should be added for an 18-gal fuel tank?

19. **Automotive Trades**   A Chevy Cruze Eco costs $10.36 per 100 miles to drive, while a conventional Chevy Volt plug-in hybrid electric costs $6.63 per 100 miles to drive. How much more would the Chevy Cruze cost to drive 580 miles? (Note: Costs per mile are based on California fuel prices in the summer of 2015.)

20. **Automotive Trades**   The air-to-fuel ratio of an engine helps determine how efficiently the engine is running. In most cases, a ratio of 14.7:1 is ideal. A larger ratio indicates that the engine is running *lean,* while a smaller ratio indicates that it is running *rich.* Suppose a certain engine draws 160 lb of air in burning 12 lb of fuel. Find the air-to-fuel ratio, and state whether the engine is running lean or rich.

21. **Automotive Trades**   In problem 20, it was stated that the ideal air-to-fuel ratio for an engine is 14.7:1. If a vehicle burns 9 lb of fuel, how many pounds of air should it draw to achieve the ideal ratio? (Round to the nearest pound.)

22. **Sports and Leisure**   Power hitter Sammy Sockitome strikes out too much to suit manager Sparky Spittoon. Sammy is promised an incentive bonus of a million dollars if he can reduce his strikeout–to–home run ratio to 3:1. If Sammy hits 56 home runs, what is the maximum number of strikeouts he can have and still earn the bonus?

23. **Landscaping**   A landscape architect is seeding a 6000-sq-ft lawn. If the seed manufacturer recommends using 6 lb of seed per 1500 sq ft, how many pounds of seed will be needed?

24. **Masonry**   A mason needs to purchase mortar mix for a retaining wall consisting of 640 blocks. The guidelines for this mix suggest that 12 bags are required for every 100 blocks. Assuming that he cannot purchase part of a bag, how many bags will he need?

25. **Allied Health**   Medication Q is available only in a liquid form with a concentration of 30 µg/mL (micrograms per milliliter). Determine how much of the liquid should be given to a patient when the following amounts of medication Q are ordered by the physician:

(a) 15 µg   (b) 10 µg   (c) 200 µg

(Round to two decimal digits if necessary.)

26. **Construction**   A rich mix of concrete requires 1.5 cu ft of sand and 3 cu ft of gravel for every bag of cement. This will produce 3.5 cu ft of concrete. If a job requires 42 cu ft of concrete, what amounts of cement, sand, and gravel are needed?

27. **Allied Health**   Each tablet of medication Z contains 50 micrograms (µg) of drug. Determine how many tablets (or fractions of tablets) should be given when the following amounts of medication Z are ordered by the physician:

(a) 100 µg   (b) 75 µg   (c) 230 µg

(Round to the nearest half-tablet.)

28. **Construction**   A building code requires one square foot (sq ft) of net-free vent area (NFVA) for every 300 sq ft of attic space. How many square feet of NFVA are required for a 1620-sq-ft attic?

29. **Agriculture**   A 12-inch-wide conveyor unloads 5000 bushels of wheat per hour. At this speed, how many bushels per hour will an 18-inch-wide conveyor unload?

30. **Trades Management**   A business owner is currently renting her shop space for $9450 per month. Her lease agreement states that her rent will be adjusted each July according to the CPI (Consumer Price Index). Specifically, the ratio of this June's CPI to the previous June's CPI will be multiplied by the current monthly rent to determine the new monthly rent. If this June's CPI is 241.038 and last June's CPI was 238.638, what will be her new monthly rent beginning in July? (Round your answer to the nearest dollar.)

31. **Trades Management**   A small welding shop employs four welders and one secretary. A workers' compensation insurance policy charges a premium of $10.59 per $100 of gross wages for the welders and $1.38 per $100 of gross wages for the secretary. If each welder earns $36,000 per year and the secretary earns $28,000 per year, what is the total annual premium for this insurance?

32. **Construction**   On a crisp fall day, a builder wants to drive from his house to a work site in the mountains to pour the concrete foundations for a cabin. The minimum temperature at which the concrete will set with adequate strength is 40°F. The temperature at his house is 60°F, and the cabin is at an altitude that is 5900 ft higher than the town where the builder lives. If temperature decreases by about 4°F for every 1000 ft of altitude increase, should the builder bother to drive to the work site?

33. **General Trades**   A trades worker is considering a job offer in another city. Her current job pays $2950 per month and is in a city with a cost of living index of 98.3. The cost of living index in the new location is 128.5. If she were to maintain her present lifestyle, how much would she need to earn per month in the new job? (*Hint:* Cost of living must be directly proportional to salary).

34. **Automotive Trades**   Suppose the fuel cost of a certain gas-powered car is $12.92 more per 100 miles than the electrical cost of a battery-powered car. If the cost of the battery-powered car was $8000 more than the gas-powered car, how many miles of driving does it take to make up for this additional cost? (Round up to the next thousand.)

35. **Allied Health**   The dosages of certain medications are based on a patient's body surface area (BSA) as measured in square meters. Suppose a physician prescribes a daily dosage of 250 mg/m^2 (milligrams per square meter) of medication D for a patient. If the patient has a BSA of 1.53 m^2, how many 60-mg capsules should the patient take per day? (Round down to the nearest whole number.)

36. **Transportation**   It is estimated that shipping by rail produces 6.4 pounds of greenhouse gas emissions per ton of cargo for every 100 miles shipped. How many pounds of such emissions would a freight train produce if it shipped 8.4 tons of cargo 480 miles? (Round to the nearest pound.)

37. **Painting**   A painting contractor knows from experience that sanding trim requires 0.80–1.00 hours of labor per 100 linear feet, and that staining that same trim requires 0.50–0.75 hours of labor per 100 linear feet. What would be the minimum and maximum number of total hours of labor required for sanding and staining 360 linear feet of trim? (Round each answer to the nearest half-hour.)

Check your answers to the odd-numbered problems in the Appendix, then turn to Section 4-2.

4-2 Special Applications of Ratio and Proportion

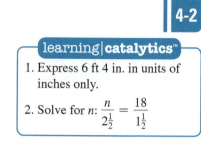

learning|catalytics™

1. Express 6 ft 4 in. in units of inches only.

2. Solve for n: $\dfrac{n}{2\frac{1}{2}} = \dfrac{18}{1\frac{1}{2}}$

In the previous section, we used the concepts of ratio and proportion to solve some simple applications. In this section we will learn about additional applications of ratio and proportion in the trades and technical areas.

Scale Drawings   Proportion equations are found in a wide variety of practical situations. For example, when a drafter makes a drawing of a machine part, building layout, or other large structure, he or she must *scale it down*. The drawing must represent the object accurately, but it must be small enough to fit on the paper. The draftsperson reduces every dimension by some fixed ratio.

EXAMPLE 1   **Drafting**   Drawings that are larger than life involve an expanded scale.

For the automobile shown, the ratio of the actual length to the scale-drawing length is equal to the ratio of the actual width to the scale-drawing width.

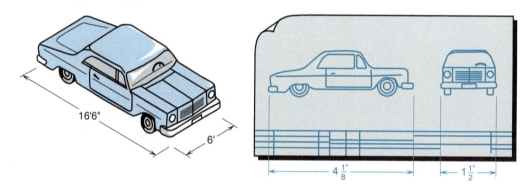

$$\frac{\text{actual length}}{\text{drawing length}} = \frac{\text{actual width}}{\text{drawing width}}$$

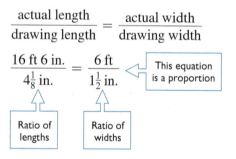

$$\frac{16 \text{ ft } 6 \text{ in.}}{4\frac{1}{8} \text{ in.}} = \frac{6 \text{ ft}}{1\frac{1}{2} \text{ in.}} \quad \Longleftarrow \quad \boxed{\text{This equation is a proportion}}$$

Ratio of lengths   Ratio of widths

Rewrite all quantities in the same units (1 ft = 12 in.):

$$\frac{198 \text{ in.}}{4\frac{1}{8} \text{ in.}} = \frac{72 \text{ in.}}{1\frac{1}{2} \text{ in.}}$$

You should notice first of all that each side of this equation is a ratio. Each side is a ratio of *like* quantities: lengths on the left and widths on the right.

Second, notice that the ratio $\dfrac{198 \text{ in.}}{4\frac{1}{8} \text{ in.}}$ is equal to $\dfrac{48}{1}$.

Divide it out: $198 \div 4\frac{1}{8} = 198 \div \dfrac{33}{8}$

$$= \frac{198}{1} \times \frac{8}{33}$$

$$= 48$$

198 ⊠ **8** ÷ **33** ⊟ →   ⟨ 48. ⟩

Notice also that the ratio $\dfrac{72 \text{ in.}}{1\frac{1}{2}\text{ in.}}$ is equal to $\dfrac{48}{1}$.

The common ratio $\dfrac{48}{1}$ is called the **scale factor** of the drawing.   ●

EXAMPLE 2   **Architecture**   In this problem, one of the dimensions is unknown.

Suppose that a rectangular room has a length of 18 ft and a width of 12 ft. An architectural scale drawing of this room is made so that, on the drawing, the length of the room is 4.5 in. What will be the width of the room on the drawing? What is the scale factor?

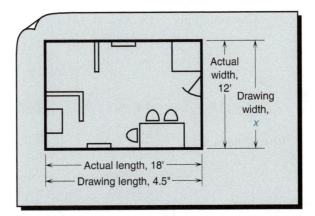

First, set up a ratio of lengths and a ratio of widths.

Length ratio $= \dfrac{\text{actual length}}{\text{drawing length}}$

$= \dfrac{18 \text{ ft}}{4.5 \text{ in.}} = \dfrac{216 \text{ in.}}{4.5 \text{ in.}}$ ← Convert 18 ft to inches so that the top and the bottom of the fraction have the same units.

Width ratio $= \dfrac{\text{actual width}}{\text{drawing width}}$

$= \dfrac{12 \text{ ft}}{x \text{ in.}} = \dfrac{144 \text{ in.}}{x \text{ in.}}$ ← Change 12 ft to 144 in.
← Let $x =$ the drawing width

Second, write a proportion equation.

$\dfrac{216}{4.5} = \dfrac{144}{x}$

Third, solve this proportion. Cross-multiply to get

$216x = (4.5)144$

$216x = 648$

$x = 3 \text{ in.}$ The width of the room on the drawing is 3 in.

For the room drawing shown, the scale factor is

Scale factor $= \dfrac{18 \text{ ft}}{4.5 \text{ in.}} = \dfrac{216 \text{ in.}}{4.5 \text{ in.}}$

$= 48$ or   48 to 1

One inch on the drawing corresponds to 48 in. or 4 ft on the actual object. We can express this as 1 in. = 4 ft. A draftsperson would divide by 4 and write it as $\frac{1}{4}$ in. = 1 ft.　●

Your Turn

Drafting　The drawing of the triangular plate shown has a height of 18 in. and a base length of 14 in.

(a) What will be the corresponding height in a reduced copy of this drawing if the base length in the copy is $2\frac{3}{16}$ in.?

(b) Find the scale factor of the reduction.

Solutions　(a)

$$\frac{18 \text{ in.}}{x} = \frac{14 \text{ in.}}{2\frac{3}{16} \text{ in.}}$$

Cross-multiply:　$18\left(2\frac{3}{16}\right) = 14x$

$$14x = 39\frac{3}{8}$$

$$x = 2\frac{13}{16} \text{ in.}$$

2 $A\frac{b}{c}$ 3 $A\frac{b}{c}$ 1 6 ✕ 1 8 ÷ 1 4 = →　[2⌴13/16] F↔D = →　[2.8125]

(b)　Scale factor = $\dfrac{18 \text{ in.}}{2\frac{13}{16} \text{ in.}}$ 　or　$\dfrac{14 \text{ in.}}{2\frac{3}{16} \text{ in.}}$

$= 6\frac{2}{5}$ 　or　6.4 to 1

1 8 ÷ 2 $A\frac{b}{c}$ 1 3 $A\frac{b}{c}$ 1 6 = →　[6⌴2/5] F↔D = →　[6.4]

Similar Figures　In general, two geometric figures that have the same shape but are not the same size are said to be **similar figures**. The blueprint drawing and the actual object are a pair of similar figures. An enlarged photograph and the smaller original are similar.

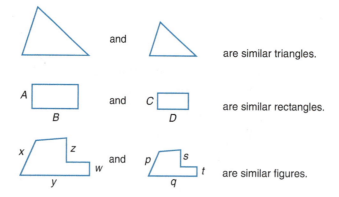

and are similar triangles.

A and C are similar rectangles.

and are similar figures.

In any similar figures, all parts of corresponding dimensions have the same scale ratio. For example, in the preceding rectangles

$$\frac{A}{C} = \frac{B}{D}$$

In the irregular figures above,

$$\frac{x}{p} = \frac{y}{q} = \frac{z}{s} = \frac{w}{t} \text{ and so on}$$

The triangles shown here are *not* similar:

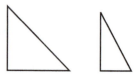

EXAMPLE 3   Suppose we know that the two cylinders pictured below are similar. To find the missing altitude H, we set up a similarity proportion and solve as follows:

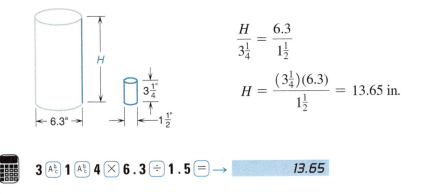

$$\frac{H}{3\frac{1}{4}} = \frac{6.3}{1\frac{1}{2}}$$

$$H = \frac{(3\frac{1}{4})(6.3)}{1\frac{1}{2}} = 13.65 \text{ in.}$$

3 $\boxed{A\frac{b}{c}}$ **1** $\boxed{A\frac{b}{c}}$ **4** $\boxed{\times}$ **6.3** $\boxed{\div}$ **1.5** $\boxed{=}$ →   *13.65*

EXAMPLE 4   **Landscaping**   A landscaper is designing a garden for a house that is not yet built. He needs to determine how long a shadow the house will cast into the garden area. To determine this, the landscaper, who is 6 ft tall, measures his shadow to be 8 ft long at a particular time on a summer afternoon.

If the roofline of the house will be 15 ft high where the landscaper was standing, how far from the edge of the house will its shadow extend at the same time of day?

To solve this problem we draw a sketch of the situation:

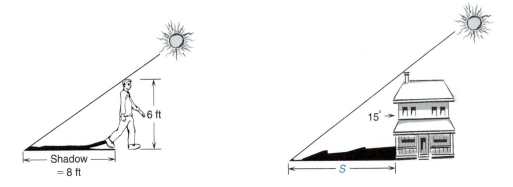

At the same time of day, the triangles formed by the objects, the sun's rays, and the shadows are similar. Therefore,

$$\frac{6 \text{ ft}}{15 \text{ ft}} = \frac{8 \text{ ft}}{S \text{ ft}}$$

$$S = \frac{15(8)}{6} = 20 \text{ ft}$$

The shadow cast by the house will extend 20 ft into the garden from the edge of the house.   ●

Your Turn   Find the missing dimension in each of the following pairs of similar figures.

(a)

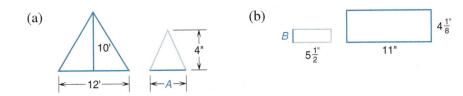

(c)

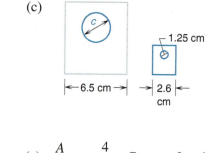

1.25 cm

|← 6.5 cm →| →| 2.6 |←
cm

(d) **Landscaping** If a 5′6″ tall person casts a 3′3″ shadow, how tall is a tree that casts a 26′ shadow at the same time of day?

Solutions (a) $\dfrac{A}{144} = \dfrac{4}{120}$ Convert ft to in.

$A = \dfrac{144(4)}{120} = 4.8$ in.

(b) $\dfrac{B}{4\frac{1}{8}} = \dfrac{5\frac{1}{2}}{11}$

$B = \dfrac{(4\frac{1}{8})(5\frac{1}{2})}{11} = 2\frac{1}{16}$ in.

(c) $\dfrac{c}{1.25} = \dfrac{6.5}{2.6}$

$c = \dfrac{1.25(6.5)}{2.6} = 3.125$ cm

(d) **First,** convert 5′6″ to 5.5′ and 3′3″ to 3.25′.

Then, set up a proportion based on Example 4 to solve for the height, H, of the tree.

$\dfrac{5.5}{H} = \dfrac{3.25}{26}$

$H = \dfrac{5.5(26)}{3.25} = 44$ ft

Direct Proportion Many trade problems can be solved by setting up a proportion involving four related quantities. But it is important that you recognize that there are *two* kinds of proportions—direct and inverse. Two quantities are said to be **directly proportional** if an increase in one quantity leads to a proportional increase in the other quantity, or if a decrease in one leads to a decrease in the other.

> **Direct proportion:**   increase → increase
> or   decrease → decrease

EXAMPLE 5   **Electrical Trades**   The electrical resistance of a wire is directly proportional to its length—the longer the wire, the greater the resistance. If 1 ft of Nichrome heater element wire has a resistance of 1.65 ohms, what length of wire is needed to provide a resistance of 19.8 ohms?

Resistance = 1.65 ohms
|← 1 ft →|

Resistance = 19.8 ohms
|←   L   →|

First, recognize that this problem involves a *direct* proportion. As the length of wire increases, the resistance increases proportionally.

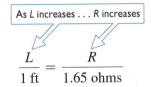

$$\frac{L}{1 \text{ ft}} = \frac{R}{1.65 \text{ ohms}}$$

Both ratios increase in size when L increases.

Second, set up a direct proportion and solve.

$$\frac{L}{1 \text{ ft}} = \frac{19.8 \text{ ohms}}{1.65 \text{ ohms}}$$

$$L = 12 \text{ ft}$$

●

Your Turn   Solve each of the following problems by setting up a direct proportion.

(a) **Painting**   If 1 gal of paint covers 825 sq ft, how much paint is needed to cover 2640 sq ft?

(b) **Transportation**   A diesel truck was driven 273 miles on 42 gal of fuel. How much fuel is needed for a trip of 600 miles? (Round to the nearest gallon.)

(c) **Industrial Technology**   A cylindrical oil tank holds 450 gal when it is filled to its full height of 8 ft. When it contains oil to a height of 2 ft 4 in., how many gallons of oil are in the tank?

Solutions   (a) A direct proportion—the more paint, the greater the area that can be covered:

$$\frac{1 \text{ gal}}{x \text{ gal}} = \frac{825 \text{ sq ft}}{2640 \text{ sq ft}}$$

$$x = 3.2 \text{ gal}$$

(b) A direct proportion—the more miles you drive, the more fuel it takes:

$$\frac{273 \text{ mi}}{600 \text{ mi}} = \frac{42 \text{ gal}}{x \text{ gal}}$$

$$x \approx 92 \text{ gallons} \quad (\textit{Reminder:} \approx \text{ means "approximately equal to"})$$

(c) A direct proportion—the volume is directly proportional to the height:

$$\frac{450 \text{ gal}}{x \text{ gal}} = \frac{8 \text{ ft}}{2\frac{1}{3} \text{ ft}} \quad \Longleftarrow \boxed{2'\ 4'' = 2\frac{4}{12}' = 2\frac{1}{3}'}$$

$$x = 131\frac{1}{4} \text{ gallons}$$

Inverse Proportion   Two quantities are said to be **inversely proportional** if an increase in one quantity leads to a proportional decrease in the other quantity, or if a decrease in one leads to an increase in the other.

> **Inverse Proportion:**   increase $\longrightarrow$ decrease
>
>   or   decrease $\longrightarrow$ increase

For example, the time required for a trip of a certain length is *inversely* proportional to the speed of travel. The faster you go (*increase* in speed), the quicker you get there (*decrease* in time).

EXAMPLE 6   If a certain trip takes 2 hours at 50 mph, how long will it take at 60 mph?

The correct proportion equation is

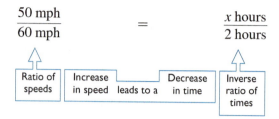

$$\frac{50 \text{ mph}}{60 \text{ mph}} = \frac{x \text{ hours}}{2 \text{ hours}}$$

| Ratio of speeds | Increase in speed leads to a Decrease in time | Inverse ratio of times |

By inverting the time ratio, we have set it up so that both sides of the equation are in balance—both ratios decrease as speed increases.

Before attempting to solve the problem, make an estimate of the answer. We expect that the time to make the trip at 60 mph will be *less* than the time at 50 mph. The correct answer should be less than 2 hours.

Now solve the proportion by cross-multiplying:

$$60x = 50 \cdot 2$$

$$x = 1\tfrac{2}{3} \text{ hours}$$ ●

Your Turn   **Automotive Trades**   In an automobile cylinder, the pressure is inversely proportional to the volume if the temperature does not change. If the volume of gas in the cylinder is 300 cu cm when the pressure is 20 psi, what is the volume when the pressure is increased to 80 psi?

Set this up as an inverse proportion and solve.

Solution   Pressure is inversely proportional to volume. If the pressure increases, we expect the volume to decrease. The answer should be less than 300 cu cm.

Set up an inverse proportion. $$\frac{P_1}{P_2} = \frac{V_2}{V_1}$$

Substitute the given information. $$\frac{20 \text{ psi}}{80 \text{ psi}} = \frac{V}{300 \text{ cu cm}}$$

Cross-multiply. $$80V = 20 \cdot 300 = 6000$$

$$V = 75 \text{ cu cm}$$

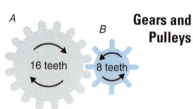

A   B
16 teeth   8 teeth

Gears and Pulleys   A particularly useful kind of inverse proportion involves the relationship between the size of a gear or pulley and the speed with which it rotates.

In the figure in the margin, B is the driving gear.

Because gear A has twice as many teeth as gear B, when B makes two turns, A makes one turn. If gear B rotates at 20 turns per second, gear A will rotate at 10 turns per second. The speed of the gear is inversely proportional to the number of teeth.

$$\frac{\text{speed of gear } A}{\text{speed of gear } B} = \frac{\text{teeth in gear } B}{\text{teeth in gear } A}$$

| This ratio has the A term on top | This ratio has the B term on top |

In this proportion, gear speed is measured in revolutions per minute, abbreviated rpm.

EXAMPLE 7 For the gear assembly shown on the previous page, if gear *A* turns at 40 rpm, what is the speed of gear *B*?

Because the relation is an inverse proportion, the smaller gear moves with the greater speed. We expect the speed of gear *B* to be faster than 40 rpm.

$$\frac{40 \text{ rpm}}{B} = \frac{8 \text{ teeth}}{16 \text{ teeth}}$$

$$8B = 40(16) = 640$$

$$B = 80 \text{ rpm}$$

On an automobile, the speed of the drive shaft is converted to rear-axle motion by the ring and pinion gear system.

$$\frac{\text{drive shaft speed}}{\text{rear-axle speed}} = \frac{\text{teeth in ring gear on axle}}{\text{teeth in pinion gear on drive shaft}}$$

| This ratio has the drive shaft term on top | This ratio has the drive shaft term on the bottom |

Again, gear speed is inversely proportional to the number of teeth on the gear.

Your Turn **Automotive Trades** If the pinion gear has 9 teeth and the ring gear has 40 teeth, what is the rear-axle speed when the drive shaft turns at 1200 rpm?

Solution $$\frac{1200 \text{ rpm}}{R} = \frac{40 \text{ teeth}}{9 \text{ teeth}}$$

Cross-multiply: $40R = 1200 \cdot 9 = 10{,}800$

$$R = 270 \text{ rpm}$$

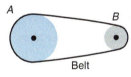

Pulleys transfer power in much the same way as gears. For the pulley system shown in the margin, the speed of a pulley is inversely proportional to its diameter.

Pulley *A* has a diameter twice that of pulley *B*. When pulley *A* makes one turn, pulley *B* will make two turns, assuming, of course, that there is no slippage of the belt.

$$\frac{\text{speed of pulley } A}{\text{speed of pulley } B} = \frac{\text{diameter of pulley } B}{\text{diameter of pulley } A}$$

| This ratio has the *A* term on top | This ratio has the *A* term on the bottom |

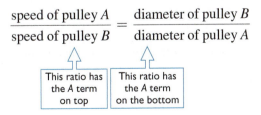

EXAMPLE 8 If pulley *B* is 16 in. in diameter and is rotating at 240 rpm, what is the speed of pulley *A* if its diameter is 20 in.?

To solve, we set up the following inverse proportion:

$$\frac{A}{240 \text{ rpm}} = \frac{16 \text{ in.}}{20 \text{ in.}}$$

Cross-multiply: $20A = 240 \cdot 16 = 3840$

$$A = 192 \text{ rpm}$$

Your Turn   Solve each of the following problems by setting up an inverse proportion.

(a) **Machine Trades**   A 9-in. pulley on a drill press rotates at 960 rpm. It is belted to a 5-in. pulley on an electric motor. Find the speed of the motor shaft.

(b) **Automotive Trades**   A 12-tooth gear mounted on a motor shaft drives a larger gear. The motor shaft rotates at 1450 rpm. If the speed of the large gear is to be 425 rpm, how many teeth must be on the large gear?

(c) **Physics**   For gases, pressure is inversely proportional to volume if the temperature does not change. If 30 cu ft of air at 15 psi is compressed to 6 cu ft, what is the new pressure?

(d) **Manufacturing**   If five assembly machines can complete a given job in 3 hours, how many hours will it take for two assembly machines to do the same job?

(e) **Physics**   The forces and lever arm distances for a lever obey an inverse proportion.

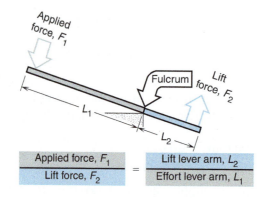

$$\frac{\text{Applied force, } F_1}{\text{Lift force, } F_2} = \frac{\text{Lift lever arm, } L_2}{\text{Effort lever arm, } L_1}$$

If a 100-lb force is applied to a 22-in. crowbar pivoted 2 in. from the end, what lift force is exerted?

Solutions   (a) $\dfrac{9 \text{ in.}}{5 \text{ in.}} = \dfrac{x}{960 \text{ rpm}}$ An inverse proportion: the larger pulley turns more slowly.

$$5x = 9 \cdot 960 = 8640$$

$$x = 1728$$

(b) $\dfrac{12 \text{ teeth}}{x \text{ teeth}} = \dfrac{425 \text{ rpm}}{1450 \text{ rpm}}$ An inverse proportion:
the larger gear turns more slowly.

$$425x = 12 \cdot 1450 = 17{,}400$$

$$x = 40.941 \ldots \qquad \text{or 41 teeth, rounded to the nearest whole number. We can't have part of a gear tooth!}$$

$$x \approx 41$$

(c) $\dfrac{30 \text{ cu ft}}{6 \text{ cu ft}} = \dfrac{P}{15 \text{ psi}}$ An inverse proportion: the higher the pressure, the smaller the volume.

$$6P = 30 \cdot 15 = 450$$

$$P = 75 \text{ psi}$$

(d) Careful on this one! An inverse proportion should be used. The *more* machines used, the *fewer* the hours needed to do the job.

$$\frac{5 \text{ machines}}{2 \text{ machines}} = \frac{x \text{ hours}}{3 \text{ hours}}$$

$$2x = 15$$

$$x = 7\tfrac{1}{2} \text{ hours} \qquad \text{Two machines will take much longer to do the job than will five machines.}$$

(e) $\dfrac{100 \text{ lb}}{F} = \dfrac{2 \text{ in.}}{20 \text{ in.}}$ If the entire bar is 22 in. long, and $L_2 = 2$ in., then $L_1 = 20$ in.

$$2F = 100(20) = 2000$$

$$F = 1000 \text{ lb}$$

Now turn to Exercises 4-2 for a set of practice problems on these special applications of ratio and proportion.

Exercises 4-2   Special Applications of Ratio and Proportion

A. Complete the following tables.

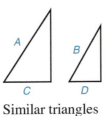
Similar triangles

1.

	A	B	C	D
(a)	$5\tfrac{1}{2}$ in.	$1\tfrac{1}{4}$ in.	$2\tfrac{3}{4}$ in.	
(b)		23.4 cm	20.8 cm	15.6 cm
(c)	12 ft		9 ft	6 ft
(d)	4.5 m	3.6 m		2.4 m

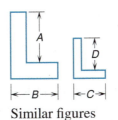
Similar figures

2.

	A	B	C	D
(a)	4 in.	5 in.	$3\tfrac{1}{2}$ in.	
(b)	5 ft		5 ft	2 ft
(c)	6.4 m	5.6 m		1.6 m
(d)		36 m	12 cm	14 cm

3.

	Number of Teeth on Gear 1	Number of Teeth on Gear 2	RPM of Gear 1	RPM of Gear 2
(a)	20	48	240	
(b)	25		150	420
(c)		40	160	100
(d)	32	40		1200

4.

	Diameter of Pulley 1	Diameter of Pulley 2	RPM of Pulley 1	RPM of Pulley 2
(a)	18 in.	24 in.	200	
(b)	12 in.		300	240
(c)		5 in.	400	640
(d)	14 in.	6 in.	300	

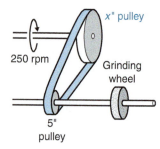

5"
pulley

Problem 1

B. Practical Applications

1. **Metalworking**   A line shaft rotating at 250 rpm is connected to a grinding wheel by the pulley assembly shown in the diagram. If the grinder shaft must turn at 1200 rpm, what size pulley should be attached to the line shaft?

2. **Architecture**   The architectural drawing of an outside deck is $3\frac{1}{2}$ in. wide by $10\frac{7}{8}$ in. long. If the deck will actually be 14 ft wide, calculate the following:
    (a) The actual length of the deck
    (b) The scale factor

3. **Automotive Trades**   Horsepower developed by an engine is directly proportional to its displacement. How many horsepower will be developed by an engine with a displacement of 240 cu in. if a 380-cu in. engine of the same kind develops 220 hp?

4. **Transportation**   The distance necessary to stop a subway train at a given speed is inversely proportional to its deceleration. If a train traveling at 30 mph requires 180 ft to stop when decelerating at 0.18 g, what is the stopping distance at the same speed when it is decelerating at 0.15 g?

5. **Physics**   A crowbar 28 in. long is pivoted 6 in. from the end. What force must be applied at the long end in order to lift a 400-lb object at the short end?

6. **Industrial Technology**   If 60 gal of oil flow through a certain pipe in 16 minutes, how long will it take to fill a 450-gal tank using this pipe?

7. **Automotive Trades**   If the alternator-to-engine drive ratio is 2.45 to 1, what rpm will the alternator have when the engine is idling at 400 rpm? (*Hint:* Use a direct proportion.)

8. **General Trades**   The length of a wrench is inversely proportional to the amount of force needed to loosen a bolt. A wrench 6 in. long requires a force of 240 lb to loosen a rusty bolt. How much force would be required to loosen the same bolt using a 10-in. wrench?

9. **Construction**   The Santa Barbara Planning Commission recently voted to restrict the size of home remodels by limiting the floor area to lot area ratio to a maximum of 0.45 to 1. Under these guidelines,
    (a) What would be the maximum allowable size of a remodel on an 11,800-sq-ft lot?
    (b) What size lot would be required in order to create a 3960-sq-ft remodel?

10. **Machine Trades**   A pair of belted pulleys have diameters of 20 in. and 16 in., respectively. If the larger pulley turns at 2000 rpm, how fast will the smaller pulley turn?

11. **Manufacturing**   A 15-tooth gear on a motor shaft drives a larger gear having 36 teeth. If the motor shaft rotates at 1200 rpm, what is the speed of the larger gear?

12. **Electronics**   The power gain of an amplifier circuit is defined as

$$\text{Power gain} = \frac{\text{output power}}{\text{input power}}$$

If the audio power amplifier circuit has an input power of 0.72 watt and a power gain of 30, what output power will be available at the speaker?

13. **Industrial Technology** If 12 assemblers can complete a certain job in 4 hours, how long will the same job take if the number of assemblers is cut back to 8?

14. **Electronics** A 115-volt power transformer has 320 turns on the primary. If it delivers a secondary voltage of 12 volts, how many turns are on the secondary? (*Hint:* Use a direct proportion.)

15. **Automotive Trades** The headlights of a car are mounted at a height of 3.5 ft. If the light beam drops 1 in. per 35 ft, how far ahead in the road will the headlights illuminate?

16. **Transportation** A truck driver covers a certain stretch of the interstate in $4\frac{1}{2}$ hours traveling at the posted speed limit of 55 mph. If the speed limit is raised to 65 mph, how much time will the same trip require?

17. **Machine Trades** It is known that a cable with a cross-sectional area of 0.60 sq in. has a capacity to hold 2500 lb. If the capacity of the cable is proportional to its cross-sectional area, what size cable is needed to hold 4000 lb?

18. **Masonry** Cement, sand, and gravel are mixed to a proportion of 1:3:6 for a particular batch of concrete. How many cubic yards of each should be used to mix 125 cubic yards of concrete?

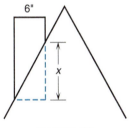

Problem 19

19. **Roofing** A cylindrical vent 6 in. in diameter must be cut at an angle to fit on a gable roof with a $\frac{2}{3}$ pitch. This means that for the vent itself the ratio of rise to run will be 2:3. Find the height x of the cut that must be made on the cylinder to make it fit the slope of the roof. (See the figure.)

20. **Architecture** A wrought-iron gate for a new house will be 18 ft 9 in. long and 9 ft 6 in. high. An architect makes a drawing of the gate using a scale factor of $\frac{1}{4}$ in. = 1 ft. What will be the dimensions of the gate on the drawing?

21. **Automotive Trades** When a tire is inflated, the air pressure is inversely proportional to the volume of the air. If the pressure of a certain tire is 28 psi when the volume is 120 cu in., what is the pressure when the volume is 150 cu in.?

22. **Masonry** A mason is using 4 in. by 8 in. by 16 in. blocks to build a wall. From prior experience, he knows that 225 blocks and 6 cu ft of mortar are needed per 100 sq ft of wall. He also knows that 7.25 hours of labor are needed per 100 sq ft of installation.
    (a) How many blocks are needed for 1760 sq ft of wall?
    (b) How many cu ft of mortar are needed for 2250 sq ft of wall?
    (c) How many hours of labor are needed for 1240 sq ft of wall?

23. **Life Skills** If you are paid $238.74 for $21\frac{1}{2}$ hours of work, what amount should you be paid for 34 hours of work at this same rate of pay?

24. **Sheet Metal Trades** If the triangular plate shown is cut into eight pieces along equally spaced dashed lines, find the height of each cut. (Round to two decimal digits.)

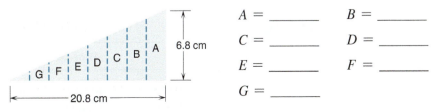

$A =$ _____ $B =$ _____

$C =$ _____ $D =$ _____

$E =$ _____ $F =$ _____

$G =$ _____

Problem 24

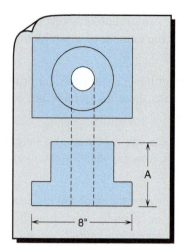

Problem 27

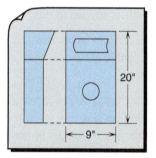

Problem 28

25. **Water/Wastewater Treatment**   The time required for an outlet pipe to empty a tank is inversely proportional to the cross-sectional area of the pipe. A pipe with a cross-sectional area of 113.0 sq in. requires 6.4 hours to empty a certain tank. If the pipe was replaced with one with a cross-sectional area of 50.25 sq in., how long would it take this pipe to empty the same tank?

26. **Physics**   A gas has a volume of 2480 cu cm at a pressure of 63.5 psi. What is the pressure when the gas is compressed to 1830 cu cm? (Round to the nearest tenth.)

27. **Machine Trades**   The base of the flange shown in the drawing is actually 8 in. Its drawing width is 3 in. (a) What will be the actual dimension A if A is 2.25 in. on the drawing? (b) Find the scale factor of the drawing.

28. **Drafting**   The drawing in the margin shows the actual length and width of a display panel. (a) What will be the corresponding drawing length of the panel if the drawing width is $2\frac{1}{4}$ in.? (b) Find the scale factor for this drawing.

29. **Plumbing**   If a 45-gal hot water tank holds 375 lb of water, what weight of water will a 55-gal tank hold? (Round to the nearest pound.)

30. **Automotive Trades**   When different-size tires are put on a vehicle, the speedometer gear must be changed so that the mechanical speedometer will continue to give the correct reading. The number of teeth on the gear is inversely proportional to the size of the tires. Suppose that a driver decides to replace 24-in. tires with $26\frac{1}{2}$-in. ones. If the old speedometer gear had 18 teeth, how many teeth should the replacement gear have? (*Remember:* You cannot have a fraction of a tooth on a gear.)

31. **Manufacturing**   A manufacturing company currently uses 24 workers to load and unload trucks. Each worker can move an average of 10 boxes per minute. A robotics firm has built a robot that can load and unload boxes at the rate of 15 boxes per minute. How many robots will it take to do the same job as the 24 humans?

When you have completed these exercises, check your answers to the odd-numbered problems in the Appendix, then turn to Section 4-3.

4-3   **Introduction to Percent**

1. Convert $\dfrac{3}{20}$ to decimal form.

2. Express 0.48 as a fraction in lowest terms.

In many practical calculations, it is helpful to be able to compare two numbers, and it is especially useful to express the comparison in terms of a percent. Percent calculations are a very important part of any work in business, technical skills, or the trades.

The word **percent** comes from the Latin phrase *per centum* meaning "by the hundred" or "for every hundred." We use the % symbol to indicate percent. A number written as a percent is being compared with a second number called the standard or **base**.

EXAMPLE 1   What part of the base length is length *A*?

Base Length	Length *A*

We could answer the question with a fraction or ratio, a decimal, or a percent. First, divide the base into 100 equal parts. Use a 100-mm metric rule to help visualize this.

A

$$\boxed{\;10\quad20\quad30\quad40\quad50\quad60\quad70\quad80\quad90\;}$$

Base = 100 mm

Then compare length A with it.

The length of A is 40 parts out of the 100 parts that make up the base.

A is $\dfrac{40\text{ mm}}{100\text{ mm}}$ or 0.40 or 40% of the base.

$\dfrac{40}{100} = 40\,\%$ 40 % means 40 parts in 100 or $\dfrac{40}{100}$

Your Turn   For the following diagram, what part of the base length is length B?

Length B

$$\boxed{\;10\quad20\quad30\quad40\quad50\quad60\quad70\quad80\quad90\;}$$ Base = 100 mm

Answer with a percent.

Solution B is $\dfrac{60}{100}$ or 60% of the base.

The compared number may be larger than the base. In such cases, the percent will be larger than 100%.

EXAMPLE 2   What percent of the base length is length C?

Length C

Base Length

In this case, divide the base into 100 parts and extend it in length.

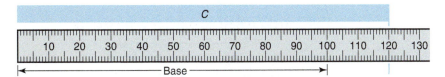

C

$$\boxed{\;10\quad20\quad30\quad40\quad50\quad60\quad70\quad80\quad90\quad100\quad110\quad120\quad130\;}$$
|← —————————— Base —————————— →|

The length of C is 120 of the 100 parts that make up the base.

C is $\dfrac{120}{100}$ or 120 % of the base.

Ratios, decimals, and percents are all alternative ways to compare two numbers. For example, in the following drawing, what fraction of the rectangle is shaded? What part of 12 is 3?

First, we can write the answer as a ratio:

$$\frac{3 \text{ shaded squares}}{12 \text{ squares}} = \frac{3}{12} = \frac{1}{4}$$

Second, by dividing 4 into 1 we can write this ratio or fraction as a decimal:

$$\frac{1}{4} = 0.25 \qquad 4\overline{)1.00}^{\,0.25}$$

Finally, we can rewrite the fraction with a denominator of 100 and then express it as a percent:

$$\frac{1}{4} = \frac{1 \times 25}{4 \times 25}$$

$$\frac{1}{4} = \frac{25}{100} \qquad \text{or} \qquad 25\%$$

We will now learn in more detail how to convert back and forth among these different comparisons.

Changing Decimal Numbers to Percents

Converting a number from fraction or decimal form to a percent is a useful skill.

> To write a decimal number as a percent, multiply the decimal number by 100%.

EXAMPLE 3 (a) $0.60 = 0.60 \times 100\% = 60\%$

Multiply by 100%

(b) $0.375 = 0.375 \times 100\% = 37.5\%$

(c) $3.4 = 3.4 \times 100\% = 340\%$

(d) $0.02 = 0.02 \times 100\% = 2\%$ ●

Learning Help Notice that multiplying by 100 is equivalent to shifting the decimal point two places to the right. For Example 3(a) it would look like this:

$$0.60 = 0.60 = 60.\% \qquad \text{or} \qquad 60\%$$

Shift decimal point two places right

Similarly, in Example 3(c) we have:

$$3.4 = 3.40 = 340\% \; ●$$

Your Turn Rewrite the following decimal numbers as percents.

(a) 0.75   (b) 1.25   (c) 0.064   (d) 3   (e) 0.05   (f) 0.004

Solutions (a) $0.75 = 0.75 \times 100\% = 75\%$:   0.75 becomes 75%

(b) $1.25 = 1.25 \times 100\% = 125\%$ 1.25 becomes 125%

(c) $0.064 = 0.064 \times 100\% = 6.4\%$ 0.064 becomes 6.4%

(d) $3 = 3 \times 100\% = 300\%$ $3 = 3.00$ or 300%

(e) $0.05 = 0.05 \times 100\% = 5\%$ 0.05 becomes 5%

(f) $0.004 = 0.004 \times 100\% = 0.4\%$ 0.004 becomes 0.4%

Changing Fractions to Percents

> To rewrite a fraction as a percent, first change it to decimal form by dividing, then multiply by 100%.

EXAMPLE 4

(a) $\frac{1}{2}$ is 1 divided by 2, or $2\overline{)1.0}$ gives 0.5 so that

$$\frac{1}{2} = 0.5 = 0.5 \times 100\% = 50\%$$

Change to a decimal → Multiply by 100%

(b) $\frac{3}{4} = 0.75 = 0.75 \times 100\% = 75\%$

(c) $\frac{3}{20} = 0.15 = 0.15 \times 100\% = 15\%$ since $\frac{3}{20} = 20\overline{)3.00}$ gives 0.15

(d) $1\frac{7}{20} = \frac{27}{20} = 1.35 = 1.35 \times 100\% = 135\%$ since $\frac{27}{20} = 20\overline{)27.00}$ gives 1.35

(e) $\frac{5}{16} = 0.3125 = 0.3125 \times 100\% = 31.25\%$

This is often written as $31\frac{1}{4}\%$. ●

In Chapter 3, we learned that some fractions cannot be converted to an exact decimal. For example, $\frac{1}{3} = 0.3333 \ldots$ is a repeating decimal. We can round to get an approximate percent:

$\frac{1}{3} \approx 0.3333 = 33.33\%$

The fraction $\frac{1}{3}$ is roughly equal to 33.33% and exactly equal to $33\frac{1}{3}\%$.

Learning Help Here is an easy way to change certain fractions into percents. If the denominator of the fraction divides exactly into 100, change the given fraction into an equivalent fraction with a denominator of 100. The new numerator is the percent. For example,

$$\frac{3}{20} = \frac{3 \times 5}{20 \times 5} = \frac{15}{100}$$ so $\frac{3}{20} = 15\%$ ●

Your Turn Rewrite the following fractions as percents.

(a) $\frac{5}{12}$ (b) $\frac{2}{3}$ (c) $3\frac{1}{8}$ (d) $\frac{4}{5}$ (e) $\frac{7}{25}$ (f) $1\frac{5}{6}$

Solutions (a) $\dfrac{5}{12}$ is $12\overline{)5.0000}$ ⟵ 0.4166... so

$$\frac{5}{12} = 0.4166\ldots \approx 41.67\% \text{ or exactly } 41\frac{2}{3}\%$$

(b) $\dfrac{2}{3} \approx 0.6667 \approx 66.67\%$ or exactly $66\frac{2}{3}\%$

(c) $\dfrac{1}{8}$ is $8\overline{)1.000}$ ⟵ 0.125   so $3\dfrac{1}{8} = 3.125$

$$3\frac{1}{8} = 3.125 = 312.5\%$$

(d) Because 5 divides exactly into 100, we can use the method of the Learning Help on p. 223.

$$\frac{4}{5} = \frac{4 \times 20}{5 \times 20} = \frac{80}{100} \quad \text{so} \quad \frac{4}{5} = 80\%$$

(e) Because 25 divides exactly into 100, let's again use the method of the Learning Help.

$$\frac{7}{25} = \frac{7 \times 4}{25 \times 4} = \frac{28}{100} \quad \text{so} \quad \frac{7}{25} = 28\%$$

(f) $1\dfrac{5}{6} = 1.833\ldots \approx 183.33\%$ or exactly $183\frac{1}{3}\%$

Practical Applications

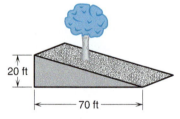

EXAMPLE 5   **Landscaping**   In landscaping or road construction, the slope of a hill or *grade* is often expressed as a percent, converted from the following fraction:

$$\text{Slope} = \frac{\text{vertical distance}}{\text{horizontal distance}}$$

The hillside in the figure has a slope of

$$\frac{20 \text{ ft}}{70 \text{ ft}} = \frac{2}{7}$$

To express this as a percent,

$$\frac{2}{7} \text{ is } 7\overline{)2.0000} \text{ ⟵ } 0.2857\ldots = 0.2857\ldots \quad \text{or} \quad 28.57\ldots\%$$

Thus, the grade is approximately 29%. ●

Your Turn   **Landscaping**   A driveway 44 ft long has a vertical rise of 2.5 ft. Express the slope of the driveway as a percent to the nearest tenth.

Solution   Expressed as a fraction, the slope is

$$\frac{2.5 \text{ ft}}{44 \text{ ft}}$$

Converting to a percent, we have

$$44\overline{)2.5000} \text{ ⟵ } 0.0568\ldots = 0.0568\ldots \approx 5.7\%$$

Changing Percents to Decimal Numbers

To use percent numbers when you solve practical problems, it is often necessary to change a percent to a decimal number.

> To change a percent to a decimal number, divide by 100%.

EXAMPLE 6

(a) $50\% = \dfrac{50\%}{100\%} = \dfrac{50}{100} = 0.5$

Divide by 100%

(b) $6\% = \dfrac{6\%}{100\%} = \dfrac{6}{100} = 0.06$

(c) $0.2\% = \dfrac{0.2\%}{100\%} = \dfrac{0.2}{100} = 0.002$

(d) If a fraction is part of the percent number, write it as a decimal number before dividing by 100%.

$8\frac{1}{2}\% = 8.5\% = \dfrac{8.5\%}{100\%} = 0.085$ ●

Learning Help

Notice that, in each of these examples, division by 100% is the same as moving the decimal point two digits to the left. For example,

$50\% = 50.\% = 0.50 \text{ or } 0.5$

Shift the decimal point two places left

You may need to insert additional zeros after shifting the decimal point.

$6\% = 06.\% = 0.06$

$0.2\% = 00.2\% = 0.002$

$8.5\% = 08.5\% = 0.085$ ●

Your Turn

Now try these. Write each percent as a decimal number.

(a) 4%   (b) 112%   (c) 0.5%

(d) $9\frac{1}{4}\%$ (e) 45%   (f) $12\frac{1}{3}\%$

Solutions

(a) $4\% = \dfrac{4\%}{100\%} = \dfrac{4}{100} = 0.04$ since $100\overline{)4.00}\;^{0.04}$

This may also be done by shifting the decimal point: $4\% = 04.\% = 0.04$.

(b) $112\% = 112.\% = 1.12$

(c) $0.5\% = 00.5\% = 0.005$

(d) $9\frac{1}{4}\% = 9.25\% = 09.25\% = 0.0925$

(e) $45\% = 45.\% = 0.45$

(f) $12\frac{1}{3}\% \approx 12.33\% = 0.1233$

A Closer Look   Did you notice that in problem (b) a percent greater than 100% gives a decimal number greater than 1?

$100\% = 1$

$200\% = 2$

$300\% = 3$ and so on ●

Changing Percents to Fractions

It is often necessary to rewrite a percent number as a fraction in lowest terms.

> To write a percent number as a fraction, form a fraction with the percent number as numerator and 100 as the denominator. Then rewrite this fraction in lowest terms.

EXAMPLE 7

(a) $40\% = \dfrac{40}{100} = \dfrac{40 \div 20}{100 \div 20} = \dfrac{2}{5}$

(b) $68\% = \dfrac{68}{100} = \dfrac{68 \div 4}{100 \div 4} = \dfrac{17}{25}$

(c) $125\% = \dfrac{125}{100} = \dfrac{125 \div 25}{100 \div 25} = \dfrac{5}{4}$ or $1\dfrac{1}{4}$ ●

Your Turn   Write each percent as a fraction in lowest terms.

(a) 8%   (b) 65%   (c) 140%

(d) 73%   (e) 0.2%   (f) $16\frac{2}{3}\%$

Solutions

(a) $8\% = \dfrac{8}{100} = \dfrac{8 \div 4}{100 \div 4} = \dfrac{2}{25}$

(b) $65\% = \dfrac{65}{100} = \dfrac{65 \div 5}{100 \div 5} = \dfrac{13}{20}$

(c) $140\% = \dfrac{140}{100} = \dfrac{140 \div 20}{100 \div 20} = \dfrac{7}{5}$ or $1\dfrac{2}{5}$

(d) $73\% = \dfrac{73}{100}$ in lowest terms

(e) $0.2\% = \dfrac{0.2}{100} = \dfrac{0.2 \times 5}{100 \times 5} = \dfrac{1}{500}$

(f) $16\frac{2}{3}\% = \dfrac{16\frac{2}{3}}{100} = \dfrac{\frac{50}{3}}{100} = \dfrac{50}{3} \times \dfrac{1}{100} = \dfrac{1}{6}$

A Closer Look   Notice in (e) that we had to multiply the numerator and denominator by 5 to avoid having a decimal number within a fraction. ●

Here is a table of percents used most often in practical applications, with their decimal and fraction equivalents. Study the table, then go to Exercises 4-3 for a set of problems involving percent conversion.

Percent Equivalents

Percent	Decimal	Fraction
5%	0.05	$\frac{1}{20}$
$6\frac{1}{4}\%$	0.0625	$\frac{1}{16}$
$8\frac{1}{3}\%$	$0.08\overline{3}$	$\frac{1}{12}$
10%	0.10	$\frac{1}{10}$
$12\frac{1}{2}\%$	0.125	$\frac{1}{8}$
$16\frac{2}{3}\%$	$0.1\overline{6}$	$\frac{1}{6}$
20%	0.20	$\frac{1}{5}$
25%	0.25	$\frac{1}{4}$
30%	0.30	$\frac{3}{10}$
$33\frac{1}{3}\%$	$0.\overline{3}$	$\frac{1}{3}$
$37\frac{1}{2}\%$	0.375	$\frac{3}{8}$
40%	0.40	$\frac{2}{5}$

Percent	Decimal	Fraction
50%	0.50	$\frac{1}{2}$
60%	0.60	$\frac{3}{5}$
$62\frac{1}{2}\%$	0.625	$\frac{5}{8}$
$66\frac{2}{3}\%$	$0.\overline{6}$	$\frac{2}{3}$
70%	0.70	$\frac{7}{10}$
75%	0.75	$\frac{3}{4}$
80%	0.80	$\frac{4}{5}$
$83\frac{1}{3}\%$	$0.8\overline{3}$	$\frac{5}{6}$
$87\frac{1}{2}\%$	0.875	$\frac{7}{8}$
90%	0.90	$\frac{9}{10}$
100%	1.00	$\frac{10}{10}$

Exercises 4-3　Introduction to Percent

A. Convert to a percent.

1. 0.32　　2. 1　　3. 0.5　　4. 2.1

5. $\frac{1}{4}$　　6. 3.75　　7. 40　　8. 0.675

9. 2　　10. 0.075　　11. $\frac{1}{2}$　　12. $\frac{1}{6}$

13. 0.335　　14. 0.001　　15. 0.005　　16. $\frac{3}{10}$

17. $\frac{3}{2}$　　18. $\frac{3}{40}$　　19. $3\frac{3}{10}$　　20. $\frac{1}{5}$

21. 0.40　　22. 0.10　　23. 0.95　　24. 0.03

25. 0.3　　26. 0.015　　27. 0.60　　28. 7.75

29. 1.2　　30. 4　　31. 6.04　　32. 9

33. $\frac{5}{4}$　　34. $\frac{1}{5}$　　35. $\frac{7}{20}$　　36. $\frac{3}{8}$

37. $\frac{5}{6}$　　38. $2\frac{3}{8}$　　39. $3\frac{7}{10}$　　40. $1\frac{4}{5}$

B. Convert to a decimal number.

1. 6%　　2. 45%　　3. 1%　　4. 33%

5. 71%　　6. 456%　　7. $\frac{1}{4}\%$　　8. 0.05%

9. $6\frac{1}{4}\%$　　10. $8\frac{3}{4}\%$　　11. 30%　　12. 2.1%

13. 800%　　14. 8%　　15. 0.25%　　16. $16\frac{1}{3}\%$

17. 7%	18. 3%	19. 56%	20. 15%
21. 1000%	22. $7\frac{1}{2}\%$	23. 90%	24. 0.3%
25. 150%	26. $1\frac{1}{2}\%$	27. $6\frac{3}{4}\%$	28. $\frac{1}{2}\%$
29. $12\frac{1}{4}\%$	30. $125\frac{1}{5}\%$	31. 1.2%	32. 240%

C. Rewrite each percent as a fraction in lowest terms.

1. 5%	2. 20%	3. 250%	4. 16%
5. 53%	6. 0.1%	7. 92%	8. 2%
9. 45%	10. 175%	11. $8\frac{1}{3}\%$	12. 37.5%
13. 24%	14. 119%	15. 0.05%	16. 15%
17. 480%	18. $83\frac{1}{3}\%$	19. 62.5%	20. 1.2%

When you have had the practice you need, check your answers to the odd-numbered problems in the Appendix, then turn to Section 4-4.

4-4　Percent Problems

1. Solve for n: $\dfrac{12}{n} = \dfrac{15}{100}$

2. What is 50% of 20?

In all your work with percent you will find that there are three basic types of problems. These three are related to all percent problems that arise in business, technology, or the trades. In this section we show you how to solve any percent problem, and we examine each of the three types of problems.

All percent problems involve three quantities:

　B, the **base** or whole or total amount, a standard used for comparison
　P, the **percentage** or part being compared with the base
　R, the **rate** or **percent**, a percent number

 Note　To avoid possible confusion between the words *percentage* and *percent*, we will refer to the percentage P as the "part." ●

For any percent problem, these three quantities are related by the proportion

$$\frac{P}{B} = \frac{R}{100}$$

For example, the proportion

$$\frac{3}{4} = \frac{75}{100}$$

can be translated to the percent statement

　　"Three is 75% of four."

Here, the part P is 3, the base B is 4, and the percent R is 75.

To solve any percent problem, we need to identify which of the quantities given in the problem is P, which is B, and which is R. Then we can write the percent proportion and solve for the missing or unknown quantity.

When *P* Is Unknown   Consider these three problems:

What is 30% of 50?

Find 30% of 50.

30% of 50 is what number?

These three questions are all forms of the same problem. They are all asking you to find *P*, the part.

EXAMPLE 1   Now let's solve these problems.

We know that 30 is the percent *R* because it has the % symbol attached to it. The number 50 is the base *B*. To solve, write the percent proportion, substituting 50 for *B* and 30 for *R*.

$$\frac{P}{50} = \frac{30}{100}$$

Now solve using cross-products and division.

$$100P = 50 \cdot 30$$

$$100P = 1500$$

$$P = \frac{1500}{100} = 15$$

15 is 30% of 50.

✓ Substitute the answer back into the proportion.

$$\frac{15}{50} = \frac{30}{100} \qquad \text{or} \qquad 15 \cdot 100 = 50 \cdot 30 \qquad \text{which is correct}$$

The answer is reasonable: 30% is roughly one-third, and 15 is roughly one-third of 50.   ●

Learning Help   The percent number *R* is easy to identify because it always has the % symbol attached to it. If you have trouble distinguishing the part *P* from the base *B*, notice that *B* is usually associated with the word *of* and *P* is usually associated with the word *is*.

What is   30%   of 50?

"is" indicates the part *P*. It is unknown in this case.

"%" indicates the percent or rate *R*.

"of" indicates the base *B*.

Therefore, the proportion $\frac{P}{B} = \frac{R}{100}$ can be thought of as $\frac{is}{of} = \frac{\%}{100}$.   ●

EXAMPLE 2   Let's try one more, step-by-step.

Find $8\frac{1}{2}\%$ of 160.

Step 1   If your mental math skills are good, estimate the answer first. In this case, $8\frac{1}{2}\%$ of 160 should be a little less than 10% of 160, or 16.

Step 2   Now identify the three quantities, *P*, *B*, and *R*. *R* is obviously $8\frac{1}{2}$. The word *is* does not appear, but *of* appears with 160, so 160 is *B*. Therefore, *P* must be the unknown quantity.

Step 3   Set up the percent proportion.

$$\frac{P}{B} = \frac{R}{100} \quad \rightarrow \quad \frac{P}{160} = \frac{8\frac{1}{2}}{100}$$

Step 4   Solve.

$$100P = 160\left(8\tfrac{1}{2}\right)$$

$$100P = 1360$$

$$P = 13.6$$

✓ **Step 5**   Check your answer by substituting it into the original proportion equation and comparing cross-products.

Notice that the calculated answer, 13.6, is a little less than 16, our rough estimate. ●

Your Turn   Solve the following percent problems.

(a) Find 2% of 140 lb.   (b) 35% of $20 is equal to what amount?

(c) What is $7\frac{1}{4}$% of $1000?   (d) Calculate $16\frac{2}{3}$% of 66.

Solutions   (a) $\dfrac{P}{140} = \dfrac{2}{100}$

$$100P = 140 \cdot 2 = 280$$

$$P = 2.8 \text{ lb}$$

✓ $\dfrac{2.8}{140} = \dfrac{2}{100}$

$280 = 280$

(b) $\dfrac{P}{\$20} = \dfrac{35}{100}$

$$100P = \$700$$

$$P = \$7$$

✓ $\dfrac{\$7}{\$20} = \dfrac{35}{100}$

$700 = 700$

(c) $\dfrac{P}{\$1000} = \dfrac{7\frac{1}{4}}{100}$

$$100P = \$1000\left(7\tfrac{1}{4}\right) = \$7250$$

$$P = \$72.50$$

✓ $\dfrac{\$72.50}{\$1000} = \dfrac{7\frac{1}{4}}{100}$

$7250 = 7250$

(d) $\dfrac{P}{66} = \dfrac{16\frac{2}{3}}{100}$ $\boxed{16\tfrac{2}{3} = \dfrac{50}{3}}$

$$100P = 66 \cdot \frac{50}{3} = \frac{\overset{22}{66}}{1} \cdot \frac{50}{\underset{1}{3}} = 22 \cdot 50$$

$$100P = 1100$$

$$P = 11 \qquad \text{Check it.}$$

16 $\boxed{A_c^b}$ 2 $\boxed{A_c^b}$ 3 $\boxed{\times}$ 66 $\boxed{\div}$ 100 $\boxed{=}$ → ⬛ *11.*

Learning Help Notice that in all problems where P is to be found, we end by dividing by 100 in the last step. Recall that the quick way to divide by 100 is to move the decimal point two places to the left. In problem (c), for example, $100P = \$7250$. The decimal point is after the zero in \$7250. Move it two places to the left:

$$\$7250 \div 100 = \$7250. = \$72.50 \bullet$$

When R Is Unknown Consider the following problems:

5 is what percent of 8?

Find what percent 5 is of 8.

What percent of 8 is 5?

Once again, these statements represent three different ways of asking the same question. In each statement the percent R is unknown. We know this because neither of the other two numbers has a % symbol attached.

EXAMPLE 3 To solve this kind of problem, first identify P and B. The word *of* is associated with the base B, and the word *is* is associated with the part P. In this case, $B = 8$ and $P = 5$.

Next, set up the percent proportion.

$$\frac{P}{B} = \frac{R}{100} \rightarrow \frac{5}{8} = \frac{R}{100}$$

Finally, solve for R.

$$8R = 5 \cdot 100 = 500$$

$$R = 62.5\%$$

$\bullet$

Note When you solve for R, remember to include the % symbol with your answer. $\bullet$

Your Turn Now try these problems for practice.

(a) What percent of 40 lb is 16 lb?

(b) 65 is what percent of 25?

(c) Find what percent \$9.90 is of \$18.00.

Solutions
(a)
$$\frac{16}{40} = \frac{R}{100}$$
$$40R = 1600$$
$$R = 40\%$$

(b)
$$\frac{65}{25} = \frac{R}{100}$$
$$25R = 6500$$
$$R = 260\%$$

(c)
$$\frac{\$9.90}{\$18.00} = \frac{R}{100}$$
$$18R = 990$$
$$R = 55\%$$

A Closer Look Notice that in problem (b) the percentage is larger than the base, resulting in a percent larger than 100%. Some students mistakenly believe that the percentage is always smaller than the base. This is not always true. When in doubt use the "is–of" method to determine P and B. $\bullet$

Learning Help In all problems where R is unknown, we end up multiplying by 100. Recall that the quick way to do this is to move the decimal point two places to the right. If there is no decimal point, attach two zeros. $\bullet$

When *B* Is Unknown   This third type of percent problem requires that you find the "whole" or the base when the percent and the part are given. Problems of this kind can be stated as

> *8.7 is 30% of what number?*
>
> *30% of what number is 8.7?*
>
> *Find a number such that 30% of it is 8.7.*

EXAMPLE 4   We can solve this problem in the same way as the previous ones.

First, identify *P*, *B*, and *R*. Because it carries the % symbol, we know that $R = 30$. Because the word *is* is associated with the number 8.7, we know that $P = 8.7$. The base *B* is unknown.

Next, set up the percent proportion.

$$\frac{P}{B} = \frac{R}{100} \rightarrow \frac{8.7}{B} = \frac{30}{100}$$

Finally, solve the proportion.

$$30B = 870$$

$$B = 29 \qquad ✓ \qquad \frac{8.7}{\boxed{29}} = \frac{30}{100}$$

$$8.7(100) = 29 \cdot 30 \quad \text{which is correct} \qquad ●$$

Your Turn   Ready for a few practice problems? Try these.

(a) 16% of what amount is equal to $5.76?

(b) $2 is 8% of the cost. Find the cost.

(c) Find a distance such that $12\frac{1}{2}$% of it is $26\frac{1}{4}$ ft.

Solutions (a) $\dfrac{\$5.76}{B} = \dfrac{16}{100}$ (b) $\dfrac{\$2}{B} = \dfrac{8}{100}$ (c) $\dfrac{26\frac{1}{4}}{B} = \dfrac{12\frac{1}{2}}{100}$

$16B = 576$ $8B = 200$ $B = \dfrac{26.25(100)}{12.5} = 210$ ft

$B = \$36$ $B = \$25$

For problem (c): $2\,6.2\,5\;\boxed{\times}\;1\,0\,0\;\boxed{\div}\;1\,2.5\;\boxed{=}\;\rightarrow$ `210.`

So far we have solved sets of problems that were all of the same type. The key to solving percent problems is to be able to identify correctly the quantities *P*, *B*, and *R*.

More Practice   Use the hints you have learned to identify these three quantities in the following problems, then solve each problem.

(a) What percent of 25 is 30?

(b) 12 is $66\frac{2}{3}$% of what amount?

(c) Find 6% of 2400.

Solutions (a) $\underbrace{\text{What percent}}_{R}$ $\underbrace{\text{of 25}}_{B}$ $\underbrace{\text{is 30?}}_{P}$ $\dfrac{30}{25} = \dfrac{R}{100}$

$$25R = 3000$$

$$R = 120\%$$

(b) $\underbrace{12}_{P}$ is $\underbrace{66\frac{2}{3}\%}_{R}$ of $\underbrace{\text{what?}}_{B}$

$$\frac{12}{B} = \frac{66\frac{2}{3}}{100}$$
$$66\frac{2}{3} \cdot B = 1200$$
$$B = 18$$

$12 \times 100 \div 66 \boxed{A^b_c} 2 \boxed{A^b_c} 3 \boxed{=} \rightarrow \quad$ *18.*

(c) $\underbrace{\text{Find}}_{P}$ $\underbrace{6\%}_{R}$ of $\underbrace{2400.}_{B}$

$$\frac{P}{2400} = \frac{6}{100}$$
$$100P = 14,400$$
$$P = 144$$

Practical Applications

When solving applied problems involving percent, the words *is* and *of* may not always appear. In these cases, identifying P, R, and B may be a bit more difficult. Always look for the rate R first. It will be accompanied by the % symbol. If you do not see the % symbol, then R will be the missing quantity. Of the remaining two quantities, try to decide logically which one is the "whole" (B) and which one is the "part" (P).

EXAMPLE 5

(a) **Allied Health** A moderately active adult weighing 175 pounds needs to consume about 2800 calories per day in order to maintain his or her weight. A nutritionist recommends that no more than 30% of those calories should come from fat. What is the maximum number of calories per day that should come from fat?

To solve, first note that the percent is given, so $R = 30\%$. We also know the total number of calories per day, so this must be the "whole." Therefore, $B = 2800$. The unknown is P—that is, the part of the 2800 total calories that should come from fat. The proportion is:

$$\frac{P}{2800} = \frac{30}{100}$$
$$100P = 84,000$$
$$P = 840 \text{ calories per day from fat}$$

(b) **Construction** A builder told his crew that the site preparation and foundation work on a house would represent about 5% of the total building time. If it took the crew members 21 days to do this preliminary work, how long will the entire construction take?

Looking for the % symbol first, we see that $R = 5\%$.

We also know that preliminary work takes 21 days and that this represents part of the unknown total building time. Therefore $P = 21$ and we must solve for B, the total building time.

21 is 5% of the total

or $\dfrac{21}{B} = \dfrac{5}{100}$

$$B = \frac{21 \cdot 100}{5} = 420 \text{ days}$$

(c) **Allied Health**   Twelve milliliters (mL) of pure acetic acid are mixed with 38 mL of water. We wish to calculate the percent concentration of the resulting solution.

In this problem, the *part* is the 12 mL of acid. The *base* is the total amount of solution, the sum of the amounts of acid and water.

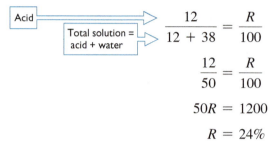

$$\frac{12}{12 + 38} = \frac{R}{100}$$

$$\frac{12}{50} = \frac{R}{100}$$

$$50R = 1200$$

$$R = 24\%$$ ●

Your Turn   (a) **Machine Trades**   During reshaping, 6 lb of metal is removed from a casting originally weighing 80 lb. What percent of the metal is removed?

(b) **Trades Management**   The $2800 actually spent on a construction job was 125% of the original estimate. What was the original estimate?

(c) **Water/Wastewater Treatment**   A water treatment operator must dose a water storage tank with 15.7 lb of chlorine. Because he must climb to the top of the tank to administer the chlorine, the operator chooses to use powdered pool chlorine instead of the much heavier liquid bleach. Pool chlorine is 70% pure chlorine. If the operator has available 5-lb containers of the powder, how many containers will he need to carry to the top of the tank?

Solutions   (a) There is no percent number given and the question begins "What percent . . . ." Therefore, R is the unknown. Of the two remaining quantities, the 6 lb removed is a part of the 80-lb whole, therefore $P = 6$ and $B = 80$.

$$\frac{6}{80} = \frac{R}{100}$$

$$80R = 600$$

$$R = 7.5\%$$

(b) The % symbol tells us that $R = 125\%$. The word *of* in the phrase "125% of the original estimate" implies that the original estimate is B, and this is the unknown. Therefore $2800 must be P; it is a percentage of the original estimate.

$$\frac{\$2800}{B} = \frac{125}{100}$$

$$B = \frac{\$2800 \cdot 100}{125} = \$2240$$

(c) **First,** determine how many pounds of actual chlorine are in 5 pounds of the powder. Find 70% of 5 lb:

$$\frac{x}{5} = \frac{70}{100}$$

$$100x = 350$$

$$x = 3.5 \text{ lb of chlorine in each container}$$

Then, use another proportion to calculate the number of containers needed to provide 15.7 lb of chlorine:

$$\frac{1 \text{ container}}{x \text{ containers}} = \frac{3.5 \text{ lb}}{15.7 \text{ lb}}$$

$$3.5x = 15.7$$

$$x \approx 4.49 \text{ containers}$$

The operator will need to use four full containers of powder and half of another one.

Now go to Exercises 4-4 for more practice on the three basic kinds of percent problems.

Exercises 4-4	**Percent Problems**

A. Solve.

1. 4 is _____ % of 5.

2. What percent of 25 is 16?

3. 20% of what number is 3?

4. 8 is what percent of 8?

5. 120% of 45 is _____.

6. 8 is _____ % of 64.

7. 3% of 5000 = _____.

8. 2.5% of what number is 2?

9. What percent of 54 is 36?

10. 60 is _____ % of 12.

11. 17 is 17% of _____.

12. 13 is what percent of 25?

13. $8\frac{1}{2}$% of $250 is _____.

14. 12 is _____ % of 2.

15. 6% of 25 is _____.

16. 60% of what number is 14?

17. What percent of 16 is 7?

18. 140 is _____ % of 105.

19. Find 24% of 10.

20. 45 is 12% of what number?

21. 30% of what number is equal to 12?

22. What is 65% of 5?

23. Find a number such that 15% of it is 750.

24. 16% of 110 is what number?

B. Solve.

1. 75 is $33\frac{1}{3}$% of _____.

2. What percent of 10 is 2.5?

3. 6% of $3.29 is _____.

4. 63 is _____ % of 35.

5. 12.5% of what number is 20?

6. $33\frac{1}{3}$% of $8.16 = _____.

7. 9.6 is what percent of 6.4?

8. $0.75 is _____ % of $37.50.

9. $6\frac{1}{4}$% of 280 is _____.

10. 1.28 is _____ % of 0.32.

11. 42.7 is 10% of _____.

12. 260% of 8.5 is _____.

13. $\frac{1}{2}$ is _____ % of 25.

14. $7\frac{1}{4}$% of 50 is _____.

15. 287.5% of 160 is _____.

16. 0.5% of _____ is 7.

17. 112% of _____ is 56.

18. $2\frac{1}{4}$% of 110 is _____.

C. Practical Applications

1. **General Interest** If you answered 37 problems correctly on a 42-question test, what percent of the problems did you answer correctly?

2. **Trades Management**   The profits from Ed's Plumbing Co. increased by $14,460, or 30%, this year. What profit did the company earn last year?

3. **Machine Trades**   A casting weighed 146 lb out of the mold. It weighed 138 lb after finishing. What percent of the weight was lost in finishing?

4. **Masonry**   On the basis of past experience, a contractor expects to find 4.5% broken bricks in every truckload. If he orders 2000 bricks, will he have enough to complete a job requiring 1900 bricks?

5. **Automotive Trades**   The weight distribution of a vehicle specifies that the front wheels carry 54% of the weight. If a certain vehicle weighs 3860 lb, what weight is carried by the front wheels? (Round to the nearest pound.)

6. **Life Skills**   If state sales tax is 6%, what is the price before tax if the tax is $7.80?

7. **General Trades**   A worker earns $28.50 per hour. If she receives a $7\frac{1}{2}\%$ pay raise, what is the amount of her hourly raise?

8. **General Trades**   If 7.65% of your salary is withheld for Social Security and Medicare, what amount is withheld from monthly earnings of $1845.00?

9. **Printing**   If an 8-in. by 10-in. photocopy is enlarged to 120% of its size, what are its new dimensions?

10. **Police Science**   Year-end statistics for a community revealed that, in 648 of the 965 residential burglaries, the burglar entered through an unlocked door or window. What percent of the entries were made in this way?

11. **Carpentry**   Preparing an estimate for a decking job, a carpenter assumes about a 15% waste factor. If the finished deck contains 1230 linear feet of boards, how many linear feet should the carpenter order? (*Hint:* If the waste is 15%, then the finished deck is 100% − 15% or 85% of the total order.)

12. **Printing**   Jiffy Print needs to produce 4000 error-free copies of a brochure. For this type of run, there is usually a 2% spoilage rate. What is the minimum number of brochures that should be printed to ensure 4000 good copies?

13. **Office Services**   The Bide-A-While Hotel manager is evaluated on the basis of the occupancy rate of the hotel, that is, the average daily percent of the available rooms that are actually occupied. During one particular week, daily occupancy totals were 117, 122, 105, 96, 84, 107, and 114. If the hotel has 140 rooms available each night, what was the occupancy rate for that week?

14. **Culinary Arts**   The manager of the Broccoli Palace restaurant needs 25 lb of fresh broccoli for an evening's business. Because of damage and spoilage, he expects to use only about 80% of the broccoli delivered by his supplier. What is the minimum amount he should order?

15. **Construction**   The BiltWell Construction Company has submitted a bid on construction of a spa deck. If its $8320 bid includes $2670 for lumber, what percent of the bid was the lumber? (Round to the nearest percent.)

16. **Flooring and Carpeting**   White oak flooring boards 3 in. wide are used to construct a hallway. If the flooring is laid at an angle, approximately 7200 linear feet of boards are needed for the job. How many linear feet

should be ordered if an allowance of 18% is made for waste? (Round to the nearest foot.) (*Hint:* If there is 18% waste, then the amount needed is 100% − 18% = 82% of the amount that should be ordered.)

17. **Construction** Quality Homes Construction Company orders only lumber that is at least 80% *clear*—free of defects, including knots and edge defects. In an order containing 65,500 linear feet, what minimum amount of lumber will be free of defects?

18. **Electrical Trades** An electrician needs fuses for a motor being installed. If the motor current is 50 amperes and the fuse must be rated at 175% of the motor current, what size fuse will the electrician need?

19. **Allied Health** A typical 35-year-old man has an age-adjusted maximum heart rate of 185 bpm (beats per minute). To maintain top conditioning, an athlete should exercise so that his or her heart rate is between 80% and 90% of his or her maximum heart rate for at least 30 minutes a day, 3 days a week. Calculate the range of heart rate (bpm) for this level of exercise.

20. **Trades Management** A masonry contractor estimates that a certain job will require eight full-time (8-hour) working days. He plans to use a crew consisting of three laborers earning $12 per hour each and three masons earning $20 per hour each. He estimates that materials will cost $3360. If he wishes to make a profit equal to 25% of his total cost for labor and materials, how much should his bid price be?

21. **Automotive Trades** The volumetric efficiency of a certain engine is 88.2%. This means that the actual airflow is 88.2% of the theoretic airflow. If the actual airflow is 268.4 cfm (cubic feet per minute), what is the theoretic airflow of this engine? (Round to the nearest 0.1 cfm.)

22. **Allied Health** If 5 oz of alcohol are mixed with 12 oz of water, what is the percent concentration of alcohol in the resulting solution? (Round to the nearest percent.) [*Hint:* See part (c) of Example 5 on page 234.]

23. **Retail Merchandising** A chain store plans to reduce its workforce by 4%. If there are currently 18,600 employees and 126 of them retire voluntarily, how many additional employees must be laid off?

24. **Electronics** A technician setting up a wireless network for an office is performing a survey to evaluate signal coverage. The technician finds that, after passing through a steel door in a brick wall, the signal power from a 62-mW wireless router is only 4 mW (milliwatts). What percent of the signal passes through the door? (Round to the nearest tenth of a percent.)

25. **Life Skills** When sales of a particular hybrid automobile reach 60,000 units, federal income tax credits begin to phase out. One year after reaching this threshold, buyers are still eligible for 25% of the original $3000 credit. What tax credit would buyers receive if they purchased the vehicle at this time?

26. **Culinary Arts** Wolfgang runs a small lunchtime cafeteria that serves about 120 customers each day, or a total of 600 meals in the normal working week. The cafeteria's weekly income is about $3680, 25% of which is set aside for taxes. An additional $1200 per week is paid for supplies, rent, and for the salary of Wolfgang's assistant. On average, what is the cafeteria's remaining profit for each meal served?

27. **Landscaping** An arborist is preparing the fuel mix for her chainsaw before climbing a tree to remove a dead limb. The two-stroke chainsaw

requires a 2% mix of engine oil to gas—that is, the amount of oil should be 2% of the amount of gas. How much engine oil should be added to a jerry can that contains 2.5 gal of gas?

28. **Automotive Trades**   An independent mechanic purchases parts from her distributor at a 20% discount from the retail price. She then makes part of her profit on the repair job by charging her customer 15% more than the retail price. If the 20% discounted price for a set of brake pads was $75.20, what is the price that the mechanic will charge her customer?

29. **Welding**   The allowable shear stress of a weld is 30% of the nominal tensile strength of the weld metal. If the nominal tensile strength is 64,000 psi, what is the allowable shear stress?

30. **Water/Wastewater Treatment**   Water and wastewater treatment formulas generally require concentrations to be expressed in parts per million (ppm). If a solution of chlorine bleach contains 5% actual chlorine, how many parts per million is this?

31. **Allied Health**   A physically active adult weighing 155 pounds needs to consume about 2790 calories per day in order to maintain his or her weight [see Example 5(a) on page 233]. A nutritionist recommends that no more than 25% of those calories should come from simple sugars. What is the maximum number of calories per day that should come from simple sugars?

32. **General Interest**   In 2015, the richest 20% of all taxpayers had an average income of $321,278 and paid an average of 25.7% of this in federal taxes. The poorest 20% of all taxpayers had an average income of $12,939 and paid an average of 3.6% in federal taxes. Determine the average after-tax income of each group to the nearest dollar. How many times higher was the after-tax income of the top 20% than the bottom 20%?

33. **Electrical Trades**   The average U.S. home uses about 11,000 kilowatt-hours (kWh) of electricity per year. Appliances and refrigeration account for about 17% of this total. If electricity costs $0.16 per kWh in a particular area, how much would the average U.S. homeowner pay annually to run appliances and refrigeration?

34. **Construction**   For a certain attic to conform to building code, 864 square inches of net-free vent area (NFVA) are required. To slightly pressurize the attic, 60% of the NFVA should be provided by the intake vents, and 40% should be provided by the exhaust vents. How many square inches of venting should be provided by the exhaust vents to slightly pressurize this attic?

35. **Retail Merchandising**   A store buyer often decides what to order for the coming year based on the previous year's sales. Last year, a clothing store recorded the following sales figures for a certain men's polo shirt:

Color	Total Sales in Thousands of $
White	24.3
Blue	16.8
Red	8.5
Green	6.6
Brown	7.2
Black	12.4

(a) To the nearest tenth, what percent of total sales were the sales of red shirts?

(b) Based on last year's figures, if the total purchase of polo shirts this year will be $48,000, what dollar amount of red shirts should be stocked?

36. **Retail Merchandising**   A store buyer purchased $6\frac{2}{3}$ dozen white T-shirts and $3\frac{1}{2}$ dozen colored T-shirts. What percent of the order consisted of white T-shirts? (Round to the nearest percent.)

37. **Allied Health**   A pharmaceutical sales representative makes a base salary of $66,200 per year. In addition, she can earn an annual bonus of up to $22,500 if she meets her prescription targets for the two medications she sells. She receives 70% of the bonus if she meets her target for medication A, and she receives 30% of the bonus if she meets her target for medication B. If she meets her target for medication A only, what would her total compensation (salary plus bonus) be for the year?

38. **Allied Health**   A pharmaceutical sales representative can earn a maximum bonus of $11,250 for meeting 100% of his annual prescription target of 4000 scrips for a certain medication. However, he can earn a portion of that bonus by meeting more than 80% of his target according to the following scale:

Percent of Target Reached	Percent of Bonus Earned
81%	5%
82%	10%
83%	15%
as this column increases by 1% . . .	this column increases by 5% . . .
98%	90%
99%	95%
100%	100%

What would the dollar amount of his bonus be for generating

(a) 3480 scrips?   (b) 3840 scrips?

39. **Water/Wastewater Treatment**   A water treatment operator must "shock" a contaminated reservoir with 12.5 lb of chlorine. How many gallons of a 5% chlorine bleach solution are required to treat the reservoir? (*Hint:* One gallon of bleach weighs 8.34 lb. Round to the nearest tenth of a gallon.)

When you have had the practice you need, check your answers to the odd-numbered problems in the Appendix, then continue with the study of some special practical applications of percent.

4-5

Special Applications of Percent Calculations

learning|catalytics™

1. Convert 6.5% to decimal form.
2. If a blueprint specifies the length of a part as 1.260 ± 0.012 in., what is the minimum allowable length for the part?

In the previous section, you learned about the three basic percent problems, and you solved a variety of practical applications. In this section, we will focus on the following special applications of percent: discount, sales tax, interest, commission, efficiency, tolerances, and percent change.

Discount   An important kind of percent problem, especially in business, involves the idea of **discount**. To stimulate sales, a merchant may offer to sell some item at less than its normal price. A discount is the amount of money by which the normal price is reduced. It is a percentage or part of the normal price. The **list price** is the normal or original price before the discount is subtracted. The **net price** is the new, reduced price. The net price is sometimes called the *discounted* price or the *sale* price. It is always less than the list price.

> Net price = list price − discount

The discount rate (R) is a percent number that enables you to calculate the discount (the percentage or part P) as part of the list price (the base B). Therefore, the percent proportion

$$\frac{\text{part}}{\text{base}} = \frac{\text{rate}}{100} \qquad \text{becomes} \qquad \frac{\text{discount } D}{\text{list price } L} = \frac{\text{discount rate } R}{100}$$

EXAMPLE 1   **General Trades**   The list price of a tool is $18.50. On a special sale, it is offered at 20% off. What is the net price?

Step 1   Write the proportion formula. $\dfrac{\text{discount}}{\text{list price}} = \dfrac{\text{discount rate}}{100}$

Step 2   Substitute the given numbers. $\dfrac{D}{\$18.50} = \dfrac{20}{100}$

Step 3   Solve for D. $100D = \$18.50 \times 20 = \$370$

$D = \$3.70$

Step 4   Calculate the net price.   Net price = list price − discount

$= \$18.50 - \$3.70$

$= \$14.80$ ●

🔍 **A Closer Look**   In Example 1 we solved for the discount and then, in Step 4, we subtracted this from the list price to find the net price. We can eliminate Step 4 if we use the following pie chart to help us write a new proportion:

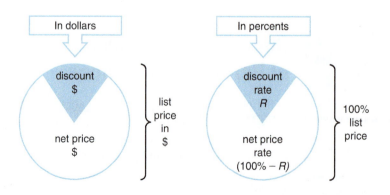

Instead of using the "discount" proportion, we can use the following "net price" proportion:

$$\frac{\text{net price } N}{\text{list price } L} = \frac{\text{net price rate } R}{100\%} \qquad \text{where net price rate} = 100\% - \text{discount rate}$$

For Example 1, the net price rate would be $100\% - 20\% = 80\%$, so the net price proportion would be:

$$\frac{N}{\$18.50} = \frac{80}{100}$$

In Step 3, we can then solve for N directly:

$$100N = 80(\$18.50) = \$1480$$

$$N = \$14.80$$

There is no need for the extra Step 4. ●

In the next example, there is a different unknown.

EXAMPLE 2   **Carpentry**   After a 25% discount, the net price of a 5-in. random orbit sander is $57. What was its list price?

We can use a pie chart diagram to help set up the correct percent proportion.

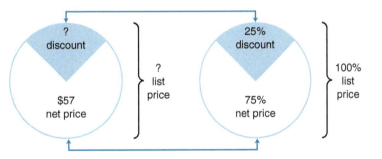

The pie chart shows that if the discount is 25% of the original list price, then the net price is $100\% - 25\%$ or 75% of the list price. Therefore, the correct proportion is

$$\frac{\text{net price}}{\text{list price}} = \frac{\text{net price rate}}{100}$$

or $$\frac{\$57}{L} = \frac{75}{100}$$

Solving for L, we have

$$75L = \$5700$$

$$L = \$76$$

The list price is $76.

$$\frac{\text{discount}}{\text{list price}} = \frac{\text{discount rate}}{100}$$

$$\frac{D}{\$76} = \frac{25}{100}$$

$$100D = \$1900$$

$$D = \$19$$

The discount is $19.

Net price = list price − discount

$$= \$76 - \$19$$

$$= \$57 \qquad \text{which is the correct net price as given in the original problem.} \qquad \bullet$$

Now let's try a problem in which the rate, or percent discount, is missing.

EXAMPLE 3   **Plumbing**   A kitchen sink that retails for $884 is offered to a plumbing contractor for $574.60. What discount rate is the contractor receiving?

First, subtract to find the discount.

Discount = list price − net price

$$= \$884 - \$574.60$$

$$= \$309.40$$

Then, set up the discount proportion.

$$\frac{D}{L} = \frac{R}{100} \qquad \text{becomes} \qquad \frac{\$309.40}{\$884} = \frac{R}{100}$$

$$884R = 30{,}940$$

$$R = 35\% \qquad \text{The contractor receives a } 35\% \text{ discount.} \qquad \bullet$$

Your Turn   Here are a few problems to test your understanding of the idea of discount.

(a) **Carpentry**   A preinventory sale advertises all tools discounted by 20%. What would be the net price of a rotary hammer drill that cost $499.95 before the sale?

(b) **Painting**   An airless sprayer is on sale for $1011.12 and is advertised as "12% off regular price." What was the regular price?

(c) **Office Services**   A set of four 205/65/15 automobile tires is on sale for $304.30. What is the discount rate if their list price is $89.50 each?

Solutions   (a) To avoid the extra subtraction step, let's use the method outlined in A Closer Look (page 240). As shown in the pie chart, the discount rate is 20%; therefore the net price rate is 80%. Use the proportion

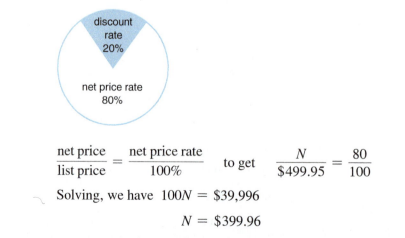

$$\frac{\text{net price}}{\text{list price}} = \frac{\text{net price rate}}{100\%} \qquad \text{to get} \qquad \frac{N}{\$499.95} = \frac{80}{100}$$

Solving, we have $100N = \$39{,}996$

$$N = \$399.96$$

(b) Pie charts will help you set up the correct proportions.

In other words, if the discount is 12% of the list price, then the net price is 100% − 12% or 88% or the list price.

Use the net price proportion, substitute, and solve.

$$\frac{\text{net price}}{\text{list price}} = \frac{\text{net price rate}}{100}$$

$$\frac{\$1011.12}{L} = \frac{88}{100}$$

$$88L = \$101,112$$

$$L = \$1149$$

(c) **First,** calculate the actual discount.

$$4(\$89.50) - \$304.30 = \$358 - \$304.30 = \$53.70$$

Then, set up the discount proportion. We use $358 as the list price (4 × $89.50).

$$\frac{D}{L} = \frac{R}{100} \quad \text{becomes} \quad \frac{\$53.70}{\$358} = \frac{R}{100}$$

$$358R = 5370$$

$$R = 15 \quad \text{The tires were discounted by 15\%.}$$

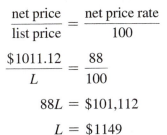

4 ⊗ 8 9 . 5 ⊖ 3 0 4 . 3 ⊜ →   53.7

⊗ 1 0 0 ⊘ 3 5 8 ⊜ →   15.

In some situations, multiple discounts are given on a purchase. These discounts are usually applied in succession so that each additional discount is applied to the newly reduced price.

EXAMPLE 4   **Retail Merchandising**   Retail buyers often negotiate two or more discounts from a vendor. Suppose a washing machine has a list price of $740, and the buyer obtains trade discounts of 30% and 10%, plus a cash discount of 3%.

To determine the final cost of the washing machine, we may apply the discounts in any order, but we must apply them in succession.

First, apply the 30% discount:

The net price rate is 100% − 30% = 70%.

$$\frac{P}{\$740} = \frac{70}{100}$$

$$P = \$518$$

Next, apply the 10% discount to this reduced price:

$$100\% - 10\% = 90\%$$

$$\frac{P}{\$518} = \frac{90}{100}$$

$$P = \$466.20$$

Finally, apply the 3% cash discount to the second reduced price:

$$100\% - 3\% = 97\%$$

$$\frac{P}{\$466.20} = \frac{97}{100}$$

$$P = \$452.214 \text{ or } \$452.21, \text{ rounded}$$

Using a calculator, we can perform all three steps as follows:

$$740 \times 70 \div 100 \times 90 \div 100 \times 97 \div 100 = \rightarrow \boxed{452.214} \quad \bullet$$

A Quick Method for Calculating Percentage   Using the proportion method for solving percent problems makes it easier to distinguish among the different types of problems. However, it seems that most practical applications involve finding the percentage (or part), and there is a quicker way to do this, especially using a calculator. Simply change the percent to a decimal number and multiply by the base.

EXAMPLE 5   Find 20% of $640.

Step 1   Change the percent to a decimal number.

$$20\% = 20.\% = 0.20 = 0.2$$

Step 2   Multiply this decimal equivalent by the base.

$$0.2 \times \$640 = \$128$$

This is the same answer you would get using a proportion.

$$\frac{P}{\$640} = \frac{20}{100} \quad \text{or} \quad P = \frac{\$640 \cdot 20}{100} = \$640(0.2) \quad \bullet$$

Your Turn   Use this quick method to solve the following problems.

(a) Find 35% of 80.   (b) What is 4.5% of 650?

(c) What is 125% of 18?   (d) Find 0.3% of 5000.

(e) **Retail Merchandising**   A set of patio furniture has a list price of $1478. A buyer obtains trade discounts of 25% and 15% and a cash discount of 5% from the supplier. What is the final cost to the buyer after all three discounts are applied?

Solutions   (a) $35\% = 35.\% = 0.35$ (b) $4.5\% = 04.5\% = 0.045$

$\qquad\qquad\quad 0.35 \times 80 = 28$ $\qquad\qquad\quad 0.045 \times 650 = 29.25$

(c) $125\% = 125.\% = 1.25$ (d) $0.3\% = 00.3\% = 0.003$

$\qquad\quad 1.25 \times 18 = 22.5$ $\qquad\quad 0.003 \times 5000 = 15$

(e) The first discount is 25%. Find 75% of the list price to apply the first reduction.

$$0.75 \times \$1478 = \$1108.50$$

The second discount is 15%. Find 85% of $1108.50 to apply the second reduction.

$$0.85 \times \$1108.50 = \$942.225 \qquad \text{(Do not round this intermediate result.)}$$

The third discount is 5%. Find 95% of $942.225 to apply the final reduction.

$$0.95 \times \$942.225 = \$895.11375$$

The final cost to the buyer, rounded to the nearest cent, is $895.11

Using a calculator, you can apply all three discounts in succession, as follows:

1478 ☒ **.75** ☒ **.85** ☒ **.95** 🟰 →   *895.11375*

Sales Tax Taxes are almost always calculated as a percent of some total amount. Property taxes are written as some fraction of the value of the property involved. Income taxes are most often calculated from complex formulas that depend on many factors. We cannot consider either income or property taxes here.

A **sales tax** is an amount calculated from the actual list price of a purchase. It is then added to the list price to determine the total cost. Retail sales tax rates in the United States are set by the individual states, counties, and cities and vary from 0 to 10% of the sales price. A sales tax of 6% is often stated as "6 cents on the dollar" because 6% of $1.00 equals 6 cents.

Using a proportion equation, we can write the formulas for sales tax and total cost as follows:

$$\frac{\text{sales tax}}{\text{list price}} = \frac{\text{tax rate}}{100}$$

$$\text{Total cost} = \text{list price} + \text{sales tax}$$

EXAMPLE 6 **Electrical Trades** If the retail sales tax rate is 7.75% in a certain city, how much would you pay for a digital multimeter with a list price of $76.50?

Using a proportion equation, we have

$$\frac{\text{sales tax}}{\text{list price}} = \frac{\text{tax rate}}{100}$$

or $$\frac{\text{sales tax } T}{\$76.50} = \frac{7.75}{100}$$

$$100T = (\$76.50)(7.75)$$

$$100T = \$592.875$$

$$T = \$5.93 \text{ rounded}$$

Using the equivalent "quick method," we have

$$\text{Sales tax} = 7.75\% \text{ of } \$76.50 \qquad 7.75\% = 07.75 = 0.0775$$

$$= (0.0775)(\$76.50)$$

$$= \$5.92875 = \$5.93 \text{ rounded}$$

$$\text{Total cost} = \text{list price} + \text{sales tax}$$

$$= \$76.50 + \$5.93$$

$$= \$82.43$$

.0775 ☒ **76.5** 🟰 →   *5.92875* ➕ **76.5** 🟰 →   *82.42875*

Many stores provide their salesclerks with computerized cash registers that calculate sales tax automatically. However, all consumers and most small business or shop owners still need to be able to do sales tax calculations. ●

Learning Help   A quicker way to calculate the total cost with tax is to add 100 to the tax rate and use this total for R. The revised proportion becomes

$$\frac{\text{total cost, } C}{\text{list price}} = \frac{100 + \text{tax rate}}{100}$$

For Example 6, we would write

$$\frac{C}{\$76.50} = \frac{100 + 7.75}{100}$$

$$\frac{C}{\$76.50} = \frac{107.75}{100}$$

$$C = \frac{(\$76.50)(107.75)}{100}$$

$$C = \$82.43, \text{ rounded}$$

Using the equivalent quick calculation method, we have

Total cost = 107.75% of $76.50   107.75% = 1.07.75 = 1.0775

= (1.0775)($76.50)

= $82.43 rounded ●

Your Turn   Here are a few problems in calculating sales tax.

General Trades   Find the total cost for each of the following items after applying the tax rate shown in parentheses.

(a) A roll of plastic electrical tape at $2.79   (5%)

(b) A random orbit sander priced at $67.50   (6%)

(c) A tape measure priced at $21.99 $(4\frac{1}{2}\%)$

(d) A new car priced at $23,550   (4%)

(e) A band saw priced at $346.50 $(6\frac{1}{2}\%)$

(f) A toggle switch priced at $10.49 $(7\frac{3}{4}\%)$

Answers

	Tax	Total Cost	Quick Calculation
(a)	$0.14	$2.93	1.05 × $2.79 = $2.9295 ≈ $2.93
(b)	$4.05	$71.55	1.06 × $67.50 = $71.55
(c)	$0.99	$22.98	1.045 × $21.99 = $22.97955 ≈ $22.98
(d)	$942	$24,492	1.04 × $23,550 = $24,492
(e)	$22.52	$369.02	1.065 × $346.50 = $369.0225 ≈ $369.02
(f)	$0.81	$11.30	1.0775 × $10.49 = $11.302975 ≈ $11.30

Interest   In modern society we have set up complex ways to enable you to use someone else's money. A *lender,* with money beyond his or her needs, supplies cash to a *borrower,* whose needs exceed his or her money. The money is called a **loan.**

Interest is the amount the lender is paid for the use of his or her money. Interest is the money you pay to use someone else's money. The more you use and the longer you use it, the more interest you must pay. **Principal** is the amount of money lent or borrowed.

When you purchase a house with a bank loan, a car or a refrigerator on an installment loan, or gasoline on a credit card, you are using someone else's money, and you pay interest for that use. If you are on the other end of the money game, you may earn interest for money you invest in a savings account or in shares of a business.

Simple interest* can be calculated using the following formulas:

$$\frac{\text{annual interest}}{\text{principal}} = \frac{\text{annual interest rate}}{100}$$

$$\text{Total interest} = \text{annual interest} \times \text{time in years}$$

EXAMPLE 7   **Life Skills**   Suppose that you have $5000 in a long-term CD (certificate of deposit) earning 1.25% interest annually in your local bank. How much interest do you receive in (a) a year? (b) 9 months?

(a) $\dfrac{\text{annual interest } I}{\$5000} = \dfrac{1.25}{100}$

$$100I = \$6250$$

$$I = \$62.50$$

The total interest for one year is $62.50.

(b) Because 9 months is $\frac{9}{12}$ or $\frac{3}{4}$ of a year, we simply multiply the annual interest amount from part (a) by $\frac{3}{4}$, or 0.75.

$$\text{Total interest} = \text{annual interest} \times \text{time in years}$$

$$= \$62.50 \times 0.75$$

$$= \$46.875 \text{ or } \$46.88, \text{ rounded}$$

The following calculator sequence shows how to use the quick calculation method to get the answers to both (a) and (b):

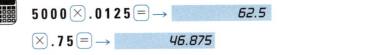

5000 ⊠ .0125 ⊟ →   ⎡ 62.5 ⎤

⊠ .75 ⊟ →   ⎡ 46.875 ⎤   ●

Your Turn   **Trades Management**   Most of us play the money game from the other side of the counter. Suppose that your small business is in need of cash and you arrange to obtain a loan from a bank. You borrow $12,000 at 9% per year for 3 months. How much interest must you pay?

Try to set up and solve this problem exactly as we did in Example 7.

---

*In many applications, compound interest is used instead of simple interest. The formula for compound interest is beyond the scope of this chapter.

Solution

$$\frac{\text{annual interest } I}{\$12{,}000} = \frac{9}{100}$$

$$I = \frac{\$12{,}000 \times 9}{100} = \$1080$$

The time of the loan is 3 months, so the time in years is

$$\frac{3 \text{ months}}{12 \text{ months}} = \frac{3}{12} \text{ yr}$$

$$= 0.25 \text{ yr}$$

Total interest = annual interest × time in years

$$= \$1080(0.25)$$

$$= \$270$$

By the quick calculation method,

$$\text{Total interest} = \$12{,}000\,(9\%)\left(\frac{3}{12} \text{ yr}\right)$$

$$= \$12{,}000(0.09)(0.25)$$

$$= \$270$$

1 2 0 0 0 ☓ **. 0 9** ☓ **3** ÷ **1 2** = → ⎯⎯⎯⎯⎯⎯ *270.*

Depending on how you and the bank decide to arrange it, you may be required to pay the total principal ($12,000) plus interest ($270) all at once, at the end of three months, or according to some sort of regular payment plan—for example, pay $4090 each month for three months.

How Does a Credit Card Work?

Many forms of small loans, such as credit card loans, charge interest by the month. These are known as *revolving credit* or *charge account* plans. (If you borrow a lot of money this way, *you* do the revolving and may run in circles for years trying to pay it back!) Essentially, you buy now and pay later. Generally, if you repay the full amount borrowed within 25 or 30 days, there is no charge for the loan.

After the first pay period of 25 or 30 days, you pay a percent of the unpaid balance each month, usually between $\frac{1}{2}$ and 2%. In addition, you must pay a yearly fee and some minimum amount each month, usually $10 or 10% of the unpaid balance, whichever is larger. Often, you are also charged a small monthly amount for insurance premiums. (The credit card company insures itself against your defaulting on the loan or disappearing, and you pay for their insurance.)

Let's see how it works. Suppose that you go on a short vacation and pay for gasoline, lodging, and meals with your Handy Dandy credit card. A few weeks later you receive a bill for $1000. You can pay it within 30 days and owe no interest, or you can pay over a longer period of time. Suppose you agree to pay $100 per month with a monthly interest rate of $1\frac{1}{2}\%$. The calculations look like this:

Month 1 $1\frac{1}{2}\%$ of $1000 = $1000 × 0.015 = $15   owe $1015
          pay $100 and carry $915 over to next month

(continued)

Month 2 $1\frac{1}{2}\%$ of \$915 = \$13.73   owe \$928.73
             pay \$100 and carry \$828.73 over to next month

. . . and so on.

Eleven months later you will have repaid the \$1000 loan and all interest.

The $1\frac{1}{2}\%$ per month interest rate seems small, but it is equivalent to between 15 and 18% annually. You pay at a high rate for the convenience of using the credit card and the no-questions-asked ease of getting the loan.

Commission   Salespeople are often paid on the basis of their success at selling and receive a **commission** or share of the sales receipts. Commission is usually described as a percent of sales income:

$$\frac{\text{commission}}{\text{sales}} = \frac{\text{commission rate}}{100}$$

EXAMPLE 8   **Life Skills**   Suppose that your job as a lumber broker pays 12% commission on all sales. How much do you earn from a sale of \$4000?

$$\frac{\text{commission } C}{\$4000} = \frac{12}{100}$$

$$C = \frac{\$4000(12)}{100}$$

$$C = \$480 \quad \text{You earn \$480 commission.}$$

By the quick calculation method,

Commission $= (\$4000)(12\%) = (\$4000)(0.12) = \$480$ ●

EXAMPLE 9   **Life Skills**   A computer salesperson is paid a salary of \$800 per month plus a 4% commission on sales. What amount of sales must he generate in order to earn \$40,000 per year in total compensation?

Salary $= \$800$ per month $\times$ 12 months

        $= \$9600$ per year

Commission $=$ total compensation $-$ salary

        $= \$40,000 - \$9600 = \$30,400$

The commission needed is \$30,400. Write the commission proportion. The sales, S, is not known.

$$\frac{\$30,400}{S} = \frac{4}{100}$$

$$S = \frac{\$30,400 \cdot 100}{4}$$

$$S = \$760,000$$

He must sell \$760,000 worth of computers to earn a total of \$40,000 per year.   ●

Your Turn   In the following two problems, concentrate on identifying each quantity in the commission proportion.

(a) **Life Skills**   A sales representative generated $360,000 in sales last year and was paid a salary of $27,000. If she were to switch to straight commission compensation, what rate of commission would be needed for her commission to match her salary?

(b) **Office Services**   At the Happy Bandit used car lot, each salesperson receives a 6% commission on sales. What would a salesperson earn if she sold a previously owned vehicle for $12,995?

Solutions   (a) Sales = $360,000   Commission = $27,000   Rate is unknown.

$$\frac{\$27{,}000}{\$360{,}000} = \frac{R}{100}$$

$$R = \frac{27{,}000 \cdot 100}{360{,}000}$$

$$R = 7.5\%$$ She needs to be paid a 7.5% rate of commission in order to match her salary.

(b) $$\frac{C}{\$12{,}995} = \frac{6}{100}$$

$$C = \frac{\$12{,}995(6)}{100}$$

$$C = \$779.70$$

or   Commission = ($12,995)(6%) = ($12,995)(0.06)

= $779.70

Small Loans

Sooner or later, most people find it necessary to borrow money. If you do, you will want to know beforehand how the loan process works. Suppose that you borrow $200 and the loan company specifies that you repay it at $25 per month plus interest at 3% per month on the unpaid balance. What interest do you actually pay?

Month 1	3% of $200 = $ 6.00	you pay	$25 + $6.00 = $ 31.00
Month 2	3% of $175 = $ 5.25	you pay	$25 + $5.25 = $ 30.25
Month 3	3% of $150 = $ 4.50	you pay	$25 + $4.50 = $ 29.50
Month 4	3% of $125 = $ 3.75	you pay	$25 + $3.75 = $ 28.75
Month 5	3% of $100 = $ 3.00	you pay	$25 + $3.00 = $ 28.00
Month 6	3% of $ 75 = $ 2.25	you pay	$25 + $2.25 = $ 27.25
Month 7	3% of $ 50 = $ 1.50	you pay	$25 + $1.50 = $ 26.50
Month 8	3% of $ 25 = $ 0.75	you pay	$25 + $0.75 = $ 25.75
	$ 27.00		$ 227.00

Total interest is $27.00

Total of eight loan payments

(continued)

The loan company might also set it up as eight equal payments of $227.00 ÷ 8 = $28.375 or $28.38 per month.

The 3% monthly interest rate seems small, but it amounts to about 20% per year.

A bank loan for $200 at 12% for 8 months would cost you

$$(12\% \text{ of } \$200)\left(\frac{8}{12}\right)$$

or $\quad (\$24)\left(\dfrac{8}{12}\right)$

or $\quad \$16 \quad$ Quite a difference for a $200 loan.

The loan company demands that you pay more, and in return they are less worried about your ability to meet the payments. For a bigger risk, they want a higher rate of interest.

Efficiency When energy is converted from one form to another in any machine or conversion process, it is useful to talk about the efficiency of the machine or the process. The **efficiency** of a process is a fraction comparing the energy or power output to the energy or power input. It is usually expressed as a percent.

$$\frac{\text{output}}{\text{input}} = \frac{\text{efficiency}}{100}$$

EXAMPLE 10 **Automotive Trades** An auto engine is rated at 175 hp and is found to deliver only 140 hp to the transmission. What is the efficiency of the process?

$$\frac{\text{output}}{\text{input}} = \frac{\text{efficiency } E}{100}$$

$$\frac{140}{175} = \frac{E}{100}$$

$$E = \frac{140 \cdot 100}{175}$$

$$E = 80\%$$

●

Your Turn **General Trades** A gasoline shop engine rated at 65 hp is found to deliver 56 hp through a belt drive to a pump. What is the efficiency of the drive system?

Solution output = 56 hp, $\quad$ input = 65 hp

$$\frac{56}{65} = \frac{E}{100}$$

$$E = \frac{56 \cdot 100}{65} = 86\% \text{ rounded}$$

56 ⊗ 1 0 0 ⊝ 6 5 ⊜ → 86.15384615

The efficiency of any practical system will always be less than 100%—you can't produce an output greater than the input!

Tolerances   On technical drawings or other specifications, measurements are usually given with a **tolerance** showing the allowed error. For example, the fitting shown in the margin has a length of

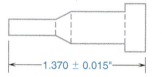

1.370 $\pm$ 0.015 in.

This means that the dimension shown must be between

1.370 + 0.015 in.   or   1.385 in.

and

1.370 − 0.015 in.   or   1.355 in.

The **tolerance limits** are 1.385 in. and 1.355 in.

Very often the tolerance is written as a percent.

$$\frac{\text{tolerance}}{\text{measurement}} = \frac{\text{percent tolerance}}{100}$$

EXAMPLE 11   To rewrite 1.370 $\pm$ 0.015 in. with a percent tolerance, substitute into the proportion equation and solve as follows:

$$\frac{0.015}{1.370} = \frac{\text{percent tolerance } R}{100}$$

$$R = \frac{(0.015)(100)}{1.370}$$

$$R = 1.0948\ldots \quad \text{or} \quad 1.09\% \text{ rounded}$$

The dimension would be written 1.370 in. $\pm$ 1.09%.   ●

Your Turn   Rewrite the following dimension with a percent tolerance:

1.426 $\pm$ 0.010 in. = _____

Solution   The tolerance is 0.010 in. and the measurement is 1.426 in. Substituting into the tolerance proportion, we have

$$\frac{0.010}{1.426} = \frac{\text{percent tolerance } R}{100}$$

$$R = \frac{(0.010)(100)}{1.426}$$

$$R = 0.70126\ldots \quad \text{or} \quad 0.70\% \text{ rounded}$$

Rewriting the dimension with a percent tolerance, we have

1.426 $\pm$ 0.010 in. = 1.426 in. $\pm$ 0.70%

If the dimension is given in metric units, the procedure is exactly the same.

EXAMPLE 12 The dimension

$$315 \pm 0.25 \text{ mm}$$

can be converted to percent tolerance.

tolerance $\longrightarrow$
measurement $\longrightarrow$
$$\frac{0.25}{315} = \frac{R}{100}$$

$$R = \frac{(0.25)(100)}{315}$$

$$R = 0.07936\ldots \qquad \text{or} \qquad 0.08\% \text{ rounded}$$

The dimension is 315 mm $\pm$ 0.08%.

.25 ⊠ 100 ⊟ 315 ⊟ → 0.079365079 •

If the tolerance is given as a percent, it is easy to use the proportion on page 252 to calculate the tolerance as a dimension value.

EXAMPLE 13 To convert 2.450 in. $\pm$ 0.20% to a dimension value, note that the measurement is 2.450 in., the percent tolerance is 0.20%, and the actual tolerance T is unknown. Therefore

$$\frac{\text{tolerance } T}{2.450} = \frac{0.20}{100}$$

$$T = \frac{(2.45)(0.2)}{100}$$

$$T = 0.0049 \text{ in.} \qquad \text{or} \qquad 0.005 \text{ in. rounded}$$

We can rewrite the dimension without a percent as follows:

$$2.450 \text{ in.} \pm 0.20\% = 2.450 \text{ in.} \pm 0.005 \text{ in.}$$

2.45 ⊠ .2 ⊟ 100 ⊟ → 0.0049 •

More Practice Now, for some practice in working with tolerances, convert the measurement tolerances below to percents, and vice versa. Round the tolerances to three decimal places and round the percents to two decimal places.

Measurement	Tolerance	Percent Tolerance
2.345 in.	±0.001 in. ±0.002 in. ±0.005 in. ±0.010 in. ±0.125 in.	
274 mm	±0.10 mm ±0.20 mm ±0.50 mm	
3.475 in.		±0.10% ±0.20%
123 mm		±0.05% ±0.15%

Answers

Measurement	Tolerance	Percent Tolerance
2.345 in.	±0.001 in.	±0.04%
	±0.002 in.	±0.09%
	±0.005 in.	±0.21%
	±0.010 in.	±0.43%
	±0.125 in.	±5.33%
274 mm	±0.10 mm	±0.04%
	±0.20 mm	±0.07%
	±0.50 mm	±0.18%
3.475 in.	±0.003 in.	±0.10%
	±0.007 in.	±0.20%
123 mm	±0.062 mm	±0.05%
	±0.185 mm	±0.15%

Percent Change   In many situations in technical work and the trades, you may need to find a **percent increase** or **percent decrease** in a given quantity. Use the following formula to calculate percent increase or decrease:

$$\frac{\text{amount of increase or decrease}}{\text{original amount}} = \frac{\text{percent of increase or decrease}}{100}$$

EXAMPLE 14   **Electrical Trades**   Suppose that the output of a certain electrical circuit is 20 amperes (A), and the output increases by 10%. What is the new value of the output?

Step 1   Write the percent change proportion.

$$\frac{\text{amount of increase}}{\text{original amount}} = \frac{\text{percent increase}}{100}$$

Step 2   Substitute. The original amount is 20 A and the percent increase is 10%.

$$\frac{\text{amount of increase } i}{20} = \frac{10}{100}$$

Step 3   Solve.

$$100i = 200$$

$$i = 2 \text{ A}$$

Step 4   Calculate the new value.

New value = original amount + increase

$$= 20 \text{ A} + 2 \text{ A} = 22 \text{ A} \qquad \bullet$$

Learning Help   As we learned with sales tax, a quicker way to solve the problem is to add 100 to the percent increase and use this total for the rate R. The revised proportion becomes

$$\frac{\text{new value } V}{\text{original amount}} = \frac{100 + \text{percent increase}}{100}$$

For the previous example,

$$\frac{V}{20} = \frac{100 + 10}{100}$$

$$\frac{V}{20} = \frac{110}{100}$$

$$V = \frac{20 \cdot 110}{100}$$

$$V = 22 \text{ A } \bullet$$

Percent decrease can be calculated in a similar way.

EXAMPLE 15

Manufacturing Suppose that the output of a certain machine at the Acme Gidget Co. is 600 gidgets per day. If the output decreases by 8%, what is the new output?

Percent *decrease* involves subtracting and, as with percent increase, we can subtract either at the beginning or at the end of the calculation. Subtracting at the end, we follow these steps:

Step 1 Write the percent change proportion.

$$\frac{\text{amount of decrease}}{\text{original amount}} = \frac{\text{percent decrease}}{100}$$

Step 2 Substitute: Original amount = 600; percent decrease = 8%.

$$\frac{\text{amount of decrease } d}{600} = \frac{8}{100}$$

Step 3 Solve.

$$d = \frac{600 \cdot 8}{100}$$

$$d = 48 \text{ gidgets per day}$$

Step 4 Now subtract to find the new value.

New value = original amount − amount of decrease

$$= 600 - 48 = 552 \text{ gidgets per day} \qquad \bullet$$

Learning Help

We can also solve the problem by subtracting the percent of decrease at the beginning of the problem.

$$\frac{\text{new value } V}{\text{original amount}} = \frac{100 - \text{percent decrease}}{100}$$

For Example 15,

$$\frac{\text{new value } V}{600} = \frac{100 - 8}{100}$$

$$\frac{V}{600} = \frac{92}{100}$$

$$100V = 55{,}200$$

$$V = 552 \text{ gidgets per day} \bullet$$

Your Turn   Try these problems involving percent change. In each, add or subtract the percents at the beginning of the problem.

(a) **Automotive Trades**   The markup, or profit, on a part is the amount a dealer adds to his cost to determine the retail price, or the amount that the customer pays. If an oil filter costs a dealer $7.85 and the markup is 40%, what retail price does he charge the customer?

(b) **Electrical Trades**   Normal line voltage, 115 volts, drops 3.5% during a system malfunction. What is the reduced voltage?

(c) **Trades Management**   Because of changes in job specifications, the cost of a small construction job is increased 25% from the original cost of $2475. What is the cost of the job now?

(d) **Welding**   By using automatic welding equipment, the time for a given job can be reduced by 40% from its original 30 hours. How long will the job take using the automatic equipment?

Solutions   (a) The markup is the percent increase.

$$\frac{\text{markup}}{\$7.85} = \frac{40}{100}$$

$$\text{Markup} = \frac{(\$7.85)(40)}{100} = \$3.14$$

Retail price = cost + markup = $7.85 + $3.14 = $10.99

(b) $$\frac{\text{voltage drop}}{115 \text{ volts}} = \frac{3.5}{100}$$

$$\text{Voltage drop} = \frac{115(3.5)}{100} = 4.025 \text{ volts}$$

Reduced voltage = normal voltage − voltage drop

$$= 115 \text{ V} - 4.025 \text{ V}$$

$$= 110.975 \text{ V or } 111 \text{ V, rounded}$$

115 $\times$ **3.5** $\div$ **100** $=$ $\rightarrow$ *4.025* **115** $-$ ANS $=$ $\rightarrow$ *110.975*

(c) $$\frac{\text{cost increase}}{\$2475} = \frac{25}{100}$$

$$\text{Cost increase} = \frac{\$2475(25)}{100} = \$618.75$$

New cost = $2475 + $618.75 = $3093.75

(d) $$\frac{\text{time reduction}}{30} = \frac{40}{100}$$

$$\text{Time reduction} = \frac{30 \cdot 40}{100} = 12 \text{ hours}$$

New time = 30 hours − 12 hours = 18 hours

In trade and technical work, changes in a measured quantity are often specified as a percent increase or decrease.

EXAMPLE 16   **General Trades**   If the reading on a pressure valve increases from 30 psi to 36 psi, the percent increase can be found by using the following proportion:

$$\text{The change} \rightarrow \frac{36 \text{ psi} - 30 \text{ psi}}{\text{Original value} \rightarrow \quad 30 \text{ psi}} = \frac{\text{percent increase}}{100}$$

$$\frac{6}{30} = \frac{R}{100}$$

$$R = \frac{6 \cdot 100}{30} = 20\%$$

The reading on the pressure valve increased by 20%.

$(\boxed{36} \boxed{-} \boxed{30}) \boxed{\times} \boxed{100} \boxed{\div} \boxed{30} \boxed{=} \rightarrow \boxed{\qquad 20.}$ ●

EXAMPLE 17   **Water/Wastewater Treatment**   During the recent drought in California, certain water districts were requiring residents to reduce their water usage by specific amounts. The Jones family lived in a district that required residents to reduce their monthly usage by 30% compared to their average for the prior year. Suppose their prior year average was 36,400 gallons per month, and they reduced their usage to 26,800 gallons per month for the first quarter of the current year. To determine whether or not they met the district's goal, we set up a percent decrease proportion;

$$\text{The change} \longrightarrow \frac{36,400 - 26,800}{\text{The original value} \longrightarrow \quad 36,400} = \frac{\text{percent decrease}}{100}$$

$$\frac{9600}{36,400} = \frac{R}{100}$$

$$R = \frac{100 \cdot 9600}{36,400}$$

$$R = 26.37 \ldots \quad \approx 26\%$$

Despite their 26% reduction in water usage, the Jones family fell short of the water district's 30% target.   ●

! Careful   Students are often confused about which number to use for the base in percent change problems. The difficulty usually arises when they think in terms of which is the bigger or smaller value. Always use the *original* value as the base. In percent increase problems, the original or base number will be the smaller number, and in percent decrease problems, the original or base number will be the larger number. ●

EXAMPLE 18   **Life Skills**   A house purchased for $400,000 during the housing boom decreased in value by 30% during the recent housing slump. By what percent must it now increase in value in order to be worth its original purchase price? (Round your answer to one decimal digit.)

Many people mistakenly believe that 30% is the answer to this problem: a 30% increase should make up for the 30% decrease, right? However, the percent decrease is based on the original price of $400,000, whereas the percent increase is based on the price after devaluation ($280,000). As you will see if you perform the calculation, 30% of 400,000 is quite a bit more than 30% of 280,000. Therefore, the required increase must be larger than 30%.

First, determine the dollar amount D of the decrease in value.

$$\frac{D}{\$400,000} = \frac{30}{100}$$

$$100D = 12,000,000$$

$$D = \$120,000$$

Then, find the new value V of the house.

$$V = \$400,000 - \$120,000$$

$$= \$280,000$$

Finally, note that the house must increase in value by $120,000 in order to be worth its original purchase price. But the percent P by which it must increase must be calculated using the current value of $280,000 as the base:

$$\frac{P}{100} = \frac{\$120,000}{\$280,000}$$

$$280,000P = 12,000,000$$

$$P \approx 43\%, \text{ rounded}$$ ●

A Closer Look   Here is a quick calculator method for doing Example 18:

First, multiply the original value by 0.70 to calculate the decreased value. $(100\% - 30\% = 70\% = 0.70)$

$400000 \times .7 = \rightarrow$ *280000.*

Then, divide the original value by this decreased value.

$400000 \div \boxed{\text{ANS}} = \rightarrow$ *1.428571429*

Finally, this tells us that the original value is about 1.43 times the decreased value, or about 143% of it. This means that the decreased value must increase by approximately 43% in order to return to the original price.   ●

Note   We do not actually need to know the original price in order to work this problem. The final answer will be the same no matter what the original amount is. For example, if we use $100 as the original price, then the first step becomes $100 \times .7 = \rightarrow$ *70.* The second step becomes $100 \div \boxed{\text{ANS}} = \rightarrow$ *1.428571429*, which gives us the same answer, approximately 43%.   ●

Your Turn   Now try these problems to sharpen your ability to work with percent changes.

(a) **HVAC**   A poorly insulated house was found to lose an average of 465,800 Btu per day. After insulation was added, the average daily heat loss decreased to 182,500 Btu. By what percent did the heat loss decrease after the insulation was added? (Round to the nearest percent.)

(b) **Electrical Trades**   What is the percent increase in voltage when the voltage increases from 65 volts to 70 volts?

(c) **Allied Health**   A woman who weighed 120 lb before becoming pregnant was 15% heavier after eventually giving birth. What percent of this increased weight must she lose in order to return to her original weight?

(d) **Automotive Trades**   A water pump costs a parts department $78.50. The pump was sold to a customer for $109.90. What was the percent markup of the pump?

Solutions   (a) To find the amount of decrease, subtract 182,500 from 465,800.

$$\text{Decrease in heat loss} \rightarrow \frac{465,800 - 182,500}{465,800} = \frac{\text{percent decrease}}{100}$$

$$\frac{283,300}{465,800} = \frac{R}{100}$$

$$465,800R = 28,330,000$$

$$R = 60.82 \ldots \text{ or } 61\%, \text{ rounded}$$

The heat loss decreased by about 61%.

(b) To find the amount of increase, subtract 65 volts from 70 volts.

$$\frac{70 \text{ volts} - 65 \text{ volts}}{\text{Original value} \rightarrow 65 \text{ volts}} = \frac{\text{percent increase}}{100}$$

$$\frac{5}{65} = \frac{R}{100}$$

$$R = \frac{5 \cdot 100}{65} = 7.6923 \ldots \text{ or } 7.7\% \text{ rounded}$$

The voltage increased by about 7.7%.

(c) **First,** determine the increase I in weight. $\dfrac{I}{120} = \dfrac{15}{100}$

$$100I = 1800$$

$$I = 18 \text{ lb}$$

Next, calculate the woman's new weight W. $W = 120 \text{ lb} + 18 \text{ lb} = 138 \text{ lb}$

Finally, she must lose 18 lb to return to her original weight. Using 138 lb as the new base, determine what percent R of her new weight she must lose.

$$\frac{18}{138} = \frac{R}{100}$$

$$138R = 1800$$

$$R \approx 13\%$$

The woman must lose about 13% of her increased weight.

(d) To find the amount of increase or markup, subtract the cost from the retail price.

Amount of markup $= \$109.90 - \$78.50 = \$31.40$

The percent markup is the percent increase over the *cost*.

$$\text{Amount of markup} \rightarrow \frac{\$31.40}{\$78.50} = \frac{\text{percent markup } R}{100}$$

$$R = \frac{(31.40)(100)}{78.50} = 40\%$$

Now, turn to Exercises 4-5 for a set of problems involving these special applications of percent calculations.

Exercises 4-5	Special Applications of Percent Calculations

1. **Electronics**   A CB radio is rated at 7.5 watts, and actual measurements show that it delivers 4.8 watts to its antenna. What is its efficiency?

2. **Electrical Trades**   An electric motor uses 6 kW (kilowatts) at an efficiency of 63%. How much power does it deliver?

3. **Electrical Trades**   An engine supplies 110 hp to an electric generator, and the generator delivers 70 hp of electrical power. What is the efficiency of the generator?

4. **Electronics**   Electrical resistors are rated in ohms and color coded to show both their resistance and percent tolerance. For each of the following resistors, find its tolerance limits and actual tolerance.

Resistance (ohms)	Limits (ohms)		Tolerance (ohms)
5300 ± 5%	_____	to _____	_____
2750 ± 2%	_____	to _____	_____
6800 ± 10%	_____	to _____	_____
5670 ± 20%	_____	to _____	_____

5. **Automotive Trades**   A 120-hp automobile engine delivers only 81 hp to the driving wheels of the car. What is the efficiency of the transmission and drive mechanism?

6. **Electronics**   An electrical resistor is rated at 500 ohms ± 10%. What is the highest value its resistance could have within this tolerance range?

7. **Roofing**   To estimate the number of board feet of one-inch sheathing needed for a section of roof, add 15% to the area (in square feet) of the roof. If a section of roof has an area of 1200 sq ft, how many board feet of such sheathing will be needed?

8. **Life Skills**   If you earn 12% commission on sales of $4200, what actual amount do you earn?

9. **Metalworking**   On a cutting operation, 2 sq ft of sheet steel is wasted for every 16 sq ft used. What is the percent waste?

10. **Metalworking**   Four pounds of a certain bronze alloy is one-sixth tin, 0.02 zinc, and the rest copper. Express the portion of each metal in (a) percents and (b) pounds.

11. **Metalworking**   An iron casting is made in a mold with a hot length of 16.40 in. After cooling, the casting is 16.25 in. long. What is the shrinkage as a percent?

12. **Manufacturing**   Because of friction, a pulley block system is found to be only 83% efficient. What actual load can be raised if the theoretical load is 2000 lb?

13. **Manufacturing**   A small gasoline shop engine develops 65 hp at 2000 rpm. At 2400 rpm, its power output is increased by 25%. What actual horsepower does it produce at 2400 rpm?

14. **Carpentry**   An online tool supplier offers a 15% discount on orders of $100 or more. If a carpenter orders a set of router bits priced at $99.99, a digital caliper priced at $39.99, and a seven-piece ratchet wrench set priced at $59.99, what would be the total cost after the discount is applied?

15. **Machine Trades** Specifications call for a hole in a machined part to be 2.315 in. in diameter. If the hole is measured to be 2.318 in., what is the machinist's percent error?

16. **Trades Management** A welding shop charges the customer 185% of its labor and materials costs to cover overhead and profit. If a bill totals $822.51, what is the cost of labor and materials?

17. Complete the following table:

Measurement	Tolerance	Percent Tolerance
3.425 in.	±0.001 in.	(a)
3.425 in.	±0.015 in.	(b)
3.425 in.	(c)	±0.20%

18. **General Trades** If you receive a pay increase of 12%, what is your new pay rate, assuming that your old rate was $16.80 per hour?

19. **Manufacturing** The pressure in a hydraulic line increases from 40 psi to 55 psi. What is the percent increase in pressure?

20. **Painting** The cost of the paint used in a redecorating job is $123.20. This is a reduction of 20% from its initial cost. What was the original cost?

21. **Life Skills** What sales tax, at a rate of $6\frac{1}{2}\%$, must you pay on the purchase of an internal hard disk drive costing $149.99?

22. **Interior Design** An interior designer working for a department store is paid a weekly salary of $350 plus a 10% commission on total sales. What would be his weekly pay if his total sales were $6875.00 for the week?

23. **Retail Merchandising** A one-terabyte external hard drive is on sale at a 25% discount. If the original price was $125, what is the sale price?

24. **Trades Management** Chris paid for the initial expenses in setting up her upholstery shop with a bank loan for $39,000 for 5 years at $8\frac{1}{2}\%$ annual interest.
    (a) What total interest will she pay?
    (b) If she pays off the interest only in equal monthly installments, what will be her monthly payments?

25. **Carpentry** A 7-in. medium-angle grinder is on sale for $104.30 and is marked as being 30% off. What was the original list price?

26. **Automotive Trades** Driving into a 20-mph breeze cuts gas mileage by about 12%. What will be your gas mileage in a 20-mph wind if it is 24 mi/gal with no wind?

27. **Automotive Trades** Supplier A offers a mechanic a part at 35% off the retail price of $68.40. Supplier B offers him the same part at 20% over the wholesale cost of $35.60. Which is the better deal, and by how much?

28. **Automotive Trades** An auto mechanic purchases the parts necessary for a repair for $126.40 and sells them to the customer at a 35% markup. How much does the mechanic charge the customer?

29. **Automotive Trades** A new tire had a tread depth of $\frac{3}{8}$ in. After one year of driving, the tire had $\frac{9}{32}$ in. of tread remaining. What percent of the tread was worn?

30. **Printing**   A printer agrees to give a nonprofit organization a 10% discount on a printing job. The normal price to the customer is $1020, and the printer's cost is $850. What is the printer's percent profit over cost after the discount is subtracted from the normal price?

31. **Water/Wastewater Treatment**   A treatment plant with a capacity of 20 MGD (million gallons per day) has a normal daily flow equal to 60% of capacity. When it rains the flow increases by 30% of the normal flow. What is the flow on a rainy day?

32. **Construction**   A retaining wall requires 28 cu yd of concrete. A good rule of thumb states that 10% should be added for handling loss. Finally, another 50% on top of this increased amount will determine the combined amounts of cement, sand, and gravel required for mixing the concrete. Calculate the combined amounts of the three components to the nearest cubic yard.

33. **Fire Protection**   The pressure reading coming out of a pump is 180 lb. Every 50-ft section reduces the pressure by about 3%. What will the nozzle pressure be at the end of six 50-ft sections? (*Hint:* One reduction of 18% is *not* equivalent to six reductions of 3% each. You must do this the long way.)

34. **Machine Trades**   Complete the following table of tolerances for some machine parts:

Measurement	Tolerance	Maximum	Minimum	Percent Tolerance
1.58 in.	± 0.002 in.			
	± 0.005 in.			
				± 0.15%
0.647 in.	± 0.004 in.			
	± 0.001 in.			
				± 0.20%
165.00 mm	± 0.15 mm			
	± 0.50 mm			
				± 0.05%
35.40 mm	± 0.01 mm			
	± 0.07 mm			
				± 0.05%
				± 0.08%
				± 0.10%

35. **Electrical Trades**   An electrician purchases some outdoor lighting for a customer at a wholesale price of $928.80. The customer would have paid the retail price of $1290. What was the wholesale discount?

36. **Life Skills**   In problem 35, what will the total retail price be after 7.75% tax is added?

37. **Trades Management**   A contractor takes a draw of $46,500 from a homeowner to start a remodeling job. He uses $22,450 to purchase materials and deposits the remaining money in a six-month CD paying 0.47% *annual* interest. How much interest does he earn on the remaining funds during the six-month term?

38. **Culinary Arts** A wholesale food sales rep is paid a base salary of $1200 per month plus a $3\frac{1}{2}\%$ commission on sales. How much sales must she generate in order to earn $35,000 per year in total compensation?

39. **Life Skills** A sales rep for roofing materials figures he can generate about $1,800,000 in business annually. What rate of commission does he need in order to earn $50,000?

40. **Carpentry** A portable hand planer originally selling for $149.50 is on sale for $119.60. By what percent is it discounted?

41. **Trades Management** Last year, the owner of the Fixum Auto Repair Shop paid a premium of $9.92 per $100 of gross wages for workers' compensation insurance. This year his premium will increase by 4%. If he has 8 employees averaging $22 per hour, 40 hours per week for 50 weeks, what will be his total premium for this year?

42. **Trades Management** Because it has a history of paying its bills on time, the Radius Electric Company receives an 8.25% discount on all purchases of electrical supplies from its supplier. What would the company pay for cable and fixtures originally costing $4760?

43. **Automotive Trades** When a turbocharger is added to a certain automobile engine, horsepower increases by 15%. If the non-turbocharged engine produces 156 peak horsepower, what peak horsepower does the turbocharged engine produce? (Round to the nearest whole number.)

44. **Electrical Trades** The governor on an engine generator regulates engine speed and frequency. When it operates in **droop** mode, the speed and frequency are regulated so that both are 3% higher with no load than they are under full load. Suppose that a certain engine has a speed of 1800 rpm and a frequency of 60 Hz at full load. What would be the speed and frequency under no load when the governor operates in droop mode?

45. **HVAC** The following table shows the recommended cooling capacity (in Btu per hour) of a room air conditioner based on the area (in square feet) of the room. If more than two people regularly occupy the room, add 600 Btu/hr per additional person; if the room is particularly sunny, add 10%; and if the room gets very little sun, subtract 10% from the listed capacity.

Area (sq ft)	Capacity (Btu/hr)
250–300	7,000
300–350	8,000
350–400	9,000
400–450	10,000

Determine the recommended cooling capacity of a room air conditioner for:
(a) A sunny 320-sq-ft room normally occupied by one or two people.
(b) A shady 440-sq-ft room normally occupied by four people.

46. **Sports and Leisure** In 1976 new world records in the high jump were set for both men and women. Dwight Stones set the men's record at 2.32 m, and Rosemarie Ackerman set the women's record at 1.96 m. The current record is 2.45 m for men (Javier Sotomayor) and 2.09 m for women (Stefka Kostadinova). Which record, men's or women's, has increased by the larger percent since 1976? Find the percent increase for each to the nearest tenth.

47. **Allied Health**   For maximum benefit during exercise, one's heart rate should increase by 50% to 85% above resting heart rate.
   (a) For a woman with a resting heart rate of 64 bpm (beats per minute), what would her heart rate be if she increased it by 75% during exercise?
   (b) A man with a resting heart rate of 70 bpm increased it to 105 bpm during exercise. By what percent did he increase his resting heart rate?

48. **Transportation**   An aerodynamically designed truck can improve fuel economy by as much as 45% over older, square-nosed rigs. If a trucker currently gets 4.5 mi/gal in his old truck, what mileage could he potentially achieve in an aerodynamically designed model? (Round to the nearest tenth.)

49. **Office Services**   A water district has been charging its customers a flat rate of $3.75 per HCF. Because of a drought emergency, it is instituting a block rate of $3.90 for the first 20 HCF and $4.15 for the next 40 HCF. If a family uses an average of 52 HCF per month, by what percent would their average bill increase under the block system? (Round to the nearest tenth.)

50. **General Trades**   Due to an economic downturn, a trades worker had to take a 10% pay cut from her $20 per hour wage. By what percent must her wage now increase in order to regain the original level? (Round your answer to one decimal digit.)

51. **Printing**   A poster was reduced in size by 25%. By what percent must it then be increased in size in order to restore it to its original size? (*Hint:* See the Note on page 258. Pick any arbitrary number to represent the original size and then work the problem.)

52. **Hydrology**   During a drought year, a customer's normal monthly water allocation of 4600 cubic feet was reduced by 15%. When the drought is over, by what percent must the allocation be increased to restore it to its original level?

53. **Allied Health**   A patient originally weighing 180 lb took a course of medication with a side effect that caused his weight to increase by 15%.
   (a) How much weight did he gain?
   (b) What percent of this resulting weight must he now lose in order to return to his original weight?

54. **General Interest**   In the year 2000, the median annual earnings among 18–34-year-old Americans were $37,355. In 2013, their median annual earnings were $33,883. By what percent did the median earnings of this age group decrease between 2000 and 2013? (Round to the nearest tenth.)

55. **Retail Merchandising**   Last quarter, sales of women's sandals in a retail store were $16,440. The goal for this quarter is to increase sales by 5%. What is the goal for this quarter in dollars?

56. **Carpentry**   A certain bench-top lathe had an initial speed of 502 rpm. With 12 lb of pressure applied on the faceplate, the speed was reduced to 366 rpm. What percent of its initial speed was lost when this pressure was applied? (Round to the nearest percent.)

57. **Retail Merchandising**   A buyer plans to sell a certain group of shirts in his store for $24 each. He needs to achieve a 45% markup over his cost. What should he pay his supplier per shirt in order to meet his goal? (Round to the nearest cent.)

58. **Retail Merchandising** A buyer for a retail store negotiates discounts of 30%, 15%, and 5% on a certain mattress. If the mattress has an initial price of $984, what price will the buyer actually pay after the discounts are applied?

59. **Meteorology** Scientists studying global warming have recently measured the reflection of radiation from glaciers and snowfields back into space. This is known as the **albedo** effect. In 1979, ice and snow in the Northern Hemisphere was reflecting 3.53 watts of solar energy per square meter. By 2008, the amount was 3.08 watts per square meter. By what percent was the albedo effect reduced over this 29-year period of time? (Round to the nearest tenth.) (*Source:* Mark Flanner, University of Michigan)

60. **Culinary Arts** A major newspaper recently challenged several restaurant chefs to reduce the number of calories in one of their menu items by at least 25%. The difficult part was to do this without compromising either the taste or the size of the item. Here are the results:

Restaurant	Original Calories	Reduced Calories
A	559	322
B	276	200
C	864	674

Calculate the percent reductions for each restaurant and determine which restaurants achieved the 25% goal.

61. **Life Skills** A recent study has found that, even in occupations that require only a high school diploma, employees who are college graduates earn more than those who are not. The following table shows expected lifetime earnings data for five sample occupations:

Occupation	Lifetime Earnings: High School Diploma Only	Lifetime Earnings: College Degree
Engineering technicians	$1,900,000	$2,200,000
Computer scientists	2,200,000	3,000,000
Firefighters/fire inspectors	2,100,000	2,700,000
Construction managers	2,000,000	3,100,000
Food-service managers	1,200,000	1,800,000

(a) For each of the listed occupations, determine the percent increase in lifetime earnings for college graduates compared to those with only a high school diploma. (Round to the nearest percent.)

(b) Which occupation showed the largest percent increase for a college degree?

(c) Which occupation showed the smallest percent increase?

(*Source:* Anthony P. Carnevale, Stephen J. Rose, and Ban Cheah, 2011: The College Payoff: Education, Occupations, Lifetime Earnings. The Georgetown University Center on Education and the Workforce.)

When you have completed these exercises, check your answers to the odd-numbered problems in the Appendix, then turn to Problem Set 4 on page 268 for a set of practice problems on ratio, proportion, and percent. If you need a quick review of the topics in this chapter, visit the chapter Summary first.

CHAPTER 4
SUMMARY

Ratio, Proportion, and Percent

Objective	Review
Calculate ratios. (p. 194)	A ratio is a comparison, using division, of two quantities of the same kind, both expressed in the same units. Final answers should be expressed either as a fraction in lowest terms or as a comparison to the number 1 using the word "to" or a colon (:). **Example:** Find the compression ratio C for an engine with maximum cylinder volume of 520 cu cm and minimum cylinder volume of 60 cu cm. $$C = \frac{520 \text{ cu cm}}{60 \text{ cu cm}} = \frac{26}{3} \quad \text{or} \quad 8\frac{2}{3} \text{ to 1} \quad \text{or} \quad 8\frac{2}{3} : 1$$
Solve proportions. (p. 198)	A proportion is an equation stating that two ratios are equal: $$\frac{a}{b} = \frac{c}{d}$$ To solve a proportion, use the cross-product rule: $$a \cdot d = b \cdot c$$ **Example:** Solve: $\dfrac{8}{x} = \dfrac{12}{15}$ $$12x = 120$$ $$x = 10$$
Solve problems involving proportions. (pp. 201, 208, 213)	Determine whether the problem involves a direct or an inverse proportion. Then set up the ratios accordingly and solve. **Example:** (a) An architectural drawing of a living room is $5\frac{1}{2}$ in. long and $4\frac{1}{2}$ in. wide. If the actual length of the room is 22 ft, find the actual width. Use a direct proportion. $$\frac{5\frac{1}{2}''}{4\frac{1}{2}''} = \frac{22'}{x}$$ $$5\frac{1}{2} \cdot x = 99$$ $$x = 18'$$ (b) A large production job is completed by four machines working for 6 hours. The same size job must be finished in 2 hours the next time it is ordered. How many machines should be used to accomplish this task? Use an inverse proportion. $$\frac{4 \text{ machines}}{x \text{ machines}} = \frac{2 \text{ hr}}{6 \text{ hr}}$$ $$2x = 24$$ $$x = 12 \text{ machines}$$
Change decimal numbers to percents. (p. 222)	Multiply the decimal number by 100%, or shift the decimal point two places to the right. **Example:** (a) $5 = 5 \times 100\% = 500\%$ (b) $0.075 = 0.075 = 7.5\%$

Objective	Review
Change fractions to percents. (p. 223)	First change the fraction to decimal form. Then proceed as in the previous objective. **Example:** $\frac{1}{4} = 0.25 = 0.25 = 25\%$
Change percents to decimal numbers. (p. 225)	Divide by 100%, or shift the decimal point two places to the left. **Example:** (a) $35\% = \dfrac{35\%}{100\%} = 0.35$ (b) $0.25\% = 00.25\% = 0.0025$
Change percents to fractions. (p. 226)	Write a fraction with the percent number as the numerator and 100 as the denominator. Then rewrite this fraction in lowest terms. **Example:** $112\% = \dfrac{112}{100} = \dfrac{28}{25}$
Solve problems involving percent. (p. 228)	Set up the proportion. $$\frac{\text{part }(P)}{\text{base }(B)} = \frac{\text{rate or percent }(R)}{100} \quad \text{or} \quad \frac{is}{of} = \frac{\%}{100}$$ Then solve for the missing quantity. **Example:** (a) Find 120% of 45. $$\frac{P}{45} = \frac{120}{100}$$ $$100P = 5400$$ $$P = 54$$ (b) 12 is 16% of what number? $$\frac{12}{B} = \frac{16}{100}$$ $$16B = 1200$$ $$B = 75$$
Solve applications involving percent. Specific applications may involve discount (p. 240), sales tax (p. 245), interest (p. 246), commission (p. 249), efficiency (p. 251), tolerances (p. 252), or percent change (p. 254).	In each case, read the problem carefully to identify P, B, and R. Use *is*, *of*, and the percent symbol (%) to help you. Then proceed as in the previous objective. **Example:** (a) The paint needed for a redecorating job originally cost $144.75, but it is now discounted by 35%. What is the discounted price? $$\frac{\text{net price}}{\$144.75} = \frac{100 - 35}{100} \longrightarrow \frac{N}{\$144.75} = \frac{65}{100}$$ $$100N = \$9408.75$$ $$N \approx \$94.09$$ (b) If the voltage in a circuit is increased from 70 volts to 78 volts, what is the percent increase in voltage? $$\frac{78 - 70}{70} = \frac{R}{100} \longrightarrow \frac{8}{70} = \frac{R}{100}$$ $$70R = 800$$ $$R \approx 11.4\%$$

PROBLEM SET 4   Ratio, Proportion, and Percent

Answers to odd-numbered problems are given in the Appendix.

A. Complete the following tables.

1.

	Diameter of Pulley A	Diameter of Pulley B	Pulley Ratio, $\frac{A}{B}$
(a)	12 in.		2 to 5
(b)	95 cm	38 cm	
(c)		15 in.	5 to 3

2.

	Teeth on Driven Gear A	Teeth on Driving Gear B	Gear Ratio, $\frac{A}{B}$
(a)	20	60	
(b)		10	6.5
(c)	56		4 to 3

B. Solve the following proportions.

1. $\dfrac{5}{6} = \dfrac{x}{42}$

2. $\dfrac{8}{15} = \dfrac{12}{x}$

3. $\dfrac{x}{12} = \dfrac{15}{9}$

4. $\dfrac{3\frac{1}{2}}{x} = \dfrac{5\frac{1}{4}}{18}$

5. $\dfrac{1.6}{5.2} = \dfrac{4.4}{x}$

6. $\dfrac{x}{12.4} = \dfrac{4 \text{ ft } 6 \text{ in.}}{6 \text{ ft } 3 \text{ in.}}$

C. Write each number as a percent.

1. 0.72   2. 0.06   3. 0.6   4. 0.358   5. 1.3

6. 3.03   7. 4   8. $\frac{7}{10}$ 9. $\frac{1}{6}$ 10. $2\frac{3}{5}$

D. Write each percent as a decimal number.

1. 4%   2. 37%   3. 11%   4. 94%

5. $1\frac{1}{4}\%$ 6. 0.09%   7. $\frac{1}{5}\%$ 8. 1.7%

9. $3\frac{7}{8}\%$ 10. 8.02%   11. 115%   12. 210%

E. Write each of the following percents as a fraction in lowest terms.

1. 28%   2. 375%   3. 81%   4. 4%

5. 0.5%   6. 70%   7. 14%   8. $41\frac{2}{3}\%$

F. Solve.

1. 3 is _____ % of 5.

2. 5% of $120 is _____.

3. 25% of what number is 1.4?

4. 16 is what percent of 8?

5. 105% of 40 is _____.

6. 1.38 is _____ % of 1.15.

7. 72% of _____ is $2.52.

8. 250% of 50 is _____.

9. 0.05% of _____ is 4.

10. $8\frac{1}{4}\%$ of 1.2 is _____.

Name

_____

Date

_____

Course/Section

_____

G. Solve.

1. **Metalworking**   Extruded steel rods shrink 12% in cooling from furnace temperature to room temperature. If a standard tie rod is 34 in. exactly when it is formed, how long will it be after cooling?

2. **Metalworking**   Cast iron contains up to 4.5% carbon, and wrought iron contains up to 0.08% carbon. How much carbon is in a 20-lb bar of each metal?

3. **Life Skills**
   (a) A real estate saleswoman sells a house for $664,500. Her company pays her 70% of the 1.5% commission on the sale. How much does she earn on the sale?
   (b) All salespeople in the Ace Department Store receive $360 per week plus a 6% commission. If you sold $3975 worth of goods in a week, what would be your total income?
   (c) A salesman at the Wasteland TV Company sold five identical LED TV sets last week and earned $509.70 in commissions. If his commission rate is 6%, what does one of these models cost?

4. **Carpentry**   What is the net price of a 10-in. band saw with a list price of $299.95 if it is on sale at a 35% discount?

5. **Life Skills**   If the retail sales tax in your state is 6%, what would be the total cost of each of the following items?
   (a) An $18.99 pair of pliers
   (b) A $17.49 adjustable wrench
   (c) 69 cents worth of washers
   (d) A $118.60 textbook
   (e) A $299.99 five-drawer mobile shop cabinet rollaway

6. **Retail Merchandising**   A computer printer sells for $376 after a 12% discount. What was its original or list price?

7. **Construction**   How many running feet of matched 1-in. by 6-in. boards will be required to lay a subfloor in a room that is 28 ft by 26 ft? Add 20% to the area to allow for waste and matching.

8. **Construction**   Limestone weighs approximately 165 pounds per cubic foot. Assuming a 7% waste, how many pounds are needed for a job requiring 120 cu ft of limestone?

9. **Machine Trades**   Complete the following table:

Measurement	Tolerance	Percent Tolerance
1.775 in.	±0.001 in.	(a)
1.775 in.	±0.05 in.	(b)
1.775 in.	(c)	±0.50%
310 mm	±0.1 mm	(d)
310 mm	(e)	±0.20%

10. **Masonry**   A mason must purchase enough mortar mix to lay 1820 bricks. The guidelines for this mix suggest that about 15 bags are required for every 400 bricks.
    (a) Assuming that he cannot purchase a fraction of a bag, how many bags will he need?
    (b) At $6.40 per bag, what will be the total cost of the mix?

11. **Manufacturing**   A production job is bid at $6275, but cost overruns amount to 15%. What is the actual job cost?

12. **Metalworking**   When heated, a metal rod expands 3.5%, to 15.23 cm. What is its cold length?

13. **Electrical Trades**   An electrical resistor is rated at 4500 ohms $\pm$ 3%. Express this tolerance in ohms and state the actual range of resistance.

14. **Automotive Trades**   A 140-hp automobile engine delivers only 126 hp to the driving wheels of the car. What is the efficiency of the transmission and drive mechanism?

15. **Metalworking**   A steel casting has a hot length of 26.500 in. After cooling, the length is 26.255 in. What is the shrinkage expressed as a percent? (Round to one decimal place.)

16. **Automotive Trades**   The parts manager for an automobile dealership can buy a part at a 25% discount off the retail price. If the retail price is $46.75, how much does he pay?

17. **Manufacturing**   An electric shop motor rated at 2.5 hp is found to deliver 1.8 hp to a vacuum pump when it is connected through a belt drive system. What is the efficiency of the drive system?

18. **Trades Management**   To purchase a truck for his mobile welding service, Jerry arranges a 36-month loan of $45,000 at 4.75% annual interest.
    (a) What total interest will he pay?
    (b) What monthly payment will he make? (Assume equal monthly payments on interest and principal.)

19. **HVAC**   The motor to run a refrigeration system is rated at 12 hp, and the system has an efficiency of 76%. What is the effective cooling power output of the system?

20. **Interior Design**   An interior designer is able to purchase a sofa from a design center for $1922. His client would have paid $2480 for the same sofa at a furniture store. What rate of discount was the designer receiving off the retail price?

21. **Life Skills**   The commission rate paid to a manufacturer's rep varies according to the type of equipment sold. If his monthly statement showed $124,600 in sales and his total commission was $4438, what was his average rate of commission to the nearest tenth of a percent?

22. **Masonry**   Six square feet of a certain kind of brick wall contains 78 bricks. How many bricks are needed for 150 sq ft?

23. **Transportation**   If 60 mph (miles per hour) is equivalent to 88 fps (feet per second), express
    (a) 45 mph in feet per second.
    (b) 22 fps in miles per hour.

24. **Automotive Trades**   The headlights on a car are mounted at the height of 28 in. The light beam must illuminate the road for 400 ft. That is, the beam must hit the ground 400 ft ahead of the car.
    (a) What should be the drop ratio of the light beam in inches per foot?
    (b) What is this ratio in inches per 25 ft?

25. **Printing**   A photograph measuring 8 in. wide by 12 in. long must be reduced to a width of $3\frac{1}{2}$ in. to fit on a printed page. What will be the corresponding length of the reduced photograph?

26. **Machine Trades** If 250 ft of wire weighs 22 lb, what will be the weight of 100 ft of the same wire?

27. **Agriculture** A sprayer discharges 600 cc of herbicide in 45 sec. For what period of time should the sprayer be discharged in order to apply 2000 cc of herbicide?

28. **Automotive Trades** A Honda Accord hybrid costs $5.21 per 100 miles to drive, while a conventional Honda Accord costs $9.26 per 100 miles to drive. How much more would the conventional Accord cost to drive 480 miles? (*Note:* Costs per mile are based on California fuel prices in the fall of 2016.)

29. **Machine Trades** Six steel parts weigh 1.8 lb. How many of these parts are in a box weighing 142 lb if the box itself weighs 7 lb?

30. **Manufacturing** Two belted pulleys have diameters of 24 in. and 10 in.
    (a) Find the pulley ratio.
    (b) If the larger pulley turns at 1500 rpm, how fast will the smaller pulley turn?

31. **Printing** If six printing presses can do a certain job in $2\frac{1}{2}$ hours, how long will it take four presses to do the same job? (*Hint:* Use an inverse proportion.)

32. **Drafting** Drafters usually use a scale of $\frac{1}{4}$ in. $= 1$ ft on their architectural drawings. Find the actual length of each of the following items if their blueprint length is given:
    (a) A printing press $2\frac{1}{4}$ in. long
    (b) A building $7\frac{1}{2}$ in. long
    (c) A rafter $\frac{11}{16}$ in. long
    (d) A car $1\frac{7}{8}$ in. long

33. **Construction** A certain concrete mix requires one sack of cement for every 650 lb of concrete. How many sacks of cement are needed for 3785 lb of concrete? (*Remember:* You cannot buy a fraction of a sack.)

34. **Automotive Trades** The rear axle of a car turns at 320 rpm when the drive shaft turns at 1600 rpm. If the pinion gear has 12 teeth, how many teeth are on the ring gear?

35. **Physics** If a 120-lb force is applied to a 36-in. crowbar pivoted 4 in. from the end, what lift force is exerted?

36. **Machine Trades** How many turns will a pinion gear having 16 teeth make if a ring gear having 48 teeth makes 120 turns?

37. **Electrical Trades** The current in a certain electrical circuit is inversely proportional to the line voltage. If a current of 0.40 A is delivered at 440 volts, what is the current of a similar system operating at 120 volts? (Round to the nearest 0.01 A.)

38. **Printing** A printer needs 12,000 good copies of a flyer for a particular job. Previous experience has shown that she can expect no more than a 5% spoilage rate on this type of job. How many should she print to ensure 12,000 clean copies?

39. **Construction** A medium mix of concrete requires 2.5 cu ft of sand and 5 cu ft of gravel for every bag of cement. This will produce 5.4 cu ft of concrete. If a job requires 32 cu ft of concrete, what amounts of cement, sand, and gravel are needed? (Round to the nearest whole number.)

40. **Manufacturing** Two belted pulleys have diameters of $8\frac{3}{4}$ in. and $5\frac{5}{8}$ in. If the smaller pulley turns at 850 rpm, how fast will the larger one turn?

41. **Trades Management** Each production employee in a plant requires an average of 120 sq ft of work area. How many employees will be able to work in an area that measures $8\frac{1}{16}$ in. by $32\frac{5}{8}$ in. on a blueprint if the scale of the drawing is $\frac{1}{4}$ in. = 1 ft?

42. **General Trades** Find the missing dimension in the following pair of similar figures:

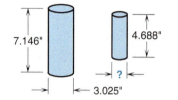

43. **Physics** In a closed container, the pressure is inversely proportional to the volume when the temperature is held constant. Find the pressure of a gas compressed to 0.386 cu ft if the pressure is 12.86 psi at 2.52 cu ft.

44. **Automotive Trades** The mechanical efficiency of a certain engine is specified to be 82.5%. This means that 82.5% of its indicated or theoretical horsepower becomes brake horsepower. If the indicated horsepower of the engine is 172, what is the brake horsepower? (Round to the nearest whole number.)

45. **Automotive Trades** An automobile engine loses approximately 3% of its power for every 1000 ft of elevation gain. A certain engine generates 180 horsepower (hp) at sea level. How much power will it generate at an elevation of 4000 ft? (Round to the nearest whole number.)

46. **Allied Health** Calculate the percent concentration of a solution created by adding 15 mL of a medication to 160 mL of water. (Round your answer to the nearest tenth of a percent.)

47. **General Interest** The following table shows the average starting salaries for three different engineering majors graduating in 2015 with either Bachelor's degrees (BA) or Master's degrees (MA). (Source: National Association of Colleges and Employers.)

Field	Average Starting Salary with BA	Average Starting Salary with MA
Petroleum engineering	$72,063	$82,983
Computer engineering	$64,155	$83,347
Geological/geophysical engineering	$63,565	$102,984

For each type of engineer, calculate the percent increase in starting salary for those with Master's degrees compared to those with Bachelor's degrees.

48. **Automotive Trades** A headlight costs an auto service parts department $269.25. If the markup is 60%, what retail price will the customer be charged?

49. **Automotive Trades** During a compression check, the cylinder with the highest compression was measured at 156 psi (pounds per square inch), while the cylinder with the lowest compression was measured at 136 psi. If the maximum allowable reduction from the highest compression is 15%, does this engine fall within the allowable limit?

50. **Allied Health** A pre-op antiseptic solution is to be used to deliver 0.3 mg of atropine to the patient. There is, on hand, a solution of atropine containing 0.4 mg per 0.5 mL.
    (a) How much of the solution should be used to obtain the amount of atropine needed?
    (b) Suppose the only measuring device available is calibrated in minims (1 mL = 16 minims). How much of the solution is needed in minims?

51. **Sports and Leisure** Use the following five steps to determine the NFL passing rating for quarterback Bubba Grassback.
    (a) Bubba completed 176 passes in 308 attempts. Calculate his pass completion rate as a percent. Subtract 30 from this number and multiply the result by 0.05.
    (b) He has 11 touchdown passes in 308 attempts. Calculate the percent of his attempts that resulted in touchdowns. Multiply that number by 0.2.
    (c) He has thrown 13 interceptions. Calculate the percent of his 308 attempts that have resulted in interceptions. Multiply this percent by 0.25, and then subtract the result from 2.375.
    (d) Bubba passed for 2186 yards. Calculate the average number of yards gained per passing attempt. Subtract 3 from this number, and then multiply the result by 0.25.
    (e) Add the four numbers obtained in parts (a), (b), (c), and (d).
    (f) Divide the total in part (e) by 6 and then multiply by 100. This is Bubba's NFL quarterback passing rating.
    (*Note:* Ratings for starting quarterbacks usually range from a low of about 50 to a high of about 110.)

52. **Allied Health** Medication Q is available only in tablets containing 30 mg of the drug. How many tablets (or fractions of tablets) should be administered if the physician orders 75 mg of medication Q?

53. **Allied Health** Medication TTQ is available only in a liquid formulation with a concentration of 25 µg/mL (micrograms per milliliter). How much of this liquid formulation should be administered if the following amounts are ordered by the physician?
    (a) 300 µg   (b) 150 µg   (c) 20 µg

54. **Transportation** For every 1-mph reduction of average speed below 65 mph, a trucker will save 1.5% in fuel consumption. If a trucker reduces her average speed from 65 mph to 60 mph, how much will she save if she currently spends $4000 per week on fuel?

55. **Life Skills** A working person who needs an emergency loan, but who has no home or business to offer as security, may require the services of a company offering short-term paycheck advances. One such company offers a 14-day maximum advance of $255 for a fee of $45.
    (a) What percent of the loan amount does the fee represent? (Round to the nearest hundredth of a percent.)
    (b) Use a proportion to calculate the annual (365-day) percentage rate of this fee. (Round to the nearest percent.)

56. **Allied Health** Over the 22-year span from 1990 to 2012, the average life expectancy in the United States increased from 75.2 years to 78.8 years. By what percent did the life expectancy increase? (Round to the nearest tenth.)

57. **Automotive Trades** The following table shows U.S. annual sales comparisons among hybrid cars in both 2010 and 2014. Study the table and then answer the questions that follow. (Source: IHS Automotive.)

Hybrid Car	2010 Sales (in thousands)	2014 Sales (in thousands)
Toyota Prius	140.3	207.6
Toyota Camry	14.6	39.7
Ford Fusion	21.0	47.0
All Other Hybrids	98.0	205.0

(a) What percent of the total hybrid market did the Prius have in 2010?

(b) What percent of the total hybrid market did the Ford Fusion have in 2014?

(c) Which of the three named models had the highest percent increase in sales between 2010 and 2014?

58. **Allied Health**   In recent years, an increasing number of young people have been choosing to have weight-loss surgery. In 2010, approximately 5% of the statewide weight-loss surgeries in New York were performed on patients under the age of 25. If there were about 9400 total statewide weight-loss surgeries, how many of these were performed in 2010 on patients younger than 25 in New York?

59. **Construction**   A contractor specializing in environmentally friendly (or green) construction proposes that he exceed the code requirements to install energy-saving upgrades in a new home. His computer software program estimates annual energy costs of $2140 without the upgrades and $960 with the upgrades. What percent of the $2140 will the residents save if they install the energy-saving upgrades? (Round to the nearest tenth.)

60. **Plumbing**   A homeowner purchases a new hot-water system to use with her newly installed solar panels. The cost of the system is $8400, but tax credits and incentives bring the cost down to $4800. What percent of the original cost do the credits and incentives save her? (Round to the nearest tenth.)

61. **Retail Merchandising**   The men's department reported the following sales totals, by size, for a particular shirt last year:

Small	$9400	Large	$19,200
Medium	$22,600	Extra Large	$12,800

If the store buyer has $40,000 to invest in the shirt this year, how much should she spend on each size if the amounts are in the same proportion as last year's sales?

62. **Water/Wastewater Treatment**   To "shock" the water in a contaminated pipe, an operator needs to treat the water with 9.75 pounds of chlorine. The operator has available containers of liquid bleach that are 5% chlorine. If each gallon of the bleach solution weighs 8.34 pounds, how many gallons are needed to shock the water in the pipe? (Round to the nearest tenth.)

63. **Water/Wastewater Treatment**   Water and wastewater formulas generally require concentrations to be expressed in parts per million (ppm). If a container of powdered pool chlorine contains 70% actual chlorine, how many parts per million is this equivalent to?

64. **Allied Health**   A physician prescribes a daily dosage of 600 mg of medication G for a certain patient. Medication G is available only in an oral solution with a concentration of 100 mg/5 mL (milligrams per 5 milliliters).

How many milliliters of the oral solution should the patient consume per day?

65. **Allied Health**   Medication X comes in an oral solution with a concentration of 5 mg/mL (milligrams per milliliter). A doctor prescribes a daily dosage of 225 mg/m² (milligrams per square meter) for a particular patient. If the patient has a BSA of 1.42 m², how many milliliters of the solution should the patient consume per day? (See problem C-35 in Exercises 4-1. Round to the nearest milliliter.)

66. **Painting**   A certain type of paint, used for porches and decks, covers 378 sq ft per gallon when applying the first coat, 540 sq ft per gallon with the second coat, and 576 sq ft per gallon with the third coat. How many gallons of this paint are needed to apply three coats to 1868 sq ft of porches and decks? (Assume that you cannot buy a fraction of a gallon.)

67. **Transportation**   It is estimated that a truck produces 16.7 pounds of greenhouse gas emissions per ton of cargo for every 100 miles shipped. How many pounds of such emissions would a truck produce if it shipped 2.4 tons of cargo 860 miles? (Round to the nearest pound.)

68. **Masonry**   A mason is using 12 in. by 8 in. by 16 in. blocks to build a wall. From prior experience, he knows that 112.5 blocks and 7.5 cu ft of mortar are needed per 100 sq ft of wall. He also knows that 7.25 hours of labor are needed per 100 sq ft of installation.
    (a) How many blocks are needed for 840 sq ft of wall?
    (b) How many cubic feet of mortar are needed for 1260 sq ft of wall?
    (c) How many hours of labor are needed for 320 sq ft of wall?

CASE STUDY: Hot Wheels

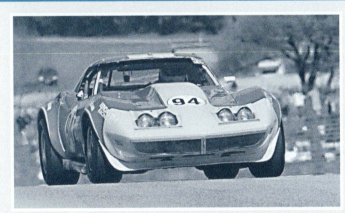

Hot Wheels toy cars have been manufactured by Mattel since 1968. The original 16 car models were produced in a variety of colors with custom features. One of the original 16 car models was a custom Corvette using the specifications from General Motors. The Hot Wheels 1:64 Corvette model was made before the 1:1 full-size car was available for purchase. Designed with the intent of being fast to race, engineers and designers from the auto industry in Detroit were recruited to create the Hot Wheels toy line.

1. The spec sheet for the actual 1968 Corvette shows the front frame dimension to be $32\frac{7}{8}$ inches. What would this measurement be, in both fractional and decimal form, on the Hot Wheels 1:64 model? Which form of the measurement would be easier to work with?

2. The Hot Wheels 1968 custom models were also produced in both a 1:43 scale model and a 1:24 scale model.

    (a) If the full-scale 1968 Corvette car was 182.1-in. long, what was the length of the 1:43 scale model? (Round to the nearest hundredth.)
    (b) The overall length of a full-scale 1968 Corvette was 182.1 in. If 1 in. = 2.54 cm, how long was the car in centimeters?

continued...

(c) What was the length of the full-scale 1968 Camaro if the 1:24 Hot Wheels model was 195.5 mm? If 1 m = 1000 mm, what is this length in meters?

3. The 1968 white-enamel Custom Camaro Hot Wheels model is an extremely rare find today. The toy originally sold for 98 cents. Today, collectors are known to pay at least $2,500 for the toy. By what percent has this Hot Wheels Camaro increased in value?

4. General Motors produced 28,566 Corvettes in 1968. If 48% of them are still being driven, how many 1968 Corvettes are still in use?

5. Different racetrack materials provide different speeds for the toy cars. The speed of the Hot Wheels toy is inversely proportional to the material rating of the track. If a track with a material rating of 1.241 allows for a maximum speed of 3.37 feet per second, what is the maximum speed on a track with a material rating of 1.068?

6. When the initial Hot Wheels were released for sale, demand far exceeded Mattel's projections and additional manufacturing sites were added to meet demand. If one production unit can produce 1.3 million Hot Wheels cars, how many production units are needed to produce at least 15 million cars?

References:
NCHWA.com. (2016). Hot Wheels history. Retrieved from http://nchwa.com/hwhistory.html
Oldride.com. (2016). 1968 Chevrolet Corvette. Retrieved from http://www.oldride.com/library/1968_chevrolet_corvette.html
Zarnock, M. (2010). Top 10 Hot Wheels cars of all time. Antique Trader, August 17, 2010. Retrieved from http://www.antiquetrader.com/features/top_10_hot_wheels_cars_of_all_time

Measurement

Objective	Sample problems		For help, go to page

When you finish this chapter, you will be able to:

1. Determine the precision, accuracy, and greatest possible error of measurement numbers.

Determine the precision of each of the following measurement numbers. Which one is most precise?
(a) 4.27 psi   (b) 6758 psi
(c) 350 psi

(a) _____
(b) _____
(c) _____

282

Determine the number of significant digits in each of the following measurement numbers.
(d) 9.6 kg   (e) 458 kg
(f) 6000 kg

(d) _____
(e) _____
(f) _____

280

Determine the greatest possible error of each measurement.
(g) 450 psi   (h) 8.6 kg
(i) $3\frac{1}{4}$ in.

(g) _____
(h) _____
(i) _____

283

2. Add and subtract measurement numbers. (Round to the correct precision.)

(a) 38.26 sec + 5.9 sec =

(b) 8.37 lb − 2.435 lb =

(a) _____
(b) _____

284

3. Multiply and divide measurement numbers. (Round to the proper accuracy.)

(a) 16.2 ft × 5.8 ft =

(b) 480 mi ÷ 32.5 mph =

(c) 240 lb/sq in. × 52.5 sq in. =

(a) _____
(b) _____
(c) _____

286

Name _____

Date _____

Course/Section _____

Objective	Sample problems	For help, go to page
4. Convert units within the U.S. customary system. (If necessary, round as directed.)	(a) 65.0 lb = _____ oz	294
	(b) 8200 yd = _____ mi (nearest tenth)	
	(c) 3.50 sq ft = _____ sq in.	
5. Think metric.	(a) Estimate the weight in pounds of a casting weighing 45 kg. (a) _____	314
	(b) Estimate the volume in quarts of 80 liters. (b) _____	
6. Convert units within the metric system.	(a) 6500 mg = _____ g	316
	(b) 0.45 km = _____ m	
	(c) 0.285 ha = _____ sq m	
7. Convert between U.S. customary and metric system units. (If necessary, round as directed.)	(a) 48.0 liters = _____ gal (nearest tenth)	326
	(b) 45 mi/hr = _____ km/h (nearest whole number)	
	(c) 42°C = _____ °F (nearest whole number)	
8. Use common technical measuring instruments.	Correctly read a length rule, micrometer, vernier caliper, protractor, and various meters and dial gauges.	338

(Answers to these preview problems are given in the Appendix. Also, worked solutions to many of these problems appear in the chapter Summary.)

If you are certain that you can work *all* these problems correctly, turn to page 368 for a set of practice problems. If you cannot work one or more of the preview problems, turn to the page indicated to the right of the problem. For those who wish to master this material with the greatest success, turn to Section 5-1 and begin work there.

Measurement

Most of the numbers used in practical or technical work come from measurements that often lead to other measurements. Trades workers and technicians must make measurements, do calculations based on measurements, round measurement numbers, and convert measurements from one unit to another. In this chapter, you will learn how to do all of these things using both the U.S. customary and the metric systems of measurement.

CASE STUDY: Mr. Ferris's Wheel

When a World's Fair or Expo takes place, units of measurement will often be expressed in both the metric and the U.S. customary system. Participants and visitors from all countries can enhance their experience by being familiar with both systems and by being able to convert from one system to another. In 1893, the very first Ferris wheel was built for the World's Fair in Chicago, Illinois. In this chapter's Case Study, "Mr. Ferris's Wheel," we will learn some facts about this Ferris wheel and perform calculations and conversions using both systems of measurement.

5-1   Working with Measurement Numbers

Units   A measurement number usually has units associated with it, and the unit name for that measurement number *must* be included when you use that number, write it, or talk about it. For example, we know immediately that 20 is a number, 20 feet (ft) is a length, 20 pounds (lb) is a weight or force, and 20 seconds (sec) is a time interval.

learning|**catalytics**™

1. What is half of 0.1?
2. Round 2568.3 to the nearest ten.

Suppose that the length of a piece of pipe is measured to be 8 ft. Two bits of information are given in this measurement: the size of the measurement (8) and the units of measurement (ft). The measurement unit compares the size of the quantity being measured to some standard.

EXAMPLE 1   If the piece of pipe is measured to be 8 ft in length, it is 8 times the standard 1-ft length.

Length = 8 ft = 8 × 1 ft

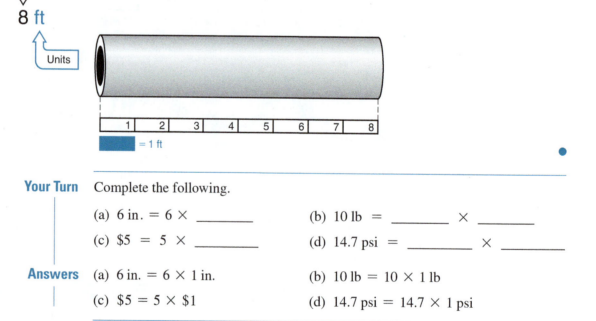

Your Turn   Complete the following.

(a) 6 in. = 6 × _____   (b) 10 lb = _____ × _____

(c) \$5 = 5 × _____   (d) 14.7 psi = _____ × _____

Answers   (a) 6 in. = 6 × 1 in.   (b) 10 lb = 10 × 1 lb

(c) \$5 = 5 × \$1   (d) 14.7 psi = 14.7 × 1 psi

Significant Digits   Every measuring instrument and every person making measurements has limitations. Measurement numbers—numbers used in technical work and in the trades—have a built-in uncertainty; this uncertainty is not stated explicitly but is expressed by the number of digits used to communicate the measurement information.

When we perform calculations with measurement numbers, we must often round our answers. For most trades workers, answers are rounded using common-sense rules that may depend on factors such as how the result will be used or the precision of their measuring instruments. Those students who are training for more technical careers, such as engineering, will often be expected to round their answers according to agreed-upon rules involving *significant digits*.

The **significant digits** in a number are those that represent an actual measurement. To determine the number of significant digits in a measurement number, follow these rules:

Rule 1 Digits other than zero are always significant.

Rule 2 A zero *is* significant when it
    (a) appears between two significant digits,
    (b) is at the right end of a decimal number,
    (c) is marked as significant with an overbar.

Rule 3 A zero is *not* significant when it
    (a) is at the right end of a whole number,
    (b) is at the left end of a number.

EXAMPLE 2 **General Trades**

- A tachometer reading of 84.2 rpm has three significant digits. All three digits in the number are nonzero. (Rule 1)

- A pressure reading of 3206 torr has four significant digits. The zero is significant because it appears between two nonzero digits. (Rule 2a)

- A voltage reading of 240 volts has two significant digits. The zero is not a significant digit because it is at the right end of a whole number. (Rule 3a)

- A voltage reading of 24$\overline{0}$ volts has three significant digits. The overbar tells us that the zero marked is significant. (Rule 2c) The person making the measurement had an instrument that could measure to the nearest unit.

- A voltage reading of 24.0 volts has three significant digits. The zero is significant because it appears at the right end of a decimal number. (Rule 2b)

- A current reading of 0.025 amp has two significant digits. The zeros are not significant because they are at the left end of the number. (Rule 3b) ●

Your Turn Determine the number of significant digits in each of the following measurement numbers.

(a) 3.04 lb   (b) 38,000 mi   (c) 15.730 sec

(d) 0.005 volt   (e) 18,4$\overline{0}$0 gal   (f) 200.0 torr

Answers (a) Three significant digits. The zero is significant because it is between two nonzero digits.

(b) Two significant digits. The zeros are not significant because they are at the right end of a whole number.

(c) Five significant digits. The zero is significant because it is at the right end of a decimal number.

(d) One significant digit. The zeros are not significant because they are at the left end of the number.

(e) Four significant digits. The zero with the overbar is significant, but the right-end zero is not significant.

(f) Four significant digits. The right-end digit is significant because it is at the right end of a decimal number. The middle two zeros are significant because they are between two significant digits.

Precision and Accuracy   Because measurement numbers are never exact, we need to be careful that the significant digits we specify in the number actually reflect the situation.

Look at this line:

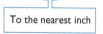

If we measured the length of this line using a rough ruler, we would write that its length is

Length = 2 in.

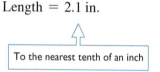

To the nearest inch

But with a better measuring device, we might be able to say with confidence that its length is

Length = 2.1 in.

To the nearest tenth of an inch

Using a caliper, we might find that the length is

Length = 2.134 in.

To the nearest thousandth of an inch

We could go on using better and better measuring devices until we found ourselves looking through a microscope trying to decide where the ink line begins. If someone told us that the length of that line was

Length = 2.1340756 in.

To the nearest 0.0000001 inch

we would know immediately that the length had been measured with incredible precision—they couldn't get a number like that by using a carpenter's rule!

To work with measurement numbers, we define precision and accuracy* as follows:

> The **precision** of a measurement number is indicated by the place value of its right-most significant digit.
>
> The **accuracy** of a measurement number refers to the number of significant digits that it contains.

EXAMPLE 3   State the precision and accuracy of each of the following measurement numbers using the rules given.

(a) 4.01 sec   (b) 40 sec   (c) 420.1 sec

---

*In engineering, *accuracy* is the degree of closeness between the measured value of some quantity and its true, or nominal, value. It has to do with the ability of an instrument to make measurements with small uncertainty. *Precision* is the degree of consistency and agreement among independent measurements of a quantity under the same conditions. For practical purposes, we define them as shown in the box.

Measurement (a) is precise to the nearest hundredth of a second. It is accurate to three significant digits.

In (b) the zero is not significant. Therefore the number is precise to the nearest 10 seconds. It is accurate to one significant digit.

Measurement (c) is precise to the nearest tenth of a second. It is accurate to four significant digits.

Notice that measurement (a) is considered the most precise because it has the most decimal digits. However, measurement (c) is considered the most accurate because it contains the most significant digits. ●

Your Turn   Determine the precision and accuracy of each of the following measurement numbers. Then indicate which is the most precise and which is the most accurate.

   (a) 34.1 lb   (b) 0.568 lb   (c) 2735 lb   (d) 5.09 lb   (e) 400 lb

Answers

Measurement	Precise to the nearest . . .	Accurate to this number of significant digits
(a)	tenth of a pound	3
(b)	thousandth of a pound	3
(c)	pound	4
(d)	hundredth of a pound	3
(e)	hundred pounds	1

Measurement (b) is the most precise and measurement (c) is the most accurate.

Greatest Possible Error   The **greatest possible error** of a measurement is defined as half the precision of that measurement. If you are reading the measurement yourself with a measuring tool, the greatest possible error is half the smallest division on the scale of the measuring tool.

EXAMPLE 4   (a) Suppose the weight of a steel bar is given as 14.6 lb. The weight of the bar has been measured to the nearest tenth of a pound, so the precision of this measurement is 0.1 lb. The greatest possible error is half the precision: $0.5(0.1\ \text{lb}) = 0.05\ \text{lb}$. This means that the actual weight of the bar is 14.6 ± 0.05 lb, or somewhere between 14.55 and 14.65 lb. (The symbol $\pm$ is read "plus or minus.")

    (b) A tachometer shows an engine speed to be 4700 rpm. The precision of this measurement is 100 rpm, so the greatest possible error is half of 100, or 50 rpm. This means that the actual speed of the engine is 4700 ± 50, or somewhere between 4650 and 4750 rpm.

    (c) Suppose that you are measuring the lengths of two pieces of tubing with a ruler, as shown in the figure.

One piece is slightly longer than $2\frac{1}{8}$ in., while the other is slightly shorter than $2\frac{1}{8}$ in. But because the smallest division on the ruler is $\frac{1}{8}$ of an inch, we would express both lengths as $2\frac{1}{8}$ in. because that is as precisely as our ruler can measure. In fact, any length that is up to $\frac{1}{16}$ in. shorter than this would be read as $2\frac{1}{8}$ in., and any length that is up to $\frac{1}{16}$ in. longer than this would also be read as $2\frac{1}{8}$ in. This illustrates why the greatest possible error of measuring instrument is half the smallest scale division on the instrument. In this case, the greatest possible error is $\frac{1}{2}\left(\frac{1}{8}\text{ in.}\right) = \frac{1}{16}$ in.   ●

Your Turn

(a) The length of a strip of metal is given as 1.42 m (meters). What is the greatest possible error of this measurement? What are the minimum and maximum limits of its actual length?

(b) You are weighing a solution on a scale that is calibrated to the nearest half-ounce. What is the greatest possible error of the weight of your solution?

Solutions

(a) The precision is 0.01 m. The greatest possible error is half of this, or 0.005 m. The minimum limit of its actual length is 1.42 m − 0.005 m, or 1.415 m. The maximum limit is 1.42 m + 0.005 m, or 1.425 m.

(b) The precision is 0.5 oz. The greatest possible error is half of this, or 0.25 oz.

Tolerance

Very often the technical measurements and numbers shown in drawings and specifications have a **tolerance** attached so that you will know the precision needed. For example, if a dimension on a machine tool is given as 2.835 ± 0.004 in., this means that the dimension should be 2.835 in. to within 4 thousandths of an inch. In other words, the dimension may be no more than 2.839 in. and no less than 2.831 in. These are the limits of size that are allowed or tolerated.

Addition and Subtraction of Measurement Numbers

When measurement numbers are added or subtracted, the resulting answer cannot be more precise than the least precise measurement number in the calculation. Therefore, use the following rule when adding or subtracting measurement numbers:

> Always round a sum or difference to agree in precision with the least precise measurement number being added or subtracted.

EXAMPLE 5

To add

4.1 sec + 3.5 sec + 1.27 sec = _____?_____

First, check to be certain that all numbers have the same units. In the given problem, all numbers have units of seconds. If one or more of the numbers are given in other units, you will need to convert so that all have the same units.

Second, add the numerical parts.

$$
\begin{array}{r}
4.1 \\
3.5 \\
+\ 1.27 \\
\hline
8.87
\end{array}
$$

Notice that we line up the decimal points and add as we would with any decimal numbers.

4.1 ⊕ 3.5 ⊕ 1.27 ⊜ → ░░░░░ *8.87*

Third, attach the common units to the sum.

8.87 sec

Fourth, round the sum to agree in precision with the least precise number in the calculation. We must round 8.87 sec to the nearest tenth, or 8.9 seconds. Our answer agrees in precision with 4.1 sec and 3.5 sec, the least precise measurements in the calculation. ●

Your Turn   Perform the following subtraction and round your answer to the correct precision.

$$7.425 \text{ in.} - 3.5 \text{ in.} = \underline{\qquad ? \qquad}$$

Solution   **First,** check to be certain that both numbers have the same units. In this problem both numbers have inch units. If one of the numbers was given in other units, we would convert so that they have the same units. (You'll learn more about converting measurement units later in this chapter.)

Second, subtract the numerical parts.

$$\begin{array}{r} 7.425 \\ -\ 3.500 \\ \hline 3.925 \end{array} \quad \Longleftarrow \text{Attach zeros if necessary}$$

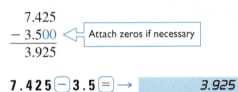 7.425 ⊖ 3.5 ⊜ →   *3.925*

Third, attach the common units to the difference. 3.925 in.

Fourth, round the answer to agree in precision with the least precise number in the original problem. The least precise number is 3.5 in., which is precise to the nearest tenth. We should round our answer, 3.925 in., to the nearest tenth, or 3.9 in.

EXAMPLE 6   Here are a few more examples to help fix the concept in your mind.

(a) 4.6 gal + 2.145 gal = 6.745 gal ≈ 6.7 gal

  Round to the nearest tenth to agree with the least precise measurement number, 4.6 gal, which is stated to the nearest tenth. (Recall from Chapter 3 that ≈ means "approximately equal to.")

(b) 0.24 mi − 0.028 mi = 0.212 mi ≈ 0.21 mi

  Round the answer to the nearest hundredth to agree with the least precise number, 0.24 mi.

(c) 2473 lb + 321.2 lb = 2794.2 lb ≈ 2794 lb

  Round to the nearest whole number to agree with the least precise measurement, 2473 lb. ●

More Practice   Try these problems for practice in adding and subtracting measurement numbers. Be careful to round your answers correctly.

(a) 10.5 in. + 8.72 in.

(b) 2.46 cu ft + 3.517 cu ft + 4.8 cu ft

(c) 47.5 sec − 21.844 sec

(d) 56 lb − 4.8 lb

(e) 325 sq ft + 11.8 sq ft + 86.06 sq ft

(f) 0.027 volt + 0.18 volt + 0.009 volt

Answers

(a) 10.5 in. + 8.72 in. = 19.22 in. ≈ 19.2 in.

(b) 2.46 cu ft + 3.517 cu ft + 4.8 cu ft = 10.777 cu ft ≈ 10.8 cu ft

(c) 47.5 sec − 21.844 sec = 25.656 sec ≈ 25.7 sec

(d) 56 lb − 4.8 lb = 51.2 lb ≈ 51 lb

(e) 325 sq ft + 11.8 sq ft + 86.06 sq ft = 422.86 sq ft ≈ 423 sq ft

(f) 0.027 volt + 0.18 volt + 0.009 volt = 0.216 volt ≈ 0.22 volt

---

✎ Note   If some of the numbers to be added are in fraction form, you should convert them to decimal form before adding. For example, the sum $3.2 \text{ lb} + 1\frac{3}{4} \text{ lb} + 2.4 \text{ lb}$ becomes

$$
\begin{array}{r}
3.2 \ \text{lb} \\
1.75 \ \text{lb} \\
+2.4 \ \text{lb} \\
\hline
7.35 \ \text{lb}
\end{array}
$$

If we do not know the true precision of the fractional measurement, the usual procedure is to round to the precision of the original *decimal* measurements. The least precise of these, 3.2 lb and 2.4 lb, are both given to the nearest tenth of a pound. Our answer, 7.35 lb, should be rounded to 7.4 lb. ●

Multiplication and Division of Measurement Numbers   When multiplying or dividing measurement numbers, we need to pay special attention to both the proper units and the proper rounding of our final answer. First, let's consider the units.

With both addition and subtraction of measurement numbers, the answer to the arithmetic calculation has the same units as the numbers being added or subtracted. This is not true when we multiply or divide measurement numbers.

EXAMPLE 7   To multiply

3 ft × 2 ft = __?__

First, multiply the numerical parts: 3 × 2 = 6

Second, multiply the units: 1 ft × 1 ft = 1 ft² or 1 sq ft

The answer is 6 ft² or 6 sq ft—both indicate 6 square feet, which is an *area*.   ●

👆 Learning Help   It may help if you remember that this multiplication is actually

$$3 \text{ ft} \times 2 \text{ ft} = (3 \times 1 \text{ ft}) \times (2 \times 1 \text{ ft})$$
$$= (3 \times 2) \times (1 \text{ ft} \times 1 \text{ ft})$$
$$= 6 \times 1 \text{ sq ft} = 6 \text{ sq ft}$$

The rules of arithmetic say that we can do the multiplications in any order we wish, so we multiply the units part separately from the number part.

Notice that the product 1 ft × 1 ft can be written either as 1 ft² or as 1 sq ft. ●

EXAMPLE 8   To multiply

2 in. × 3 in. × 5 in. = __?__

First, multiply the numbers: 2 × 3 × 5 = 30

Second, multiply the units: 1 in. × 1 in. × 1 in. = 1 in.³ or 1 cu in. (1 cubic inch).

The answer is 30 cu in. This is a *volume*.   ●

Visualizing Units

It is helpful to have a visual understanding of measurement units and the "dimension" of a measurement.

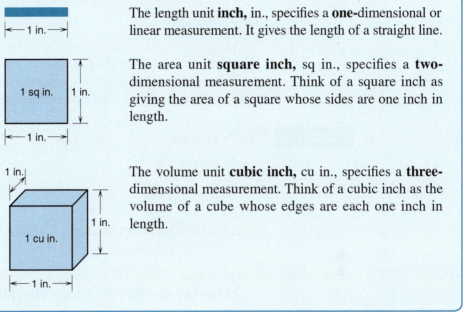

The length unit **inch,** in., specifies a **one**-dimensional or linear measurement. It gives the length of a straight line.

The area unit **square inch,** sq in., specifies a **two**-dimensional measurement. Think of a square inch as giving the area of a square whose sides are one inch in length.

The volume unit **cubic inch,** cu in., specifies a **three**-dimensional measurement. Think of a cubic inch as the volume of a cube whose edges are each one inch in length.

Division with measurement numbers requires that you take the same care with the units that is needed in multiplication.

EXAMPLE 9

To divide

$$24 \text{ mi} \div 1.5 \text{ hr} = \underline{\ ?\ }$$

First, divide the numerical parts as usual.

$$24 \div 1.5 = 16$$

Second, divide the units.

$$1 \text{ mi} \div 1 \text{ hr} = \frac{1 \text{ mi}}{1 \text{ hr}}$$

In this case, the unit $\frac{1 \text{ mi}}{1 \text{ hr}}$ is usually written mi/hr or mph—miles per hour—and the answer to the problem is 16 mph. Remember that

$$24 \text{ mi} \div 1.5 \text{ hr} = \frac{24 \times 1 \text{ mi}}{1.5 \times 1 \text{ hr}} = \left(\frac{24}{1.5}\right) \times \left(\frac{1 \text{ mi}}{1 \text{ hr}}\right)$$

$$= 16 \times 1 \text{ mph} = 16 \text{ mph}$$

Units such as sq ft or mph, which are made up of a combination of simpler units, are called **compound units.** ●

Note Both h and hr are acceptable abbreviations for the unit *hour.* The standard metric abbreviation is h, and hr is in common usage in the United States. ●

Your Turn $36 \text{ lb} \div 1.5 \text{ sq in.} = \underline{\hspace{2cm}}$

Solution **First,** divide the numbers: $36 \div 1.5 = 24$

Second, divide the units: 1 lb ÷ 1 sq in. = $\dfrac{1 \text{ lb}}{1 \text{ sq in.}}$ = 1 lb/sq in. = 1 psi

$\left(\text{The unit } psi \text{ means "pounds per square inch" or } \dfrac{\text{lb}}{\text{sq in.}}.\right)$

The answer is 24 psi.

When multiplying or dividing quantities having compound units, it is sometimes possible to "cancel" or divide out common units. This procedure is identical to the cancellation of common numerical factors when multiplying fractions.

EXAMPLE 10   (a)   To multiply

45 mph × 4 hr = ___?___

set it up this way:

$$45\,\frac{\text{mi}}{\text{hr}} \times 4 \text{ hr} = (45 \times 4) \times \left(\frac{\text{mi}}{\cancel{\text{hr}}} \times \cancel{\text{hr}}\right)$$

$$= 180 \text{ mi}$$

Notice that the "hr" units are cancelled just like common numerical factors.

(b)   To divide

42 sq yd ÷ 6 yd = _____?_____

work with the numbers and the units separately as follows.

$$\left(\frac{42}{6}\right) \times \left(\frac{\text{sq yd}}{\text{yd}}\right) = \left(\frac{42}{6}\right) \times \left(\frac{\text{yd} \times \cancel{\text{yd}}}{\cancel{\text{yd}}}\right) = 7 \text{ yd}$$ ●

More Practice   Before we discuss rounding in multiplication and division, try the following practice problems. Be careful with units.

(a)   5 ft × 4 ft   (b)   4 ft ÷ 2 sec

(c)   2 yd × 6 yd × 0.5 yd   (d)   24 sq in. ÷ 3 in.

(e)   30 mph × 2 hr   (f)   70 ft ÷ 35 ft/sec

Answers   (a)   20 sq ft   (b)   2 ft/sec   (c)   6 cu yd

(d)   8 in.   (e)   60 mi   (f)   2 sec

A Closer Look   Did you have difficulty with (d), (e), or (f)? Do them this way:

(d)   24 sq in. ÷ 3 in. = $\left(\dfrac{24}{3}\right) \times \left(\dfrac{\text{sq in.}}{\text{in.}}\right)$

$$= 8 \times \left(\frac{\text{in.}^2}{\text{in.}}\right) = 8 \times \left(\frac{\text{in.} \times \cancel{\text{in.}}}{\cancel{\text{in.}}}\right) = 8 \text{ in.}$$

(e)   30 mph × 2 hr = $30\,\dfrac{\text{mi}}{\text{hr}} \times 2 \text{ hr}$

$$= (30 \times 2) \times \left(\frac{\text{mi}}{\cancel{\text{hr}}} \cdot \cancel{\text{hr}}\right) = 60 \text{ mi}$$

(f) $70 \text{ ft} \div 35 \text{ ft/sec} = \left(\dfrac{70}{35}\right) \times \left(\text{ft} \div \dfrac{\text{ft}}{\text{sec}}\right)$

$= 2 \times \left(\text{ft} \times \dfrac{\text{sec}}{\text{ft}}\right)$ ⟵ To divide by a fraction, multiply by its reciprocal.

$= 2 \text{ sec}$ ●

Rounding Products and Quotients

When measurement numbers are multiplied or divided, the resulting answer cannot be expressed with greater accuracy than the least accurate measurement number in the calculation. The result of a calculation should not be written to reflect greater accuracy than any number used in that calculation.

> Always round a product or quotient to the same number of significant digits as the least accurate number used in the calculation.

EXAMPLE 11   Multiply: $4.35 \text{ ft} \times 3.6 \text{ ft}$.

First, multiply the numerical factors: $4.35 \times 3.6 = 15.66$

Second, multiply the units: $1 \text{ ft} \times 1 \text{ ft} = 1 \text{ ft}^2$ or 1 sq ft

Third, round the answer to agree in accuracy with the least accurate number in the calculation.

4.35 ft is accurate to three significant digits.

3.6 ft is accurate to two significant digits.

The least accurate factor has two significant digits, so the answer must be rounded to two significant digits.

$15.66 \text{ sq ft} \approx 16 \text{ sq ft}$ ●

EXAMPLE 12   Divide: $375 \text{ mi} \div 65 \text{ mph}$.

First, divide the numerical parts: $375 \div 65 = 5.7692\ldots$

Second, divide the units:

$$\text{mi} \div \dfrac{\text{mi}}{\text{hr}} = \text{mi} \times \dfrac{\text{hr}}{\text{mi}} = \text{hr}$$

Third, round the answer to two significant digits to agree in accuracy with 65, the least accurate measurement number in the calculation.

$375 \text{ mi} \div 65 \text{ mph} \approx 5.8 \text{ hr}$ ●

✎ **Note**   Numbers that result from simple counting should be considered exact. They do not affect the accuracy of the calculation. For example, if a flywheel turns 48 rotations in 1.25 minutes, the speed of the flywheel is calculated as $48 \div 1.25 = 38.4 \text{ rpm}$. We should assume that the 48 is an exact number—rotations are counted directly—so we express the answer as 38.4 with three significant digits to agree with 1.25, the least accurate *measurement* number in the calculation. ●

Your Turn   Perform the following calculations. Round each answer to the proper accuracy.

(a) $7.2 \text{ in.} \times 0.48 \text{ in.}$ (b) $0.18 \text{ ft} \times 3.15 \text{ ft} \times 20 \text{ ft}$

(c) $82.45 \text{ ft} \div 3.65 \text{ sec}$ (d) $28 \text{ sq yd} \div 1.25 \text{ yd}$

(e) $2450 \text{ tons/sq ft} \times 3178 \text{ sq ft}$ (f) $148 \text{ mi} \div 15 \text{ mph}$

Answers   (a) 7.2 in. $\times$ 0.48 in. $=$ 3.456 sq in. $\approx$ 3.5 sq in.

(b) 0.18 ft $\times$ 3.15 ft $\times$ 20 ft $=$ 11.34 cu ft $\approx$ 10 cu ft

(c) 82.45 ft $\div$ 3.65 sec $=$ 22.58904 . . . ft/sec $\approx$ 22.6 ft/sec

(d) 28 sq yd $\div$ 1.25 yd $=$ 22.4 yd $\approx$ 22 yd

(e) 2450 tons/sq ft $\times$ 3178 sq ft $=$ 7,786,100 tons $\approx$ 7,790,000 tons

(f) 148 mi $\div$ 15 mph $=$ 9.8666 . . . hr $\approx$ 9.9 hr

---

! **Careful**   When dividing by hand it is important that you carry the arithmetic to one digit more than will be retained after rounding. ●

Practical Applications

EXAMPLE 13   **Machine Trades**   Each inch of cold-rolled steel weighs 0.22 lb. A certain job calls for 27.5 inches of the steel. How much would this amount weigh?

To solve, we must multiply the weight per inch by the number of inches:

Total weight $=$ (0.22 lb/in.) $\times$ (27.5 in.) $=$ 6.05 lb

For multiplication, we must round our answer to the same number of significant digits as the least accurate factor, which is 0.22 lb/in. Therefore, we must round our product to two significant digits, giving us a total weight of 6.1 lb.   ●

Your Turn   (a) **Industrial Technology**   The volume of a rectangular storage bin can be calculated by multiplying the area of the base times the height. If a bin is designed to have a volume of 12.4 cubic meters and if the base area needs to be 3.8 square meters, what must the height of the bin be?

(b) **Agriculture**   A crate full of tomatoes weighs 39.8 lb. The manufacturer lists the weight of the crate itself as 4.25 lb. What is the weight of the tomatoes alone?

Solutions   (a) Base area times height equals volume, so the height H equals volume divided by base area, or

$$H = (12.4 \text{ cu m}) \div (3.8 \text{ sq m}) = 3.263 \ldots \text{ m}$$

With division, we must round our answer to the same number of significant digits as the least accurate number in the calculation, which is 3.8 square meters. Rounded to two significant digits, the height of the bin must be 3.3 meters.

(b) To determine the weight of the tomatoes alone, we must subtract the weight of the empty crate from the weight of the packed crate. Therefore,

$$\text{Weight of tomatoes} = 39.8 \text{ lb} - 4.25 \text{ lb} = 35.55 \text{ lb}$$

With subtraction, we must round our result to the precision of the least precise measurement in the calculation, 39.8 lb. Therefore the weight of the tomatoes, rounded to the nearest tenth, is 35.6 lb.

✎ **Note**   As we stated earlier, the rounding rules presented in this section are primarily for those who are training for technical, scientific, and engineering careers. In deference to those students who are not preparing for these types of careers, we will give rounding instructions where necessary in future sections and chapters of the text. ●

Decimal Equivalents Some of the measurements made by technical workers in a shop are made in fractions. But on many shop drawings and specifications, the dimensions may be given in decimal form, as shown in the figure. (*Note:* The " symbol stands for "inches.")

The steel rule used in the shop or in other technical work may be marked in 8ths, 16ths, 32nds, or even 64ths of an inch. A common problem is to rewrite the decimal number to the nearest 32nd or 64th of an inch and to determine how much error is involved in using the fraction number rather than the decimal number. In Section 3-3, we learned how to find the closest fractional equivalent to a given decimal using a table. Here we will learn a mathematical technique for doing this when a table is not available. The following example illustrates how to do this.

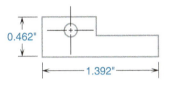

EXAMPLE 14 **Machine Trades** In the preceding drawing, (a) find the fraction, to the nearest 32nd of an inch, equivalent to 0.462 in., and (b) find the error involved in using the fraction number instead of the decimal number.

(a) Multiply the decimal number by the fraction $\frac{32}{32}$.

$$0.462 \text{ in.} = 0.462 \text{ in.} \times \frac{32}{32}$$

$$= \frac{0.462 \text{ in.} \times 32}{32}$$

$$= \frac{14.784}{32} \text{ in.}$$

Round to the nearest 32nd of an inch by rounding the numerator to the nearest whole number.

$$\approx \frac{15}{32} \text{ in.}$$

(b) The error involved is the difference between

$$\frac{15}{32} \text{ in.} \quad \text{and} \quad \frac{14.784}{32} \text{ in.,} \quad \text{or} \quad \frac{15 - 14.784}{32} \text{ in.} = \frac{0.216}{32} \text{ in.}$$

$$= 0.00675 \text{ in.}$$

$$\approx 0.0068 \text{ in.} \quad \text{rounded}$$

The error involved in using $\frac{15}{32}$ in. instead of 0.462 in. is about 68 ten-thousandths of an inch. In problems of this kind, we usually round the error to the nearest ten-thousandth.

Fraction: **.462** ⊗ **32** ⊜ → 14.784 Round to $\frac{15}{32}$.

Error: **15** ⊖ [ANS] ⊜ ⊘ **32** ⊜ → 0.00675 about 0.0068 in. ●

🔍 **A Closer Look** Pressing the [ANS] key retrieves the last answer—in this case, 14.784. ●

EXAMPLE 15 (a) Change 1.392 in. to a fraction expressed in 64ths of an inch, and (b) find the error involved in using the fraction instead of the decimal.

(a) $1.392 \text{ in.} = 1.392 \text{ in.} \times \dfrac{64}{64}$

$= \dfrac{1.392 \times 64}{64} \text{ in.}$

$= \dfrac{89.088}{64} \text{ in.}$

$\approx \dfrac{89}{64} \text{ in.}$ ⟵ Rounded to the nearest unit

$\approx 1\dfrac{25}{64} \text{ in.}$

(b) The error in using the fraction rather than the decimal is

$\dfrac{89.088}{64} \text{ in.} - \dfrac{89}{64} \text{ in.} = \dfrac{0.088}{64} \text{ in.}$

$\approx 0.0014 \text{ in.}$ or about 14 ten-thousandths of an inch.

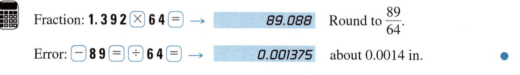

Fraction: **1 . 3 9 2** ✕ **6 4** = → ⬚ **89.088** Round to $\dfrac{89}{64}$.

Error: − **8 9** = ÷ **6 4** = → ⬚ **0.001375** about 0.0014 in.   ●

Your Turn For each of the following decimal measurements, find (a) the closest fractional equivalent using the given denominator, and (b) the error to the nearest ten-thousandth if the fraction is used in place of the decimal number.

(1) 0.438 in. (16ths)   (2) 0.30 in. (16ths)

(3) 0.720 in. (32nds)   (4) 2.285 in. (32nds)

(5) 0.047 in. (64ths)   (6) 1.640 in. (64ths)

Answers (1) (a) $\dfrac{7}{16}$ in. (b) 0.0005 in.   (2) (a) $\dfrac{5}{16}$ in. (b) 0.0125 in.

(3) (a) $\dfrac{23}{32}$ in. (b) 0.0013 in.   (4) (a) $2\dfrac{9}{32}$ in. (b) 0.0038 in.

(5) (a) $\dfrac{3}{64}$ in. (b) 0.0001 in.   (6) (a) $1\dfrac{41}{64}$ in. (b) 0.0006 in.

For a set of practice problems covering this section of Chapter 5, turn to Exercises 5-1.

Exercises 5-1 Working with Measurement Numbers

A. State (a) the precision, (b) the accuracy, and (c) the greatest possible error of each of the following measurement numbers.

   1. 4.7 gal   2. 0.008 ft

   3. 562 psi   4. 3400 rpm

   5. 7.024 lb   6. 360 volts

   7. 19,$\overline{0}$00 tons   8. 19.65 oz

   9. 10.0 in.   10. 0.20 sec

B. Add or subtract as shown. Round your answer to the correct precision.

1. 7 in. + 14 in.
2. 628 lb + 400 lb
3. 3.25 in. + 1.7 in. + 4.6 in.
4. 16.2 psi + 12.64 psi
5. 0.28 gal + 1.625 gal
6. 64.5 mi − 12.72 mi
7. 1.565 oz − 0.38 oz
8. 12.372 sec − 2.1563 sec
9. 28.3 psi + 16 psi + 5.9 psi
10. 6.4 tons + 8.26 tons − 5.0 tons

C. Multiply or divide as shown. Round to the appropriate number of significant digits.

1. 6 ft × 7 ft
2. 3.1 in. × 1.7 in.
3. 3.0 ft × 2.407 ft
4. 215 mph × 1.2 hr
5. $1\frac{1}{4}$ ft × 1.5 ft
6. 48 in. × 8.0 in.
7. 2.1 ft × 1.7 ft × 1.3 ft
8. 18 sq ft ÷ 2.1 ft
9. 458 mi ÷ 7.3 hr
10. $1.37 ÷ 3.2 lb
11. $40\frac{1}{2}$ mi ÷ 25 mph
12. 2.0 sq ft ÷ 1.073 ft

D. Find the closest fractional equivalent for each of the following decimal measurements using the given denominator.

1. 0.921 in. (16ths)
2. 2.55 in. (16ths)
3. 3.69 in. (16ths)
4. 0.306 in. (16ths)
5. 1.90 in. (32nds)
6. 2.350 in. (32nds)
7. 2.091 in. (32nds)
8. 0.685 in. (32nds)
9. 0.235 in. (64ths)
10. 1.80 in. (64ths)
11. 0.645 in. (64ths)
12. 1.935 in. (64ths)

E. For each of the following decimal measurements, (a) find the closest fractional equivalent using the given denominator, and (b) find the error, to the nearest ten-thousandth if the fraction is used in place of the decimal number.

1. 1.80 in. (16ths)
2. 1.571 in. (16ths)
3. 0.666 in. (32nds)
4. 0.285 in. (32nds)
5. 2.420 in. (64ths)
6. 3.175 in. (64ths)

F. Practical Applications

1. **Machine Trades** Specifications call for drilling a hole 0.637 ± 0.005 in. in diameter. Will a $\frac{41}{64}$-in. hole be within the required tolerance?

2. **Metalworking** According to standard American wire size tables, 3/0 wire is 0.4096 in. in diameter. Write this size to the nearest 32nd of an inch. What error is involved in using the fraction rather than the decimal?

3. **Automotive Trades** An auto mechanic converts a metric part to 0.473 in. The part comes only in fractional sizes given to the nearest 64th of an inch. What would be the closest size?

4. **General Trades** What size wrench, to the nearest 32nd of an inch, will fit a bolt with a 0.748-in. head?

5. **Drafting** Find the missing distance x in the drawing.

6. **Carpentry** A truckload of knotty pine paneling weighs 180 lb. If this kind of wood typically has a density of 38 lb/cu ft, find the volume of wood in the load. (*Hint:* Divide the weight by the density.)

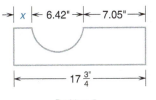

Problem 5

7. **Automotive Trades**   The wear on a cylinder has increased its bore to 3.334 in. The most likely original bore was the nearest 32nd-in. size less than this. What was the most likely original bore?

8. **Interior Design**   If the rectangular floor of a room has an area of 455 sq ft and a length of 26 ft, what is its width? (*Hint:* Area = length × width.)

9. **Transportation**   A truck, moving continuously, averages 52.5 mph for 6.75 hr. How far does it travel during this time?
(*Hint:* Distance = speed × time.)

10. **Industrial Technology**   The volume of a cylindrical tank can be calculated as its base area multiplied by its height. If the tank has a capacity of 2760 cu ft of liquid and has a base area of 120 sq ft, what is its height?

11. **Electronics**   If an electromagnetic signal moves at the speed of light, about 186,282 miles per second, how long will it take this signal to travel from New Jersey to California, approximately 2500 miles? (*Hint:* Time = distance ÷ speed.)

12. **Aviation**   What speed must an airplane average in order to travel 1260 miles in 4.8 hours? (*Hint:* Speed = distance ÷ time.)

13. **Agriculture**   A crate full of oranges weighs 28.4 lb. If the empty crate weighs 3.75 lb, what do the oranges alone weigh?

14. **Water/Wastewater Treatment**   After repairing a section of a water main, a water treatment specialist must disinfect the section with chlorine. To determine the amount of chlorine needed in pounds, the specialist performs the following calculation:

Pounds of chlorine = (0.002349 million gallons)(450 mg/L)(8.34 lb/gal)

How many pounds of chlorine are needed to disinfect the section of pipe?

15. **Sheet Metal Trades**   Four pieces of sheet metal with thicknesses 0.148 in., 0.24 in., 0.375 in., and 0.42 in. are bolted together. What is the total resulting thickness?

16. **General Trades**   Find the greatest possible error for each of the following situations:

   (a) The length of a steel part is measured to be $3\frac{3}{4}$ in. with a ruler that is divided into sixteenths of an inch.

   (b) A liquid solution is measured at $2\frac{1}{3}$ cups with a cup measure that is calibrated to the nearest $\frac{1}{3}$ of a cup.

When you have completed these problems, check your answers to the odd-numbered problems in the Appendix, then turn to Section 5-2 for some information on units and unit conversion.

| 5-2   U.S. Customary Units and Unit Conversion

learning|**catalytics**™

1. How many feet are in 5 yards?
2. Use cancellation to simplify and then multiply:
$$\frac{8}{1} \times \frac{1}{12} \times \frac{15}{1}$$

When you talk about a 4-ft by 8-ft wall panel, a $1\frac{1}{4}$-in. bolt, 3 lb of solder, or a gallon of paint, you are comparing an object with a *standard*. The units or standards used in science, technology, or the trades have two important characteristics: They are convenient in size, and they are standardized. Common units such as the foot, pound, gallon, or hour were originally chosen because they were related either to natural quantities or to body measurements.

A pinch of salt, a drop of water, a handful of sand, or a tank of gas are all given in *natural* units. The width of a finger, the length of a foot, or a stride length are all distances given in *body-related* units. In either case, there is a need to define and standardize these units before they are useful for business, science, or technology.

In ancient times, the *digit* was a length unit defined as the width of a person's finger. Four digits was called a *hand,* a unit still used to measure horses. Four hands was a *cubit,* the distance from fingertip to elbow, about 18 in. One *foot* was defined as the length of a man's foot. A *yard* was the distance from nose to tip of outstretched arm. Of course all these lengths depend on whose finger, hand, arm, or foot is used.

Standardization in the English system of units began in the fifteenth century, and one yard was defined as the length of a standard iron bar kept in a London vault. One foot was defined as exactly one-third the length of the bar, and one inch was one-twelfth of a foot. (The word *inch* comes from Latin and Anglo-Saxon words meaning "one-twelfth.") Today the inch is legally defined worldwide in terms of metric or SI units.

The English system of units was adopted for use in the United States, and it is commonly known as the U.S. customary system of units.

Because there are a great many units available for writing the same quantity, it is necessary that you be able to convert a given measurement from one unit to another. For example, a length of exactly one mile can be written as

$$
\begin{aligned}
1 \text{ mile} &= 1760 \text{ yd, for the traffic engineer} \\
&= 5280 \text{ ft, for a landscape architect} \\
&= 63{,}360 \text{ in., in a science problem} \\
&= 320 \text{ rods, for a surveyor} \\
&= 8 \text{ furlongs, for a horse racing fan} \\
&= 1609.344 \text{ meters, in metric units}
\end{aligned}
$$

Here are some basic conversion factors for length, weight, and liquid capacity in the U.S. customary system of units:

Length Units

1 foot (ft) = 12 inches (in.)
1 yard (yd) = 3 ft = 36 in.
1 rod (rd) = $16\frac{1}{2}$ ft
1 mile (mi) = 5280 ft = 1760 yd

Liquid Capacity Units

1 tablespoon (T) = 3 teaspoons (t)
1 fluid ounce (fl oz) = 2 tablespoons (T)
1 measuring cup = 8 fl oz
1 pint (pt) = 2 cups = 16 fl oz
1 quart (qt) = 2 pt = 32 fl oz
1 gallon (gal) = 4 qt = 128 fl oz

Weight Units

1 pound (lb) = 16 ounces (oz)
1 ton = 2000 lb

❗ Careful Household tableware does not come in standard sizes and therefore should *not* be used for measuring teaspoons and tablespoons. Instead use a proper measuring device specifically designed to measure these amounts. ●

In this section we will show you a quick and mistake-proof way to convert measurements from one unit to another. Whatever the units given, if a new unit is defined, you should be able to convert the measurement to the new units.

EXAMPLE 1   Consider the following simple unit conversion. Convert 3 yards to feet.

3 yd = _____ ft

From the table, we know that 1 yard is defined as exactly 3 feet; therefore,

3 yd = 3 × 3 ft = 9 ft

Easy.   ●

Your Turn   Try another, more difficult, conversion. Surveyors use a unit of length called a *rod*, defined as exactly $16\frac{1}{2}$ ft.

Convert: 11 yd = _____ rods.

Solution   The correct answer is 2 rods.

You might have thought first to convert 11 yd to 33 ft and then noticed that 33 ft divided by $16\frac{1}{2}$ equals 2. There are 2 rods in a 33-ft length.

Unity Fractions   Most people find it difficult to reason through a problem in this way. To help you, we have devised a method of solving *any* unit conversion problem quickly with no chance of error. This is the **unity fraction method**.

The first step in converting a number from one unit to another is to set up a unity fraction from the definition linking the two units.

EXAMPLE 2   To convert a distance from yard units to foot units, take the equation relating yards to feet:

1 yd = 3 ft

and form the fractions $\frac{1\ yd}{3\ ft}$ and $\frac{3\ ft}{1\ yd}$.

These fractions are called **unity fractions** because they are both equal to 1. Any fraction whose top and bottom are equal has the value 1.   ●

Your Turn   Practice this step by forming pairs of unity fractions from each of the following definitions.

(a) 12 in. = 1 ft   (b) 1 lb = 16 oz   (c) 4 qt = 1 gal

Solutions   (a) 12 in. = 1 ft. The unity fractions are $\frac{12\ in.}{1\ ft}$ and $\frac{1\ ft}{12\ in.}$.

(b) 1 lb = 16 oz. The unity fractions are $\frac{1\ lb}{16\ oz}$ and $\frac{16\ oz}{1\ lb}$.

(c) 4 qt = 1 gal. The unity fractions are $\frac{4\ qt}{1\ gal}$ and $\frac{1\ gal}{4\ qt}$.

The second step in converting units is to multiply the original number by one of the unity fractions. Always choose the unity fraction that allows you to cancel out the units you do not want in your answer. The units you do not want will be in the denominator of the appropriate unity fraction.

EXAMPLE 3 To convert the distance 4.0 yd to foot units, multiply this way:

$$4.0 \text{ yd} = (4.0 \times 1 \text{ y̶d̶}) \times \left(\frac{3 \text{ ft}}{1 \text{ y̶d̶}} \right) \quad \longleftarrow \boxed{\begin{array}{l}\text{Choose the unity fraction that cancels}\\ \text{yd and replaces it with ft}\end{array}}$$

$$= 4.0 \times 3 \text{ ft}$$

$$= 12 \text{ ft}$$

You must choose one of the two fractions. Multiply by the fraction that allows you to cancel out the *yard* units you do not want and to keep the *feet* units you do want.   ●

Did multiplying by a fraction give you any trouble? Remember that any number A can be written as $\frac{A}{1}$, so that we have

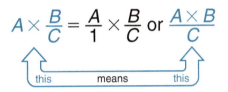

If you need to review the multiplication of fractions, turn back to page 86.

The next example illustrates two ways to deal with a remainder.

EXAMPLE 4 To convert 44 oz to pounds, remember that 1 lb = 16 oz.

Use the equation 1 lb = 16 oz to set up the unity fractions $\frac{1 \text{ lb}}{16 \text{ oz}}$ and $\frac{16 \text{ oz}}{1 \text{ lb}}$.

Multiply by the first fraction to convert ounces to pounds.

$$44 \text{ oz} = (44 \text{ o̶z̶}) \times \left(\frac{1 \text{ lb}}{16 \text{ o̶z̶}} \right) \quad \longleftarrow \boxed{\begin{array}{l}\text{Choose the unity fraction that cancels}\\ \text{oz and replaces it with lb.}\end{array}}$$

$$= \frac{44 \times 1 \text{ lb}}{16}$$

$$= \frac{44}{16} \text{ lb}$$

Because 44 is not exactly divisible by 16, we can simplify our answer in one of three ways:

As a mixed number: $\dfrac{44}{16} \text{ lb} = \dfrac{11}{4} \text{ lb} = 2\dfrac{3}{4} \text{ lb}$

As a decimal number: $= 2.75 \text{ lb}$

Or in pounds and ounces: $\dfrac{44}{16} \text{ lb} = 2r12 = 2 \text{ lb } 12 \text{ oz}$

(*Reminder:* 2r12 means "a quotient of two with a remainder of 12." The units of the remainder are always the same as the units of the numerator or dividend, which in this case was ounces.)

It may happen that several unity fractions are needed in the same problem.   ●

EXAMPLE 5   If you know that

$$1 \text{ rod} = 16\tfrac{1}{2} \text{ ft} \quad \text{and} \quad 1 \text{ yd} = 3 \text{ ft}$$

then to convert 11 yd to rods, multiply as follows:

| First cancel yd and keep ft | | Then cancel ft and keep rod |

$$11 \text{ yd} = (11 \text{ yd}) \times \left(\frac{3 \text{ ft}}{1 \text{ yd}} \right) \times \left(\frac{1 \text{ rod}}{16\tfrac{1}{2} \text{ ft}} \right)$$

$$= \frac{11 \times 3 \times 1 \text{ rod}}{16\tfrac{1}{2}}$$

$$= \frac{33}{16\tfrac{1}{2}} \text{ rod}$$

$$11 \text{ yd} = 2 \text{ rods}$$

To work with unity fractions on a calculator, simply multiply by any quantity in a numerator and divide by any quantity in a denominator. For this example, we would enter

 1 1 ⊠ **3** ÷ **1 6 . 5** ▭ → ⬛ 2. •

📝 **Note**   When a measurement is converted from one unit to another, the result should be rounded to the same number of significant digits as contained in the original measurement. •

Your Turn   Try these practice problems. Use unity fractions to convert.

(a) 6.25 ft = _____ in.   (b) $5\tfrac{1}{4}$ yd = _____ ft

(c) 2.1 mi = _____ yd   (d) 12 gallons = _____ pints

(e) 24 cups = _____ quarts   (f) 87 in. = _____ ft _____ in.

(g) 11 lb 14 oz = _____ oz   (h) 0.75 fl oz = _____ teaspoons

Solutions   (a) $6.25 \text{ ft} = (6.25 \text{ ft}) \times \left(\dfrac{12 \text{ in.}}{1 \text{ ft}} \right)$ (b) $5\tfrac{1}{4} \text{ yd} = (5\tfrac{1}{4} \text{ yd}) \times \left(\dfrac{3 \text{ ft}}{1 \text{ yd}} \right)$

$\qquad\qquad\qquad = 6.25 \times 12 \text{ in.}$ $= 5\tfrac{1}{4} \times 3 \text{ ft}$

$\qquad\qquad\qquad = 75 \text{ in.}$ $= 15\tfrac{3}{4} \text{ ft}$

(c) $2.1 \text{ mi} = (2.1 \text{ mi}) \times \left(\dfrac{1760 \text{ yd}}{1 \text{ mi}} \right)$ (d) $12 \text{ gal} = (12 \text{ gal}) \times \left(\dfrac{4 \text{ qt}}{1 \text{ gal}} \times \dfrac{2 \text{ pt}}{1 \text{ qt}} \right)$

$\qquad\qquad\quad = 2.1 \times 1760 \text{ yd}$ $= 12 \times 4 \times 2 \text{ pt}$

$\qquad\qquad\quad = 3696 \text{ yd}$ $= 96 \text{ pt}$

$\qquad\qquad\quad \approx 3700 \text{ yd}$, rounded to two
$\qquad\qquad\qquad$ significant digits

(e) $24 \text{ cups} = 24 \text{ cups} \times \left(\dfrac{1 \text{ pt}}{2 \text{ cups}} \times \dfrac{1 \text{ qt}}{2 \text{ pt}} \right)$

$$= \dfrac{24}{4} \text{ qt} = 6 \text{ qt}$$

(f) $87 \text{ in.} = (87 \text{ in.}) \times \left(\dfrac{1 \text{ ft}}{12 \text{ in.}} \right)$

$$= \dfrac{87}{12} \text{ ft} = 7r3 \text{ ft} = 7 \text{ ft } 3 \text{ in.}$$

(g) **First,** convert the pounds to ounces:

$$11 \text{ lb} = (11 \text{ lb}) \times \left(\dfrac{16 \text{ oz}}{1 \text{ lb}} \right)$$

$$= 11 \times 16 \text{ oz} = 176 \text{ oz}$$

Then, add the ounces portion of the original measurement:

$$11 \text{ lb } 14 \text{ oz} = 176 \text{ oz} + 14 \text{ oz} = 190 \text{ oz}$$

(h) $0.75 \text{ fl oz} = 0.75 \text{ fl oz} \times \left(\dfrac{2 \text{ T}}{1 \text{ fl oz}} \times \dfrac{3 \text{ t}}{1 \text{ T}} \right)$

$$= 0.75 \times 2 \times 3 \text{ t} = 4.5 \text{ t}$$

Practical Applications

EXAMPLE 6   **Allied Health**   The average American consumes about 44.7 gallons of soft drinks per year. How many 12-fluid-ounce cans would this be?

To answer this question, we first use a series of three unity fractions to convert gallons to ounces:

$$44.7 \text{ gal} = (44.7 \text{ gal}) \times \left(\left(\dfrac{4 \text{ qt}}{1 \text{ gal}} \right) \left(\dfrac{2 \text{ pt}}{1 \text{ qt}} \right) \left(\dfrac{16 \text{ fl oz}}{1 \text{ pt}} \right) \right)$$

$$= 44.7 \times 4 \times 2 \times 16 \text{ fl oz} = 5721.6 \text{ fl oz}$$

Then, we convert fluid ounces to cans. Use the equation $1 \text{ can} = 12 \text{ fl oz}$ to set up a unity fraction that cancels ounces and replaces them with cans:

$$5721.6 \text{ fl oz} = (5721.6 \text{ fl oz}) \times \left(\dfrac{1 \text{ can}}{12 \text{ fl oz}} \right) = \dfrac{5721.6}{12} \text{ cans} = 476.8 \text{ cans}$$

Therefore, 44.7 gal is equivalent to about 477 12-fl-oz cans of soft drinks.   ●

EXAMPLE 7   **Automotive Trades**   The circumference of each tire on a truck is approximately 98 in., which means that the truck travels 98 in. with each revolution of the tire. How many miles will the truck travel if the wheels make 8500 revolutions (rev)?

First, convert revolutions to inches:

$$8500 \text{ rev} = (8500 \text{ rev}) \times \left(\dfrac{98 \text{ in.}}{1 \text{ rev}} \right)$$

$$= 8500 \times 98 \text{ in.}$$

$$= 833,000 \text{ in.}$$

Then, convert inches to miles:

$$833{,}000 \text{ in.} = (833{,}000 \text{ in.}) \times \left(\frac{1 \text{ ft}}{12 \text{ in.}}\right) \times \left(\frac{1 \text{ mi}}{5280 \text{ ft}}\right)$$

$$= \frac{833{,}000}{12 \times 5280} \text{ mi}$$

$$\approx 13.147 \text{ mi}$$

$$\approx 13 \text{ mi, rounded} \qquad \bullet$$

Your Turn   (a) **Allied Health**   A patient is directed to take 2 teaspoons of a liquid medication per day. How many days will a 16-fluid-ounce bottle of this medication last?

(b) **Construction**   Sandstone weighs approximately 150 pounds per cubic foot. How many tons are needed for a job requiring 1600 cu ft of sandstone?

Solutions   (a) We will use three unity fractions to perform this calculation: one to convert fluid ounces (fl oz) to tablespoons (T), one to convert tablespoons to teaspoons (t), and one using teaspoons per day to convert our answer to days.

$$16 \text{ fl oz} = 16 \text{ fl oz} \times \frac{2 \text{ T}}{1 \text{ fl oz}} \times \frac{3 \text{ t}}{1 \text{ T}} \times \frac{1 \text{ day}}{2 \text{ t}} = \frac{16 \times 2 \times 3}{2} \text{ days} = 48 \text{ days}$$

(b) 150 pounds per cubic foot can be expressed in fraction form as $\dfrac{150 \text{ lb}}{1 \text{ cu ft}}$.

Now multiply as follows:

$$1600 \text{ cu ft} \times \frac{150 \text{ lb}}{1 \text{ cu ft}} \times \frac{1 \text{ ton}}{2000 \text{ lb}} = 120 \text{ tons}$$

Compound Units   In the U.S. customary system of units, the basic units for length, weight, and time are named with a single word or abbreviation: ft, lb, and sec. We often define units for other similar quantities in terms of these. For example, 1 mile = 5280 ft, 1 ton = 2000 lb, 1 hr = 3600 sec, and so on.

We may also define units for more complex quantities using these basic units. For example, the common unit of speed is miles per hour or mph, and it involves both distance (miles) and time (hour) units. Density is the quotient of a weight divided by a volume—for example, pounds per cubic foot. Part (b) of the previous Your Turn gave the density of sandstone. Units named using the product or quotient of two or more simpler units are called **compound units**.

EXAMPLE 8   If a car travels 75 miles at a constant rate in 1.5 hours, it is moving at a speed of

$$\frac{75 \text{ miles}}{1.5 \text{ hours}} = \frac{75 \times 1 \text{ mi}}{1.5 \times 1 \text{ hr}}$$

$$= \frac{75}{1.5} \times \frac{1 \text{ mi}}{1 \text{ hr}}$$

$$= 50 \text{ mi/hr}$$

We write this speed as 50 mi/hr or 50 mph, and read it as "50 miles per hour." As shown by this example, speed is a compound unit because it is the quotient of a distance unit and a time unit. $\qquad \bullet$

Your Turn   Which of the following are expressed in compound units?

(a) Diameter, 4.65 in.

(b) Density, 12 lb/cu ft

(c) Gas usage, mi/gal or mpg

(d) Current, 4.6 amperes

(e) Rotation rate, revolution/min or rpm

(f) Pressure, lb/sq in. or psi

(g) Cost ratio, cents/lb

(h) Length, 5 yd 2 ft

(i) Area, sq ft

(j) Volume, cu in.

Answers   Compound units are used in quantities (b), (c), (e), (f), (g), (i), and (j). Two different units are used in (h), but they have not been combined by multiplying or dividing.

In Chapters 8 and 9, we study the geometry needed to calculate the area and volume of the various plane and solid figures used in technical work. Here we take a first look at the many different compound units used to measure area and volume.

Area   When the area of a surface is found, the measurement or calculation is given in "square units." If the lengths are measured in inches, the area is given in square inches or sq in.; if the lengths are measured in feet, the area is given in square feet or sq ft.

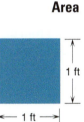

Area = 1 ft × 1 ft

= 1 sq ft

To see the relation between square feet area units and square inch area units, rewrite the above area in inch units.

Area = 12 in. × 12 in.

= 144 × 1 in. × 1 in.

= 144 sq in.

Therefore,

1 sq ft = 144 sq in.

Note   The area units "sq ft," "sq in.," and so on are often expressed as ft^2, $in.^2$, and so on. The reason for this will become clear when we study exponents in Chapter 6. ●

EXAMPLE 9   Let's use this information to do the following conversion:

15.25 sq ft = _____ sq in.

$$= (15.25 \times 1 \text{ sq ft}) \times \left(\frac{144 \text{ sq in.}}{1 \text{ sq ft}} \right)$$

This unity fraction cancels sq ft and keeps sq in.

= 15.25 × 144 sq in.

= 2196 sq in.

15.25 ⊠ 144 ⊟ → 2196.

Learning Help   In Example 9, if you do not remember the conversion number, 144, try it this way:

$$15.25 \text{ sq ft} = (15.25 \times 1 \text{ ft} \times 1 \text{ ft}) \times \left(\frac{12 \text{ in.}}{1 \text{ ft}}\right) \times \left(\frac{12 \text{ in.}}{1 \text{ ft}}\right)$$

$$= 15.25 \times 1 \times 1 \times 12 \text{ in.} \times 12 \text{ in.}$$

$$= 2196 \text{ sq in.} \quad \bullet$$

A number of convenient area units have been defined.

Area Units

1 square foot (sq ft or ft^2)	= 144 square inches (sq in. or in.^{2})
1 square yard (sq yd or yd^2)	= 9 sq ft = 1296 sq in.
1 square rod (sq rod)	= 30.25 sq yd
1 acre = 160 sq rod	= 4840 sq yd = 43,560 sq ft
1 sq mile (sq mi or mi^2)	= 640 acres

EXAMPLE 10   Suppose we need to convert 4800 acres to square miles. From the table, we find that 1 sq mi = 640 acres. Using the appropriate unity fraction, we have

$$4800 \text{ acres} = (4800 \text{ acres}) \times \left(\frac{1 \text{ sq mi}}{640 \text{ acres}}\right)$$

$$= \frac{4800}{640} \text{ sq mi} = 7.5 \text{ sq mi} \qquad \bullet$$

Your Turn   Solve the following conversion problems.

(a) 2.50 sq yd = _____ sq in.

(b) 1210 sq yd = _____ acre

Solutions   (a) $2.50 \text{ sq yd} = (2.50 \times 1 \text{ sq yd}) \times \left(\frac{1296 \text{ sq in.}}{1 \text{ sq yd}}\right)$

$$= 3240 \text{ sq in.}$$

(b) $1210 \text{ sq yd} = (1210 \times 1 \text{ sq yd}) \times \left(\frac{1 \text{ acre}}{4840 \text{ sq yd}}\right)$

$$= \frac{1210}{4840} \text{ acre} = 0.25 \text{ acre}$$

Volume   Volume units are given in cubic units. If the lengths are measured in inches, the volume is given in cubic inches or cu in.; if the lengths are measured in feet, the volume is given in cubic feet or cu ft.

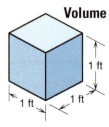

Volume = 1 ft × 1 ft × 1 ft = 1 cu ft

In inch units, this same volume is 12 in. × 12 in. × 12 in. = 1728 cu in. This means that 1 cu ft = 1728 cu in.

Note   Volume units such as cu ft and cu in. can also be written as ft^3 and in.^{3}, respectively. ●

EXAMPLE 11   To convert 0.250 cu ft to cu in., proceed as follows:

$$0.250 \text{ cu ft} = (0.250 \text{ cu ft}) \times \left(\frac{1728 \text{ cu in.}}{1 \text{ cu ft}} \right)$$

$$= (0.250 \times 1728) \text{ cu in.}$$

$$= 432 \text{ cu in.} \qquad ●$$

A very large number of special volume units have been developed for various uses.

Volume Units

1 cubic foot (cu ft or ft^3)	= 1728 cubic inches (cu in. or in.^{3})
1 cubic yard (cu yd or yd^3)	= 27 cu ft = 46,656 cu in.
1 gallon (gal)	= 231 cu in. or 1 cu ft ≈ 7.48 gal
1 bushel (bu)	= 2150.42 cu in.
1 fluid ounce (fl oz)	= 1.805 cu in.
1 pint (pt)	= 28.875 cu in. (liquid measure)

 Note   You may have noticed that the volume units of the gallon, the fluid ounce, and the pint were all defined earlier as liquid capacity units (see p. 295). Think of liquid capacity as the amount of liquid that a container can hold. As indicated in the table, these units can also be defined in terms of cubic units, which is the amount of three-dimensional space taken up by each given amount of liquid. For example, a gallon container holds 128 fluid ounces (liquid capacity), and this amount of liquid also takes up 231 cubic inches of space (volume). ●

EXAMPLE 12   To convert 10,500 cubic inches to bushels, set up the following calculation:

$$10,500 \text{ cu in.} = (10,500 \text{ cu in.}) \times \left(\frac{1 \text{ bu}}{2150.42 \text{ cu in.}} \right)$$

$$= \left(\frac{10,500}{2150.42} \right) \text{ bu}$$

$$= 4.882 \ldots \text{ bu} \approx 4.88 \text{ bu} \qquad ●$$

Your Turn   Convert the following volume measurements for practice. Round all answers to the nearest tenth.

(a) 94 cu ft = _____ cu yd   (b) 1.255 gal = _____ cu in.

(c) 282.5 fl oz = _____ cu in.   (d) 150 cu in. = _____ pt

Answers   (a) 3.5 cu yd   (b) 289.9 cu in.   (c) 509.9 cu in.   (d) 5.2 pt

Applications of Area and Volume

EXAMPLE 13 **Agriculture** Barley has a density of 38.4 lb/cu ft. One bushel of barley weighs 48 lb. How many bushels of barley are contained in 3.25 cu yd?

First, calculate the number of bushels in one cubic yard. Set up unity fractions using the following conversion equations: 1 cu ft = 38.4 lb, 1 bushel (bu) = 48 lb, 1 cu yd = 27 cu ft.

$$\frac{38.4 \text{ lb}}{1 \text{ cu ft}} \times \frac{1 \text{ bu}}{48 \text{ lb}} \times \frac{27 \text{ cu ft}}{1 \text{ cu yd}} = 21.6 \text{ bu/cu yd}$$

Then, multiply by 3.25 cu yd to find the total number of bushels:

$$\frac{21.6 \text{ bu}}{1 \text{ cu yd}} \times 3.25 \text{ cu yd} = 70.2 \text{ bu}$$

●

Your Turn **Flooring and Carpeting** Carpeting used to be sold strictly by the square yard, but now many companies are pricing it by the square foot. Suppose Carpet City offered a certain carpet at $42.95 per square yard, and a competitor, Carpeterium, offered a comparable carpet at $4.95 per square foot. Which is the better buy?

Solution To compare prices, we will convert Carpeterium's price to cost per square yard.

$$\$4.95 \text{ per sq ft} = \left(\frac{\$4.95}{1 \text{ sq ft}}\right) \times \left(\frac{9 \text{ sq ft}}{1 \text{ sq yd}}\right) = \frac{\$44.55}{1 \text{ sq yd}} = \$44.55 \text{ per sq yd}$$

Carpeterium's cost is $44.55 per square yard compared to Carpet City's cost of $42.95 per square yard. Carpet City has the better deal. Notice that we could have also converted Carpet City's price to cost per square foot and reached the same conclusion.

Lumber Measure Carpenters and workers in the construction trades use a special unit of volume to measure the amount of lumber. The volume of lumber is measured in **board feet**. One board foot (bf or bd f or fbm) of lumber is a piece having an area of 1 sq ft and a thickness of 1 in. or less.

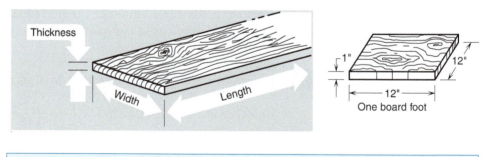

Number of board feet = thickness in inches × width in feet × length in feet

To find the number of board feet in a piece of lumber, multiply the length in feet by the width in feet by the thickness in inches. A thickness of less than 1 in. should be counted as 1 in. If the lumber is dressed or finished, use the full size or rough stock dimension to calculate board feet.

EXAMPLE 14 **Construction** A 2-by-4 used in framing a house would actually measure $1\frac{1}{2}$ in. by $3\frac{1}{2}$ in., but we would use the dimensions 2 in. by 4 in. in calculating board feet. A 12-ft length of framing lumber that is 2 in. thick by 4 in. wide would have a volume of

$2 \text{ in.} \times \frac{4}{12} \text{ ft} \times 12 \text{ ft} = 8 \text{ board ft or } 8 \text{ bf}$

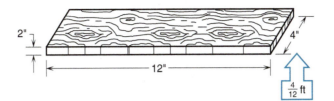

Note   In lumber measure the phrase "per foot" means "per board foot." The phrase "per running foot" means "per foot of length." ●

EXAMPLE 15   **Flooring and Carpeting**   To floor a small building requires 245 boards, each $1\frac{1}{2}$ in. by 12 in. by 12 ft long. To determine how many board feet should be ordered,

First, find the volume of each board:

$2 \text{ in.} \times \frac{12}{12} \text{ ft} \times 12 \text{ ft} = 24 \text{ bf}$

Then, for 245 boards, we must order

$245 \times 24 \text{ bf} = 5880 \text{ bf}$ ●

Your Turn   1. **Construction**   Find the number of board feet in each of the following pieces of lumber. Be sure to convert the middle dimension, width, to feet.

  (a) $\frac{3}{4}$ in. × 6 in. × 4 ft   (b) 2 in. × 12 in. × 16 ft

  (c) 1 in. × 8 in. × 12 ft   (d) 4 in. × 4 in. × 8 ft

  2. **Construction**   How many board feet are there in a shipment containing 80 boards, 2 in. by 6 in., each 16 ft long?

Answers   1. (a) 2 bf   (b) 32 bf   (c) 8 bf   (d) $10\frac{2}{3}$ bf

2. $2\,\boxed{\times}\,6\,\boxed{\div}\,12\,\boxed{\times}\,16\,\boxed{\times}\,80\,\boxed{=}\; \rightarrow \qquad$ *1280.*

There are 1280 board feet.

A Closer Look   In part 1(a), remember to round the thickness up to 1 in. ●

Speed   As illustrated in Example 8 on p. 300, speed is usually specified by a compound unit that consists of a distance unit divided by a time unit. For example, a speed given as 60 mph (miles per hour) can be written as

$60 \text{ mi/hr or } \dfrac{60 \text{ mi}}{1 \text{ hr}}$

To convert speed units within the U.S. customary system of units, use the length units from page 295 and the following time conversion factors:

> **Time Units**
>
> 1 minute (min) = 60 seconds (sec)
>
> 1 hour (hr or h) = 60 min = 3600 sec
>
> 1 day = 24 hr or 24 h

Converting speeds often involves converting both the distance and the time units. This requires two separate unity fractions.

EXAMPLE 16   To convert 55 mph to ft/sec, use two conversion factors: 1 mi = 5280 ft (to convert miles to feet) and 1 hr = 3600 sec (to convert hours to seconds). Set up the unity fractions like this:

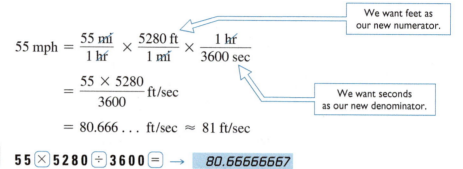

$$55 \text{ mph} = \frac{55 \text{ mi}}{1 \text{ hr}} \times \frac{5280 \text{ ft}}{1 \text{ mi}} \times \frac{1 \text{ hr}}{3600 \text{ sec}}$$

We want feet as our new numerator.

$$= \frac{55 \times 5280}{3600} \text{ ft/sec}$$

We want seconds as our new denominator.

$$= 80.666\ldots \text{ ft/sec} \approx 81 \text{ ft/sec}$$

55 ⊠ 5280 ⊡ 3600 ⊟ → **80.66666667**

EXAMPLE 17   To convert 3.8 in./sec to ft/min, set up unity fractions as follows:

This replaces in. with ft in the numerator.

$$3.8 \text{ in./sec} = \frac{3.8 \text{ in.}}{1 \text{ sec}} \times \frac{1 \text{ ft}}{12 \text{ in.}} \times \frac{60 \text{ sec}}{1 \text{ min}}$$

This replaces sec with min in the denominator.

$$= 19 \text{ ft/min}$$

Your Turn   Convert the following speed measurements.

(a) 98.5 ft/sec = _____ mi/hr   (b) 160 rpm = _____ rev/sec

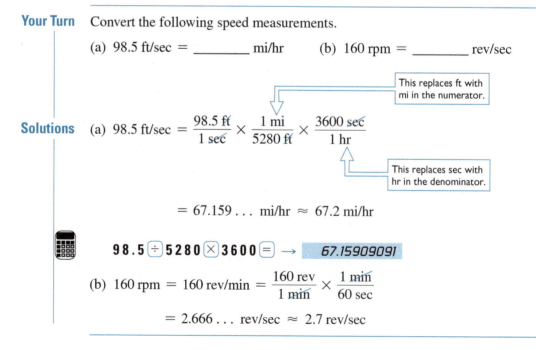

Solutions   (a) $98.5 \text{ ft/sec} = \dfrac{98.5 \text{ ft}}{1 \text{ sec}} \times \dfrac{1 \text{ mi}}{5280 \text{ ft}} \times \dfrac{3600 \text{ sec}}{1 \text{ hr}}$

This replaces ft with mi in the numerator.

This replaces sec with hr in the denominator.

$$= 67.159\ldots \text{ mi/hr} \approx 67.2 \text{ mi/hr}$$

98.5 ⊡ 5280 ⊠ 3600 ⊟ → **67.15909091**

(b) $160 \text{ rpm} = 160 \text{ rev/min} = \dfrac{160 \text{ rev}}{1 \text{ min}} \times \dfrac{1 \text{ min}}{60 \text{ sec}}$

$$= 2.666\ldots \text{ rev/sec} \approx 2.7 \text{ rev/sec}$$

More Applications   We can use the cancellation of units method to solve problems involving unique conversion factors stated in the problem.

EXAMPLE 18

Automotive Trades The 2017 Toyota Camry 4-cylinder hybrid has an EPA fuel economy rating of 40 mi/gal (miles per gallon; combined city and highway driving). The comparable 4-cylinder nonhybrid has an average fuel economy of 27 mi/gal, but it costs $3720 less. If gas costs an average of $2.40 per gallon, we can calculate the amount of money the hybrid would save over a 160,000-mi lifespan.

First, use the cancellation of units method to calculate the cost of gas over the life of each car.

For the hybrid: $\dfrac{160{,}000 \text{ mi}}{1 \text{ lifespan}} \times \dfrac{1 \text{ gal}}{40 \text{ mi}} \times \dfrac{\$2.40}{1 \text{ gal}} = \$9600$

For the nonhybrid: $\dfrac{160{,}000 \text{ mi}}{1 \text{ lifespan}} \times \dfrac{1 \text{ gal}}{27 \text{ mi}} \times \dfrac{\$2.40}{1 \text{ gal}} \approx \$14{,}222$

Then, subtract to find the savings on gas.

The hybrid would save $14,222 − $9600 = $4622 on gas.

Finally, because the hybrid costs $3720 more than the nonhybrid, we subtract this from the gas savings to find the total savings.

The hybrid saves a total of $4622 − $3720 = $902 over a 160,000-mi lifespan.

 160000 ÷ 27 × 2.4 − 160000 ÷ 40 × 2.4 − 3720 = → *902.2222222*

Gas cost for nonhybrid   Gas cost for hybrid   Price difference

If you happened to calculate the gas costs in the opposite order, you would have ended up with the same number preceded by a negative sign. ●

Your Turn **Life Skills** An employee who lives 50 miles from her place of work (100-mile round trip) can either drive her car to work each day or take a bus. To compare the relative costs of these two options, use the following information:

1. A monthly bus pass costs $125, and she can walk to the bus stop.

2. Her car gets 25 miles per gallon of gas.

3. The estimated cost of gas is $2.40 per gallon.

4. She works an average of 21 days per month.

How much would the employee save per month if she took the bus?

Solution In this problem, we need to calculate the total cost of driving per month and compare it to the $125 cost of a bus pass. The conversion equations are 1 month = 21 round trips, 1 round trip = 100 miles, 1 gallon = 25 miles, and 1 gallon = $2.40. To end up with dollars per month, multiply the following unity fractions:

$$\dfrac{21 \text{ round-trips}}{1 \text{ month}} \times \dfrac{100 \text{ mi}}{1 \text{ round-trip}} \times \dfrac{1 \text{ gal}}{25 \text{ mi}} \times \dfrac{\$2.40}{1 \text{ gal}} = \$201.60 \text{ per month}$$

Finally, subtract the cost of the monthly bus pass from this result to determine how much the employee will save by taking the bus:

$$\$201.60 - \$125 = \$76.60 \text{ saved per month by taking the bus}$$

The following table summarizes all U.S. customary conversion factors given in this section.

Length		Weight	
1 ft = 12 in.		1 lb = 16 oz	
1 yd = 3 ft = 36 in.		1 ton = 2000 lb	
1 rod (rd) = $16\frac{1}{2}$ ft			
1 mi = 5280 ft = 1760 yd			
Liquid Capacity		**Area**	
1 tablespoon (T) = 3 teaspoons (t)		1 sq ft = 144 sq in.	
1 fluid ounce (fl oz) = 2 T		1 sq yd = 9 sq ft = 1296 sq in.	
1 measuring cup = 8 fl oz		1 sq rod = 30.25 sq yd	
1 pint (pt) = 2 cups = 16 fl oz		1 acre = 160 sq rod = 4840 sq yd	
1 qt = 2 pt = 32 fl oz		1 acre = 43,560 sq ft	
1 gal = 4 qt = 128 fl oz		1 sq mi = 640 acres	
Volume		**Time**	
1 cu ft = 1728 cu in.		1 min = 60 sec	
1 gal = 231 cu in.		1 hr or 1 h = 60 min = 3600 sec	
1 bu = 2150.42 cu in.		1 day = 24 hr	
1 pt = 28.875 cu in.			
1 cu yd = 27 cu ft = 46,656 cu in.			
1 cu ft ≈ 7.48 gal			
1 fl oz = 1.805 cu in.			

Now, turn to Exercises 5-2 for a set of problems involving unit conversion in the U.S. customary system.

Exercises 5-2   U.S. Customary Units and Unit Conversion

A. Convert the units as shown. (Rounding instructions are provided where needed.)

1. 4.25 ft = _____ in.

2. 33 yd = _____ ft

3. 3.4 mi = _____ ft (nearest thousand)

4. 17.3 lb = _____ oz (whole number)

5. 126 gal = _____ qt

6. 8.5 mi = _____ yd (thousand)

7. 5.50 gal = _____ cu in. (ten)

8. 7.5 lb = _____ oz

9. 18.4 pt = _____ gal

10. 33.0 in. = _____ ft

11. 3960 ft = _____ mi

12. 22 yd = _____ rd

13. 126 ft = _____ yd

14. 230 oz = _____ lb _____ oz

15. 3465 cu in. = _____ gal

16. 351 in. = _____ ft _____ in.

17. 1408 yd = _____ mi

18. 0.85 cu ft = _____ gal

19. 6 ft 11 in. = _____ in.

20. 13.5 lb = _____ oz

21. 72 qt = _____ gal

22. 4.5 gal = _____ pt

23. 326 oz = _____ lb _____ oz

24. 29 ft = _____ yd _____ ft

25. 0.75 pt = _____ fl oz

26. 12 cups = _____ pt

27. 18 t = _____ T

28. $2\frac{1}{3}$ T = _____ t

29. 8.0 rd = _____ yd

30. 6.5 cups = _____ fl oz

31. 4.0 gal = _____ cups

32. 5.0 T = _____ fl oz

33. 2.5 qt = _____ T

34. 12 t = _____ cups

B. Convert the units as shown. (Rounding instructions are provided where needed.)

1. 6.50 sq ft = _____ sq in.

2. 2.50 acre = _____ sq ft (thousand)

3. $17\frac{1}{3}$ cu yd = _____ cu ft

4. 387 sq ft = _____ sq yd

5. 972.0 cu in. = _____ cu ft

6. 330 rev/min = _____ rev/sec

7. $660 \dfrac{lb}{cu\ ft}$ = _____ $\dfrac{lb}{cu\ in.}$ (hundredth)

8. $30\overline{0}$ cu ft = _____ cu in. (thousand)

9. 14,520 sq ft = _____ acre

10. $2500 \dfrac{ft}{min}$ = _____ $\dfrac{in.}{sec}$

11. 20.0 sq mi = _____ acre

12. 65 mph = _____ ft/sec (whole number)

13. 2.1 sq yd = _____ sq in. (hundred)

14. 42.0 sq yd = _____ sq ft

15. 14,000 cu in. = _____ cu yd (hundredth)

16. $8\frac{1}{3}$ cu yd = _____ cu ft

17. 64.0 ft/sec = _____ mi/hr (tenth)

18. 3.80 rev/sec = _____ rpm

19. 12.5 in./sec = _____ ft/min

20. 360 sq in. = _____ sq ft

21. 2304 acres = _____ sq mi

22. 648 sq in. = _____ sq yd

23. 1.50 lb/sq in. = _____ lb/sq ft

24. 12.5 cu ft = _____ gal

25. 8.00 bu = _____ cu in. (hundred)

26. 24 fl oz = _____ cu in. (whole number)

27. 4650 gal = _____ cu ft (whole number)

28. 2.7 cu in. = _____ fl oz (tenth)

C. Solve:

1. **Sports and Leisure** The Manitou Incline, which leads to the top of Pike's Peak in Colorado, climbs from an elevation of 6574 feet to 8585 feet above sea level. The fastest reported time for running up the incline is 16 minutes and 42 seconds, by Mark Fretta. How many vertical feet per second did Mark climb during his record-breaking performance? (Round to 1 decimal digit.)

2. **Carpentry**   Find the number of board feet in each of the following quantities of lumber:

   (a) 24 pieces of 2 in. × 10 in. × 16 ft

   (b) 36 pieces of 1 in. × 8 in. × 10 ft

   (c) 50 pieces of 2 in. × 4 in. × 12 ft

   (d) 72 pieces of $\frac{7}{8}$ in. × 6 in. × 8 ft

3. **Construction**   The hole for a footing needs to be 5 ft deep. If it is currently 2 ft 8 in. deep, how much deeper does it need to be dug? Give the answer in inches.

4. **Construction**   A contractor is excavating for the basement of a new house. The finished first-floor line is at an elevation of 148.25 feet. The basement floor is to be 8.00 feet below the finished first floor, and an additional 9 inches must be dug for the floor and cinder fill. What is the final elevation of the excavation?

5. **General Interest**   The earliest known unit of length to be used in a major construction project is the "megalithic yard" used by the builders of Stonehenge in southwestern Britain about 2600 B.C. If 1 megalithic yard = 2.72 ± 0.05 ft, convert this length to inches. (Round to the nearest tenth of an inch.)

6. **General Interest**   In the Bible (Genesis, Chapter 7), Noah built an ark 300 cubits long, 50 cubits wide, and 30 cubits high. In I Samuel, Chapter 17, it is reported that the giant Goliath was "six cubits and a span" in height. If 1 cubit = 18 in. and 1 span = 9 in.:

   (a) What were the dimensions of the ark in feet?

   (b) How tall was Goliath in feet and inches?

7. **Flooring and Carpeting**   A homeowner needing carpeting has calculated that $140\frac{5}{9}$ sq yd of carpet must be ordered. However, carpet is sold by the square foot. Make the necessary conversion for her.

8. **Metalworking**   Cast aluminum has a density of 160.0 lb/cu ft. Convert this density to units of oz/cu in. and round to two decimal places.

9. **Machine Trades**   A $\frac{5}{16}$-in. twist drill with a periphery speed of 50.0 ft/min has a cutting speed of 611 rpm (revolutions per minute). Convert this speed to rps (revolutions per second) and round to one decimal place.

10. **Welding**   How many pieces $8\frac{3}{8}$ in. long can be cut from a piece of 20-ft stock? Allow for $\frac{3}{16}$ in. kerf (waste due to the width of a cut).

11. **Water/Wastewater Treatment**   A sewer line has a flow rate of 1.25 MGD (million gallons per day). Convert this to cfs (cubic feet per second). (Round to two decimal places.)

12. **Hydrology**   A reservoir has a capacity of 9000 cu ft. How long will it take to fill the reservoir at the rate of 250 gallons per minute?

13. **Sheet Metal Trades**   A machinist must cut 12 strips of metal that are each 3 ft $2\frac{7}{8}$ in. long. What is the total length needed in inches?

14. **Allied Health**   A hospitalized patient must drink $\frac{1}{2}$ pint of liquid every 2 hours. How many quarts of liquid will the patient drink in 3 days?

15. **Culinary Arts**   A recipe for gazpacho soup calls for 2 tablespoons of lemon juice and $\frac{1}{2}$ cup of olive oil. The given recipe serves 4 people, but a cook wants to make a larger batch that serves 40.

    (a) How many cups of lemon juice will the chef need for the larger batch?

    (b) How many pints of olive oil will the chef need for the larger batch?

16. **Carpentry**   A Metabol Steb 135 Plus top-handle jigsaw weighs 6.20 lb. A Porter-Cable PC600JS model weighs 5.40 lb. How many *ounces* lighter is the Porter-Cable model?

17. **Culinary Arts**   A restaurant manager has developed time guidelines for his kitchen staff. One such guideline requires a food prep worker to prepare 24 containers of granola in 7 minutes 40 seconds or less. What is the maximum time allowed per container? (Round down to the nearest second.)

18. **Marine Technology**   In the petroleum industry, a barrel is defined as 42 gal. A scientist wishes to treat a 430,000-barrel oil spill with a bacterial culture. If the culture must be applied at the rate of 1 oz of culture per 100 cu ft of oil, how many ounces of culture will the scientist need? (Round to the nearest thousand.)

19. **Painting**   A painter must thin some paint for use in a sprayer. If the recommended rate is $\frac{1}{2}$ pint of water per gallon of paint, how many total gallons will there be after thinning 8 gal of paint?

20. **Interior Design**   A designer is pricing a certain wool carpet for a new house. She finds it listed for $139.50 per square yard at Carpet City and $16.65 per square foot at Carpetorium. (a) Which is the better buy? (b) How much will she save on 166 sq yd?

21. **Landscaping**   A 2800-sq-ft house is built on a flat and bare $\frac{1}{2}$-acre parcel of land. If an additional 600 sq ft are needed for a driveway, how many square feet remain for landscaping?

22. **Transportation**   A car at rest begins traveling at a constant acceleration of 18 feet per second per second (18 ft/sec^2). After 6.0 sec, the car is traveling at 108 ft/sec. Convert this final speed to miles per hour. (Round to the nearest whole number.)

23. **Automotive Trades**   If the circumference of each tire on a vehicle is approximately 88 in., how many miles will the vehicle travel when the tires make 12,000 revolutions? (See Example 7 on page 299. Round to the nearest mile.)

24. **Automotive Trades**   A drum of oil contains 55 gal and costs $164.80. What is the cost per quart?

25. **Automotive Trades**   Suppose a hybrid automobile has an average fuel economy of 40 mi/gal, and a comparable nonhybrid averages 25 mi/gal. If the hybrid model costs $6000 more than the nonhybrid, and gasoline averages $2.50 per gallon, which model would be less expensive over a lifespan of 125,000 miles? (See Example 18 on page 307.)

26. **Architecture**   Calculate the height of the church tower shown on the next page. (Recall that the ′ symbol stands for "feet," so that 18′5″ means 18 feet 5 inches.)

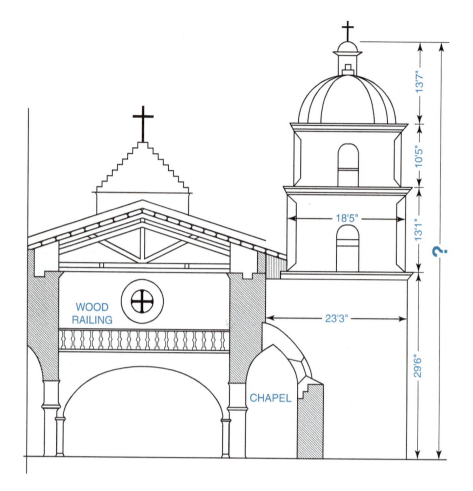

27. **Hydrology** Hydrological engineers often measure very large amounts of water in acre-feet. 1 acre-ft of water fills a rectangular volume of base 1 acre and height 1 ft. Convert 1 acre-ft to

    (a) _____ cu ft   (b) _____ gal   (c) _____ cu yd

28. **Hydrology** The average daily usage of water from Bradbury Dam at Lake Cachuma during a particular week in the summer was 134.2 acre-feet. To the nearest thousand, how many gallons is this? (See problem 27.)

29. **Automotive Trades** The 2017 Toyota Prius is EPA-rated at 54 mi/gal (miles per gallon) for city driving and 50 mi/gal for highway driving. The 2017 Toyota Yaris is rated at 30 mi/gal city and 36 mi/gal highway. How many gallons of gas, to the nearest tenth, would each vehicle consume for:

    (a) 300 mi of city driving plus 100 mi of highway driving.

    (b) 100 mi of city driving plus 300 mi of highway driving.

30. **Life Skills** A trades worker has a job in a large city. She can either rent an apartment in the city for $1400 per month or a comparable one in the sub-urbs for $1200 per month. If she lives in the city, she can walk or bike to work. If she lives in the suburbs, she must drive 60 miles per day round trip an average of 22 days per month. Her car averages 24 mi/gal and gas costs $2.50 per gallon. If she decides to live in the suburbs, would the total monthly cost of gas be more or less than the extra rent?

31. **General Interest** Researchers have "cracked the code" of the Aztec sys-tem of measurement. Their basic unit of length was the rod, which is

approximately 8 ft. (Notice that this is not the same as the modern definition of a rod.) Other units are defined in terms of the rod as follows:

5 hands = 3 rods   3 arms = 1 rod

2 arrows = 1 rod   5 bones = 1 rod

5 hearts = 2 rods

(*Source:* Barbara J. Williams and Maria del Carmen Jorge y Jorge, "Aztec Arithmetic Revisited: Land-Area Algorithms and Acolhua Congruence Arithmetic," *Science* 320, pp. 72–77, 4 April 2008.)

Use these definitions to calculate

(a) the number of hands in 15 rods.

(b) the number of rods in 35 hearts.

(c) the number of bones in 8 arrows.

(d) the number of hands in 12 arms.

(e) the approximate number of feet in 20 hearts.

32. **Agriculture**   Oats have a density of 25.6 lb/cu ft. One bushel of oats weighs 32 lb. How many cubic yards of oats are there in 4200 bushels? (Round to the nearest whole number.)

33. **Life Skills**   An employee who lives 35 miles from her place of work can either drive or take the subway. Use the following information to compare the relative costs of these two options:

1. A monthly subway pass costs $65.

2. Her car gets 24 miles per gallon of gas.

3. The estimated cost of gas is $2.60 per gallon.

4. She works an average of 22 days per month.

How much would the employee save per month if she takes the subway? (Round to the nearest dollar.)

34. **Allied Health**   The average American consumes approximately 44.7 gallons of soft drinks per year. Many states are in the process of levying a special tax on these beverages to discourage their consumption and help pay for health-related obesity costs. In a state with a population of 20 million people, how much tax revenue would be generated by a one-tenth of a cent ($0.001) per ounce tax on soft drinks?

35. **Manufacturing**   A conveyor belt 18 in. wide is moving at a linear speed of 66 ft/min. The belt is moving boxes that are 6 in. wide and 18 in. long. Using the full width of the belt, how many boxes can be unloaded off the belt in an hour? Assume that the boxes are placed end to end with no space in between them.

36. **Life Skills**   A potential car buyer is comparing fuel costs of two different models. Model A has a combined rating of 22 mi/gal and uses regular gas. Model B has a combined fuel rating of 24 mi/gal but requires premium grade. If we assume that regular gas averages $2.60 per gallon and premium gas averages $2.85 per gallon, which model will have the higher fuel cost? (*Hint:* Select an arbitrary number of miles, such as 1000, and determine each car's fuel cost over that distance. The answer will be the same no matter what total number of miles you use for the calculation.)

37. **Automotive Trades**   The 2017 BMW 3 Series comes in a gas-powered model and a diesel engine model. The gas-powered model, which requires

premium fuel, has an average fuel economy of 29 mi/gal, while the diesel model averages 37 mi/gal. The base price of the diesel model is $7100 more than that of the gas-powered model. A potential buyer notes that the current cost of diesel fuel is $2.45 per gallon, and the current cost of premium fuel is $2.75 per gallon. If the buyer opts for the diesel model, would the savings in total fuel cost over 100,000 miles be enough to make up for the extra initial cost of the diesel model?

38. **Life Skills**   When gas prices are high, people sometimes drive some distance to fill their tank with less expensive gas. But is it really worth it? Suppose your local gas station is selling your grade for $2.79 per gallon, but a station 5 miles away sells it for $2.69. Suppose that your car averages 23 miles per gallon, and you plan to purchase 18 gallons of gas. Would the cost of gas for the 10-mile round trip be less or more than the money you would save on the 18-gallon purchase at the less-expensive station? Assume that the extra gas is valued at $2.69 per gallon.

39. **Construction**   A certain grade of sheet glass weighs 45.5 ounces per square foot. How many *pounds* would 10 sheets weigh if they were each 5 square feet in size? (Round to the nearest pound.)

Check your answers to the odd-numbered problems in the Appendix, then continue in Section 5-3 with the study of metric units.

5-3   Metric Units

For all technical and many trades workers, the ability to use metric units is important. Familiarity with metric units is also important for people who travel internationally or who interact with companies or other people who use the metric system. In many ways, working with meters, kilograms, and liters is easier and more logical than using our traditional units of feet, pounds, or quarts. In this section we define and explain the most useful metric units, explain how to convert from one metric unit to another, and provide practice problems designed to help you use the metric system and to "think metric."

The most important common units in the metric system are those for length or distance, speed, weight, volume, area, and temperature. Time units—year, day, hour, minute, second—are the same in the metric as in the U.S. customary system.

Length   The basic unit of length in the metric system is the **meter**, pronounced *meet-ur* and abbreviated *m*. (The word is spelled *metre* in British English but is pronounced exactly the same.) One meter is roughly equal to one yard.

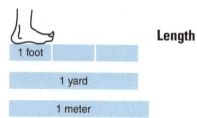

The meter is the appropriate unit to use in measuring your height, the width of a room, length of lumber, or the height of a building.

Your Turn   Estimate the following lengths in meters:

(a) The length of the room in which you are now sitting = _____ m.

(b) Your height = _____ m.

(c) The length of a ping-pong table = _____ m.

(d) The length of a football field = _____ m.

Guess as closely as you can before looking at the solution.

Answers
  (a) A small room might be 3 or 4 m long, and a large one might be 6 or 8 m long.

  (b) Your height is probably between $1\frac{1}{2}$ and 2 m. Note that 2 m is roughly 6 ft 7 in.

  (c) A ping-pong table is a little less than 3 m in length.

  (d) A football field is about 100 m long.

All other length units used in the metric system are defined in terms of the meter, and these units differ from one another by powers of ten. For example, the **centimeter**, pronounced *cent-a-meter* and abbreviated *cm*, is defined as exactly one-hundredth of a meter. The **kilometer**, pronounced *kill-AH-meter* and abbreviated *km*, is defined as exactly 1000 meters. The **millimeter**, abbreviated *mm*, is defined as exactly $\frac{1}{1000}$ meter.

Because metric units increase or decrease in powers of ten, they may be named using prefixes attached to a basic unit. The following table summarizes the most commonly used metric prefixes.

Metric Prefix	Multiplier	Common Example
tera	1,000,000,000,000 (10^{12})	terabyte: one trillion bytes
giga	1,000,000,000 (10^{9})	gigahertz: one billion hertz
mega	1,000,000 (10^{6})	megawatt: one million watts
kilo	1,000 (10^{3})	kilopascal: one thousand pascal
centi	0.01 (10^{-2})	centimeter: one hundredth of a meter
milli	0.001 (10^{-3})	milliliter: one thousandth of a liter
micro	0.000001 (10^{-6})	microgram: one millionth of a gram
nano	0.000000001 (10^{-9})	nanosecond: one billionth of a second
pico	0.000000000001 (10^{-12})	picofarad: one trillionth of a farad

Note   The prefixes "hecto" (multiplier of 100), "deca" (multiplier of 10), and "deci" (multiplier of one-tenth) were not included in the table because they are rarely used in the trades or even in technical work. ●

The following length conversion factors are especially useful for workers in the trades.

> **Metric Length Units**
>
> 1 centimeter (cm) = 10 millimeters (mm)
>
> 1 meter (m) = 100 cm = 1000 mm
>
> 1 kilometer (km) = 1000 m

The millimeter (mm) unit of length is very often used on technical drawings, and the centimeter (cm) is handy for shop measurements.

1 cm is roughly the width of a large paper clip.

1 mm is roughly the thickness of the wire in a paper clip.

1 meter is roughly 10% more than a yard.

Metric to Metric Length Conversion

One very great advantage of the metric system is that we can convert easily from one metric unit to another. There are no hard-to-remember conversion factors such as 36 in. in a yard, 5280 ft in a mile, 220 yd in a furlong, or whatever.

For example, with U.S. customary units, if we convert a length of 137 in. to feet or yards, we find

$$137 \text{ in.} = 11\tfrac{5}{12} \text{ ft} = 3\tfrac{29}{36} \text{ yd}$$

Here we must divide by 12 and then by 3 to convert the units. But in the metric system we simply shift the decimal point. Using unity fractions is also easy because we are always multiplying or dividing by powers of 10, such as 10 or 100 or 1000.

EXAMPLE 1   Convert 348 cm to meters like this:

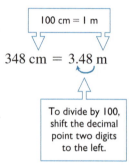

100 cm = 1 m

$$348 \text{ cm} = 3.48 \text{ m}$$

To divide by 100, shift the decimal point two digits to the left.

Of course, we may also use unity fractions:

$$348 \text{ cm} = 348 \text{ cm} \times \left(\frac{1 \text{ m}}{100 \text{ cm}} \right)$$

Choose the unity fraction that cancels cm and replaces it with m.

$$= \frac{348}{100} \text{ m}$$

$$= 3.48 \text{ m}$$

EXAMPLE 2   Convert 7 cm to millimeters like this:

1 cm = 10 mm

$$7 \text{ cm} = 7.0 \text{ mm} = 70 \text{ mm}$$

To multiply by 10, shift the decimal point one digit to the right.

Or, use a unity fraction to cancel out cm and replace it with mm:

$$7 \text{ cm} = 7 \text{ cm} \times \left(\frac{10 \text{ mm}}{1 \text{ cm}} \right)$$

$$= 70 \text{ mm}$$

Your Turn   Complete the following unit conversions.

(a) 3.1 m = _____ cm   (b) 2.1 km = _____ m

(c) 307 m = _____ km   (d) 20.5 mm = _____ cm

Solutions   (a) $3.1 \text{ m} = 3.1 \text{ m} \times \left(\dfrac{100 \text{ cm}}{1 \text{ m}} \right)$

$= 3.1 \times 100 \text{ cm}$

$= 310 \text{ cm}$ The decimal point shifts two places right.

(b) $2.1 \text{ km} = 2.1 \text{ km} \times \left(\dfrac{1000 \text{ m}}{1 \text{ km}} \right)$

$= 2100 \text{ m}$ The decimal point shifts three places right.

(c) $307 \text{ m} = 307 \text{ m} \times \left(\dfrac{1 \text{ km}}{1000 \text{ m}} \right)$

$= 0.307 \text{ km}$ The decimal point shifts three places left.

(d) $20.5 \text{ mm} = 20.5 \text{ mm} \times \left(\dfrac{1 \text{ cm}}{10 \text{ mm}} \right)$

$= 2.05 \text{ cm}$ The decimal point shifts one place left.

Weight   The weight of an object is the gravitational pull of the earth exerted on that object. The mass of an object is related to how it behaves when pushed or pulled. For scientists, the difference between mass and weight can be important. For all practical purposes, they are the same.

The basic unit of mass or metric weight in the International Metric System (SI) is the **kilogram,** pronounced *kill-o-gram* and abbreviated *kg.* The kilogram is defined as the mass of a standard platinum–iridium cylinder kept at the International Bureau of Weights and Measures near Paris. By law in the United States, the pound is defined as exactly equal to the weight of 0.45359237 kg, so that 1 kg weighs about 2.2046 lb.

One kilogram weighs about 10% more than 2 lb.

The kilogram was originally designed so that it was almost exactly equal to the weight of one liter of water.

Three smaller metric units of mass are also defined. The **gram,** abbreviated *g,* is equal to 0.001 kg; the **milligram,** abbreviated *mg,* is equal to 0.001 g; and the **microgram,** abbreviated *μg,* is equal to 0.001 mg. One larger unit, the metric ton, is equal to 1000 kg, and it is abbreviated *t.*

The following conversion factors will make it easy for you to convert weight units within the metric system.

Metric Weight/Mass Units

1 milligram (mg) $=$ 1000 micrograms (μg)

1 gram (g) $=$ 1000 milligrams (mg)

1 kilogram (kg) $=$ 1000 g

1 metric ton (t) $=$ 1000 kg

The gram is a very small unit of weight, equal to roughly $\frac{1}{500}$ of a pound, or the weight of a paper clip. The milligram is often used by druggists and nurses to measure very small amounts of medication or other chemicals, and it is also used in scientific measurements.

Metric to Metric Weight Conversion   To convert weights within the metric system, use the conversion factors in the table to form unity fractions.

EXAMPLE 3   To convert 4500 mg to grams, use the conversion factor 1 g = 1000 mg.

$$4500 \text{ mg} = 4500 \text{ mg} \times \frac{1 \text{ g}}{1000 \text{ mg}} = 4.5 \text{ g}$$

EXAMPLE 4   To convert 0.0062 kg to milligrams, we need to use two conversion factors: 1 kg = 1000 g and 1 g = 1000 mg

$$0.0062 \text{ kg} = 0.0062 \text{ kg} \times \frac{1000 \text{ g}}{1 \text{ kg}} \times \frac{1000 \text{ mg}}{1 \text{ g}} = 6200 \text{ mg}$$

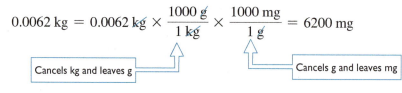

Cancels kg and leaves g   Cancels g and leaves mg

Your Turn   Convert the following measurements as shown.

(a) 21.4 g = _____ mg   (b) 156,000 mg = _____ kg

(c) 0.56 kg = _____ g   (d) 467 μg = _____ mg

Answers   (a) 21,400 mg   (b) 0.156 kg   (c) 560 g   (d) 0.467 mg

Area   Because an area is essentially the product of two lengths, it is specified by a compound unit. When the area of a surface is calculated, the units will be given in "square length units." If the lengths are measured in feet, the area will be given in square feet; if the lengths are measured in meters, the area will be given in square meters.

For example, the area of a carpet 2 m wide and 3 m long would be

2 m × 3 m = 6 square meters

= 6 sq m

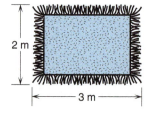

Areas roughly the size of carpets and room flooring are usually measured in square meters. Larger areas are usually measured in *hectares,* a metric surveyor's unit. One hectare (ha) is the area of a square 100 m on each side. The hectare is roughly $2\frac{1}{2}$ acres, or about double the area of a football field.

> **Metric Area Units**
>
> 1 hectare (ha) = 100 m × 100 m
>
> = 10,000 sq m (or m²)

EXAMPLE 5   Convert 456,000 sq m to hectares.

Using a unity fraction,

$$456,000 \text{ sq m} = 456,000 \text{ sq m} \times \frac{1 \text{ ha}}{10,000 \text{ sq m}} = 45.6 \text{ ha}$$

Your Turn Convert 0.85 hectare to square meters.

Solution $0.85 \text{ ha} = 0.85 \, \cancel{\text{ha}} \times \dfrac{10{,}000 \text{ sq m}}{1 \, \cancel{\text{ha}}} = 8500 \text{ sq m}$

Area conversions within the metric system, such as square meters to square kilometers, are seldom used in the trades and we will not cover them here. We will learn much more about area in Chapter 8.

Volume and Liquid Capacity

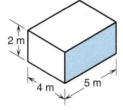

Recall from Section 5-2 that when the volume of an object is calculated or measured, the units will normally be given in "cubic length units." If the lengths are measured in feet, the volume will be given in cubic feet; if the lengths are measured in meters, the volume will be given in cubic meters.

For example, the volume of a box 2 m high by 4 m wide by 5 m long would be

$2 \text{ m} \times 4 \text{ m} \times 5 \text{ m} = 40 \text{ cu m or } 40 \text{ m}^3$

The cubic meter is an appropriate unit for measuring the amount of three-dimensional space in a room, in a pile of sand, in a concrete slab, and so on. For liquid volume or capacity, the most useful metric unit is the **liter,** pronounced *leet-ur*. You may find this volume unit spelled either *liter* in the United States or *litre* elsewhere in the world.

One liter (L) is defined as 1000 cubic centimeters (cu cm), the volume of a cube 10 cm on each edge (see the figure in the margin). To help you visualize a liter, think of it as being roughly equal to a quart.

Because 1000 cubic centimeters are needed to make up 1 liter, a cubic centimeter is $\dfrac{1}{1000}$ of a liter, or a **milliliter** (see the figure in the margin). Liquid volumes smaller than 1 liter are usually measured in units of cubic centimeters or milliliters. To help you visualize a milliliter, think of it as the amount of liquid that would fill a standard eye-dropper. You might even have a measuring cup at home that is calibrated in milliliters, and this will also help you visualize this amount of liquid.

We will cover the subject of volume in depth in Chapter 9, but for now, we will learn some basic volume unit conversions.

Metric to Metric Volume Conversions The following table summarizes the most useful conversion factors for metric volume and liquid capacity.

Metric Units for Volume and Liquid Capacity

1 cubic centimeter (cu cm or cc or cm^3) = 1 milliliter (mL)

1 milliliter (mL) = 1000 microliters (μL)

1 liter (L) = 1000 mL = 1000 cu cm

EXAMPLE 6 To convert 0.24 liter to cubic centimeters, use the conversion factor 1 liter = 1000 cu cm.

Using a unity fraction,

$0.24 \text{ liter} = 0.24 \, \cancel{\text{liter}} \times \dfrac{1000 \text{ cu cm}}{1 \, \cancel{\text{liter}}} = 240 \text{ cu cm}$

Your Turn   (a) Convert 5780 mL to liters.

(b) Convert 0.45 mL to microliters.

Solutions   (a) We use the conversion factor 1 L = 1000 mL to set up a unity fraction.

$$5780 \text{ mL} = 5780 \text{ mL} \times \frac{1 \text{ liter}}{1000 \text{ mL}} = 5.78 \text{ L}$$

(b) We use the conversion factor 1 mL = 1000 μL.

$$0.45 \text{ mL} = 0.45 \text{ mL} \times \frac{1000 \text{ μL}}{1 \text{ mL}} = 450 \text{ μL}$$

Speed   In the metric system, ordinary highway speeds are measured in kilometers per hour, abbreviated km/h. To "think metric" while you drive, remember that

> 100 km/h is approximately equal to 62 mph.

To convert speed units within the metric system, use the appropriate combination of time and distance unity fractions.

EXAMPLE 7   To convert 1360 cm/sec to m/min,

First, determine the necessary conversion factors:

For time, 1 min = 60 sec. For distance, 1 m = 100 cm.

Then, multiply by the appropriate unity fractions:

$$1360 \text{ cm/sec} = \frac{1360 \text{ cm}}{1 \text{ sec}} \times \frac{60 \text{ sec}}{1 \text{ min}} \times \frac{1 \text{ m}}{100 \text{ cm}} = 816 \text{ m/min}$$ ●

Your Turn   (a) Convert 27.5 km/hr to cm/sec. (Round to the nearest whole number.)

(b) Convert 30.0 m/sec to km/hr.

Solutions   (a) For time, 1 hr = 3600 sec. For distance, 1 km = 1000 m and 1 m = 100 cm. Setting up unity fractions and multiplying, we have:

$$\frac{27.5 \text{ km}}{1 \text{ hr}} \times \frac{1 \text{ hr}}{3600 \text{ sec}} \times \frac{1000 \text{ m}}{1 \text{ km}} \times \frac{100 \text{ cm}}{1 \text{ m}} = 763.888\ldots \text{ cm/sec} \approx 764 \text{ cm/sec}$$

27.5 ÷ **3600** × **1000** × **100** = → ⬛ 763.8888889

$$\text{(b)} \quad \frac{30.0 \text{ m}}{1 \text{ sec}} \times \frac{3600 \text{ sec}}{1 \text{ hr}} \times \frac{1 \text{ km}}{1000 \text{ m}} = 108 \text{ km/hr}$$

Temperature   The metric temperature scale gives the temperature in degrees Celsius (°C). On the Celsius scale, water boils at 100°C and freezes at 0°C. Because this scale is much simpler than the Fahrenheit scale, the Celsius scale is used almost everywhere in the world except the United States. There are no conversions to perform within the Celsius scale, but in the next section of this chapter, and again in Chapter 7, we will teach you ways to convert between the Celsius and Fahrenheit scales.

Applications of Metric Units   Because most medications are measured in metric units, dosage calculations often involve metric conversions.

EXAMPLE 8   **Allied Health**   If a patient takes 60 mg of a medication three times a day, how many grams will the patient take in four days?

To solve this problem, **first** determine how many milligrams the patient will take in four days:

In four days the patient will take $\dfrac{60 \text{ mg}}{1 \text{ dose}} \times \dfrac{3 \text{ doses}}{1 \text{ day}} \times 4 \text{ days} = 720 \text{ mg}$

Then, convert milligrams to grams:

$$720 \text{ mg} = 720 \text{ mg} \times \dfrac{1 \text{ g}}{1000 \text{ mg}} = 0.72 \text{ g}$$

Your Turn   **Allied Health**   A certain medication is available only in 625 μg tablets. If a patient needs to take 4.5 mg per day, how many tablets should she take?

Solution   **First,** convert 625 μg to mg:

$$625 \text{ μg} = 625 \text{ μg} \times \dfrac{1 \text{ mg}}{1000 \text{ μg}} = 0.625 \text{ mg per tablet}$$

Then, calculate the number of tablets needed:

$$\dfrac{4.5 \text{ mg}}{1 \text{ day}} \times \dfrac{1 \text{ tablet}}{0.625 \text{ mg}} = 7.2 \text{ tablets per day}$$

This would normally be rounded down to 7 tablets per day.

EXAMPLE 9   **Machine Trades**   The length of a steel component is given as 8.6 cm. What is the greatest possible error of this measurement in millimeters?

To solve this, recall from Section 5-1 that the greatest possible error of a measurement is half the precision. In this case, the precision is 0.1 cm, so the greatest possible error is 0.5(0.1 cm) = 0.05 cm. Converting this to millimeters, we have

$$0.05 \text{ cm} = 0.05 \text{ cm} \times \dfrac{10 \text{ mm}}{1 \text{ cm}} = 0.5 \text{ mm}$$

Here is an example involving speed.

EXAMPLE 10   **Aviation**   The cruising speed of a Boeing 787 is Mach 0.85, meaning 0.85 times the speed of sound. If the speed of sound at cruising altitude is 295 meters per second, what is the cruising speed of the 787 in kilometers per hour?

To solve this problem, **first** convert Mach 0.85 to meters per second:

$$0.85 \times 295 \text{ m/s} = 250.75 \text{ m/s}$$

Then, convert the speed to kilometers per hour:

$$\dfrac{250.75 \text{ m}}{1 \text{ sec}} \times \dfrac{1 \text{ km}}{1000 \text{ m}} \times \dfrac{3600 \text{ sec}}{1 \text{ hr}} = 902.7 \text{ km/hr} \approx 903 \text{ km/hr, rounded}$$

We can enter all of this on a calculator as follows:

.85 ✕ 295 ÷ 1000 ✕ 3600 = →   *902.7*

Your Turn (a) **Allied Health** The volume of a solution of medication is given as 1.5 liters (L). What is the greatest possible error of this measurement in milliliters?

(b) **Metalworking** The density of copper is 8940 kg/cu m. Express this in grams per cubic centimeter. (*Hint:* 1 cu m = 1,000,000 cu cm.)

Solutions (a) Precision is 0.1 L. Greatest possible error is half of this, or 0.05 L.

$$0.05 \text{ L} = 0.05 \, \cancel{L} \times \frac{1000 \text{ mL}}{1 \, \cancel{L}} = 50 \text{ mL}$$

(b) $8940 \text{ kg/cu m} = \dfrac{8940 \, \cancel{kg}}{1 \, \cancel{cu\,m}} \times \dfrac{1000 \text{ g}}{1 \, \cancel{kg}} \times \dfrac{1 \, \cancel{cu\,m}}{1,000,000 \text{ cu cm}} = 8.94 \text{ g/cu cm}$

Sometimes you must work with unique conversion factors to solve a problem. The next three examples illustrate this type of situation.

EXAMPLE 11 **Allied Health** A very active adult who wishes to maintain her weight of 128 pounds is on a diet of 2300 calories per day. A nutritionist advises her to limit her intake of saturated fat to no more than 7% of her total calories. Fat contains 9 calories per gram. To determine the maximum number of grams of saturated fat she should consume per day, follow these steps:

Step 1 Use a percent proportion to calculate the number of calories per day from saturated fat.

$$\frac{P}{2300} = \frac{7}{100}$$

$$P = \frac{2300 \times 7}{100} = 161 \text{ calories per day from saturated fat}$$

Step 2 Use unity fractions to calculate the number of grams per day of saturated fat.

$$\frac{161 \, \cancel{cal}}{1 \text{ day}} \times \frac{1 \text{ gram}}{9 \, \cancel{cal}} = 17.88 \ldots \text{ grams per day}$$

Step 3 Because this represents her maximum intake of saturated fat, we would round down to 17 grams per day. ●

EXAMPLE 12 **Water/Wastewater Treatment** The most common unit for measuring contaminants in water is milligrams per liter (mg/L). In the field of water treatment, this is defined as being equivalent to parts per million (ppm)—that is, 1 mg/L = 1 ppm. When very small amounts are measured, they are reported in units of micrograms per liter (μg/L) or parts per billion (ppb): 1 μg/L = 1 ppb.

Suppose a particular contaminant was measured at a level of 0.21 ppm, and we needed to report this in micrograms per liter. According to the definition, 0.21 ppm is equivalent to 0.21 mg/L. To convert this to μg/L, we need only convert milligrams to micrograms:

$$\frac{0.21 \, \cancel{mg}}{1 \text{ L}} \times \frac{1000 \, \mu g}{1 \, \cancel{mg}} = 210 \, \mu g/L$$ ●

EXAMPLE 13 **Allied Health** In working with intravenous (IV) infusion, nurses must often measure flow rates in drops or microdrops per minute while the solution itself is measured in milliliters. Suppose that a 750-mL IV must be given to a patient over a period of 6 hours, and the nurse must set the flow rate in drops per minute. If the "drop factor" for this solution is 15 drops per milliliter, the nurse must perform the following conversion:

$$\frac{750 \text{ mL}}{6 \text{ hr}} \times \frac{15 \text{ drops}}{1 \text{ mL}} \times \frac{1 \text{ hr}}{60 \text{ min}} = 31.25 \text{ drops/min}$$

For practical purposes, this would be rounded to 31 drops per minute.   ●

Here are several similar problems for you to try.

Your Turn   (a) **Allied Health**   A nutritionist recommends a diet of 3000 calories per day for a patient. The nutritionist also specifies that 20% of the calories should come from protein. If protein contains 4 calories per gram, how many grams of protein should the patient consume per day?

(b) **Water/Wastewater Treatment**   Suppose a contaminant was measured at a level of 46 parts per billion. How would this be expressed in milligrams per liter?

(c) **Allied Health**   A nurse sets a 400-mL solution to infuse intravenously at a rate of 20 drops per minute. If the drop factor is 10 drops per milliliter, how many hours will it take to administer the entire solution?

Solutions   (a) 150 grams/day   (b) 0.046 mg/L

(c) We are starting with milliliters and we want to end up with hours. Here's how we do it with unity fractions:

$$400 \text{ mL} \times \frac{10 \text{ drops}}{1 \text{ mL}} \times \frac{1 \text{ min}}{20 \text{ drops}} \times \frac{1 \text{ hr}}{60 \text{ min}} = 3\frac{1}{3} \text{ hr, or 3 hr 20 min}$$

The first unity fraction cancelled mL and left us with drops. The second fraction cancelled drops and left us with minutes. And the third fraction cancelled minutes and left us with hours.

The following table summarizes the conversion factors we have covered within the metric system. You can refer to this table as you work on Exercises 5-3, which follow the table.

Length	Weight
1 cm = 10 mm	1 mg = 1000 μg
1 m = 100 cm = 1000 mm	1 g = 1000 mg
1 km = 1000 m	1 kg = 1000 g
	1 metric ton (t) = 1000 kg

Area	Volume/Liquid Capacity
1 ha = 10,000 sq m (m²)	1 mL = 1000 μL
	1 cu cm (cm³) = 1 mL
	1 liter (L) = 1000 cu cm
	1 L = 1000 mL

Exercises 5-3   Metric Units

A. **Think Metric**   For each problem, circle the measurement closest to the first one given. No calculations are needed.

Remember: (1) A meter is a little more (about 10%) than a yard.
(2) A kilogram is a little more (about 10%) than 2 pounds.
(3) A liter is a little more (about 6%) than a quart.

1. 30 cm   (a) 30 in.   (b) 75 in.   (c) 1 ft
2. 5 ft   (a) 1500 cm   (b) 1.5 m   (c) 2 m
3. 1 yd   (a) 90 cm   (b) 110 cm   (c) 100 cm
4. 2 m   (a) 6 ft 6 in.   (b) 6 ft   (c) 2 yd
5. 3 km   (a) 3 mi   (b) 2 mi   (c) 1 mi
6. 200 km   (a) 20 mi   (b) 100 mi   (c) 120 mi
7. 50 km/h   (a) 30 mph   (b) 50 mph   (c) 60 mph
8. 55 mph   (a) 30 km/h   (b) 60 km/h   (c) 90 km/h
9. 100 m   (a) 100 yd   (b) 100 ft   (c) 1000 in.
10. 400 lb   (a) 800 kg   (b) 180 kg   (c) 250 kg
11. 6 oz   (a) 1.7 g   (b) 17 g   (c) 170 g
12. 5 kg   (a) 2 lb   (b) 5 lb   (c) 10 lb
13. 50 liters   (a) 12 gal   (b) 120 gal   (c) 50 gal
14. 6 qt   (a) 2 liters   (b) 3 liters   (c) 6 liters
15. 212°F   (a) 100°C   (b) 400°C   (c) 50°C
16. 0°C   (a) 100°F   (b) 32°F   (c) −30°F

B. **Think Metric**   Choose the closest estimate.

1. Diameter of a penny   (a) 3 cm   (b) 1.5 cm   (c) 5 cm
2. Length of a man's foot   (a) 3 m   (b) 3 cm   (c) 30 cm
3. Tank of gasoline   (a) 50 liters (b) 5 liters   (c) 500 liters
4. Volume of a wastebasket   (a) 10 liters (b) 100 liters (c) 10 cu cm
5. One-half gallon of milk   (a) 1 liter   (b) 2 liters   (c) $\frac{1}{2}$ liter
6. Hot day in Phoenix, Arizona (a) 100°C   (b) 30°C   (c) 45°C
7. Cold day in Minnesota   (a) −80°C   (b) −10°C   (c) 20°C
8. 200-lb barbell   (a) 400 kg   (b) 100 kg   (c) 40 kg
9. Length of paper clip   (a) 10 cm   (b) 35 mm   (c) 3 mm
10. Your height   (a) 17 m   (b) 17 cm   (c) 1.7 m
11. Your weight   (a) 80 kg   (b) 800 kg   (c) 8 kg
12. Boiling water   (a) 100°C   (b) 212°C   (c) 32°C

C. Perform the following metric–metric conversions.

1. 5.6 cm   = _____ mm   2. 0.08 g   = _____ mg
3. 0.25 ha   = _____ sq m   4. 64,600 cu cm   = _____ liters
5. 0.045 kg   = _____ g   6. 12.5 km/h   = _____ cm/sec
7. 1.25 liters = _____ cu cm   8. 720,000 g   = _____ kg
9. 4.2 m   = _____ cm   10. 1026 mm   = _____ cm
11. 0.16 m/sec   = _____ cm/sec   12. 598 cm   = _____ m
13. 9.62 km   = _____ m   14. 785 mm   = _____ m

15. 56,500 m   = _____ km   16. 78,000 sq m = _____ ha

17. 9500 mg   = _____ g   18. 870̄ m/min = _____ km/h

19. 580 ml   = _____ liters   20. 42.0 cm/sec = _____ m/min

21. 1400 μL   = _____ mL   22. 42.5 mL   = _____ μL

23. 650 cu cm = _____ liters   24. 27 km   = _____ m

25. 246 km/h = _____ m/min   26. 0.2 L/kg   = _____ mL/g

27. 0.75 m   = _____ cm   28. 0.16 m   = _____ mm

29. 550 μg   = _____ mg   30. 25 mg/mL   = _____ g/L

31. 2.75 metric tons = _____ kg   32. 47,500 kg = _____ t

33. 1.65 g/L = _____ μg/mL

D. Practical Applications

1. **Electronics**   Convert the following quantities as shown:

    (a) 18,000 volts = _____ kilovolts

    (b) 435 millivolts = _____ volts

2. **Sports and Leisure**   By changing the rear derailleur of a racing bike, a mechanic is able to reduce the overall weight of the bike by 60 g. If the bike weighs 7.50 kg before changing the derailleur, what is its weight in kilograms after the derailleur is changed?

3. **Water/Wastewater Treatment**   A contaminant in a reservoir was measured at a level of 0.15 mg/L. How would this be expressed in parts per billion?

4. **Allied Health**   If a patient takes 25 mg of a medication twice a day, how many grams will he take in 14 days?

5. **Allied Health**   A certain medication is available only in 500 μg tablets. If the patient needs to take 3.0 mg per day, how many tablets must she take?

6. **Construction**   The Taipei 101 in Taipei, Taiwan, is 509 m tall. Its elevators are the fastest in the world, rising at 60.48 km/hr. How many seconds would it take for an elevator to rise the entire height of the building, nonstop?

7. **Allied Health**   A moderately active man weighing 175 pounds should consume no more than 840 calories per day from fat sources. If fat contains 9 calories per gram, what is the maximum number of grams of fat he should consume?

8. **Allied Health**   A pharmacist gave the weight of a certain medication as 420 mg. What is the greatest possible error of this weight in grams?

9. **Allied Health**   A wound was measured to be 0.8 cm in length. What is the greatest possible error of this measurement in millimeters?

10. **Metalworking**   The density of zinc is 7.13 g/cu cm. What is this density expressed in kilograms per cubic meter? (*Hint*: 1 cu m = 1,000,000 cu cm. Round to the nearest whole number.)

11. **Allied Health**   An intravenous solution has a drop factor of 60 microdrops per milliliter. If a patient must receive 500 mL of an IV solution in 3 hours, what should be the flow rate in microdrops per minute? (Round to the nearest whole number.)

12. **Aviation**   In October of 1967, the North American X-15 reached a world-record speed of Mach 6.70. What would this speed be in kilometers per hour? (See Example 10. Round to the nearest 10 km/hr.)

13. **Allied Health**   A nurse sets a 750-mL solution to infuse intravenously at a rate of 15 drops per minute. If the drop factor is 20 drops per milliliter, how many hours and minutes will it take to administer the solution?

14. **General Interest**   Environmental scientists have estimated that the global warming potential of methane is 25 times that of carbon dioxide. For example, 1 kilogram of methane would be equivalent to 25 kilograms of carbon dioxide. In 2015, a methane gas leak in Southern California was serious enough to cause a long-term evacuation of the Porter Ranch subdivision. In a $2\frac{1}{2}$-month period, 84 million kilograms of methane had leaked from a damaged well. What would be the equivalent amount of carbon dioxide in metric tons?

When you have completed these exercises, check your answers to the odd-numbered problems in the Appendix, then turn to Section 5-4 to learn how to convert between U.S. customary and metric measurements.

5-4 Metric–U.S. Customary Conversions

In Section 5-2, we reviewed the most useful units of measurement in the U.S. customary system and how to convert units within that system. In Section 5-3, we did the same with the metric system. But in this globally connected world, we often need to convert units of measurement from one system to the other. That is the subject of this section.

Metric–U.S. Customary Length Conversion

Because metric units are the only international units, all U.S. customary units are defined in terms of the metric system. Use the following table for metric—U.S. customary length conversions.

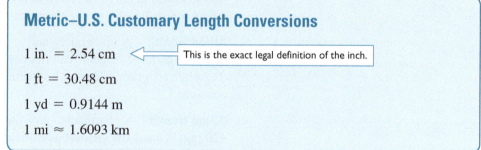

Metric–U.S. Customary Length Conversions

1 in. = 2.54 cm   ← This is the exact legal definition of the inch.

1 ft = 30.48 cm

1 yd = 0.9144 m

1 mi ≈ 1.6093 km

To shift from U.S. customary to metric units or from metric to U.S. customary units, use either unity fractions or the conversion factors given in the table.

EXAMPLE 1   Convert 4.3 yd to meters.

From the conversion table, 1 yd = 0.9144 m.

Using unity fractions,

$$4.3 \text{ yd} \times \frac{0.9144 \text{ m}}{1 \text{ yd}} = 3.93192 \text{ m} \approx 3.9 \text{ m, rounded}$$

Sometimes more than one unity fraction is needed. This is demonstrated in the next example.

EXAMPLE 2 Convert 2.50 in. to millimeters.

The conversion table does not list an inch–millimeter conversion factor.

Therefore we must use two factors: 1 in. = 2.54 cm and 1 cm = 10 mm

$$2.50 \text{ in.} = 2.50 \text{ in.} \times \frac{2.54 \text{ cm}}{1 \text{ in.}} \times \frac{10 \text{ mm}}{1 \text{ cm}} = 63.5 \text{ mm}$$

This unity fraction cancels in. and leaves cm.

This unity fraction cancels cm and leaves mm.

2.5 ⊠ **2.54** ⊠ **10** ⊟ → **63.5**

Your Turn Convert the following measurements as shown. Round answers to the nearest tenth of a unit.

(a) 16.0 in. = _____ cm   (b) 6.0 km = _____ mi

(c) 1.2 m = _____ ft   (d) 4.2 m = _____ yd

(e) 140 cm = _____ ft   (f) 18 ft 6 in. = _____ m

Answers (a) 40.6 cm   (b) 3.7 mi   (c) 3.9 ft

(d) 4.6 yd   (e) 4.6 ft   (f) 5.6 m

A Closer Look In problem (c), $1.2 \text{ m} \times \dfrac{1 \text{ yd}}{0.9144 \text{ m}} \times \dfrac{3 \text{ ft}}{1 \text{ yd}} = 3.937 \text{ ft} \ldots \approx 3.9 \text{ ft.}$

1.2 ⊟ **.9144** ⊠ **3** ⊟ → **3.937007874** ●

Dual Dimensioning

Some companies involved in international trade use "dual dimensioning" on their technical drawings and specifications. With dual dimensioning, both inch and metric dimensions are given. For example, a part might be labeled like this:

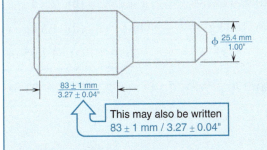

Notice that the metric measurement is written first or on top of the fraction bar. Diameter dimensions are marked with the symbol ɸ.

Metric–U.S. Customary Weight Conversions   To convert units of weight between the metric system and the U.S. customary system, use the conversion factors in the following table.

Metric–U.S. Customary Weight Conversions

1 ounce (oz) ≈ 28.35 grams (g)

1 pound (lb) ≈ 0.4536 kilogram (kg)

1 ton (T) ≈ 907.2 kilograms (kg) ≈ 0.9072 metric tons (t)

! Careful   The abbreviation for ton, T, is the same as the abbreviation for tablespoon, and the abbreviation for metric ton, t, is the same as the abbreviation for teaspoon. Whenever it is not clear if you are working with weight or liquid capacity, always spell out the desired unit. ●

EXAMPLE 3   (a) To convert 75 g to ounces, use the conversion factor 1 ounce = 28.35 grams to form a unity fraction.

$$75 \text{ g} = 75 \text{ g} \times \frac{1 \text{ oz}}{28.35 \text{ g}} = 2.6455 \dots \approx 2.6 \text{ oz, rounded}$$

(b) Convert 1.40 T to kg as follows:

$$1.40 \text{ T} = 1.40 \text{ T} \times \frac{907.2 \text{ kg}}{1 \text{ T}}$$

$$= 1.40 \times 907.2 \text{ kg}$$

$$\approx 1270 \text{ kg}$$

(c) To convert 0.075 pound to grams, use two unity fractions:

$$0.075 \text{ lb} = 0.075 \text{ lb} \times \frac{0.4536 \text{ kg}}{1 \text{ lb}} \times \frac{1000 \text{ g}}{1 \text{ kg}}$$

$$= 34.02 \text{ g} \approx 34 \text{ g, rounded}$$

📟 .075 ⊗ .4536 ⊗ 1000 ⊜ →   *34.02*   ●

Your Turn   Convert the following weight measurements as shown. (Round to the nearest whole number unless otherwise instructed in parentheses.)

(a) 6.0 oz = _____ g (b) 2.3 T = _____ t (nearest tenth)

(c) 100̄ kg = _____ lb (d) 1.5 kg = _____ oz

(e) 15,000 g = _____ lb (f) 4 lb 5 oz = _____ kg (nearest hundredth)

Answers   (a) 170 g (b) 2.1 t (c) 220 lb (d) 53 oz (e) 33 lb (f) 1.96 kg

---

🔍 A Closer Look   In problem (d) $1.5 \text{ kg} \times \dfrac{1 \text{ lb}}{0.4536 \text{ kg}} \times \dfrac{16 \text{ oz}}{1 \text{ lb}} = 52.91 \dots \approx 53 \text{ oz}$

Alternatively, $1.5 \text{ kg} \times \dfrac{1000 \text{ g}}{1 \text{ kg}} \times \dfrac{1 \text{ oz}}{28.35 \text{ g}} \approx 53 \text{ oz}$

There are also two ways to do problem (f):

$$\text{Method 1: 4 lb 5 oz} = 4\frac{5}{16}\text{ lb} = \frac{69}{16}\text{lb} \times \frac{0.4536\text{ kg}}{1\text{ lb}} = 1.95615\text{ kg} \approx 1.96\text{ kg}$$

$$\text{Method 2: 4 lb 5 oz} = 64\text{ oz} + 5\text{ oz} = 69\text{ oz} \times \frac{28.35\text{ g}}{1\text{ oz}} \times \frac{1\text{ kg}}{1000\text{ g}} \approx 1.96\text{ kg} \bullet$$

Metric–U.S. Customary Area Conversion

As we learned earlier in this chapter, area is a compound unit because it is the product of *two* length units. As a result, most areas are expressed in "square" units. Use the following table to convert area units between the metric and U.S. customary systems.

> ## Metric–U.S. Customary Area Conversions
>
> 1 square inch (sq in. or in.^{2}) $\approx$ 6.452 sq cm (or cm^2)
>
> 1 square foot (sq ft or ft^2) $\approx$ 0.0929 sq m (or m^2)
>
> 1 square yard (sq yd or yd^2) $\approx$ 0.836 sq m
>
> 1 acre $\approx$ 0.405 hectare (ha)

EXAMPLE 4

(a) To convert 15 square feet to square meters, multiply as follows;

$$15\text{ sq ft} = 15\text{ sq ft} \times \frac{0.0929\text{ sq m}}{1\text{ sq ft}} = 1.3935\text{ sq m} \approx 1.4\text{ sq m, rounded}$$

(b) To convert 5.80 hectares to acres, multiply as follows:

$$5.80\text{ ha} = 5.80\text{ ha} \times \frac{1\text{ acre}}{0.405\text{ ha}} \approx 14.3\text{ acres, rounded} \qquad\qquad \bullet$$

Your Turn

Use this information to solve the following problems. Round to the nearest tenth.

(a) 16 acres = _____ ha   (b) 0.21 sq m = _____ sq ft

(c) 11 sq cm = _____ sq in.   (d) 78.0 sq yd = _____ sq m

Answers   (a) 6.5 ha   (b) 2.3 sq ft   (c) 1.7 sq in.   (d) 65.2 sq m

Volume and Liquid Capacity Conversions

The following table summarizes the most useful metric–U.S. customary conversion factors for volume and liquid capacity.

> ## Metric–U.S. Customary Volume and Liquid Capacity Conversions
>
> 1 cubic inch (cu in. or in.^{3}) $\approx$ 16.387 cubic centimeters (cu cm or cm^3)
>
> 1 cubic yard (cu yd or yd^3) $\approx$ 0.765 cubic meters (cu m or m^3)
>
> 1 teaspoon (t) = 5 milliliters (mL)
>
> 1 tablespoon (T) = 15 mL
>
> 1 fluid ounce (fl oz) $\approx$ 29.574 mL (or cu cm)
>
> 1 quart (qt) $\approx$ 0.946 liter (L)*
>
> 1 gallon (gal) $\approx$ 3.785 L

*In the United States, 1 quart is legally defined as 0.94635295 liter. This means that a volume of 1 liter is slightly larger than 1 qt.

EXAMPLE 5   (a) To convert 5.5 gal to liters, use the conversion factor 1 gal ≈ 3.785 L. Using unity fractions,

$$5.5 \text{ gal} \times \frac{3.785 \text{ L}}{1 \text{ gal}} \approx 20.8175 \text{ L} \approx 21 \text{ liters}$$

(b) To convert 125 mL to fluid ounces, use the conversion factor 1 fl oz ≈ 29.574 mL to form a unity fraction and multiply as follows:

$$125 \text{ mL} = 125 \text{ mL} \times \frac{1 \text{ fl oz}}{29.574 \text{ mL}} = 4.2266 \ldots \approx 4.23 \text{ fl oz}$$ ●

Your Turn   Convert the following measurements as shown. (Round to the nearest tenth unless otherwise directed.)

(a) 13.4 gal = _____ L   (b) 9.10 fl oz = _____ mL (nearest whole number)

(c) 16.0 cu yd = _____ cu m   (d) 5.1 liters = _____ qt

(e) 6 t ≈ _____ mL   (f) 75 mL ≈ _____ T

Answers   (a) 50.7 L   (b) 269 mL   (c) 12.2 cu m   (d) 5.4 qt   (e) 30 mL   (f) 5 T

Speed   We know that speed is a compound unit consisting of a distance divided by a time. To convert speed units between the U.S. customary and metric system, use the appropriate combination of distance (page 326) and time unity fractions.

EXAMPLE 6   To convert 2.4 ft/sec to m/min, use the following conversion factors:

For converting time in the denominator, use 1 min = 60 sec.

For converting distance in the numerator, use 1 ft = 30.48 cm and 1 m = 100 cm.

Using unity fractions,

$$2.4 \text{ ft/sec} = \frac{2.4 \text{ ft}}{1 \text{ sec}} \times \frac{30.48 \text{ cm}}{1 \text{ ft}} \times \frac{1 \text{ m}}{100 \text{ cm}} \times \frac{60 \text{ sec}}{1 \text{ min}}$$

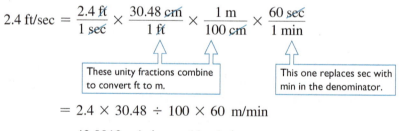

These unity fractions combine to convert ft to m.

This one replaces sec with min in the denominator.

$$= 2.4 \times 30.48 \div 100 \times 60 \text{ m/min}$$

$$= 43.8912 \text{ m/min} \approx 44 \text{ m/min}$$ ●

EXAMPLE 7   To convert 285 km/hr to mi/hr, note that the time units are the same. We need only convert the distance units using 1 mi ≈ 1.6093 km.

$$285 \text{ km/hr} = \frac{285 \text{ km}}{1 \text{ hr}} \times \frac{1 \text{ mi}}{1.6093 \text{ km}} = 177.09 \ldots \approx 177 \text{ mi/hr}$$ ●

Your Turn   Convert the following speed units as indicated. (Round as directed.)

(a) 35 mph = _____ km/h (whole number)

(b) 26.5 m/min = _____ ft/sec (hundredth)

(c) 1240 cm/sec = _____ mph (tenth)

(d) 18 in/sec = _____ cm/sec (whole number)

Answers (a) 56 km/h (b) 1.45 ft/sec

(c) 27.7 mph (d) 46 cm/sec

Temperature The Fahrenheit temperature scale is commonly used in the United States for weather reports, cooking, and other practical work. On this scale, water boils at 212°F and freezes at 32°F. The metric or *Celsius* temperature scale is a simpler scale originally designed for scientific work but now used worldwide for most temperature measurements. On the Celsius scale, water boils at 100°C and freezes at 0°C.

Notice that the range from freezing to boiling is covered by 180 degrees on the Fahrenheit scale and 100 degrees on the Celsius scale. The size and meaning of a temperature degree is different for the two scales. A temperature of zero does not mean "no temperature"—it is simply another point on the scale. Temperatures less than zero are possible, and they are labeled with negative numbers.

Because zero does not correspond to the same temperature on both scales, converting Celsius to Fahrenheit or Fahrenheit to Celsius is not as easy as converting length or weight units. In Chapter 7, we will learn how to use algebraic formulas to perform these conversions. For now, we will use the following chart to convert from one scale to the other.

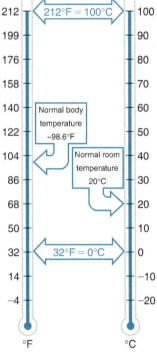

TEMPERATURE CONVERSION CHART
F = Fahrenheit   C = Celsius

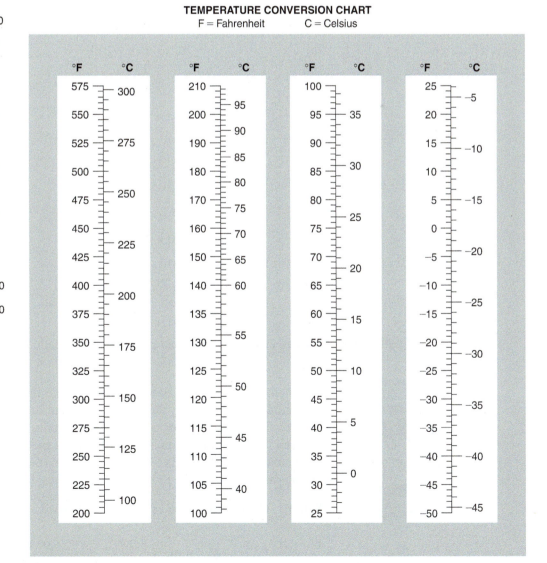

EXAMPLE 8   (a) On this chart, equal temperatures are placed side by side. To convert 50°F to Celsius:

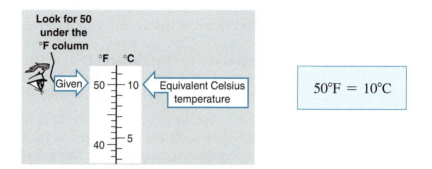

Look for 50 under the °F column

Given   50 — 10   Equivalent Celsius temperature

$$50°F = 10°C$$

(b) To convert 125°C to Fahrenheit, locate 125°C in the °C column of the far-left chart. This Celsius mark does not coincide exactly with one of the Fahrenheit divisions, but it is closest to the second mark above 250°F. Because there are 5 divisions between 250°F and 275°F, each division represents 5°F, so the second mark represents an additional 10 degrees. Therefore, 125°C ≈ 260°F.   ●

Your Turn   Now use this temperature conversion chart to determine the following temperatures. Estimate your answers to the nearest "tick" mark.

(a) 375°F = _____ °C   (b) −20°F = _____ °C

(c) 14°C = _____ °F   (d) 80°C = _____ °F

(e) 37°C = _____ °F   (f) 68°F = _____ °C

(g) −40°C = _____ °F   (h) 525°F = _____ °C

(i) 80°F = _____ °C   (j) −14°C = _____ °F

Answers   (a) 190°C   (b) −29°C

(c) 57°F   (d) 176°F

(e) 99°F   (f) 20°C

(g) −40°F   (h) 275°C

(i) 27°C   (j) 7°F

Applications   Here is an application involving a length conversion.

EXAMPLE 9   **Automotive Trades**   The size of an automobile tire is given as P205/55R16. The first number indicates that the width, W, of the tire (from sidewall to sidewall) is 205 mm. The second number, called the **aspect ratio,** indicates that the height, H, of the sidewall is 55% of the width. The third number indicates that the tire is designed to be mounted on a wheel with a 16-inch diameter. (See the figure.) In order to find the overall diameter, D, of the mounted tire in inches, follow these steps:

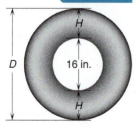

Step 1.   Convert the width of the tire to inches. Use 1 in. ≈ 25.4 mm

$$W = 205 \text{ mm} \times \frac{1 \text{ in.}}{25.4 \text{ mm}} = 8.070 \ldots \approx 8.07 \text{ in.}$$

Step 2. Find 55% of this result to determine the height of the sidewall.

$$\frac{H}{8.07} = \frac{55}{100}$$

$$100H = 443.85$$

$$H = 4.4385 \text{ in.} \approx 4.44 \text{ in.}$$

Step 3. Double the result of Step 2 to get the total height of both sidewall sections, and then add the diameter of the wheel. This will give you the overall diameter of the mounted tire.

$$D = 2(4.44 \text{ in.}) + 16 \text{ in.} \approx 24.88 \text{ in.}$$

The overall diameter of the mounted tire is about $24\frac{7}{8}$ in.   ●

Now let's try one requiring conversion of liquid capacity units.

EXAMPLE 10 **Allied Health** Suppose a patient was given 30 mL of a liquid medication twice a day when he was in the hospital. If he continues this dosage at home, how many tablespoons per day will the patient take?

In one day, the patient will take 30 mL × 2 or 60 mL.

Converting this to tablespoons, we have:

$$60 \text{ mL} = 60 \text{ mL} \times \frac{1 \text{ T}}{15 \text{ mL}} = 4 \text{ T}$$ ●

The next example requires us to convert both weight and area units as well as a unique conversion factor given in the problem.

EXAMPLE 11 **Agriculture** For 2016/2017, the European Union projected an average wheat yield of 5.83 metric tons (t) per hectare (ha). The U.S. projection for the same time period was 51.3 bushels per acre, where one bushel of wheat weighs approximately 60 pounds. (Source: USDA) To compare these projected yields, we will convert the European projection to U.S. customary units.

We know that 1 metric ton = 1000 kg, 1 lb ≈ 0.4536 kg, and 1 acre ≈ 0.405 ha. To convert 5.83 metric tons per hectare to bushels per acre, we must use four different unity fractions:

$$\frac{5.83 \text{ t}}{1 \text{ ha}} \times \frac{0.405 \text{ ha}}{1 \text{ acre}} \times \frac{1000 \text{ kg}}{1 \text{ t}} \times \frac{1 \text{ lb}}{0.4536 \text{ kg}} \times \frac{1 \text{ bu}}{60 \text{ lb}} \approx 86.7 \text{ bu/acre}$$

The European projected yield is approximately 35.4 bushels per acre more than the projected U.S. yield. This is about a 69% increase over the U.S. yield.   ●

For those who work with metals, building materials, and other substances, the concept of *density* is useful. **Density** is defined as the amount of mass per unit of volume of a material.

EXAMPLE 12 **Manufacturing** The density of steel is 7850 kilograms per cubic meter (kg/cu m). In the U.S. customary system, we would usually express this in pounds per cubic foot (lb/cu ft). We will need the following factors to perform this conversion: 1 lb ≈ 0.4536 kg, 1 cu yd ≈ 0.765 cu m, and 1 cu yd = 27 cu ft. We can now set up unity fractions as follows:

$$\frac{7850 \text{ kg}}{1 \text{ cu m}} \times \frac{1 \text{ lb}}{0.4536 \text{ kg}} \times \frac{0.765 \text{ cu m}}{1 \text{ cu yd}} \times \frac{1 \text{ cu yd}}{27 \text{ cu ft}} = 490.33 \ldots \approx 490 \text{ lb/cu ft}$$

Using a calculator, we have:

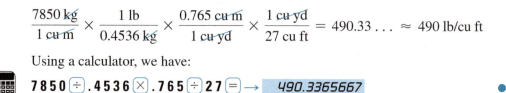

Here are some applied problems for you to try.

Your Turn   (a) **General Interest**   A room in a European hotel features a bed that measures 160 cm by 200 cm. Which of the following American beds is closest in size to this one?

A. Full: 54 in. by 75 in.   B. Queen: 60 in. by 80 in.   C. King: 76 in. by 80 in.

(b) **Allied Health**   If a patient is taking two teaspoons of a liquid medication per dosage, how many doses of the medication are in a 500 mL bottle?

(c) **Machine Trades**   The cutting speed for a soft steel part in a lathe is 165 ft/min. Express this in cm/sec.

(d) **Manufacturing**   A U.S. company orders steel rectangular tubing from a European distributor. The distributor lists the weight of the tubing as 2.92 kg/m^2. The U.S. company has on hand tubing that weighs 0.625 lb/ft^2. Is the European tubing heavier or lighter than this?

Solutions   (a) Using 1 in. = 2.54 cm, we have

$$160 \text{ cm} = 160 \text{ cm} \times \frac{1 \text{ in.}}{2.54 \text{ cm}} = 62.99 \ldots \approx 63 \text{ in}$$

$$200 \text{ cm} = 200 \text{ cm} \times \frac{1 \text{ in.}}{2.54 \text{ cm}} = 78.74 \ldots \approx 79 \text{ in}$$

This European bed is closest in size to the American queen-size bed (choice B).

(b) **First,** convert two teaspoons to milliliters:

$$2 \text{ t} = 2 \text{ t} \times \frac{5 \text{ mL}}{1 \text{ t}} = 10 \text{ mL}$$

Then, calculate the number of doses in 500 mL:

$$500 \text{ mL} \times \frac{1 \text{ dose}}{10 \text{ mL}} = 50 \text{ doses}$$

(c) Using 1 min = 60 sec and 1 ft = 30.48 cm, we have:

$$\frac{165 \text{ ft}}{1 \text{ min}} \times \frac{1 \text{ min}}{60 \text{ sec}} \times \frac{30.48 \text{ cm}}{1 \text{ ft}} = 83.82 \text{ cm/sec} \approx 83.8 \text{ cm/sec, rounded}$$

(d) To solve this problem, convert the European weight to pounds per square foot.

$$2.92 \text{ kg/m}^2 = \frac{2.92 \text{ kg}}{1 \text{ m}^2} \times \frac{1 \text{ lb}}{0.4536 \text{ kg}} \times \frac{0.0929 \text{ m}^2}{1 \text{ ft}^2} = 0.598 \text{ lb/ft}^2$$

The European tubing is slightly lighter.

The following table summarizes the conversion factors we have covered between the metric system and the U.S. customary system. You can refer to these as you work on the set of exercises that follow the table.

Metric–U.S. Customary Conversion Factors

Length	Area
1 in. = 2.54 cm	1 sq in. (in.^{2}) ≈ 6.452 sq cm (cm^2)
1 ft = 30.48 cm	1 sq ft (ft^2) ≈ 0.0929 sq m (m^2)
1 yd = 0.9144 m	1 sq yd (yd^2) ≈ 0.836 m^2
1 mi ≈ 1.6093 km	1 acre ≈ 0.405 hectare (ha)
Weight/Mass	**Volume and Liquid Capacity**
1 oz ≈ 28.35 g	1 cu in. (in.^{3}) ≈ 16.387 cu cm (cm^3)
1 lb ≈ 0.4536 kg	1 cu yd (yd^3) ≈ 0.765 cu m (m^3)
1 ton (T) ≈ 907.2 kg	1 teaspoon (t) = 5 mL
1 T ≈ 0.9072 metric tons (t)	1 tablespoon (T) = 15 mL
	1 fluid ounce (fl oz) ≈ 29.574 mL
	1 qt ≈ 0.946 liter (L)
	1 gal ≈ 3.785 L

Exercises 5-4 | Metric–U.S. Customary Conversions

A. Convert to the units shown. Round to the nearest tenth unless otherwise directed.

1. 478 cm = _____ ft
2. 4.20 ft = _____ cm (whole number)

3. 2.0 mi = _____ km
4. 20 ft 6 in. = _____ m

5. $9\frac{1}{4}$ in. = _____ cm
6. 33.0 kg = _____ lb _____ oz

7. 152 lb = _____ kg
8. 3 lb 4 oz = _____ kg

9. 8.50 kg = _____ lb
10. 10.4 oz = _____ g (whole number)

11. 3.15 gal = _____ L
12. 2.5 liters = _____ qt

13. 6.5 qt = _____ L
14. 61.0 mph = _____ km/h

15. 8.0 km/h = _____ mph
16. 230 °F = _____ °C

17. 10 °C = _____ °F
18. 35.0 in./sec = _____ cm/sec

19. 63.0 ft/sec = _____ m/sec
20. 2.0 m = _____ ft

21. 2.85 sq in. = _____ sq cm
22. 512 cu in. = _____ cu cm (nearest ten)

23. 2.8 m^3 = _____ cu yd
24. 46.2 sq cm = _____ sq in. (hundredth)

25. 687 km = _____ mi (whole number)
26. 260 mL = _____ fl oz

27. 2468 sq ft = _____ sq m
28. 62.0 m/sec = _____ ft/sec (whole number)

29. 31 kg/sq m = _____ lb/sq ft
30. 4.50 metric tons = _____ lb (nearest ten)

31. 28 L = _____ gal
32. 7.20 fl oz = _____ mL (whole number)

33. 94.6 cm = _____ in.
34. 120 g = _____ oz

35. 6800 lb = _____ metric tons
36. 365 cu cm = _____ cu in.

37. 8.7 cu yd   = _____ cu m   38. 56 cm/sec   = _____ mph

39. 1.08 g/cu cm = _____ lb/cu ft   40. 165.0 lb/cu ft = _____ kg/cu m
                                                        (whole number)

41. 2.80 mi/hr   = _____ m/min   42. 180 m/min   = _____ ft/sec

B. Practical Applications. Round as indicated.

1. **Welding**   Welding electrode sizes are presently given in inches. Convert the following set of electrode sizes to millimeters. (Round to the nearest one-thousandth of a millimeter.)

in.	mm		in.	mm		in.	mm
0.030			$\frac{1}{16}$			$\frac{5}{32}$	
0.035			$\frac{5}{64}$			$\frac{3}{16}$	
0.040			$\frac{3}{32}$			$\frac{3}{8}$	
0.045			$\frac{1}{8}$			$\frac{11}{64}$	

2. **Hydrology**   One cubic foot of water weighs 62.4 lb. (a) What is the weight in pounds of 1 liter of water? (b) What is the weight in kilograms of 1 cu ft of water? (c) What is the weight in kilograms of 1 liter of water? (Round to one decimal place.)

3. **Sports and Leisure**   At the 2016 Olympic Games in Rio, Eliud Kipchoge won the 26.2-mile men's marathon race in 2 hr 8 min 44 sec. What was his average speed in mph and km/h? (Round to the nearest tenth.)

4. **Physics**   The metric unit of pressure is the *pascal,* where

   1 lb/sq in. (psi) ≈ 6894 pascal (Pa), and 1 lb/sq ft (psf) ≈ 47.88 Pa

   For very large pressures, the kilopascal (kPa) is the appropriate unit.

   1 kPa = 1000 Pa and therefore

   1 psi ≈ 6.894 kPa and 1 psf ≈ 0.04788 kPa

   Convert the following pressures to metric units:

   (a) 15.0 psi = _____ Pa   (b) $10\overline{0}$ psi = _____ kPa

   (c) 40 psf = _____ Pa   (d) 2500 psf = _____ kPa

5. Find the difference between

   (a) 1 mile and 1 kilometer   _____ km

   (b) 3 miles and 5000 meters   _____ ft

   (c) 120 yards and 110 meters   _____ ft

   (d) 1 quart and 1 liter   _____ qt

   (e) 10 pounds and 5 kilograms   _____ lb

6. **Flooring and Carpeting**   Standard American carpeting comes in 12-ft widths, but imported European carpeting comes in 4-m widths. What would this European width be in feet and inches? (Round to the nearest half-inch.)

7. **Automotive Trades**   Which is performing more efficiently: a car getting 12 km per liter of gas or one getting 25 miles per gallon of gas?

8. **Painting**   A gallon of exterior enamel sells for $38.29. At this rate, what would 1 liter cost?

9. **Roofing** In roofing, the unit "one square" is sometimes used to mean 100 sq ft. Convert this unit to the metric system. (Round to one decimal place.)

   1 square = _____ sq m

10. **General Interest** A *cord* of wood is a volume of wood 8 ft long, 4 ft wide, and 4 ft high.

    (a) What are the equivalent metric dimensions for a "metric cord" of wood?

    (b) What is the volume of a cord in cubic meters? (Round to the nearest tenth.) (*Hint:* volume = length × width × height.)

11. **General Interest** Translate these well-known phrases into metric units.

    (a) A miss is as good as _____ km. (1 mile = _____ km)

    (b) An _____ cm by _____ cm sheet of paper. ($8\frac{1}{2}$ in. = _____ cm; 11 in. = _____ cm)

    (c) He was beaten within _____ cm of his life. (1 in. = _____ cm)

    (d) Race in the Indy _____ km. (500 miles = _____ km)

    (e) Take it with _____ grams of salt. (*Hint:* 437.5 grains = 1 oz)

    (f) _____ grams of prevention is worth _____ grams of cure.

      (1 oz = _____ g; 1 lb = _____ g)

    (g) Peter Piper picked _____ liters of pickled peppers.

      (*Hint:* 1 peck = 8 quarts)

    (h) A cowboy in a _____ liter hat. (10 gal = _____ liter)

12. **Automotive Trades** In an automobile, cylinder displacement is measured in either cubic inches or liters. What is the cylinder displacement in liters of a 400-cu in. engine? (Round to the nearest liter.)

13. **Sports and Leisure** In the United States, distance markers on golf courses are given in yards, but in Europe they are given in meters. A European player, golfing in the United States, has a shot that measures 150 yards. He wants his caddy to give him the distance in meters. What should the caddy tell him?

14. **Culinary Arts** Fried crickets, served as an appetizer, sell for 5400 Cambodian riels per kg. If 4060 riels are equal to one dollar, how much will 5 lb of crickets cost in U.S. dollars?

15. **Sports and Leisure** Nutrition scientists have determined that for long-distance running in the desert, a drink containing 5.0 g of glycerol per 100 mL of water is ideal. Convert this to ounces per quart. (Round to two significant digits.)

16. **Automotive Trades** A BMW X5 has a fuel tank capacity of 85.0 liters. How many gallons does the fuel tank hold? (Round to the nearest tenth.)

17. **Culinary Arts** A British recipe for Beef Wellington requires a cooking temperature of 170°C. Determine the equivalent of this temperature in °F.

18. **General Interest** An American family is looking for an apartment rental in Europe. They would like one that is at least 1500 sq ft in size. A website advertises an apartment with a total area of 130 square meters. Does this meet the family's requirement?

19. **Aviation** The normal cruising speed of the Airbus A380 is about 250 m/sec. What is this speed in miles per hour?

20. **Allied Health** For a baby's milk formula, 1 oz of formula should be mixed with 0.5 qt of water. Express this baby formula concentration in grams per liter (g/L).

21. **Culinary Arts**   A caterer for a wedding reception wants to use a Hungarian recipe for eggplant salad that recommends using 1.5 kg of eggplant for 8 guests. However, in her area, eggplant is sold only by the pound. If 150 guests are expected, how many pounds of eggplant should the caterer purchase? (Round to the nearest pound.)

22. **Culinary Arts**   The standard recipe for chocolate chip cookies calls for 12 oz of chocolate chips to make approximately 60 cookies. An American visiting Europe wants to make 200 cookies for a Christmas party. Because chocolate chips are difficult to find in Europe, he decides to chop up some chocolate bars to replace the chips. How many 100-g chocolate bars will he need to purchase? (Assume that you cannot buy a fraction of a bar.)

23. **Automotive Trades**   The size of an automobile tire is given as P215/70R15. Find the overall diameter of the mounted tire. (See Example 9. Round to the nearest eighth of an inch.)

24. **Sports and Leisure**   In the 2016 Olympics, Usain Bolt of Jamaica ran the 100 meters in 9.81 sec and the 200 meters in 19.78 sec. Determine his average speed in miles per hour for (a) the 100-m run, and (b) the 200-m run. (Round your answer to two decimal digits.)

25. **Agriculture**   For 2016/2017, the USDA projected the U.S. soybean yield to be 48.9 bushels per acre. If a bushel of soybeans weighs 60 lb, how does this compare to a European yield of 3.16 metric tons per hectare?

26. **Metalworking**   The density of aluminum is 2712 kg/cu m. Express this in pounds per cubic foot.

27. **Aviation**   The cruising speed of a Boeing 737 is approximately Mach 0.74, meaning 0.74 times the speed of sound. If the speed of sound at cruising altitude is approximately 295 m/s, what is the cruising speed of the 737 in miles per hour?

28. **Automotive Trades**   An automobile tire mounted on a 16-in. wheel has an overall diameter of 24.68 in. and an aspect ratio of 45%. Determine the size of the tire in standard form. (*Hint:* Perform the steps of Example 9 in reverse order.)

29. **Welding**   The wire feed controls the amperage during the welding process. A control knob labeled in metric units shows a maximum wire-feed speed of 350 mm/sec. What is this maximum speed in inches per minute? (Round to two significant digits.)

30. **Allied Health**   If a patient must take three tablespoons of a liquid medication per day, how many days will a 750 mL supply last?

When you have completed these exercises, check your answers to the odd-numbered problems in the Appendix, then turn to Section 5-5 to learn about direct measurements.

5-5   Direct Measurements

The simplest kind of measuring device to use is one in which the output is a digital display. For example, in a digital clock or digital voltmeter, the measured quantity is translated into electrical signals, and each digit of the measurement number is displayed.

A digital display is easy to read, but most technical work involves direct measurement where the numerical reading is displayed on some sort of scale. Reading the scale correctly requires skill and the ability to estimate accurately between scale divisions.

Length Measurements

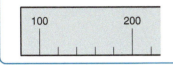
Length measurement involves many different instruments that produce either fraction or decimal number readouts. Carpenters, electricians, plumbers, roofers, and machinists all use rulers, yardsticks, or other devices to make length measurements. Each instrument is made with a different scale that determines the precision of the measurement that can be made. The more precise the measurement has to be, the smaller the scale divisions needed.

For example, on the carpenter's rule shown, scale A is graduated in eighths of an inch. Scale B is graduated in sixteenths of an inch.

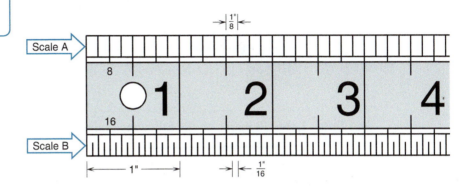

A carpenter or roofer may need to measure no closer than $\frac{1}{8}$ or $\frac{1}{16}$ in., but a machinist will often need a scale graduated in 32nds or 64ths of an inch. The finely engraved steel rule shown is a machinist's rule.

Some machinists' rules are marked in tenths or hundredths of an inch on one side and may be marked in metric units on the other side.

Notice that, on all these rules, only a few scale markings are actually labeled with numbers. It would be easy to read a scale marked like this:

Any useful scale has divisions so small that this kind of marking is impossible. An expert in any technical field must be able to identify the graduations on the scale correctly even though the scale is *not* marked. To use any scale, first determine the size of the smallest division by counting the number of spaces in a 1-in. interval.

EXAMPLE 1

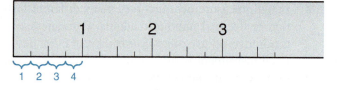

There are four spaces in the first 1-in. interval. The smallest division is therefore $\frac{1}{4}$ in., and we can label the scale like this:

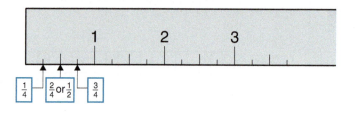

Your Turn   Label all marks on the following scales. The units are inches.

(a)

Smallest division = _____

(b)

Smallest division = _____

(c)

Smallest division = _____

(d)

Smallest division = _____

Answers   (a) Smallest division = $\frac{1}{8}$ in.

(b) Smallest division = $\frac{1}{5}$ in.

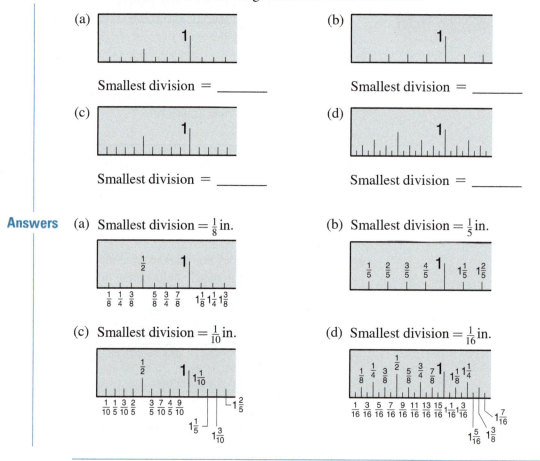

(c) Smallest division = $\frac{1}{10}$ in.

(d) Smallest division = $\frac{1}{16}$ in.

To use the rule once you have mentally labeled all scale markings, first count inches, then count the number of smallest divisions from the last whole-inch mark.

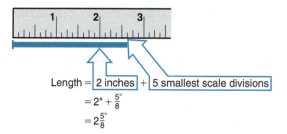

Length = 2 inches + 5 smallest scale divisions

$= 2" + \frac{5}{8}"$

$= 2\frac{5}{8}"$

If the length of the object being measured falls between scale divisions, record the nearest scale division.

EXAMPLE 2 In this measurement:

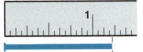

the endpoint of the object falls between $1\frac{3}{16}$ and $1\frac{4}{16}$ in. Here is an enlarged view:

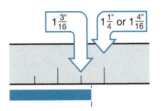

To the nearest scale division, the length is $1\frac{3}{16}$ in. When a scale with $\frac{1}{16}$-in. divisions is used, we usually write the measurement to the nearest $\frac{1}{16}$ in., but we may estimate it to the nearest half of a scale division, or $\frac{1}{32}$ in. Thus, we might estimate this length as $1\frac{7}{32}$ in. ●

Your Turn For practice in using length rulers, find the dimensions given on the following inch rules. Express all fractions in lowest terms.

(a)

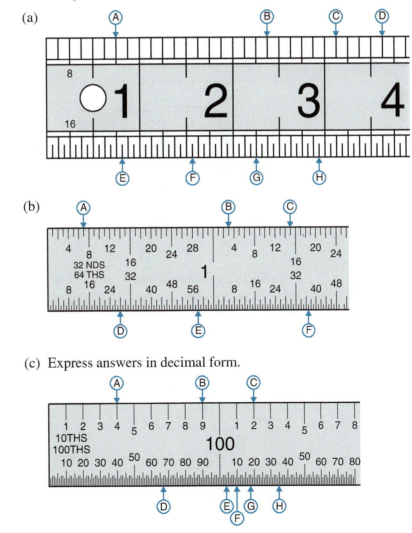

(b)

(c) Express answers in decimal form.

Answers   (a) A. $\frac{3}{4}$ in.   B. $2\frac{3}{8}$ in.   C. $3\frac{1}{8}$ in.   D. $3\frac{5}{8}$ in.

E. $\frac{13}{16}$ in.   F. $1\frac{9}{16}$ in.   G. $2\frac{1}{4}$ in.   H. $2\frac{15}{16}$ in.

(b) A. $\frac{7}{32}$ in.   B. $1\frac{3}{32}$ in.   C. $1\frac{15}{32}$ in.

D. $\frac{28}{64}$ in. $= \frac{7}{16}$ in.   E. $\frac{58}{64}$ in. $= \frac{29}{32}$ in.   F. $1\frac{37}{64}$ in.

(c) A. $\frac{4}{10}$ in. $= 0.4$ in.   B. $\frac{9}{10}$ in. $= 0.9$ in.   C. $1\frac{2}{10}$ in. $= 1.2$ in.

D. $\frac{67}{100}$ in. $= 0.67$ in.   E. $1\frac{4}{100}$ in. $= 1.04$ in.   F. $1\frac{10}{100}$ in. $= 1.10$ in.

G. $1\frac{18}{100}$ in. $= 1.18$ in.   H. $1\frac{35}{100}$ in. $= 1.35$ in.

Micrometers   A steel rule can be used to measure lengths of $\pm\frac{1}{32}$ in. or $\pm\frac{1}{64}$ in., or as fine as $\pm\frac{1}{100}$ in. To measure lengths with greater accuracy than this, more specialized instruments are needed. A **micrometer** can be used to measure lengths to ±0.001 in., one one-thousandth of an inch.

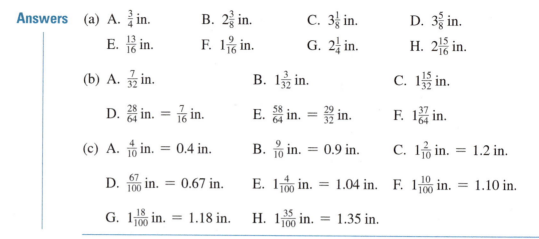

A modern **digital micrometer** is accurate to 0.0001 in., and the measurement is given as a digital readout in inches or millimeters.

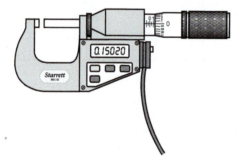

Two scales are used on a micrometer. The first scale, marked on the sleeve, records the movement of a screw machined accurately to 40 threads per inch. The smallest divisions on this scale record movements of the spindle of one-fortieth of an inch, or 0.025 in.

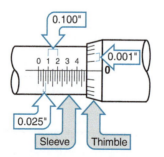

The second scale is marked on the rotating spindle. One complete turn of the thimble advances the screw one turn, or $\frac{1}{40}$ in. The thimble scale has 25 divisions, so that each mark on the thimble scale represents a spindle movement of $\frac{1}{25} \times \frac{1}{40}$ in. $= \frac{1}{1000}$ in., or 0.001 in.

To read a micrometer, follow these steps:

Step 1 Read the largest numeral visible on the sleeve and multiply this number by 0.100 in.

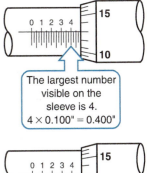

The largest number visible on the sleeve is 4.
$4 \times 0.100" = 0.400"$

Step 2 Read the number of additional scale spaces visible on the sleeve. (This will be 0, 1, 2, or 3.) Multiply this number by 0.025 in.

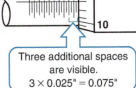

Three additional spaces are visible.
$3 \times 0.025" = 0.075"$

Step 3 Read the number on the thimble scale opposite the horizontal line on the sleeve and multiply this number by 0.001 in.

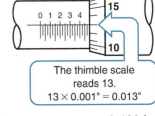

The thimble scale reads 13.
$13 \times 0.001" = 0.013"$

Step 4 Add these three products to find the measurement.

$$\begin{array}{r} 0.400 \text{ in.} \\ 0.075 \text{ in.} \\ + 0.013 \text{ in.} \\ \hline 0.488 \text{ in.} \end{array}$$

⚠ **Careful** To count as a "space" in step 2, the right end marker must be visible. ●

Here are more examples.

EXAMPLE 3

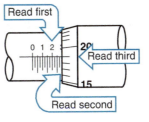

Step 1 The largest number completely visible on the sleeve is 2.

2×0.100 in. $= 0.200$ in.

Step 2 Three additional spaces are visible on the sleeve.

3×0.025 in. $= 0.075$ in.

Step 3 The thimble scale reads 19.

19×0.001 in. $= 0.019$ in.

Step 4 Add:

$$\begin{array}{r} 0.200 \text{ in.} \\ 0.075 \text{ in.} \\ 0.019 \text{ in.} \\ \hline \text{Length} = 0.294 \text{ in.} \end{array}$$

●

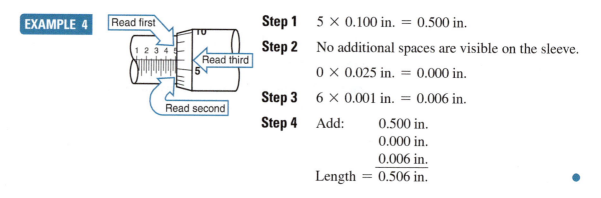

EXAMPLE 4

Step 1	5×0.100 in. $= 0.500$ in.
Step 2	No additional spaces are visible on the sleeve.
	0×0.025 in. $= 0.000$ in.
Step 3	6×0.001 in. $= 0.006$ in.
Step 4	Add:

$$\begin{aligned} &0.500 \text{ in.}\\ &0.000 \text{ in.}\\ &\underline{0.006 \text{ in.}}\\ \text{Length} = &0.506 \text{ in.} \end{aligned}$$

Your Turn   For practice, read the following micrometers.

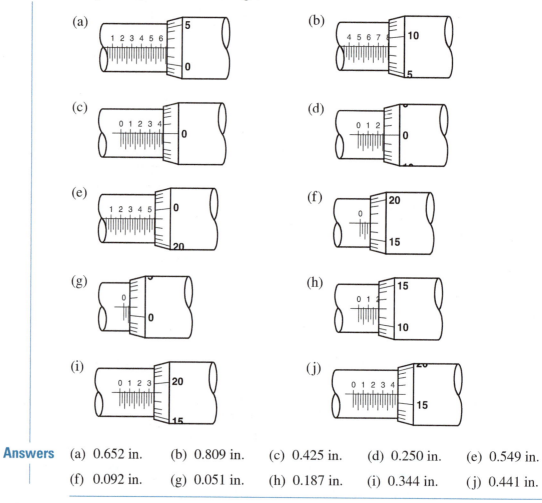

Answers   (a) 0.652 in.   (b) 0.809 in.   (c) 0.425 in.   (d) 0.250 in.   (e) 0.549 in.

(f) 0.092 in.   (g) 0.051 in.   (h) 0.187 in.   (i) 0.344 in.   (j) 0.441 in.

Metric micrometers are easier to read. Each small division on the sleeve represents 0.5 mm. The thimble scale is divided into 50 spaces; therefore, each division is $\frac{1}{50} \times 0.5$ mm $= \frac{1}{100}$ mm or 0.01 mm.

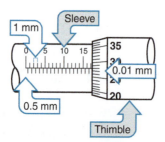

Turning the thimble scale of a metric micrometer by one scale division advances the spindle one one-hundredth of a millimeter, or 0.001 cm.

To read a metric micrometer, follow these steps.

Step 1   Read the largest mark visible on the sleeve and multiply this number by 1 mm.

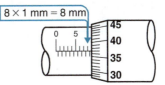

Step 2   Read the number of additional half spaces visible on the sleeve. (This number will be either 0 or 1.) Multiply this number by 0.5 mm.

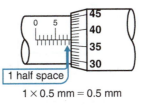

Step 3   Read the number on the thimble scale opposite the horizontal line on the sleeve and multiply this number by 0.01 mm.

The thimble scale reads 37.
37×0.01 mm $= 0.37$ mm

Step 4   Add these three products to find the measurement value.

8.00 mm
0.50 mm
0.37 mm
————
8.87 mm

EXAMPLE 5

17×1 mm $= 17$ mm

1×0.5 mm $= 0.5$ mm

21×0.01 mm $= 0.21$ mm

Length $= 17$ mm $+ 0.5$ mm $+ 0.21$ mm $= 17.71$ mm ●

Your Turn   Read these metric micrometers.

(a)

(b)

(c)

(d)

Answers   (a) 11.16 mm   (b) 15.21 mm

(c) 7.96 mm   (d) 21.93 mm

Vernier Micrometers   When more accurate length measurements are needed, the **vernier micrometer** can be used to provide an accuracy of ± 0.0001 in.

On the vernier micrometer, a third or vernier scale is added to the sleeve. This new scale appears as a series of ten lines parallel to the axis of the sleeve above the usual scale.

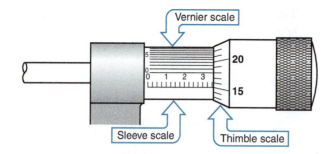

The ten vernier marks cover a distance equal to nine thimble scale divisions—as shown on this spread-out diagram of the three scales:

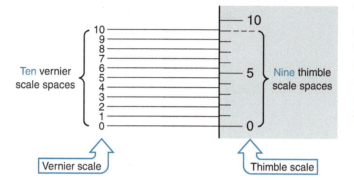

The difference between a vernier scale division and a thimble scale division is one-tenth of a thimble scale division, or $\frac{1}{10}$ of 0.001 in. = 0.0001 in.

To read the vernier micrometer, follow the first three steps given for the ordinary micrometer, then:

Step 4 Find the number of the line on the vernier scale that lines up exactly with any line on the thimble scale. Multiply this number by 0.0001 in. and add this amount to the other three distances.

EXAMPLE 6 For the vernier micrometer shown:

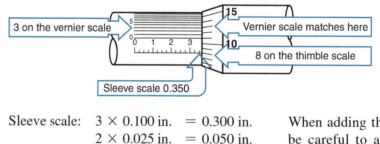

Sleeve scale: 3×0.100 in. $= 0.300$ in.
2×0.025 in. $= 0.050$ in.
Thimble scale: 8×0.001 in. $= 0.008$ in.
Vernier scale: 3×0.0001 in. $= 0.0003$ in.
Length $= 0.3583$ in.

When adding these four lengths, be careful to align the decimal digits correctly.

The vernier scale provides a way of measuring accurately the distance to an additional tenth of the finest scale division. ●

Your Turn Now for some practice in using the vernier scale, read each of the following.

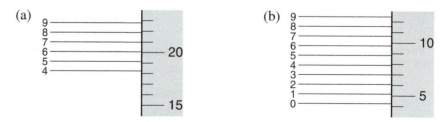

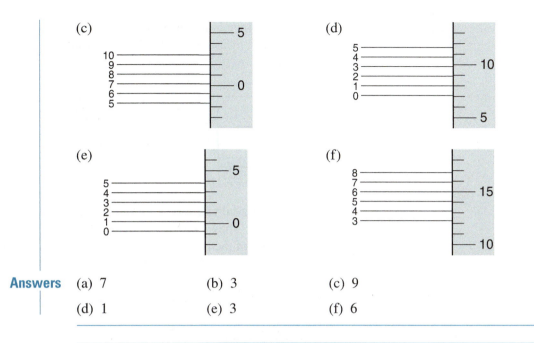

Answers
(a) 7	(b) 3	(c) 9
(d) 1	(e) 3	(f) 6

More Practice Now read the following vernier micrometers.

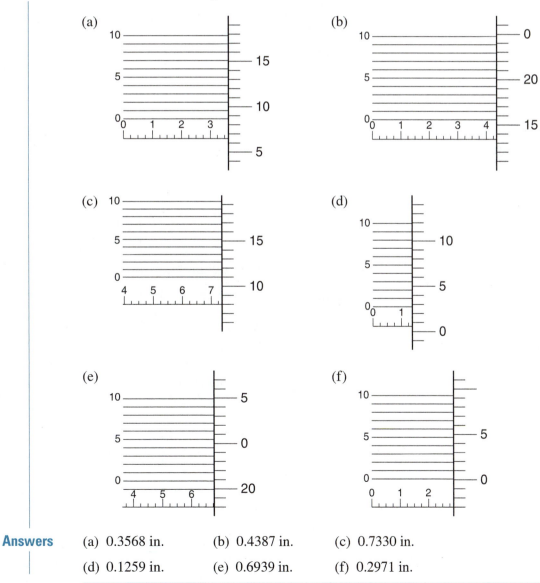

Answers
(a) 0.3568 in.	(b) 0.4387 in.	(c) 0.7330 in.
(d) 0.1259 in.	(e) 0.6939 in.	(f) 0.2971 in.

Vernier Calipers   The **vernier caliper** is another length-measuring instrument that uses the vernier principle and can measure lengths to ± 0.001 in.

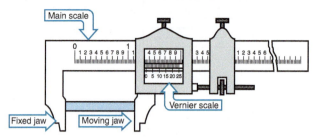

Modern direct **digital calipers,** battery or solar-powered, give measurements accurate to 0.0001 in. They convert automatically between U.S. customary and metric units. If the jaws of the measuring device cannot access both sides of the part being measured, an ultrasonic digital readout thickness gauge will do the job.

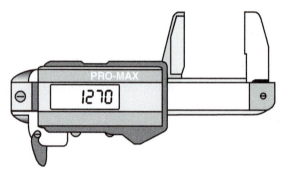

Notice that, on the manual instrument, each inch of the main scale is divided into 40 parts so that each smallest division on the scale is 0.025 in. Every fourth division or tenth of an inch is numbered. The numbered marks represent 0.100 in., 0.200 in., 0.300 in., and so on.

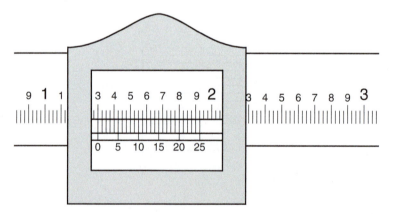

The sliding vernier scale is divided into 25 equal parts numbered by 5s. These 25 divisions on the vernier scale cover the same length as 24 divisions on the main scale.

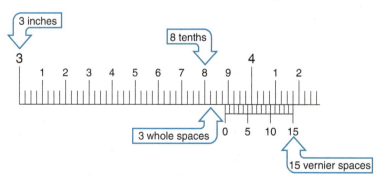

To read the vernier caliper just shown, follow these steps:

Step 1 Read the number of whole-inch divisions on the main scale to the left of the vernier zero. This gives the number of whole inches.

3 in. = 3.000 in.

Step 2 Read the number of tenths on the main scale to the left of the vernier zero. Multiply this number by 0.100 in.

8 tenths
8×0.100 in. = 0.800 in.

Step 3 Read the number of additional whole spaces on the main scale to the left of the vernier zero. Multiply this number by 0.025 in.

3 spaces
3×0.025 in. = 0.075 in.

Step 4 On the sliding vernier scale, find the number of the line that lines up exactly with any line on the main scale. Multiply this number by 0.001 in.

15 lines up exactly.
15×0.001 in. = 0.015 in.

Step 5 Add the four numbers.

3.000 in.
0.800 in.
0.075 in.
+ 0.015 in.
———————
3.890 in.

EXAMPLE 7

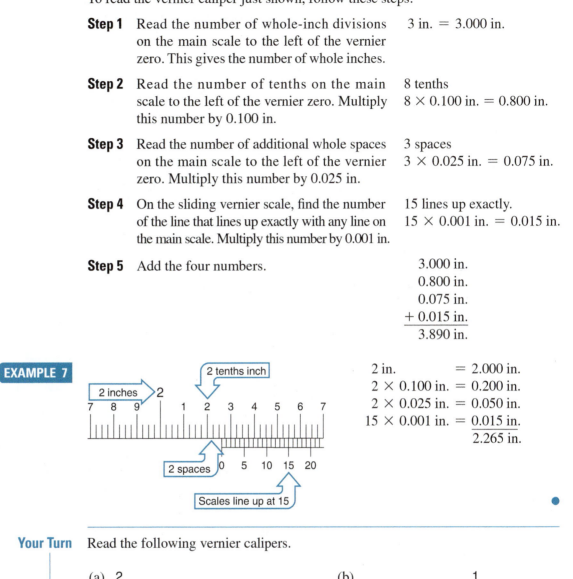

2 in.   = 2.000 in.
2×0.100 in. = 0.200 in.
2×0.025 in. = 0.050 in.
15×0.001 in. = 0.015 in.
———————
2.265 in.

Your Turn Read the following vernier calipers.

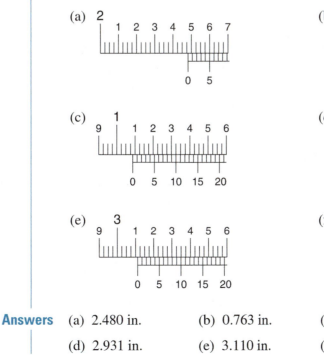

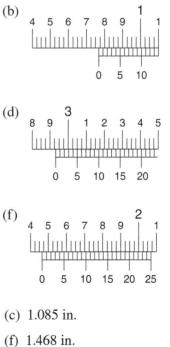

Answers   (a) 2.480 in.   (b) 0.763 in.   (c) 1.085 in.

(d) 2.931 in.   (e) 3.110 in.   (f) 1.468 in.

Protractors   A **protractor** is used to measure and draw angles. The simplest kind of protractor, shown here, is graduated in degrees with every tenth degree labeled with a number.

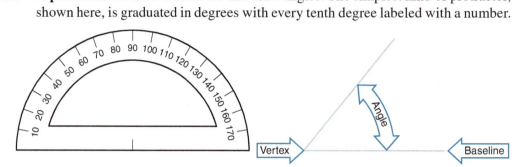

In measuring or drawing angles with a protractor, be certain that the vertex of the angle is exactly at the center of the protractor base. One side of the angle should be placed along the 0°–180° baseline.

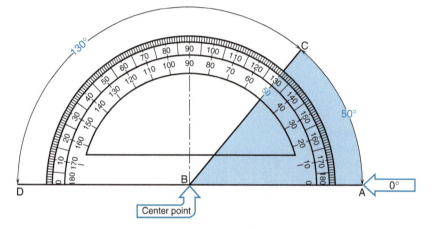

Notice that the protractor can be read from the 0° point on either the right or left side. Measuring from the right side, angle ABC is read as 50° on the inner scale. Measuring from the left side, angle DBC is 130° on the outer scale.

The **bevel protractor** is very useful in shopwork because it can be adjusted with an accuracy of ±0.5°. The angle being measured is read directly from the movable scale.

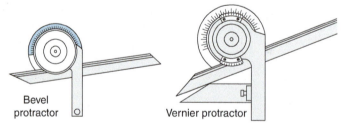

Bevel protractor

Vernier protractor

For very accurate angle measurements, the **vernier protractor** can be used with an accuracy of $\pm\frac{1}{60}$ of a degree, or 1 minute, because 60 minutes = 1 degree.

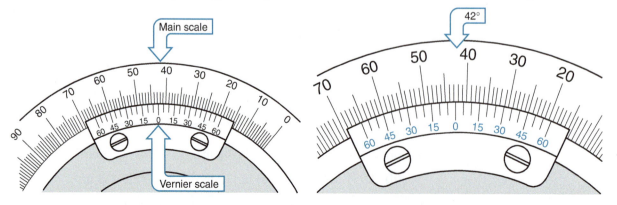

EXAMPLE 8 To read the vernier protractor shown, follow these steps:

Step 1 Read the number of whole degrees on the main scale between the zero on the main scale and the zero on the vernier scale.   The mark on the main scale just to the right of 0° is 42°.

Step 2 Find the vernier scale line on the left of the vernier zero that lines up exactly with a main scale marking. The vernier protractor in the example reads 42°45′.   The vernier scale lines up with a mark on the main scale at 45. This gives the fraction of a degree in minutes.   ●

Your Turn Read the following vernier protractors. The non-zero numbers in blue indicate where the vernier scales line up.

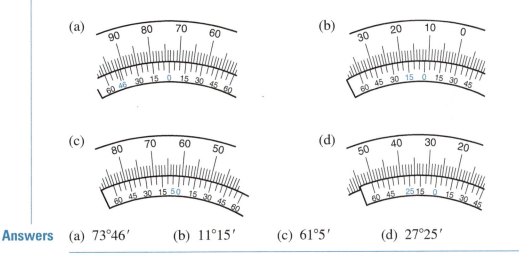

(a) (b)

(c) (d)

Answers (a) 73°46′   (b) 11°15′   (c) 61°5′   (d) 27°25′

Meters with Uniform Scales Many measuring instruments convert the measurement into an electrical signal and then display that signal by means of a pointer and a decimal scale. Instruments of this kind, called **meters**, are used to measure electrical quantities such as voltage, current, or power, and other quantities such as air pressure, flow rates, or speed. As with length scales, not every division on the meter scale is labeled, and practice is required in order to read meters quickly and correctly.

The **range** of a meter is its full-scale or maximum reading. The **main divisions** of the scale are the numbered divisions. The **small divisions** are the smallest marked portions of a main division. On a **uniform scale**, all marks are equally spaced, and all divisions represent the same quantity.

EXAMPLE 9 **Electrical Trades**   The following voltmeter has a uniform scale. The *range* is 10 volts;

each *main* division is 1 volt;
each *small* division is 0.5 volt.

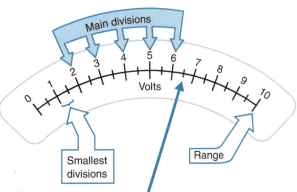

A **dual-scale meter** has an upper and lower scale and allows the user to switch ranges as needed.

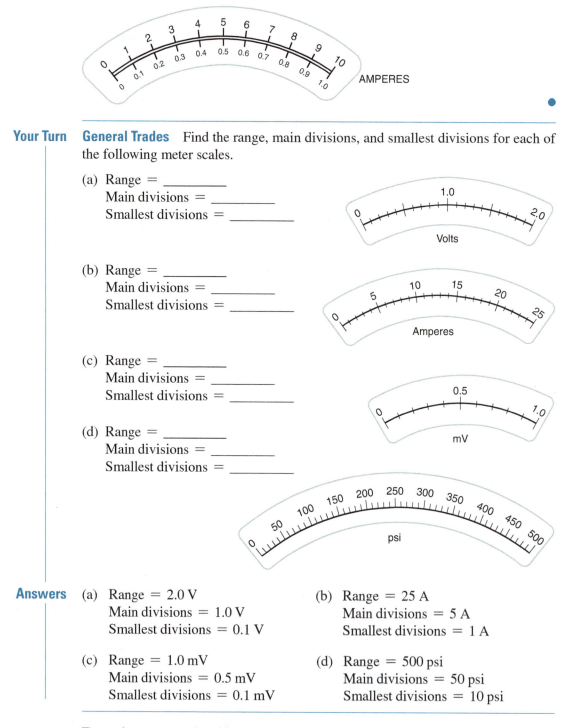

Your Turn　**General Trades**　Find the range, main divisions, and smallest divisions for each of the following meter scales.

(a)　Range = _____
　　Main divisions = _____
　　Smallest divisions = _____

(b)　Range = _____
　　Main divisions = _____
　　Smallest divisions = _____

(c)　Range = _____
　　Main divisions = _____
　　Smallest divisions = _____

(d)　Range = _____
　　Main divisions = _____
　　Smallest divisions = _____

Answers　(a)　Range = 2.0 V
　　　Main divisions = 1.0 V
　　　Smallest divisions = 0.1 V

(b)　Range = 25 A
　　Main divisions = 5 A
　　Smallest divisions = 1 A

(c)　Range = 1.0 mV
　　Main divisions = 0.5 mV
　　Smallest divisions = 0.1 mV

(d)　Range = 500 psi
　　Main divisions = 50 psi
　　Smallest divisions = 10 psi

To read a meter scale, either we can choose the small scale division nearest to the pointer, or we can estimate between scale markers.

EXAMPLE 10　**Electrical Trades**　Look at the following two meters:

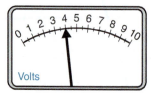

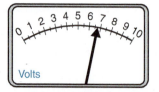

The one on the left reads exactly 4.0 volts, but the one on the right reads 6.5 volts to the nearest scale marker, or about 6.7 volts if we estimate between scale markers. The smallest division is 0.5 volt.

On the following meter scale

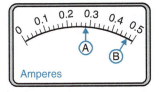

the smallest scale division is $\frac{1}{5}$ of 0.1 ampere, or 0.02 ampere. The reading at A is therefore 0.2 plus 4 small divisions, or 0.28 ampere. The reading at B is 0.4 plus $3\frac{1}{2}$ small divisions, or 0.47 ampere. ●

Your Turn   **General Trades**   Read the following meters. Estimate between scale divisions where necessary.

(a)

(b)

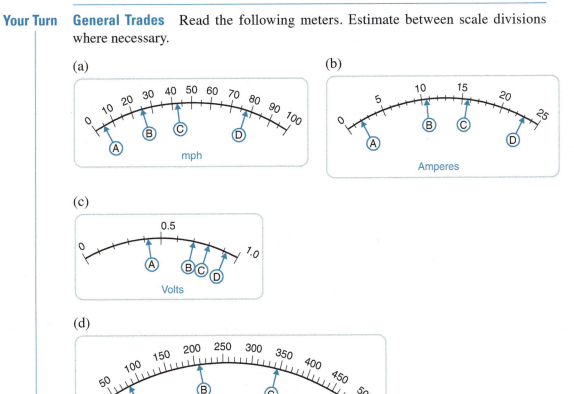

(c)

(d)

(e)

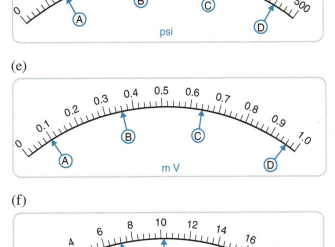

(f)

Answers   Note: Your estimated answers may differ slightly from ours.

(a) A. 5   B. 25   C. 43   D. 78 mph

(b) A. 2   B. 10.6   C. 15.8   D. 23 amperes

(c) A. 0.42   B. 0.7   C. 0.8   D. 0.93 volt

(d) A. 82   B. 207   C. 346   D. 458 psi

(e) A. 0.1   B. 0.36   C. 0.64   D. 0.96 mV

(f) A. 2.8   B. 7.2   C. 10.2   D. 15.8 volts

A Closer Look   Notice in meter (f) that the smallest division is $\frac{1}{5}$ of 2 volts, or 0.4 volt. If your answers differ slightly from ours, the difference is probably due to the difficulty of estimating the position of the arrows on the diagrams. ●

Meters with Non-uniform Scales   On a non-uniform scale, spacing between the divisions changes at various places along the scale, and the number of units represented by each small division also changes. The following ohmmeter scale is an example of a non-uniform scale:

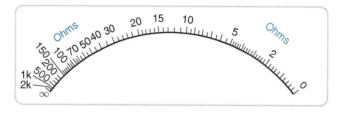

Notice that between 0 and 5 ohms (Ω), each one-ohm range contains five subdivisions, so that each of these small divisions represents 0.2 Ω. However, between 5 and 10 Ω, each one-ohm range contains just two subdivisions; therefore, each of these small divisions represents 0.5 Ω. As you move from right to left, the scale continues to compress, with each small division representing a larger number of units. The following table gives the value of the subdivisions for each part of the ohm scale.

Range on the Scale	Value of Each Subdivision
0–5 Ω	0.2 Ω
5–10 Ω	0.5 Ω
10–20 Ω	1 Ω
20–30 Ω	2 Ω
30–100 Ω	5 Ω
100–150 Ω	10 Ω
200–500 Ω	100 Ω

EXAMPLE 11   **Electrical Trades**

(a) On the ohmmeter in Figure (a), the needle points to the second small division between 3 and 4, so the reading is 3 Ω + 2(0.2Ω), or 3.4 Ω.

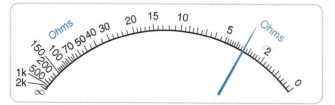

Figure (a)

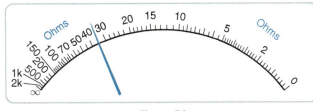

Figure (b)

(b) On the ohmmeter in Figure (b), the needle points to the only small division between 30 and 40, so the reading is 30 Ω + 5 Ω = 35 Ω. ●

Your Turn   **Electrical Trades**   Read the following ohmmeters.

(a)

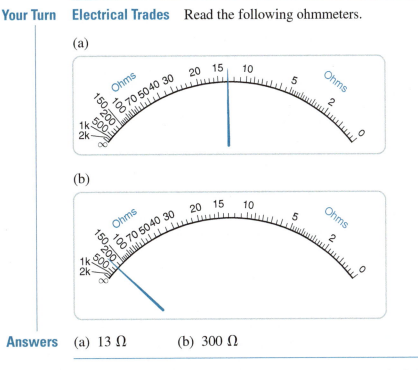

(b)

Answers   (a) 13 Ω   (b) 300 Ω

Meters with Circular Scales   Some meters have two or more circular dials, and each dial represents a different unit. To read these meters, you must pair the reading on each dial with its unit value and then combine the results into a single measurement.

The following water meter measures the number of cubic feet of water use. It contains six dials, each labeled with a power of ten ranging from 1 to 100,000. To determine the number of cubic feet, read the dials in descending order, beginning with the 100,000 dial. If the needle is between two numbers, always read the smaller one. If it's between 9 and 0, read the 9. Example 12 demonstrates a quick and easy way to write the number of cubic feet shown.

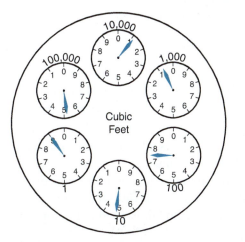

EXAMPLE 12   **General Trades**   To read the number of cubic feet indicated on the previous meter, **first,** create six blanks that will represent the six-digit number of cubic feet indicated.

___ ___ ___ , ___ ___ ___

The first blank represents the hundred-thousands place, the second blank represents the ten-thousands place, and so on.

Then, fill in each blank with the reading of the appropriate dial.

• The needle on the 100,000 dial is between 5 and 6, so we read it as 5 and fill in the hundred-thousands place of our number with a 5:

_5_ ___ ___ , ___ ___ ___

• The needle on the 10,000 dial is between 1 and 2, so we read it as 1 and fill in the ten-thousands place of our number with a 1:

_5_ _1_ ___ , ___ ___ ___

• The needle on the 1,000 dial is between 0 and 1, so we read it as 0 and fill in the thousands place with a 0:

_5_ _1_ _0_ , ___ ___ ___

• Continuing in this manner, we read the 100 dial as 7, the 10 dial as 4, and the 1 dial as 9. Filling in the remaining places in our six-digit number, we have:

_5_ _1_ _0_ , _7_ _4_ _9_

Finally, we see that the meter reads 510,749 cu ft.   ●

Your Turn   **General Trades**   Read the following water meters.

(a)   (b)

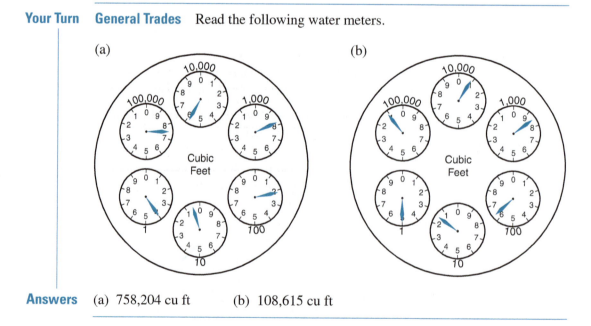

Answers   (a) 758,204 cu ft   (b) 108,615 cu ft

Many electric meters also contain circular dials representing the number of kilowatt-hours consumed. As with water meters, the dials represent powers of ten. On the following electric meter, the four dials represent the thousands, hundreds, tens, and ones digits of a four-digit number of kilowatt-hours. Read these meters using the same method illustrated in Example 12.

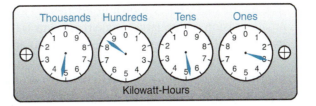

EXAMPLE 13   **General Trades**   To read the number of kilowatt-hours (kWh) shown on the preceding meter, **first,** draw four blanks representing the thousands place, the hundreds place, the tens place, and the ones place of a four-digit number: ___ ___ ___ ___

Next, read each dial as you did with water meters, again choosing the smaller number whenever the needle is between two numbers.

- The needle on the thousands dial is between 4 and 5, so we read it as 4 and write a 4 in the thousands place of our number:

   4 ___ ___ ___

- The needle on the hundreds dial is between 8 and 9, so we read it as 8 and write an 8 in the hundreds place of our number:

   4   8 ___ ___

- Continuing, we read 5 on the tens dial and 3 on the ones dial. We finish filling in our slots as follows:

   4   8   5   3

Finally, we read our meter as indicating 4853 kWh.   ●

Your Turn   **General Trades**   Read the following electric meters.

(a)

(b)

Answers   (a) 7021 kWh   (b) 3992 kWh

Dial gauges, also known as **dial indicators,** are used to measure very small distances. They are useful to detect a difference in height along a flat surface, to center cylindrical stock, to monitor the depth of drilled holes, to determine if the edges of a rectangular surface are parallel, and to perform many other precise measurements in machine tool technology, manufacturing, woodworking, designing, and automotive technology.

The metric dial gauges shown in Figures (c) and (d) have two circular scales. Each small division on the larger scale represents 0.01 mm. Each number on the smaller scale indicates the number of complete revolutions made by the larger needle, and each revolution represents 1.00 mm. The larger needle rotates either clockwise (to the right) or counterclockwise (to the left). A clockwise rotation is considered to be a positive deflection of the needle, indicating that the object being measured is larger than a standard or desired measurement. A counterclockwise rotation is considered to be a negative deflection of the needle, indicating that the object being measured is smaller than a standard or desired measurement. The arrows on the upper portion of the larger scales have been drawn in to indicate in which direction the needle deflected.

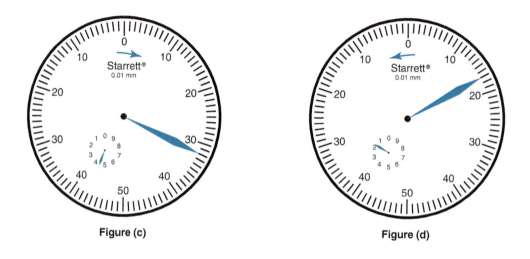

Figure (c)   Figure (d)

Note   When using an actual dial gauge, you must watch the needle to determine whether the deflection is positive or negative. ●

EXAMPLE 14   **General Trades**

(a) To read the dial gauge in Figure (c):

First, notice that the arrow at the top of the dial points to the right, indicating a positive deflection. This means that the measurement will be a positive number and that the larger dial will be read in a clockwise direction.

Next, notice that the needle on the smaller scale is between 4 and 5. This indicates 4 complete revolutions of the larger needle, which represents 4×1.00 mm, or 4.00 mm.

Then, because the deflection is positive, we read the larger scale in a clockwise direction. The larger needle therefore reads 32, representing 32×0.01 mm, or 0.32 mm.

Finally, we add this to the 4.00 mm indicated on the smaller scale and attach a plus sign to indicate a positive deflection. The resulting measurement is +4.32 mm.

(b) To read the dial gauge in Figure (d):

First, notice that the arrow at the top of the dial points to the left, indicating a negative deflection. This means that the measurement will be a negative number and that the larger dial will be read in a counterclockwise direction.

Next, notice that the needle on the smaller scale is between 1 and 2. This indicates one complete revolution of the larger needle, giving us 1.00 mm so far.

Then, because the deflection is negative, we read the larger scale in a counter-clockwise direction. This means that, as we move past 50, the number 40 actually represents 60×0.01 mm, or 0.60 mm; the number 30 represents 70×0.01 mm, or 0.70 mm; and so on. Thus, the needle on this dial is indicating 83×0.01 mm, or 0.83 mm.

Finally, we add this to the 1.00 mm from the previous step and attach a minus sign to indicate a negative deflection. The resulting measurement is -1.83 mm. ●

Your Turn   **General Trades**   Read the following dial gauges.

(a)   (b)

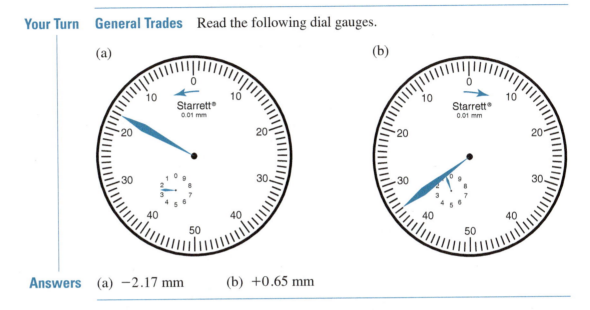

Answers   (a) -2.17 mm   (b) $+0.65$ mm

A U.S. customary dial gauge measures distances in inches. As with metric dial gauges, the smaller dial records the number of complete revolutions of the larger dial, with each revolution representing 0.100 in., or one-tenth of an inch. Each small division on the larger scale represents 0.001 in., or one-thousandth of an inch. As with the metric gauges, a clockwise rotation represents a positive deflection [Figure (e)], and a counterclockwise rotation represents a negative deflection [Figure (f)].

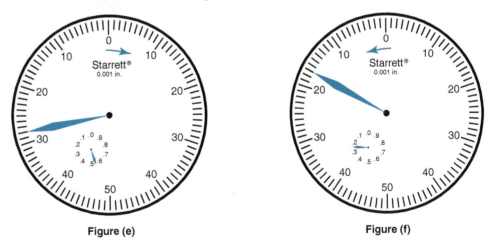

Figure (e)   Figure (f)

EXAMPLE 15   **General Trades**

(a) To read the dial gauge in Figure (e):

Step 1   Notice that the arrow at the top points to the right, indicating a positive deflection.

Step 2   The needle on the smaller dial is between 5 and 6, indicating 5 complete rotations of the larger dial. This gives us a reading of 5×0.100 in., or 0.500 in.

Step 3   With a positive deflection, we read the larger scale in a clockwise direction, so the larger needle gives us a reading of 72 × 0.001 in., or 0.072 in.

Step 4   Adding the two measurements together and attaching a plus sign, we have a final reading of +0.572 in.

(b) To read the dial gauge in Figure (f):

Step 1   Notice that the arrow at the top points to the left, indicating a negative deflection.

Step 2   The needle on the smaller dial is between 2 and 3, indicating 2 complete rotations of the larger dial. This gives us a reading of 2 × 0.100 in., or 0.200 in.

Step 3   With a negative deflection, we read the larger scale in a counterclockwise direction, so the larger needle gives us a reading of 17 × 0.001 in., or 0.017 in.

Step 4   Adding the two measurements together and attaching a negative sign, we obtain a final reading of −0.217 in.   ●

Your Turn   General Trades   Read the following U.S. customary dial gauges.

(a)   (b)

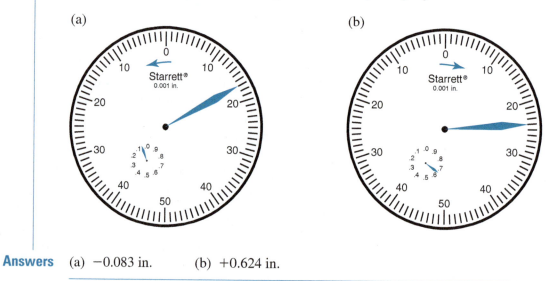

Answers   (a) −0.083 in.   (b) +0.624 in.

📝 **Note**   Many other types of dial gauges that differ in some way from the ones shown here are used in the trades. Some measure distances to a different degree of precision than the gauges pictured in our examples. Also, some have scales that are numbered in a clockwise direction from 0 to 100 rather than from 0 to 50 in both directions.   ●

More Practice   General Trades   Read the following meters and gauges.

(a)

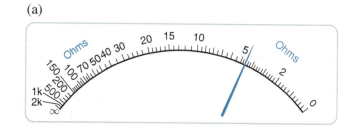

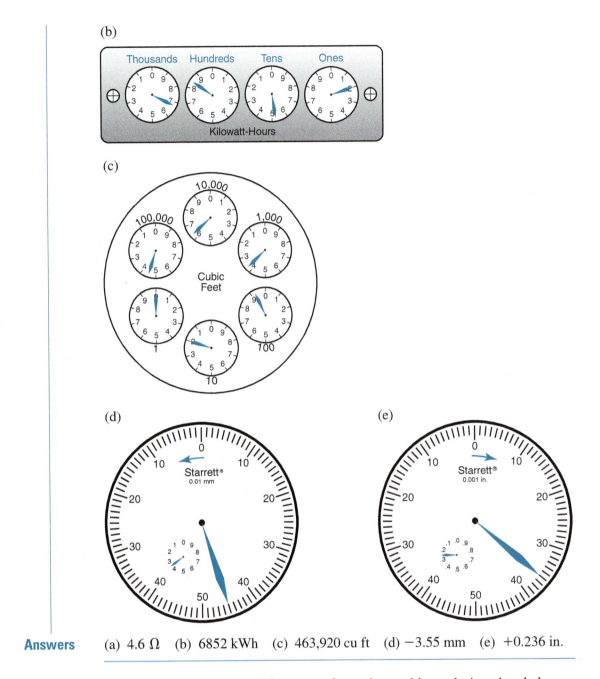

(b)

(c)

(d)

(e)

Answers   (a) 4.6 Ω   (b) 6852 kWh   (c) 463,920 cu ft   (d) −3.55 mm   (e) +0.236 in.

Now turn to Exercises 5-5 for a set of practice problems designed to help you become more proficient in making direct measurements.

Exercises 5-5 | Direct Measurements

A. Find the lengths marked on the following rules.

1.

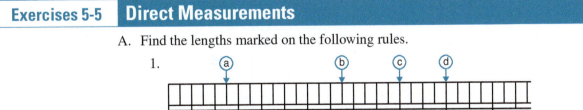

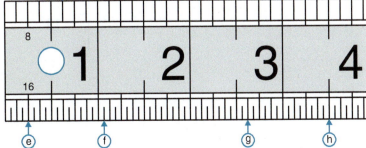

2.

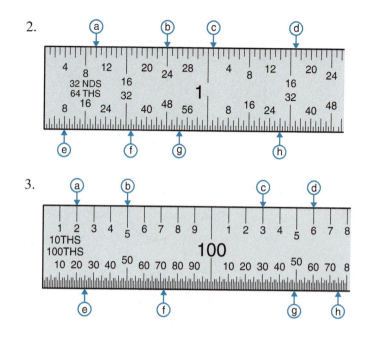

3.

B. Read the following micrometers.

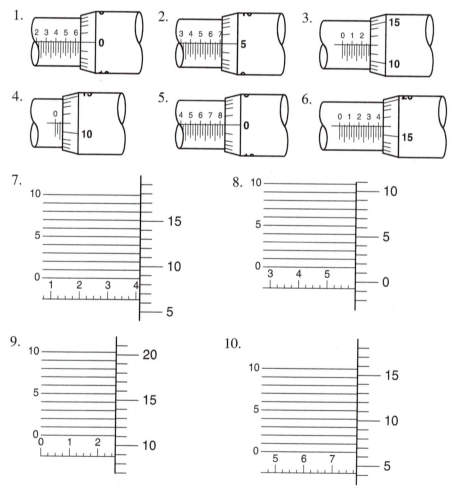

11.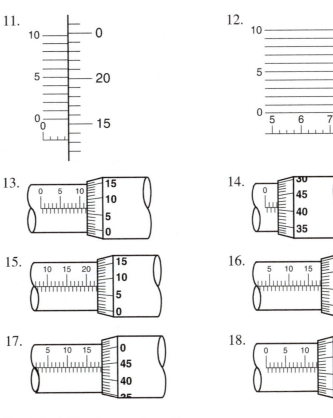

12.

13.

14.

15.

16.

17.

18.

C. Read the following vernier calipers.

1.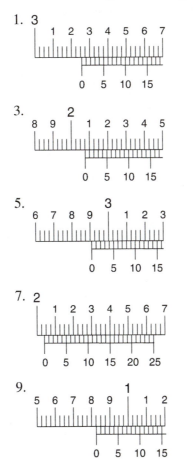

2.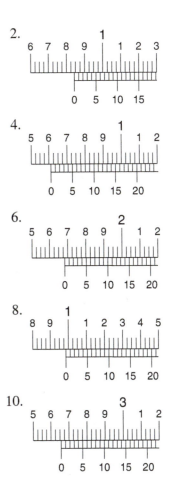

3.

4.

5.

6.

7.

8.

9.

10.

D. Read the following vernier protractors.

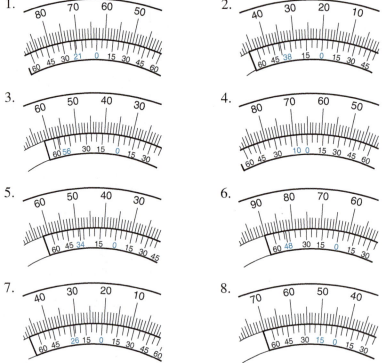

E. Read the following meters. Estimate between scale divisions where necessary. Be sure to include units with your answers.

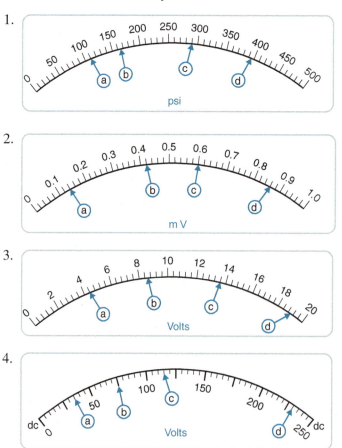

5.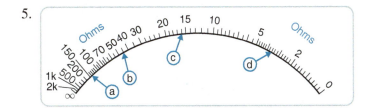

F. Read the following water meters, electric meters, and dial gauges.

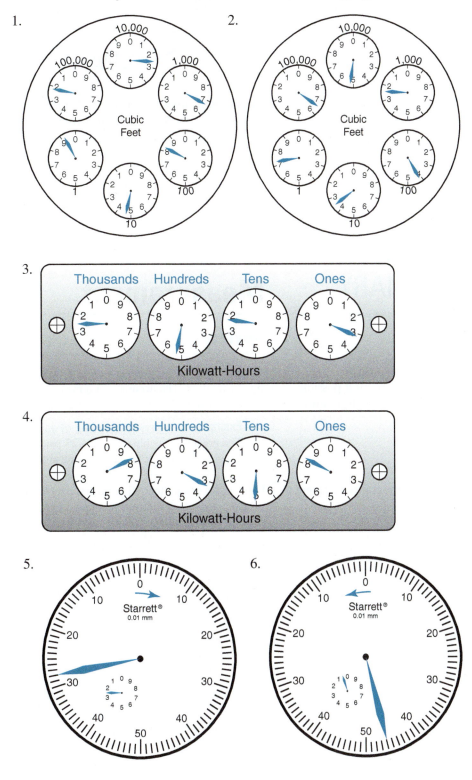

7.   8.

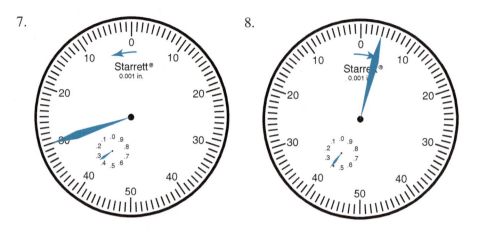

Check your answers to the odd-numbered problems in the Appendix, then turn to Problem Set 5 on page 368 for practice on working with measurement numbers. If you need a quick review of the topics in this chapter, visit the chapter Summary first.

CHAPTER 5
SUMMARY

Measurement

Objective	Review
Determine the precision, accuracy, and greatest possible error of measurement numbers. (p. 282)	The precision of a measurement number is indicated by the place value of its right-most significant digit. The accuracy of a measurement number refers to the number of significant digits it contains. The greatest possible error of a measurement number is half of its precision. **Example:** A pressure reading of 4.27 psi is precise to the nearest hundredth. The greatest possible error of 4.27 psi is 0.005 psi. The measurement of a mass given as 6000 kg is accurate to one significant digit.
Add and subtract measurement numbers. (p. 284)	**First,** make certain that all numbers have the same units. **Then,** add or subtract the numerical parts and attach the common units. **Finally,** round the answer to agree in precision with the least precise number in the problem. **Example:** 38.26 sec + 5.9 sec = 44.16 sec = 44.2 sec, rounded The final answer has been rounded to one decimal digit to agree with 5.9 sec, the least precise number in the sum.
Multiply and divide measurement numbers. (p. 286)	Multiply or divide the given numbers, then round the answer to the same number of significant digits as the least accurate number used in the calculation. The units of the answer will be the product or quotient of the units in the original problem. **Example:** 16.2 ft × 5.8 ft = 93.96 sq ft = 94 sq ft, rounded The final answer is rounded to two significant digits to agree with 5.8 ft, the least accurate of the original factors. Note that ft × ft = sq ft.

Objective	Review
Convert units: (a) within the U.S. customary system (p. 295), (b) within the metric system (p. 315), and (c) between the U.S. customary and metric systems (p. 326)	Use the basic conversion factors given in the text, along with the unity fraction method, for all conversions. Round your answer to the same number of significant digits as the original number. **Example:** (a) $65.0 \text{ lb} = (65.0 \text{ lb}) \times \left(\dfrac{16 \text{ oz}}{1 \text{ lb}} \right) = 1040 \text{ oz}$ (b) $0.45 \text{ km} = (0.45 \text{ km}) \times \left(\dfrac{1000 \text{ m}}{1 \text{ km}} \right) = 450 \text{ m}$ (c) $45 \text{ mph} = \dfrac{45 \text{ mi}}{1 \text{ hr}} \times \dfrac{1.6093 \text{ km}}{1 \text{ mi}} = 72.4185 \text{ km/h} \approx 72 \text{ km/h}$ The final answer has been rounded to two significant digits to agree in accuracy with the original speed, 45 mph.
Think metric. (p. 314)	Use appropriate conversion factors to perform U.S. customary to metric or metric to U.S. customary conversions mentally. **Example:** Estimate the weight, in pounds, of a casting weighing 45 kg. Remember that a kilogram is about 10% more than 2 lb. Our first estimate is that 45 kg is about 2×45 or 90 lb. For a better estimate add 10% to 90: $90 + 9 = 99$, or about 100 lb.
Use common technical measuring instruments. (p. 338)	Correctly read a length rule, micrometer, vernier caliper, protractor, various meters, and dial gauges.

PROBLEM SET 5   Measurement

Answers to odd-numbered problems are given in the Appendix.

A. State (a) the accuracy, (b) the precision, and (c) the greatest possible error of each of the following measurement numbers.

1. 8.3 lb   2. 960 ft   3. 3.50 in.   4. 8775 gal

5. 0.04 kg   6. 17 psi   7. 4000 sq yd   8. 1.375 liters

B. Perform the following calculations with measurement numbers. Round to the correct precision or accuracy.

1. 5.75 sec − 2.3 sec

2. 3.1 ft × 2.2 ft

3. 4.56 sq in. ÷ 6.1 in.

4. 3.2 cm × 1.26 cm

5. $7\frac{1}{2}$ in. + $6\frac{1}{4}$ in. + $3\frac{1}{8}$ in.

6. 9 lb 6 oz − 5 lb 8 oz

7. 765 mi ÷ 36.2 gal

8. 1.5 cm + 2.38 cm + 5.8 cm

9. 34.38 psi − 16.732 psi

10. $7\frac{1}{4}$ ft − 4.1 ft

11. 7.064 in. − 3.19 in.

12. 22.0 psi × 18.35 sq in.

13. 8 ft 7 in. + 11 ft 11 in. + 9 in.

14. 2.6 in. × 3.85 in. × 14 in.

C. Find the closest fractional equivalent to each of the following decimal measurements using the given denominator. Then calculate the error if the fraction is used in place of the decimal.

1. 0.628 in. (16ths)   2. 3.902 in. (16ths)   3. 2.44 in. (32nds)

4. 0.179 in. (32nds)   5. 0.588 in. (64ths)   6. 1.31 in. (64ths)

D. Convert the following measurement numbers to new units as shown. Round all units marked with an asterisk (*) to the nearest whole number. Otherwise, round to the nearest tenth if necessary.

1. 6.25 in. = _____ cm

2. 54.0 mph = _____ ft/sec

3. 22°C = _____ °F*

4. 2.6 L = _____ qt

5. 0.82 kg = _____ g

6. 1.62 cm = _____ mm

7. 0.265 cu ft = _____ cu in.*

8. 41°F = _____ °C*

9. 14.3 kg = _____ lb

10. 65.2 cm = _____ in.

11. 24.0 km/h = _____ cm/sec*

12. 127,000 sq m = _____ ha

13. $14\frac{1}{2}$ mi = _____ km

14. 356.0 cu in. = _____ cu cm*

15. $14\frac{1}{3}$ cu yd = _____ cu ft

16. 165 sq ft = _____ sq m

17. 1400 mL = _____ L

18. 5652 m/min = _____ km/h

19. 24,700 cu in. = _____ bu

20. 6.9 L = _____ gal

21. 6.875 sq ft = _____ sq in.

22. $12\frac{1}{4}$ qt = _____ L

23. 0.095 g = _____ mg

24. 596,000 m = _____ km

25. 3.40 ft = _____ cm*

26. 15.2 in./sec = _____ cm/sec

27. 1.85 L = _____ cu cm

28. 4,000,000 mg = _____ g

Name

_____

Date

_____

Course/Section

_____

29. 65.0 km/h = _____ mph   30. 28.8 lb = _____ kg

31. 1.5 qt = _____ fl oz   32. 525 mL = _____ fl oz

33. 12.4 m = _____ ft   34. 425.0 yd = _____ m

35. 465 g = _____ oz   36. 26.8 km = _____ mi

37. 560 cm = _____ m   38. 6.28 metric tons = _____ kg

39. 0.85 L = _____ mL   40. 27.5 oz = _____ g*

41. 1200 μL = _____ mL   42. 94 mm = _____ cm

43. 1.875 metric tons = _____ lb*   44. 16.0 gal = _____ L

45. 1.85 km = _____ m   46. 4.45 fl oz = _____ mL*

47. 8600 ft = _____ mi   48. 12.5 ft = _____ in.

49. 3.75 lb = _____ oz   50. 2.8 tons = _____ lb

51. 280 rpm = _____ rev/sec   52. 97.5 ft = _____ yd

E.   Think metric. Choose the closest estimate.

1. Weight of a pencil   (a) 4 kg   (b) 4 g   (c) 40 g

2. Size of a house   (a) 6 sq m   (b) 200 sq m   (c) 1200 sq m

3. Winter in New York   (a) −8°C   (b) −40°C   (d) 20°C

4. Volume of a 30-gal trash can (a) 0.1 cu m   (b) 2.0 cu m   (c) 5.0 cu m

5. Height of an oak tree   (a) 500 mm   (b) 500 cm   (c) 50 km

6. Capacity of a 50-gal water   (a) 25 liters (b) 100 liters (c) 180 liters
   heater

7. Distance you can walk   (a) 500 cm   (b) 50 m   (c) 5 km
   in an hour

8. Highway speed of a car   (a) 8 km/hr   (b) 80 km/hr   (c) 800 km/hr

F.   Read the following measuring devices.

1. Rulers.

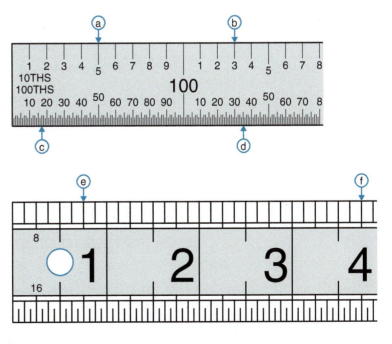

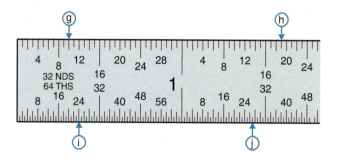

2. Micrometers. Problems (a) and (b) are U.S. customary, (c) and (d) are metric, and (e) and (f) are U.S. customary with a vernier scale.

(a)

(b)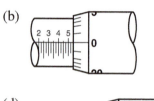

(c)

(d)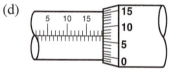

(e)

(f)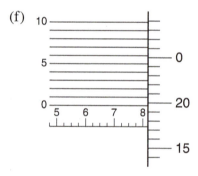

3. Vernier calipers.

(a)

(b)

(c)

(d)

(e)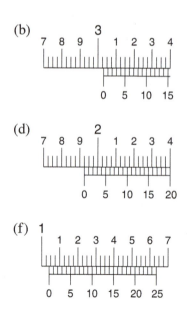

(f)

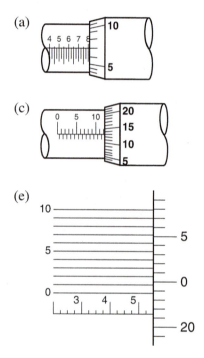

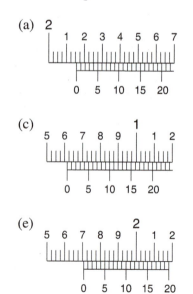

4. Meters.

(a)

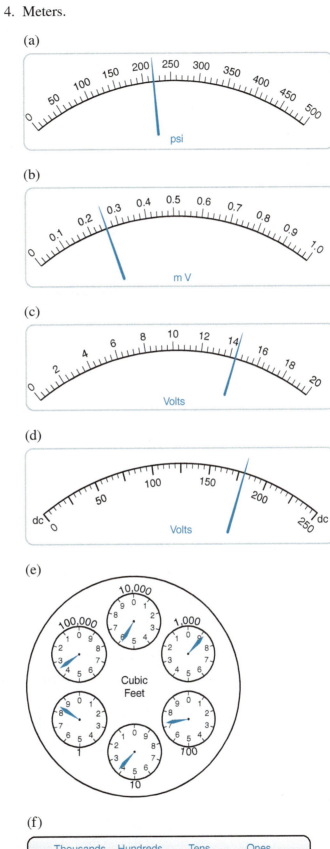

(b)

(c)

(d)

(e)

(f)

5. Dial gauges.

(a)

(b)

(c)

(d)

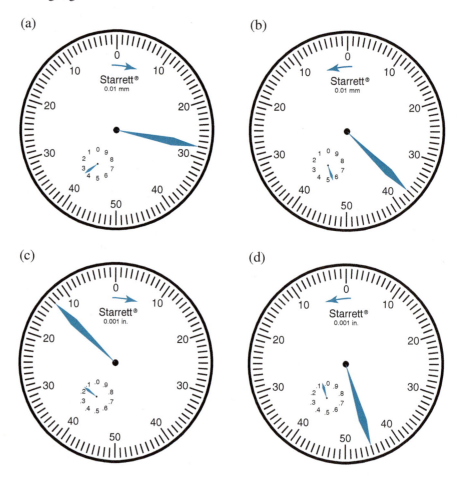

G. **Practical Applications. If necessary, round to the nearest tenth unless otherwise directed.**

1. **Machine Trades**   A piece $6\frac{1}{16}$ in. long is cut from a steel bar $28\frac{5}{8}$ in. long. How much is left?

2. **Life Skills**   An American traveling in Italy notices a road sign saying 85 km/h. What is this speed in miles per hour rounded to the nearest whole number?

3. **Automotive Trades**   An automotive technician testing cars finds that a certain bumper will prevent damage up to 9.0 ft/sec. Convert this to miles per hour.

4. **Construction**   Marble weighs approximately 170 pounds per cubic foot. How many tons are needed for a job requiring 1460 cu ft of marble? (Round to the nearest ton.)

5. **Carpentry**   What size wrench, to the nearest 32nd of an inch, will fit a 0.455-in. bolt head?

6. **Metalworking**   The melting point of a casting alloy is 260°C. What temperature is this on the Fahrenheit scale?

7. **Sheet Metal Trades**   What area of sheet metal in square inches is needed to construct a vent with a surface area of 6.50 sq ft?

8. **Life Skills**   How many gallons of gas are needed to drive a car exactly 350 miles if the car normally averages 18.5 mi/gal?

9. **Metalworking**   A certain alloy is heated to a temperature of 560°F. What Celsius temperature does this correspond to? (Round to the nearest degree.)

10. **Automotive Trades**   A Nissan 370Z has a 3.7-L engine. What would this be in cubic inches? (Round to the nearest cubic inch.)

11. **Sports and Leisure**   The radar gun at a baseball game measures the speed of a pitcher's fastball as 92.6 mph. What is the greatest possible error of this measurement converted to feet per second? (Round to three decimal digits.)

12. **Plumbing**   The Uniform Plumbing Code is being converted to the metric system. Convert the following measurements and round to two decimal digits.
    (a) The amount of gas used by a furnace (136 cu ft/hr) to cubic meters per hour
    (b) The length of a gas line (25.0 ft) to meters
    (c) The diameter of a water pipe (2.50 in.) to centimeters

13. **Metalworking**   Wrought iron has a density of 480 lb/cu ft. Convert this to pounds per cubic inch rounded to the nearest hundredth.

14. **Hydrology**   A cubic foot of water contains 7.48 gal. What volume is this in liters?

15. **Carpentry**   What are the dimensions of a metric "2-by-4" (a piece of wood $1\frac{1}{2}$ in. by $3\frac{1}{2}$ in.) in centimeters?

16. **Architecture**   A common scale used by architects and drafters in making blueprints is $\frac{1}{4}$ in. = 1 ft. What actual length in centimeters would correspond to a drawing dimension of $1\frac{1}{4}$ in.?

17. **Forestry**   The world's largest living organism is a fungus known as the Honey Mushroom, growing in the Malheur National Forest in Oregon. The fungus covers approximately 2200 acres. How many square miles is this?

18. **Automotive Trades**   A mechanic spends 45 min on an oil change. Convert this to hours and determine the labor charge given a rate of $62 per hour. Do not round the answer.

19. **Automotive Trades**   The circumference of each tire on a motor vehicle is approximately 81.7 in. If the tires make 10,500 revolutions, how many miles will the vehicle travel?

20. **Automotive Trades**   An automobile tire is designated as P205/65R15. Find the overall diameter of the tire in inches.

21. **Automotive Trades**   An auto repair shop allots 1.2 hr to replace a fan belt. The mechanic actually spends 1 hr 20 min doing the job. By how many minutes did the mechanic exceed the allotted time?

22. **Automotive Trades**   A customer buying a new car has narrowed her decision down to two finalists: a hybrid selling for $30,000 and an economical nonhybrid selling for $25,000. The average fuel economy of the hybrid is 48 mi/gal, compared to 30 mi/gal for the nonhybrid. By purchasing the hybrid, the customer will receive a one-time tax credit of $1500 and will save about $500 over the life of the vehicle on discounted auto insurance. Assume that the customer will keep each car for 120,000 miles and that the cost of fuel will average $2.60 per gallon over that time period.
    (a) What will be the total cost of fuel for the nonhybrid for 120,000 miles?
    (b) To find the comparative cost of the nonhybrid, add the answer to part (a) to the purchase price.

(c) What is the total cost of fuel for the hybrid for 120,000 miles?

(d) To find the comparative cost of the hybrid, first add the answer to part (c) to the purchase price. Then subtract the tax credit and the insurance savings.

(e) If total cost were the only consideration, which vehicle would the customer choose?

23. **Construction**   A remodeling job requires the following quantities of lumber: 36 boards, 2 in. by 10 in., each 8 ft long; and 48 boards, 2 in. by 4 in., each 12 ft long. What is the total number of board feet required?

24. **Hydrology**   The average daily evaporation from Bradbury Dam at Lake Cachuma during a particular week in the summer was 63.6 acre-feet. To the nearest thousand, how many gallons is this? (See Exercises 5-2, problem C-27.)

25. **Automotive Trades**   Europeans commonly express gas mileage in "liters per 100 km" (L/100 km), whereas Americans express mileage in miles per gallon (mi/gal). If a European car has a mileage rating of 6.5 L per 100 km, how would this be expressed in the United States? (Round to the nearest whole number.)

26. **Automotive Trades**   Tire pressure is expressed in pounds per square inch (psi) in the United States and in kiloPascals (kPa) in countries using the metric system. What is the pressure in pounds per square inch of a tire with a pressure of 220 kPa? (Use 1 psi = 6.895 kPa)

27. **Sports and Leisure**   Kenenisa Bekele (Ethiopia) holds the world record for both the men's 5000-m run and the 10,000-m run. His record time for 5000 m is 12 min 37.35 sec. His record time for 10,000 m is 26 min 17.53 sec. To solve the following problems, convert his times from minutes and seconds to only seconds, and round each answer to two decimal places.

(a) What was Bekele's average speed (in meters per second) during his record-breaking 5000-m run?

(b) What was Bekele's average speed (in meters per second) during his record-breaking 10,000-m run?

(c) If Bekele could maintain his average 10,000-m speed for an entire marathon (42,195 m), what would his total time be for the marathon? (Express the final answer in hours, minutes, seconds.)

28. **Allied Health**   If a patient must take two tablespoons of a liquid medication twice a day, how many days will a 600-mL supply last?

29. **Manufacturing**   A conveyor belt 12 in. wide is moving at a linear speed of 120 ft/min. The belt is moving boxes that are 6 in. wide and 24 in. long. Using the full width of the belt, how many boxes can be unloaded off the belt in 8 hours of operation? (Assume that the boxes are placed end to end with no space between them.)

30. **Electronics**   In computer technology, the **bit rate** can be determined by multiplying the number of bits per symbol by the **baud rate**, which is the number of symbols per second. If there are 8 bits per symbol, and the baud rate is 4000 symbols per second, find the bit rate and label your answer with the proper units.

31. **Allied Health**   A doctor prescribes a medication for a patient at a dosage of 25 milligrams per kilogram of body weight. If the patient weighs 172 pounds, how many milligrams per dose should he be given? (Round to the nearest milligram.)

32. **Water/Wastewater Treatment** The level of a particular contaminant is given as 525 µg/L. Express this in units of parts per million.

33. **Metalworking** The density of copper is 8940 kg/cu m. Express this in pounds per cubic foot rounded to the nearest whole number.

34. **Allied Health** A certain medication is available only in 750-µg tablets. If a patient needs to take 3.0 mg per day, how many tablets should she take?

35. **Aviation** The cruising speed of a Cessna TR182 is approximately 232 km/hr. What is this speed in miles per hour?

36. **Allied Health** An adult weighing 185 lb is on a diet of 3500 calories per day. He wishes to limit his intake of saturated fat to no more than 10% of his total calories. Fat contains 9 calories per gram. What is the daily maximum number of grams of saturated fat that he should consume? (Round down to the nearest whole number.)

37. **Machine Trades** The length of a steel part is given as 28.4 mm. What is the greatest possible error in centimeters?

38. **Allied Health** A 540-mL IV solution must be given to a patient over a period of 3 hours. If the drop factor for this solution is 10 drops per milliliter, what should the flow rate be in drops per minute?

39. **Agriculture** A European country has a wheat yield of 7.25 metric tons per hectare. How would this compare to a yield in the U.S. of 116 bushels per acre? (*Remember:* One bushel of wheat weighs 60 pounds.)

CASE STUDY: Mr. Ferris's Wheel

In the summer of 1893, the first Ferris wheel was displayed at the World's Fair Columbian Exposition in Chicago, Illinois. The Columbian Exposition was in commemoration of the 400 years since the European discovery of the Americas. The structure was designed by George Washington Gale Ferris, Jr., a 32-year-old Pittsburgh steel engineer. The Ferris wheel was designed to be a showstopper like the Eiffel Tower was at the 1889 Exposition in Paris. Because World's Fairs and Expos are international events, those involved in the design, organization, and setup of the various components must be proficient in both the metric and the U.S. customary systems of measurement.

Mr. Ferris's wheel was 80 meters tall, had a diameter of 76.2 meters, and weighed 2,079,884 pounds when empty. The main axle alone weighed 71 tons. The wheel had 36 passenger cars mounted on the rim, and each car could hold up to 60 riders. Each ride consisted of two revolutions and lasted 20 minutes. During the first revolution, the cars were unloaded and then filled with new passengers. Once loaded, the second full, uninterrupted revolution lasted 9 minutes.

The Ferris wheel operated from June 21 through November 6 for up to 16 hours every day. Many times the wheel was at full capacity every ride. A total of $726, 805 was collected in ticket sales from 1,453,611 riders.

Since that time, hundreds of Ferris wheels, or "observation wheels," have been built around the world. The current recordholder as the world's largest Ferris wheel is the High Roller in Las Vegas, with a diameter of 550 feet.

continued...

1. What was the cost of a ticket to ride the original Ferris wheel?

2. Compare the diameter of the Columbian Exposition Ferris wheel to that of the High Roller. What is the difference in length between the two diameters in feet?

3. Express the weight of the main axle of the 1893 Ferris wheel in kilograms.

4. What was the maximum number of riders the original Ferris wheel could hold at any one time?

5. If each rider weighed an average of 150 pounds, what was the total weight (riders and structure) of the original Ferris wheel at full capacity? Express your answer in both (a) pounds and (b) kilograms.

6. a. What was the maximum daily capacity of the 1893 Ferris wheel during a 16-hour day?
    b. What was the actual average number of riders per day? (*Hint:* Use a calendar to count the number of days from June 21 through November 6.)
    c. What percent of the maximum capacity was the actual average capacity?

7. One revolution of the Columbian Exposition Ferris wheel covered a circular distance or circumference of 239.4 meters. Calculate the speed of the wheel during an uninterrupted revolution. Express your answer in both (a) meters per minute and (b) miles per hour.

8. The average Ferris wheel rotates at 6.9 miles per hour. What circular distance, in feet, does the average Ferris wheel cover in a 5-minute ride?

9. Today, a handicap access ramp is required for loading many Ferris wheels. Wheelchair ramps must have a slope ratio of rise to run of 1:12. If the loading height to a standard Ferris wheel is 3 feet, what is the length, in meters, of the horizontal base of the loading ramp?

References

Malanowski, J. (2015). The brief history of the Ferris wheel. Retrieved from http://www.smithsonianmag. com/history/history-ferris-wheel-180955300/

Meehan, P. (2015). Hyde Park Historical Society. Retrieved from http://www.hydeparkhistory. org/2015/04/27/ferris-wheel-in-the-1893-chicago-worlds-fair/

Objective	Sample problems		For help, go to page
When you finish this chapter, you will be able to:			
1. Understand the meaning of signed numbers.	(a) Which is smaller, -12 or -14?	_____	380
	(b) Mark the following points on a number line: $-5\frac{1}{4}, +1.0, -0.75, +3\frac{1}{2}$.	_____	
	(c) Represent a loss of $250 as a signed number.	_____	
2. Add signed numbers.	(a) $(-3) + (-5)$	_____	381
	(b) $18\frac{1}{2} + \left(-6\frac{3}{4}\right)$	_____	
	(c) $-21.7 + 14.2$	_____	
	(d) $(-3) + 15 + 8 + (-20) + 6 + (-9)$	_____	
3. Subtract signed numbers.	(a) $4 - 7$	_____	390
	(b) $-11 - (-3)$	_____	
	(c) $6.2 - (-4.7)$	_____	
	(d) $-\frac{5}{8} - 1\frac{1}{4}$	_____	

Name _____

Date _____

Course/Section _____

Objective	Sample problems	For help, go to page
4. Multiply and divide signed numbers.	(a) $(-12) \times 6$	396
	(b) $(-54) \div (-9)$	
	(c) $(-4.73)(-6.5)$	
	(d) $\frac{3}{8} \div \left(-2\frac{1}{2}\right)$	
	(e) $60 \div (-0.4) \times (-10)$	
5. Work with exponents.	(a) 4^3	401
	(b) $(2.05)^2$	
6. Use the order of operations.	(a) $2 + 3^2$	404
	(b) $(4 \times 5)^2 + 4 \times 5^2$	
	(c) $6 + 8 \times 2^3 \div 4$	
7. Find square roots.	(a) $\sqrt{169}$	406
	(b) $\sqrt{14.5}$ (to two decimal places)	

(Answers to these preview problems are given in the Appendix. Worked solutions to many of these problems appear in the chapter Summary.)

If you are certain that you can work *all* these problems correctly, turn to page 414 for a set of practice problems. If you cannot work one or more of the preview problems, turn to the page indicated to the right of the problem. Those who wish to master this material should turn to Section 6-1 and begin work there.

Pre-Algebra

Before you can master basic algebra, you must become familiar with signed numbers, exponents, and square roots. In this chapter, we will cover these topics as well as the expanded order of operations.

CASE STUDY: Making Accurate Job Estimates

The Case Study at the end of this chapter illustrates how signed numbers can be used in **trades management**. In this particular application, the owner of a classic car body shop is refurbishing a 1960 T-Bird. She decides to use positive and negative numbers to assess the accuracy of her estimates and to try to avoid losing money on future jobs.

6-1   Addition of Signed Numbers

The Meaning of Signed Numbers

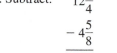
What kind of number would you use to name each of the following: a golf score two strokes below par, a $2 loss in a poker game, a debt of $2, a loss of 2 yards on a football play, a temperature 2 degrees below zero?

The answer is that they are all *negative numbers,* all equal to −2.

We can better understand the meaning of negative numbers if we can picture them on a number line. First we will display the familiar whole numbers as points along a number line:

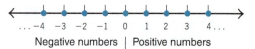

Whole numbers: 0, 1, 2, 3, 4, . . .

Notice that the numbers become greater as you move to the right along the number line, and they become lesser as you move to the left. The numbers to the *right* of zero, *greater* than zero, are considered to be **positive numbers**. The **negative numbers** are all the numbers *less than* zero, and we must extend the number line to the left in order to picture them:

$$\ldots -4 \quad -3 \quad -2 \quad -1 \quad 0 \quad 1 \quad 2 \quad 3 \quad 4 \ldots$$

Negative numbers | Positive numbers

Think of the negative part of the number line as a reflection of the positive part through an imaginary mirror placed through zero. This means that ordering the negative numbers is the opposite of ordering the positive numbers.

To indicate the ordering or relative size of numbers, we often use the following symbols:

The symbol $>$ means "greater than." For example, $a > b$ signifies that "a is greater than b."

The symbol $<$ means "less than." For example, $c < d$ signifies that "c is less than d."

EXAMPLE 1   (a) On the number line, 4 is to *the right* of 1 so that 4 is *greater than* 1, or $4 > 1$.

(b) On the negative side, −4 is to *the left* of −1 so that −4 is *less than* −1, or $-4 < -1$.   ●

Your Turn   Make each statement true by replacing the ☐ with either the > or the < symbol.

(a) $-5 \ \square \ -3$ (b) $9 \ \square \ 6$ (c) $-15 \ \square \ -22$

(d) $12 \ \square \ 31$ (e) $-8 \ \square \ 2$ (f) $43 \ \square \ -6$

Answers   (a) $-5 < -3$ (b) $9 > 6$ (c) $-15 > -22$

(d) $12 < 31$ (e) $-8 < 2$ (f) $43 > -6$

 Note   In algebra, we also use the symbol $\geq$ to signify "greater than or equal to," and the symbol $\leq$ to signify "less than or equal to."   ●

To help in solving the preceding problems, you may extend the number line in both directions. This is perfectly legitimate because the number line is infinitely long. The numbers we have marked, . . . −4, −3, −2, −1, 0, 1, 2, 3, 4, . . . , are called **integers**. The positions of fractions and decimals, both positive and negative, can be located between the integers.

Your Turn   Try marking the following points on a number line.

(a) $-3\frac{1}{2}$ (b) $\frac{1}{4}$ (c) -2.4 (d) 1.75 (e) -0.6

Answers

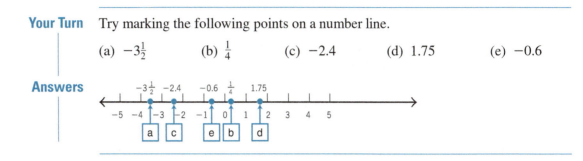

When we work with both positive and negative numbers and need to use names to distinguish them from each other, we often refer to them as **signed numbers**. Signed numbers are a bookkeeping concept. They were used in bookkeeping by the ancient Greeks, Chinese, and Hindus more than 2000 years ago. Chinese merchants wrote positive numbers in black and negative numbers in red in their account books. (Being "in the red" still means losing money or having a negative income.) The Hindus used a dot or circle to show that a number was negative. We use a minus sign (−) to show that a number is negative.

For emphasis, the "+" symbol is sometimes used to denote a positive number, although the absence of any symbol automatically implies that a number is positive. Because the + and − symbols are used also for addition and subtraction, we will use parentheses when writing problems involving addition and subtraction of signed numbers.

More Practice   Before we attempt these operations, test your understanding of signed numbers by representing each of the following situations with the appropriate signed number.

(a) Two seconds before liftoff

(b) A profit of eight dollars

(c) An elevator going up six floors

(d) Diving 26 feet below the surface of the ocean

(e) A debt of $50

(f) Twelve degrees above zero

Answers   (a) −2 sec   (b) +$8   (c) +6 floors

(d) −26 ft   (e) −$50   (f) +12°

Addition of Signed Numbers   Addition of signed numbers may be pictured using the number line. A positive number may be represented on the line by an arrow to the right or positive direction. Think of a positive arrow as a "gain" of some quantity, such as money. The sum of two numbers is the net "gain."

EXAMPLE 2   The sum of $4 + 3$ or $(+4) + (+3)$ is

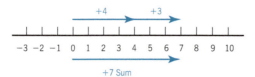

The numbers $+4$ and $+3$ are represented by arrows directed to the right. The $+3$ arrow begins where the $+4$ arrow ends. The sum $4 + 3$ is the arrow that begins at the start of the $+4$ arrow and ends at the tip of the $+3$ arrow. In terms of money, a "gain" of \$4 followed by a "gain" of \$3 produces a net gain of \$7.   ●

A negative number may be represented on the number line by an arrow to the left or negative direction. Think of a negative arrow as a loss of some quantity.

EXAMPLE 3   The sum $(-4) + (-3)$ is -7.

The numbers -4 and -3 are represented by arrows directed to the left. The -3 arrow begins at the tip of the -4 arrow. Their sum, -7, is the arrow that begins at the start of the -4 arrow and ends at the tip of the -3 arrow. A loss of \$4 followed by another loss of \$3 makes a total loss of \$7.   ●

Your Turn   Try setting up number line diagrams for the following two sums.

(a) $4 + (-3)$ (b) $(-4) + 3$

Solutions   (a) $4 + (-3) = 1$

The number 4 is represented by an arrow directed to the right. The number -3 is represented by an arrow starting at the tip of $+4$ and directed to the left. A gain of \$4 followed by a loss of \$3 produces a net gain of \$1.

(b) $(-4) + 3 = -1$

The number -4 is represented by an arrow directed to the left. The number 3 is represented by an arrow starting at the tip of -4 and directed to the right. A loss of \$4 followed by a gain of \$3 produces a net loss of \$1.

Another Way to Picture Addition of Signed Numbers

If you are having trouble with the number line, you might find it easier to picture addition of signed numbers using ordinary poker chips. Suppose that we designate white chips to represent positive integers and blue chips to represent negative integers. The sum of $(+4) + (+3)$ can be found by taking four white chips and combining them with three more white chips. The result is seven white chips or $+7$:

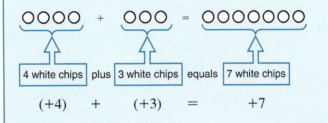

4 white chips	plus	3 white chips	equals	7 white chips
$(+4)$	$+$	$(+3)$	$=$	$+7$

Similarly, $(-4) + (-3)$ can be illustrated by taking four *blue* chips and combining them with three additional *blue* chips. We end up with seven blue chips or -7:

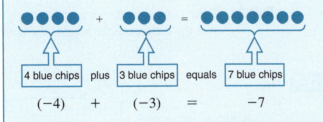

4 blue chips	plus	3 blue chips	equals	7 blue chips
(-4)	$+$	(-3)	$=$	-7

To illustrate $4 + (-3)$, take four white chips and three blue chips. Realizing that a white and a blue will "cancel" each other out $[(+1) + (-1) = 0]$, we rearrange them as shown below, and the result is one white chip or $+1$:

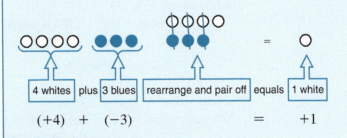

4 whites	plus	3 blues	rearrange and pair off	equals	1 white
$(+4)$	$+$	(-3)		$=$	$+1$

Finally, to represent $(-4) + 3$, picture the following:

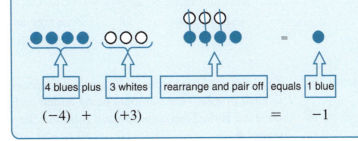

4 blues	plus	3 whites	rearrange and pair off	equals	1 blue
(-4)	$+$	$(+3)$		$=$	-1

Number lines and poker chips provide a good way to see what is happening when we add signed numbers, but we need a nice easy rule that will allow us to do the addition more quickly.

Absolute Value

To add or subtract signed numbers you should work with their absolute value. The **absolute value** of a number is the distance along the number line between that number and the zero point. We indicate the absolute value by enclosing the number with vertical lines. The absolute value of any number is always a nonnegative number. For example,

$$|4| = 4$$

The distance between 0 and 4 is 4 units, so the absolute value of 4 is 4.

$$|-5| = 5$$

The distance between 0 and −5 is 5 units, so the absolute value of −5 is 5.

To add two signed numbers, first find the absolute value of each number. Then apply the appropriate rule:

Rule 1   If the two numbers have the *same sign, add* their absolute values and give the sum the sign of the original numbers.

Rule 2   If the two numbers have *opposite signs, subtract* the absolute values and give the difference the sign of the number with the larger absolute value.

EXAMPLE 4

(a)
$$9 + 3 = +12$$
Add absolute values: $9 + 3 = 12$
Absolute value = 9
Absolute value = 3
Sign is positive because both numbers are positive

(b)
$$(-6) + (-9) = -15$$
Add absolute values: $6 + 9 = 15$
Absolute value = 6
Absolute value = 9
Sign is negative because both numbers are negative

(c)
$$7 + (-12) = -5$$
Subtract absolute values: $12 - 7 = 5$
Absolute value = 7
Absolute value = 12
Same sign as the number with the larger absolute value (−12)

(d)
$$22 + (-14) = +8$$
Subtract absolute values: $22 - 14 = 8$
Absolute value = 22
Absolute value = 14
Same sign as the number with the larger absolute value (+22)   ●

A Closer Look   To summarize the results of Example 4: In parts (a) and (b), we used Rule 1 because in each problem the numbers being added had the same sign. In parts (c) and (d), we used Rule 2 because in each problem the numbers being added had opposite signs. ●

Your Turn Use this method to do the following calculations.

(a) $7 + (-2)$ (b) $-6 + (-14)$ (c) $(-15) + (-8)$

(d) $23 + 13$ (e) $-5 + 8 + 11 + (-21) + (-9) + 12$

Solutions (a) $7 + (-2) = +5$ Signs are opposite, so apply Rule 2: Subtract absolute values. Answer has the sign of the number with the larger absolute value (7).

(b) $-6 + (-14) = -20$ Signs are the same, so apply Rule 1: Add absolute values. Answer has the sign of the original numbers.

(c) $(-15) + (-8) = -23$ Signs are the same; add absolute values. Answer has the sign of the original numbers.

(d) $23 + 13 = +36$ Signs are the same; add absolute values. Answer has the sign of the original numbers.

(e) When more than two numbers are being added, rearrange them so that numbers with like signs are grouped together:

$$-5 + 8 + 11 + (-21) + (-9) + 12 = (8 + 11 + 12) + [(-5) + (-21) + (-9)]$$

Rearrange: ⬆ Positives   ⬆ Negatives

Find the positive total: $8 + 11 + 12 = +31$

Find the negative total: $(-5) + (-21) + (-9) = -35$

Add the positive total to the negative total using Rule 2: $31 + (-35) = -4$

The procedure is exactly the same if the numbers being added are fractions or decimal numbers.

EXAMPLE 5 (a) To add $6\frac{3}{8} + \left(-14\frac{3}{4}\right)$, **first** note that the signs are opposite. Apply Rule 2.

We must subtract absolute values:
$$\begin{array}{r} 14\frac{3}{4} \\ -6\frac{3}{8} \\ \hline \end{array}$$

The common denominator is 8.
$$\begin{array}{r} 14\frac{6}{8} \\ -6\frac{3}{8} \\ \hline 8\frac{3}{8} \end{array}$$

$\frac{3}{4} = \frac{3 \times 2}{4 \times 2} = \frac{6}{8}$

The answer has the sign of the number with the larger absolute value $\left(-14\frac{3}{4}\right)$;   Answer: $-8\frac{3}{8}$

(b) To add $-21.6 + (-9.8)$, notice that the signs are the same. Apply Rule 1.

We must add absolute values:
$$\begin{array}{r} 21.6 \\ + 9.8 \\ \hline 31.4 \end{array}$$

The original numbers were both negative.   Answer: -31.4

Your Turn   Find each sum.

(a) $-1\frac{2}{3} + \left(-2\frac{3}{4}\right)$ (b) $-7.2 + 11.6$

Solutions   (a) The signs are the same so we must add absolute values (Rule 1).

The common denominator is 12:

$$1\frac{2}{3} \longrightarrow 1\frac{8}{12} \quad \Longleftarrow \boxed{\frac{2\times 4}{3\times 4}}$$

$$+2\frac{3}{4} \longrightarrow +2\frac{9}{12} \quad \Longleftarrow \boxed{\frac{3\times 3}{4\times 3}}$$

$$3\frac{17}{12} = 3 + 1\frac{5}{12} = 4\frac{5}{12}$$

The original numbers were both negative.   Answer: $-4\frac{5}{12}$

(b) The signs are opposite, so we must subtract absolute values (Rule 2).

$$\begin{array}{r} 11.6 \\ -\ 7.2 \\ \hline 4.4 \end{array}$$

The answer is $+4.4$ because the number with the larger absolute value ($+11.6$) is positive.

Using a Calculator   To work with signed numbers on a calculator, we need only one new key, the $\boxed{(-)}$ key. This key allows us to enter a negative number on our display.

For example, to enter -4, simply press $\boxed{(-)}$ followed by 4.

EXAMPLE 6   To add $(-587) + 368$, enter the following sequence:

$\boxed{(-)}\,\mathbf{587}\,\boxed{+}\,\mathbf{368}\,\boxed{=} \rightarrow$ $-219.$ ●

More Practice   Use a calculator to do the following addition problems.

(a) $(-2675) + 1437$ (b) $(-6.975) + (-5.2452)$

(c) $0.026 + (-0.0045)$ (d) $(-683) + (-594) + 438 + (-862)$

(e) $(-\frac{5}{8}) + 2\frac{2}{3} + (-3\frac{1}{2})$

Solutions   (a) $\boxed{(-)}\,\mathbf{2675}\,\boxed{+}\,\mathbf{1437}\,\boxed{=} \rightarrow$ $-1238.$

(b) $\boxed{(-)}\,\mathbf{6.975}\,\boxed{+}\,\boxed{(-)}\,\mathbf{5.2452}\,\boxed{=} \rightarrow$ -12.2202

(c) $\mathbf{.026}\,\boxed{+}\,\boxed{(-)}\,\mathbf{.0045}\,\boxed{=} \rightarrow$ 0.0215

(d) $\boxed{(-)}\,\mathbf{683}\,\boxed{+}\,\boxed{(-)}\,\mathbf{594}\,\boxed{+}\,\mathbf{438}\,\boxed{+}\,\boxed{(-)}\,\mathbf{862}\,\boxed{=} \rightarrow$ $-1701.$

(e) $\boxed{(-)}\,\mathbf{5}\,\boxed{A^{b}_{c}}\,\mathbf{8}\,\boxed{+}\,\mathbf{2}\,\boxed{A^{b}_{c}}\,\mathbf{2}\,\boxed{A^{b}_{c}}\,\mathbf{3}\,\boxed{+}\,\boxed{(-)}\,\mathbf{3}\,\boxed{A^{b}_{c}}\,\mathbf{1}\,\boxed{A^{b}_{c}}\,\mathbf{2}\,\boxed{=} \rightarrow$ $-1 \sqcup 11/24$

Practical Applications

EXAMPLE 7   **Electrical Trades**   Excess electricity generated by solar panels can often be sold back to a power company. Over the course of a year, one particular homeowner using solar energy consumed the following monthly amounts of electricity according to her electric bill: $-112, -42, 86, 108, 123, -65, 144, -122, -186, -114, 22,$ and 65 kWh (kilowatt-hours). A positive number means she purchased electricity, while a negative number means she sold electricity back to the power company. We can calculate the net amount of electricity bought or sold during this period by adding the 12 monthly amounts. Rearranging the numbers into a positive group and a negative group, we have:

$$\text{Net amount} = (86 + 108 + 123 + 144 + 22 + 65) + [(-112) + (-42) +$$
$$(-65) + (-122) + (-186) + (-114)]$$

$$= (+548) + (-641) = -93 \text{ kWh}$$

Because the net amount was a negative number, the electric company had to pay the homeowner for 93 kWh of electricity for the year.   ●

Your Turn   **Business and Finance**   A group of sales reps receive bonuses if they exceed their monthly quotas for unit sales. Over a particular six-month period, their monthly comparisons to their quotas were: $-24, -16, +33, +12, -19,$ and $+8$ units.

A positive number means they exceeded their quota, and a negative number means they fell short of their quota. How far above or below the quotas were they for the entire six-month period?

Answer   They were a total of 6 units below their quotas.

For more practice in addition of signed numbers, continue with Exercises 6-1 for a set of problems.

Exercises 6-1   Addition of Signed Numbers

A. Make each statement true by replacing the □ symbol with either the $>$ or $<$ symbol.

1. $-4 \;\square\; 7$ 2. $-17 \;\square\; -14$ 3. $23 \;\square\; 18$

4. $13 \;\square\; -15$ 5. $-9 \;\square\; -10$ 6. $-31 \;\square\; 41$

7. $-65 \;\square\; -56$ 8. $47 \;\square\; 23$ 9. $-47 \;\square\; -23$

10. $-3 \;\square\; 0$ 11. $6 \;\square\; 0$ 12. $-1.4 \;\square\; -1.39$

13. $2.7 \;\square\; -2.8$ 14. $-3\frac{1}{4} \;\square\; -3\frac{1}{8}$ 15. $-11.4 \;\square\; -11\frac{1}{4}$

16. $-0.1 \;\square\; -0.001$

B. For each problem, make a separate number line and mark the points indicated.

1. $-8.7, -9.4, -8\frac{1}{2}, -9\frac{1}{2}$ 2. $-1\frac{3}{4}, +1\frac{3}{4}, -\frac{1}{4}, +\frac{1}{4}$

3. $-0.2, -1\frac{1}{2}, -2.8, -3\frac{1}{8}$ 4. $+4.1, -1.4, -3.9, +1\frac{1}{4}$

C. Represent each of the following situations with a signed number.

1. A temperature of six degrees below zero

2. A debt of three hundred dollars

3. An airplane rising to an altitude of 12,000 feet

4. The year A.D. 240

5. A golf score of nine strokes below par

6. A quarterback getting sacked for a seven-yard loss

7. Diving eighty feet below the surface of the ocean

8. Winning fifteen dollars at poker

9. Taking five steps backward

10. Ten seconds before takeoff

11. A profit of six thousand dollars

12. An elevation of thirty feet below sea level

13. A checking account overdrawn by $17.50

14. A 26-point drop in the NAZWAC stock index

D. Add.

1. $-1 + 6$

2. $-11 + 7$

3. $-5 + (-8)$

4. $9 + 4$

5. $-13 + (-2)$

6. $-6 + (-19)$

7. $31 + (-14)$

8. $23 + (-28)$

9. $21 + 25$

10. $12 + 19$

11. $-8 + (-8)$

12. $-7 + (-9)$

13. $-34 + 43$

14. $-56 + 29$

15. $45 + (-16)$

16. $22 + (-87)$

17. $-18 + (-12)$

18. $-240 + 160$

19. $-3.6 + 2.7$

20. $5.7 + (-7.9)$

21. $-12.43 + (-15.66)$

22. $-90.15 + (-43.69)$

23. $-0.05 + (-0.09)$

24. $-0.6 + 0.4$

25. $2\frac{5}{8} + (-3\frac{7}{8})$

26. $-4\frac{1}{4} + 6\frac{3}{4}$

27. $-9\frac{1}{2} + (-4\frac{1}{2})$

28. $-6\frac{2}{3} + (-8\frac{2}{3})$

29. $-2\frac{1}{2} + 1\frac{3}{4}$

30. $-5\frac{5}{8} + 8\frac{3}{4}$

31. $9\frac{7}{16} + (-7\frac{7}{8})$

32. $-13\frac{3}{10} + (-8\frac{4}{5})$

33. $6\frac{5}{6} + (-11\frac{1}{4})$

34. $(-4\frac{3}{8}) + (-3\frac{7}{12})$

35. $-6 + 4 + (-8)$

36. $22 + (-38) + 19$

37. $-54 + 16 + 22$

38. $11 + (-6) + (-5)$

39. $-7 + 12 + (-16) + 13$

40. $23 + (-19) + 14 + (-21)$

41. $-36 + 45 + 27 + (-18)$

42. $18 + (-51) + (-8) + 26$

43. $-250 + 340 + (-450) + (-170) + 320 + (-760) + 880 + (-330)$

44. $-1 + 3 + (-5) + 7 + (-9) + 11 + (-13) + (15) + (-17) + 19$

45. $23 + (-32) + (-14) + 55 + (-66) + 44 + 28 + (-31)$

46. $-12.64 + 22.56 + 16.44 + (-31.59) + (-27.13) + 44.06 + 11.99$

47. $(-2374) + (5973)$

48. $(6.875) + (-11.765)$

49. $(-46,720) + (-67,850) + 87,950$

50. $(0.0475) + (-0.0875) + (-0.0255)$

E. **Practical Applications**

1. **Printing**   The Jiffy Print Shop has a daily work quota that it tries to meet. During the first two weeks of last month, its daily comparisons to this quota showed the following: $+\$30, -\$27, -\$15, +\$8, -\$6, +\$12, -\$22, +\$56, -\$61, +\$47$. (In each case "$+$" means the shop exceeded the quota and "$-$" means it was below.) How far below or above its quota was the two-week total?

2. **Sports and Leisure**   During a series of plays in a football game, a running back carried the ball for the following yardages: $7, -3, 2, -1, 8, -5,$ and -2. Find his total yards and his average yards per carry.

3. **Meteorology**   On a cold day in Fairbanks the temperature stood at 12 degrees below zero $(-12°)$ at 2 P.M. Over the next 5 hours the temperature rose 4 degrees, dropped 2 degrees, rose 5 degrees, dropped 1 degree, and finally dropped 3 degrees. What was the temperature at 7 P.M.?

4. **Life Skills**   Pedro's checking account contained $489 on June 1. He then made the following transactions: deposited $150, withdrew $225, withdrew $34, deposited $119, withdrew $365, and deposited $750. What was his new balance after these transactions?

5. **Business and Finance**   A struggling young company reported the following results for the four quarters of the year: a loss of $257,000, a loss of $132,000, a profit of $87,000, a profit of $166,000. What was the net result for the year?

6. **Plumbing**   In determining the fitting allowance for a pipe configuration, a plumber has negative allowances of $-2\frac{1}{2}$ in. and $-1\frac{7}{16}$ in. and two positive allowances of $\frac{1}{2}$ in. each. What is the net fitting allowance for the job?

7. **Electronics**   What is the sum of currents flowing into Node N in the circuit below?

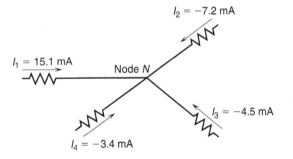

8. **Electrical Trades**   As explained in Example 7, excess energy generated by solar panels can often be sold back to the power company. Over a 6-month period, a household consumed $212, 150, -160, -178, -148,$ and 135 kWh (kilowatt-hours) of electricity from the power company. What was the net amount of electricity consumed by the household? (Express your answer as a signed number.)

When you have completed these exercises, check your answers to the odd-numbered problems in the Appendix, then turn to Section 6-2 to learn how to subtract signed numbers.

6-2 Subtraction of Signed Numbers

Opposites

When the number line is extended to the left to include the negative numbers, every positive number can be paired with a negative number whose absolute value, or distance from zero, is the same. These pairs of numbers are called **opposites**. For example, -3 is the opposite of $+3$, $+5$ is the opposite of -5, and so on.

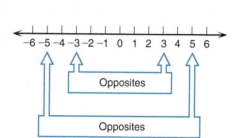

📝 **Note**   Zero has no opposite because it is neither positive nor negative. ●

The concept of opposites will help us express a rule for subtraction of signed numbers.

Subtraction of Signed Numbers

We know from basic arithmetic that $5 - 2 = 3$. From the rules for addition of signed numbers, we know that $5 + (-2) = 3$ also. Therefore,

$$(+5) - (+2) = (+5) + (-2) = 3$$

This means that subtracting $(+2)$ is the same as adding its opposite, (-2). Our knowledge of arithmetic and signed numbers verifies that subtracting and adding the opposite are equivalent for all the following cases:

$$5 - 5 = 5 + (-5) = 0$$
$$5 - 4 = 5 + (-4) = 1$$
$$5 - 3 = 5 + (-3) = 2$$
$$5 - 2 = 5 + (-2) = 3$$
$$5 - 1 = 5 + (-1) = 4$$
$$5 - 0 \qquad\qquad = 5 \quad \text{(zero has no opposite)}$$

But what happens with a problem like $5 - 6 = ?$ If we rewrite this subtraction as addition of the opposite like the others, we have

$$5 - 6 = 5 + (-6) = -1$$

Notice that this result fits right into the pattern established by the previous results. The answers *decrease* as the numbers being subtracted *increase*: $5 - 3 = 2$, $5 - 4 = 1, 5 - 5 = 0$, so it seems logical that $5 - 6 = -1$.

What happens if we subtract a negative? For example, how can we solve the problem $5 - (-1) = ?$ If we rewrite this subtraction as addition of the opposite, we have

$$5 - (-1) = 5 + (+1) = 6$$

subtracting -1 adding the opposite of -1 (or $+1$)

Notice that this too fits the pattern established by the previous results. The answers *increase* as the numbers being subtracted *decrease:* $5 - 2 = 3, 5 - 1 = 4,$ $5 - 0 = 5$, so again it seems logical that $5 - (-1) = 6$.

We can now state a rule for subtraction of signed numbers. To subtract two signed numbers, follow this two-step rule:

Step 1 Rewrite the subtraction as an addition of the opposite.

Step 2 Use the rules for addition of signed numbers to add.

EXAMPLE 1

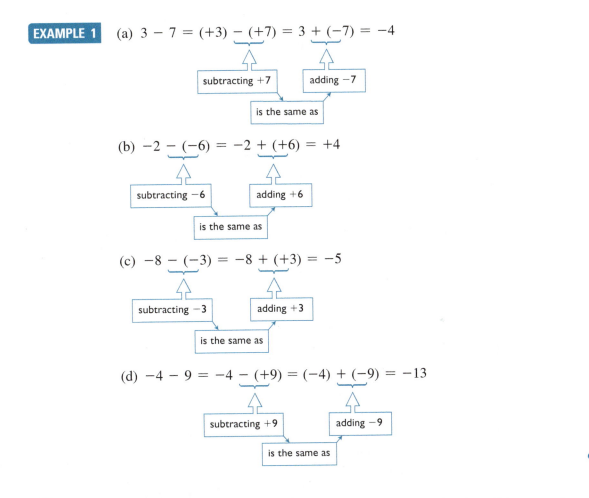

(a) $3 - 7 = (+3) - (+7) = 3 + (-7) = -4$

subtracting $+7$ ⇒ is the same as ⇐ adding -7

(b) $-2 - (-6) = -2 + (+6) = +4$

subtracting -6 ⇒ is the same as ⇐ adding $+6$

(c) $-8 - (-3) = -8 + (+3) = -5$

subtracting -3 ⇒ is the same as ⇐ adding $+3$

(d) $-4 - 9 = -4 - (+9) = (-4) + (-9) = -13$

subtracting $+9$ ⇒ is the same as ⇐ adding -9

! Careful Never change the sign of the first number. You change the sign of the *second* number only when changing subtraction to addition. ●

Your Turn Try these:

(a) $-7 - 5$ (b) $8 - 19$ (c) $4 - (-4)$ (d) $-12 - (-8)$

(e) $\left(-\dfrac{3}{4}\right) - \left(\dfrac{5}{16}\right)$ (f) $-2 - (-3) + (-1)$

(g) $-3 - 9 - (-7) + (-8) + (+2)$

(h) $(-26.75) - (-44.38) - 16.96$

Solutions (a) $-7 - 5 = -7 + (-5) = -12$

(b) $8 - 19 = (+8) + (-19) = -11$

(c) $4 - (-4) = (+4) + (+4) = +8$

(d) $-12 - (-8) = -12 + (+8) = -4$

(e) $\left(-\dfrac{3}{4}\right) - \left(\dfrac{5}{16}\right) = \left(-\dfrac{3}{4}\right) + \left(-\dfrac{5}{16}\right)$

$$= \left(-\dfrac{12}{16}\right) + \left(-\dfrac{5}{16}\right)$$

$$= -\dfrac{17}{16} = -1\dfrac{1}{16}$$

(f) $-2 - (-3) + (-1) = -2 + (+3) + (-1)$

$$= (+1) + (-1)$$

$$= 0$$

(g) $-3 - 9 - (-7) + (-8) + (+2) = -3 + (-9) + (+7) + (-8) + (+2) = -11$

> These last two operations are additions, so they remain unchanged when the problem is rewritten.

(h) $\boxed{(-)}\,26.75\,\boxed{-}\,\boxed{(-)}\,44.38\,\boxed{-}\,16.96\,\boxed{=} \rightarrow$ ⟨ 0.67 ⟩

! **Careful**   When using a calculator, make sure to use the $\boxed{-}$ key only for subtraction and the $\boxed{(-)}$ key only to enter a negative number. Using the incorrect key will result in either an "Error" message or a wrong answer. ●

Subtraction with Chips

If you are having trouble visualizing subtraction of signed numbers, try the poker chip method that we used to show addition. Subtraction means "taking away" chips. First, a simple example: $7 - 3$ or $(+7) - (+3)$.

| Begin with 7 white (positive) chips. | Take away 3 white chips. | Four white chips remain. |

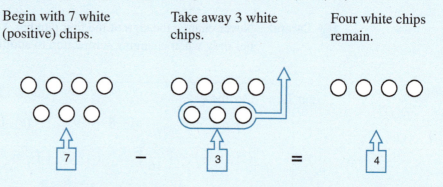

What about the problem: $(+3) - (+7) = ?$ How can you remove 7 white chips if you have only 3 white chips? Here we must be very clever. Remember that a positive chip and a negative chip "cancel" each other $[(+1) + (-1) = 0]$. So if we add both a white (positive) and a blue (negative) chip to a collection, we do not really change its value. Do $(+3) - (+7)$ this way:

(continued)

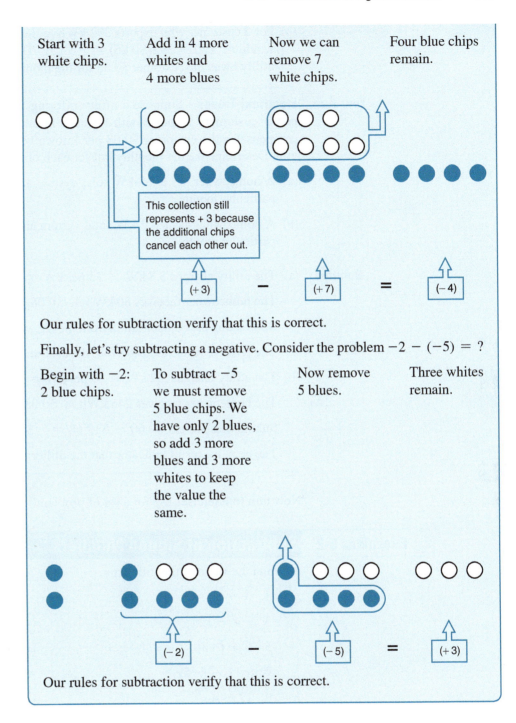

Start with 3 white chips.

Add in 4 more whites and 4 more blues

Now we can remove 7 white chips.

Four blue chips remain.

This collection still represents + 3 because the additional chips cancel each other out.

$$(+3) \quad - \quad (+7) \quad = \quad (-4)$$

Our rules for subtraction verify that this is correct.

Finally, let's try subtracting a negative. Consider the problem $-2 - (-5) = ?$

Begin with -2: 2 blue chips.

To subtract -5 we must remove 5 blue chips. We have only 2 blues, so add 3 more blues and 3 more whites to keep the value the same.

Now remove 5 blues.

Three whites remain.

$$(-2) \quad - \quad (-5) \quad = \quad (+3)$$

Our rules for subtraction verify that this is correct.

Practical Applications

EXAMPLE 2 **Electrical Trades** Customers who install solar power systems can usually sell their excess electricity back to the utility company. This is called "net energy metering," and it may result in a "negative" monthly bill if the compensation for excess electricity exceeds any monthly fees to operate and upgrade the grid. Suppose a utility charges solar power customers a fixed monthly fee of $20 but compensates the customer $0.06 per kilowatt-hour (kWh) exported to the grid. To determine the customer's net monthly bill, we multiply the number of kilowatt-hours exported to the grid by $0.06 and subtract this from $20.

(a) For a customer who exports 280 kWh to the grid in a particular month, the bill would be $20 - 280(\$0.06) = +\$3.20$. The positive result means that the customer owes the utility $3.20 for the month.

(b) For a customer who exports 392 kWh to the grid in a particular month, the bill would be $20 − 392($0.06) = −$3.52. The negative result indicates that the utility owes the customer $3.52 for the month. •

Your Turn   Electrical Trades   Suppose a utility offering net energy metering charges solar power customers $3 per month for each kilowatt (kW) of installed capacity but compensates the customer $0.08 per kilowatt-hour of electricity exported to the grid. Determine the net monthly bill for each of the following situations.

(a) A homeowner has a 3.8-kW solar system and exports 80 kWh to the grid in a particular month.

(b) A homeowner has a 3.2-kW solar system and exports 246 kWh to the grid in a particular month.

Solution   (a) The utility charges 3.8 kW × $3 per kW = $11.40.

The homeowner receives 80 kWh × $0.08 per kWh = $6.40.

Subtracting the compensation from the fee, we get $11.40 − $6.40 = +$5.00.

The positive result indicates that the customer owes the utility $5.

(b) The utility charges 3.2 kW × $3 per kW = $9.60.

The homeowner receives 246 kWh × $0.08 = $19.68.

Subtracting, we get $9.60 − $19.68 = −$10.08.

The negative result indicates that the utility owes the customer $10.08.

Now turn to Exercises 6-2 for a set of problems on subtraction of signed numbers.

Exercises 6-2   Subtraction of Signed Numbers

A. Find the value of the following.

1. $3 - 5$	2. $7 - 20$
3. $-4 - 12$	4. $-9 - 15$
5. $6 - (-3)$	6. $14 - (-7)$
7. $-11 - (-5)$	8. $-22 - (-10)$
9. $12 - 4$	10. $23 - 9$
11. $-5 - (-16)$	12. $-14 - (-17)$
13. $12 - 30$	14. $30 - 65$
15. $-23 - 8$	16. $-18 - 11$
17. $2 - (-6)$	18. $15 - (-12)$
19. $44 - 18$	20. $70 - 25$
21. $-34 - (-43)$	22. $-27 - (-16)$
23. $0 - 19$	24. $0 - 26$
25. $-17 - 0$	26. $-55 - 0$
27. $0 - (-8)$	28. $0 - (-14)$
29. $-56 - 31$	30. $-45 - 67$

31. $22 - 89$

32. $17 - 44$

33. $-12 - 16 + 18$

34. $23 - (-11) - 30$

35. $19 + (-54) - 7$

36. $-21 + 6 - (-40)$

37. $-2 - 5 - (-7) + 6$

38. $5 - 12 + (-13) - (-7)$

39. $-3 - 4 - (-5) + (-6) + 7$

40. $2 - (-6) - 10 + 4 + (-11)$

41. $\frac{2}{9} - \frac{7}{9}$

42. $\frac{5}{12} - \left(-\frac{7}{12}\right)$

43. $\left(-\frac{3}{8}\right) - \frac{1}{4}$

44. $2\frac{1}{2} - \left(-3\frac{1}{4}\right)$

45. $6\frac{9}{16} - 13\frac{1}{16}$

46. $\left(-\frac{5}{6}\right) - \left(-\frac{2}{3}\right)$

47. $\left(-2\frac{3}{4}\right) - 5\frac{5}{6}$

48. $\left(-1\frac{7}{8}\right) - \left(-3\frac{1}{6}\right)$

49. $6.5 - (-2.7)$

50. $(-0.275) - 1.375$

51. $(-86.4) - (-22.9)$

52. $0.0025 - 0.075$

53. $22.75 - 36.45 - (-54.72)$

54. $(-7.35) - (-4.87) - 3.66$

55. $3\frac{1}{2} - 6.6 - \left(-2\frac{1}{4}\right)$

56. $(-5.8) - \left(-8\frac{1}{8}\right) - 3.9$

B. **Practical Applications**

1. **General Interest**   The height of Mt. Rainier is $+14{,}410$ ft. The lowest point in Death Valley is -288 ft. What is the difference in altitude between these two points?

2. **Meteorology**   The highest temperature ever recorded on earth was $+136°$F at Libya, North Africa. The coldest was $-127°$F at Vostok in Antarctica. What is the difference in temperature between these two extremes?

3. **Meteorology**   At 8 P.M. the temperature was $-2°$F in Anchorage, Alaska. At 4 A.M. it was $-37°$F. By how much did the temperature drop in that time?

4. **General Interest**   A vacation trip involved a drive from a spot in Death Valley 173 ft below sea level (-173 ft) to the San Bernardino Mountains 4250 ft above sea level ($+4250$ ft). What was the vertical ascent for this trip?

5. **Trades Management**   Dave's Auto Shop was $2465 "in the red" ($-\$2465$) at the end of April—they owed this much to their creditors. At the end of December they were only $648 "in the red." How much money did they make in profit between April and December to achieve this improvement?

6. **Sports and Leisure**   A golfer had a total score of six strokes over par going into the last round of a tournament. He ended with a total score of three strokes under par. What was his score relative to par during the final round?

7. **Meteorology**   The greatest recorded rise in temperature within a 12-hour period occurred in Granville, North Dakota, on February 21, 1918. After a morning low of $-33°$F, the temperature rose to a high of $50°$F. How many degrees of warming occurred during this 12-hour period?

8. **Meteorology**   The fastest 24-hour chill ever recorded occurred in Browning, Montana, in January 1916, when the temperature plunged from $44°$F to $-54°$F. By how many degrees did the temperature fall?

9. **Trades Management**   Pacific Roofing was $3765 "in the red" (see problem 5) at the end of June. At the end of September they were $8976 "in the red." Did they make or lose money between June and September? How much?

10. **Sheet Metal Trades**   To establish the location of a hole relative to a fixed zero point, a machinist must make the following calculation:

$$y = 5 - (3.750 - 0.500) - 2.375$$

Find y.

11. **Electronics**   The peak-to-peak voltage of an ac circuit is found by subtracting the negative peak voltage from the positive peak voltage. Calculate the peak-to-peak voltage V in the diagram shown.

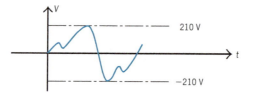

12. **Automotive Trades**   During an alignment, the caster angle on a vehicle was adjusted from $-0.25°$ to $+1.75°$. What was the amount of the adjustment?

13. **General Interest**   The James Webb Telescope, which NASA hopes to launch by the year 2018, has a sun shield that protects the telescope by dissipating heat. The temperature on the light side of the telescope is 185°F, while the temperature on the dark side is $-388°F$. Find the difference in temperature between the two sides.

14. **Meteorology**   The hottest official temperature ever recorded in the United States was 134°F in Death Valley, California. The coldest official temperature ever recorded was $-80°F$ in Prospect Creek, Alaska. What is the difference between these two extremes?

15. **Electrical Trades**   Suppose a utility charges solar power customers $3.50 per month for each kilowatt (kW) of installed capacity but compensates the customer $0.09 per kilowatt-hour of electricity exported to the grid. Determine the net monthly bill for each of the following situations. (See "Your Turn" on p. 394.)

    (a) A homeowner has a 3.5-kW solar system and exports 108 kWh to the grid in a particular month.

    (b) A homeowner has a 4.6-kW solar system and exports 258 kWh to the grid in a particular month.

Check your answers to the odd-numbered problems in the Appendix, then turn to Section 6-3 to learn about multiplication and division of signed numbers.

6-3   Multiplication and Division of Signed Numbers

Multiplication of Signed Numbers

As we learned earlier, multiplication is the shortcut for repeated addition. For example,

$$5 \times 3 = \underline{3 + 3 + 3 + 3 + 3} = 15$$

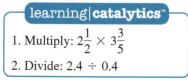

| Five times three | means | three added five times |

Notice that the *product of two positive numbers* is always a *positive* number.

We can use repeated addition to understand multiplication involving negative numbers. Consider the problem $(+5) \times (-3)$.

$$(+5) \times (-3) = \underbrace{(-3) + (-3) + (-3) + (-3) + (-3)}_{} = -15$$

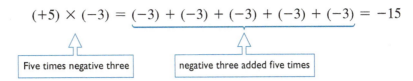

Five times negative three negative three added five times

Notice that the product of a *positive number* and a *negative number* is a *negative number*. The order in which you multiply two numbers does not matter, so a negative times a positive is also a negative. For example,

$$(-5) \times (+3) = (+3) \times (-5) = (-5) + (-5) + (-5) = -15$$

To understand the product of a negative times a negative, look at the pattern in the following sequence of problems:

$$(-5) \times (+3) = -15$$
$$(-5) \times (+2) = -10$$
$$(-5) \times (+1) = -5$$
$$(-5) \times 0 = 0$$
$$(-5) \times (-1) = ?$$

Each time we *decrease* the factor on the right by 1, we *increase* the product by 5. So to preserve this pattern, the product of $(-5) \times (-1)$ must be $+5$. In other words, *a negative times a negative is a positive.*

Rules for Multiplication In summary, to multiply two signed numbers, follow this two-step rule:

Step 1 Find the product of their absolute values.

Step 2 Attach a sign to the answer:

- If both numbers have the *same* sign, the answer is *positive*.

$$(+) \times (+) = + \qquad (-) \times (-) = +$$

- If the two numbers have *opposite* signs, the answer is *negative*.

$$(+) \times (-) = - \qquad (-) \times (+) = -$$

EXAMPLE 1 Multiply:

(a) $(+3) \times (-2)$

Step 1 The product of their absolute values is $3 \times 2 = 6$

Step 2 The original numbers have opposite signs, so the product is negative. $(+3) \times (-2) = -6$

(b) $(-4) \times (-8)$

Step 1 The product of their absolute values is $4 \times 8 = 32$

Step 2 The original numbers have the same sign, so the product is positive. $(-4) \times (-8) = +32$

(c) $\left(-\dfrac{1}{4}\right) \times \left(+\dfrac{2}{3}\right)$

Step 1 The product of their absolute values is $\dfrac{1}{\overset{}{\underset{2}{4}}} \times \dfrac{\overset{1}{2}}{3} = \dfrac{1}{6}$

Step 2 The original fractions have opposite signs, so the product is negative. $\left(-\dfrac{1}{4}\right) \times \left(+\dfrac{2}{3}\right) = -\dfrac{1}{6}$ ●

Your Turn   Use these rules to work the following problems.

(a) $(-4) \times (+7)$ (b) $(-8) \times (-6)$ (c) $(+5) \times (+9)$

(d) $(+10) \times (-3)$ (e) $(-3) \times (-3)$ (f) $(-2) \times (-2) \times (-2)$

(g) $(-2) \times (-3) \times (+4) \times (-5) \times (-6)$ (h) $\left(-2\frac{1}{3}\right) \times \left(\frac{9}{14}\right)$

(i) $(-3.25) \times (-1.40)$

Solutions   (a) $(-4) \times (+7) = -28$ (b) $(-8) \times (-6) = +48$

| opposite signs | answer is negative | | same sign | answer is positive |

(c) $+45$ (same sign, answer is positive)

(d) -30 (opposite signs, answer is negative)

(e) $+9$ (f) -8 (g) $+720$ (h) $-1\frac{1}{2}$

(i) $\boxed{(-)}\, 3\,.\,2\,5\, \boxed{\times}\, \boxed{(-)}\, 1\,.\,4\, \boxed{=} \rightarrow$ *4.55*

A Closer Look   Did you discover an easy way to do problems (f) and (g), where more than two signed numbers are being multiplied? You could go left to right, multiply two numbers, then multiply the answer by the next number. You would then need to apply the sign rules for each separate product. However, because every *two* negative factors results in a positive, an *even* number of negative factors gives a positive answer, and an *odd* number of negative factors gives a negative answer. In a problem with more than two factors, count the total number of negative signs to determine the sign of the answer.

In problem (f) there are three negative factors—an odd number—so the answer is negative. Multiply the absolute values as usual to get the numerical part of the answer, 8. In problem (g) there are four negative factors—an even number—so the answer is positive. The product of the absolute values of all five numbers gives us the numerical part of the answer, 720. ●

Rules for Division   Recall that division is the reverse of the process of multiplication. The division

$$24 \div 4 = \square \quad \text{can be written as the multiplication} \quad 24 = 4 \times \square$$

and in this case the number $\square$ is 6. Notice that *a positive divided by a positive equals a positive*. We can use this "reversing" process to verify that the rest of the sign rules for division are exactly the same as those for multiplication. To find $(-24) \div (+4)$, rewrite the problem as follows:

$$(-24) \div (+4) = \square \qquad -24 = (+4) \times \square$$

From our knowledge of signed number multiplication, $\square = -6$. Therefore, *a negative divided by a positive equals a negative*. Similarly,

$$(+24) \div (-4) = \square \quad \text{means} \quad (+24) = (-4) \times \square \quad \text{so that} \quad \square = -6$$

A positive divided by a negative is also a negative. Finally, let's try a negative divided by a negative:

$$(-24) \div (-4) = \square \quad \text{means} \quad (-24) = (-4) \times \square \quad \text{so that} \quad \square = +6$$

A negative divided by a negative equals a positive.

When dividing two signed numbers, determine the sign of the answer the same way you do for multiplication:

- If both numbers have the *same* sign, the answer is *positive*.

$$(+) \div (+) = + \qquad (-) \div (-) = +$$

- If the two numbers have *opposite* signs, the answer is *negative*.

$$(+) \div (-) = - \qquad (-) \div (+) = -$$

In summary, to divide two signed numbers, follow this two-step rule:

> **Step 1**   Divide the absolute values of the numbers to get the numerical value of the quotient.
>
> **Step 2**   Attach the proper sign to the answer.

EXAMPLE 2   Divide:

(a) $(+28) \div (-4)$

> **Step 1**   The quotient of their absolute values is $28 \div 4 = 7$
>
> **Step 2**   The original numbers have opposite signs, so the quotient is negative. $(+28) \div (-4) = -7$

(b) $(-45) \div (-5)$

> **Step 1**   The quotient of their absolute values is $45 \div 5 = 9$
>
> **Step 2**   The original numbers have the same sign, so the quotient is positive. $(-45) \div (-5) = +9$

(c) $\left(-\dfrac{2}{5}\right) \div \left(+\dfrac{7}{10}\right)$

> **Step 1**   The quotient of their absolute values is $\dfrac{2}{5} \div \dfrac{7}{10} = \dfrac{2}{\overset{}{\underset{1}{5}}} \times \dfrac{\overset{2}{\cancel{10}}}{7} = \dfrac{4}{7}$
>
> **Step 2**   The original fractions have opposite signs, so the quotient is negative. $\left(-\dfrac{2}{5}\right) \div \left(+\dfrac{7}{10}\right) = -\dfrac{4}{7}$ ●

Your Turn   Try these division problems:

(a) $(-36) \div (+3)$ 　　　　　　　 (b) $(-72) \div (-9)$

(c) $(+42) \div (-6)$ 　　　　　　　 (d) $(-5.4) \div (-0.9)$

(e) $\left(\dfrac{3}{8}\right) \div \left(-\dfrac{1}{4}\right)$ 　　　　　 📟 (f) $(-3.75) \div (6.4)$
(Round to two decimal places.)

Answers   (a) -12 　　 (b) $+8$ 　　 (c) -7 　　 (d) $+6$ 　　 (e) $-1\tfrac{1}{2}$ 　　 (f) -0.59

Enter (f) this way on a calculator:

📟 $(-)\,3.75\,\div\,6.4\,= \rightarrow$ 　 -0.5859375

For more practice in multiplication and division of signed numbers, turn to Exercises 6-3 for a set of problems.

Exercises 6-3 | Multiplication and Division of Signed Numbers

A. Multiply or divide as indicated.

1. $(-7) \times (9)$ 2. $8 \times (-5)$ 3 $(-7) \times (-11)$

4. $(-4) \times (-9)$ 5. $28 \div (-7)$ 6. $(-64) \div 16$

7. $(-72) \div (-6)$ 8. $(-44) \div (-11)$ 9. $(-3.4) \times (-1.5)$

10. $2.3 \times (-1.5)$ 11. $(-1.6) \times (5.0)$ 12. $(-4) \div (16)$

13. $(-5) \div (-25)$ 14. $(-10.8) \div (-1.2)$ 15. $(-8.4) \div 2.1$

16. $\frac{2}{3} \times (-\frac{3}{4})$ 17. $(-\frac{5}{8}) \times (-\frac{4}{15})$ 18. $(-\frac{5}{6}) \div (-\frac{3}{4})$

19. $8\frac{1}{2} \div (-3\frac{1}{3})$ 20. $(-13) \times (-4)$ 21. $15 \times (-6)$

22. $(-18) \times 3$ 23. $(-12) \times (-7)$ 24. $(-120) \div 6$

25. $140 \div (-5)$ 26. $(-200) \div (-10)$ 27. $(-350) \div (-70)$

28. $(-6\frac{2}{5}) \times (-1\frac{1}{4})$ 29. $(-4\frac{1}{2}) \times (-2\frac{3}{4})$ 30. $(-2) \times (-5\frac{7}{8})$

31. $(-3) \times (6\frac{1}{2})$ 32. $(-2.25) \div 0.15$ 33. $3.2 \times (-1.4)$

34. $(-0.07) \times (-1.1)$ 35. $(-4.8) \div (-12)$ 36. $(-0.096) \div 1.6$

37. $(-5) \times (-6) \times (-3)$ 38. $(-8) \times 4 \times (-7)$

39. $(-24) \div (-6) \times (-5)$ 40. $35 \div (-0.5) \times (-10)$

41. $(-12) \times (-0.3) \div (-9)$ 42. $0.2 \times (-5) \div (-0.02)$

B. Multiply or divide using a calculator and round as indicated.

1. $(-3.87) \times (4.98)$ (tenth)

2. $(-5.8) \times (-9.75)$ (whole number)

3. $(-0.075) \times (-0.025)$ (ten-thousandths)

4. $(648) \times (-250)$ (whole number)

5. $(-4650) \div (1470)$ (hundredth)

6. $(-14.5) \div (-3.75)$ (hundredth)

7. $(-0.58) \div (-2.5)$ (hundredth)

8. $0.0025 \div (-0.084)$ (thousandth)

9. $(-4.75) \times 65.2 \div (-6.8)$ (whole number)

10. $78.4 \div (-8.25) \times (-22.6)$ (whole number)

11. $(-2400) \times (-450) \times (-65)$ (whole number)

12. $(-65,000) \div (2.75) \times 360$ (hundred thousands)

C. Practical Applications

1. **Meteorology**   To convert 17°F to degrees Celsius, it is necessary to perform the calculation

$$°C = \frac{5}{9} \times (°F - 32) = \frac{5}{9} \times (17 - 32)$$

What is the Celsius equivalent of 17°F?

2. **Business and Finance**   In the first four months of the year, a certain country had an average monthly trade deficit (a *negative* quantity) of $16.5 billion.

During the final eight months of the year, the country had an average monthly trade surplus (a *positive* quantity) of $5.3 billion. What was the overall trade balance for the year? (Use the appropriate sign to indicate an overall deficit or surplus.)

3. **Industrial Technology**   Due to the relative sizes of two connected pulleys A and B, pulley A turns $1\frac{1}{2}$ times more than pulley B and in the opposite direction. If positive numbers are used to indicate counterclockwise revolutions and negative numbers are used for clockwise revolutions, express as a signed number the result for:
   (a) Pulley A when pulley B turns 80 revolutions counterclockwise.
   (b) Pulley B when pulley A turns 60 revolutions clockwise.

4. **Machine Trades**   A long metal rod changes in length by 0.4 mm for every 1°C change in temperature. If it expands when heated and contracts when cooled, express its change in length as a signed number as the temperature changes from 12°C to 4°C.

5. **Aviation**   An airplane descends from 42,000 ft to 18,000 ft in 12 minutes. Express its rate of change in altitude in feet per minute as a signed number.

6. **Meteorology**   To convert $-8°C$ to °F, you must do the following calculation:

$$°F = \frac{9 \times °C}{5} + 32 = \frac{9 \times (-8)}{5} + 32$$

What is the Fahrenheit equivalent of $-8°C$?

Check your answers to the odd-numbered problems in the Appendix, then turn to Section 6-4 to learn about exponents and square roots.

| **6-4** | # Exponents and Square Roots |

Exponents   When the same number appears many times in a multiplication, writing the product may become monotonous, tiring, and even inaccurate. It is easy, for example, to miscount the twos in

$$32{,}768 = 2 \times 2 \times 2 \times 2 \times 2 \times 2 \times 2 \times 2 \times 2 \times 2 \times 2 \times 2 \times 2 \times 2 \times 2$$

or the tens in

$$100{,}000{,}000{,}000 = 10 \times 10 \times 10 \times 10 \times 10 \times 10 \times 10 \times 10 \times 10 \times 10 \times 10$$

Products of this sort are usually written in a shorthand form as 2^{15} and 10^{11}. In this **exponential form** the raised integer 15 shows the number of times that 2 is to be used as a factor in the multiplication.

EXAMPLE 1

$2 \times 2 = 2^2$ Product of <u>two</u> factors of 2

$2 \times 2 \times 2 = 2^3$ Product of <u>three</u> factors of 2

$\boxed{2 \times 2 \times 2 \times 2} = 2^4$ Product of <u>four</u> factors of 2

Four 2s

Your Turn   Write $3 \times 3 \times 3 \times 3 \times 3$ in exponential form.

$3 \times 3 \times 3 \times 3 \times 3 = \underline{\hspace{2cm}}$

Solution   Counting 3s,

$$3 \times 3 \times 3 \times 3 \times 3 = 3^5$$

Five factors of 3

In this expression, 3 is called the **base**, and the integer 5 is called the **exponent**. The exponent 5 tells you how many times the base 3 must be used as a factor in the multiplication.

EXAMPLE 2   Find the value of $4^3 = $ _____

The exponent 3 tells us to use 4 as a factor 3 times.

$$4^3 = 4 \times 4 \times 4 = (4 \times 4) \times 4 = 16 \times 4 = 64$$ ●

Note   It is important that you be able to read exponential form correctly.

2^2 is read "two to the second power" or "two squared".

2^3 is read "two to the third power" or "two cubed".

2^4 is read "two to the fourth power".

2^5 is read "two to the fifth power" and so on.

Students studying basic electronics, or those going on to another mathematics or science class, will find that it is important to understand exponential notation.   ●

More Practice   Do the following problems for practice in using exponents.

(a) Write in exponential form.

$5 \times 5 \times 5 \times 5$	= ____	base = ____	exponent = ____
7×7	= ____	base = ____	exponent = ____
$10 \times 10 \times 10 \times 10 \times 10$	= ____	base = ____	exponent = ____
$3 \times 3 \times 3 \times 3 \times 3 \times 3 \times 3$	= ____	base = ____	exponent = ____
$9 \times 9 \times 9$	= ____	base = ____	exponent = ____

(b) Write as a product of factors and multiply out.

2^6	= _____	= ____	base = ____	exponent = ____
10^7	= _____	= ____	base = ____	exponent = ____
$(-3)^4$	= _____	= ____	base = ____	exponent = ____
5^2	= _____	= ____	base = ____	exponent = ____
$(-6)^3$	= _____	= ____	base = ____	exponent = ____
4^5	= _____	= ____	base = ____	exponent = ____
1^5	= _____	= ____	base = ____	exponent = ____

Solutions

(a)

$5 \times 5 \times 5 \times 5$	= 5^4	base = 5	exponent = 4
7×7	= 7^2	base = 7	exponent = 2
$10 \times 10 \times 10 \times 10 \times 10$	= 10^5	base = 10	exponent = 5
$3 \times 3 \times 3 \times 3 \times 3 \times 3 \times 3$	= 3^7	base = 3	exponent = 7
$9 \times 9 \times 9$	= 9^3	base = 9	exponent = 3

(b) $2^6 = 2 \times 2 \times 2 \times 2 \times 2 \times 2$

$\qquad\qquad\qquad\qquad\qquad = \underline{\ \ 64\ \ }$ base $= \underline{\ \ 2\ \ }$ exponent $= \underline{\ \ 6\ \ }$

$10^7 = 10 \times 10 \times 10 \times 10 \times 10 \times 10 \times 10 = 10,000,000$

$\qquad\qquad\qquad\qquad\qquad\qquad\quad$ base $= \underline{\ \ 10\ \ }$ exponent $= \underline{\ \ 7\ \ }$

$(-3)^4 = (-3) \times (-3) \times (-3) \times (-3)$

$\qquad\qquad\qquad\qquad = \underline{\ +81\ }$ base $= \underline{\ -3\ }$ exponent $= \underline{\ \ 4\ \ }$

$5^2 = 5 \times 5 \qquad\qquad\qquad = \underline{\ \ 25\ \ }$ base $= \underline{\ \ 5\ \ }$ exponent $= \underline{\ \ 2\ \ }$

$(-6)^3 = (-6) \times (-6) \times (-6)$

$\qquad\qquad\qquad = \underline{\ -216\ }$ base $= \underline{\ -6\ }$ exponent $= \underline{\ \ 3\ \ }$

$4^5 = 4 \times 4 \times 4 \times 4 \times 4 = \underline{1024}$ base $= \underline{\ \ 4\ \ }$ exponent $= \underline{\ \ 5\ \ }$

$1^5 = 1 \times 1 \times 1 \times 1 \times 1 = \underline{\ \ 1\ \ }$ base $= \underline{\ \ 1\ \ }$ exponent $= \underline{\ \ 5\ \ }$

A Closer Look Notice that $(-3)^4 = +81$, but $(-6)^3 = -216$. The rules for multiplying signed numbers tell us that a negative number raised to an **even** power is positive, but a negative number raised to an **odd** power is negative. ●

! Careful Be careful when there is a negative sign in front of the exponential expression. For example,

$3^2 = 3 \times 3 = 9 \quad$ and $\quad (-3)^2 = (-3) \times (-3) = 9$

but

$-3^2 = -(3^2) = -(3 \times 3) = -9$

To calculate $-(-2)^3$ first calculate

$(-2)^3 = (-2) \times (-2) \times (-2) = -8$

Then

$-(-2)^3 = -(-8) = 8$. A useful way of remembering this: exponents act only on what they touch. ●

Special Bases and Powers Here are some useful facts about special bases and powers:

The number 1 raised to any power is equal to 1, of course.

$1^2 = 1 \times 1 = 1$

$1^3 = 1 \times 1 \times 1 = 1$

$1^4 = 1 \times 1 \times 1 \times 1 = 1$ and so on

Notice that when the base is ten, the product is easy to find.

$10^2 = 10 \times 10 = 100$

$10^3 = 10 \times 10 \times 10 = 1000$

$10^4 = 10,000$

$10^5 = 100,000$

$\qquad\quad\underbrace{}_{5\ \text{zeros}}$

A whole number exponent is always exactly equal to the number of zeros in the final product when the base is 10.

Continue the pattern to find the value of 10^0 and 10^1.

$10^5 = 100,000$ $10^2 = 100$

$10^4 = 10,000$ $10^1 = 10$

$10^3 = 1000$ $10^0 = 1$

For any base, powers of 1 or 0 are easy to find.

$2^1 = 2$ $2^0 = 1$

$3^1 = 3$ $3^0 = 1$

$4^1 = 4$ $4^0 = 1$ and so on

Order of Operations with Exponents   In Chapter 1 we discussed the order of operations when you add, subtract, multiply, or divide. To evaluate an arithmetic expression containing exponents, first calculate the value of the exponential factor, then perform the other operations using the order of operations already given. If the calculations are to be performed in any other order, parentheses are used to show the order. The revised order of operations is as follows:

> **First,** simplify all operations within **parentheses.**
>
> **Second,** simplify all **exponents,** left to right.
>
> **Third,** perform all **multiplications** and **divisions,** left to right.
>
> **Finally,** perform all **additions** and **subtractions,** left to right.

Learning Help   If you think you will have trouble remembering this order of operations, try memorizing this phrase: **P**lease **E**xcuse **M**y **D**ear **A**unt **S**ally. The first letter of each word should remind you of **P**arentheses, **E**xponents, **M**ultiply/**D**ivide, **A**dd/**S**ubtract. ●

Here are some examples showing how to apply this expanded order of operations.

EXAMPLE 3   (a) To calculate $12 + 54 \div 3^2$

First, simplify the exponent factor. $= 12 + 54 \div 9$
Second, divide. $= 12 + 6$
Finally, add. $= 18$

(b) To calculate $237 - (2 \times 3)^3$

First, work within parentheses. $= 237 - (6)^3$
Second, simplify the exponent term. $= 237 - 216$
Finally, subtract. $= 21$ ●

Your Turn   Calculate.

(a) $2^4 \times 5^3$ (b) $6 + 3^2$ (c) $(6 + 3)^2$

(d) $(24 \div 6)^3$ (e) $2^2 + 4^2 \div 2^3$ (f) $4 + 3 \times 2^4 \div 6$

Solutions   (a) $2^4 \times 5^3 = 16 \times 125 = 2000$ (b) $6 + 3^2 = 6 + 9 = 15$

(c) $(6 + 3)^2 = (9)^2 = 81$ (d) $(24 \div 6)^3 = (4)^3 = 64$

(e) $2^2 + 4^2 \div 2^3 = 4 + 16 \div 8 = 4 + 2 = 6$ (Do the division before adding.)

(f) $4 + 3 \times 2^4 \div 6 = 4 + 3 \times 16 \div 6 = 4 + 48 \div 6 = 4 + 8 = 12$

Finding Powers on a Calculator To square a number using a calculator, enter the number, then press the x^2 key followed by the $=$ key.

EXAMPLE 4 The value of 38^2 can be found by entering

$3\ 8\ \boxed{x^2}\ \boxed{=} \rightarrow$ ⟦ *1444.* ⟧ ●

To raise a number to any other power, enter the base number, press the power key (labeled $\boxed{\wedge}$ or $\boxed{y^x}$ or $\boxed{x^y}$), then enter the exponent number, and finally press the $=$ key.

EXAMPLE 5 To calculate 3.8^5, enter

$3\ .\ 8\ \boxed{\wedge}\ 5\ \boxed{=} \rightarrow$ ⟦ *792.35168* ⟧ ●

✎ **Note** Many calculators have an $\boxed{x^3}$ key for raising a number to the third power. ●

To raise a *negative* number to a power, you must enclose the number in parentheses.

EXAMPLE 6 (a) To calculate $(-25.4)^2$, enter

$\boxed{(}\ \boxed{(-)}\ 2\ 5\ .\ 4\ \boxed{)}\ \boxed{x^2}\ \boxed{=} \rightarrow$ ⟦ *645.16* ⟧

(b) To calculate $(-0.85)^5$, enter

$\boxed{(}\ \boxed{(-)}\ .\ 8\ 5\ \boxed{)}\ \boxed{\wedge}\ 5\ \boxed{=} \rightarrow$ ⟦ *−0.443705313* ⟧ ●

Your Turn A bit of practice will help. Calculate the following:

(a) 68^2 (b) 65^3 (c) $(-38)^2$

(d) 2.75^4 (Round to the nearest tenth.)

(e) $(-0.475)^5$ (Round to four decimal places.)

Solutions (a) $6\ 8\ \boxed{x^2}\ \boxed{=} \rightarrow$ ⟦ *4624.* ⟧

(b) $6\ 5\ \boxed{\wedge}\ 3\ \boxed{=} \rightarrow$ ⟦ *274625.* ⟧

(c) $\boxed{(}\ \boxed{(-)}\ 3\ 8\ \boxed{)}\ \boxed{x^2}\ \boxed{=} \rightarrow$ ⟦ *1444.* ⟧

(d) $2\ .\ 7\ 5\ \boxed{\wedge}\ 4\ \boxed{=} \rightarrow$ ⟦ *57.19140625* ⟧ or 57.2 rounded

(e) $\boxed{(}\ \boxed{(-)}\ .\ 4\ 7\ 5\ \boxed{)}\ \boxed{\wedge}\ 5\ \boxed{=} \rightarrow$ ⟦ *−0.024180654* ⟧ or −0.0242 rounded

Applications of Exponents

EXAMPLE 7 **Electronics** The percent resolution of an 8-bit digital-to-analog converter is given by the calculation

$$\frac{1}{2^8 - 1} \times 10^2$$

To calculate this resolution,

First, simplify exponents: $\qquad\qquad \dfrac{1}{2^8 - 1} \times 10^2 = \dfrac{1}{256 - 1} \times 100$

Next, simplify the denominator: $\qquad\qquad = \dfrac{1}{255} \times 100$

Finally, multiply and divide: $\qquad\qquad = \dfrac{100}{255} = 0.3921\ldots\%$

Rounded to three decimal places, the percent resolution is 0.392%.

Using a calculator, we enter:

$1 \div (2 \wedge 8 - 1) \times 10 \, x^2 = \rightarrow$ *0.392156863*

Your Turn   **Electronics**   To convert the binary number 1101101 to a decimal number, the following calculation must be performed:

$$2^6 + 2^5 + 2^3 + 2^2 + 2^0$$

Calculate this as a decimal number.

Solution $2^6 + 2^5 + 2^3 + 2^2 + 2^0 = 64 + 32 + 8 + 4 + 1 = 109$

Square Roots   What is interesting about the numbers

1, 4, 9, 16, 25, 36, 49, 64, 81, 100, . . . ?

Do you recognize them?

These numbers are the squares or second powers of the counting numbers,

$1^2 = 1$

$2^2 = 4$

$3^2 = 9$

$4^2 = 16$, and so on. 1, 4, 9, 16, 25, . . . are called **perfect squares**.

Perfect Squares			
$1^2 = 1$	$6^2 = 36$	$11^2 = 121$	$16^2 = 256$
$2^2 = 4$	$7^2 = 49$	$12^2 = 144$	$17^2 = 289$
$3^2 = 9$	$8^2 = 64$	$13^2 = 169$	$18^2 = 324$
$4^2 = 16$	$9^2 = 81$	$14^2 = 196$	$19^2 = 361$
$5^2 = 25$	$10^2 = 100$	$15^2 = 225$	$20^2 = 400$

If you have memorized the multiplication table for one-digit numbers, you will recognize them immediately. The number 3^2 is read "three squared." What is "square" about $3^2 = 9$? The name comes from an old Greek idea about the nature of numbers. Ancient Greek mathematicians called certain numbers "square numbers" or "perfect squares" because they could be represented by a square array of dots:

4   9   16   and so on

The number of dots along the side of the square was called the "root" or origin of the square number. We call it the **square root**. When you square this number, or multiply it by itself, you obtain the original number. The symbol $\sqrt{\ }$ is used to indicate the square root of any number written inside of the symbol.

EXAMPLE 8   (a) The square root of 16 is 4, because 4^2 or $4 \times 4 = 16$. Using the square root symbol, we write $\sqrt{16} = 4$.

(b) $\sqrt{64} = 8$ because 8^2 or $8 \times 8 = 64$.

(c) $\sqrt{169} = 13$ In words, we read this as "the square root of 169 is 13."   •

Your Turn   Find $\sqrt{81} =$ _____ $\sqrt{361} =$ _____ $\sqrt{289} =$ _____

Try using the table of perfect squares if you do not recognize these.

Solutions $\sqrt{81} = 9$ ✓ $9 \times 9 = 81$

$\sqrt{361} = 19$ ✓ $19 \times 19 = 361$

$\sqrt{289} = 17$ ✓ $17 \times 17 = 289$

Always check your answer as shown.

🖉 **Note**   Every positive number actually has two square roots—one positive and one negative. For example, both $+9$ and -9 are considered to be square roots of 81 because $(+9)^2 = 81$ and $(-9)^2 = 81$. However, the $\sqrt{}$ symbol asks for only the positive or *principal* square root of the number inside the symbol.

If we wanted to ask for the negative square root of 81, we would write $-\sqrt{81}$.

If we wanted to ask for both square roots of 81, we would write $\pm\sqrt{81}$.

In trades work, we rarely, if ever, use the negative square root of a number. In Chapter 11, we will cover some higher-level applications that require both square roots.   •

How do you find the square root of any number? Using a calculator is the easiest way to find the square or other power of a number, or to find a square roots.

To find the square root of a number, simply press the $\boxed{\sqrt{}}$ key and then enter the number followed by the $\boxed{=}$ key. On some calculators, $\boxed{\sqrt{}}$ is a second function of the $\boxed{x^2}$ key.

EXAMPLE 9   To calculate $\sqrt{237}$, enter

$\boxed{\sqrt{}}$ **2 3 7** $\boxed{=}$ → ▮ 15.39480432 ▮

You will usually need to round your answer. Rounded to the nearest tenth, $\sqrt{237} \approx 15.4$.   •

🖉 **Note**   Some calculators automatically open parentheses with the $\boxed{\sqrt{}}$ function, and our key sequences will be based on ones that do. However, unless there are additional calculations after the square root is taken, it is not necessary to close parentheses before pressing $\boxed{=}$.   •

Your Turn   Perform the following calculations. (Round as indicated.)

(a) $\sqrt{760}$ (one decimal digit)   (b) $\sqrt{5.74}$ (two decimal digits)

Answers   (a) $\boxed{\sqrt{}}$ **7 6 0** $\boxed{=}$ → ▮ 27.5680975 ▮ $\sqrt{760} \approx 27.6$

(b) $\boxed{\sqrt{}}$ **5 . 7 4** $\boxed{=}$ → ▮ 2.39582971 ▮ $\sqrt{5.74} \approx 2.40$

! **Careful**   Notice that you cannot take the square root of a negative number. There is no number $\square$ such that $\square \times \square = -81$. If $\square$ is positive, then $\square \times \square$ is positive, and if $\square$ is negative, then $\square \times \square$ is also positive. You cannot multiply a number by itself and get a negative result.

(To perform higher-level algebra, mathematicians have invented imaginary numbers, which are used to represent the square roots of negative numbers. Students who go on to higher-level courses will learn about these numbers later.) ●

Order of Operations with Square Roots   Square roots have the same place in the order of operations as exponents. Simplify square roots before performing multiplication, division, addition, or subtraction.

The expanded order of operations now reads:

> **First,** simplify all operations within **parentheses.**
>
> **Second,** simplify all **exponents and roots,** left to right.
>
> **Next,** perform all **multiplications and divisions,** left to right.
>
> **Finally,** perform all **additions and subtractions,** left to right.

EXAMPLE 10   (a) $3\sqrt{49} = 3 \times \sqrt{49} = 3 \times 7 = 21$

Note that $3\sqrt{49}$ is a shorthand way of writing $3 \times \sqrt{49}$.

(b) $2 + \sqrt{64} = 2 + 8 = 10$

(c) $\dfrac{\sqrt{81}}{3} = \dfrac{9}{3} = 3$

(d) If an arithmetic operation appears under the square root symbol, perform this operation before taking the square root. For example,

$$\sqrt{9 + 16} = \sqrt{25} = 5$$

! **Careful**   Notice $\sqrt{9 + 16}$ is not the same as $\sqrt{9} + \sqrt{16} = 3 + 4 = 7$. Treat whatever is under the square root symbol as if it is enclosed in parentheses. Simplify these operations first. ●

Your calculator is programmed to follow the order of operations. When square roots are combined with other operations, you may enter the calculation in the order that it is written, but you must be aware of the correct use of parentheses. If your calculator automatically opens parentheses with square root, then you must close parentheses if another operation follows the square root. Remember that our key sequences are based on models that do automatically open parentheses.

EXAMPLE 11   (a) To calculate $\dfrac{38^2\sqrt{3}}{4}$, enter the following sequence:

3 8 $\boxed{x^2}$ $\boxed{\times}$ $\boxed{\sqrt{}}$ **3** $\boxed{)}$ $\boxed{\div}$ **4** $\boxed{=}$ → **625.2703415**

If your calculator does not open parentheses with square root, then do not press $\boxed{)}$ after entering 3.

(b) To calculate $\sqrt{17^2 - 15^2}$, enter

$\boxed{\sqrt{}}$ **1 7** $\boxed{x^2}$ $\boxed{-}$ **1 5** $\boxed{x^2}$ $\boxed{=}$ → **8.**

On this problem, if your calculator does not automatically open parentheses with square root, then you must press ⊙ after √ in order to take the square root of the entire expression. If you wish to avoid parentheses, use this alternate key sequence instead:

17 x^2 ⊖ **15** x^2 ⊜ √ (ANS) ⊜ → ⬛ *8.*

Recall that "ANS" stands for "last answer," so that √ (ANS) takes the square root of the answer to $17^2 - 15^2$.   ●

Your Turn   Calculate:

(a) $100 - \sqrt{100}$ (b) $\dfrac{4^2 \times \sqrt{36}}{3}$ (c) $\sqrt{169 - 144}$ (d) $\sqrt{64} + \sqrt{121}$

(e) $6 + 4\sqrt{25}$ (f) $49 - 3^2\sqrt{4}$ (g) $\dfrac{\sqrt{225} - \sqrt{81}}{\sqrt{36}}$ (h) $\sqrt{\dfrac{225 - 81}{36}}$

(i) $\sqrt{12.4^2 + 21.6^2}$ (j) $\dfrac{3(8.4)^2\sqrt{3}}{2}$ (Round (i) and (j) to the nearest tenth.)

Solutions   (a) $100 - \sqrt{100} = 100 - 10 = 90$

(b) $\dfrac{4^2 \times \sqrt{36}}{3} = \dfrac{16 \times (6)}{3} = \dfrac{96}{3} = 32$

(c) $\sqrt{169 - 144} = \sqrt{25} = 5$

(d) $\sqrt{64} + \sqrt{121} = 8 + 11 = 19$

(e) $6 + 4\sqrt{25} = 6 + 4 \times 5 = 6 + 20 = 26$

(f) $49 - 3^2\sqrt{4} = 49 - 9\sqrt{4} = 49 - 9 \times 2 = 49 - 18 = 31$

(g) $\dfrac{\sqrt{225} - \sqrt{81}}{\sqrt{36}} = \dfrac{15 - 9}{6} = \dfrac{6}{6} = 1$

(h) $\sqrt{\dfrac{225 - 81}{36}} = \sqrt{\dfrac{144}{36}} = \sqrt{4} = 2$

(i) √ **12.4** x^2 ⊕ **21.6** x^2 ⊜ → *24.90622412* ≈ 24.9

or **12.4** x^2 ⊕ **21.6** x^2 ⊜ √ (ANS) ⊜ → *24.90622412*

(j) **3** ⊗ **8.4** x^2 ⊗ √ **3** ⊙ ⊘ **2** ⊜ → *183.3202575* ≈ 183.3

Applications of Square Roots

EXAMPLE 12   **Agriculture**   A grower wishes to plant flowers in a square plot containing 1500 sq ft. What should be the length of each side of the plot?

To find the side length of a square, we simply take the square root of the area. Using a calculator,

√ **1500** ⊜ → *38.72983346*

Rounded to the nearest tenth, the length of each side of the flower plot should be approximately 38.7 ft.   ●

Your Turn   **Manufacturing**   A particular flat-screen television has a width w of 40 in. and a height h of 22.5 in. However, the size of a television is always given by the diagonal length of the screen. The formula for the diagonal length is $\sqrt{w^2 + h^2}$. Determine the diagonal length for this television rounded to the nearest inch.

Solution   Diagonal length $= \sqrt{(40 \text{ in.})^2 + (22.5 \text{ in.})^2} = 45.89\ldots$ in. ≈ 46 in.

$\sqrt{\ }$ **4 0** $\boxed{x^2}$ **+** **2 2.5** $\boxed{x^2}$ **=** → **45.89389938**

Now go to Exercises 6-4 for a set of problems on exponents and square roots.

Exercises 6-4   ## Exponents and Square Roots

A. Find the value of these.

1. 2^4	2. 3^2	3. 4^3
4. 5^3	5. 10^3	6. 7^2
7. 2^8	8. 6^2	9. 8^3
10. 3^4	11. 5^4	12. 10^5
13. $(-2)^3$	14. $(-3)^5$	15. 9^3
16. 6^0	17. 5^1	18. 1^4
19. 2^5	20. 8^2	21. $(-14)^2$
22. $(-21)^2$	23. 15^3	24. 16^2
25. $2^2 \times 3^3$	26. $2^6 \times 3^2$	27. $3^2 \times 5^3$
28. $2^3 \times 7^3$	29. $6^2 \times 5^2 \times 3^1$	30. $2^{10} \times 3^2$
31. $2 + 4^2$	32. $14 - 2^3$	33. $(8 - 2)^3$
34. $(14 \div 2)^2$	35. $3^3 - 8^2 \div 2^3$	36. $12 - 5 \times 2^5 \div 2^4$
37. $5 \times 6^2 - 4^3$	38. $64 \div 4^2 + 18$	39. $(28 \div 7)^2 - 14$
40. $3(8^2 - 3^3)$	41. $26 + 40 \div 2^3 \times 3^2$	42. $18 + 4 \times 5^2 - 3^4$

B. Calculate. Round to two decimal places if necessary.

1. $\sqrt{81}$	2. $\sqrt{144}$	3. $\sqrt{36}$
4. $\sqrt{16}$	5. $\sqrt{25}$	6. $\sqrt{9}$
7. $\sqrt{256}$	8. $\sqrt{400}$	9. $\sqrt{225}$
10. $\sqrt{49}$	11. $\sqrt{324}$	12. $\sqrt{121}$
13. $\sqrt{4.5}$	14. $\sqrt{500}$	15. $\sqrt{12.4}$
16. $\sqrt{700}$	17. $\sqrt{210}$	18. $\sqrt{321}$
19. $\sqrt{810}$	20. $\sqrt{92.5}$	21. $\sqrt{1000}$
22. $\sqrt{2000}$	23. $\sqrt{25000}$	24. $\sqrt{2500}$
25. $\sqrt{150}$	26. $\sqrt{300}$	27. $\sqrt{3000}$
28. $\sqrt{30000}$	29. $\sqrt{1.25}$	30. $\sqrt{1.008}$
31. $3\sqrt{20}$	32. $4 + 2\sqrt{5}$	33. $55 - \sqrt{14}$
34. $\dfrac{3^2\sqrt{24}}{3 \times 2}$	35. $\sqrt{2^5 - 3^2}$	36. $5.2^2 - 3\sqrt{52 - 5^2}$

37. $\sqrt{105^2 + 360^2}$ 38. $\dfrac{8}{3}(68.5 + 27.4 + \sqrt{68.5 \times 27.4})$

C. Practical Applications

1. **General Trades** Find the length of the side of a square whose area is 184.96 sq ft. (*Hint:* The area of a square is equal to the square of one of its sides. The side of a square equals the square root of its area.)

2. **Construction** A square building covers an area of 1269 sq ft. What is the length of each side of the building? (Round to the nearest tenth.)

3. **Mechanical Engineering** The following calculation must be performed to find the stress (in pounds per square inch) on a plunger. Find this stress to the nearest thousand psi.

$$\frac{3000\sqrt{34}}{0.196}$$

4. **Electronics** The resistance (in ohms) of a certain silicon diode is found using the expression

$$\sqrt{15 \times 10^6 \times 24}$$

Calculate this resistance to the nearest thousand ohms.

5. **Fire Protection** The velocity (in feet per second) of water discharged from a hose with a nozzle pressure of 62 psi is given by

$$12.14\sqrt{62}$$

Calculate and round to the nearest whole number.

6. **Police Science** Skid marks can be used to determine the maximum speed of a car involved in an accident. If a car leaves a skid mark of 175 ft on a road with a 35% coefficient of friction, its speed (in miles per hour) can be estimated using

$$\sqrt{30 \times 0.35 \times 175}$$

Find this speed to the nearest whole number.

7. **Allied Health** The Fit-4-U weight loss center calculates the body mass index (BMI) for each prospective client.

$$\text{BMI} = \frac{703 \times (\text{weight in pounds})}{(\text{height in inches})^2}$$

A person with a BMI greater than 30 is considered to be obese, and a person with a BMI over 25 is considered overweight. Calculate the BMI of a person 5 ft 6 in. tall weighing 165 lb.

8. **Roofing** The length L of a rafter on a shed roof is calculated using the formula

$$L = \sqrt{(H_2 - H_1)^2 + W^2}$$

Suppose that H_2, the maximum roof height, is 13 ft; H_1, the minimum wall height, is 7.5 ft; and W, the width of the shed, is 12 ft. Calculate the rafter length, not including allowance for overhang. (Round to the nearest tenth.)

9. **Construction** The Golden Gate bridge is about 1 mi long. On a warm day it expands about 2 ft in length. If there were no expansion joints to compensate for the expansion, how high a bulge would this produce? Use the following formula for your calculation, and round to the nearest foot.

$$\text{Height} \atop (\text{in ft}) = \sqrt{\left(\frac{5280}{2} + 1\right)^2 - \left(\frac{5280}{2}\right)^2}$$

10. **Electronics**   In computer technology, a **bus** is a set of connections between two or more devices that allows them to communicate with each other. The following calculation gives the binary bandwidth of a certain bus in units of mebibytes per second (MBps):

$$\frac{2^6 \times (15.625) \times 10^6}{8 \times 2^{20}}$$

Calculate this bandwidth to one decimal place.

11. **Allied Health**   As explained in Section 4-1 (see Exercises 4-1, problem C-35), the dosage of certain medications are sometimes based on a patient's body surface area (BSA). The BSA (in square meters) of a patient 6 feet tall and weighing 174 pounds is given by the expression

$$\sqrt{\frac{(174)(72)}{3131}}$$

Calculate the BSA to the nearest hundredth of a square meter.

12. **Electrical Trades**   The electrical resistance of a given length of wire is inversely proportional to the square of the diameter of the wire:

$$\frac{\text{resistance of wire } A}{\text{resistance of wire } B} = \frac{(\text{diameter of } B)^2}{(\text{diameter of } A)^2}$$

If a certain length of wire with a diameter of 34.852 mils has a resistance of 8.125 ohms, what is the resistance of the same length of the same composition wire with a diameter of 45.507 mils?

13. **Business and Finance**   Suppose you deposit $8000 in an investment account that pays 2% annual interest compounded quarterly. Perform the following calculation to determine the value of the account after 3 years:

$$\$8000(1.005)^{12}$$

Check your answers to the odd-numbered problems in the Appendix, then turn to Problem Set 6 on page 414 for practice problems on Chapter 6. If you need a quick review of the topics in this chapter, visit the chapter Summary first.

CHAPTER 6
SUMMARY   **Pre-Algebra**

Objective | Review

Objective	**Review**
Understand the meaning of signed numbers. (p. 380)	Negative numbers are used to represent real-world situations such as debts, losses, temperatures below zero, and locations below sea level. The negative part of the number line is a reflection of the positive part of the line. Therefore, the order of the negative numbers is the opposite of the order of the positive numbers.

Example: 14 is greater than 12 ($14 > 12$), but -14 is less than -12 ($-14 < -12$). A loss of $250 can be represented as the signed number $-\$250$.

Objective	Review
Add signed numbers. (p. 381)	To add two signed numbers, follow these steps.

Step 1 Find the absolute value of each number.

Step 2 If the two numbers have the same sign, add their absolute values and give the sum the sign of the original numbers.

Step 3 If the two numbers have opposite signs, subtract the absolute values and give the difference the sign of the number with the larger absolute value.

Example: (a) $(-3) + (-5) = -8$ ← Add absolute values: $3 + 5 = 8$
Absolute value $= 3$
Absolute value $= 5$
Sign of the sum is negative because both numbers are negative.

(b) $-21.7 + 14.2 = -7.5$ ← Subtract absolute values: $21.7 - 14.2 = 7.5$
Same sign as number with the larger absolute value (-21.7)

| **Subtract signed numbers.** (p. 390) | Rewrite the subtraction as an addition of the opposite signed number. |

Example: (a) $4 - 7 = 4 + (-7) = -3$

(b) $-11 - (-3) = -11 + (+3) = -8$

| **Multiply and divide signed numbers.** (p. 396) | To multiply or divide two signed numbers, first find the product or quotient of their absolute values. If the numbers have the same sign, the answer is positive. If they have opposite signs, the answer is negative. |

Example: (a) $(-12) \times 6 = -72$ Opposite signs, answer is negative

(b) $(-54) \div (-9) = +6$ Same sign, answer is positive

| **Work with exponents.** (p. 401) | In exponential form, the exponent indicates the number of times the base is multiplied by itself. |

Example: Exponent
Base ⟶ $4^3 = 4 \times 4 \times 4 = 64$
4 multiplied by itself 3 times

| **Use the order of operations.** (p. 404) | 1. Simplify all operations within **parentheses**. |

1. Simplify all operations within **parentheses**.
2. Simplify all **exponents**, left to right.
3. Perform all **multiplications** and **divisions**, left to right.
4. Perform all **additions** and **subtractions**, left to right.

Example: Simplify: $(4 \times 5)^2 + 4 \times 5^2$

First, simplify within parentheses. $= (20)^2 + 4 \times 5^2$

Second, simplify exponents. $= 400 + 4 \times 25$

Third, multiply. $= 400 + 100$

Finally, add. $= 500$

| **Find square roots.** (p. 406) | The square root of a positive number is that number such that when it is squared (multiplied by itself), it produces the original number. Square roots occupy the same place in the order of operations as exponents. |

Example: $\sqrt{169} = 13$ because 13^2 or $13 \times 13 = 169$

PROBLEM SET 6   Pre-Algebra

Answers to odd-numbered problems are given in the Appendix.

A. Rewrite each group of numbers in order from smallest to largest.

1. $-13, 7, 4, -8, -2$

2. $-0.6, 0, 0.4, -0.55, -0.138, -0.96$

3. $-150, -140, -160, -180, -120, 0$

4. $-\dfrac{5}{8}, \dfrac{3}{4}, \dfrac{1}{2}, -\dfrac{1}{2}, -\dfrac{3}{4}, \dfrac{3}{8}$

B. Add or subtract as indicated.

1. $-16 + 6$	2. $-5 + (-21)$
3. $7 - 13$	4. $-4 - (-4)$
5. $23 + (-38)$	6. $-13 - 37$
7. $20 - (-5)$	8. $11 + (-3)$
9. $-2.6 + 4.9$	10. $2\frac{1}{4} - 4\frac{3}{4}$
11. $-\$5.26 - \$3.89$	12. $-2\frac{3}{8} + \left(-7\frac{1}{2}\right)$

13. $-9 - 6 + 4 + (-13) - (-8)$

14. $-12.4 + 6\frac{1}{2} - \left(-1\frac{3}{4}\right) + (-18) - 5.2$

C. Multiply or divide as indicated.

1. $6 \times (-10)$	2. $(-7) \times (-5)$
3. $-48 \div (-4)$	4. $(-320) \div 8$
5. $\left(-1\frac{1}{8}\right) \times \left(-6\frac{1}{4}\right)$	6. $\left(-5\frac{1}{4}\right) \div 7$
7. $6.4 \div (-16)$	8. $(-2.4) \times 15$
9. $(-4) \times (-5) \times (-6)$	10. $72 \div (-4) \times (-3)$

11. $(2.4735) \times (-6.4)$

12. $(-0.675) \div (-2.125)$ (round to three decimal places)

D. Find the numerical value of each expression.

1. $12.5 \times (-6.3 + 2.8)$	2. $(-3) \times 7 - 4 \times (-9)$
3. $\dfrac{14 - 46}{(-4) \times (-2)}$	4. $\left(-4\frac{1}{2} - 3\frac{3}{4}\right) \div 8$
5. 2^6	6. 3^5
7. 17^2	8. 23^3
9. 0.5^2	10. 1.2^3
11. 0.02^2	12. 0.03^3
13. $(-5)^4$	14. 10^6
15. 4.02^2	16. $(-3)^3$
17. $(-2)^4$	18. $\sqrt{1369}$

Name

_____

Date

_____

Course/Section

_____

19. $\sqrt{784}$ 20. $\sqrt{4.41}$

21. $\sqrt{0.16}$ 22. $\sqrt{5.29}$

23. $5.4 + 3.3^2$ 24. $2.1^2 + 3.1^3$

25. 4.5×5.2^2 26. $4.312 \div 1.4^2$

27. $(3 \times 6)^2 + 3 \times 6^2$ 28. $28 - 4 \times 3^2 \div 6$

29. $2 + 2\sqrt{16}$ 30. $5\sqrt{25} - 4.48$

31. $\sqrt{0.49} - \sqrt{0.04}$ 32. $10 - 3\sqrt{1.21}$

Round to two decimal digits.

33. $\sqrt{80}$ 34. $\sqrt{106}$

35. $\sqrt{310}$ 36. $\sqrt{1.8}$

37. $\sqrt{4.20}$ 38. $\sqrt{3.02}$

39. $\sqrt{1.09}$ 40. $\sqrt{0.08}$

41. $7 - 3\sqrt{2}$ 42. $\sqrt{6} - \sqrt{5}$

43. $3\sqrt{2} - 2\sqrt{3}$ 44. $4\sqrt{10} + 1.5^2$

45. $\sqrt{9.65^2 - 5.73^2}$ 46. $\frac{4}{3}(13^2 + 18^2 + \sqrt{13^2 \cdot 18^2})$

E. Practical Applications

1. **Electrical Trades**   The current in a circuit changes from -2 A (amperes) to 5 A. What is the change in current?

2. **General Trades**   Find the side of a square whose area is 225 sq ft. (The side of a square is the square root of its area.)

3. **Meteorology**   In Hibbing, Minnesota, the temperature dropped 8°F overnight. If the temperature read -14°F before the plunge, what did it read afterward?

4. **Trades Management**   Morgan Plumbing had earnings of $3478 in the last week of January and expenses of $4326. What was the net profit or loss? (Indicate the answer with the appropriate sign.)

5. **Electrical Trades**   The current capacity (in amperes) of a service using a three-phase system for 124 kW of power and a line voltage of 440 volts is given by the calculation

$$\frac{124,000}{440\sqrt{3}}$$

Determine this current to the nearest whole number.

6. **Fire Protection**   The flow rate (in gallons per minute) of water from a $1\frac{1}{2}$-in. hose at 65 psi is calculated from the expression

$$29.7 \times (1.5)^2\sqrt{65}$$

Find the flow rate to the nearest gallon per minute.

7. **Construction**   The crushing load (in tons) for a square pillar 8 in. thick and 12 ft high is given by the formula

$$L = \frac{25 \times (8)^4}{12^2}$$

Find this load to the nearest hundred tons.

8. **Trades Management**   An auto shop was $14,560 "in the red" at the beginning of the year and $47,220 "in the black" at the end of the year. How much profit did the shop make during the year?

9. **Electronics**   In determining the bandwidth of a high-fidelity amplifier, a technician would first perform the following calculation to find the rise time of the signal:

$$T = \sqrt{3200^2 - 22.5^2}$$

Calculate this rise time to the nearest hundred. The time will be in nanoseconds.

10. **Electronics**   The following calculation is used to determine one of the voltages for a sweep generator:

$$V = 10 - 9.5 \div 0.03$$

Calculate this voltage. (Round to the nearest hundred.)

11. **Meteorology**   Use the formula

$$°C = \frac{5}{9} \times (°F - 32)$$

to convert the following Fahrenheit temperatures to Celsius, rounded to the nearest degree:
(a) −4°F   (b) 24°F

12. **Meteorology**   Use the formula

$$°F = \frac{9 \times °C}{5} + 32$$

to convert the following Celsius temperatures to Fahrenheit:
(a) −21°C   (b) −12°C

13. **Electrical Trades**   The value of a 1000-ohm resistor decreases by 0.2 ohm for every degree Celsius of temperature decrease. By how many ohms does the resistance decrease if the temperature drops from 125°C to −40°C?

14. **Meteorology**   The fastest temperature drop in recorded history in the United States occurred on January 10, 1911, in Rapid City, South Dakota. Between 7 A.M. and 7:15 A.M., the temperature plummeted from 55°F to 8°F. To the nearest hundredth, what was the average change in temperature per minute? Express the answer as a signed number.

15. **Electronics**   The percent resolution of a 12-bit digital-to-analog converter is given by the calculation

$$\frac{1}{2^{12} - 1} \times 10^2$$

Calculate this resolution to three decimal places.

16. **Allied Health**   The following expression gives the body surface area (BSA), in square meters, of a patient 185 cm tall and weighing 96 kg:

$$\sqrt{\frac{(96)(185)}{3600}}$$

Calculate this BSA to the nearest hundredth of a square meter.

17. **Electrical Trades**   Suppose an electric utility charges solar power customers a fixed fee of $20 per month but compensates the customer

$0.07 per kilowatt-hour of electricity exported to the grid. Determine the net monthly bill for each of the following situations: (See Example 2 on p. 393.)

(a) A homeowner exports 166 kWh to the grid in a particular month.

(b) A homeowner exports 358 kWh to the grid in a particular month.

18. **Electrical Trades** If a certain length of wire with diameter 42.653 mils has a resistance of 7.375 ohms, what is the resistance of the same length of the same composition of wire with a diameter of 30.497 mils? (*Hint:* Use the formula in problem 12 on p. 412.)

19. **Business and Finance** Suppose you have invested $6400 in an investment account paying 3% annual interest compounded monthly. Determine the value of the account after 4 years by performing the following calculation:

$$\$6400(1.0025)^{48}$$

CASE STUDY: Making Accurate Job Estimates

Allison, owner of Custom Car Design, specializes in the restoration of classic vehicles. Last year, Allison had been losing money because she had been underestimating the cost of some of her jobs. This year, Allison decided that, for each job, she would prepare a spreadsheet comparing her original estimate to the actual costs. She hoped this would help her understand how to prepare more accurate estimates in the future. To simplify her calculations, she decided to use positive numbers to represent areas where the estimate ended up being more than the cost (profit) and negative numbers to represent areas where the estimate ended up being less than the actual cost (loss).

Custom Car Design Budget vs. Actual Comparison			
Invoice #	48931		
Date	December 21		
Customer	Jim Smith		
Master Technician	Mike Brown		
	Estimate	Actual	Difference
Parts, Engine			
Alternator	$495.19		
Alternator Pulley	$34.96		
Generator	$79.47		
Voltage Regulator	$74.81		
Parts, Body			
Convertible Top	$379.35		
Convertible Top Pad	$56.72		
Fender Vent	$0.00		
Paint			
Paint, $8.70/can	$69.60		
Prepping Supplies	$58.00		
Labor			
Master Technician, $25.50/hour	$306.00		
Auto Body Specialist, $20.75/hour	$114.13		
Paint Specialist, $18.50/hour	$277.50		
Total			

A customer came in with a 1960 Ford Thunderbird and three requests: (1) Make some needed repairs to the charging system; (2) change the hard top to a convertible top; and (3) repaint the car a different color.

1. When the master technician estimated the cost of the four parts for the engine, he forgot to add in an extra 7.5% for sales tax. Determine the actual costs with the sales tax added in. Enter these in the spreadsheet in the "actual" column. Then fill in the differences using signed numbers.

2. The original estimate for the convertible top was $379.35, but the customer ended up choosing a black top, which cost only $329.80. The convertible top pad came in exactly at the estimated cost. However, when the body was being buffed, one of the fender vents was damaged and needed to be replaced at a cost of $95.84. Use this information to fill in the actual costs and the differences for the body parts section of the spreadsheet.

continued...

3. When the car was repainted, 11 cans of paint were used instead of the original amount estimated. Was this more or less than the original estimate? By how much? There was no change in the amount of prepping supplies required. Fill in the actual costs and differences for the paint section of the spreadsheet.

4. The employees working on the job actually put in the following number of hours: Master Technician, 14.75 hours; Auto Body Specialist, 5 hours; Paint Specialist, 18 hours. Which employees worked more than originally estimated? By how many hours? Which employees worked less than originally estimated? By how many hours? Use this information to fill in the actual costs and differences for the labor section of the spreadsheet.

5. Find the total of each of the three columns. Verify the total difference by finding the difference between the total estimate and total actual costs. Did the job come in over or under budget?

6. Based on the results of this job, what steps do you think Allison and her employees can take to ensure they do not lose money on future jobs due to inaccurate estimates?

Extension: Set up this spreadsheet in Excel or Google Sheets using formulas. In order to get a basic understanding of how to set up formulas in a spreadsheet, use a search engine to find resources on "How do you create formulas in a spreadsheet?"

7 Basic Algebra

Objective	Sample problems	For help, go to page

When you finish this chapter, you will be able to:

1. Evaluate formulas and literal expressions.

If $x = 2$, $y = 3$, $a = 5$, $b = 6$, find the value of

(a) $2x + y$ _____   426

(b) $\dfrac{1 + x^2 + 2a}{y}$ _____

(c) $A = x^2y$ $A =$ _____

(d) $T = \dfrac{2(a + b + 1)}{3x}$ $T =$ _____

(e) $P = abx^2$ $P =$ _____

2. Perform the basic algebraic operations.

(a) $3ax^2 + 4ax^2 - ax^2$ $=$ _____   443

(b) $5x - 3y - 8x + 2y$ $=$ _____

(c) $5x - (x + 2)$ $=$ _____   445

(d) $3(x^2 + 5x) - 4(2x - 3)$ $=$ _____   446

Name _____

Date _____

Course/Section _____

Objective	Sample problems	For help, go to page
3. Solve linear equations in one unknown and solve formulas.	(a) $3x - 4 = 11$ $x =$ _____	459
	(b) $2x = 18$ $x =$ _____	451
	(c) $2x + 7 = 43 - x$ $x =$ _____	475
	(d) Solve the following formula for N: $S = \dfrac{N}{2} + 26$ $N =$ _____	478
	(e) Solve the following formula for A: $M = \dfrac{(A - B)L}{8}$ $A =$ _____	
4. Translate simple English phrases and sentences into algebraic expressions and equations.	Write each phrase as an algebraic expression or equation.	
	(a) Four times the area. _____	484
	(b) Current squared times resistance. _____	
	(c) Efficiency is equal to 100 times the output divided by the input. _____	
	(d) Resistance is equal to 12 times the length divided by the diameter squared. _____	
5. Solve word problems.	(a) **Trades Management** An electrician collected $1265.23 for a job that included $840 labor, which was not taxed, and the rest for parts, which were taxed at 6%. Determine how much of the total was tax. _____	489
	(b) **General Trades** A plumber's helper earns $18 per hour plus $27 per hour overtime. If he works 40 regular hours during a week, how much overtime does he need in order to earn $1000? _____	
6. Multiply and divide simple algebraic expressions.	(a) $2y \cdot 3y$ _____	498
	(b) $(6x^4y^2)(-2xy^2)$ _____	
	(c) $3x(y - 2x)$ _____	

Objective	Sample problems		For help, go to page

$$\text{(d) } \frac{10x^7}{-2x^2}$$

_____   500

$$\text{(e) } \frac{6a^2b^4}{18a^5b^2}$$

_____

$$\text{(f) } \frac{12m^4 - 9m^3 + 15m^2}{3m^2}$$

_____

7. Use scientific notation.

Write in scientific notation.

(a) 0.000184

_____   505

(b) 213,000

_____

Calculate.

(c) $(3.2 \times 10^{-6}) \times (4.5 \times 10^2)$

_____   506

(d) $(1.56 \times 10^{-4}) \div (2.4 \times 10^3)$

_____

(Answers to these preview problems are given in the Appendix. Also, worked solutions to many of these problems appear in the chapter Summary.)

If you are certain that you can work *all* these problems correctly, turn to page 516 for the set of practice problems. If you cannot work one or more of the preview problems, turn to the page indicated to the right of the problem. Those who wish to master this material should turn to Section 7-1 and begin work there.

Basic Algebra

In this chapter you will study algebra, but not the very formal algebra that deals with theorems, proofs, sets, and abstract problems. Instead, we shall study practical or applied algebra as actually used by technical and trades workers.

CASE STUDY: Investigating the Basal Metabolic Rate

In the Case Study at the conclusion of the chapter, you will learn how these algebraic skills can help you work with several formulas for the **basal metabolic rate (or BMR).** These formulas are useful for doctors and nutritionists who are assisting patients in weight management.

7-1 Algebraic Language and Formulas

learning|**catalytics**™

1. Calculate:

$\frac{1}{3}(12)(9 + 4 + \sqrt{9 \times 4})$

2. If you travel for 3 hours at an average speed of 50 mph, how far would you go?

The most obvious difference between algebra and arithmetic is that, in algebra, letters are used to replace or to represent numbers. A mathematical statement in which letters are used to represent numbers is called a **literal expression**. Algebra is the arithmetic of literal expressions—a kind of symbolic arithmetic.

Any letters will do, but in practical algebra we use the normal lower- and uppercase letters of the English alphabet. It is helpful in practical problems to choose the letters to be used on the basis of their memory value: t for time, D for diameter, C for cost, A for area, and so on. The letter used reminds you of its meaning.

Multiplication   Most of the usual arithmetic symbols have the same meaning in algebra that they have in arithmetic. For example, the addition (+) and subtraction (−) signs are used in exactly the same way. However, the multiplication sign (×) of arithmetic looks like the letter x, so to avoid confusion we have other ways to show multiplication in algebra. The product of two algebraic quantities a and b, "a times b," may be written using

a raised dot: $a \cdot b$

parentheses: $a(b)$ or $(a)b$ or $(a)(b)$

nothing at all: ab

Obviously, this last way of showing multiplication won't do in arithmetic; we cannot write "two times four" as "24"—it looks like twenty-four. But it is a quick and easy way to write a multiplication in algebra.

Placing two quantities side by side to show multiplication is not new and it is not only an algebra gimmick; we use it every time we write 20 cents or 4 feet.

20 cents = 20 × 1 cent

4 feet   = 4 × 1 foot

Your Turn   Write the following multiplications using no multiplication symbols.

(a) 8 times a = _____   (b) m times p = _____

(c) 2 times s times t = _____   (d) 3 times x times x = _____

Answers   (a) $8a$ (b) mp (c) $2st$ (d) $3x^2$

A Closer Look   Did you notice in problem (d) that powers are written just as in arithmetic?

$x \cdot x = x^2$

$x \cdot x \cdot x = x^3$

$x \cdot x \cdot x \cdot x = x^4$ and so on. ●

Parentheses   Parentheses ( ) are used in arithmetic to show that some complicated quantity is to be treated as a unit. For example,

$$2(13 + 14 - 6)$$

means that the number 2 multiplies *all* of the quantity in the parentheses.

In exactly the same way in algebra, parentheses ( ), brackets [ ], or braces { } are used to show that whatever is enclosed in them should be treated as a single quantity. An expression such as

$$(3x^2 - 4ax + 2by^2)^2$$

should be thought of as (something)^2. The expression

$$(2x + 3a - 4) - (x^2 - 2a)$$

should be thought of as (first quantity) − (second quantity). Parentheses are the punctuation marks of algebra. Like the period, comma, or semicolon in regular sentences, they tell you how to read an expression and get its correct meaning.

Division   In arithmetic we would write "48 divided by 2" as

$$2\overline{)48} \quad \text{or} \quad 48 \div 2 \quad \text{or} \quad \frac{48}{2}$$

But the first two ways of writing division are used very seldom in algebra. Division is usually written as a fraction.

"x is divided by y" is written $\frac{x}{y}$

"$(2n + 1)$ divided by $(n - 1)$" is written $\frac{(2n + 1)}{(n - 1)}$ or $\frac{2n + 1}{n - 1}$

Your Turn   Write the following using algebraic notation.

(a) 8 times $(2a + b)$ = _____   (b) $(a + b)$ times $(a - b)$ = _____

(c) x divided by y^2 = _____   (d) $(x + 2)$ divided by $(2x - 1)$ = _____

Answers   (a) $8(2a + b)$ (b) $(a + b)(a - b)$ (c) $\frac{x}{y^2}$ (d) $\frac{x + 2}{2x - 1}$

Algebraic Expressions   The word *expression* is used very often in algebra. An **expression** is a general name for any collection of numbers and letters connected by arithmetic signs. For example,

$$x + y \qquad 2x^2 + 4 \qquad 3(x^2 - 2ab)$$

$$\frac{D}{T} \qquad \sqrt{x^2 + y^2} \qquad \text{and} \qquad (b - 1)^2$$

are all **algebraic expressions**.

If the algebraic expression has been formed by multiplying quantities, each multiplier is called a **factor** of the expression.

Expression	Factors
ab	a and b
$2x(x + 1)$	$2x$ and $(x + 1)$ or 2, x, and $(x + 1)$
$(R - 1)(2R + 1)$	$(R - 1)$ and $(2R + 1)$

The algebraic expression can also be a sum or difference of simpler quantities. Quantities that are added or subtracted are called **terms**.

Expression	Terms
$x + 4y$	x and $4y$
$2x^2 + xy$	$2x^2$ and xy
$A - R$	A and R

The first term is x
The second term is $4y$

Your Turn Now let's check to see if you understand the difference between terms and factors. In each algebraic expression below, tell whether the portion being named is a term or a factor.

(a) $2x^2 - 3xy$ $3xy$ is a _____

(b) $7x - 4$ $7x$ is a _____

(c) $4x(a + 2b)$ $4x$ is a _____

(d) $-2y(3y - 5)$ $(3y - 5)$ is a _____

(e) $3x^2y + 8y^2 - 9$ $3x^2y$ is a _____

Answers   (a) term   (b) term   (c) factor   (d) factor   (e) term

---

Evaluating Formulas   One of the most useful algebraic skills for any technical or practical work involves finding the value of an algebraic expression when the letters are given numerical values. A **formula** is a rule for calculating the numerical value of one quantity from the values of the other quantities. The formula or rule is usually written in mathematical form because algebra gives a brief, convenient to use, and easy to remember form for the rule. Here are a few examples of rules and formulas used in the trades:

1. **Rule:** Ohm's law: The voltage V (in volts) across a simple resistor is equal to the product of the current i (in amperes) through the resistor and the value of its resistance R (in ohms).

    Formula: $V = iR$

2. **Rule:** The detention time T in a basin is equal to 24 times the basin volume V, in cubic feet, divided by the flow F, in gallons per day.

    Formula: $T = \dfrac{24V}{F}$

3. **Rule:** The number of standard bricks N needed to build a wall is about 21 times the volume of the wall.

    Formula: $N = 21LWH$

Evaluating a formula or algebraic expression means to find its value by substituting numbers for the letters in the expression.

To be certain you do it correctly, follow this two-step process:

Step 1   Place the numbers being substituted in parentheses, and then substitute them in the formula.

Step 2   Do the arithmetic carefully *after* the numbers are substituted.

The order of operations for arithmetic calculations must be used when evaluating formulas. Remember:

> **First,** do any operations **inside parentheses**.
>
> **Second,** find all **powers** and **roots**.
>
> **Next,** do all **multiplications** or **divisions** left to right.
>
> **Finally,** do all **additions** or **subtractions** left to right.

EXAMPLE 1 **Construction** The following formula is used to calculate the number of joists J needed to cover a distance L when spaced 16 in. on center:

$$J = \tfrac{3}{4}L + 1$$

Suppose you know that $L = 44$ ft. You can determine J as follows:

First, substitute 44 for L. $\qquad\qquad\qquad\qquad J = \tfrac{3}{4}(44) + 1$

Then, do the arithmetic according to the order of operations.

$$\text{Multiply:} \qquad J = 33 + 1$$

$$\text{Then add:} \qquad J = 34$$

34 joists will be needed for the given length. ●

Your Turn (a) **Automotive Trades** Automotive engineers use the following formula to calculate the SAE rating (R) of an engine

$$R = \frac{D^2N}{2.5} \qquad \text{where } D \text{ is the diameter of a cylinder in inches and } N \text{ is the number of cylinders.}$$

Find R when $D = 3\tfrac{1}{2}$ in. and $N = 6$.

(b) **Allied Health** The basal metabolic rate, or BMR, is the minimum amount of energy needed in a 24-hour period to keep the body functioning. The BMR is based on a person's gender, age, weight, and height. One formula used to calculated the BMR for a woman is

$$\text{BMR} = 9.247w + 3.098h - 4.330a + 447.593$$

where w is the woman's weight in kilograms, h is her height in centimeters, a is her age in years, and the resulting BMR is in units of kilocalories per day.

Calculate the BMR for a woman who weighs 145 lb, has a height of 5 ft 6 in., and is 55 years old. Round to the nearest whole number. (Be careful of units!)

Solutions (a) **Step 1** Substitute the given values, in parentheses, into the formula.

$$R = \frac{(3.5)^2(6)}{2.5}$$

Step 2 Perform the calculation using the order of operations.

First, find the power: $R = \dfrac{(12.25)(6)}{2.5}$

$$\text{Next, multiply:} = \frac{73.5}{2.5}$$

$$\text{Finally, divide:} = 29.4$$

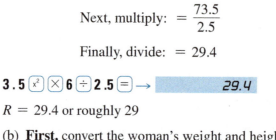

3.5 x^2 $\times$ **6** $\div$ **2.5** $=$ → `29.4`

$R = 29.4$ or roughly 29

(b) **First,** convert the woman's weight and height to metric units.

$$w = 145 \text{ lb} = 145 \text{ lb} \times \frac{0.4536 \text{ kg}}{1 \text{ lb}} = 65.772 \text{ kg}$$

$$h = 5 \text{ ft } 6 \text{ in.} = 66 \text{ in.} = 66 \text{ in.} \times \frac{2.54 \text{ cm}}{1 \text{ in.}} = 167.64 \text{ cm}$$

Then, substitute into the formula.

$$\text{BMR} = 9.247(65.772) + 3.098(167.64) - 4.330(55) + 447.593$$

Using a calculator, we have

9.247 $\times$ **65.772** $+$ **3.098** $\times$ **167.64** $-$ **4.33** $\times$ **55** $+$ **447.593**

$=$ → `1336.985404`

The woman's BMR is approximately 1337 kcal per day.

Distance, Rate, and Time   Suppose you drive a car for 2 hours at 60 miles per hour, and you want to know how far you have traveled. A lot of people would instinctively know to multiply 60 by 2 to come up with the answer of 120 miles. They are using one of the most basic and useful formulas in real-world applications, namely, "distance equals rate times time," or $D = RT$. Notice that the word "speed" is often used instead of "rate." Two other variations of this formula are

$$R = \frac{D}{T} \quad \text{and} \quad T = \frac{D}{R}$$

EXAMPLE 2   **Transportation**   The developers of a high-speed rail project in California have guaranteed that their high-speed train (or "bullet train") will cover the 400 miles from Los Angeles to San Jose in no more than 2 hours 40 minutes. What speed, in miles per hour, must the train average in order to fulfill this guarantee?

In this problem, we are given distance and time, and we must find the rate. But before we substitute the time into the formula, we need to convert 40 minutes to a fraction of an hour so that time is expressed in hours:

$$40 \text{ min} = 40 \text{ min} \times \frac{1 \text{ hr}}{60 \text{ min}} = \frac{40}{60} \text{ hr} = \frac{2}{3} \text{ hr}$$

Therefore, 2 hr 40 min $= 2\frac{2}{3}$ hr $= \frac{8}{3}$ hr

Next, substitute into the rate formula:

$$R = \frac{D}{T} = \frac{(400 \text{ mi})}{\frac{8}{3} \text{ hr}} = 400 \times \frac{3}{8} = 150 \text{ mi/hr}$$

The bullet train would have to average at least 150 mi/hr in order to meet the guaranteed time limit.   ●

In the next example, we must use two of these formulas to solve the problem.

EXAMPLE 3   **Sports and Leisure**   In the 2016 Summer Olympic Games, Usain Bolt of Jamaica won the men's 100-meter dash in a time of 9.81 seconds. In the 1964 Games, Bob Hayes of the United States won the same event in a time of 10.0 seconds. Suppose these two champions could have run against each other at those same speeds. How far behind, in meters, would Hayes be as Bolt crossed the finish line? To solve this problem,

First, determine Hayes's speed using the formula for R:

$$R = \frac{D}{T} = \frac{(100 \text{ m})}{(10.0 \text{ sec})} = 10 \text{ m/sec}$$

Next, calculate how far Hayes would have gone in the time it took Bolt to finish the race. Here we use the formula for D:

$$D = RT = (10 \text{ m/sec})(9.81 \text{ sec})$$

$$= \frac{10 \text{ m}}{1 \text{ sec}} \cdot \frac{9.81 \text{ sec}}{1} = 98.1 \text{ m}$$

Finally, subtract the distance that Hayes ran from the 100 meters that Bolt ran to determine how far behind Hayes would have been:

$$100 \text{ m} - 98.1 \text{ m} = 1.9 \text{ m}$$

Bob Hayes would have been 1.9 meters behind Bolt as Bolt crossed the finish line.

The solution can be entered on a calculator as follows:

100 ÷ **10** × **9.81** = → | 98.1 | **100** − ANS = → | 1.9 | ●

Your Turn   (a) **Transportation**   A truck driver drove for 80 miles at an average speed of 50 mi/hr and then drove an additional 210 miles at an average speed of 60 mi/hr. What was her total time for the trip?

(b) **Sports and Leisure**   In the 2016 Olympics, Elaine Thompson of Jamaica won the gold medal in the women's 200 meters in a time of 21.78 seconds. In the 1960 Olympics, Wilma Rudolph of the United States won the same race in a time of 24.13 seconds. If these two champions could have raced against each other at their gold-medal speeds, how far behind would Wilma Rudolph have been as Elaine Thompson crossed the finish line? (Round to the nearest tenth.)

Solutions   (a) There are two parts to the trip, so we must use the time formula twice.

For the first part: $T = \dfrac{D}{R} = \dfrac{(80 \text{ mi})}{(50 \text{ mi/hr})} = 1.6 \text{ hr}$

For the second part: $T = \dfrac{D}{R} = \dfrac{(210 \text{ mi})}{(60 \text{ mi/hr})} = 3.5 \text{ hr}$

Adding the two times together, we have:

Total time $= 1.6 \text{ hr} + 3.5 \text{ hr} = 5.1 \text{ hr}$, or 5 hr 6 min

(a) **First,** calculate Wilma Rudolph's speed:

$$R = \frac{D}{T} = \frac{(200 \text{ m})}{(24.13 \text{ sec})} \approx 8.288 \text{ m/sec}$$

Next, determine how far Rudolph would have gone in the time it took Elaine Thompson to finish the race (21.78 sec):

$$D = RT \approx (8.288 \text{ m/sec})(21.78 \text{ sec}) \approx 180.5 \text{ m}$$

Finally, subtract this from 200 m to determine how far behind Wilma Rudolph would have been when Elaine Thompson crossed the finish line:

$$200 \text{ m} - 180.5 \text{ m} \approx 19.5 \text{ m}$$

Using a calculator, we have:

2 0 0 ⊕ **2 4 . 1 3** ⊗ **2 1 . 7 8** ⊜ → ☐ 180.5221716
2 0 0 ⊖ (ANS) ⊜ → ☐ 19.47782843

When negative numbers must be substituted into a formula, it is especially important to place them in parentheses.

EXAMPLE 4   To evaluate the formula $B = x^2 - y$ for $x = 4$ and $y = -3$, follow these steps:

Step 1   Substitute the numbers in parentheses. $B = (4)^2 - (-3)$

Step 2   Do the arithmetic. $= 16 - (-3)$

$$= 16 + 3 = 19$$

Using parentheses in this way may seem like extra work for you, but it is the key to avoiding mistakes when evaluating formulas.   ●

Temperature Conversion   In Chapter 5 we used a chart to convert temperatures from Fahrenheit to Celsius and from Celsius to Fahrenheit. More precise conversions can be performed using two algebraic formulas.

To convert Fahrenheit F to Celsius C temperature, use the following formula:

$$°C = \frac{5(°F - 32)}{9}$$

To convert Celsius C to Fahrenheit F temperature, use this formula:

$$°F = \frac{9°C}{5} + 32$$

EXAMPLE 5   (a) To convert 85°F to Celsius, use the first of the preceding formulas:

Step 1   Substitute 85 for °F.

$$°C = \frac{5(°F - 32)}{9} = \frac{5(85 - 32)}{9}$$

Step 2   Calculate.

$$= \frac{5(53)}{9}$$

$$= \frac{265}{9}$$

$$\approx 29.4°C$$

(b) To convert $-18°C$ to Fahrenheit, use the second formula:

Step 1 Substitute -18 for $°C$.

$$°F = \frac{9°C}{5} + 32 = \frac{9(-18)}{5} + 32$$

Step 2 Calculate.

$$= \frac{-162}{5} + 32$$

$$= -32.4 + 32$$

$$= -0.4°F$$

More Practice

1. Evaluate the following formulas using the standard order of operations.

   (a) $I = \$500RT$ for $R = 0.03$ and $T = 6$ years

   (b) $V = 3.14r^2h$ for $r = 7.50$ cm and $h = 11.2$ cm

   (c) $T = \dfrac{H(A + B)}{2}$ for $H = 2\frac{1}{2}$ in., $A = 3\frac{1}{4}$ in., and $B = 4\frac{3}{4}$ in.

   (d) $L = 2\sqrt{2RH - H^2}$ for $R = 6.5$ ft and $H = 4$ ft

2. Use the temperature conversion formulas to find the following temperatures. Round to the nearest tenth.

   (a) $650°F =$ _____ $°C$ (b) $-160°C =$ _____ $°F$

   (c) $-10°F =$ _____ $°C$ (d) $52.5°C =$ _____ $°F$

Solutions

1. (a) $I = \$500(0.03)(6) = \$90$

   (b) $V = 3.14(7.50)^2(11.2) = 3.14(56.25)(11.2) = 1978.2$ cu cm

   (c) $T = \dfrac{\dfrac{5}{2}\left(\dfrac{13}{4} + \dfrac{19}{4}\right)}{2} = \dfrac{\dfrac{5}{2}\left(\dfrac{32}{4}\right)}{2} = \dfrac{20}{2} = 10$ sq in.

   (d) $L = 2\sqrt{2(6.5)(4) - 4^2} = 2\sqrt{2(6.5)(4) - 16}$
   $= 2\sqrt{52 - 16} = 2\sqrt{36} = 12$ ft

2. (a) $°C = \dfrac{5(650 - 32)}{9}$

   $= \dfrac{5(618)}{9}$

   $= \dfrac{3090}{9}$

   $\approx 343.3°C$

   (b) $°F = \dfrac{9(-160)}{5} + 32$

   $= \dfrac{-1440}{5} + 32$

   $= -288 + 32$

   $= -256°F$

   (c) $°C = \dfrac{5[(-10) - 32]}{9}$

   $= \dfrac{5[-42]}{9}$

   $= \dfrac{-210}{9}$

   $\approx -23.3°C$

   (d) $°F = \dfrac{9(52.5)}{5} + 32$

   $= \dfrac{472.5}{5} + 32$

   $= 94.5 + 32$

   $= 126.5°F$

On a calculator, do 2(a) like this:

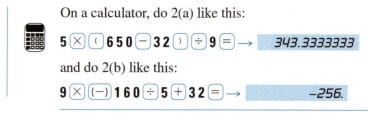

5 ✕ (6 5 0 − 3 2) ÷ 9 = → 　343.3333333

and do 2(b) like this:

9 ✕ (−) 1 6 0 ÷ 5 + 3 2 = → 　−256.

! Careful   Avoid the temptation to combine steps when you evaluate formulas. Take it slowly and carefully, follow the standard order of operations, and you will arrive at the correct answer. If you rush through problems like these, you may make mistakes.   ●

Now turn to Exercises 7-1 for a set of practice problems in evaluating formulas.

Exercises 7-1   Algebraic Language and Formulas

A. Write the following using algebraic notation.

1. 6 times y times x 2. $2x$ divided by 5

3. $(a + c)$ times $(a - c)$ 4. $4x$ times $(x + 6)$

5. $(n + 5)$ divided by $4n$ 6. $7y$ divided by $(y - 4)$

B. For each algebraic expression, tell whether the portion named is a term or a factor.

1. $5x$ x is a _____

2. $6x^2 + 4$ 4 is a _____

3. $5n^3$ n^3 is a _____

4. $2y^2 - 3y + 6$ $3y$ is a _____

5. $4abc$ b is a _____

6. $5m(m + n)$ $(m + n)$ is a _____

7. $2y(3y^2 - 6)$ $2y$ is a _____

8. $4x^2y + 6xy^2 + 2y^3$ $6xy^2$ is a _____

C. Evaluate each of the following formulas. Round to the nearest whole number if necessary.

1. $A = 3.14(R^2 - r^2)$ for $R = 42$ mm and $r = 33$ mm

2. $V = (L + W)(2L + W)$ for $L = 7.5$ ft, $W = 5.0$ ft

3. $I = PRT$ for $P = \$150, R = 0.05, T = 2$ years

4. $H = 2(a^2 + b^2)$ for $a = 2$ cm, $b = 1$ cm

5. $T = \dfrac{(A + B)H}{2}$ for $A = 3.26$ m, $B = 7.15$ m,
                                                          $H = 4.4$ m

6. $V = \dfrac{\pi D^2 H}{4}$ for $\pi = 3.14, D = 6.25$ in., $H = 7.2$ in.

7. $P = \dfrac{NR(T + 273)}{V}$ for $N = 5, R = 0.08, T = 27, V = 3$

8. $W = D(AB - \pi R^2)H$ for $D = 9$ lb/in.^{3}, $A = 6.3$ in., $B = 2.7$ in.,
                                                          $\pi = 3.14, R = 2$ in., $H = 1.0$ in.

9. $V = LWH$ for $L = 16.25$ m, $W = 3.1$ m, $H = 2.4$ m

10. $V = \pi R^2 A$ for $\pi = 3.14, R = 3.2$ ft, $A = 0.425$ ft

11. $P = \dfrac{V^2}{R}$ for $V = 12, R = 3.0$

12. $L = \dfrac{1}{2}(P_1 + P_2)s$ for $P_1 = 24$ ft, $P_2 = 36$ ft, $s = 8.0$ ft

13. $V = \dfrac{4\pi r^3}{3}$ for $\pi = 3.14, r = 12.50$ m

14. $A = \dfrac{3a^2\sqrt{3}}{4}$ for $a = 6.0$ in.

15. $f = \dfrac{1}{2\pi C X_c}$ for $\pi = 3.14, C = 0.000005,$ and $X_c = 50.0$

16. $V = \dfrac{1}{3}h\left(A_1 + A_2 + \sqrt{A_1 A_2}\right)$ for $h = 12$ in., $A_1 = 100$ sq in., $A_2 = 196$ sq in.

17. $A = \dfrac{1}{2}b\sqrt{a^2 - \left(\dfrac{b}{2}\right)^2}$ for $a = 12$ cm, $b = 10$ cm

18. $C = \$38.615 + \$0.186(T - 382)$ for $T = 502$

19. $A = \sqrt{s(s - a)(s - b)(s - c)}$ for $a = 14\dfrac{1}{2}$ in., $b = 13\dfrac{1}{4}$ in., $c = 12\dfrac{1}{4}$ in., $s = 20$ in.

20. $Z = \sqrt{R^2 + (X_L - X_C)^2}$ for $R = 300\ \Omega, X_L = 76.6\ \Omega, X_C = 482\ \Omega$

D. Practical Applications

1. **Retail Merchandising**   A store planner uses the following formula to determine how much space to give to a particular department:

$$A = \frac{V}{S}$$

where A is the amount of space in square feet, V is the projected annual sales volume, and S is the estimated sales per square foot.

If the shoe department of a store is projected to have annual sales of \$880,000, and the estimated sales per square foot is \$760 per year, how much space should be given to the department? (Round to the nearest ten square feet.)

2. **Electronics**   The current in amperes in a simple electrical circuit is given by Ohm's law $i = \frac{V}{R}$, where V is the voltage and R is the resistance of the circuit. Find the current in a circuit whose resistance is 10 ohms and that is connected across a 120-volt power source.

3. **Electrical Trades**   Find the power used in an electric light bulb, $P = i^2 R$, if the current $i = 0.80$ ampere and the resistance $R = 150$ ohms. P will be in watts.

4. **Electrical Trades**   When charging a battery with solar cells connected in series, the number N of solar cells required is determined by the following formula:

$$N = \frac{\text{voltage of the battery}}{\text{voltage of each cell}}$$

How many solar cells, providing 0.5 volt each, must be connected in series to provide sufficient voltage to charge a 12-volt battery?

5. **Meteorology**   Use the appropriate formula on page 430 to find the Fahrenheit temperature when the Celsius temperature is 40°.

6. **Welding**   In forge welding, metal pieces are heated to around 2200°F. Convert this to degrees Celsius. (See the formulas on page 430 and round to the nearest 100 degrees.)

7. **Welding**   One way in which welded parts shrink in groove joints is called **transverse shrinkage.** The amount of transverse shrinkage S_T in millimeters is given by the formula

$$S_T = \frac{A}{5T} + 0.05d$$

   where A is the cross-sectional area of the weld in square millimeters
   T is the thickness of the plates in millimeters
   d is the root opening in millimeters

   Suppose a plate is 16 mm thick, the cross-sectional area of the weld is 102 sq mm, and the root opening is 3.0 mm. Calculate the amount of transverse shrinkage. Round to the nearest tenth.

8. **Machine Trades**   The volume of a round steel bar depends on its length L and diameter D according to the formula $V = \frac{\pi D^2 L}{4}$. Find the volume of a bar 20.0 in. long and 3.0 in. in diameter. Use $\pi \approx 3.14$ and round to the nearest 10 cu in.

9. **Construction**   In framing a house, a contractor estimates the load L (in pounds) on a header board using the formula

$$L = \frac{THF}{2}$$

   where T is the length of the tail beam in feet, H is the length of the header in feet, and F is the floor load in pounds per square foot (lb/ft²). Find the load on a header if the tail beam is 15 feet long, the header is 8.5 feet long, and the floor load is 55 lb/ft². (Round to the nearest hundred pounds.)

10. **Electronics**   The total resistance R of two resistances a and b connected in parallel is $R = \frac{ab}{a + b}$. What is the total resistance if $a = 200$ ohms and $b = 300$ ohms?

11. **HVAC**   One way to save on energy costs is to have an air-tight house. One measure of air-tightness is the ACH (air changes per hour) of a house using a blower-door test. The ACH is given by the formula

$$\text{ACH} = \frac{60L}{V}$$

   where L is the air leakage in cubic feet per minute (cfm) and V is the volume of the house in cubic feet.

   Suppose that a blower-door test resulted in an air-leakage reading of 538 cfm in a house with a volume of 22,400 cu ft. Calculate the ACH to the nearest hundredth.

12. **Plumbing**   Use the formula $T = \frac{1}{2}(D - d)$ to find the wall thickness T of tubing having the following dimensions. (See the figure.)

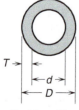

Problem 12

D, outside diameter	d, inside diameter
(a) 2.125 in.	1.500 in.
(b) $\frac{15}{16}$ in.	$\frac{5}{8}$ in.

13. **Roofing** The number of squares S of shingles required for a section of roof with area A and valley length V can be calculated using the following formula:

$$S = \frac{1.1A + V}{100}$$

Calculate the number of squares required for a 2280-sq-ft roof with a valley length of 60 ft. (Round up to the nearest whole number.)

14. **HVAC** The hourly heat loss H (in Btu) through a uniform slab of material can be calculated using the following formula:

$$H = \frac{A(T_i - T_o)}{R}$$

where A is the area of the slab in square feet, T_i and T_o are the inside and outside temperatures in degrees Fahrenheit, and R is the thermal resistance of the slab.

Calculate the hourly heat loss through 672 sq ft of walls with an R-value of 19.5 if T_i is 68°F and T_o is 42°F.

15. **Construction** The total load T in pounds carried by a single floor joist is given by

$$T = (W_D + W_L)dL$$

where W_D is the dead load of the floor, W_L is the live load, d is the center-to-center distance between the joists, and L is the length of the floor. Find T when $W_D = 10$ lb/ft², $W_L = 40$ lb/ft², $d = 2.0$ ft, and $L = 15$ ft.

16. **Construction** For a rectangular beam, the vertical shear force V is given by the formula

$$V = \frac{wL}{2}$$

where w is the weight in pounds per inch that each joist must support and L is the length of the unsupported span in inches. The horizontal shear stress h in pounds per square inch (psi) is given by the formula

$$h = \frac{3V}{2bd}$$

where b and d are the breadth and depth of the joists in inches and V is the vertical shear force as given by the first formula.

Use the following information to calculate V from the first formula and then find h using the second formula: $w = 12$ lb/in., $L = 120$ in., $b = 4.0$ in., and $d = 7.0$ in. (Round your final answer to the nearest whole number).

17. **Electrical Trades** An electrician uses a bridge circuit to locate a ground in an underground cable several miles long. The formula

$$\frac{R_1}{L - x} = \frac{R_2}{x} \quad \text{or} \quad x = \frac{R_2 L}{R_1 + R_2}$$

is used to find x, the distance to the ground. Find x if $R_1 = 750$ ohms, $R_2 = 250$ ohms, and $L = 4000$ ft.

18. **Machine Trades**   The cutting speed of a lathe is the rate, in feet per minute, that the revolving workpiece travels past the cutting edge of the tool. Machinists use the following formula to calculate cutting speed:

$$\text{Cutting speed, } C = \frac{\pi D N}{12}$$

where $\pi \approx 3.1416$, D is the diameter in inches of the work, and N is the turning rate in rpm. Find the cutting speed if a steel shaft 3.25 in. in diameter is turned at 210 rpm. (Round to the nearest whole number.)

19. **Electronics**   Find the power load P in kilowatts of an electrical circuit that takes a current I of 12.0 amperes at a voltage V of 220 volts if

$$P = \frac{VI}{1000}$$

20. **Aviation**   A jet engine developing T pounds of thrust and driving an airplane at V mph has a thrust horsepower, H, given approximately by the formula

$$H = \frac{TV}{375} \qquad \text{or} \qquad V = \frac{375H}{T}$$

Find the airspeed V if $H = 16{,}000$ hp and $T = 10{,}000$ lb.

21. **Electronics**   The resistance R of a conductor is given by the formula

$$R = \frac{PL}{A} \qquad \text{or} \qquad L = \frac{AR}{P}$$

where $P =$ coefficient of resistivity
   $L =$ length of conductor
   $A =$ cross-sectional area of the conductor

Find the length in cm of No. 16 nichrome wire needed to obtain a resistance of 8.0 ohms. For this wire, $P = 0.000113$ ohm-cm and $A = 0.013$ cm^2. (Round to the nearest centimeter.)

22. **Automotive Trades**   The pressure P and total force F exerted on a piston of diameter D are approximately related by the equation

$$P = \frac{4F}{\pi D^2} \qquad \text{or} \qquad F = \frac{\pi P D^2}{4}$$

Find the total force on a piston of diameter 3.25 in. if the pressure exerted on it is 150 lb/sq in. Use $\pi \approx 3.14$ and round to the nearest 50 lb.

23. **Water/Wastewater Treatment**   The operating time T, in hours, for a zeolite softener is

$$T = \frac{24W}{F}$$

where W is the amount of water treated, in gallons, and F is the average flow in gallons per day. If 265,000 gallons of water are treated, and the average flow is 175,000 gallons per day, find the operating time of the softener. (Round to the nearest tenth of an hour.)

24. **Electronics**   In this circuit, the shunt resistance (R_s) has been selected to pass a current of 0.5 mA. Determine the value of R_s using the formula given. (See the figure.)

$$R_s = \frac{R_m \cdot I_m}{I_s}$$

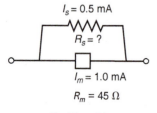

$I_s = 0.5$ mA
$R_s = ?$
$I_m = 1.0$ mA
$R_m = 45\ \Omega$

Problem 24

25. **Automotive Trades**   The overall valve lift L_v for a vehicle is given by $L_v = L_c R$, where L_c is the lift at the cam and R is the rocker-arm ratio. Find the overall valve lift if the lift at the cam is 0.350 in. and the rocker arm ratio is 1.50 to 1.

26. **Plumbing**   To determine the number N of smaller pipes that provide the same total flow as a larger pipe, use the formula

$$N = \frac{D^2}{d^2}$$

where D is the diameter of the larger pipe and d is the diameter of each smaller pipe. How many $1\frac{1}{2}$-in. pipes will it take to produce the same flow as a $2\frac{1}{2}$-in. pipe?

27. **Automotive Trades**   The braking distance d, in meters, of a vehicle traveling at a velocity v, in meters per second, is given by the formula

$$d = \frac{v^2}{2\mu g}$$

where μ is the coefficient of friction and g is the acceleration due to gravity. On Earth, $g \approx 9.8 \text{ m/s}^2$.

   (a) Determine the braking distance of a car traveling at 55 km/hr on dry pavement ($\mu = 0.80$). (Be sure to convert the speed to meters per second and round your answer to the nearest meter.)

   (b) Determine the braking distance of a car traveling at 65 mi/hr on wet pavement ($\mu = 0.45$). Give the final answer in feet rounded to the nearest ten feet.

28. **Machine Trades**   How many minutes will it take a lathe to make 17 cuts, each 24.5 in. in length, on a steel shaft if the tool feed F is 0.065 in. per revolution and the shaft turns at 163 rpm? Use the formula

$$T = \frac{LN}{FR}$$

where T is the cutting time (min), N is the number of cuts, L is the length of cut (in.), F is the tool feed rate (in./rev), and R is the rpm rate of the workpiece. (Round to the nearest minute.)

29. **Electrical Trades**   The resistance R of a wire (in ohms) is given by

$$R = \frac{KL}{d^2}$$

where K is the specific resistance of the material (in ohms), L is the length of the wire (in feet), and d is the diameter of the wire in thousandths of an inch (or mils).

   (a) Find R if $L = 1850$ feet, $K = 10.4$ ohms (copper), and $d = 162$ mils.

   (b) Find R if $L = 925$ feet, $K = 17.2$ ohms (aluminum), and $d = 289$ mils.

30. **Automotive Trades**   For modern automotive engines, horsepower (hp) is calculated as

$$\text{hp} = \frac{\text{torque (in lb} \cdot \text{ft)} \times \text{engine speed (in rpm)}}{5252}$$

In the 2017 Ford Mustang, at 4750 rpm the engine produces 429 lb · ft of torque. Calculate the horsepower at this engine speed. (Round to the nearest whole number.)

31. **Automotive Trades**   The theoretic airflow of an engine is given by

$$T = \frac{DS}{3456}$$

where D is the displacement in cubic inches and S is the engine speed in rpm (revolutions per minute). The theoretic airflow will be in cubic feet per minute, or cfm. Find the theoretic airflow if the displacement is 190.0 in.^{3} and the engine is running at 3800 rpm. (Round to the nearest tenth.)

32. **Business and Finance**   A company's net profit margin M, expressed as a percent, is determined using the formula

$$M = \frac{100I}{R}$$

where I represents net income and R represents net revenue. Calculate the net profit margin for a company with a net income of $254,000 and a net revenue of $7,548,000. (Round to the nearest hundredth of a percent.)

33. **Electronics**   For a sinusoidal voltage, the root mean square voltage, V_{rms}, is defined as $V_{rms} = \dfrac{V_{peak}}{\sqrt{2}}$. If V_{peak} is 2.50 volts what is V_{rms}? (Round to the nearest hundredth.)

34. **Allied Health**   The body mass index (BMI), which is used to determine whether a person is overweight, underweight, or in the normal range, is calculated with metric units as

$$BMI = \frac{\text{weight in kg}}{(\text{height in m})^2}$$

An adult with a BMI of less than 18.5 is considered to be underweight, and an adult with a BMI above 25.0 is considered to be overweight. A BMI between 18.5 and 25.0 indicates weight in the normal range. Find the BMI for each person described and determine whether they are overweight, underweight, or in the normal range.

(a) A person 1.87 m tall and weighing 95.0 kg.

(b A person 1.58 m tall and weighing 55.0 kg.

(c) A person 1.70 m tall and weighing 52.0 kg.

35. **Retail Merchandising**   A measure called **weeks of supply** is used to estimate the lead time on purchases in the retail industry. The formula for this measure is

$$\text{Weeks of supply} = \frac{(\text{on-hand stock} + \text{on-order stock})}{\text{week-to-date sales}}$$

Suppose that a store had $4200 of an item on hand, $5800 on order, and week-to-date sales were $2500. How many weeks of supply are available?

36. **HVAC**   We can estimate the cost C_N (in $ per million Btu) of a natural gas heater using the formula

$$C_N = \frac{10T}{E}$$

where T is the cost per therm of natural gas and E is the appliance efficiency.

Suppose that natural gas costs an average of $0.75 per therm, and a certain heater operates at 84% efficiency (use 0.84), what would the cost be for using *five* million Btu? (Round to the nearest cent.)

37. **Manufacturing**   The capacity C of a conveyor belt in metric tons per hour is given by the formula

$$C = 3.6SDA$$

where S is the belt speed in meters per second
         D is the density of the load material in kilograms per cubic meter
         A is the cross-sectional area of the load in square meters

Find the capacity of a belt if it is moving at 2.5 m/s and if the load has a density of 2000 kg/cu m with a cross-sectional area of 0.25 sq m.

38. **Water/Wastewater Treatment**   A tank with a leak or a drain at the bottom initially empties faster due to the pressure of the liquid, and then more slowly as it gradually empties and the pressure diminishes. **Torricelli's law** is a formula that gives the volume V of liquid remaining at any time t. For a tank that holds 50 gallons of water and empties in 20 minutes with an open drain in the bottom, the formula is

$$V = 50\left(1 - \frac{t}{20}\right)^2$$

where V is in gallons and t is in minutes. Find the remaining amount of water after 15 minutes if the drain is open. (Round to the nearest tenth.)

39. **Forestry**   The American Forests Organization has created the National Big Tree Program to locate the largest trees in the country. Each tree is given a numerical rating R using the following formula:

$$R = C + H + \tfrac{1}{4}S$$

where C is the circumference of the tree in inches, H is the height in feet, and S is the average crown spread in feet. (The circumference is measured at a height of 4.5 feet.) The champion giant sequoia tree in Sequoia National Park has a circumference of 1020 inches, a height of 274 feet, and an average crown spread of 107 feet. Calculate its rating rounded to the nearest whole number.

40. **Transportation**   A truck driver covers the first 150 miles of a 400-mile run at an average speed of 50.0 mi/hr. What average speed must he maintain during the rest of the run in order to finish in a *total* time of 7 hours?

41. **Aviation**   If an airplane pilot cruises at an average speed of 220 mi/hr for 3 hours 15 minutes, what distance does she fly?

42. **Sports and Leisure**   In the 1936 Summer Olympic Games, Jesse Owens won the 100-meter dash in a time of 10.3 seconds. If he could have raced at that speed against Usain Bolt in the 2016 Olympic Games, how far behind, to the nearest tenth of a meter, would Owens have been when Bolt finished the race? (See Example 3 on page 429.)

43. **General Interest**   Suppose that a steel band was placed tightly around the earth at the equator. If the temperature of the steel is raised 1°F, the metal expands 0.000006 in. each inch. How much space D would there be between the earth and the steel band if the temperature was raised 1°F? Use the formula

$$D \text{ (in ft)} = \frac{(0.000006)(\text{diameter of the earth})(5280)}{\pi}$$

where the diameter of the earth $= 7917$ miles. Use $\pi \approx 3.1415927$ and round to the nearest foot.

44. **Sheet Metal Trades**   The length of a chord of a circle is given by the formula $L = 2\sqrt{2RH - H^2}$, where R is the radius of the circle and H is the height of the arc above the chord. If you have a portion of a circular disk for which $R = 5\frac{1}{2}$ in. and $H = 4\frac{1}{4}$ in., what is the length of the chord? (See the figure and round to the nearest tenth of an inch.)

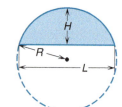

Problem 44

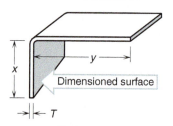

Problem 45

45. **Sheet Metal Trades**   To make a right-angle inside bend in sheet metal, the length of sheet used is given by the formula $L = x + y + \frac{1}{2}T$. Find L when $x = 6\frac{1}{4}$ in., $y = 11\frac{7}{8}$ in., $T = \frac{1}{4}$ in. (See the figure.)

46. **Manufacturing**   A millwright uses the following formula as an approximation to find the required length of a pulley belt. (See the figure.)

$$L = 2C + 1.57(D + d) + \frac{(D + d)}{4C}$$

Find the length of belt needed if

$C = 36$ in. between pulley centers
$D = 24$-in. follower
$d = 4.0$-in. driver

(Round to the nearest 10 in.)

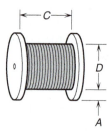

Problem 46

47. **Construction**   The rope capacity of a drum is given by the formula $L = ABC(A + D)$. How many feet of $\frac{1}{2}$-in. rope can be wound on a drum where $A = 6.0$ in., $C = 30.0$ in., $D = 24.0$ in., and $B = 1.05$ for $\frac{1}{2}$-in. rope? (Round to the nearest 100 in.) (See the figure.)

48. **Plumbing**   When a cylindrical container is lying on its side, the following formula can be used for calculating the volume of liquid in the container:

$$V = \frac{4}{3}h^2L \sqrt{\frac{d}{h} - 0.608}$$

Use this formula to calculate the volume of water in such a tank 6.0 ft long (L), 2.0 ft in diameter (d), and filled to a height (h) of 0.75 ft. (Round to the nearest tenth.)

Problem 47

49. **Meteorology**   The following formula is used to calculate the wind chill factor W in degrees Celsius:

$$W = 33 - \frac{(10.45 + 10\sqrt{V} - V)(33 - T)}{22.04}$$

where T is the air temperature in degrees Celsius and V is the wind speed in meters per second. Determine the wind chill for the following situations:

(a) Air temperature 7°C and wind speed 20 meters per second.

(b) Air temperature 28°F and wind speed 22 mph. (*Hint:* Use the conversion factors from Section 5-3 and the temperature formulas given on page 430 in this chapter to convert to the required units. Convert your final answer to degrees Fahrenheit.)

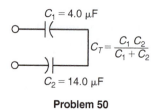

Problem 50

50. **Electronics**   Calculate the total capacitance C_T in the circuit shown in the margin. Round to the nearest 0.1 μF (microfarad).

51. **Allied Health** The *vital capacity* is the amount of air that can be exhaled after one deep inhalation. Vital capacity is usually measured directly, but it can be estimated using the following formula:

    $$V = (21.78 - 0.101a)h$$

    where a is the patient's age in years, h is his height in cm, and V is his vital capacity in cc (cubic centimeters).

    If Bob is 185 cm tall and 60.5 years old, calculate his vital capacity. (Round to the nearest ten cc.)

52. **Sheet Metal Trades** A sheet metal worker uses the following formula for calculating bend allowance, BA, in inches:

    $$BA = N(0.01743R + 0.0078T)$$

    where $N =$ number of degrees in the bend
                 $R =$ inside radius of the bend, in inches
                 $T =$ thickness of the metal, in inches

    Find BA for each of the following situations. (Round as directed.)

N	R	T
(a) 50°	$1\frac{1}{4}$ in.	0.050 in. (nearest hundredth)
(b) 40°	1.025 in.	0.0856 in. (nearest thousandth)

53. **Automotive Trades** The fuel economy F of a certain hybrid electric car at speeds greater than or equal to 50 mph is given by the formula

    $$F = 0.005V^2 - 1.25V + 96$$

    where F is in miles per gallon and V is the speed in miles per hour. What is the fuel economy at a speed of (a) 50 mph? (b) 80 mph? (Round to the nearest whole number.)

54. **Business and Finance** If an amount of money P is invested at $r\%$ annual interest compounded n times per year, then the total value V of the money after t years is given by

    $$V = P\left(1 + \frac{r}{100n}\right)^{nt}$$

    Find the value of an initial investment of $10,000 after five years if

    (a) it is invested at 4% annual interest compounded quarterly.

    (b) it is invested at 2% annual interest compounded yearly.

    (c) it is invested at 6% annual interest compounded monthly.

    (d) it is invested at 1% annual interest compounded daily (use $n = 360$).

55. **Sports and Leisure** Major League Baseball general managers are depending more and more on specialized mathematical formulas known as "analytics" to evaluate players and make personnel decisions. One such formula is known as BABIP, or "Batting Average on Balls in Play." The formula is defined as follows:

    $$\text{BABIP} = \frac{H - HR}{AB - K - HR + SF}$$

where H = hits, HR = home runs, AB = at bats, K = strikeouts, and SF = sacrifice flies. In 2015, Matt Duffy of the Giants and Kris Bryant of the Cubs were contending for Rookie of the Year honors. Here are their stats:

	At Bats	Hits	Home Runs	Strikeouts	Sacrifice Flies
Duffy	573	169	12	96	2
Bryant	559	154	26	199	5

Determine each player's BABIP rounded to three decimal places.

56. **Sports and Leisure**   The following formulas are used to calculate the number of points awarded for individual events in the men's decathlon:

For running events: Points $= A(B - P)^C$

For field events: Points $= A(P - B)^C$

P is the time or distance of the performance, and A, B, and C are constants as shown in the following table:

Event	A	B	C	P	Event Type
100 m	25.437	18.0	1.81	sec	Running
Long jump	0.14354	220	1.40	cm	Field
Shot put	51.39	1.5	1.05	m	Field
High jump	0.8465	75	1.42	cm	Field
400 m	1.53775	82	1.81	sec	Running
110 m hurdles	5.74352	28.5	1.92	sec	Running
Discus throw	12.91	4.0	1.1	m	Field
Pole vault	0.2797	100	1.35	cm	Field
Javelin throw	10.14	7.0	1.08	m	Field
1500 m	0.03768	480	1.85	sec	Running

Note that the P-column indicates the units that must be used in the formulas.

Use the two formulas and the values in the table to calculate the points awarded for each of the following performances. (Round to the nearest whole number and be sure to convert units where necessary!)

(a) A 100-meter run of 10.42 sec   (b) A long jump of 7.68 m

(c) A shot put of 15.38 m   (d) A 400-meter run of 47.25 sec

(e) A discus throw of 43.86 m   (f) A 1500-meter run of 4 min 31.68 sec

When you have completed these exercises, check your answers to the odd-numbered problems in the Appendix, then turn to Section 7-2 to learn how to add and subtract algebraic expressions.

7-2   Adding and Subtracting Algebraic Expressions

learning|catalytics™

1. True or false:
   $a(b + c) = ab + ac$
2. Calculate: $16 - (9 - 12)$

In the preceding section, we learned how to find the value of an algebraic expression after substituting numbers for the letters. There are other useful ways of using algebra where we must manipulate algebraic expressions *without* first substituting numbers for letters. In this section we learn how to simplify algebraic expressions by adding and subtracting terms.

Combining Like Terms   Two algebraic terms are said to be **like terms** if they contain exactly the same literal part. For example, the terms

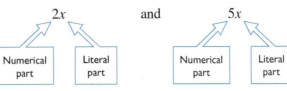

are like terms. The literal part, x, is the same for both terms.

The number multiplying the letters is called the **numerical coefficient** of the term.

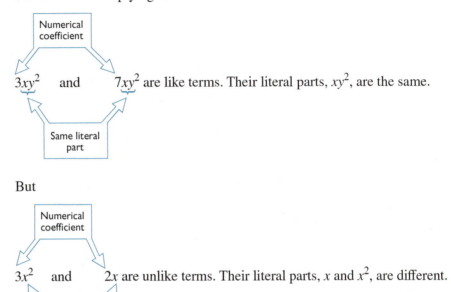

$3xy^2$ and $7xy^2$ are like terms. Their literal parts, xy^2, are the same.

But

Numerical coefficient

$3x^2$ and $2x$ are unlike terms. Their literal parts, x and x^2, are different.

Different literal parts

You can add and subtract like terms but not unlike terms. To add or subtract like terms, add or subtract their numerical coefficients and keep the same literal part.

EXAMPLE 1   (a) Add $2x + 3x$ like this:

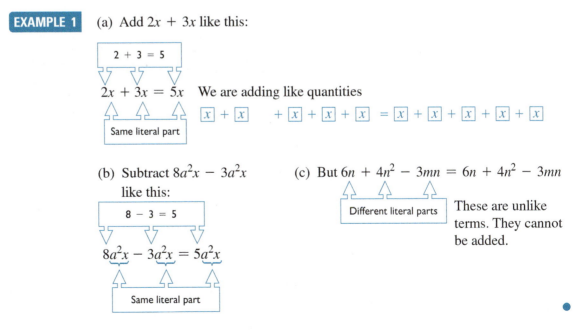

$2 + 3 = 5$

$2x + 3x = 5x$ We are adding like quantities

Same literal part

(b) Subtract $8a^2x - 3a^2x$ like this:

$8 - 3 = 5$

$8a^2x - 3a^2x = 5a^2x$

Same literal part

(c) But $6n + 4n^2 - 3mn = 6n + 4n^2 - 3mn$

Different literal parts

These are unlike terms. They cannot be added.

 Note   If a term does not have a numerical coefficient in front of it, the coefficient is 1. For example, x means $1x$. ●

Your Turn   Try these problems for practice.

(a) $12d^2 + 7d^2$ = _____   (b) $2ax - ax - 5ax =$ _____

(c) $3(y + 1) + 9(y + 1) =$ _____   (d) $8x^2 + 2xy - 2x^2 =$ _____

(e) $x - 6x + 2x$ = _____   (f) $4xy - xy + 3xy =$ _____

Answers   (a) $12d^2 + 7d^2 = 19d^2$

(b) $2ax - ax - 5ax = -4ax$ The term ax is equal to $1 \cdot ax$.

(c) $3(y + 1) + 9(y + 1) = 12(y + 1)$

(d) $8x^2 + 2xy - 2x^2 = 6x^2 + 2xy$

We cannot combine the x^2-term and the xy-term because the literal parts are not the same. They are unlike terms.

(e) $x - 6x + 2x = 1x - 6x + 2x = -3x$

(f) $4xy - xy + 3xy = 4xy - 1xy + 3xy = 6xy$

In general, to simplify a series of terms being added or subtracted, first group together like terms, then add or subtract.

EXAMPLE 2 $3x + 4y - x + 2y + 2x - 8y$

becomes

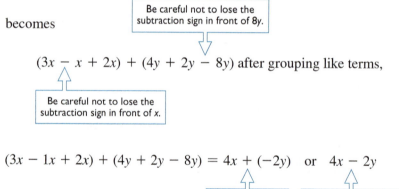

$(3x - x + 2x) + (4y + 2y - 8y)$ after grouping like terms,

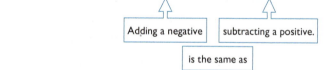

$(3x - 1x + 2x) + (4y + 2y - 8y) = 4x + (-2y)$ or $4x - 2y$

It is simpler to write the final answer as a subtraction rather than as an addition.   ●

Your Turn   Simplify the following expressions by adding and subtracting like terms.

(a) $5x + 4xy - 2x - 3xy$

(b) $3ab^2 + a^2b - ab^2 + 3a^2b - a^2b$

(c) $x + 2y - 3z - 2x - y + 5z - x + 2y - z$

(d) $17pq - 9ps - 6pq + ps - 6ps - pq$

(e) $4x^2 - x^2 + 2x + 2x^2 + x$

Solutions   (a) $(5x - 2x) + (4xy - 3xy) = 3x + xy$

(b) $(3ab^2 - ab^2) + (a^2b + 3a^2b - a^2b) = 2ab^2 + 3a^2b$

(c) $(x - 2x - x) + (2y - y + 2y) + (-3z + 5z - z) = -2x + 3y + z$

(d) $(17pq - 6pq - pq) + (-9ps + ps - 6ps) = 10pq - 14ps$

(e) $(4x^2 - x^2 + 2x^2) + (2x + x) = 5x^2 + 3x$

Expressions with Parentheses   Parentheses* are used in algebra to group together terms that are to be treated as a unit. Adding and subtracting expressions usually involves working with parentheses. There are three main rules for removing parentheses.

| **Rule 1**   If the parenthesis has a plus sign in front, simply remove the parentheses. |

EXAMPLE 3   (a) $1 + (3x + y) = 1 + 3x + y$

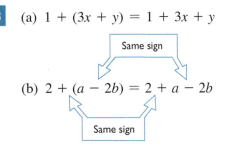

(b) $2 + (a - 2b) = 2 + a - 2b$

| **Rule 2**   If the parenthesis has a negative sign in front, change the sign of each term inside, then remove the parentheses. |

EXAMPLE 4

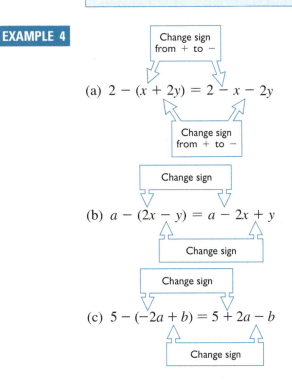

(a) $2 - (x + 2y) = 2 - x - 2y$

(b) $a - (2x - y) = a - 2x + y$

(c) $5 - (-2a + b) = 5 + 2a - b$

✎ **Note**   In using Rule 2 you are simply rewriting subtraction as addition of the opposite. However, you must add the opposite of *all* terms inside parentheses.   ●

*The word "parentheses" is plural, referring to a pair: one to open, and one to close. The word "parenthesis" is singular, referring to just one of the pair.

Your Turn   Simplify the following expressions by using these two rules to remove parentheses.

(a) $x + (2y - a^2)$ (b) $4 - (x^2 - y^2)$

(c) $-(x + 1) + (y + a)$ (d) $ab - (a - b)$

(e) $(x + y) - (p - q)$ (f) $-(-x - 2y) - (a + 2b)$

Answers   (a) $x + 2y - a^2$ (b) $4 - x^2 + y^2$

(c) $-x - 1 + y + a$ (d) $ab - a + b$

(e) $x + y - p + q$ (f) $x + 2y - a - 2b$

A third rule is needed when a multiplier is in front of the parentheses.

> **Rule 3**   If the parenthesis has a multiplier in front, multiply each term inside the parentheses by the multiplier.

Note   Rule 3 is often called the "distributive property" because you are *distributing* the factor in front to all terms inside parentheses. ●

EXAMPLE 5   (a) $+2(a + b) = +2a + 2b$

Think of this as $(+2)(a + b) = (+2)a + (+2)b$
$$= +2a + 2b$$

Each term inside the parentheses is multiplied by $+2$.

(b) $-2(x + y) = -2x - 2y$

Think of this as $(-2)(x + y) = (-2)x + (-2)y$

Each term inside the parentheses is multiplied by -2.

(c) $-(x - y) = (-1)(x - y)$
$$= (-1)(x) + (-1)(-y) = -x + y$$

(d) $2 - 5(3a - 2b) = 2 + (-5)(3a - 2b) = 2 + (-5)(3a) + (-5)(-2b)$
$$= 2 + (-15a) + (+10b) \quad \text{or} \quad 2 - 15a + 10b$$ ●

Careful   In the last example it would be incorrect to first subtract the 5 from the 2. The order of operations requires that we multiply before we subtract. ●

When you multiply negative numbers, you may need to review the arithmetic of negative numbers starting on page 380.

Notice that we must multiply *every* term inside the parentheses by the number outside the parentheses.

Your Turn   Simplify the following expressions by removing parentheses.

(a) $2(2x - 3y)$ (b) $1 - 4(x + 2y)$ (c) $a - 2(b - 2x)$

(d) $x^2 - 3(x - y)$ (e) $p - 2(-y - 2x)$ (f) $3(x - y) - 2(2x^2 + 3y^2)$

Solutions

(a) $2(2x - 3y) = 2(2x) + 2(-3y) = 4x + (-6y)$ or $4x - 6y$

(b) $1 - 4(x + 2y) = 1 + (-4)(x) + (-4)(2y) = 1 + (-4x) + (-8y)$
 or $1 - 4x - 8y$

(c) $a - 2(b - 2x) = a + (-2)(b) + (-2)(-2x) = a + (-2b) + (+4x)$
 or $a - 2b + 4x$

(d) $x^2 - 3(x - y) = x^2 + (-3)(x) + (-3)(-y) = x^2 + (-3x) + (+3y)$
 or $x^2 - 3x + 3y$

(e) $p - 2(-y - 2x) = p + (-2)(-y) + (-2)(-2x) = p + (+2y) + (+4x)$
 or $p + 2y + 4x$

(f) $3(x - y) - 2(2x^2 + 3y^2) = 3(x) + 3(-y) + (-2)(2x^2) + (-2)(3y^2)$
$$= 3x + (-3y) + (-4x^2) + (-6y^2)$$
 or $3x - 3y - 4x^2 - 6y^2$

Once you can simplify expressions by removing parentheses, it is easy to add and subtract like terms.

EXAMPLE 6 $(3x - y) - 2(x - 2y) = 3x - y - 2x + 4y$ Simplify by removing parentheses.

$$= \underbrace{3x - 2x} \ \underbrace{- y + 4y} \qquad \text{Group like terms.}$$

$$= x + 3y \qquad\qquad\qquad \text{Combine like terms.} \qquad \bullet$$

More Practice   Try these problems for practice.

(a) $(3y + 2) + 2(y + 1)$ (b) $(2x + 1) + 3(4 - x)$

(c) $(a + b) - (a - b)$ (d) $2(a + b) - 2(a - b)$

(e) $2(x - y) - 3(y - x)$ (f) $-2(3x - 5) - 4(x - 1)$

Answers

(a) $5y + 4$ (b) $-x + 13$ (c) $2b$

(d) $4b$ (e) $5x - 5y$ (f) $-10x + 14$

Some step-by-step solutions:

(a) $(3y + 2) + 2(y + 1) = 3y + 2 + 2y + 2$
$$= 3y + 2y + 2 + 2$$
$$= 5y + 4$$

(b) $(2x + 1) + 3(4 - x) = 2x + 1 + 12 - 3x$
$$= 2x - 3x + 1 + 12$$
$$= -x + 13$$

(c) $(a + b) - (a - b) = a + b - a - (-b)$
$$= a + b - a + b$$
$$= a - a + b + b = 0 + 2b = 2b$$

(f) $-2(3x - 5) - 4(x - 1) = -2(3x) - 2(-5) - 4x - 4(-1)$
$$= -6x + 10 - 4x + 4$$
$$= -6x - 4x + 10 + 4 = -10x + 14$$

Now continue with Exercises 7-2 for additional practice in addition and subtraction of algebraic expressions.

Exercises 7-2 | Adding and Subtracting Algebraic Expressions

A. Simplify by adding or subtracting like terms.

1. $3y + y + 5y$

2. $4x^2y + 5x^2y$

3. $E + 2E + 3E$

4. $ax - 5ax$

5. $9B - 2B$

6. $3m - 3m$

7. $3x^2 - 5x^2$

8. $4x + 7y + 6x + 9y$

9. $6R + 2R^2 - R$

10. $1.4A + 0.05A - 0.8A^2$

11. $x - \frac{1}{2}x - \frac{1}{4}x - \frac{1}{8}x$

12. $x + 2\frac{1}{2}x - 5\frac{1}{2}x$

13. $2 + W - 4.1W - \frac{1}{2}$

14. $q - p - 1\frac{1}{2}p$

15. $2xy + 3x + 4xy$

16. $ab + 5ab - 2ab$

17. $x^2 + x^2y + 4x^2 + 3x$

18. $1.5p + 0.3pq + 3.1p$

B. Simplify by removing parentheses and, if possible, combining like terms.

1. $3x^2 + (2x - 5)$

2. $6 + (-3a + 8b)$

3. $8m + (4m^2 + 2m)$

4. $9x + (2x - 5x^2)$

5. $2 - (x + 5y)$

6. $7x - (4 + 2y)$

7. $3a - (8 - 6b)$

8. $5 - (w - 6z)$

9. $4x - (10x^2 + 7x)$

10. $12m - (6n + 4m)$

11. $15 - (3x - 8)$

12. $-12 - (5y - 9)$

13. $-(x - 2y) + (2x + 6y)$

14. $-(3 + 5m) + (11 - 4m)$

15. $-(14 + 5w) - (2w - 3z)$

16. $-(16x - 8) - (-2x + 4y)$

17. $3(3x - 4y)$

18. $4(5a + 6b)$

19. $-8(7m + 6)$

20. $-2(6x - 3)$

21. $x - 5(3 + 2x)$

22. $4y - 2(8 + 3y)$

23. $9m - 7(-2m + 6)$

24. $w - 5(4w - 3)$

25. $3 - 4(2x + 3y)$

26. $6 - 2(3a - 5b)$

27. $12 - 2(3w - 8)$

28. $2 - 11(7 + 5a)$

29. $2(3x + 4y) - 4(6x^2 - 5y^2)$

30. $-4(5a - 6b) + 7(2ab - 4b^2)$

31. $(22x - 14y) + 3(8y - 6x)$

32. $8(3w - 5z) + (9z - 6w)$

33. $6(x + y) - 3(x - y)$

34. $-5(2x - 3x^2) + 2(9x^2 + 8x)$

35. $4(x^2 - 6x + 8) - 6(x^2 + 3x - 5)$

36. $2(3a - 5b + 6ab) - 8(2a + b - 4ab)$

When you have completed these exercises, check your answers to the odd-numbered problems in the Appendix, then turn to Section 7-3 to learn how to solve algebraic equations.

7-3 Solving Simple Equations

An arithmetic equation such as $3 + 2 = 5$ means that the number named on the left $(3 + 2)$ is the same as the number named on the right (5).

An algebraic equation such as $x + 3 = 7$ is a statement that the sum of some number x and 3 is equal to 7. If we choose the correct value for x, the number $x + 3$ will be equal to 7.

x is a **variable**, a symbol that stands for a number in an equation, a blank space to be filled. Many numbers might be put in the space, but only one makes the equation a true statement.

Your Turn Find the missing numbers in the following arithmetic equations.

(a) $37 +$ _____ $= 58$ (b) _____ $- 15 = 29$

(c) $4 \times$ _____ $= 52$ (d) $28 \div$ _____ $= 4$

Answers (a) $37 + 21 = 58$ (b) $44 - 15 = 29$

(c) $4 \times 13 = 52$ (d) $28 \div 7 = 4$

We could have written these equations as follows, with variables instead of blanks:

$$37 + A = 58 \qquad B - 15 = 29 \qquad 4C = 52 \qquad \frac{28}{D} = 4$$

Of course, any letters would do in place of A, B, C, and D in these algebraic equations.

How did you solve these equations? You probably "eyeballed" them—mentally juggled the other information in the equation until you found a number that made the equation true. Solving algebraic equations is very similar except that we can't "eyeball" it entirely. We need certain and systematic ways of solving the equation that will produce the correct answer quickly every time.

In this section you will learn first what a solution to an algebraic equation is—how to recognize it if you stumble over it in the dark—then how to solve linear equations.

Solution Each value of the variable that makes an equation true is called a **solution** of the equation.

EXAMPLE 1 (a) The solution of $x + 3 = 7$ is $x = 4$.

$(4) + 3 = 7$

(b) The solution of the equation $2x - 9 = 18 - 7x$ is $x = 3$.

$$2(3) - 9 = 18 - 7(3)$$
$$6 - 9 = 18 - 21$$
$$-3 = -3$$

For certain equations, more than one value of the variable may make the equation true.

EXAMPLE 2 ✓ The equation $x^2 + 6 = 5x$ is true for $x = 2$,

$$(2)^2 + 6 = 5(2)$$
$$4 + 6 = 5 \cdot 2$$
$$10 = 10$$

and it is also true for $x = 3$.

✓
$$(3)^2 + 6 = 5(3)$$
$$9 + 6 = 5 \cdot 3$$
$$15 = 15$$

●

More Practice   Determine whether each given value of the variable is a solution to the equation.

(a) For $4 - x = -3$ is $x = -7$ a solution?

(b) For $3x - 8 = 22$ is $x = 10$ a solution?

(c) For $-2x + 15 = 5 + 3x$ is $x = -2$ a solution?

(d) For $3(2x - 8) = 5 - (6 - 4x)$ is $x = 11.5$ a solution?

(e) For $3x^2 - 5x = -2$ is $x = \dfrac{2}{3}$ a solution?

Answers   (a) No—the left side equals 11.

(b) Yes.   (c) No—the left side equals 19, the right side equals -1.

(d) Yes.   (e) Yes.

✓   Here is the check for (e): $3\left(\dfrac{2}{3}\right)^2 - 5\left(\dfrac{2}{3}\right) = -2$

$$3\left(\dfrac{4}{9}\right) - 5\left(\dfrac{2}{3}\right) = -2$$

$$\dfrac{4}{3} - \dfrac{10}{3} = -2$$

$$-\dfrac{6}{3} = -2$$

$$-2 = -2$$

Equivalent Equations   Equations with the exact same solution are called **equivalent equations**. The equations $2x + 7 = 13$ and $3x = 9$ are equivalent because substituting the value 3 for x makes them both true.

We say that an equation with the variable x is *solved* if it can be put in the form

$x = \square$ where $\square$ is some number

For example, the solution to the equation

$2x - 1 = 7$ is $x = 4$

because $2(4) - 1 = 7$

or $8 - 1 = 7$

is a true statement.

Solving Equations by Adding or Subtracting

Equations as simple as the previous one are easy to solve by guessing, but guessing is not a very dependable way to do mathematics. We need some sort of rule that will enable us to rewrite the equation to be solved ($2x - 1 = 7$, for example) as an equivalent solution equation ($x = 4$).

The general rule is to treat every equation as a balance of the two sides.

$2x = 6$	$\dfrac{2x \quad = \quad 6}{\blacktriangle}$
$3x - 4 = 8 - x$	$\dfrac{3x - 4 \quad = \quad 8 - x}{\blacktriangle}$

Any changes made in the equation must not disturb this balance.

Any operation performed on one side of the equation must also be performed on the other side.

Two kinds of balancing operations may be used.

1. Adding or subtracting a number on both sides of the equation does not change the balance.

Original equation: $a = b$

$a + 2 = b + 2$

$a - 2 = b - 2$

2. Multiplying or dividing both sides of the equation by a number (but not zero) does not change the balance.

$2 \cdot a = 2 \cdot b$

$\dfrac{a}{3} = \dfrac{b}{3}$

The trick to solving an equation is to choose the balancing operation or operations that will isolate the variable only on one side of the equation.

EXAMPLE 3

Let's work through an example.

Solve: $x - 4 = 2$.

Step 1 We want to change this equation to an equivalent equation with only x on the left, so we add 4 to each side of the equation.

$$x - 4 + 4 = 2 + 4$$

Step 2 Combine terms.

$$x \underbrace{- 4 + 4}_{0} = \underbrace{2 + 4}_{6}$$

$$x = 6 \quad \text{Solution}$$

We chose the operation (addition) that isolated x on one side. But to maintain the balance, we also performed the same operation on the other side of the equation. ●

Your Turn   Use a balancing operation to solve the equation

$9 + x = 14$

Solution   **Solve:** $9 + x = 14$

Step 1   We want to change this equation to an equivalent equation with only x on the left, so we subtract 9 from each side of the equation.

$9 + x \boxed{- 9} = 14 \boxed{- 9}$

Step 2   Combine terms.

$\underbrace{x + 9 - 9}_{0} = \underbrace{14 - 9}_{5}$ where $9 + x = x + 9$

The solution is: $x = 5$

☑ $9 + (5) = 14$

$14 = 14$

Learning Help   When adding or subtracting numbers on both sides of an equation, you may find it easier to show the work underneath the original equation. Here's what this would look like:

For Example 3		**For Your Turn**	
Solve:	$x - 4 = 2$	Solve:	$9 + x = 14$
Add 4 to both sides:	$\underline{+4 \quad +4}$	Subtract 9 from both sides:	$\underline{-9 \qquad -9}$
Combine terms:	$x + 0 = 6$	Combine terms:	$0 + x = 5$
Solution:	$x = 6$	Solution:	$x = 5$ ●

More Practice   Solve these in the same way.

(a) $x - 7 = 10$ (b) $12 + x = 27$ (c) $x + 6 = 2$

(d) $8.4 = 3.1 + x$ (e) $6.7 + x = 0$ (f) $\frac{1}{4} = x - \frac{1}{2}$

Solutions   (a) **Solve:** $x - 7 = 10$

Add 7 to each side. $x - 7 \boxed{+ 7} = 10 \boxed{+ 7}$

Combine terms. $\underbrace{x - 7 + 7}_{0} = \underbrace{10 + 7}_{17}$

$x = 17$ Solution

☑ $(17) - 7 = 10$

$10 = 10$

(b) **Solve:** $12 + x = 27$

Subtract 12 from each side. $12 + x \boxed{- 12} = 27 \boxed{- 12}$

Combine terms. $\underbrace{x + 12 - 12}_{0} = \underbrace{27 - 12}_{15}$
(Note that $12 + x = x + 12$.)

$x = 15$ Solution

☑ $12 + (15) = 27$

$27 = 27$

(c) **Solve:** $x + 6 = 2$

Subtract 6 from each side. $\qquad\qquad x + 6 - 6 = 2 - 6$

Combine terms. $\qquad\qquad\qquad\quad \underbrace{x + 6 - 6}_{0} = \underbrace{2 - 6}_{-4}$

The solution is a negative number. $\qquad x = -4$
Remember, any number, positive or
negative, may be the solution of an
equation.

$$(-4) + 6 = 2$$
$$2 = 2$$

(d) **Solve:** $8.4 = 3.1 + x$

The variable x is on the right side of this equation.

Subtract 3.1 from each side. $\qquad\qquad 8.4 - 3.1 = 3.1 + x - 3.1$

Combine terms. $\qquad\qquad\qquad\qquad\quad 8.4 - 3.1 = x$

Decimal numbers often appear $\qquad\qquad\quad 5.3 = x$
in practical problems.

$5.3 = x$ is the same as $x = 5.3$. $\qquad\qquad\qquad x = 5.3$

The check is left to you.

(e) **Solve:** $6.7 + x = 0$

Subtract 6.7 from each side. $\qquad\qquad 6.7 + x - 6.7 = 0 - 6.7$

$$x = -6.7$$

Be sure to check your answer.

(f) **Solve:** $\dfrac{1}{4} = x - \dfrac{1}{2}$

Add $\dfrac{1}{2}$ to each side. $\qquad \underbrace{\dfrac{1}{4} + \dfrac{1}{2}}_{} = x - \underbrace{\dfrac{1}{2} + \dfrac{1}{2}}_{0} \qquad\qquad \dfrac{1}{4} + \dfrac{1}{2} = \dfrac{3}{4}$

Solution: $\qquad\qquad\qquad \dfrac{3}{4} = x \qquad \text{or} \qquad x = \dfrac{3}{4}$

$$\dfrac{1}{4} = \left(\dfrac{3}{4}\right) - \dfrac{1}{2}$$
$$\dfrac{1}{4} = \dfrac{1}{4}$$

Learning Help Notice that, to solve these equations, we always do the opposite of whatever operation is being performed on the variable. If a number is being added to x in the equation, we must subtract that number in order to solve. If a number is being subtracted from x in the equation, we must add that number in order to solve. Finally, whatever we do to one side of the equation, we must also do to the other side of the equation. ●

Solving Equations by Multiplying or Dividing In the equations solved so far, we needed to add or subtract to solve because numbers were added to or subtracted from the variable. In some equations the variable may be multiplied or divided by a number, so that it appears as $2x$ or $\dfrac{x}{3}$, for example. To solve such an equation, do the operation opposite of the one being performed on the variable. If the variable appears as $2x$, divide by 2; if the variable appears as $\frac{x}{3}$, multiply by 3, and so on. Be sure to perform the same operation on both sides of the equation.

EXAMPLE 4   To solve the equation $27 = -3x$, follow these steps:

$27 = -3x$ Notice that the x-term is on the right side of the equation. It is x multiplied by -3. To change $-3x$ into $1x$ or x, we must perform the opposite operation, that is, divide by -3.

Step 1   Divide each side by -3. $\dfrac{27}{-3} = \dfrac{-3x}{-3}$

Step 2   Simplify. $-9 = \left(\dfrac{-3}{-3}\right)x$

Solution: $-9 = x$ or $x = -9$

✓ $27 = -3(-9)$
$27 = 27$

EXAMPLE 5   Here is a slightly different problem to solve: $\dfrac{x}{3} = 13$

In this equation the variable x is divided by 3. We want to change this equation to an equivalent equation with x alone on the left. Therefore, we must perform the opposite operation, multiplying by 3.

Step 1   Multiply both sides by 3. $\dfrac{x}{3} \cdot 3 = 13 \cdot 3$

Step 2   Simplify. $x = 39$

$$\boxed{\frac{x}{3} \cdot \frac{3}{1} = \frac{3x}{3} = \left(\frac{3}{3}\right)x = 1x = x}$$

Be sure to check your answer.

EXAMPLE 6   To solve $-\dfrac{x}{4} = 5$

Step 1   Multiply both sides by -4. $\left(-\dfrac{x}{4}\right)(-4) = 5 \, (-4)$

Step 2 $\dfrac{(-x)(-4)}{4} = -20$ because $-\dfrac{x}{4} = \dfrac{-x}{4}$

$\dfrac{4x}{4} = -20$ because $(-x)(-4) = +4x$

$x = -20$ because $\dfrac{4x}{4} = \left(\dfrac{4}{4}\right)x = 1x = x$

✓ $-\dfrac{(-20)}{4} = 5$
$-(-5) = 5$
$5 = 5$

🔍 **A Closer Look** When a negative sign precedes a fraction, we may move it either to the numerator or to the denominator, but not to both. In the previous problem, we placed the negative sign in the numerator in Step 2. If we had moved it to the denominator instead, we would have the following:

$$\left(\frac{x}{-4}\right)(-4) = -20 \quad \text{or} \quad \frac{-4x}{-4} = -20 \quad \text{or} \quad x = -20$$

which is exactly the same solution. ●

Your Turn Try the following problems for more practice with one-step multiplication and division equations.

(a) $-5x = -25$ (b) $\dfrac{x}{6} = 6$ (c) $-16 = 2x$

(d) $7x = 35$ (e) $-\dfrac{x}{2} = 14$ 🖩 (f) $2.4x = 0.972$

Solutions (a) **Solve:** $-5x = -25$ The variable x appears multiplied by -5.

     Step 1 Divide both sides by -5. $\dfrac{-5x}{-5} = \dfrac{-25}{-5}$

     Step 2 Simplify: $\dfrac{-5x}{-5} = \left(\dfrac{-5}{-5}\right)x = 1x = x$ $x = 5$

✓ $-5(5) = -25$

    $-25 = -25$

   (b) **Solve:** $\dfrac{x}{6} = 6$ The variable x appears divided by 6.

     Step 1 Multiply both sides by 6. $\left(\dfrac{x}{6}\right)(6) = 6(6)$

     Step 2 Simplify: $\left(\dfrac{x}{6}\right)(6) = x\left(\dfrac{6}{6}\right) = x(1) = x$ $x = 36$

✓ $\dfrac{(36)}{6} = 6$

    $6 = 6$

   (c) **Solve:** $-16 = 2x$ $-16 = 2x$

     Step 1 Divide each side by 2. $\dfrac{-16}{2} = \dfrac{2x}{2}$

     Step 2 Simplify: $\dfrac{2x}{2} = \left(\dfrac{2}{2}\right)x = 1x = x$ $-8 = x$ or $x = -8$

✓ $-16 = 2(-8)$
    $-16 = -16$

   (d) **Solve:** $7x = 35$ $7x = 35$

     Step 1 Divide both sides by 7. $\dfrac{7x}{7} = \dfrac{35}{7}$

     Step 2 Simplify: $\dfrac{7x}{7} = \left(\dfrac{7}{7}\right)x = x$ $x = 5$

    Be sure to check your answer.

(e) **Solve:** $-\dfrac{x}{2} = 14$

$$-\dfrac{x}{2} = 14$$

Step 1   Multiply both sides by -2.

$$\left(-\dfrac{x}{2}\right)(-2) = 14\,(-2)$$

Step 2   Simplify:

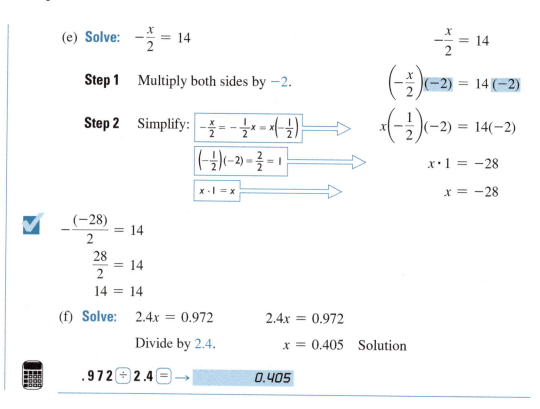

$$x\left(-\dfrac{1}{2}\right)(-2) = 14(-2)$$

$$x \cdot 1 = -28$$

$$x = -28$$

✓ $-\dfrac{(-28)}{2} = 14$

$\dfrac{28}{2} = 14$

$14 = 14$

(f) **Solve:** $2.4x = 0.972$ $\qquad$ $2.4x = 0.972$

$\qquad\qquad$ Divide by 2.4. $\qquad$ $x = 0.405$ Solution

$.972 \boxed{\div} 2.4 \boxed{=} \rightarrow \qquad 0.405$

We have now covered the four basic single-operation equations. Before moving on, try the following set of problems covering these four operations. For each problem, think carefully whether you must add, subtract, multiply, or divide in order to make the equation read $x =$ _____ or _____ $= x$.

More Practice   **Solve:**

(a) $14 = x - 8$ (b) $\dfrac{y}{6} = 3$ $\qquad$ (c) $n + 11 = 4$ (d) $3x = 33$

(e) $7 = -\dfrac{a}{5}$ (f) $z - 2 = -10$ (g) $-8 = -16x$ (h) $26 = y + 5$

Solutions   Each solution states the operation you should have performed on both sides and then gives the final answer.

(a) Add 8 to get $x = 22$. $\qquad$ (b) Multiply by 6 to get $y = 18$.

(c) Subtract 11; $n = -7$. $\qquad$ (d) Divide by 3; $x = 11$.

(e) Multiply by -5; $a = -35$. $\qquad$ (f) Add 2; $z = -8$.

(g) Divide by -16; $x = \dfrac{1}{2}$. $\qquad$ (h) Subtract 5; $y = 21$.

Practical Applications

EXAMPLE 7   **Manufacturing**   The capacity C of a conveyor belt in metric tons per hour is given by the formula

$$C = 3.6SDA$$

where S is the belt speed in meters per second
$\qquad$ D is the density of the load material in kilograms per cubic meter
$\qquad$ A is the cross-sectional area of the load in square meters

Suppose that a certain load material has a density of 750 kg/cu m and a cross-sectional area of 0.083 sq m. What belt speed is needed to move 1000 metric tons per hour?

To answer this question, **first** substitute the given numbers into the formula. We know that $D = 750, A = 0.083$, and $C = 1000$. This gives us the equation

$$1000 = (3.6)S(750)(0.083)$$

Next, simplify the right side by multiplying the numerical quantities:

$$1000 = (224.1)S$$

Finally, divide both sides by 224.1 to solve for S:

$$S = 4.462\ldots \text{m/s} \approx 4.5 \text{ m/s, rounded}$$ ●

Your Turn   **Roofing**   The formula $A = LR$ gives the relationship among the number of linear feet L of roof vents, the rating R of the vents, and the area A of net-free vent area (NFVA) in the attic. If a certain ridge vent has a rating of 9.0 sq in. of NFVA per linear foot, and an attic has 432 sq in. of NFVA, how many linear feet of ridge vents are needed?

Solution   **First**, substitute $R = 9.0$ and $A = 432$ into the formula:

$$432 = L(9.0)$$

Then, divide both sides by 9 to solve for L:

$$L = 48 \text{ linear feet}$$

Now turn to Exercises 7-3 for practice solving simple equations.

Exercises 7-3	**Solving Simple Equations**

A. Solve the following equations.

1. $x + 4 = 13$ 　　　　　　　　2. $23 = A + 6$

3. $41 = x - 17$ 　　　　　　　　4. $z - 18 = 29$

5. $6 = a - 2\frac{1}{2}$ 　　　　　　　6. $73 + x = 11$

7. $y - 16.01 = 8.65$ 　　　　　8. $11.6 - R = 3.7$

9. $-39 = 3x$ 　　　　　　　　10. $-9y = 117$

11. $13a = 0.078$ 　　　　　　　12. $\frac{x}{3} = 7$

13. $\frac{z}{1.3} = 0.5$ 　　　　　　　14. $\frac{N}{2} = \frac{3}{8}$

15. $m + 18 = 6$ 　　　　　　　16. $-34 = x - 7$

17. $-6y = -39$ 　　　　　　　18. $\frac{a}{-4} = 9$

19. $-5.9 = -6.6 + Q$ 　　　　20. $79.2 = 2.2y$

21. $12 + x = 37$ 　　　　　　　22. $\frac{5}{4} = \frac{Z}{8}$

23. $\frac{K}{0.5} = 8.48$ 　　　　　　24. $-12z = 3.6$

B. Practical Applications

1. **Electrical Trades**   Ohm's law is often written in the form

$$I = \frac{E}{R}$$

where I is the current in amperes (A), E is the voltage, and R is the resistance in ohms. What is the voltage necessary to push a 0.80-A current through a resistance of 450 ohms?

2. **Aviation**   The formula

$$D = RT$$

is used to calculate the distance D traveled by an object moving at a constant average speed R during an elapsed time T. How long would it take a pilot to fly 1240 miles at an average speed of 220 mph?

3. **Water/Wastewater Treatment**   A wastewater treatment operator uses the formula

$$A = 8.34FC$$

to determine the amount A of chlorine in pounds to add to a basin. The flow F through the basin is in millions of gallons per day, and the desired concentration C of chlorine is in parts per million.

If 1800 pounds of chlorine is added to a basin with a flow rate of 7.5 million gallons per day, what would be the resulting concentration?

4. **Sheet Metal Trades**   The water pressure P in pounds per square foot is related to the depth of water D in feet by the formula

$$P = 62.4D$$

where 62.4 lb is the weight of 1 cu ft of water. If the material forming the bottom of a tank is made to withstand 800 pounds per square foot of pressure, what is the maximum safe height for the tank?

5. **Physics**   The frequency f (in waves per second) and the wavelength w (in meters) of a sound traveling in air are related by the equation

$$fw = 343$$

where 343 m/s is the speed of sound in air. Calculate the wavelength of a musical note with a frequency of 200 waves per second.

6. **Physics**   At a constant temperature, the pressure P (in psi) and the volume V (in cu ft) of a particular gas are related by the equation

$$PV = 1080$$

If the volume is 60 cu ft, calculate the pressure.

7. **Electrical Trades**   For a particular transformer, the voltage E in the circuits is related to the number of windings W of wire around the core by the equation

$$E = 40W$$

How many windings will produce a voltage of 840 V?

8. **Metalworking**   The surface speed S in fpm (feet per minute) of a rotating cylindrical object is

$$S = \frac{\pi dn}{12}$$

where d is the diameter of the object in inches and n is the speed of rotation in rpm (revolutions per minute). If an 8-in. grinder must have a surface speed of 6000 fpm, what should the speed of rotation be? Use $\pi \approx 3.14$, and round to the nearest hundred rpm.

9. **Construction** The amount of lumber in board feet (bf) can be expressed by the formula

$$bf = \frac{TWL}{12}$$

where T is the thickness of a board in inches, W is its width in inches, and L is its length in feet. What total length of 1-in. by 6-in. boards is needed for a total of 16 bf?

10. **Automotive Trades** For modern automotive engines, horsepower is defined as

$$hp = \frac{\text{torque (in lb} \cdot \text{ft)} \times \text{engine speed (in rpm)}}{5252}$$

If an engine has a peak horsepower of 285 hp at 4800 rpm, what is the torque of the engine at this speed? (Round to the nearest whole number.)

11. **Business and Finance** The formula for a company's net profit margin M expressed as a percent is given by

$$M = \frac{100I}{R}$$

where I represents net income and R represents net revenue.

If a company had a net profit margin of 2.85% on net revenue of $4,625,000, what was its net income? (Round to the nearest hundred dollars.)

12. **Allied Health** In problem 34 on page 438, the metric formula for BMI (body mass index) was given as

$$BMI = \frac{\text{weight in kg}}{(\text{height in m})^2}$$

If a patient has a BMI of 23.8 and is 1.77 m tall, how much does the patient weigh?

13. **Construction** On a construction site, a winch is used to wind cable onto the drum of the winch. For the first layer of cable, the line pull P of the winch is related to the diameter D of the drum by the formula

$$P = \frac{k}{D}$$

where k is a constant. If $P = 5400$ lb and $D = 4$ in., find the value of the constant k in foot-pounds. (*Hint:* Convert D to feet first.)

When you have completed these exercises, check your answers to the odd-numbered problems in the Appendix, then continue to Section 7-4.

7-4 Solving Two-Operation Equations

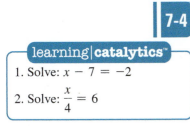

1. Solve: $x - 7 = -2$

2. Solve: $\dfrac{x}{4} = 6$

Solving many simple algebraic equations involves both kinds of operations: addition/subtraction and multiplication/division.

EXAMPLE 1 **Solve:** $2x + 6 = 14$

Step 1   We want to change this equation to an equivalent equation with only x or terms that include x on the left.

Subtract 6 from both sides. $2x + 6 \boxed{-6} = 14 \boxed{-6}$

Combine terms. (Be careful. You cannot add $2x$ and 6—they are unlike terms.)

$$2x + \underbrace{6 - 6}_{0} = 14 - 6$$

Now this is an equivalent equation with only an x-term on the left. $2x = 8$

Step 2   Divide both sides of the equation by 2. $\dfrac{2x}{2} = \dfrac{8}{2}$

$$\dfrac{2x}{2} = x \qquad\qquad x = 4$$

✓ $2(4) + 6 = 14$

$8 + 6 = 14$

$14 = 14$ ●

Your Turn   Try this one to test your understanding of the process.

Solve: $3x - 7 = 11$

Solution **Step 1**   Add 7 to each side. $3x - 7 \boxed{+7} = 11 \boxed{+7}$

$$3x \underbrace{- 7 + 7}_{0} = 11 + 7$$

$$3x = 18$$

Step 2   Divide both sides of the equation by 3. $\dfrac{3x}{3} = \dfrac{18}{3}$

Solution: $x = 6$

✓ $3(6) - 7 = 11$

$18 - 7 = 11$

$11 = 11$

EXAMPLE 2   Here is another two-operation equation.

Solve: $23 = 9 - \dfrac{y}{3}$

Step 1   Notice that the variable y is on the right side of the equation. We must first eliminate the 9 by subtracting 9 from both sides. Be sure to keep the negative sign in front of the $\dfrac{y}{3}$ term.

$$23 - 9 = 9 - \dfrac{y}{3} - 9$$

$$\boxed{9 - 9 = 0}$$

$$14 = -\dfrac{y}{3}$$

Step 2   Multiply both sides by -3.

$$(-3)(14) = (-3)\left(-\frac{y}{3}\right)$$

$$(-3)(14) = (-3)\left(-\frac{1}{3}\right)y \quad \longleftarrow \quad \boxed{-\frac{y}{3} = -\frac{1}{3}y}$$

$$-42 = 1y \quad \longleftarrow \quad \boxed{(-3)\left(-\frac{1}{3}\right) = 1}$$

Solution: $y = -42$

$$23 = 9 - \frac{(-42)}{3}$$

$$23 = 9 - (-14)$$

$$23 = 23$$

Your Turn   Solve: $\dfrac{y}{2} - 6 = 4$

Solution   **Step 1**   Add 6 to each side.

$$\frac{y}{2} \underbrace{- 6 + 6}_{0} = 4 + 6$$

$$\frac{y}{2} = 10$$

Step 2   Multiply both sides by 2.

$$2\left(\frac{y}{2}\right) = 2 \cdot 10$$

Solution: $y = 20$

In Step 2, $2\left(\dfrac{y}{2}\right) = \dfrac{\overset{1}{2}}{1} \cdot \dfrac{y}{\underset{1}{2}} = 1y = y$.

Be sure to check your answer.

Note   In these two-operation problems, we did the addition or subtraction in Step 1 and the multiplication or division in Step 2. We could have reversed this order and arrived at the correct solution, but the problem might have become more complicated to solve. Always add or subtract first, and when you multiply or divide, do so to *all* terms.

If more than one variable term appears on the same side of an equation, combine these like terms before performing any operation to both sides.

EXAMPLE 3   Solve: $2x + 5 + 4x = 17$

Step 1   Combine the x-terms on the left side. $(2x + 4x) + 5 = 17$

$$6x + 5 = 17$$

Step 2   Subtract 5 from each side. $6x + 5 - 5 = 17 - 5$

$$6x = 12$$

Step 3   Divide each side by 6. $\dfrac{6x}{6} = \dfrac{12}{6}$

$$x = 2$$

✓ Substitute 2 for x in each x-term.

$$2(2) + 5 + 4(2) = 17$$
$$4 + 5 + 8 = 17$$
$$17 = 17$$

●

Your Turn   Now you try one.

Solve: $32 = x - 12 - 5x$

Solution   **Step 1**   Combine the x-terms on the right. $32 = (x - 5x) - 12$
$$32 = -4x - 12$$

Step 2   Add 12 to both sides. $32 + 12 = -4x - 12 + 12$
$$44 = -4x$$

Step 3   Divide both sides by -4. $\dfrac{44}{-4} = \dfrac{-4x}{-4}$

$$-11 = x \qquad \text{or} \qquad x = -11$$

✓ $32 = (-11) - 12 - 5(-11)$

$32 = -11 - 12 + 55$

$32 = -23 + 55$

$32 = 32$

The next example involves the two operations of multiplication and division.

EXAMPLE 4   **Solve:** $\dfrac{5x}{3} = 25$

Step 1   Multiply both sides by 3. $\dfrac{5x}{3} \cdot 3 = 25 \cdot 3$

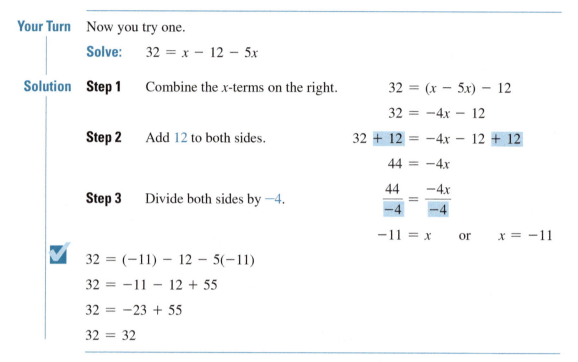

$5x = 75$

Step 2   Divide both sides by 5. $\dfrac{5x}{5} = \dfrac{75}{5}$

$$x = 15$$ ●

🔍 **A Closer Look**   In the previous example, you could solve for x in one step if you multiply both sides by the fraction $\frac{3}{5}$.

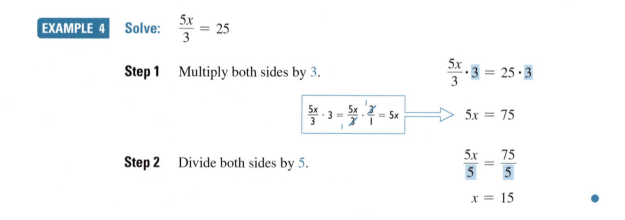

$$x = 15 \; ●$$

More Practice   Here are a few more two-operation equations for practice.

(a) $7x + 2 = 51$ (b) $18 - 5x = 3$ (c) $15.3 = 4x - 1.5$

(d) $\dfrac{x}{5} - 4 = 6$ (e) $11 - x = 2$ (f) $5 = 7 - \dfrac{x}{4}$

(g) $\dfrac{x + 2}{3} = 4$ (h) $2x - 9.4 = 0$ (i) $2.75 = 14.25 - 0.20x$

(j) $3x + x = 18$ (k) $12 = 9x + 4 - 5x$ (l) $\dfrac{3x}{4} = -15$

Solutions   (a) **Solve:** $7x + 2 = 51$

Change this equation to an equivalent equation with only an x-term on the left.

Step 1   Subtract 2 from each side. $7x + 2 \boxed{-\ 2} = 51 \boxed{-\ 2}$

Combine like terms. $7x + \underbrace{2 - 2}_{0} = 51 - 2$

$7x = 49$

Step 2   Divide both sides by 7. $\dfrac{7x}{\boxed{7}} = \dfrac{49}{\boxed{7}}$

$x = \dfrac{49}{7}$

$x = 7$

✔ $7(7) + 2 = 51$

$49 + 2 = 51$

$51 = 51$

(b) **Solve:** $18 - 5x = 3$

Step 1   Subtract 18 from each side. $18 - 5x \boxed{-\ 18} = 3 \boxed{-\ 18}$

Rearrange terms and combine like terms. $-5x + \underbrace{18 - 18}_{0} = 3 - 18$

$-5x = -15$

Step 2   Divide both sides by -5. $\dfrac{-5x}{\boxed{-5}} = \dfrac{-15}{\boxed{-5}}$

$x = 3$

✔ $18 - 5(3) = 3$

$18 - 15 = 3$

$3 = 3$

(c) **Solve:** $15.3 = 4x - 1.5$

Step 1   Add 1.5 to each side and combine like terms. $15.3 \boxed{+\ 1.5} = 4x \underbrace{-\ 1.5 \boxed{+\ 1.5}}_{0}$

$16.8 = 4x$

or $4x = 16.8$

Step 2   Divide both sides by 4. $\dfrac{4x}{4} = \dfrac{16.8}{4}$

$$x = 4.2$$

Decimal number solutions are common in practical problems.

$1\,5\,.\,3\;\boxed{+}\;1\,.\,5\;\boxed{=}\;\boxed{\div}\;4\;\boxed{=}\;\rightarrow\;\boxed{\qquad\qquad 4.2}$

Be sure to check your answer.

(d) **Solve:** $\dfrac{x}{5} - 4 = 6$

Step 1   Add 4 to both sides. $\dfrac{x}{5} - 4 \boxed{+ 4} = 6 \boxed{+ 4}$

$$\dfrac{x}{5} = 10$$

Step 2   Multiply both sides by 5. $\left(\dfrac{x}{5}\right) \cdot \boxed{5} = 10 \cdot \boxed{(5)}$

$$x = 50$$

✓ $\dfrac{(50)}{5} - 4 = 6$

$$10 - 4 = 6$$

$$6 = 6$$

(e) **Solve:** $11 - x = 2$

Step 1   Subtract 11 from each side. $11 - x \boxed{-\ 11} = 2 \boxed{-\ 11}$

Combine terms. Be sure to keep the negative sign in front of the x. $-x + \underbrace{11 \boxed{-\ 11}}_{0} = 2 \boxed{-\ 11}$

Step 2   Multiply each side by -1. $-x = -9$

$$x = 9$$

Be sure to check the solution.

(f) **Solve:** $5 = 7 - \dfrac{x}{4}$

Step 1   Subtract 7 from both sides. $5 \boxed{-\ 7} = 7 - \dfrac{x}{4} \boxed{-\ 7}$

$\boxed{7 - \dfrac{x}{4} - 7 = (7 - 7) - \dfrac{x}{4} = -\dfrac{x}{4}}$ $-2 = -\dfrac{x}{4}$

Step 2   Multiply both sides by -4. $-2\,\boxed{(-4)} = \left(-\dfrac{x}{4}\right)(-4)$

$\boxed{\left(-\dfrac{x}{4}\right)(-4) = (-x)\left(\dfrac{-4}{4}\right) = (-x)(-1) = x}$ $8 = x$

Solution: $x = 8$

✓ $5 = 7 - \dfrac{(8)}{4}$

$$5 = 7 - 2$$

$$5 = 5$$

(g) **Solve:** $\dfrac{x + 2}{3} = 4$

Multiply both sides by 3.

$$\dfrac{x + 2}{3} \cdot 3 = 4 \cdot 3$$

This eliminates the fraction on the left side.

$$\dfrac{x + 2}{{}_1\cancel{3}} \cdot \dfrac{{}^1\cancel{3}}{1} = 4 \cdot 3$$

$$x + 2 = 12$$

Subtract 2 from both sides.

$$x + 2 - 2 = 12 - 2$$

$$x = 10$$

✓ $\dfrac{10 + 2}{3} = 4$

$\dfrac{12}{3} = 4$

$4 = 4$

(h) **Solve:** $2x - 9.4 = 0$

Step 1 Add 9.4 to each side.

$$2x - 9.4 + 9.4 = 0 + 9.4$$

$$2x = 9.4$$

Step 2 Divide each side by 2.

$$x = 4.7$$

Check your solution.

(i) **Solve:** $2.75 = 14.25 - 0.20x$

Step 1 Subtract 14.25 from both sides.

$$2.75 - 14.25 = 14.25 - 0.20x - 14.25$$

$$-11.5 = -0.20x$$

Step 2 Divide both sides by -0.20.

$$\dfrac{-11.5}{-0.20} = \dfrac{-0.20x}{-0.20}$$

$$57.5 = x$$

or $x = 57.5$

2.75 ⊟ 14.25 ⊟ ÷ (−) .2 ⊟ → 57.5

You may also check your answer with a calculator.

(j) **Solve:** $3x + x = 18$

Step 1 Combine like terms on the left side.

$$4x = 18$$

Step 2 Divide each side by 4.

$$\dfrac{4x}{4} = \dfrac{18}{4}$$

$$x = 4.5$$

✓ $3(4.5) + (4.5) = 18$

$13.5 + 4.5 = 18$

$18 = 18$

(k) **Solve:** $12 = 9x + 4 - 5x$

Step 1 Combine like terms.

$$12 = 4x + 4$$

Step 2 Subtract 4 from each side.

$$12 - 4 = 4x + 4 - 4$$

$$8 = 4x$$

Step 3 Divide each side by 4.

$$\dfrac{8}{4} = \dfrac{4x}{4}$$

$$2 = x \quad \text{or} \quad x = 2$$

The check is left to you.

(l) **Solve:** $\dfrac{3x}{4} = -15$

> **Step 1**   Multiply both sides by 4.
>
> $$\dfrac{3x}{4} \cdot 4 = -15 \cdot 4$$
>
> $$3x = -60$$
>
> **Step 2**   Divide both sides by 3.
>
> $$\dfrac{3x}{3} = \dfrac{-60}{3}$$
>
> $$x = -20$$

Practical Applications

EXAMPLE 5   **HVAC**   We can estimate the annual cost of operating a fuel oil–fired boiler (C_F, in dollars per million Btu) using the following formula:

$$C_F = \dfrac{7.25F}{E}$$

where F is the cost per gallon of fuel oil, and E is the operating efficiency of the boiler.

The annual cost of operating a ground-source heat pump (C_G, also in dollars per million Btu) is given by the formula

$$C_G = \dfrac{293K}{R}$$

where K is the cost per kilowatt-hour of electricity, and R is the efficiency rating of the heat pump.

Assume the following:

- Fuel oil costs $2.60 per gallon.

- An existing boiler is 84% efficient. (Use 0.84 in the formula.)

- A new ground-source heat pump would have an efficiency rating of 3.3.

What would the cost of electricity have to be to make the two annual costs the same?

To solve this problem, **first** substitute the given numbers into the two formulas:

$$C_F = \dfrac{7.25(\$2.60)}{0.84} \qquad C_G = \dfrac{293K}{3.3}$$

Then, set the formulas equal to each other and solve for K:

$$\dfrac{7.25(\$2.60)}{0.84} = \dfrac{293K}{3.3}$$

Because this is in the form of a proportion $\left(\dfrac{a}{b} = \dfrac{c}{d}\right)$, we can cross-multiply ($ad = bc$):

$$7.25(\$2.60)(3.3) = (0.84)(293K)$$

Multiplying and switching sides, we have:

$$246.12K = \$62.205$$

Divide both sides by 246.12 to obtain:

$$K = \$0.2527\ldots$$

Rounded to the nearest cent, the cost of electricity must be $0.25 per kWh to make the annual costs the same.   ●

Your Turn   **Electronics**   The resistance R in ohms (Ω) of a particular circuit element is given by the formula

$$R = 0.02T + 5.00$$

where T is the temperature of the element in degrees Celsius (°C). At what temperature is the resistance 5.48 Ω?

Solution   **First**, substitute 5.48 for R:

$$5.48 = 0.02T + 5.00$$

Then, solve:

Subtract 5.00 from both sides.

$$5.48 - 5.00 = 0.02T + 5.00 - 5.00$$

$$0.48 = 0.02T$$

Divide both sides by 0.02.

$$\frac{0.48}{0.02} = \frac{0.02T}{0.02}$$

$$24 = T$$

The resistance is 5.48 Ω at a temperature of 24°C.

Be sure to check your solution.

EXAMPLE 6   **General Interest**   A student's first four semester test scores have been 82, 85, 74, and 73. What score must the student receive on the fifth test in order to achieve an average score of 80?

To solve this problem, recall from Chapter 3 the formula for the average test score:

$$\text{Average} = \frac{\text{sum of the test scores}}{\text{number of test scores}}$$

If we let $x =$ the score needed on the fifth test, we can substitute into the formula to set up the following equation:

$$80 = \frac{82 + 85 + 74 + 73 + x}{5}$$

Simplifying the numerator on the right side, we have

$$80 = \frac{314 + x}{5}$$

Multiplying both sides of the equation by 5 gives us

$$400 = 314 + x$$

Subtracting 314 from both sides gives us x:

$$86 = x$$

The student must earn a score of 86 on the fifth test in order to achieve an overall average of 80.   ●

Your Turn   **Allied Health**   For the first 11 months of the year, sales of a particular medication have averaged 984 units per month. What must sales be for December in order for the monthly average to increase to 1000 units?

Solution   **First**, write the formula for average monthly sales for the year.

$$\text{Average monthly sales} = \frac{\text{total annual sales}}{12}$$

Next, calculate the total units sold during the first 11 months:

Total units sold = 11 months × 984 units/month = 10,824 units

If we let $x =$ the required number of sales for December, then the total annual sales would be $x + 10,824$. Substituting into the formula, we now have

$$1000 = \frac{x + 10,824}{12}$$

Finally, solve for x:

| Multiply both sides by 12: | $12,000 = x + 10,824$ |
| Subtract 10,824 from both sides: | $1176 = x$ |

To achieve a monthly average of 1000 units for the entire year, sales for December must be 1176 units.

Now turn to Exercises 7-4 for more practice in solving these two-operation equations.

| **Exercises 7-4** | **Solving Two-Operation Equations** |

A. Solve.

1. $2x - 3 = 17$ 2. $4x + 6 = 2$

3. $2\dfrac{x}{5} = 7$ 4. $-8y + 12 = 32$

5. $4.4m - 1.2 = 9.8$ 6. $\dfrac{3}{4} = 3x + \dfrac{1}{4}$

7. $14 - 7n = -56$ 8. $38 = 58 - 4a$

9. $3z - 5z = 12$ 10. $17 = 7q - 5q - 3$

11. $23 - \dfrac{x}{4} = 11$ 12. $9m + 6 + 3m = -60$

13. $-15 = 12 - 2n + 5n$ 14. $2.6y - 19 - 1.8y = 1$

15. $\dfrac{1}{2} + 2x = 1$ 16. $3x + 16 = 46$

17. $-4a + 45 = 17$ 18. $\dfrac{x}{2} + 1 = 8$

19. $-3Z + \dfrac{1}{2} = 17$ 20. $2x + 6 = 0$

21. $1 = 3 - 5x$ 22. $23 = 17 - \dfrac{x}{4}$

23. $-5P + 18 = 3$ 24. $5x - 2x = 24$

25. $6x + 2x = 80$

26. $x + 12 - 6x = -18$

27. $27 = 2x - 5 + 4x$

28. $-13 = 22 - 3x + 8x$

29. $\dfrac{x - 5}{4} = -2$

30. $7 = \dfrac{x + 6}{2}$

31. $\dfrac{2x}{3} = -4$

32. $6 = -\dfrac{5x}{4}$

B. Practical Applications

1. **Office Services**   A repair service charges $45 for a house call and an additional $60 per hour of work. The formula

   $$T = 45 + 60H$$

   represents the total charge T for H hours of work. If the total bill for a customer was $375, how many hours of actual labor were there?

2. **Meteorology**   The air temperature T (in degrees Fahrenheit) at an altitude h (in feet) above a particular area can be approximated by the formula

   $$T = -0.002h + G$$

   where G is the temperature on the ground directly below. If the ground temperature is 76°F, at what altitude will the air temperature drop to freezing, 32°F?

3. **Sports and Leisure**   Physical fitness experts sometimes use the following formula to approximate the maximum target heart rate R during exercise based on a person's age A:

   $$R = -0.8A + 176$$

   At what age should the heart rate during exercise not exceed 150 beats per minute?

4. **Life Skills**   The formula $A = p + prt$ is used to determine the total amount of money A in a bank account after an amount p is invested for t years at a rate of interest r. What rate of interest is needed for $8000 to grow to $12,000 after 5 years? Be sure to convert your decimal answer to a percent and round to the nearest tenth of a percent.

5. **Automotive Trades**   The formula

   $$P + 2T = C$$

   gives the overall diameter C of the crankshaft gear for a known pitch diameter P of the small gear and the height T of teeth above the pitch diameter circle. If $P = 2.875$ in. and $C = 3.125$ in., find T.

6. **Sheet Metal Trades**   The allowance A for a Pittsburgh lock is given by

   $$A = 2w + \frac{3}{16}\text{ in.}$$

   where w is the width of the pocket. If the allowance for a Pittsburgh lock is $\frac{11}{16}$ in., what is the width of the pocket?

7. **Trades Management**   A plumber's total bill A can be calculated using the formula

   $$A = RT + M$$

where R is his hourly rate, T is the total labor time in hours, and M is the cost of materials. A plumber bids a particular job at \$3400. If materials amount to \$985, and his hourly rate is \$84 per hour, how many hours should the job take for the estimate to be accurate?

8. **Machine Trades**   The formula

$$L = 2d + 3.26(r + R)$$

can be used under certain conditions to approximate the length L of belt needed to connect two pulleys of radii r and R if their centers are a distance d apart. How far apart can two pulleys be if their radii are 8 in. and 6 in. and the total length of the belt connecting them is 82 in.? (Round to the nearest inch.)

9. **Forestry**   The fire damage potential D of a class 5 forest is given by the formula

$$D = 2A + 5$$

where A is the average age (in years) of the brush in the forest. If $D = 20$, find the average age of the brush.

10. **Machine Trades**   Suppose that, on the average, 3% of the parts produced by a particular machine have proven to be defective. Then the formula

$$N - 0.03N = P$$

will give the number of parts N that must be produced in order to manufacture a total of P nondefective ones. How many parts should be produced by this machine in order to end up with 7500 nondefective ones?

11. **Automotive Trades**   The 2017 Mercedes AMG E63S generates 603 hp at 5750 rpm. Use the following formula to solve for the torque created at this engine speed:

$$\text{hp} = \frac{\text{torque (in lb} \cdot \text{ft)} \times \text{engine speed (in rpm)}}{5252}$$

(Round to the nearest whole number.)

12. **Automotive Trades**   The total engine displacement D (in inches) is given by

$$D = \frac{\pi B^2 SN}{4}$$

where B is the bore, S is the stroke, and N is the number of cylinders. What is the stroke of a 6-cylinder engine with a bore of 3.50 in. and a total displacement of 216 in.^{3}? Use $\pi \approx 3.14$.

13. **Water/Wastewater Treatment**   The pressure P (in pounds per square inch or psi) on a submerged body in wastewater is given by

$$P = 0.520D + 14.7$$

where D is the depth in feet. At what depth is the pressure 25.0 psi?

14. **HVAC**   We can estimate the cost C_N (in dollars per million Btu) of a natural gas–powered heating system using the formula

$$C_N = \frac{10T}{E}$$

where T is the cost per therm of natural gas and E is the operating efficiency of the system. Similarly, we can estimate the cost C_A (also in dollars per million Btu) of an air-source heat pump using the formula

$$C_A = \frac{1000K}{R}$$

where K is the cost per kilowatt-hour of electricity and R is the efficiency rating of the pump. Assume the following:

- An existing natural-gas heater is 92% efficient. (Use 0.92.)

- A new air-source heat pump would have an efficiency rating of 9.5.

- Electricity costs $0.09 per kilowatt-hour.

What would the cost of natural gas have to be to make the two annual costs the same?

15. **HVAC** As explained in problem 11 on page 434, the ACH of a house is a measure of its air-tightness. The formula for ACH is

$$\text{ACH} = \frac{60L}{V}$$

where L is the air leakage in cubic feet per minute (cfm) and V is the volume of the house in cubic feet. If a house has a volume of 25,600 cu ft, what is the maximum amount of air leakage that would result in an ACH of no more than 1.25?

16. **General Interest** A student has a test average of 89.4 for the first five tests of the semester. What score would the student need on the sixth test in order to achieve an average of 90 for all six tests?

17. **Retail Merchandising** A salesperson has a goal of averaging $2000 worth of sales per day. During the first four days of a week, she has achieved the following daily totals: $1860, $2228, $452, and $3448. What amount of sales does she need on the fifth day in order to achieve her goal for the week?

18. **Electronics** Use the formula

$$R = 0.02T + 5.00 \quad \text{(see "Your Turn" on page 467)}$$

to find the temperature T at which the resistance R of the circuit element is 5.24 Ω.

When you have completed these exercises, check your answers to the odd-numbered problems in the Appendix, then continue in Section 7-5.

| 7-5 | Solving More Equations and Solving Formulas

Parentheses in Equations In Section 7-2 you learned how to deal with algebraic expressions involving parentheses. Now you will learn to solve equations containing parentheses by using these same skills.

EXAMPLE 1 **Solve:** $2(x + 4) = 27$

Step 1 Use Rule 3 on page 446 (the distributive property). Multiply each term inside the parentheses by 2.

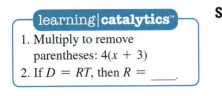

$$2x + 8 = 27$$

Now solve this equation using the techniques of the previous section.

Step 2   Subtract 8 from both sides. $2x + 8 - 8 = 27 - 8$

$$2x = 19$$

Step 3   Divide both sides by 2. $$\frac{2x}{2} = \frac{19}{2}$$

Solution: $x = 9.5$

$2[(9.5) + 4] = 27$

$2(13.5) = 27$

$27 = 27$

Your Turn   Try this similar example. Solve for y.

$$-3(y - 4) = 36$$

Solution   **Solve:** $-3(y - 4) = 36$

Step 1   Multiply each term inside the parentheses by -3.

$-3(y - 4) = -3(y) + (-3)(-4)$ $\Longrightarrow$ $-3y + 12 = 36$

Step 2   Subtract 12 from each side. $-3y + 12 - 12 = 36 - 12$

$$-3y = 24$$

Step 3   Divide each side by -3. $$\frac{-3y}{-3} = \frac{24}{-3}$$

Solution: $y = -8$

$-3[(-8) - 4] = 36$

$-3(-12) = 36$

$36 = 36$

A Closer Look   In each of the last two problems, you could have first divided both sides by the number in front of the parentheses.

Here is how each solution would have looked:

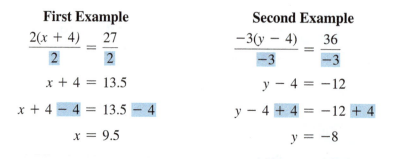

First Example	**Second Example**
$\dfrac{2(x + 4)}{2} = \dfrac{27}{2}$	$\dfrac{-3(y - 4)}{-3} = \dfrac{36}{-3}$
$x + 4 = 13.5$	$y - 4 = -12$
$x + 4 - 4 = 13.5 - 4$	$y - 4 + 4 = -12 + 4$
$x = 9.5$	$y = -8$

Some students may find this technique preferable, especially when the right side of the equation is exactly divisible by the number in front of the parentheses.

Here is a more difficult equation involving parentheses.

EXAMPLE 2   **Solve:** $5x - (2x - 3) = 27$

Step 1   Use Rule 2 on page 445. Change the sign of each term inside the parentheses and then remove them.

$-(2x - 3) = -2x + 3$ ⟹ $5x - 2x + 3 = 27$

Step 2   Combine the like terms on the left.

$5x - 2x = 3x$ ⟹ $3x + 3 = 27$

Step 3   Subtract 3 from both sides. $3x + 3 - 3 = 27 - 3$

$3x = 24$

Step 4   Divide both sides by 3. $\dfrac{3x}{3} = \dfrac{24}{3}$

$x = 8$

✓ $5(8) - [2(8) - 3] = 27$

$40 - (16 - 3) = 27$

$40 - 13 = 27$

$27 = 27$ ●

Your Turn   Try this problem.

Solve: $8 - 3(2 - 3x) = 34$

Solution   **Step 1**   Be very careful here. Some students are tempted to subtract the 3 from the 8. However, the order of operations rules on page 408 specify that multiplication must be performed before addition or subtraction. Therefore, your first step is to multiply the expression in parentheses by -3.

$8 - 3(2 - 3x) = 8 + (-3)(2) + (-3)(-3x)$ $8 - 6 + 9x = 34$

Step 2   Combine the like terms on the left in Step 1. $2 + 9x = 34$

Step 3   Subtract 2 from each side. $2 + 9x - 2 = 34 - 2$

$9x = 32$

Step 4   Divide each side by 9. $\dfrac{9x}{9} = \dfrac{32}{9}$

$x = 3\tfrac{5}{9}$

Be sure to check your answer.

More Practice   Here are more equations with parentheses for you to solve.

(a) $4(x - 2) = 26$ (b) $11 = -2(y + 5)$

(c) $23 = 6 - (3n + 4)$ (d) $4x - (6x - 9) = 41$

(e) $7 + 3(5x + 2) = 38$ (f) $20 = 2 - 5(9 - a)$

(g) $(3m - 2) - (5m - 3) = 19$ (h) $2(4x + 1) + 5(3x + 2) = 58$

Answers
(a) $x = 8.5$ (b) $y = -10.5$ (c) $n = -7$ (d) $x = -16$

(e) $x = 1\frac{2}{3}$ (f) $a = 12.6$ (g) $m = -9$ (h) $x = 2$

Here are worked solutions to (b), (d), (f), and (h). The checks are left to you.

(b) **Solve:** $11 = -2(y + 5)$

Step 1	Multiply each term inside the parentheses by -2.	$11 = -2y - 10$
Step 2	Add 10 to each side.	$11 + 10 = -2y - 10 + 10$
		$21 = -2y$
Step 3	Divide each side by -2.	$\dfrac{21}{-2} = \dfrac{-2y}{-2}$
		$-10.5 = y$ or $y = -10.5$

(d) **Solve:** $4x - (6x - 9) = 41$

Step 1	Change the sign of each term inside the parentheses and then remove the parentheses.	$4x - 6x + 9 = 41$
Step 2	Combine like terms on the left side.	$-2x + 9 = 41$
Step 3	Subtract 9 from each side.	$-2x + 9 - 9 = 41 - 9$
		$-2x = 32$
Step 4	Divide each side by -2.	$\dfrac{-2x}{-2} = \dfrac{32}{-2}$
		$x = -16$

(f) **Solve:** $20 = 2 - 5(9 - a)$

Step 1   Multiply both terms inside the parentheses by -5 and then remove the parentheses.

$$\boxed{\begin{array}{l} 2 - 5(9 - a) \\ = 2 + (-5)(9) + (-5)(-a) \end{array}} \longrightarrow 20 = 2 - 45 + 5a$$

Step 2	Combine like terms on the right side.	$20 = -43 + 5a$
Step 3	Add 43 to both sides.	$20 + 43 = -43 + 5a + 43$
		$63 = 5a$
Step 4	Divide both sides by 5.	$\dfrac{63}{5} = \dfrac{5a}{5}$
		$12.6 = a$ or $a = 12.6$

(h) **Solve:** $2(4x + 1) + 5(3x + 2) = 58$

Step 1   Multiply each term inside the first parentheses by 2 and each term inside the second parentheses by 5.

$$8x + 2 + 15x + 10 = 58$$

Step 2   Combine both pairs of like terms on the left side. $23x + 12 = 58$

Step 3 Subtract 12 from both sides. $23x + 12 - 12 = 58 - 12$

$$23x = 46$$

Step 4 Divide both sides by 23. $$\frac{23x}{23} = \frac{46}{23}$$

$$x = 2$$

Variable on Both Sides In all the equations we have solved so far, the variable has been on only one side of the equation. Sometimes it is necessary to solve equations with variable terms on both sides.

EXAMPLE 3 To solve

$$3x - 4 = 8 - x$$

First, move all variable terms to the left side by adding x to both sides.

$3x + x = 4x$
$-x + x = 0$

$3x - 4 + x = 8 - x + x$

$4x - 4 = 8$

Next, proceed as before.

Add 4 to both sides. $4x - 4 + 4 = 8 + 4$

$$4x = 12$$

Divide both sides by 4. $$\frac{4x}{4} = \frac{12}{4}$$

$$x = 3$$

☑ **Finally**, check your answer.

$$3(3) - 4 = 8 - (3)$$

$$9 - 4 = 5$$

$$5 = 5$$

●

Your Turn Ready to attempt one yourself? Solve this equation for y, then check it with our solution.

$$5y - 21 = 8y$$

Solution Did you move the variable terms to the left? Here you can save a step by moving them to the right instead of to the left.

Solve: $5y - 21 = 8y$

Step 1 Subtract $5y$ from both sides. $5y - 21 - 5y = 8y - 5y$

$$-21 = 3y$$

Step 2 Divide both sides by 3. $$\frac{-21}{3} = \frac{3y}{3}$$

$$-7 = y \quad \text{or} \quad y = -7$$

☑ $5(-7) - 21 = 8(-7)$

$$-35 - 21 = -56$$

$$-56 = -56$$

Learning Help   If a variable term is already by itself on one side of the equation, move all variable terms to this side. As in the last example, this will save a step. ●

More Practice   Now try these problems for practice in solving equations in which the variable appears on both sides.

(a) $x - 6 = 3x$ (b) $5(x - 2) = x + 4$

(c) $2(x - 1) = 3(x + 1)$ (d) $4x + 9 = 7x - 15$

(e) $2A = 12 - A$ (f) $6y = 4(2y + 7)$

(g) $8m - (2m - 3) = 3(m - 4)$ (h) $-2(3x - 5) = 7 - 5(2x + 3)$

Solutions   (a) **Solve:** $x - 6 = 3x$

Subtract x from each side, so that x-terms will appear only on the right. $x - 6 - x = 3x - x$

Combine like terms. $-6 = 2x$

Write with the x-term on the left. $2x = -6$

Divide by 2. $x = -3$

☑ $(-3) - 6 = 3(-3)$

$-9 = -9$

(b) **Solve:** $5(x - 2) = x + 4$

Multiply each term inside the parentheses by 5. $5x - 10 = x + 4$

Subtract x from each side. $5x - 10 - x = x + 4 - x$

Combine terms. $4x - 10 = 4$

Add 10 to each side. $4x - 10 + 10 = 4 + 10$

$4x = 14$

Divide each side by 4. $x = 3\frac{1}{2}$

☑ $5(3\frac{1}{2} - 2) = (3\frac{1}{2}) + 4$

$5(1\frac{1}{2}) = 7\frac{1}{2}$

$7\frac{1}{2} = 7\frac{1}{2}$

(c) **Solve:** $2(x - 1) = 3(x + 1)$

Remove the parentheses by multiplying. $2x - 2 = 3x + 3$

Subtract $3x$ from each side. $2x - 2 - 3x = 3x + 3 - 3x$

Combine like terms. $-2 - x = 3$

Add 2 to each side. $-2 - x + 2 = 3 + 2$

$-x = 5$

Multiply both sides by -1.
Solve for x, not $-x$. $x = -5$

☑ $2(-5 - 1) = 3(-5 + 1)$

$2(-6) = 3(-4)$

$-12 = -12$

(d) **Solve:** $4x + 9 = 7x - 15$

Subtract $7x$ from each side. $\qquad 4x + 9 - 7x = 7x - 15 - 7x$

Combine like terms. $\qquad -3x + 9 = -15$

Subtract 9 from each side. $\qquad -3x + 9 - 9 = -15 - 9$

$$-3x = -24$$

Divide each side by -3. $\qquad x = 8$

The check is left to you.

(e) **Solve:** $2A = 12 - A$

Add A to each side. $\qquad 2A + A = 12 - A + A$

The A-terms now appear only on the left. $\qquad 3A = 12$

Divide by 3. $\qquad \dfrac{3A}{3} = \dfrac{12}{3}$

$$A = 4$$

☑ $\quad 2(4) = 12 - (4)$

$\quad 8 = 8$

(f) **Solve:** $6y = 4(2y + 7)$

Multiply to remove the parentheses. $\qquad 6y = 8y + 28$

Subtract $8y$ from both sides. $\qquad 6y - 8y = 8y + 28 - 8y$

Combine like terms. $\qquad -2y = 28$

Divide by -2. $\qquad \dfrac{-2y}{-2} = \dfrac{28}{-2}$

$$y = -14$$

Check the solution.

(g) **Solve:** $8m - (2m - 3) = 3(m - 4)$

Change signs to remove the parentheses on the left. $\qquad 8m - 2m + 3 = 3(m - 4)$

Multiply by 3 to remove the parentheses on the right. $\qquad 8m - 2m + 3 = 3m - 12$

Combine like terms. $\qquad 6m + 3 = 3m - 12$

Subtract $3m$ from both sides. $\qquad 6m + 3 - 3m = 3m - 12 - 3m$

$$3m + 3 = -12$$

Subtract 3 from both sides. $\qquad 3m + 3 - 3 = -12 - 3$

Combine like terms. $\qquad 3m = -15$

Divide by 3. $\qquad \dfrac{3m}{3} = \dfrac{-15}{3}$

$$m = -5$$

Check the solution.

(h) **Solve:** $-2(3x - 5) = 7 - 5(2x + 3)$

Multiply by -2 to remove the parentheses on the left. $\qquad -6x + 10 = 7 - 5(2x + 3)$

Multiply by -5 to remove the parentheses on the right. (Leave the 7 alone!) $\qquad -6x + 10 = 7 - 10x - 15$

Combine like terms on the right.	$-6x + 10 = -8 - 10x$
Add $10x$ to both sides.	$-6x + 10 + 10x = -8 - 10x + 10x$
Combine like terms.	$4x + 10 = -8$
Subtract 10 from both sides.	$4x + 10 - 10 = -8 - 10$
Combine like terms.	$4x = -18$
Divide by 4.	$x = -4.5$

Check it.

Remember:

1. Do only legal operations: Add or subtract the same quantity from both sides of the equation; multiply or divide both sides of the equation by the same nonzero quantity.

2. Remove all parentheses carefully.

3. Combine like terms when they are on the same side of the equation.

4. Use legal operations to change the equation so that you have only the variable by itself on one side of the equation and a number on the other side of the equation.

5. To avoid errors, be sure to check your answer.

Solving Formulas   To **solve a formula** for some letter means to rewrite the formula as an equivalent formula with that letter isolated on the left of the equals sign.

For example, the area of a triangle is given by the formula

$$A = \frac{BH}{2}$$ where A is the area, B is the length of the base, and H is the height.

Solving for the base B gives the equivalent formula

$$B = \frac{2A}{H}$$

Solving for the height H gives the equivalent formula

$$H = \frac{2A}{B}$$

Solving formulas is a very important practical application of algebra. Very often a formula is not written in the form that is most useful. To use it you may need to rewrite the formula, solving it for the letter whose value you need to calculate.

To solve a formula, use the same balancing operations that you used to solve equations. You may add or subtract the same quantity on both sides of the formula and you may multiply or divide both sides of the formula by the same nonzero quantity.

EXAMPLE 4   To solve the formula

$$S = \frac{R + P}{2} \qquad \text{for } R$$

First, multiply both sides of the equation by 2 to eliminate or "clear" the fraction.

$$2 \cdot S = 2 \cdot \left(\frac{R + P}{2} \right)$$

$$2S = R + P$$

Second, subtract P from both sides of the equation to isolate R on one side.

$$2S \underbrace{- P = R + P - P}_{0}$$

$$2S - P = R$$

This formula can be reversed to read

$R = 2S - P$ We have solved the original formula for R.   ●

Your Turn   Solve the following formulas for the variable indicated.

(a) $V = \dfrac{3K}{T}$ for K

(b) $Q = 1 - R + T$ for R

(c) $V = \pi R^2 H - AB$ for H

(d) $P = \dfrac{T}{A - B}$ for A

Solutions   (a) $V = \dfrac{3K}{T}$

First, multiply both sides by T to get $VT = 3K$

Second, divide both sides by 3 to get $\dfrac{VT}{3} = K$

Solved for K, the formula is $K = \dfrac{VT}{3}$

(b) $Q = 1 - R + T$

First, add R to both sides to get $R + Q = 1 + T$

Second, subtract Q from both sides to get $R = 1 + T - Q$

(c) $V = \pi R^2 H - AB$

First, add AB to both sides to get $V + AB = \pi R^2 H$

Second, divide both sides by πR^2 to get $\dfrac{V + AB}{\pi R^2} = H$

Notice that we divide *all* of the left side by πR^2.

Solved for H, the formula is $H = \dfrac{V + AB}{\pi R^2}$

(d) $P = \dfrac{T}{A - B}$

First, we need to "clear" or eliminate the fraction. To do so, we multiply both sides by the denominator $(A - B)$ to get $P(A - B) = T$

Second, multiply to remove the parentheses. $PA - PB = T$

Next, add PB to each side. $PA = T + PB$

Finally, divide each side by P. $A = \dfrac{T + PB}{P}$

! Careful   Remember, when using the multiplication/division rule, you must multiply or divide *all* of both sides of the formula by the same quantity.   ●

More Practice   Practice solving formulas with the following problems.

Solve:

(a) $P = 2A + 3B$ for A (b) $E = MC^2$ for M

(c) $S = \dfrac{A - RT}{1 - R}$ for A (d) $S = \dfrac{1}{2}gt^2$ for g

(e) $A = \dfrac{2V - W}{R}$ for V (f) $F = \dfrac{9C}{5} + 32$ for C

(g) $A = \dfrac{\pi R^2 S}{360}$ for S (h) $P = \dfrac{t^2 dN}{3.78}$ for d

(i) $C = \dfrac{AD}{A + 12}$ for D (j) $V = \dfrac{\pi L T^2}{6} + 2$ for L

Answers   (a) $A = \dfrac{P - 3B}{2}$ (b) $M = \dfrac{E}{C^2}$

(c) $A = S - SR + RT$ (d) $g = \dfrac{2S}{t^2}$

(e) $V = \dfrac{AR + W}{2}$ (f) $C = \dfrac{5F - 160}{9}$

(g) $S = \dfrac{360A}{\pi R^2}$ (h) $d = \dfrac{3.78P}{t^2 N}$

(i) $D = \dfrac{CA + 12C}{A}$ (j) $L = \dfrac{6V - 12}{\pi T^2}$

Note   The equations you learned to solve in this chapter are all *linear* equations. The variable appears only to the first power—no x^2 or x^3 terms appear in the equations. You will learn how to solve more difficult algebraic equations in Chapter 11. ●

Now turn to Exercises 7-5 for a set of practice problems on solving equations and formulas.

Exercises 7-5   **Solving More Equations and Solving Formulas**

A. Solve the following equations.

  1. $5(x - 3) = 30$ 2. $22 = -2(y + 6)$

  3. $3(2n + 4) = 41$ 4. $-6(3a - 7) = 21$

  5. $2 - (x - 5) = 14$ 6. $24 = 5 - (3 - 2m)$

  7. $6 + 2(y - 4) = 13$ 8. $7 - 11(2z + 3) = 18$

  9. $8 = 5 - 3(3x - 4)$ 10. $7 + 9(2w + 3) = 25$

11. $6c - (c - 4) = 29$ 12. $9 = 4y - (y - 2)$

13. $5x - 3(2x - 8) = 31$ 14. $6a + 2(a + 7) = 8$

15. $9t - 3 = 4t - 2$ 16. $7y + 5 = 3y + 11$

17. $12x = 4x - 16$ 18. $22n = 16n - 18$

19. $8y - 25 = 13 - 11y$ 20. $6 - 2p = 14 - 4p$

21. $9x = 30 - 6x$

22. $12 - y = y$

23. $2(3t - 4) = 10t + 7$

24. $5A = 4(2 - A)$

25. $2 - (3x - 20) = 4(x - 2)$

26. $2(2x - 5) = 6x - (5 - x)$

27. $8 + 3(6 - 5x) = 11 - 10x$

28. $2(x - 5) - 3(2x - 8) = 16 - 6(4x - 3)$

B. Solve the following formulas for the variable shown.

1. $S = LW$ for L

2. $A = \dfrac{1}{2}BH$ for B

3. $V = IR$ for I

4. $H = \dfrac{D - R}{2}$ for D

5. $S = \dfrac{W}{2}(A + T)$ for T

6. $V = \pi R^2 H$ for H

7. $P = 2A + 2B$ for B

8. $H = \dfrac{R}{2} + 0.05$ for R

9. $T = \dfrac{RP}{R + 2}$ for P

10. $I = \dfrac{E + V}{R}$ for V

C. Practical Applications

Problem 1

1. **Sheet Metal Trades**   The length of the arc of a sector of a circle is given by the formula

$$L = \frac{2\pi Ra}{360}$$

R is the radius of the circle and a is the central angle in degrees (see figure at left).

(a) Solve for a.

(b) Solve for R.

(c) Find L when $R = 10$ in. and $a = 30°$. Use $\pi \approx 3.14$.

2. **Sheet Metal Trades**   The area of the sector shown in problem 1 is

$$A = \frac{\pi R^2 a}{360}.$$

(a) Solve this formula for a.

(b) Find A if $R = 12$ in., $a = 45°$, $\pi \approx 3.14$. (Round to the nearest whole number. The answer will be in square inches.)

3. **Machine Trades**   The linear speed L of a grinding wheel in feet per minute (fpm) is given by the formula

$$L = \frac{\pi dr}{12}$$

where d is the diameter of the wheel in inches and r is the rotational speed in revolutions per minute (rpm). What rotational speed is needed for a 6.0-inch grinding wheel to generate a linear speed of 4200 fpm? (Round to the nearest hundred rpm.)

4. **Machine Trades**   Machinists use a formula known as Pomeroy's formula to determine roughly the power required by a metal punch machine.

$$P \approx \frac{t^2 dN}{3.78}$$

where P = power needed, in horsepower
t = thickness of the metal being punched
d = diameter of the hole being punched
N = number of holes to be punched at one time

(a) Solve this formula for N.

(b) Find the power needed to punch six 2-in.-diameter holes in a sheet $\frac{1}{8}$ in. thick. (Round to the nearest hundredth.)

5. **Physics**   When a gas is kept at constant temperature and the pressure on it is changed, its volume changes in accord with the pressure–volume relationship known as Boyle's law:

$$\frac{V_1}{V_2} = \frac{P_2}{P_1}$$

where P_1 and V_1 are the beginning pressure and volume and P_2 and V_2 are the final pressure and volume.

(a) Solve for V_1.

(b) Solve for V_2.

(c) Solve for P_1.

(d) Solve for P_2.

(e) Find P_1 when V_1 = 10 cu ft, V_2 = 25 cu ft, and P_2 = 120 psi.

6. **Sports and Leisure**   The volume of a football is roughly $V = \frac{\pi L T^2}{6}$, where L is its length and T is its thickness.

(a) Solve for L.

(b) Solve for T^2.

7. **Allied Health**   Nurses use a formula known as Young's rule to determine the amount of medicine to give a child under 12 years of age when the adult dosage is known.

$$C = \frac{AD}{A + 12}$$

C is the child's dose; A is the age of the child in years; D is the adult dose.

(a) Work backward and find the adult dose in terms of the child's dose. Solve for D.

(b) Find D if C = 0.05 gram and A = 7.

8. **Carpentry**   The projection or width P of a protective overhang of a roof is determined by the height T of the window, the height H of the header above the window, and a factor F that depends on the latitude of the construction site.

$$P = \frac{T + H}{F}$$

Solve this equation for the header height H.

9. **Electronics**   For a current transformer, $\dfrac{i_L}{i_S} = \dfrac{T_P}{T_S}$

where i_L = line current
i_S = secondary current
T_P = number of turns of wire in the primary coil
T_S = number of turns of wire in the secondary coil

(a) Solve for i_L.

(b) Solve for i_S.

(c) Find i_L when $i_S = 1.5$ amperes, $T_P = 1500$, $T_S = 100$.

10. **Electronics** The electrical power P dissipated in a circuit is equal to the product of the current I and the voltage V, where P is in watts, I is in amperes, and V is in volts.

(a) Write an equation giving P in terms of I and V.

(b) Solve for I.

(c) Find V when $P = 15{,}750$ watts, $I = 42$ amperes.

11. **Roofing** The formula

$$L = U(R + H)$$

is used to determine rafter length (L) of a roof, where R is the run and H is the overhang. The quantity U represents unit line length. When the run and the overhang are in feet, the rafter length will be in inches. If a rafter 247 in. long is used on a roof with a run of 16.5 ft and $U = 13.0$, how long will the overhang be?

12. **Automotive Trades** The compression ratio R of a cylinder in a diesel engine is given by

$$R = \frac{S + C}{C}$$

where S is the swept volume of the cylinder and C is the clearance volume of the cylinder.

(a) Solve for S.

(b) Find the swept volume if the clearance volume is 6.46 in.^{3} and the compression ratio is 17.4 to 1. (Round to the nearest cubic inch.)

13. **Water/Wastewater Treatment** The detention time in a settling basin is the amount of time that entering water will be held before being drawn out for use. If V is the volume of the basin in gallons and F is the flow rate in gallons per day, then the detention time T is given by the formula

$$T = \frac{24V}{F}$$

(a) Solve this formula for V.

(b) Solve this formula for F.

(c) Find the flow rate if a basin holding 7480 gal has a detention time of 3.0 hr. (Round to the nearest thousand.)

(d) Find the volume of a basin if a flow rate of 42,000 gal/day results in a detention time of 3.5 hr. (Round to the nearest hundred gallons.)

14. **Life Skills** The formula

$$A = P(1 + rt)$$

is used to find the total amount A of money in an account when an original amount or principal P is invested at a rate of simple interest r for t years. How long would it take $8000 to grow to $10,000 at 4% simple interest?

15. **Construction**   The formula

$$I = 0.000014L(T - t)$$

gives the expansion I of a particular highway of length L at a temperature of T degrees Fahrenheit. The variable t stands for the temperature at which the highway was built. If a 2-mile stretch of highway was built at an average temperature of 60°F, what is the maximum temperature it can withstand if expansion joints allow for 7.5 ft of expansion? (*Hint:* The units of L must be the same as the units of I.)

16. **Electrical Trades**   The length L of wire (in feet) that would have resistance R (in ohms) is given by

$$L = \frac{Rd^2}{K}$$

where d is the diameter of the wire in thousandths of an inch (or mils) and K is the specific resistance of the material.

(a) Solve this formula for R.

(b) Substitute $\dfrac{E}{I}$ for R in your answer to part (a) and then solve for d.

(c) Use your answer to part (b) to find the diameter of copper wire ($K = 10.4$ ohms) 1620 ft long if the current I is 16.0 amperes and the voltage drop E is 12.0 volts. (Round to the nearest whole number.)

When you have completed these exercises, check your answers to the odd-numbered problems in the Appendix, then turn to Section 7-6 to learn about word problems.

7-6   Solving Word Problems

Translating English to Algebra   Algebra is a very useful tool for solving real-world problems. But in order to use it, you may find it necessary to translate simple English sentences and phrases into mathematical expressions or equations. In technical work especially, the formulas to be used are often given in the form of English sentences, and they must be rewritten as algebraic formulas before they can be used.

EXAMPLE 1   **Transportation**   The statement

The horsepower required to overcome vehicle air resistance is equal to the cube of the vehicle speed in miles per hour multiplied by the frontal area in square feet divided by 150,000

translates to the formula

$$hp = \frac{mph^3 \cdot area}{150,000}$$

or $P = \dfrac{v^3 A}{150,000}$ in algebraic form, where v is the vehicle speed.   ●

> **learning|catalytics™**
>
> 1. Write as an algebraic expression: "The sum of 4 and twice x equals 11."
> 2. Solve: $x + 2x + 3x = 72$

In the next few pages of this chapter we show you how to translate English statements into algebraic formulas. To begin, try the following problem.

Your Turn   **Automotive Trades**   An automotive technician found the following statement in a manual:

The pitch diameter of a cam gear is twice the diameter of the crank gear.

Translate this sentence into an algebraic equation.

Answer   The equation is $P = 2C$ where P is the pitch diameter of the cam gear and C is the diameter of the crank gear.

You may use any letters you wish, of course, but we have chosen letters that remind you of the quantities they represent: P for *pitch* and C for *crank*.

Notice that the phrase "twice C" means "two times C" and is written as $2C$ in algebra.

Certain words and phrases appear again and again in statements to be translated. They are signals alerting you to the mathematical operations to be used. Here is a handy list of the *signal words* and their mathematical translations.

Signal Words

English Term	Math Translation	Example
Equals	$=$	$A = B$
Is, is equal to, was, are, were		
The same as . . .		
What is left is . . .		
The result is . . .		
Gives, makes, leaves, having		
Plus, sum of	$+$	$A + B$
Increased by, more than		
A and B		
Minus B, subtract B	$-$	$A - B$
Less B		
Decreased by B, take away B		
Reduced by B, diminished by B		
B less than A		
B subtracted from A		
Difference between A and B		
Times, multiply, of	$\times$	AB
Multiplied		
Product of		
Divide, divided by B	$\div$	$A \div B$ or $\dfrac{A}{B}$
Quotient of		
Twice, twice as much	$\times 2$	$2A$
Double		
Squared, square of A		A^2
Cubed, cube of A		A^3

Your Turn   Translate the phrase *length plus 3 inches* into an algebraic expression. Try it, then check your answer with ours.

Solution   **First**, make a word equation by using parentheses.

(length)   (plus)   (3 inches)

Second, substitute mathematical symbols.

(length)   (plus)   (3 inches)
   ↓   ↓   ↓
   L $+$ 3   or $L + 3$

Notice that signal words, such as *plus,* are translated directly into math symbols. Unknown quantities are represented by letters of the alphabet, chosen to remind you of their meaning.

More Practice   **General Trades**   Translate the following phrases into math expressions.

(a) Weight divided by 13.6   (b) $6\frac{1}{4}$ in. more than the width

(c) One-half of the original torque   (d) The sum of the two lengths

(e) The voltage decreased by 5   (f) Five times the gear reduction

(g) 8 in. less than twice the height

Solutions   (a) (weight)   (divided by)   (13.6)
                      ↓   ↓   ↓
                     W $\div$ 13.6   or $\dfrac{W}{13.6}$

(b) $\left(6\frac{1}{4}\text{ in.}\right)$ (more than)   (the width)
        ↓   ↓   ↓
       $6\frac{1}{4}$ $+$ W or $6\frac{1}{4} + W$

(c) (one-half)   (of)   (the original torque)
        ↓   ↓   ↓
       $\dfrac{1}{2}$ $\times$ T or $\dfrac{1}{2}T$ or $\dfrac{T}{2}$

(d) (the sum of)   (the two lengths)
         ↓   ↓   ↓
              $L_1 + L_2$

(e) (the voltage)   (decreased by)   (5)
        ↓   ↓   ↓
        V $-$ 5   or $V - 5$

(f) (five)   (times)   (the gear reduction)
       ↓   ↓   ↓
       5       $\times$ G or $5G$

(g) 8 in. less than twice the height means

    (twice the height)   (less)   (8)
       ↓   ↓   ↓   ↓
       2   H $-$ 8   or $2H - 8$

Of course, any letters could be used in place of the ones used above.

! Careful The phrase *less than* has a meaning very different from the word *less*.

"8 *less* 5" means 8 − 5

"8 *less than* 5" means 5 − 8 ●

Translating Sentences to Equations So far we have translated only phrases, pieces of sentences, but complete sentences can also be translated. An English phrase translates into an algebraic expression, and an English sentence translates into an algebraic formula or equation.

EXAMPLE 2 **HVAC** The sentence

The height of the duct   is equal to   its width

translates to H $=$ W or $H = W$

Each word or phrase in the sentence becomes a mathematical term, letter, number, expression, or symbol. ●

Your Turn **Machine Trades** Translate the following sentence into algebraic form as we did above.

The size of a drill for a tap is equal to the tap diameter minus the depth.

Solution $S = T - D$

Follow these steps when you must translate an English sentence into an algebraic equation or formula.

Step 1 Cross out all unnecessary words.

The size ~~of a drill for a tap~~ is equal to ~~the~~ tap diameter minus ~~the~~ depth.

Step 2 Make a word equation using parentheses.

(Size) (is equal to) (tap diameter) (minus) (depth)

Step 3 Substitute a letter or an arithmetic symbol for each phrase in parentheses.

S $=$ T $-$ D

Step 4 Combine and simplify.

$S = T - D$

In most formulas the units for the quantities involved must be given. In the formula above, T and D are in inches.

← Learning Help Translating English sentences or verbal rules into algebraic formulas requires that you read the sentences very differently from the way you read stories or newspaper articles. Very few people are able to write out the math formula after reading the problem only once. You should expect to read it several times, and you'll want to read it slowly. No speed reading here! ●

The ideas in technical work and formulas are usually concentrated in a few key words, and you must find them. If you find a word you do not recognize, stop reading and look it up in a dictionary, textbook, or manual. It may be important. Translating and working with formulas is one of the skills you must have if you are to succeed at any technical occupation.

EXAMPLE 3   **Electrical Trades**   Here is another example of translating a verbal rule into an algebraic formula:

The electrical resistance of a length of wire is equal to the resistivity of the metal times the length of the wire divided by the square of the wire diameter.

Step 1   Eliminate all but the key words.

"~~The electrical~~ resistance ~~of a length of wire~~ is equal to ~~the~~ resistivity ~~of the metal~~ times ~~the~~ length ~~of the wire~~ divided by ~~the~~ square of ~~the wire~~ diameter."

Step 2   Make a word equation.   (Resistance) (is equal to) (resistivity) (times) (length) (divided by) (square of diameter)

Step 3   Substitute letters and symbols. $R = \dfrac{rL}{D^2}$

If the resistivity r has units of ohms times inches, L and D will be in inches. The resistance R will be in ohms.   ●

More Practice   The more translations you do, the easier it gets. Translate each of the following technical statements into algebraic formulas.

(a) **Sheet Metal Trades**   A sheet metal worker measuring a duct cover finds that the width is $8\frac{1}{2}$ in. less than the height.

(b) **General Trades**   One-quarter of a job takes $3\frac{1}{2}$ days.

(c) **Electrical Trades**   One-half of a coil of wire weighs $16\frac{2}{3}$ lb.

(d) **Industrial Technology**   The volume of an elliptical tank is approximately 0.7854 times the product of its height, length, and width.

(e) **Automotive Trades**   The engine speed is equal to 168 times the overall gear reduction multiplied by the speed in miles per hour and divided by the rolling radius of the tire.

(f) **Transportation**   The air resistance force in pounds acting against a moving vehicle is equal to 0.0025 times the square of the speed in miles per hour times the frontal area of the vehicle.

(g) **Electrical Trades**   Two pieces of wire have a combined length of 24 in. The longer piece is five times the length of the shorter piece. (*Hint:* Write two separate equations.)

Answers   (a) $W = H - 8\frac{1}{2}$

(b) $\frac{1}{4}J = 3\frac{1}{2}$ or $\frac{J}{4} = 3\frac{1}{2}$

(c) $\frac{1}{2}C = 16\frac{2}{3}$ or $\frac{C}{2} = 16\frac{2}{3}$

(d) $V = 0.7854HLW$

(e) $S = \dfrac{168Gv}{R}$ where v is in mph.

(f) $R = 0.0025v^2A$ where v is in mph.

(g) $24 = L + S$ and $L = 5S$

General Word Problems   Now that you can translate English phrases and sentences into algebraic expressions and equations, you should be able to solve many practical word problems.

The following six-step method will help you to solve any word problem.

1. **Read** the problem carefully, noting what information is given and what is unknown.

2. **Assign a variable** to represent the unknown. If there are additional unknowns, try to represent these with an expression involving the same variable.

3. **Write an equation** using the unknown(s) and some given fact about the unknown(s) in the problem statement.

4. **Solve** the equation.

5. **Answer** the original question; be careful to state the value of all unknown quantities asked for in the problem.

6. **Check** your solution.

EXAMPLE 4   **Construction**   A total of 66 cu yd of concrete is required for a pour. The combined amounts of cement, sand, and gravel should be $1\frac{1}{2}$ times the amount of concrete needed. For this particular job, cement, sand, and gravel must be combined in the ratio of 1:2:3. What volume of each component will be necessary?

To solve this problem, **first** calculate the total amount A of cement, sand, and gravel needed:

$$A = 1.5(66 \text{ cu yd}) = 99 \text{ cu yd}$$

Next, let $x =$ the amount of cement needed. The required ratios tell us that $2x$ would then represent the amount of sand needed, and $3x$ would represent the amount of gravel needed.

Finally, set up an equation and solve:

$$\text{Cement} + \text{Sand} + \text{Gravel} = 99 \text{ cu yd}$$
$$x + 2x + 3x = 99$$

Combine like terms: $6x = 99$

Divide by 6: $x = 16.5 \text{ cu yd}$

Therefore, 16.5 cubic yards of cement are needed for the job. Now find $2x$ and $3x$:

$$2x = 2(16.5) = 33 \text{ cu yd}$$
$$3x = 3(16.5) = 49.5 \text{ cu yd}$$

The job also requires 33 cu yd of sand and 49.5 cu yd of gravel.

✔ $x + 2x + 3x = 16.5 + 33 + 49.5 = 99$ ●

EXAMPLE 5   **Electrical Trades**   The Z, G, & E utility charges customers who generate solar energy a monthly fee of $15. Because they offer net energy metering, they compensate the customers $0.04 per kilowatt-hour (kWh) of net electricity returned to the grid. The Ready Edison utility charges solar customers a maintenance fee of $20 per month, and they compensate the customers $0.06 per kWh of net electricity returned to the grid. Suppose we wish to determine the amount of net return for which the two plans would result in the same monthly bill. If we let $x =$ the amount of net return we are looking for, then for Z, G, & E the monthly bill would be the monthly fee minus the compensation, or $15 - $0.04x$. Similarly, the monthly bill for Ready Edison would be $20 - $0.06x$. To determine the unknown number of kWh, simply set these two expressions equal to each other and solve for x.

$$\$15 - \$0.04x = \$20 - \$0.06x$$

First, add $0.06x$ to both sides to get

$$\$15 + \$0.02x = \$20$$

Next, subtract $15 from both sides to get

$$\$0.02x = \$5$$

Finally, divide both sides by $0.02 to get

$$x = 250$$

For customers returning 250 kWh to the grid, the two plans would result in the same bill.

We can go one step further and determine what that common bill would be.

Substituting $x = 250$ in the left side of the original equation, we have

$$\$15 - \$0.04(250) = \$15 - \$10 = \$5$$

If we substitute in the right side of the original equation, we should get the same result:

$$\$20 - \$0.06(250) = \$20 - \$15 = \$5$$

Although it was not necessary to substitute our answer into both sides, doing so serves as a way of checking our answer.   ●

Your Turn   Use your knowledge of algebra to translate the following problems to algebraic equations and then solve them.

(a) **Electrical Trades**   For customers generating their own solar power, an electric utility charges a $20 per month maintenance fee but refunds them $0.05 per kilowatt-hour of net electricity returned to the grid. How many kilowatt-hours per month would they need to return to the grid in order to break even?

(b) **Life Skills**   Bob's small appliance repair service charges a fixed cost of $65 plus $50 per hour for the total repair time. Frank's small appliance repair service charges a simple rate of $70 per hour.

   (1) After how many hours would the total repair bill be the same for Bob and Frank?

   (2) Who would be less expensive for a repair job lasting 2 hours?

Solutions   (a) To "break even," we want the amount we pay to be equal to the amount we earn back. If we let $x =$ the number of kilowatt-hours needed to break even, then we can set up the following equation:

$$\$20 = \$0.05x$$

To solve, divide both sides by $0.05 to get

$$400 = x \quad \text{or} \quad x = 400$$

The customers would need to return 400 kilowatt-hours to the grid in order to break even on their monthly bill.

(b) (1) There is one unknown:

   Let $x =$ the repair time for which the two shops would charge the same amount.

Use the fact that the two amounts would be the same to write an equation:

Bob's charge   Frank's charge

$$65 + 50x = 70x$$

To solve, subtract $50x$ from both sides. $65 = 20x$

Then divide each side by 20. $3.25 = x$

Bob and Frank would charge the same amount for a repair job involving 3.25 hours of work.

(2) For a repair job of 2 hours:

Bob's charge is $\$65 + \$50(2) = \$165$
Frank's charge is $\$70(2) = \$140$

Frank would be less expensive. (This would be the case for any job involving up to 3.25 hours, after which Bob would be less expensive.)

Some of the more difficult percent problems can be simplified using algebraic equations. One such problem, backing the tax out of a total, is commonly encountered by those who own their own businesses.

EXAMPLE 6   **Trades Management**   Suppose that an auto mechanic has collected $1468.63 for parts, including 6% tax. For accounting purposes he must determine exactly how much of the total is sales tax. If we let x stand for the dollar amount of the parts before tax was added, then the tax is 6% of that, or $0.06x$, and we have

$$x \quad + \quad 0.06x \quad = \quad \$1468.63$$

Cost of the parts   6% tax on the parts   Total amount

To solve the equation $x + 0.06x = \$1468.63$

First, note that $x = 1x = 1.00x$. $1.00x + 0.06x = \$1468.63$

Next, combine like terms. $1.06x = \$1468.63$

Now, divide by 1.06 to get $x = \$1385.50$

x is the cost of the parts. To calculate the tax,

multiply x by 0.06 $0.06x = 0.06(\$1385.50) = \$83.13$

or subtract x from $1468.63 $\$1468.63 - \$1385.50 = \$83.13$ •

EXAMPLE 7   **Automotive Trades**   When fuel costs increase, more people consider the purchase of a hybrid car, and they need to make informed decisions about whether such a purchase will save them money in the long run. Consider a vehicle that is available in a hybrid and a nonhybrid model. The following table shows us several pertinent facts about the two models:

Vehicle	Purchase Price	Combined Mileage Rating	Tax Credit
Hybrid	$29,500	28 mi/gal	$2200
Nonhybrid	$25,000	20 mi/gal	none

Let's assume that the average price of gas is $2.50 per gallon. We wish to determine how many miles of driving it will take before the fuel savings of the hybrid will make up for its greater initial cost.

First, determine the difference in their initial costs. The tax credit on the hybrid reduces its initial cost to $29,500 − $2200, or $27,300, so the actual difference in their initial costs is $27,300 − $25,000, or $2300.

Next, determine the cost per mile for each of the two vehicles. To do so, divide cost per gallon by miles per gallon. Note that

$$\frac{\$}{\text{gal}} \div \frac{\text{mi}}{\text{gal}} = \frac{\$}{\text{gal}} \times \frac{\text{gal}}{\text{mi}} = \frac{\$}{\text{mi}}$$

In our example,

Cost per mile of the hybrid: $2.50 per gallon ÷ 28 mi/gal ≈ $0.089 per mile

Cost per mile of the nonhybrid: $2.50 per gallon ÷ 20 mi/gal = $0.125 per mile

Next, set up an equation. Let $x =$ the number of miles one must drive to make up for the difference in the initial cost: At what value of x will the difference between the cost of driving the nonhybrid ($0.125x$) and the cost of driving the hybrid ($0.089x$) be equal to the difference in the initial cost ($2300)?

$$\$0.125x - \$0.089x = \$2300$$

Finally, solve the equation:

Combine like terms. $0.036x = $2300

Divide by $0.036. $x = 63,888.8 \ldots$

Therefore, it will take approximately 64,000 miles of driving before the fuel savings of the hybrid will make up for its greater initial cost.   ●

✎ Note   While the method of the previous example gives us a fairly accurate picture of the payback time for a hybrid, there may be other cost differences involved, such as insurance or maintenance, that may also affect the cost comparison. ●

More Practice   Practice makes perfect. Work the following set of word problems.

(a) **Trades Management**   Geeks R Us, a computer repair service, charges $100 for the first hour of labor and $85 for each additional hour. The total labor charge for a repair job was $610. Set up an equation to find the total repair time, then solve the equation.

(b) **HVAC**   A family is currently using an oil-fired boiler to heat their house. The family wishes to replace this old system with a more energy-efficient ground-source heat pump (GSHP). The boiler costs about $1254 per year to operate, while a GSHP would cost about $792 per year. If the initial cost of the GSHP is $10,000, how many years will it take for the savings in energy cost to pay back the cost of installing the GSHP? (Round up to the next whole number.)

(c) **General Trades**   A tech is paid $22 per hour plus $33 per hour for overtime (weekly time in excess of 40 hours). How much total time must the tech work to earn $1078?

(d) **Life Skills**   A couple prepares a budget that allows for $220 per month to be spent on gas. Suppose that gas costs $2.50 per gallon and the couple's vehicles have an average gas mileage of 25 mi/gal. Determine the maximum number of miles, m, they can drive and stay within this budget.

(e) **Trades Management** Mike and Jeff are partners in a small manufacturing firm. Because Mike provided more of the initial capital for the business, they have agreed that Mike's share of the profits should be one-fourth greater than Jeff's. The total profit for the first quarter of this year was $17,550. How should they divide it?

(f) **Trades Management** A plumber collected $2784.84 during the day: $1650 for labor, which is not taxed, and the rest for parts, which includes 5% sales tax. Determine the total amount of tax that was collected.

(g) **Printing** A printer knows from past experience that about 2% of a particular run of posters will be spoiled. How many should she print in order to end up with 1500 usable posters?

(h) **Construction** A contractor needs to order some gravel for a driveway. Acme Building Materials sells it for $61 per cubic yard plus a $62 delivery charge. The Stone Yard sells it for $55 per cubic yard plus an $83 delivery charge. (1) For an order of 2 cu yd, which supplier is less expensive? (2) For an order of 5 cu yd, which is less expensive? (3) For what size order would both suppliers charge the same amount?

Solutions (a) Let $H =$ the total number of hours

then $H - 1 =$ the number of additional hours charged at $85 per hour

$$\$100 + \$85(H - 1) = \$610$$

| The cost of the first hour | | The cost of $H - 1$ additional hours |

Multiply to remove parentheses. $\$100 + \$85H - \$85 = \$610$
Combine like terms. $\$85H + \$15 = \$610$
Subtract $15 from both sides. $\$85H = \$595$
Divide both sides by $85. $H = 7$

The total repair time was 7 hours.

(b) Let $t =$ the number of years required for payback. By switching to a GSHP, the amount saved each year is $1254 - $792 = $462.

Therefore, in t years the amount saved is $462t. We wish to know when the total savings is equal to the initial cost of $10,000, so the equation is simply:

$$\$462t = \$10,000$$

Dividing both sides by $462, we have

$$t = 21.645 \ldots$$

Rounding up to the next whole number, we find that after 22 years the family will have paid back the initial investment and will be starting to save money. Although this may seem like a long time to wait for payback, another factor to consider is that the older oil-fired boiler may need repairs or replacement within those 22 years.

(c) Let $t =$ total time worked
then $t - 40 =$ hours of overtime worked

For the first 40 hours, the tech's pay is

$\$22(40) = \$880$

For $t - 40$ overtime hours, the tech's pay is

$\$33(t - 40)$

Therefore, the equation for total pay is

$\$880 + \$33(t - 40) = \$1078$

Solve for t. (We can eliminate the $ signs to simplify the equation.):

$$880 + 33t - 1320 = 1078$$
$$33t - 440 = 1078$$
$$33t = 1518$$
$$t = 46$$

The tech must work 46 total hours to earn $1078.

(d) **First,** set up and solve an equation to determine the number of gallons, g, the couple can purchase.

$\$2.50g = \$220$
$g = 88$ gal

Then, multiply this result by their gas mileage.

$m = (88 \text{ gal})(25 \text{ mi/gal})$

Notice that the gallons unit cancels out.

$m = \dfrac{88 \text{ gal}}{1 \text{ month}} \times \dfrac{25 \text{ mi}}{1 \text{ gal}}$

Finally, this leaves us with the maximum number of miles they can drive in a month.

$m = \dfrac{2200 \text{ mi}}{1 \text{ month}}$

or 2200 miles per month

(e) Let Jeff's share $= J$
then Mike's share $= J + \frac{1}{4}J$

The equation is

$\$17{,}550 = J + (J + \frac{1}{4}J)$

Combine terms; remember $J = 1 \cdot J$.

$\$17{,}550 = (2\frac{1}{4})J$

Write the mixed number as an improper fraction.

$\$17{,}550 = \dfrac{9}{4}J = \dfrac{9J}{4}$

or

$\dfrac{9J}{4} = \$17{,}550$

Multiply by 4.

$9J = \$70{,}200$

Divide by 9.

$J = \$7800$ Jeff's share

Mike's share $= \$7800 + \dfrac{\$7800}{4} = \$7800 + \$1950 = \$9750$

(f) Let $x =$ cost of the parts
then $0.05x =$ amount of the tax

Total cost with tax

Subtract the labor to get the total cost of the parts.

The equation is $x + 0.05x = \$2784.84 - \$1650$

Combine terms. $1.05x = \$1134.84$ (Remember: $x = 1x = 1.00x$)

Divide by 1.05. $x = \$1080.80$ This is the cost of the parts.

The amount of the tax is $0.05(\$1080.80) = \$54.04$.

(g) Let $x =$ number of posters she needs to print
then $0.02x =$ number that will be spoiled

$$x \quad - \quad 0.02x \quad = \quad 1500$$

The number printed   minus   the number spoiled   must total 1500.

We can rewrite this equation as $1.00x - 0.02x = 1500$

Now subtract like terms. $0.98x = 1500$

Divide by 0.98. $x = 1531$ rounded

She must print approximately 1531 posters to end up with 1500 unspoiled ones.

(h) (1) For 2 cu yd
Acme price: $61(2) + $62 = $184
Stone Yard price: $55(2) + $83 = $193
Acme is less expensive.

(2) For 5 cu yd
Acme price: $61(5) + $62 = $367
Stone Yard price: $55(5) + $83 = $358
Stone Yard price is less expensive.

(3) Let x = the amount of gravel for which each supplier's price will be the same. Then if Acme's price = Stone Yard's price

$$\$61x + \$62 = \$55x + \$83$$

Subtract $55x$. $6x + 62 = 83$

Subtract 62. $6x = 21$

Divide by 6. $x = 3.5$ cu yd

The total cost for each supplier is the same for an order of 3.5 cu yd.

Now turn to Exercises 7-6 for a set of word problems.

Exercises 7-6   Solving Word Problems

A. Translate the following into algebraic equations:

1. **Water/Wastewater Treatment**   The height of the tank is 1.4 times its width.

2. **General Trades**   The sum of the two weights is 167 lb.

3. **General Trades**   The volume of a cylinder is equal to $\frac{1}{4}$ of its height times π times the square of its diameter.

4. **General Trades**   The weight in kilograms is equal to 0.454 times the weight in pounds.

5. **Machine Trades**   The volume of a solid bar is equal to the product of the cross-sectional area and the length of the bar.

6. **General Trades**   The volume of a cone is $\frac{1}{3}$ times π times the height times the square of the radius of the base.

7. **Machine Trades**   The pitch diameter D of a spur gear is equal to the number of teeth on the gear divided by the pitch.

8. **Machine Trades**   The cutting time for a lathe operation is equal to the length of the cut divided by the product of the tool feed rate and the revolution rate of the workpiece.

9. **Machine Trades**   The weight of a metal cylinder is approximately equal to 0.785 times the height of the cylinder times the density of the metal times the square of the diameter of the cylinder.

10. **Electrical Trades**   The voltage across a simple circuit is equal to the product of the resistance of the circuit and the current flowing in the circuit.

B. Set up equations and solve.

1. **Water/Wastewater Treatment**   Two tanks must have a total capacity of 400 gallons. If one tank needs to be twice the size of the other, how many gallons should each tank hold?

2. **Machine Trades**   Two metal castings weigh a total of 84 lb. One weighs 12 lb more than the other. How much does each one weigh?

3. **Plumbing**   The plumber wants to cut a 24-ft length of pipe into three sections. The largest piece should be twice the length of the middle piece, and the middle piece should be twice the length of the smallest piece. How long should each piece be?

4. **Electrical Trades**   For customers primarily using their own solar power, a certain electric utility charges an $18 per month maintenance fee but refunds to the customer $0.06 per kilowatt-hour of net electricity that is returned to the grid. Following the method of Example 5, set up an equation and solve for the amount of electricity a customer would need to return to the grid in order to break even.

5. **Masonry**   There are 156 concrete blocks available to make a retaining wall. The bottom three rows will all have the same number of blocks. The next six rows will each have two blocks fewer than the row below it. How many blocks are in each row?

6. **Construction**   A total of 18.6 cu yd of concrete is required for pouring a foundation. The ratio of cement to sand to gravel for this application is 1:2:4, and the total amount of these components must be $1\frac{1}{2}$ times the volume of concrete needed. Use the method of Example 4 to determine how many cubic yards of each component are needed.

7. **Trades Management**   Jo and Ellen are partners in a painting business. Because Jo is the office manager, she is to receive one-third more than Ellen when the profits are distributed. If their profit is $85,400, how much should each of them receive?

8. **Construction**   A total of $1,120,000 is budgeted for constructing a roadway. The rule of thumb for this type of project is that the pavement costs twice the amount of the base material, and the sidewalk costs one-fourth the amount of the pavement. Using these figures, how much should each item cost?

9. **Photography**   A photographer has enough liquid toner for fifty 5-in. by 7-in. prints. How many 8-in. by 10-in. prints will this amount cover? (*Hint:* The toner covers the *area* of the prints: Area = length × width.)

10. **Construction**   A building foundation has a length of 82 ft and a perimeter of 292 ft. What is the width? (*Hint:* Perimeter = 2 × length + 2 × width.)

11. **General Trades**   A roofer earns $24 per hour plus $36 per hour overtime. If he puts in 40 hours of regular time during a certain week and he wishes to earn $1200, how many hours of overtime should he work?

12. **Agriculture**   An empty avocado crate weighs 4.2 kg. How many avocados weighing 0.3 kg each can be added before the total weight of the crate reaches 15.0 kg?

13. **Machine Trades**   One-tenth of the parts tooled by a machine are rejects. How many parts must be tooled to ensure 4500 acceptable ones?

14. **Office Services** An auto mechanic has total receipts of $1861.16 for a certain day. He determines that $1240 of this is labor. The remaining amount is for parts plus 6% sales tax on the parts only. Find the amount of the sales tax he collected.

15. **Office Services** A travel agent receives a 7% commission on the base fare of a customer—that is, the fare *before* sales tax is added. If the total fare charged to a customer comes to $753.84 *including* 5.5% sales tax, how much commission should the agent receive?

16. **Office Services** A newly established carpenter wishes to mail out letters advertising her services to the families of a small town. She has a choice of mailing the letters first class or obtaining a bulk-rate permit and mailing them at the cheaper bulk rate. The bulk-rate permit costs $215 and each piece of bulk mail then costs 29 cents. If the first-class rate is 47 cents, how many pieces would she have to mail in order to make bulk rate the cheaper way to go?

17. **Trades Management** An appliance repair shop charges $95 for the first hour of labor and $80 for each additional hour. The total labor charge for an air-conditioning system installation was $655. Write an equation that allows you to solve for the total repair time; then solve the equation.

18. **Trades Management** Four partners in a plumbing business decide to sell the business. Based on their initial investments, Al will receive five shares of the selling price, Bob will receive three shares, and Cody and Dave will receive two shares each. If the business sells for $840,000, how much is each share worth?

19. **Automotive Trades** An automotive technician has two job offers. Shop A would pay him a straight salary of $720 per week. Shop B would pay him $20 per hour to work flat rate, where his hours would vary according to how many jobs he completes. How many hours would he have to log during a week at Shop B in order to match the salary at Shop A?

20. **Culinary Arts** Laurie runs a small cafeteria. She must set aside 25% of her revenue for taxes and spends another $1300 per week on other expenses. How much total revenue must she generate in order to clear a profit of $2000 per week?

21. **Life Skills** Suppose you budgeted $2800 for fuel expenses for the year. How many miles could you drive if gas were $2.70 per gallon and your vehicle averaged 28 mi/gal? (Round to the nearest thousand.)

22. **Automotive Trades** A certain vehicle has both a hybrid and a nonhybrid model. Their purchase prices, combined mileage ratings, and eligible tax credits are shown in the following table:

Vehicle	Purchase Price	Combined Mileage Rating	Tax Credit
Hybrid	$32,000	34 mi/gal	$500
Nonhybrid	$28,000	23 mi/gal	none

How many miles of driving will it take until the savings on gas makes up for the higher initial cost of the hybrid if the average price of gas is $2.70 per gallon? (Round to the nearest hundred miles.) (See Example 7 on page 491.)

23. **Automotive Trades** In problem 22, suppose the average price of gas has decreased to $2.40 per gallon by the time you purchase your vehicle.

(a) How many miles would you have to drive in order to offset the higher initial cost of the hybrid? (Round to the nearest hundred.)

(b) Choose the correct word to fill in the underlined portion of the quote: "The higher the price of gas, the (more/fewer) miles it takes for the savings on gas to make up for the higher the initial cost of the hybrid."

24. **Landscaping**   A landscaper must order some topsoil for a job. The Dirt Yard sells it for $60 per cubic yard plus a $75 delivery charge. Mud City sells it for $68 per cubic yard plus a $55 delivery charge. Set up and solve an equation to determine the number of cubic yards at which both suppliers would charge the same total amount. [See problem (h) on page 493.]

25. **Construction**   A contractor estimates that it would take $16,800 to make a home 50% more energy-efficient than a code-built house. She projects that the code-built house would have annual heating costs of $1848, while the more energy-efficient house would have annual heating costs of $806. How long would it take for the annual savings to pay off the cost of the upgrades? (Round up to the next whole number.)

26. **Automotive Trades**   A couple owns two cars: an electric car and a gas-powered car. The gas-powered car averages 25 miles per gallon, and gas currently costs $2.50 per gallon. To charge the electric car, they first paid $1000 to have a home charger installed and then paid about $0.05 per mile in electricity costs. After how many miles of driving would the electric car save money compared to the gas-powered car if both were driven equally?

27. **Manufacturing**   A conveyor belt 24 in. wide is moving boxes that are 12 in. wide and 18 in. long. Using the full width of the belt, how fast, in *feet* per minute, would the belt need to move in order to unload 6400 boxes in an hour? (Assume that the boxes are placed end to end with no space in between them.)

Check your answers to the odd-numbered problems in the Appendix, then turn to Section 7-7 to learn more algebra.

7-7   Multiplying and Dividing Algebraic Expressions

Multiplying Simple Factors   Earlier we learned that only like algebraic terms, those with the same literal part, can be added and subtracted. However, any two terms, like or unlike, can be multiplied.

1. True or false: $2^3 \times 2^2 = 2^5$
2. True or false: $2^6 \div 2^2 = 2^3$

To multiply two terms such as $2x$ and $3xy$, first remember that $2x$ means 2 times x. Second, recall from arithmetic that the order in which you do multiplications does not make a difference. For example, in arithmetic

$$2 \cdot 3 \cdot 4 = (2 \cdot 4) \cdot 3 = (3 \cdot 4) \cdot 2$$

and in algebra

$$a \cdot b \cdot c = (a \cdot c) \cdot b = (c \cdot b) \cdot a$$

or $2 \cdot x \cdot 3 \cdot x \cdot y = 2 \cdot 3 \cdot x \cdot x \cdot y$

Finally, remember that $x \cdot x = x^2$, $x \cdot x \cdot x = x^3$, and so on. Therefore,

$$2x \cdot 3xy = 2 \cdot 3 \cdot x \cdot x \cdot y = 6x^2y$$

The following examples show how to multiply two terms.

EXAMPLE 1

(a) $a \cdot 2a = a \cdot 2 \cdot a$

$\qquad = 2 \cdot \underbrace{a \cdot a}$ Group like factors together.

$\qquad = 2 \cdot a^2$

$\qquad = 2a^2$

(b) $2x^2 \cdot 3xy = 2 \cdot x \cdot x \cdot 3 \cdot x \cdot y$

$\qquad = \underbrace{2 \cdot 3} \cdot \underbrace{x \cdot x \cdot x} \cdot y$ Group like factors together.

$\qquad = 6 \cdot x^3 \cdot y$

$\qquad = 6x^3y$

(c) $3x^2yz \cdot 2xy = 3 \cdot x^2 \cdot y \cdot z \cdot 2 \cdot x \cdot y$

$\qquad = \underbrace{3 \cdot 2} \cdot \underbrace{x^2 \cdot x} \cdot \underbrace{y \cdot y} \cdot z$

$\qquad = 6 \cdot x^3 \cdot y^2 \cdot z$

$\qquad = 6x^3y^2z$

Remember to group like factors together before multiplying.

If you need to review exponents, return to page 401.   ●

Your Turn Now try the following problems.

(a) $x \cdot y$ $\qquad =$ _____   (b) $2x \cdot 3x$ $\qquad =$ _____

(c) $2x \cdot 5xy$ $\qquad =$ _____   (d) $4a^2b \cdot 2a$ $\qquad =$ _____

(e) $3x^2y \cdot 4xy^2$ $\qquad =$ _____   (f) $3x \cdot 2y^2 \cdot 2x^2y$ $\qquad =$ _____

(g) $2x^2(x + 3x^2)$ $\qquad =$ _____   (h) $-2a^2b(a^2 - 3b^2)$ $\qquad =$ _____

Answers (a) xy (b) $6x^2$ (c) $10x^2y$ (d) $8a^3b$

$\qquad$ (e) $12x^3y^3$ (f) $12x^3y^3$ (g) $2x^3 + 6x^4$ (h) $-2a^4b + 6a^2b^3$

A Closer Look Were the last two problems tricky for you? Recall from Rule 3 on page 446 (the distributive property) that when there is a multiplier in front of parentheses, every term inside must be multiplied by this factor. Try it this way:

(g) $2x^2(x + 3x^2) = (2x^2)(x) + (2x^2)(3x^2)$

$\qquad = 2 \cdot \underbrace{x^2 \cdot x} + \underbrace{2 \cdot 3} \cdot \underbrace{x^2 \cdot x^2}$

$\qquad = 2 \cdot x^3 + 6 \cdot x^4$

$\qquad = 2x^3 + 6x^4$

(h) $-2a^2b(a^2 - 3b^2) = (-2a^2b)(a^2) + (-2a^2b)(-3b^2)$

$\qquad = (-2) \cdot \underbrace{a^2 \cdot a^2} \cdot b + \underbrace{(-2) \cdot (-3)} \cdot \underbrace{a^2 \cdot b \cdot b^2}$

$\qquad = -2 \cdot a^4 \cdot b + 6 \cdot a^2 \cdot b^3$

$\qquad = -2a^4b + 6a^2b^3$ ●

Rule for Multiplication   As you did the preceding problems, you may have noticed that when you multiply like factors together, you actually add their exponents. For example,

$$x^2 \cdot x^3 = x \cdot x \cdot x \cdot x \cdot x = x^5$$

In general, we can state the following rule:

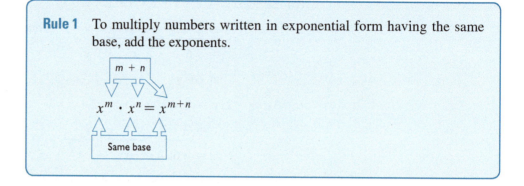

> **Rule 1**   To multiply numbers written in exponential form having the same base, add the exponents.
>
> $$x^m \cdot x^n = x^{m+n}$$

EXAMPLE 2   (a) $x^2 \cdot x^3 = x^{2+3} = x^5$ (b) $3^5 \times 3^2 = 3^{5+2} = 3^7$

(c) $2a^6 \cdot 8a \cdot a^5 = 2 \cdot 8 \cdot a^6 \cdot a^1 \cdot a^5 = 16a^{6+1+5} = 16a^{12}$ ●

! Careful   1.   Notice in part (b) that the like bases remain unchanged when you multiply—that is, $3^5 \times 3^2 = 3^7$, not 9^7.

2.   Exponents are added during multiplication, but numerical factors are multiplied as usual. For example,

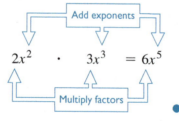

Add exponents

$$2x^2 \quad \cdot \quad 3x^3 \quad = 6x^5$$

Multiply factors   ●

Your Turn   Use Rule 1 to simplify.

(a) $m^4 \cdot m^3$ (b) $5^3 \cdot 5^6$ (c) $4y^4 \cdot 6y \cdot y^2$

Solutions   (a) $m^4 \cdot m^3 = m^{4+3} = m^7$ (b) $5^3 \cdot 5^6 = 5^{3+6} = 5^9$

(c) $4y^4 \cdot 6y \cdot y^2 = 4 \cdot 6 \cdot y^4 \cdot y^1 \cdot y^2 = 24y^{4+1+2} = 24y^7$

Dividing Simple Factors   To divide a^5 by a^2, think of it this way:

$$\frac{a^5}{a^2} = \frac{\cancel{a} \cdot \cancel{a} \cdot a \cdot a \cdot a}{\cancel{a} \cdot \cancel{a}} = a^3$$

Cancel common factors

Notice that the final exponent, 3, is the *difference* between the original two exponents, 5 and 2. Remember that multiplication and division are reverse operations. It makes sense that if you add exponents when multiplying, you would subtract exponents when dividing. We can now state the following rule:

> **Rule 2**   To divide numbers written in exponential form having the same base, subtract the exponents.
>
> $$\frac{x^m}{x^n} = x^{m-n}$$

EXAMPLE 3   (a) $\dfrac{x^7}{x^3} = x^{7-3} = x^4$ (b) $\dfrac{4^5}{4^2} = 4^{5-2} = 4^3$

(c) $\dfrac{6a^6}{2a^3} = \dfrac{6}{2} \cdot \dfrac{a^6}{a^3} = 3a^3$ (d) $\dfrac{24m^7n^4}{-6m^2n} = \dfrac{24}{-6} \cdot \dfrac{m^7}{m^2} \cdot \dfrac{n^4}{n^1} = -4m^5n^3$ ●

! Careful   1. As with multiplication, the like bases remain unchanged when dividing with exponents. This is especially important to keep in mind when the base is a number, as in part (b) of Example 3.

2. When dividing expressions that contain both exponential expressions and numerical factors, *subtract* the exponents but *divide* the numerical factors. For Example 3(c),

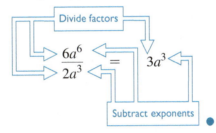

●

Your Turn   Try the division rule on these problems:

(a) $\dfrac{x^6}{x^3}$ (b) $\dfrac{6a^4}{2a}$ (c) $\dfrac{-12m^3n^5}{4m^2n^2}$ (d) $\dfrac{5^8}{5^2}$ (e) $\dfrac{4x^5 - 8x^4 + 6x^3}{2x^2}$

Solutions   (a) $\dfrac{x^6}{x^3} = x^{6-3} = x^3$ (b) $\dfrac{6a^4}{2a} = \dfrac{6}{2} \cdot \dfrac{a^4}{a^1} = 3a^3$

(c) $\dfrac{-12m^3n^5}{4m^2n^2} = \dfrac{-12}{4} \cdot \dfrac{m^3}{m^2} \cdot \dfrac{n^5}{n^2} = -3mn^3$ (d) $\dfrac{5^8}{5^2} = 5^{8-2} = 5^6$

(e) $\dfrac{4x^5 - 8x^4 + 6x^3}{2x^2} = \dfrac{4x^5}{2x^2} - \dfrac{8x^4}{2x^2} + \dfrac{6x^3}{2x^2} = 2x^3 - 4x^2 + 3x$

🔍 A Closer Look   Did you have trouble with (e)? Unlike the other problems, which have only one term in the numerator (dividend), problem (e) has *three* terms in the numerator. Just imagine that the entire numerator is enclosed in parentheses and divide each of the three terms separately by the denominator (divisor). ●

✎ Note   The rule for division confirms the fact stated in Section 6-4 that $a^0 = 1$. We know from arithmetic that any quantity divided by itself is equal to 1, so

$$\frac{a^n}{a^n} = 1 \qquad (\text{for } a \neq 0)$$

But according to the division rule,

$$\frac{a^n}{a^n} = a^{n-n} = a^0 \qquad \text{Therefore, } a^0 \text{ must be equal to 1.} \quad ●$$

Negative Exponents   If we apply the division rule to a problem in which the divisor has a larger exponent than the dividend, the answer will contain a negative exponent. For example,

$$\frac{x^4}{x^6} = x^{4-6} = x^{-2}$$

If we use the cancellation method to do this problem, we have

$$\frac{x^4}{x^6} = \frac{\cancel{x} \cdot \cancel{x} \cdot \cancel{x} \cdot \cancel{x}}{\cancel{x} \cdot \cancel{x} \cdot \cancel{x} \cdot \cancel{x} \cdot x \cdot x} = \frac{1}{x^2} \qquad \text{Therefore,} \quad x^{-2} = \frac{1}{x^2}.$$

In general, we define negative exponents as follows:

$$x^{-n} = \frac{1}{x^n}$$

It is usually preferable to write an answer in fraction form rather than with negative exponents.

EXAMPLE 4   (a) To divide $-\dfrac{12a}{4a^5}$

Think of this as the product of two separate quotients:

$$-\frac{12a}{4a^5} = -\frac{12}{4} \cdot \frac{a^1}{a^5} \qquad \boxed{a = a^1}$$

First, divide the numerical coefficients:

$$= -3 \cdot \frac{a^1}{a^5}$$

Second, divide the a's by subtracting exponents:

$$= -3 \cdot a^{-4}$$

Next, apply the definition of negative exponents to write the answer with positive exponents:

$$= \frac{-3}{1} \cdot \frac{1}{a^{+4}} \qquad \boxed{a^{-4} = \frac{1}{a^4}}$$

Finally, combine the results into a single fraction:

$$= -\frac{3}{a^4}$$

(b) To divide $\dfrac{2m^6n^2}{8m^3n^3}$

Think of this as the product of three separate quotients:

$$\frac{2m^6n^2}{8m^3n^3} = \frac{2}{8} \cdot \frac{m^6}{m^3} \cdot \frac{n^2}{n^3}$$

Simplify each quotient separately:

$$= \frac{1}{4} \cdot m^3 \cdot n^{-1}$$

Eliminate negative exponents:

$$= \frac{1}{4} \cdot m^3 \cdot \frac{1}{n} \qquad \boxed{\begin{array}{l} n^{-1} = \frac{1}{n^{+1}} \\ = \frac{1}{n} \end{array}}$$

Combine the three factors into a single fraction:

$$= \frac{1}{4} \cdot \frac{m^3}{1} \cdot \frac{1}{n} = \frac{m^3}{4n} \quad \bullet$$

Your Turn   Practice using the definition of negative exponents by giving two answers for each of the following problems—one with negative exponents and one using fractions to eliminate the negative exponents.

(a) $\dfrac{a^2}{a^6}$ (b) $\dfrac{8x^2}{-2x^3}$ (c) $\dfrac{6^3}{6^5}$ (d) $\dfrac{5x^4y^2}{15xy^7}$ (e) $\dfrac{12x^6 - 9x^3 + 6x}{3x^2}$

Solutions   (a) $\dfrac{a^2}{a^6} = a^{2-6} = a^{-4} = \dfrac{1}{a^4}$ (b) $\dfrac{8x^2}{-2x^3} = \dfrac{8}{-2} \cdot \dfrac{x^2}{x^3} = -4x^{-1} = \dfrac{-4}{x} = -\dfrac{4}{x}$

(c) $\dfrac{6^3}{6^5} = 6^{3-5} = 6^{-2} = \dfrac{1}{6^2} = \dfrac{1}{36}$

(d) $\dfrac{5x^4y^2}{15xy^7} = \dfrac{5}{15}\cdot\dfrac{x^4}{x}\cdot\dfrac{y^2}{y^7} = \dfrac{1}{3}x^3y^{-5} = \dfrac{1}{3}\cdot\dfrac{x^3}{1}\cdot\dfrac{1}{y^5} = \dfrac{x^3}{3y^5}$

(e) $\dfrac{12x^6 - 9x^3 + 6x}{3x^2} = \dfrac{12x^6}{3x^2} - \dfrac{9x^3}{3x^2} + \dfrac{6x}{3x^2} = 4x^4 - 3x + 2x^{-1} = 4x^4 - 3x + \dfrac{2}{x}$

Now turn to Exercises 7-7 for a set of problems on multiplication and division of algebraic expressions.

Exercises 7-7   Multiplying and Dividing Algebraic Expressions

A. Simplify by multiplying.

1. $5\cdot 4x$ 2. $(a^2)(a^3)$

3. $-3R(-2R)$ 4. $3x\cdot 3x$

5. $5^6\cdot 5^4$ 6. $3^2\cdot 3^7$

7. $10^2\cdot 10^{-5}$ 8. $2^{-6}\cdot 2^9$

9. $p^2\cdot p^4$ 10. $3\cdot 2x^2$

11. $5x^3\cdot x^4$ 12. $4x\cdot 2x^2\cdot x^3$

13. $(4x^2y)(-2xy^3)$ 14. $2x\cdot 2x\cdot 2x$

15. $0.4a\cdot 1.5a$ 16. $x\cdot x\cdot A\cdot x\cdot 2x\cdot A^2$

17. $2y^5\cdot 5y^2$ 18. $2a^2b\cdot 2ab^2\cdot a$

19. $\frac{1}{2}Q\cdot\frac{1}{2}Q\cdot\frac{1}{2}Q\cdot\frac{1}{2}Q$ 20. $(pq^2)(\frac{1}{2}pq)(2.4p^2q)$

21. $(-2M)(3M^2)(-4M^3)$ 22. $xy\cdot x^2\cdot xy^3$

23. $2x(3x - 1)$ 24. $x(x + 2)$

25. $-2(1 - 2y)$ 26. $ab(a^2 - b^2)$

27. $p^2(p + 2p)$ 28. $xy(x + y)$

29. $-3x^2(2x^2 - 5x^3)$ 30. $6x^3(4x^4 + 3x^2)$

31. $2ab^2(3a^2 - 5ab + 7b^2)$ 32. $-5xy^3(4x^2 - 6xy + 8y^2)$

B. Divide as indicated. Express all answers using positive exponents.

1. $\dfrac{4^5}{4^3}$ 2. $\dfrac{10^9}{10^3}$

3. $\dfrac{x^4}{x}$ 4. $\dfrac{y^6}{y^2}$

5. $\dfrac{10^4}{10^6}$ 6. $\dfrac{m^5}{m^{10}}$

7. $\dfrac{8a^8}{4a^4}$ 8. $\dfrac{-15m^{12}}{5m^6}$

9. $\dfrac{16y^2}{-24y^3}$ 10. $\dfrac{-20t}{-5t^3}$

11. $\dfrac{-6a^2b^4}{-2ab}$ 12. $\dfrac{36x^4y^5}{-9x^2y}$

13. $\dfrac{48m^2n^3}{-16m^6n^4}$

14. $\dfrac{15c^6d}{20c^2d^3}$

15. $\dfrac{-12a^2b^2}{20a^2b^5}$

16. $\dfrac{100xy}{-10x^2y^4}$

17. $\dfrac{6x^4 - 8x^2}{2x^2}$

18. $\dfrac{9y^3 + 6y^2}{3y}$

19. $\dfrac{12a^7 - 6a^5 + 18a^3}{-6a^2}$

20. $\dfrac{-15m^5 + 10m^2 + 5m}{5m}$

When you have completed these exercises, check your answers to the odd-numbered problems in the Appendix, then turn to Section 7–8 to learn about scientific notation.

| 7-8 | Scientific Notation

learning | catalytics™

1. Write 1,000,000 as a power of 10.
2. What must you multiply 4.62 by to get 4620?

In technical work you will often deal with very small and very large numbers that may require a lot of space to write. For example, a certain computer function may take 5 nanoseconds (0.000000005 s), and in electricity, 1 kilowatt-hour (kWh) is equal to 3,600,000 joules (J) of work. Rather than take the time and space to write all the zeros, we often use *scientific notation* to express such numbers.

Definition of Scientific Notation

In scientific notation, 5 nanoseconds would be written as 5×10^{-9} s, and 1 kWh would be 3.6×10^6 J. As you can see, a number written in **scientific notation** is a product of a number between 1 and 10 and a power of 10.

> ### Scientific Notation
>
> A number is expressed in scientific notation when it is in the form
>
> $P \times 10^k$
>
> where P is a number less than 10 and greater than or equal to 1, and k is an integer.

Numbers that are powers of ten are easy to write in scientific notation:

$1000 = 1 \times 10^3$

$100 = 1 \times 10^2$

$10 = 1 \times 10^1$ and so on

In general, to write a positive decimal number in scientific notation, first write it as a number between 1 and 10 times a power of 10, then write the power of ten using an exponent.

EXAMPLE 1   (a) $3,600,000 = 3.6 \times 1,000,000$

A power of 10

A number between 1 and 10

$= 3.6 \times 10^6$

(b) $0.0004 = 4 \times 0.0001 = 4 \times \dfrac{1}{10,000} = 4 \times \dfrac{1}{10^4} = 4 \times 10^{-4}$

Note   If you need to review negative exponents, see page 502. ●

Converting to Scientific Notation   For a shorthand way of converting to scientific notation, use the following four-step procedure illustrated by Example 2 (a) and (b):

EXAMPLE 2   **Steps for Converting to Scientific Notation**   (a)   (b)

Step 1   If the number is given without a decimal point, rewrite it with a decimal point.

$57,400$
$= 57,400.$ 0.0038

Decimal point   Decimal point already shown

Step 2   Place a mark $\wedge$ after the first nonzero digit.

$5 \wedge 7400.$ $0.003 \wedge 8$

Step 3   Count the number of digits from the mark to the decimal point.

$5 \wedge 7400.$ $0.003 \wedge 8$

4 digits   3 digits

Step 4   Place the decimal point in the marked position. Use the resulting number as the multiplier P. Use the number of digits from the mark $\wedge$ to the original decimal point as the exponent. If the shift is to the right, the exponent is positive; if the shift is to the left, the exponent is negative.

5.74×10^4 3.8×10^{-3}

| Exponent is $+4$ because decimal point shifts 4 digits *right*. | Exponent is -3 because decimal point shifts 3 digits *left*. |

Here are several more examples.

EXAMPLE 3   (a) $150,000 = 1 \wedge 50000. = 1.5 \times 10^5$

Shift 5 digits right. Exponent is $+5$.

(b) $0.00000205 = 0.000002 \wedge 05 = 2.05 \times 10^{-6}$

Shift 6 digits left. Exponent is -6.

(c) $47 = 4 \wedge 7. = 4.7 \times 10^1$

Shift 1 digit right.   ●

Your Turn   For practice, write the following numbers in scientific notation.

(a) 2900   (b) 0.006   (c) 1,960,100

(d) 0.0000028   (e) 600   (f) 0.0001005

Solutions   (a) $2900 = 2 \wedge 900. = 2.9 \times 10^3$

Shift 3 digits right.

(b) $0.006 = 0.006 \wedge = 6 \times 10^{-3}$

Shift 3 digits left.

(c) $1,960,100 = 1 \wedge 960100. = 1.9601 \times 10^6$

Shift 6 digits right.

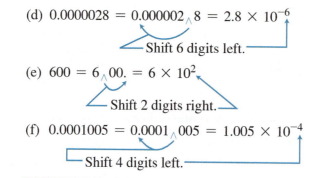

(d) $0.0000028 = 0.000002\,8 = 2.8 \times 10^{-6}$

Shift 6 digits left.

(e) $600 = 6\,00. = 6 \times 10^{2}$

Shift 2 digits right.

(f) $0.0001005 = 0.0001\,005 = 1.005 \times 10^{-4}$

Shift 4 digits left.

Learning Help   When a positive number greater than or equal to 10 is written in scientific notation, the exponent is positive. When a positive number less than 1 is written in scientific notation, the exponent is negative. When a number between 1 and 10 is written in scientific notation, the exponent is 0. ●

Converting from Scientific Notation to Decimal Form   To convert a number from scientific notation to decimal form, shift the decimal point as indicated by the power of 10—to the right for a positive exponent, to the left for a negative exponent. Attach additional zeros as needed.

EXAMPLE 4   (a)

Attach additional zeros.

$3.2 \times 10^{4} = 3\,2000. = 32{,}000$

Shift the decimal point 4 places right.

For a positive exponent, shift the decimal point to the right.

(b)

Attach additional zeros.

$2.7 \times 10^{-3} = 0.002\,7 = 0.0027$

Shift the decimal point 3 places left.

For a negative exponent, shift the decimal point to the left. ●

Your Turn   Write each of the following numbers in decimal form.

(a) 8.2×10^{3} (b) 1.25×10^{-6}

(c) 2×10^{-4} (d) 5.301×10^{5}

Solutions   (a) $8.2 \times 10^{3} = 8\,200. = 8200$

(b) $1.25 \times 10^{-6} = 0.000001\,25 = 0.00000125$

(c) $2 \times 10^{-4} = 0.0002 = 0.0002$

(d) $5.301 \times 10^{5} = 5\,30100. = 530{,}100$

Multiplying and Dividing in Scientific Notation   Scientific notation is especially useful when we must multiply or divide very large or very small numbers. Although most calculators have a means of converting decimal numbers to scientific notation and can perform arithmetic with scientific notation, it is important that you be able to do simple calculations of this kind quickly and accurately without a calculator.

To multiply or divide numbers given in exponential form, use the rules given in Section 7-7. (See pages 500–501.)

$$a^m \times a^n = a^{m+n}$$

$$a^m \div a^n = \frac{a^m}{a^n} = a^{m-n}$$

To multiply numbers written in scientific notation, work with the decimal and exponential parts separately.

$$(A \times 10^B) \times (C \times 10^D) = (A \times C) \times 10^{B+D}$$

EXAMPLE 5 $26,000 \times 3,500,000 = ?$

Step 1   Rewrite each number in scientific notation.

$$= (2.6 \times 10^4) \times (3.5 \times 10^6)$$

Step 2   Regroup to work with the decimal and exponential parts separately.

$$= (2.6 \times 3.5) \times (10^4 \times 10^6)$$

Step 3   Multiply using the rule for multiplying exponential numbers.

$$= 9.1 \times 10^{4+6}$$

$$= 9.1 \times 10^{10}$$

When dividing numbers written in scientific notation, it may help to think of the division as a fraction.

$$(A \times 10^B) \div (C \times 10^D) = \frac{A \times 10^B}{C \times 10^D} = \frac{A}{C} \times \frac{10^B}{10^D}$$

Therefore,

$$(A \times 10^B) \div (C \times 10^D) = (A \div C) \times 10^{B-D}$$

EXAMPLE 6 $45,000 \div 0.0018 = ?$

Step 1   Rewrite.

$$= \frac{4.5 \times 10^4}{1.8 \times 10^{-3}}$$

Step 2   Split into two fractions.

$$= \frac{4.5}{1.8} \times \frac{10^4}{10^{-3}}$$

Step 3   Divide using the rules for exponential division.

$$\boxed{4.5 \div 1.8} \quad = 2.5 \times 10^{4-(-3)} \quad \boxed{4 - (-3) = 4 + 3 = 7}$$

$$= 2.5 \times 10^7$$

If the result of the calculation is not in scientific notation, that is, if the decimal part is greater than 10 or less than 1, rewrite the decimal part so that it is in scientific notation.

EXAMPLE 7

$$0.0000072 \div 0.0009 = \frac{7.2 \times 10^{-6}}{9 \times 10^{-4}}$$

$$= \frac{7.2}{9} \times \frac{10^{-6}}{10^{-4}}$$

$$= 0.8 \times 10^{-6-(-4)}$$

$$= 0.8 \times 10^{-2} \qquad \boxed{-6-(-4)=-6+4=-2}$$

But 0.8 is not a number between 1 and 10, so write it as

$$0.8 = 8 \times 10^{-1}.$$

Then, $0.8 \times 10^{-2} = 8 \times 10^{-1} \times 10^{-2}$

$$= 8 \times 10^{-3} \quad \text{in scientific notation} \qquad \bullet$$

Your Turn   Perform the following calculations using scientific notation, and write the answer in scientific notation.

(a) $1600 \times 350{,}000$ (b) $64{,}000 \times 250{,}000$

(c) 2700×0.0000045 (d) $15{,}600 \div 0.0013$

(e) $0.000348 \div 0.087$ (f) $0.00378 \div 540{,}000{,}000$

Solutions

(a) $(1.6 \times 10^3) \times (3.5 \times 10^5) = (1.6 \times 3.5) \times (10^3 \times 10^5)$

$$= 5.6 \times 10^{3+5}$$

$$= 5.6 \times 10^8$$

(b) $(6.4 \times 10^4) \times (2.5 \times 10^5) = (6.4 \times 2.5) \times (10^4 \times 10^5)$

$$= 16 \times 10^{4+5}$$

$$= 16 \times 10^9$$

$$\boxed{16 = 1.6 \times 10^1} \Rightarrow = 1.6 \times 10^1 \times 10^9$$

$$= 1.6 \times 10^{10}$$

(c) $(2.7 \times 10^3) \times (4.5 \times 10^{-6}) = (2.7 \times 4.5) \times (10^3 \times 10^{-6})$

$$= 12.15 \times 10^{3+(-6)} \qquad \boxed{3+(-6)=-3}$$

$$= 12.15 \times 10^{-3}$$

$$= 1.215 \times 10^1 \times 10^{-3}$$

$$= 1.215 \times 10^{-2}$$

(d) $(1.56 \times 10^4) \div (1.3 \times 10^{-3}) = \frac{1.56}{1.3} \times \frac{10^4}{10^{-3}}$

$$= 1.2 \times 10^{4-(-3)} \qquad \boxed{4-(-3)=4+3=7}$$

$$= 1.2 \times 10^7$$

(e) $(3.48 \times 10^{-4}) \div (8.7 \times 10^{-2}) = \frac{3.48}{8.7} \times \frac{10^{-4}}{10^{-2}}$

$$= 0.4 \times 10^{-4-(-2)} \qquad \boxed{-4-(-2)=-4+2=-2}$$

$$= 0.4 \times 10^{-2}$$

Rewrite 0.4 as 4×10^{-1}: $= 4 \times 10^{-1} \times 10^{-2}$

$= 4 \times 10^{-3}$

(f) $(3.78 \times 10^{-3}) \div (5.4 \times 10^8) = \dfrac{3.78}{5.4} \times \dfrac{10^{-3}}{10^8}$

$= 0.7 \times 10^{-3-8}$

$= 0.7 \times 10^{-11}$

Rewrite 0.7 as 7×10^{-1}: $= 7 \times 10^{-1} \times 10^{-11}$

$= 7 \times 10^{-12}$

Calculators and Scientific Notation

If a very large or very small number contains too many digits to be shown on the display of a scientific calculator, it will be converted to scientific notation. For example, if you enter the product

6 4 8 0 0 0 0 $\times$ **7 5 0 0 0** $=$

on a calculator, the answer will be displayed in one of the following ways:

4.86×10^{11} or $4.86^{\,11}$ or $4.86\ E\ 11$

Interpret all of these as

$6{,}480{,}000 \times 75{,}000 = 4.86 \times 10^{11}$

Similarly, the division

. 0 0 0 0 0 6 $\div$ **4 8 0 0 0 0 0 0** $=$

gives one of the following displays:

1.25×10^{-13} or $1.25^{\,-13}$ or $1.25\ E -13$

Therefore, $0.000006 \div 48{,}000{,}000 = 1.25 \times 10^{-13}$

You may also enter numbers in scientific notation directly into your calculator using a key labeled $\boxed{\text{EE}}$, $\boxed{\text{EXP}}$, or $\boxed{\times 10^x}$. For the calculation,

$0.000006 \div 48{,}000{,}000$ or $(6 \times 10^{-6}) \div (4.8 \times 10^7)$

enter

6 $\boxed{\text{EE}}$ $\boxed{(-)}$ **6** $\boxed{\div}$ **4 . 8** $\boxed{\text{EE}}$ **7** $=$

and your calculator will again display 1.25×10^{-13}

If you wish to see all of your answers in scientific notation, look for a key marked $\boxed{\text{SCI/ENG}}$ or use the $\boxed{\text{SET UP}}$ or $\boxed{\text{MODE}}$ menus to switch to scientific mode.

Practical Applications

EXAMPLE 8 **Electrical Trades** At a solar energy plant being built in the Mojave Desert, power is generated using 173,500 heliostats. Each heliostat has a luminance value of 1,340,000,000. To calculate the total luminance generated at the plant, we multiply the number of heliostats by the luminance value of each heliostat. Because this calculation involves very large numbers, we use scientific notation to simplify the multiplication:

$$(173,500)(1,340,000,000) = (1.735 \times 10^5)(1.34 \times 10^9)$$
$$= (1.735 \times 1.34)(10^5 \times 10^9)$$
$$= 2.3249 \times 10^{14} \text{ lumens}$$

Your Turn   **Electronics**   Hard-disk storage capacities are often given in units of terabytes (TB). One TB is equal to 10^{12} bytes (B). The latest auxiliary storage units have much larger capacities, which are now measured in units of petabytes (PB). One PB is equal to 10^{15} B. How many terabytes are there in 250 PB? (Give your answer in scientific notation.)

Solution   One way to solve this is to use unity fractions.

First, convert 250 PB to bytes:

$$\frac{250 \text{ PB}}{1} \times \frac{10^{15} \text{ B}}{1 \text{ PB}} = 250 \times 10^{15} \text{ B}$$

Then, convert this result to terabytes:

$$(250 \times 10^{15} \text{ B}) \times \frac{1 \text{ TB}}{10^{12} \text{ B}} = 250 \times 10^3 \text{ TB}$$
$$= (2.5 \times 10^2)(10^3) \text{ TB}$$
$$= 2.5 \times 10^5 \text{ TB}$$

Now turn to Exercises 7-8 for a set of problems on scientific notation.

Exercises 7-8   Scientific Notation

A. Rewrite each number in scientific notation.

1. 5000	2. 450	3. 90	4. 40,700
5. 0.003	6. 0.071	7. 0.0004	8. 0.0059
9. 6,770,000	10. 38,200	11. 0.0292	12. 0.009901
13. 1001	14. 0.0020	15. 0.000107	16. 810,000
17. 31.4	18. 0.6	19. 125	20. 0.74

21. Young's modulus for the elasticity of steel: 29,000,000 lb/in.^{2}

22. Thermal conductivity of wood: 0.00024 cal/(cm·s°C)

23. Power output: 95,500,000 watts

24. Speed of light: 670,600,000 mph

B. Rewrite each number in decimal form.

1. 2×10^5	2. 7×10^6	3. 9×10^{-5}	4. 3×10^{-4}
5. 1.7×10^{-3}	6. 3.7×10^{-2}	7. 5.1×10^4	8. 8.7×10^2
9. 4.05×10^4	10. 7.01×10^{-6}	11. 3.205×10^{-3}	12. 1.007×10^3
13. 2.45×10^6	14. 3.19×10^{-4}	15. 6.47×10^5	16. 8.26×10^{-7}

C. Rewrite each of the following in scientific notation and calculate the answer in scientific notation. If necessary, convert your answer to scientific notation. Round to one decimal place if necessary.

1. $2000 \times 40,000$

2. 0.0037×0.0000024

3. $460,000 \times 0.0017$

4. $0.0018 \times 550,000$

5. $0.0000089 \div 3200$

6. $0.000125 \div 5000$

7. $45,500 \div 0.0091$

8. $12,450 \div 0.0083$

9. $2,240,000 \div 16,000$

10. $25,500 \div 1,700,000$

11. $0.000045 \div 0.00071$

12. $0.000086 \div 0.000901$

13. $9,501,000 \times 2410$

14. 9800×0.000066

15. $0.0000064 \div 80,000$

16. 1070×0.0000055

17. $\dfrac{0.000056}{0.0020}$

18. $\dfrac{0.0507}{43,000}$

19. $\dfrac{0.00602 \times 0.000070}{72,000}$

20. $\dfrac{2,780,000 \times 512,000}{0.000721}$

21. $64,000 \times 2800 \times 370,000$

22. $0.00075 \times 0.000062 \times 0.014$

23. $\dfrac{0.0517}{0.0273 \times 0.00469}$

24. $\dfrac{893,000}{5620 \times 387,000}$

D. Solve.

1. **Allied Health** The levels of some blood components are usually expressed using scientific notation. For example, red blood cells are commonly expressed as 10^{12} cells/liter (abbreviated as 10^{12}/L), and white blood cells are commonly expressed as 10^{9} cells/liter. Write the following amounts without using scientific notation:

   (a) A red blood cell value of 5.3×10^{12}/L.

   (b) A white blood cell value of 9.1×10^{9}/L.

   Write the following amounts using scientific notation:

   (c) A red blood cell value of 3,600,000,000,000/L.

   (d) A white blood cell value of 6,700,000,000/L.

2. **Marine Technology** Migrating gray whales swim 7000 miles north from the Sea of Cortez to the Alaskan coast in the spring and then return in the fall. Use scientific notation to calculate the total number of miles swum by a gray whale in an average 80-year life span. Give the answer both in scientific notation and as a whole number.

3. **HVAC** A brick wall 15 m by 25 m is 0.48 m thick. Under particular temperature conditions, the rate of heat flow through the wall, in calories per second, is given by the expression

   $$H = (1.7 \times 10^{-4}) \times \left(\frac{1500 \times 2500}{48} \right)$$

   Calculate the value of this quantity to the nearest tenth.

4. **Physics** If an atom has a diameter of 4×10^{-8} cm and its nucleus has a diameter of 8×10^{-13} cm, find the ratio of the diameter of the nucleus to the diameter of the atom.

5. **Electrical Trades**   The capacitance of a certain capacitor (in microfarads) can be found using the following expression. Calculate this capacitance.

$$C = (5.75 \times 10^{-8}) \times \left(\frac{8.00}{3.00 \times 10^{-3}} \right)$$

6. **Physics**   The energy (in joules) of a photon of visible light with a wavelength of 5.00×10^{-7} m is given by the following expression. Calculate this energy.

$$E = \frac{(6.63 \times 10^{-34}) \times (3.00 \times 10^{8})}{5.00 \times 10^{-7}}$$

7. **Welding**   The longitudinal shrinkage S_L (in mm) that occurs during welding is given by the formula

$$S_L = \frac{305IL}{t} \times 10^{-7}$$

where I is the welding current in amperes
L is the length of the weld in millimeters
t is the thickness of the plate in millimeters

Suppose an electrode is welded at 125 amp, the length of the weld is 270 mm, and the plate is 16 mm thick. Calculate the longitudinal shrinkage that occurs during welding. (Round to the nearest thousandth.)

8. **General Interest**   In 1977, the *Voyager I* spacecraft was launched from Cape Canaveral, Florida. Its mission was to explore the outer solar system, and in August 2013, it left the solar system and traveled into interstellar space. The spacecraft is so far from the earth that scientists measure its distance in **astronomical units** (**AU**), where 1 AU is equal to 93 million miles—that is, the average distance from the earth to the sun. When the *Voyager I* reached interstellar space, it was approximately 170 AU from the earth. Express this distance in miles and in scientific notation.

9. **General Interest**   Glacial geologists have recently developed 3-D technologies enabling them to measure the mass of glaciers over time from old photographs. They have estimated that from 1900 to 2010, the Greenland ice sheet has lost 9013 gigatons of mass. Given that a gigaton is 10^{12} metric tons, and a metric ton is 1000 kilograms, convert this shrinkage to U.S. customary tons. Express your answer in scientific notation to three decimal places. (Source: Kurt Kjaer, Natural History Museum of Denmark.

Check your answers to the odd-numbered problems in the Appendix, then turn to Problem Set 7 on page 516 for some practice problems on the work of this chapter. If you need a quick review of the topics in this chapter, visit the chapter Summary first.

CHAPTER 7
SUMMARY

Basic Algebra

Objective	Review

Evaluate formulas and literal expressions. (p. 426)

Step 1   Place the numbers being substituted within the parentheses, then substitute them in the formula being evaluated.

Step 2   After the numbers are substituted, do the arithmetic carefully. Be careful to use the proper order of operations.

Example: If $x = 2$, $y = 3$, $a = 5$, and $b = 6$, then evaluate

(a) $T = \dfrac{2(a + b + 1)}{3x} = \dfrac{2[(5) + (6) + 1]}{3(2)}$

$= \dfrac{2(12)}{3(2)} = \dfrac{24}{6} = 4$

(b) $A = x^2y = (2)^2(3) = 4(3) = 12$

Combine like terms. (p. 443)

Only like terms, those with the same literal parts, may be added and subtracted.

Example: $5x - 3y - 8x + 2y = (5x - 8x) + (-3y + 2y)$

$= -3x + (-1y)$

$= -3x - y$

Remove parentheses properly. (p. 445)

Follow these rules for removing parentheses:

Rule 1   If the left parenthesis immediately follows a plus sign, simply remove the parentheses.

Rule 2   If the left parenthesis immediately follows a negative sign, change the sign of each term inside the parentheses, then remove the parentheses.

Rule 3   If the left parenthesis immediately follows a multiplier, multiply each term inside the parentheses by the multiplier.

Example: (a) $4 + (x - 2) = 4 + x - 2 = 2 + x$

(b) $5x - (x + 2) = 5x - x - 2 = 4x - 2$

(c) $3(x^2 + 5x) - 4(2x - 3) = 3x^2 + 15x - 8x + 12$

$= 3x^2 + 7x + 12$

Solve linear equations in one unknown and solve formulas. (p. 449–480)

First, remove the parentheses as shown in the preceding objective, then combine like terms that are on the same side of the equation. Finally, use the two kinds of balancing operations (add/subtract and multiply/divide) to isolate the desired variable on one side of the equation. The solution will appear on the other side.

Example: (a)   To solve the equation, $7 + 3(5x + 2) = 38$

Remove the parentheses (Rule 3). $7 + 15x + 6 = 38$

Combine like terms. $13 + 15x = 38$

Subtract 13 from each side. $15x = 25$

Divide each side by 15. $x = 1\frac{2}{3}$

Objective	**Review**

	(b) Solve for N. $\qquad\qquad\qquad\qquad\qquad\qquad\qquad S = \dfrac{N}{2} + 26$
	Subtract 26 from each side. $\qquad\qquad\qquad S - 26 = \dfrac{N}{2}$
	Multiply each side by 2. $\qquad\qquad\qquad 2(S - 26) = N$
	This can also be written as: $\qquad\qquad\qquad N = 2S - 52$

Translate English phrases and sentences into algebraic expressions and equations, and solve word problems. (p. 484)	To solve any word problem:

1. **Read** the problem carefully, noting what is given and what is unknown.

2. **Assign a variable** to represent the unknown. Use an expression involving the same variable to represent additional unknowns.

3. **Write an equation** using the unknown(s) and some given fact about the unknown(s). Look for signal words to help you translate words to symbols.

4. **Solve** the equation.

5. **Answer** the original question.

6. **Check** your solution.

Example: An electrician collected $1265.23 for a job that included $840 for labor, which was not taxed, and the remainder for parts, which were taxed at 6%. Determine how much of the total was tax.

Solution: Let x = cost of parts before tax. Then $0.06x$ is the amount of tax on the parts. The cost of the parts plus the tax on the parts is equal to the total cost minus the labor.

$$x + 0.06x = \$1265.23 - \$840$$

$$1.06x = \$425.23$$

$$x = \$401.16, \text{ rounded. This is the cost of the parts.}$$

The amount of tax is: $0.06x = 0.06(\$401.16) = \$24.07$

Or $\$425.23 - \$401.16 = \$24.07$

The tax on the parts was $24.07.

Multiply algebraic expressions. (p. 498)	Group like factors together and use Rule 1:

$$x^m \cdot x^n = x^{m+n}$$

If there is a factor multiplying an expression in parentheses, be careful to multiply every term inside the parentheses by this factor. That is,

$$a(b + c) = a \cdot b + a \cdot c$$

Example: (a) $(6x^4y^2)(-2xy^2) = 6(-2) \cdot x^4 \cdot x \cdot y^2 \cdot y^2$

$$= -12 \cdot x^5 \cdot y^4 = -12x^5y^4$$

(b) $3x(y - 2x) = 3x \cdot y - 3x \cdot 2x = 3xy - 6x^2$

Objective	Review
Divide algebraic expressions. (p. 500)	Divide numerical factors and use Rule 2 to divide exponential expressions. $$\frac{x^m}{x^n} = x^{m-n}$$ If a negative exponent results, use the following rule to convert to a positive exponent: $$x^{-n} = \frac{1}{x^n}$$ **Example:** (a) $\dfrac{6a^2b^4}{18a^5b^2} = \dfrac{6}{18}\cdot\dfrac{a^2}{a^5}\cdot\dfrac{b^4}{b^2} = \dfrac{1}{3}\cdot a^{-3}\cdot b^2$ $$= \frac{1}{3}\cdot\frac{1}{a^3}\cdot\frac{b^2}{1} = \frac{b^2}{3a^3}$$ (b) $\dfrac{12m^4 - 9m^3 + 15m^2}{3m^2} = \dfrac{12m^4}{3m^2} - \dfrac{9m^3}{3m^2} + \dfrac{15m^2}{3m^2}$ $$= 4m^2 - 3m + 5$$
Convert between decimal notation and scientific notation. (p. 504)	A number expressed in scientific notation has the form $P \times 10^k$, where P is less than 10 and greater than or equal to 1 and k is an integer. To convert between decimal notation and scientific notation, use the method explained on pages 505–507 and illustrated here. **Example:** (a) $0.000184 = 0.0001{}_\wedge 84 = 1.84 \times 10^{-4}$ 4 digits left (b) $213,000 = 2{}_\wedge 13000. = 2.13 \times 10^5$ 5 digits right
Multiply and divide in scientific notation. (p. 506)	$(A \times 10^B) \times (C \times 10^D) = (A \times C) \times 10^{B+D}$ $(A \times 10^B) \div (C \times 10^D) = (A \div C) \times 10^{B-D}$ Convert your final answer to proper scientific notation. **Example:** (a) $(3.2 \times 10^{-6}) \times (4.5 \times 10^2) = (3.2 \times 4.5) \times (10^{-6+2})$ $= 14.4 \times 10^{-4}$ $= 1.44 \times 10^1 \times 10^{-4}$ $= 1.44 \times 10^{-3}$ (b) $(1.56 \times 10^{-4}) \div (2.4 \times 10^3) = \dfrac{1.56}{2.4} \times 10^{-4-3}$ $= 0.65 \times 10^{-7}$ $= 6.5 \times 10^{-1} \times 10^{-7}$ $= 6.5 \times 10^{-8}$

PROBLEM SET 7   Basic Algebra

Answers to odd-numbered problems are given in the Appendix.

A. **Find the value of each of the following. Round to two decimal places when necessary.**

1. $L = 2W - 3$ for $W = 8$

2. $M = 3x - 5y + 4z$ for $x = 3, y = 5, z = 6$

3. $I = PRt$ for $P = 800, R = 0.06, t = 3$

4. $I = \dfrac{V}{R}$ for $V = 220, R = 0.0012$

5. $V = LWH$ for $L = 3\frac{1}{2}, W = 2\frac{1}{4}, H = 5\frac{3}{8}$

6. $N = (a + b)(a - b)$ for $a = 7, b = 12$

7. $L = \dfrac{s(P + p)}{2}$ for $s = 3.6, P = 38, p = 26$

8. $f = \dfrac{1}{8N}$ for $N = 6$

9. $t = \dfrac{D - d}{L}$ for $D = 12, d = 4, L = 2$

10. $V = \dfrac{gt^2}{2}$ for $g = 32.2, t = 4.1$

B. **Simplify.**

1. $4x + 6x$

2. $6y + y + 5y$

3. $3xy + 9xy$

4. $6xy^3 + 9xy^3$

5. $3\frac{1}{3}x - 1\frac{1}{4}x - \frac{5}{8}x$

6. $8v - 8v$

7. $0.27G + 0.78G - 0.65G$

8. $7y - 8y^2 + 9y$

9. $3x \cdot 7x$

10. $(4m)(2m^2)$

11. $(2xy)(-5xyz)$

12. $3x \cdot 3x \cdot 3x$

13. $5^7 \cdot 5^3$

14. $(6a^2b)(4a^3b^4c)$

15. $3(4x - 7)$

16. $2ab(3a^2 - 5b^2)$

17. $(4x + 3) + (5x + 8)$

18. $(7y - 4) - (2y + 3)$

19. $3(x + y) + 6(x - y)$

20. $(x^2 - 6) - 3(2x^2 - 5)$

21. $\dfrac{6^8}{6^4}$

22. $\dfrac{x^3}{x^8}$

23. $\dfrac{-12m^{10}}{4m^2}$

24. $\dfrac{6y}{-10y^3}$

25. $\dfrac{-16a^2b^2}{-4ab^3}$

26. $\dfrac{8x^4 - 4x^3 + 12x^2}{4x^2}$

Name

_____

Date

_____

Course/Section

_____

C. **Solve the following equations. Round your answer to two decimal places if necessary.**

1. $x + 5 = 17$

2. $m + 0.4 = 0.75$

3. $e - 12 = 32$

4. $a - 2.1 = -1.2$

5. $12 = 7 - x$

6. $4\frac{1}{2} - x = -6$

7. $5x = 20$

8. $-4m = 24$

9. $\frac{1}{2}y = 16$

10. $\frac{n}{6} = 11$

11. $22 = 8 - 4x$

12. $2x + 7 = 13$

13. $-4x + 11 = 35$

14. $0.75 - 5f = 6\frac{1}{2}$

15. $0.5x - 16 = -18$

16. $5y + 8 + 3y = 24$

17. $3g - 12 = g + 8$

18. $7m - 4 = 11 - 3m$

19. $3(x - 5) = 33$

20. $2 - (2x + 3) = 7$

21. $5x - 2(x - 1) = 11$

22. $2(x + 1) - x = 8$

23. $12 - 7y = y - 18$

24. $2m - 6 - 4m = 4m$

25. $9 + 5n + 11 = 10n - 25$

26. $6(x + 4) = 45$

27. $5y - (11 - 2y) = 3$

28. $4(3m - 5) = 8m + 20$

Solve the following formulas for the variable shown.

29. $A = bH$ for b

30. $R = S + P$ for P

31. $P = 2L + 2W$ for L

32. $P = \dfrac{w}{F}$ for F

33. $S = \frac{1}{2}gt - 4$ for g

34. $V = \pi R^2 H - AB$ for A

D. Rewrite in scientific notation.

1. 7500

2. 12,800

3. 0.041

4. 0.000236

5. 0.00572

6. 0.00000482

7. 447,000

8. 2,127,000

9. 80,200,000

10. 46,710

11. 0.00000705

12. 0.001006

Rewrite as a decimal.

13. 9.3×10^5

14. 6.02×10^3

15. 2.9×10^{-4}

16. 3.05×10^{-6}

17. 5.146×10^{-7}

18. 6.203×10^{-3}

19. 9.071×10^4

20. 4.006×10^6

Calculate using scientific notation, and write your answer in standard scientific notation. Round to one decimal place if necessary.

21. $45,000 \times 1,260,000$

22. $625,000 \times 12,000,000$

23. 0.0007×0.0043

24. 0.0000065×0.032

25. $56,000 \times 0.0000075$

26. 1020×0.00055

27. $0.0074 \div 0.00006$

28. $0.000063 \div 0.0078$

29. $96,000 \div 3,400,000$

30. $26,500,000 \div 12,000$

31. $0.00089 \div 37,000$

32. $123,500 \div 0.00077$

E. Practical Applications

1. **Machine Trades**   The linear speed L of a grinding wheel in feet per minute (fpm) is given by the formula

$$L = \frac{\pi dr}{12}$$

where d is the diameter of the wheel in inches, and r is the rotational speed in revolutions per minute (rpm). What is the linear speed of a 6.0-inch grinding wheel rotating at 2500 rpm? (Use $\pi \approx 3.14$ and round to the nearest hundred fpm.)

2. **Construction**   Structural engineers have found that a good estimate of the crushing load for a square wooden pillar is given by the formula

$$L = \frac{25T^4}{H^2}$$

where L is the crushing load in tons, T is the thickness of the wood in inches, and H is the height of the post in feet. Find the crushing load for a 6-in.-thick post 12 ft high.

3. **Welding**   Tensile strength refers to the maximum load-carrying capability of a material, and it is given by the formula

$$T = \frac{L}{A}$$

where T is the tensile strength in pounds per square inch, L is the maximum load in pounds, and A is the original cross-sectional area in square inches. If the maximum load of a material is 12,500 lb, and the original area is 0.175 sq in., calculate the tensile strength to the nearest hundred psi.

4. **Manufacturing**   The length of belt on a roll of conveyor belt material is given by the formula

$$L = \pi H N$$

where $\pi \approx 3.14$, H is the height of the center core in meters, and N is the number of wraps on the roll. If a particular roll has 18 wraps of belt and the height of the center core is 2.20 meters, what length of belt is on the roll? (Round to the nearest meter.)

5. **Manufacturing**   For a conveyor belt, the **feeder belt capacity** C in metric tons per hour is given by the formula

$$C = \frac{W^2 DS}{1.085}$$

where W is the width of the belt in meters, D is the density of the material on the belt in kilograms per cubic meter, and S is the speed of the belt in meters per second. Find the feeder belt capacity for a 0.7-meter wide belt moving at 3.5 meters per second and carrying material with a density of 1700 kilograms per cubic meter.

6. **Plumbing**   The formula

$$N = \sqrt{\left(\frac{D}{d}\right)^5}$$

gives the approximate number N of smaller pipes of diameter d necessary to supply the same total flow as one larger pipe of diameter D. Unlike the formula in problem 26 on page 437, this formula takes into account the extra friction caused by the smaller pipes. Use this formula to determine the number of $\frac{1}{2}$-in. pipes that will provide the same flow as one $1\frac{1}{2}$-in. pipe.

7. **Sheet Metal Trades**   The length L of the stretch-out for a square vent pipe with a grooved lock-seam is given by

$$L = 4s + 3w$$

where s is the side length of the square end and w is the width of the lock. Find the length of stretch-out for a square pipe $12\frac{3}{4}$ in. on a side with a lock width of $\frac{3}{16}$ in.

8. **HVAC**   In planning a solar energy heating system for a house, a contractor uses the formula

$$Q = 8.33GDT$$

to determine the energy necessary for heating water. In this formula Q is the energy in Btu, G is the number of gallons heated per day, D is the number of days, and T is the temperature difference between tap water and the desired temperature of hot water. Find Q when G is 50 gallons, D is 30 days, and the water must be heated from 60° to 140°.

9. **Sheet Metal Trades**   The bend allowance (BA) for sheet metal is given by the formula

$$BA = N(0.01743R + 0.0078T)$$

where N is the angle of the bend in degrees, R is the inside radius of the bend in inches, and T is the thickness of the metal in inches. Find BA if N is 47°, R is 0.725 in., and T is 0.0625 in. (Round to the nearest thousandth.)

10. **Machine Trades**   To find the taper per inch of a piece of work, a machinist uses the formula

$$T = \frac{D - d}{L}$$

where D is the diameter of the large end, d is the diameter of the small end, and L is the length. Find T if $D = 4.1625$ in., $d = 3.2513$ in., and $L = 8$ in.

11. **Transportation**   A so-called "bullet train" can reach a top speed of 220 mi/hr. At this rate, how far can the train travel in 90 minutes?

12. **Sports and Leisure**   If a runner maintains an average pace of 8 mi/hr, how long will it take him to complete a marathon (26.2 miles)? Give your answer in hours, minutes, and seconds.

13. **Transportation**   A truck driver leaves a warehouse at 6:30 A.M. and needs to arrive at his destination 340 miles away by noon. What average speed must the driver maintain? (Round to the nearest whole number.)

14. **Allied Health**   Use the formula in Your Turn, part (b), on page 427, to calculate the basal metabolic rate of a 48-year-old woman who weighs 128 lb and who stands 5 ft 4 in. tall. (Round to the nearest ten calories.)

15. **Automotive Trades**   The fuel economy F of a car with a 3.5-liter diesel engine at speeds at or above 50 mph is given by the formula

$$F = -0.0007V^2 - 0.13V + 42$$

where F is in miles per gallon and V is the speed in miles per hour. What is the fuel economy at a speed of (a) 50 mph? (b) 80 mph? (Round to the nearest whole number.)

16. **Allied Health**   Body surface area (BSA) is often used to calculate dosages of drugs for patients undergoing chemotherapy or those with severe burns. Body surface area can be estimated from the patient's height and weight.

$$BSA = \sqrt{\frac{\text{weight in kg} \times \text{height in cm}}{3600}}$$

The BSA will be in square meters.
(a) Calculate the BSA for a man who weighs 95 kg and is 180 cm tall.

(b) A chemotherapy drug is given at a dose of 20 mg/m². Calculate the proper dosage for the man in part (a).

17. **Construction**   In the remodel of a two-story home, a stairwell opening, S, is determined by the formula

$$S = U_{ru}\left[\frac{H + T}{U_{ri}} - 1\right]$$

where U_{ru} is the unit run, U_{ri} is the unit rise, H is the head clearance, and T is the floor thickness. The quantity in brackets must be rounded up to the next whole number before multiplying by U_{ru}. Find the stairwell opening if $U_{ru} = 9$ in., $U_{ri} = 7\frac{1}{2}$ in., $H = 84$ in., and $T = 10\frac{3}{8}$ in.

18. **Electronics**   The inductance L in microhenrys of a coil constructed by a ham radio operator is given by the formula

$$L = \frac{R^2 N^2}{9R + 10D}$$

where $R =$ radius of the coil
   $D =$ length of the coil
   $N =$ number of turns of wire in the coil

Find L if $R = 3$ in., $D = 6$ in., and $N = 200$.

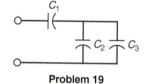

Problem 19

19. **Electronics**   Calculate the total capacitance of the series-parallel circuit shown in the figure using the formula given. Round to the nearest 0.1 μF.

$C_1 = 14\mu F$
$C_2 = 4\mu F$ $C_T = \dfrac{C_1(C_2 + C_3)}{C_1 + C_2 + C_3}$
$C_3 = 6\mu F$

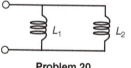

Problem 20

20. **Electronics**   Assuming that there is no mutual coupling, the total inductance in this circuit (see the figure) can be calculated from the following formula:

$$L_T = \frac{L_1 L_2}{L_1 + L_2}$$

Calculate L_T for $L_1 = 200$ mH and $L_2 = 50$ mH.

21. **Carpentry**   When selecting the dimensions of floor joists and beams, it is important to know the moment of inertia I_L of the load to be supported. We use the following formula to calculate I_L in units of in.⁴:

$$I_L = \frac{75wL^3}{16E}$$

where $w =$ the weight in pounds per inch that each joist must support
   $L =$ the length in inches of the unsupported span
   $E =$ the modulus of elasticity in pounds per square inch (psi) of the material

(a) Find I_L to the nearest whole number if $w = 12$ lb/in., $L = 180$ in., and $E = 1,500,000$ psi.

(b) The moment of inertia I_J of joists with a breadth b and a depth d, both in inches, is given by the formula

$$I_J = \frac{bd^3}{12}$$

The value of I_J must be greater than or equal to the value of I_L in order for the joist to provide sufficient support. Will a joist with a breadth of 6.0 in. and a depth of 8.0 in. be sufficient to support the load from part (a)?

22. **Electrical Trades**   Electricians use a formula that states the level of light in a room, in foot-candles, is equal to the product of the fixture rating, in lumens; the coefficient of depreciation; and the coefficient of utilization; all divided by the area, in square feet.
    (a) State this as an algebraic equation.
    (b) Find the level of illumination for four fixtures rated at 2800 lumens each if the coefficient of depreciation is 0.75, the coefficient of utilization is 0.6, and the area of the room is 120 sq ft.
    (c) Solve for area.
    (d) Use your answer to (c) to determine the size of the room in which a level of 60 foot-candles can be achieved with 10,000 lumens, given that the coefficient of depreciation is 0.8 and the coefficient of utilization is 0.5.

23. **Aviation**   In weight and balance calculations, airplane mechanics are concerned with the center of gravity (CG) of an airplane. The center of gravity may be calculated from the formula

    $$CG = \frac{100(H - x)}{L}$$

    where H is the distance from the datum to the empty CG, x is the distance from the datum to the leading edge of the mean aerodynamic chord (MAC), and L is the length of the MAC. CG is expressed as a percent of the MAC.

    (All lengths are in inches.)
    (a) Find the center of gravity if $H = 180$, $x = 155$, and $L = 80$.
    (b) Solve the formula for L, and find L when $CG = 30\%$, $H = 200$, and $x = 150$.
    (c) Solve the formula for H, and find H when $CG = 25\%$, $x = 125$, and $L = 60$.
    (d) Solve the formula for x, and find x when $CG = 28\%$, $H = 170$, and $L = 50$.

24. **Sports and Leisure**   In baseball analytics, a statistic known as FIP, or **fielding independent pitching**, is a measurement of a pitcher's performance that removes the role of defense and luck. The formula is defined as follows:

    $$FIP = \frac{13 \times HR + 3(BB + HBP) - 2 \times K}{IP} + 3.13$$

    where HR = home runs yielded, BB = number of batters walked, HBP = number of batters hit by a pitch, K = strikeouts, and IP = innings pitched. A lower FIP indicates a better pitching performance. In 2015, Jake Arrieta of the Cubs and Zack Greinke of the Dodgers were both contending for the Cy Young Award, given to the outstanding pitcher in each major league. Here are their stats:

	HR	BB	HBP	K	IP
Arrieta	10	48	6	236	229
Greinke	14	40	5	200	$222\frac{2}{3}$

Determine each pitcher's FIP rounded to two decimal places. Based on this statistic, who should have gotten the Cy Young Award?

25. **Electronics** Solve for the variable indicated in each of these electronics formulas:

(a) $R = \dfrac{KL}{A}$ Solve for L.

(b) $R_s = \dfrac{R_m \cdot I_m}{I_s}$ Solve for R_m.

(c) $X_c = \dfrac{1}{2\pi FC}$ Solve for C.

(d) $L_T = \dfrac{L_1 + L_2}{2L_m}$ Solve for L_m.

26. **Automotive Trades** The static stability factor (SSF) is a measure of the relative rollover risk associated with an automobile.

$$SSF = \frac{1}{2} \times \frac{\text{tire-track width}}{\text{height of center of gravity}}$$

An SSF of 1.04 or less indicates a 40% or higher rollover risk, and an SSF of 1.45 or more indicates a 10% or less rollover risk. Suppose a particular SUV has a tire-track width of 56 in. and a center of gravity 26 in. above the road.

(a) Calculate the SSF for this SUV.

(b) By how much would auto designers need to lower the center of gravity to improve the SSF to 1.45? Round to the nearest inch.

27. **Agriculture** Agricultural researchers developed the following formula from a study on the effects of the salinity of irrigation water on the yield potential of rice fields:

$$Y = 100 - 9.1(S - 1.9)$$

In this formula, Y is the percent yield of the rice fields, and S is the average salinity of the irrigation water in decisiemens per meter (dS/m).

(a) If $S = 5.2$ dS/m, find the percent yield of the rice fields. (Round to the nearest percent.)

(b) To achieve a yield of 80%, what must be the salinity of the irrigation water? (Round to the nearest 0.1 dS/m.)

28. **Automotive Trades** The indicated horsepower (ihp) of an engine measures the power of the engine prior to friction loss. The ihp is given by

$$\text{ihp} = \frac{(\text{MEP}) \times D \times S}{792{,}000}$$

where MEP represents the mean effective pressure of a typical cylinder in pounds per square inch (psi), D represents displacement in cubic inches (in.^{3}), and S represents engine speed in revolutions per minute (rpm).

(a) Find the ihp of a 224-in.^{3} engine operating at 3800 rpm with an MEP of 170 psi. (Round to the nearest whole number.)

(b) If the ihp of a 244-in.^{3} engine operating at 3250 rpm is 206, what is the MEP of a typical cylinder? (Round to the nearest whole number.)

29. **Electrical Trades** A certain electric utility charges residents according to a tier system. If they reach tier 3, where total usage T is between 361 and 556 kilowatt-hours, their monthly cost C for electricity can be expressed as

$$C = \$34.817 + \$0.171(T - 361)$$

How many kilowatt-hours would result in a monthly bill of $60? (Round to the nearest whole number.)

30. **Automotive Trades** When balancing a crankshaft, bobweights are used in place of the piston and connecting rod. To choose the correct bobweight B for a cylinder, the following formula is used:

$$B = L + \frac{S + P}{2}$$

where $L =$ weight of the large end of the connecting rod
$S =$ weight of the small end of the connecting rod
$P =$ weight of the piston assembly

(a) Find the bobweight if $L = 518$ grams, $S = 212$ grams, and $P = 436$ grams.
(b) Solve the formula for P.

31. **Construction** In building energy-efficient homes, many contractors are now adopting the Passive House Standard—a strict building code developed in Germany in 1996. The Passive House Standard for air infiltration is no more than 0.6 ACH (air changes per hour). The formula for calculating ACH is

$$ACH = \frac{60L}{V}$$

where L is the air leakage in cubic feet per minute (cfm) and V is the volume of the house in cubic feet. Use this formula to determine what the air leakage would have to be in order to achieve an ACH of 0.6 in a 22,400-cubic-foot house.

32. **HVAC** Air-source heat pumps (ASHPs) and ground-source heat pumps (GSHPs) are new, more efficient alternatives to traditional heating systems for a home. The following formulas are used to estimate the operating costs of the different systems. All costs are per million Btu.

Fuel oil: Cost $= 7.25 \times$ (cost per gallon) $\div$ (efficiency rating)
ASHP: Cost $= 1000 \times$ (cost per kilowatt-hour) $\div$ (efficiency rating)
GSHP: Cost $= 293 \times$ (cost per kilowatt-hour) $\div$ (efficiency rating)

(a) Find the cost per million Btu for a fuel oil–fired boiler if fuel oil is $2.40 per gallon and the boiler is 82% efficient.
(b) Find the cost per million Btu for an ASHP if the cost of electricity is $0.16 per kilowatt-hour and the efficiency rating of the unit is 9.5.
(c) Find the cost per million Btu for a GSHP if the cost of electricity is $0.16 per kilowatt-hour and the efficiency rating of the unit is 3.2.
(d) Use your answer to part (a) to determine what the cost of electricity would have to be in order for the ASHP cost to match the boiler cost.
(e) Suppose a fuel-oil boiler costs $6500 installed and a GSHP costs $10,000 installed. Use your answers to parts (a) and (c) to determine the number of Btus of consumption that it would take for the GSHP to have a lower overall cost than the fuel-oil boiler.

33. **Electrical Trades** For customers primarily generating their own solar power, a certain electric utility charges a $16 per month maintenance fee but refunds to the customer $0.08 per kilowatt-hour of net electricity that is returned to the grid. Following the method of Your Turn (a) on page 490, set up an equation and solve for the amount of electricity a customer would need to return to the grid in order to break even.

34. **Construction**   A total of 32.6 cu yd of concrete is required for pouring a foundation. The ratio of cement to sand to gravel for this application is 1:2:4, and the total amount of these components must be $1\frac{1}{2}$ times the volume of concrete needed. Use the method of Example 4 on page 489 to determine how many cubic yards of each component are needed.

35. **Agriculture**   Jean is mixing a batch of her CowVita supplement for cattle. She adds $3\frac{1}{2}$ times as much CowVita A as CowVita B, and the total amount of mixture is 120 lb. How much CowVita A is in the mix? (Round to the nearest 0.1 lb.)

36. **Trades Management**   The Zeus Electronics repair shop charges $110 for the first hour of labor and $84 for each additional hour. The total labor cost for a lighting repair job was $656. Write an equation that allows you to solve for the total repair time, then solve this equation.

37. **Life Skills**   Suppose you budgeted $1300 for annual fuel expenses and drove 12,000 miles per year. If gas cost $2.60 per gallon, what combined mileage would your vehicle need to meet your budget?

38. **Life Skills**   Roxanne needs to buy a new car. She is debating between a hybrid and a nonhybrid model. The hybrid has an average fuel economy of 34 mi/gal, and it costs $32,000, but it will provide a tax credit of $3500. The nonhybrid has an average fuel economy of 22 mi/gal and costs $26,500. If she assumes an average gasoline cost of $2.60 per gallon, how many miles of driving will it take before the fuel savings of the hybrid compensate for its higher initial cost? (Round to the nearest thousand.)

39. **Trades Management**   Eric runs a small shop. He must set aside 22% of his revenue for taxes, and he spends another $3600 per month on other expenses. How much total revenue must he generate in order to clear a profit of $6000 per month? (Round to the nearest dollar.)

40. **HVAC**   In designing a new home, a resident is considering either a ground-source heat pump (GSHP) or a gas furnace with separate air conditioner. Assume the following:
    - The installation cost of the furnace is $6200.
    - The installation cost of the GSHP is $11,800.
    - The estimated annual cost of the furnace is $1245.
    - The estimated annual cost of the GSHP is $1020.

    What is the payback time for selecting the GSHP over the gas furnace? (Round up to the next whole number.)

41. **Sheet Metal Trades**   The length of a piece of sheet metal is four times its width. The perimeter of the sheet is 80 cm. Set up an equation relating these measurements and solve the equation to find the dimensions of the sheet.

42. **Trades Management**   Sal, Martin, and Gloria are partners in a welding business. Because of the differences in their contributions, it is decided that Gloria's share of the profits will be one-and-a-half times as much as Sal's share, and Martin's share will be twice as much as Gloria's share. How should a profit of $48,400 be divided?

43. **Machine Trades**   Two shims must have a combined thickness of 0.048 in. One shim must be twice as thick as the other. Set up an equation and solve for the thickness of each shim.

44. **Agriculture**   A farmer sold 24,000 pounds of tangerines to a packing house for an average of $0.76 per pound. He has a second harvest of 16,000 pounds that he intends to sell to a different packing house. What price per pound would the farmer need to receive for his second harvest in order to achieve an overall average of $0.80 per pound?

45. **Electrical Trades**   ABC electric utility charges customers with rooftop solar panels a monthly fee of $20 per month, and they compensate the customers $0.08 per kilowatt-hour (kWh) of net electricity returned to the grid. XYZ utility charges solar customers a maintenance fee of $16 per month, and they compensate the customers $0.05 per kWh of net electricity returned to the grid. Following the method of Example 5 in Section 7-6, set up an equation and solve for the amount of net return for which the two plans would result in the same monthly bill.

46. **Electrical Engineering**   The calculation shown is used to find the velocity of an electron at an anode. Calculate this velocity (units are in meters per second).

$$V = \sqrt{\frac{2(1.6 \times 10^{-19})(2.7 \times 10^2)}{9.1 \times 10^{-31}}}$$

47. **Electronics**   To convert the binary number 0.1011 to a decimal, the following calculation must be performed:

$$2^{-1} + 2^{-3} + 2^{-4}$$

Calculate this decimal number.

48. **Electronics**   How many terabytes are in 80 petabytes? (See Your Turn on page 510.)

CASE STUDY: Investigating the Basal Metabolic Rate

The basal metabolic rate, or BMR, is the minimum amount of energy required in a day to keep the body functioning. Another way to think of it is that the BMR represents the number of calories a person would burn if he or she was completely at rest in a neutral environment for 24 hours. Doctors and nutritionists use the BMR to help determine whether a person is likely to gain weight, lose weight, or maintain his or her weight based on calorie intake and additional calorie-burning activities he or she participates in, such as exercise. In Your Turn part (b) on page 427, we introduced the revised Harris–Benedict equation for a woman's BMR:

$$BMR = 9.247w + 3.098h - 4.330a + 447.593$$

where w is the woman's weight in kilograms, h is her height in centimeters, a is her age in years, and the BMR is measured in kilocalories per day.

For men, the revised Harris–Benedict equation is

$$BMR = 13.397w + 4.799h - 5.677a + 88.362$$

continued...

1. Why do you think there are different formulas for men and women?

2. Use the first formula to calculate the BMR for a 42-year-old woman with a weight of 72.5 kg and a height of 173 cm. (Round the BMR to the nearest whole number of kilocalories throughout this Case Study.)

3. Use the second formula to calculate the BMR for a man with the same age, weight, and height as the woman in question 2.

4. For two people of different gender with the same age, weight, and height, do you think the BMR of a man will always be higher than that of a woman? See if you can find a realistic combination of age, weight, and height that results in a higher BMR for a woman.

5. If a person maintains the same weight and height, does his or her BMR increase or decrease as he or she gets older? Did you need to substitute a different number for a in the formulas, or can you answer this question by just looking at the formula? Explain.

6. If a person loses weight, what happens to his or her BMR? Did you need to substitute a different number for w in the formulas, or can you answer this question by just looking at the formula? Explain.

The original Harris–Benedict equations were introduced in 1918 and were revised in 1984. Then, in 1990, M.D. Mifflin, S.T. St. Jeor, and others introduced the following new BMR formula called the Mifflin–St. Jeor equation:

$$BMR = 10.0w + 6.25h - 5.0a + s$$

where $s = +5$ for men and $s = -161$ for women

7. Use the data in question 2 to calculate the BMR for both a man and a woman using this new formula. Does the new formula seem to result in a higher or lower BMR for men? For women?

8. All three formulas require metric inputs. Create equivalent U.S. customary formulas for all three by converting the coefficients of w to pounds and the coefficients of h to inches. For example, using the conversion factor of 1 in. = 2.54 cm, we would multiply 6.25 in the Mifflin–St. Jeor equation by 2.54 to get the revised coefficient of h that would allow us to input the height in inches.

9. Once you have completed question 8, use both formulas to determine your own BMR. Use the data for a classmate or friend of the opposite gender to calculate his or her BMR using both formulas.

10. You cannot control your age or your height, but you can control your weight. Suppose you wanted to reduce your BMR reading by 100 kcal per day. Substitute into both formulas and solve the resulting equations to determine how much weight you would have to lose to achieve this reduction.

11. By researching BMR further, try to find scientific answers to the following questions:

    (a) Why are there different formulas for men and women?
    (b) Why does BMR change as people with the same height and weight get older?
    (c) Find out exactly how you can use your BMR, combined with your exercise program, to determine the number of calories per day that you can consume and still maintain your current weight.

8 Practical Plane Geometry

Objective	Sample problems	For help, go to page

When you finish this chapter, you will be able to:

1. Measure angles with a protractor.

∡*ABC* = _____   533

2. Classify angles.

(a)

(a) ∡*DEF* is _____   534

(b) G

(b) ∡*GHI* is _____

(c) J

(c) ∡*JKL* is _____

Name   _____

Date   _____

Course/Section   _____

Objective	Sample problems	For help, go to page
3. Use simple geometric relationships involving intersecting lines and triangles.	(a) Assume that the two horizontal lines are parallel.	$\angle a =$ _____ 536 $\angle b =$ _____ $\angle c =$ _____ $\angle d =$ _____ $\angle e =$ _____
	(b)	$\angle x =$ _____ 537
4. Identify polygons, including triangles, squares, rectangles, parallelograms, trapezoids, and hexagons.	(a)	_____ 561
	(b)	_____ 545
	(c)	_____
	(d)	_____
	(e)	_____ 572
5. Use the Pythagorean theorem.	Round to one decimal place.	$x =$ _____ 562
6. Find the area and perimeter of geometric figures.	(a)	Area = _____ 566
	(b)	Area = _____ 550

Objective	Sample problems	For help, go to page

(c)

Area = _____
(Round to two decimal places.)   572

Perimeter = _____

0.40"

(d)

Area = _____   590

r = 2.70"

Circumference = _____   585

Use π ≈ 3.14 and round both answers to the nearest tenth.

7. Solve practical problems involving area and perimeter of plane figures.

(a)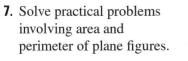

10'

21'

24'

12'

Flooring and Carpeting   If carpet costs $4.75 per square foot, how much will it cost to purchase carpet for the room in the figure?

_____   579

(b) **Landscaping**   A circular flowerbed, with a radius of 8.5 ft, is surrounded by a circular ring of lawn 12 ft wide. At $9.50 per square foot, how much will it cost to replace the lawn with artificial turf?

_____   592

(Answers to these preview problems are given in the Appendix. Also, worked solutions to many of these problems appear in the chapter Summary.)

If you are certain that you can work *all* these problems correctly, turn to page 602 for a set of practice problems. If you cannot work one or more of the preview problems, turn to the page indicated to the right of the problem. Those who wish to master this material with the greatest success should turn to Section 8-1 and begin work there.

CHAPTER

8

Practical Plane Geometry

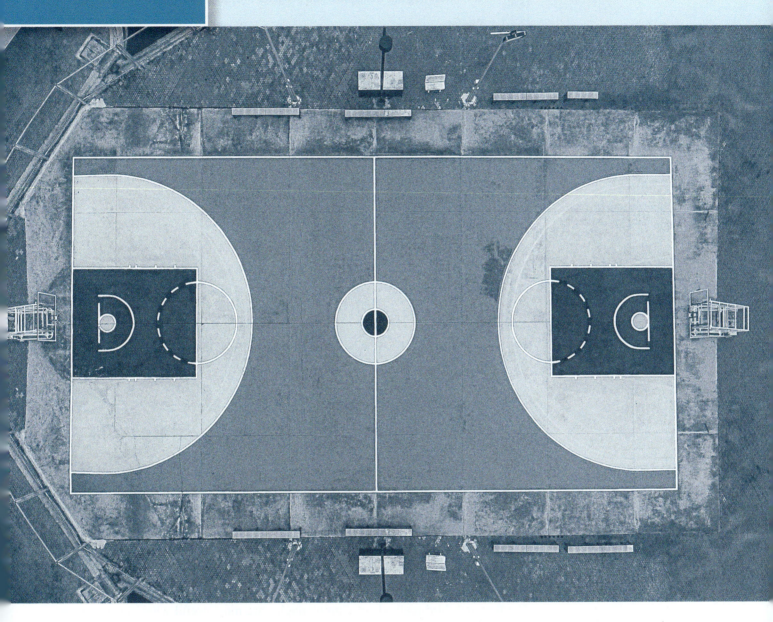

Geometry is one of the oldest branches of mathematics. It involves the study of the properties of points, lines, plane surfaces, and solid figures. Ancient Egyptian engineers used these properties when they built the pyramids, and modern trades workers use them when cutting sheet metal, installing plumbing, building cabinets, laying flooring and carpeting, and performing countless other tasks. In this chapter, we concentrate on the applications of plane geometry, with an emphasis on angle measurement, perimeter, and area.

CASE STUDY: The Basketball Arena

The Case Study at the end of the chapter illustrates how useful area formulas can be in the **painting** and **flooring** trades. In this case, your main project will be to calculate the materials and labor costs for painting and refinishing a basketball court.

8-1 Angle Measurement

In geometry, the word **plane** is used to refer to a mathematical concept describing an infinite set of points in space. For practical purposes, a plane surface can be thought of as a perfectly flat, smooth surface such as a tabletop or a window pane.

Labeling Angles   An **angle** is a measure of the size of the opening between two intersecting lines. The point of intersection is called the **vertex**, and the lines forming the opening are called the **sides**.

learning|catalytics™

1. What fraction of a circle is this?

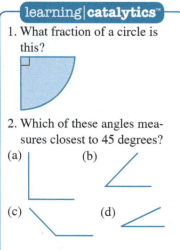

2. Which of these angles measures closest to 45 degrees?

(a)   (b)

(c)   (d)

An angle may be identified in any one of the following ways:

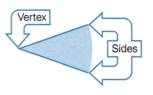

| Vertex letter in the middle | Vertex letter only | A letter or number placed inside the angle |

The angle symbol ∡ is simply a shorthand way to write the word *angle*.

For the first angle, the middle letter is the vertex letter of the angle, while the other two letters represent points on each side. Notice that upper case letters are used. The second angle is identified by the letter D near the vertex. The third angle is named by a lower case letter or number placed inside the angle.

The first and third methods of naming angles are most useful when several angles are drawn with the same vertex.

EXAMPLE 1

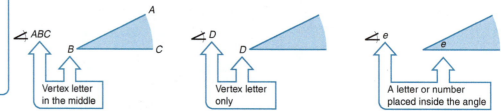

The interior of angle *POR* (or *ROP*) is shaded.

The interiors of angles *a* and 1 are shaded.   ●

Your Turn   Name the angle shown in the margin using the three different methods.

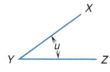

Answers   ∡ *XYZ*, ∡ *ZYX*, ∡ *Y*, or ∡ *u*

Measuring Angles Naming or labeling an angle is the first step. Measuring it may be even more important. Angles are measured according to the size of their opening. The length of the sides of the angle is not related to the size of the angle.

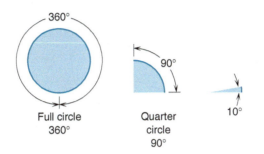 and are the same size angle.

The basic unit of measurement of angle size is the **degree**. Because of traditions going back thousands of years, we define a degree as an angle that is $\frac{1}{360}$ of a full circle, or $\frac{1}{90}$ of a quarter circle.

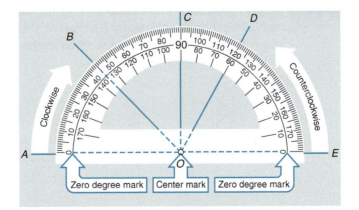

Full circle
360°

Quarter
circle
90°

The degree can be further subdivided into finer angle units known as *minutes* and *seconds* (which are not related to the time units).

1 degree, 1° = 60 minutes = 60′

1 minute, 1′ = 60 seconds = 60″

An angle of 62 degrees, 12 minutes, 37 seconds would be written 62°12′37″. Most trade work requires precision only to the nearest degree.

As you saw in Chapter 5, page 350, a protractor is a device used to measure angles. To learn to measure angles, first look at the protractor pictured.

visibleA protractor that you may cut out and use is available in the Appendix on page 883.

To measure an angle, place the protractor over it so that the zero-degree mark is lined up with one side of the angle and the center mark is on the vertex. Read the measure in degrees where the other side of the angle intersects the scale of the protractor. When reading an angle clockwise, use the upper scale, and when reading counterclockwise, use the lower scale.

EXAMPLE 2   (a) In the drawing on page 533, ⊀ *EOB* should be read counterclockwise from the 0° mark on the right. ⊀ *EOB* = 135°.

(b) ⊀ *AOD* should be read clockwise from the 0° mark on the left. ⊀ *AOD* = 120°. ●

Your Turn   Find the measure of (a) ⊀ *EOD*   (b) ⊀ *AOB*

Solutions   (a) ⊀ *EOD* = 60°   Measure counterclockwise from the right 0° mark and read the lower scale.

(b) ⊀ *AOB* = 45°   Measure clockwise from the left 0° mark and read the upper scale.

Classifying Angles   An **acute angle** is an angle *less than* 90°. Angles *AOB* and *EOD* in the previous Your Turn are both less than 90°, so both are acute angles.

Acute angles:

An **obtuse angle** is an angle *greater than* 90° and less than 180°. Angles *AOD* and *EOB* in Example 2 are both greater than 90°, so both are obtuse angles.

Obtuse angles:

A **right angle** is an angle exactly equal to 90°. Angle *EXC* in the drawing at the left is a right angle.

Two straight lines that meet in a right or 90° angle are said to be **perpendicular**. In this drawing, *CX* is perpendicular to *XE*.

C

X ⌐ *E*
Perpendicular lines

Notice that a small square is placed at the vertex of a right angle to show that the sides are perpendicular.

A **straight angle** is an angle equal to 180°.

Angle *BOA* shown here is a straight angle.   *A*   *O*   *B*

Your Turn   Use your protractor to measure each of the angles given and tell whether each is acute, obtuse, right, or straight. Estimate the size of the angle before you measure. (You will need to extend the sides of the angles to fit your protractor.)

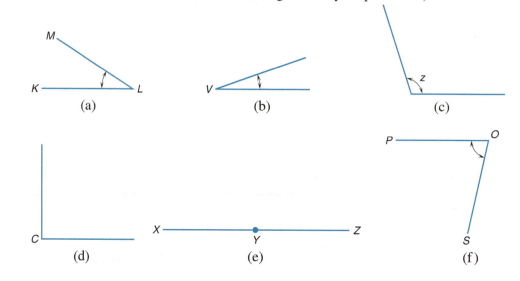

Answers (a) ∡ KLM = 33°, an acute angle   (b) ∡ V = 20°, an acute angle

(c) ∡ z = 108°, an obtuse angle   (d) ∡ C = 90°, a right angle

(e) ∡ XYZ = 180°, a straight angle   (f) ∡ POS = 78°, an acute angle

Drawing Angles In addition to measuring angles that already exist, trades workers will sometimes need to draw their own angles. To draw an angle of a given size follow these steps:

Step 1 Draw a line representing one side of the angle.

Step 2 Place the protractor over the line so that the center mark is on the vertex, and the 0° mark coincides with the other end of the line. (*Note:* We could have chosen either end of the line for the vertex.)

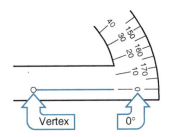

Step 3 Now place a small dot above the degree mark corresponding to the size of your angle. In this case the angle is 30°.

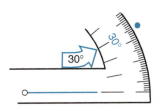

Step 4 Remove the protractor and connect the dot to the vertex.

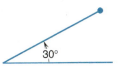

Your Turn Now you try it. Use your protractor to draw the following angles.

∡ A = 15°   ∡ B = 112°   ∡ C = 90°

Answers Here are our drawings:

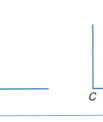

Angle Facts   There are several important geometric relationships involving angles that you should know. For example, if we measure the four angles created by the intersecting lines shown,

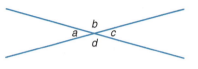

we find that

$$\angle a = 30° \qquad \angle b = 150° \qquad \angle c = 30° \qquad \text{and} \qquad \angle d = 150°$$

Do you notice that opposite angles are equal?

For any pair of intersecting lines the opposite angles are called **vertical angles.** In the preceding drawing, angles a and c are a pair of vertical angles. Angles b and d are also a pair of vertical angles.

An important geometric rule is that

> When two lines intersect, the vertical angles are always equal.

$$\angle p = \angle r$$
and $\angle q = \angle s$

Your Turn   If you remember that a straight angle $= 180°$, you can complete the following statements for the preceding drawing.

$\angle p + \angle q =$ _____ $\angle p + \angle s =$ _____

$\angle s + \angle r =$ _____ $\angle q + \angle r =$ _____

Answers $\angle p + \angle q = 180°$ $\angle p + \angle s = 180°$

$\angle s + \angle r = 180°$ $\angle q + \angle r = 180°$

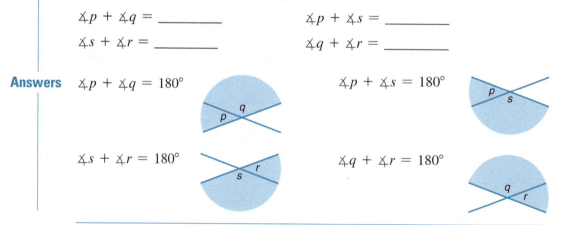

Each of the four pairs of angles in the preceding problem are called **adjacent angles.**

This leads to a second important geometry relation:

> When two lines intersect, each pair of adjacent angles sums to 180°.

Your Turn Use these relationships to answer the following questions.

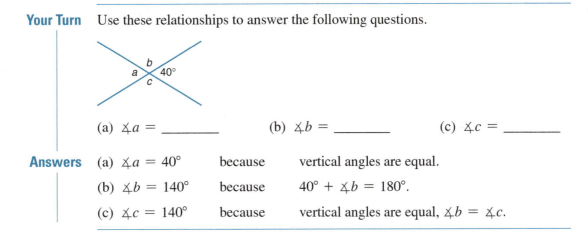

(a) $\angle a =$ _____ (b) $\angle b =$ _____ (c) $\angle c =$ _____

Answers (a) $\angle a = 40°$ because vertical angles are equal.

(b) $\angle b = 140°$ because $40° + \angle b = 180°$.

(c) $\angle c = 140°$ because vertical angles are equal, $\angle b = \angle c$.

A **triangle** is formed from three distinct line segments. The line segments are called the **sides** of the triangle, and the points where they meet are called its **vertices**. Every triangle forms three angles, and these are related by an important geometric rule.

If we use a protractor to measure each angle in the following triangle, *ABC*, we find

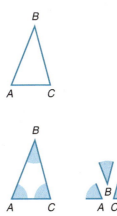

$\angle A = 69°$ $\angle B = 36°$ $\angle C = 75°$

Notice that the three angles of the triangle sum to 180°.

$69° + 36° + 75° = 180°$

Measuring angles on triangles of every possible size and shape leads to the third geometry relationship:

> The interior angles of a triangle always add to 180°.

EXAMPLE 3 We can use this fact to solve the following problems.

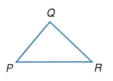

(a) If $\angle P = 50°$ and $\angle Q = 85°$, find $\angle R$.

By the relationship given, $\angle P + \angle Q + \angle R = 180°$, so

$\angle R = 180° - \angle P - \angle Q = 180° - 50° - 85° = 45°$

(b) If $\angle P = 27° \, 42'$ and $\angle Q = 76° \, 25'$, find $\angle R$.

Step 1 $\angle R = 180° - 27° \, 42' - 76° \, 25'$

Step 2 Combine the two angles that are being subtracted.

$27° \, 42' + 76° \, 25' = (27 + 76)° + (42 + 25)' = 103° \, 67'$

Step 3 To simplify $103° \, 67'$ use the fact that $60' = 1°$.

$103° \, 67' = 103° + 60' + 7' = 103° + 1° + 7' = 104° \, 7'$

Step 4 Subtract this total from 180°. First change 180° to 179° 60'.

$$180° - 104° 7' = 179° 60' - 104° 7'$$
$$= (179 - 104)° + (60 - 7)' = 75° 53'$$

●

Your Turn Find the missing angle in each problem.

(a) If $\angle X = 34°$ and $\angle Y = 132°$, find $\angle Z$.

(b) If $\angle X = 29° 51'$ and $\angle Z = 24° 34'$, find $\angle Y$.

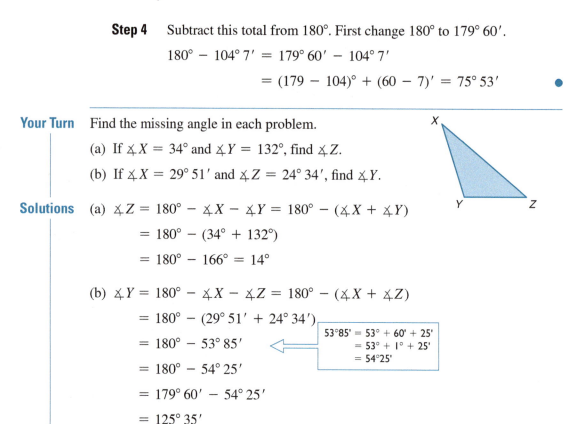

Solutions (a) $\angle Z = 180° - \angle X - \angle Y = 180° - (\angle X + \angle Y)$

$$= 180° - (34° + 132°)$$
$$= 180° - 166° = 14°$$

(b) $\angle Y = 180° - \angle X - \angle Z = 180° - (\angle X + \angle Z)$

$$= 180° - (29° 51' + 24° 34')$$
$$= 180° - 53° 85'$$
$$= 180° - 54° 25'$$
$$= 179° 60' - 54° 25'$$
$$= 125° 35'$$

> 53°85' = 53° + 60' + 25'
> = 53° + 1° + 25'
> = 54°25'

A fourth and final angle fact you will find useful involves parallel lines. Two straight lines are said to be **parallel** if they are always the same distance apart and never meet. If a pair of parallel lines is cut by a third line called a **transversal**, several important pairs of angles are formed.

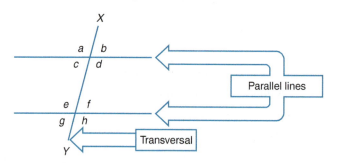

Angles c and f are called **alternate interior angles**. Angles d and e are also alternate interior angles.

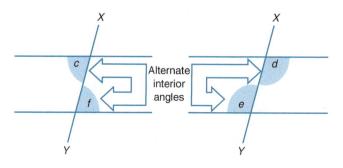

The angles are *interior* (or inside) the original parallel lines and they are on *alternate* sides of the transversal XY. The alternate interior angles are equal when the lines cut by the transversal are parallel.

The angles a and e form one of four pairs of **corresponding angles**. The others are b and f, c and g, and d and h.

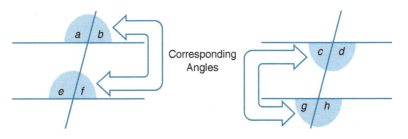

These angles are on the same side of the transversal, and they are on the same side of their respective parallel line. For example, b and f are both above their parallel lines while c and g are both below their lines. Corresponding angles are always equal when the lines cut by the transversal are parallel.

The geometric rule is as follows:

> When parallel lines are cut by a third line, the corresponding angles are equal, and the alternate interior angles are equal.

Your Turn   Use this rule and the previous ones to answer the following questions about the figure shown. Assume that the two horizontal lines are parallel.

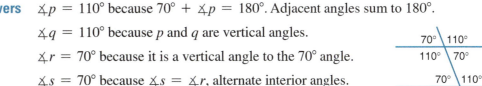

$\angle p =$ _____ $\angle q =$ _____

$\angle r =$ _____ $\angle s =$ _____

$\angle t =$ _____ $\angle u =$ _____

$\angle w =$ _____

Answers $\angle p = 110°$ because $70° + \angle p = 180°$. Adjacent angles sum to $180°$.

$\angle q = 110°$ because p and q are vertical angles.

$\angle r = 70°$ because it is a vertical angle to the $70°$ angle.

$\angle s = 70°$ because $\angle s = \angle r$, alternate interior angles.

$\angle t = 110°$ because $\angle t = \angle q$, alternate interior angles, and $\angle p = \angle t$, corresponding angles.

$\angle u = 110°$ because $\angle u = \angle q$, corresponding angles.

$\angle w = 70°$ because $\angle u + \angle w = 180°$ and $\angle w = \angle r$, corresponding angles.

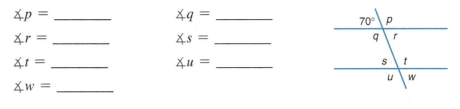

Practical Applications

EXAMPLE 4   **Plumbing**   A plumber connects the two vertical sections of pipe shown in Figure (1) with a $30°$ elbow (angle a). What angle x does the elbow make with the upper pipe?

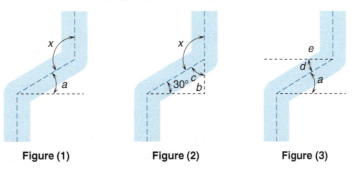

Figure (1)   **Figure (2)**   **Figure (3)**

To help us solve this problem, we first extend the horizontal and vertical lines as shown in Figure (2). These lines are perpendicular, so angle b is a right angle, or 90°. Because the interior angles of the triangle sum to 180°, angle c must be $180° - 90° - 30° = 60°$. Finally, because angles c and x are adjacent angles forming a straight line, they sum to 180°. Therefore angle $x = 180° - 60° = 120°$.

Another way to determine angle x is to first draw the two horizontal lines shown in Figure (3). These lines are parallel, so angles d and a are alternate interior angles. This means that $\angle d = \angle a = 30°$. Angle e is a right angle, and $\angle x = \angle d + \angle e$. Therefore $\angle x = 30° + 90° = 120°$.   ●

Your Turn   Welding   In the pipe flange shown, the eight bolts are equally spaced. How many degrees x separate the bolts?

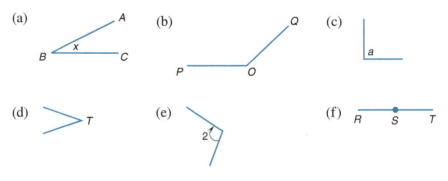

Solution   We know that a full circle contains 360°. Because the bolts are equally spaced, angle x must be $\frac{1}{8}$ of 360°, or $360° \div 8$, which is 45°.

Now, ready for some problems on angle measurement? Turn to Exercises 8-1 for practice in measuring and working with angles.

Exercises 8-1   Angle Measurement

A. Solve the following. If you need a protractor, cut out and use the one in the Appendix.

1. Name each of the following angles using the three kinds of notation shown in the text.

(a)   (b)   (c)

(d)   (e)   (f) $\underset{R \quad\ S \quad\ T}{\bullet}$

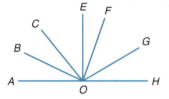

2. For the figure in the margin,

(a) Name an acute angle that has AO as a side.

(b) Name an acute angle that has HO as a side.

(c) Name an obtuse angle that has AO as a side.

(d) Name a right angle.

(e) Use your protractor to measure $\angle GOF$, $\angle BOF$, and $\angle COG$.

(f) Use your protractor to measure $\angle AOB$, $\angle HOC$, and $\angle FOH$.

(g) Based on your answers to part (f), determine the measures of $\angle HOB$, $\angle AOC$, and $\angle AOF$ without using a protractor.

3. Use your protractor to draw the following angles.

(a) $\angle LMN = 29°$ (b) $\angle 3 = 100°$ (c) $\angle Y = 152°$

(d) $\angle PQR = 137°$ (e) $\angle t = 68°$ (f) $\angle F = 48°$

4. Measure the indicated angles on the following shapes. (You will need to extend the sides of the angles to improve the accuracy of your measurement.)

(a)

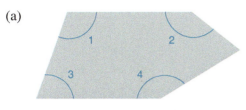

(b)

(c)

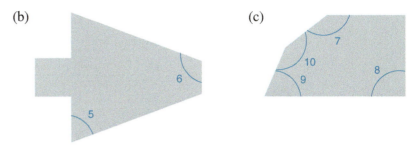

B. Use the geometry relationships to answer the following questions.

1. In each problem one angle measurement is given. Determine the others for each figure without using a protractor.

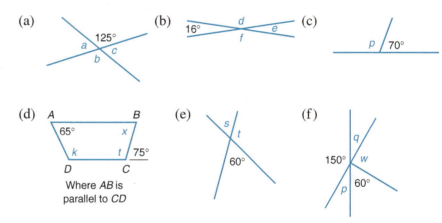

2. For each triangle shown, two angles are given. Find the third angle without using a protractor.

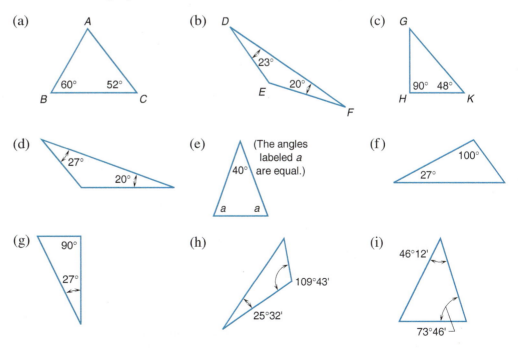

3. In each figure, two parallel lines are cut by a third line. Find the angles that are marked with letters.

(a)
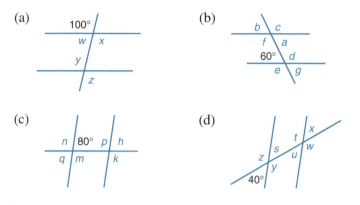

(b)

(c)

(d)

C. Applied Problems (*Note:* Figures accompany all problems.)

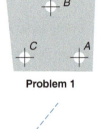

Problem 1

1. **Machine Trades**   A machinist must punch holes *A* and *B* in the piece of steel shown by rotating the piece through a certain angle from vertex *C*. Use a protractor to find angle *ACB*.

Problem 2

2. **Plumbing**   A plumber must connect the two pipes shown by first selecting the proper elbows. What angle elbows does he need?

3. **Carpentry**   A carpenter needs to build a triangular hutch that will fit into the corner of a room as shown. Measure the three angles of the hutch.

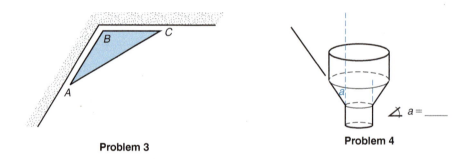

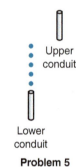

Upper conduit

Lower conduit

Problem 5

Problem 3

Problem 4

4. **Sheet Metal Trades**   A sheet metal worker must connect the two vent openings shown. At what angle must the sides of the connecting pieces flare out?

5. **Electrical Trades**   An electrician wants to connect the two conduits shown with a 45° elbow. How far up must she extend the lower conduit before they will connect at that angle? Give the answer in inches. (*Hint:* Draw the 45° connection on the upper conduit first.) Each dot represents a 1-in. extension.

6. **Roofing**   At what angle must the rafters be set to create the roof gable shown in the drawing?

7. **Machine Trades**   A machinist receives a sketch of a part he must make as shown. If the indicated angle *BAC* is precisely half of angle *BAD*, find ∡*BAD*.

Problem 6

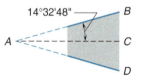

Problem 7

Problem 8

8. **Machine Trades**   The drawing illustrates a cross section of a V-thread. Find the indicated angle.

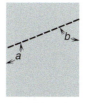

Problem 9

9. **Carpentry** At what angles must drywall board be cut to create the wall shown in the figure?

10. **Carpentry** A carpenter wishes to make a semicircular deck by cutting eight equal angular pieces of wood as shown. Measure the angle of one of the pieces. Was there another way to determine the angle measure without using a protractor?

Problem 10

Problem 11

Problem 12

11. **Carpentry** A carpenter's square is placed over a board as shown in the figure. If ∡1 measures 74°, what is the size of ∡2?

12. **Carpentry** Board 1 must be joined to board 2 at a right angle. If ∡x measures 42°, what must ∡y measure?

13. **Welding** In the pipe flange shown, the six bolt holes are equally spaced. What is the spacing of the holes, in degrees?

Problem 13

When you have finished these problems, check your answers to the odd-numbered problems in the Appendix, then continue in Section 8-2 with the study of plane figures.

8-2 Perimeter of Polygons and Area of Quadrilaterals

Polygons

A **polygon** is a closed plane figure containing three or more angles and bounded by three or more straight sides. The word *polygon* itself means "many sides." The following figures are all polygons:

A figure with a curved side is *not* a polygon.

and are not polygons

Every trades worker will find that an understanding of polygons is important. In this section you will learn how to identify the parts of a polygon, recognize some different kinds of polygons, and compute the perimeter and area of certain polygons.

In the general polygon shown, each **vertex** or corner is labeled with a letter: *A*, *B*, *C*, *D*, and *E*. The polygon is named simply "polygon *ABCDE*"—not a very fancy name, but it will do.

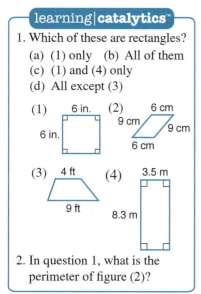

learning|catalytics™

1. Which of these are rectangles?
   (a) (1) only   (b) All of them
   (c) (1) and (4) only
   (d) All except (3)

   (1) 6 in.   (2) 6 cm
   6 in.   9 cm / 9 cm
                6 cm

   (3) 4 ft   (4) 3.5 m
       9 ft   8.3 m

2. In question 1, what is the perimeter of figure (2)?

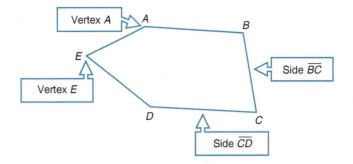

The **sides** of the polygon are named after the line segments that form each side: $\overline{AB}, \overline{BC}, \overline{CD}$, and so on. Notice that we place a bar over the letters to indicate that we are talking about a line segment. A **line segment** is a finite portion of a straight line; it has two endpoints.

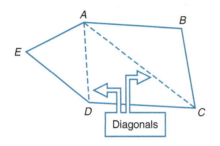

The **diagonals** of the polygon are the line segments connecting nonconsecutive vertices such as $\overline{AC}$ and $\overline{AD}$. These are shown as dashed lines in the figure. Other diagonals not shown are $\overline{EB}, \overline{EC}$, and $\overline{BD}$.

EXAMPLE 1   Consider the following polygon, *UVWXYZ:*

The sides are $\overline{UV}, \overline{VW}, \overline{WX}, \overline{XY}, \overline{YZ}$, and $\overline{ZU}$.

The vertices are *U, V, W, X, Y,* and *Z.*

There are nine diagonals; we need to be certain to find them all.

The diagonals are $\overline{UW}, \overline{UX}, \overline{UY}, \overline{VX}, \overline{VY}, \overline{VZ}, \overline{WY}, \overline{WZ}$, and $\overline{XZ}$.   ●

✎ **Note**   The letters naming a side or a diagonal can be reversed: side $\overline{VU}$ is the same as side $\overline{UV}$ and diagonal $\overline{YW}$ is the same as diagonal $\overline{WY}$. ●

Perimeter   For those who do practical work with polygons, the most important measurements are the lengths of the sides, the *perimeter,* and the *area.* The **perimeter** of any polygon is simply the distance around the outside of the polygon.

> To find the perimeter of a polygon, add the lengths of its sides.

✎ **Note**   In drawings, we will sometimes use the ' symbol for feet and the " symbol for inches. ●

EXAMPLE 2   In the polygon *KLMN* the perimeter is

3 in. + 5 in. + 6 in. + 4 in. = 18 in.

The perimeter is a length; therefore, it has length units, in this case inches.

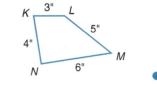

Your Turn   Find the perimeter of each of the following polygons.

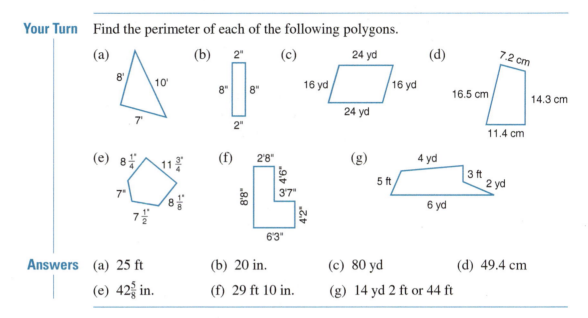

(a)   (b)   (c)   (d)

(e)   (f)   (g)

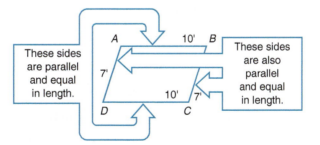

Answers   (a) 25 ft   (b) 20 in.   (c) 80 yd   (d) 49.4 cm

(e) $42\frac{5}{8}$ in.   (f) 29 ft 10 in.   (g) 14 yd 2 ft or 44 ft

---

! Careful   As you may have discovered in problem (g), it is important that all sides be expressed in the same units before calculating the perimeter. ●

Adding the lengths of the sides will always give you the perimeter of a polygon, but handy formulas are needed to enable you to calculate the area of a polygon. Before you can use these formulas you must learn to identify the different types of polygons. Let's look at a few. In this section we examine the **quadrilaterals**, or four-sided polygons.

Quadrilaterals   Figure *ABCD* is a **parallelogram**. In a parallelogram, opposite sides are parallel and equal in length.

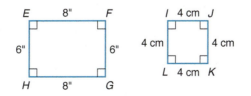

These sides are parallel and equal in length.

These sides are also parallel and equal in length.

Figure *EFGH* is a **rectangle**, a parallelogram in which the four corner angles are right angles. The ⌐ symbols at the vertices indicate right angles. Because a rectangle is a parallelogram, opposite sides are parallel and equal.

IJKL is a **square**, a rectangle in which *all* sides are the same length.

MNOP is called a **trapezoid**. A trapezoid contains two parallel sides and two nonparallel sides. $\overline{MN}$ and $\overline{OP}$ are parallel; $\overline{MP}$ and $\overline{NO}$ are not.

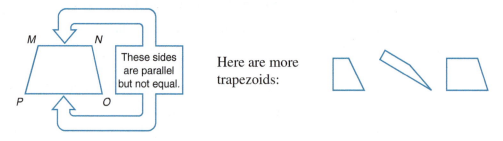

Here are more trapezoids:

Parallelograms, rectangles, squares, and trapezoids are all quadrilaterals as well. But if a four-sided polygon has none of these special features—no parallel sides—we call it simply a *quadrilateral. QRST* is a quadrilateral.

Your Turn   Name each of the following quadrilaterals. If more than one name fits, give the most specific name.

(a) 6'
6'   6'
6'

(b) *AD* is parallel to *BC*
7 cm   C
D
15 cm   18.6 cm
A 6 cm B

(c) 16"
13"   13"
16"

(d) 11 yd
16 yd
26 yd
22 yd

(e) 24"
18"   18"
24"

Answers   (a) Square (It is also a rectangle and a parallelogram, but "square" is a more specific name.)

(b) Trapezoid   (c) Parallelogram   (d) Quadrilateral

(e) Rectangle (It is also a parallelogram, but "rectangle" is a more specific name.)

Area and Perimeter of Rectangles   Once you can identify a polygon, you can use a formula to find its area. Next let's examine and use some area formulas for these quadrilaterals.

The **area** of any plane figure is the number of square units of surface within the figure.

In this rectangle,

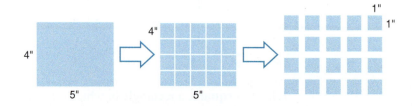

we can divide the surface into exactly 20 small squares each one inch on a side. By counting squares we can see that the area of this 4-in. by 5-in. rectangle contains 20 square inches. We abbreviate these area units as 20 sq in.

✎ **Note**   As mentioned in Chapter 5, area units may be abbreviated using exponents. For example, the area unit square inches or sq in. results from multiplying inches × inches. Using the definition of exponents from Chapter 6, this can be expressed as inches² or in.². Similarly, square feet or sq ft can be written as ft², square centimeters as cm², and so on. Both types of unit abbreviations are used by trades workers, and we will use both in the remainder of this text. ●

Of course, there is no need to draw lines in the messy way shown on the previous page. We can find the area by multiplying the two dimensions of the rectangle.

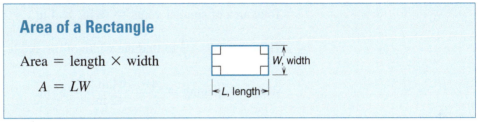

Area of a Rectangle

Area = length × width

$A = LW$

W, width

← L, length →

✎ **Note**   The dimensions "length" and "width" are sometimes called "base" and "height." ●

EXAMPLE 3   Find the area of rectangle *EFGH*.

$L = 9$ in.

$W = 5$ in.

A, area $= LW = (9\text{ in.})(5\text{ in.})$

$A = 45\text{ in.}^2$

E ——— 9" ——— F

5"

H ——————— G

●

✎ **Note**   In technical work, it is important to both state measurements and round calculations to the proper precision and accuracy. However, in the trades and in many occupations, people do not normally state quantities with the exact accuracy with which they have been measured. For example, the length of a box might very well be measured accurately to two significant digits, but a carpenter would write it as 9 in. rather than 9.0 in. To avoid confusion in calculations, either state your unrounded answer or follow the rounding instructions that accompany the problems. ●

Your Turn   Find the area of rectangle *IJKL*.

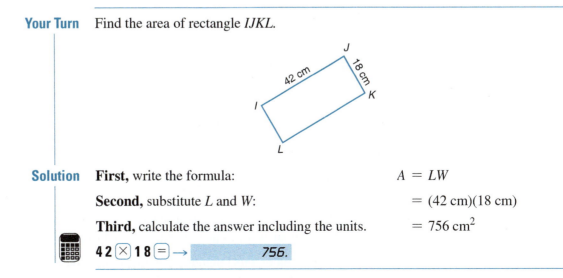

Solution   **First,** write the formula: $A = LW$

Second, substitute *L* and *W*: $= (42\text{ cm})(18\text{ cm})$

Third, calculate the answer including the units. $= 756\text{ cm}^2$

⌨ 4 2 ⊗ 1 8 ⊜ →   [756.]

Of course, the formula $A = LW$ may also be used to find L or W when the other quantities are known. As you learned in your study of basic algebra in Chapter 7, the following formulas are all equivalent:

$$A = LW \quad \text{means that} \quad L = \frac{A}{W} \quad \text{or} \quad W = \frac{A}{L}$$

EXAMPLE 4   **Sheet Metal Trades**   A rectangular opening 12 in. long must have the same total area as two smaller rectangular vents 6 in. by 4 in. and 8 in. by 5 in. What must be the width of the opening?

To solve this problem, **first** find the total area of the two smaller vents:

Total area of smaller vents $= (6 \text{ in.} \times 4 \text{ in.}) + (8 \text{ in.} \times 5 \text{ in.}) = 64 \text{ in.}^2$

Then, use $W = \dfrac{A}{L}$ to find the width of the single opening:

Width of opening $= \dfrac{A}{L} = \dfrac{64 \text{ in.}^2}{12 \text{ in.}} = 5\frac{1}{3} \text{ in.}$ ●

Another formula that you may find useful allows you to calculate the perimeter of a rectangle from its length and width.

Perimeter of a Rectangle

$$P = 2L + 2W$$

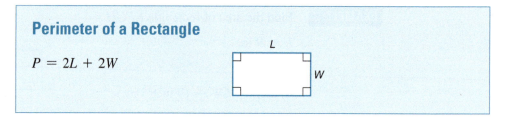

EXAMPLE 5   We can use this formula to find the perimeter of the rectangle shown here.

$$\begin{aligned}
\text{Perimeter} &= 2L + 2W \\
&= 2(7 \text{ in.}) + 2(3 \text{ in.}) \\
&= 14 \text{ in.} + 6 \text{ in.} \\
&= 20 \text{ in.}
\end{aligned}$$

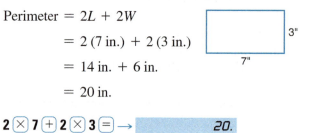

More Practice   For practice using the area and perimeter formulas for rectangles, work the following problems.

(a) Find the area and perimeter of each of the following rectangles.

(1) 8", 6"   (2) 12', 43'   (3) $9\frac{2}{3}$ yd, $2\frac{2}{3}$ yd   (4) 1.80 m, 1.80 m

(b) Find the area of a rectangle whose length is 4 ft 6 in. and whose width is 3 ft 3 in. (Give your answer in square feet in decimal form. Round to one decimal place.)

(c) Find the area of a rectangular opening 20 in. wide and 3 ft long.

Solutions   (a) (1) $A = 48$ in.^{2}, $P = 28$ in.   (2) $A = 516$ ft^2, $P = 110$ ft

(3) $A = 25\frac{7}{9}$ yd^2, $P = 24\frac{2}{3}$ yd   (4) $A = 3.24$ m^2, $P = 7.20$ m

(b) 4 ft 6 in. $= 4\frac{6}{12}$ ft $= 4.5$ ft

3 ft 3 in. $= 3\frac{3}{12}$ ft $= 3.25$ ft

4.5 ft $\times$ 3.25 ft ≈ 14.6 ft^2

(c) 720 in.^{2} or 5 ft^2

**Area and Perimeter
of Squares**

Did you notice that the rectangle in problem (a)(4) in the last set is actually a square? All sides are the same length. To save time when calculating the area or perimeter of a square, use the following formulas:

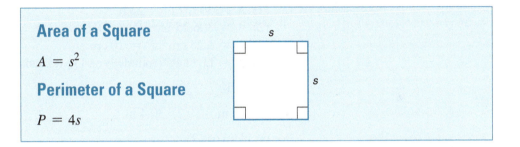

Area of a Square

$A = s^2$

Perimeter of a Square

$P = 4s$

EXAMPLE 6

We can use these formulas to find the area and perimeter of the square shown.

Area, $A = s^2$

$\qquad = (9 \text{ ft})^2$

$\qquad = (9 \text{ ft})(9 \text{ ft}) = 81 \text{ ft}^2$

Remember, s^2 means s times s, not $2s$.

Perimeter, $P = 4s$

$\qquad = 4(9 \text{ ft})$

$\qquad = 36 \text{ ft}$ ●

A calculation that is very useful in carpentry, sheet metal work, and many other trades involves finding the length of one side of a square given its area. The following formula will help:

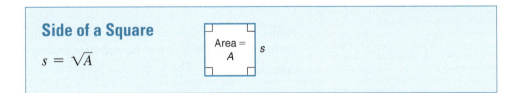

Side of a Square

$s = \sqrt{A}$

EXAMPLE 7

If a square opening is to have an area of 64 in.^{2} what must be its side length?

$s = \sqrt{A}$

$\quad = \sqrt{64 \text{ in.}^2}$

$\quad = 8 \text{ in.}$ $\left(\sqrt{\text{in.}^2} = \text{in.} \right)$ ●

More Practice

Ready for a few problems involving squares? Try these.

(a) Find the area and perimeter of each of the following squares.

(1)   6'

(2) $17\frac{1}{2}$"

(3)   2.50 cm

(4)   3'4"

(b) Find the area of a square whose side length is $2\frac{1}{3}$ yd. (Round to two decimal places.)

(c) Find the length of the side of a square whose area is (1) 144 m², (2) 927 in.² Use a calculator and round to one decimal place.

Answers

(a) (1) $A = 36$ ft²; $P = 24$ ft
    (2) $A = 306.25$ in.²; $P = 70$ in.
    (3) $A = 6.25$ cm²; $P = 10$ cm
    (4) $A \approx 11.11$ ft² (rounded) or $11\frac{1}{9}$ ft², using fractions; $P = 13$ ft 4 in.

(b) 5.44 yd²

(c) (1) 12 m   (2) 30.4 in.

For problem (c)(2),

$\boxed{\vee}$ **9 2 7** $\boxed{=}$ →   **30.4466747** or 30.4, rounded

---

Area and Perimeter of Parallelograms

You should recall that a parallelogram is a four-sided figure whose opposite pairs of sides are equal in length and parallel. Here are a few parallelograms:

To find the area of a parallelogram, use the following formula:

> ### Area of a Parallelogram
>
> $A = bh$
>
> h h Height
>
>   ←b→
>   Base

Notice that we do not use only the lengths of the sides to find the area of a parallelogram. The height h is the perpendicular distance between the base and the side opposite the base. The height h is perpendicular to the base b.

EXAMPLE 8

In parallelogram *ABCD*, the base is 7 in. and the height is 10 in. Find its area.

D   *C*
$h = 10"$ 12.2"
A   *B*
$b = 7"$

First, write the formula: $A = bh$

Second, substitute the given values: $= (7 \text{ in.})(10 \text{ in.})$

Third, calculate the area including the units: $= 70$ sq in.

Notice that we do not need to use the length of the slant side, 12.2 in., in this calculation. ●

Your Turn   Find the area of this parallelogram:

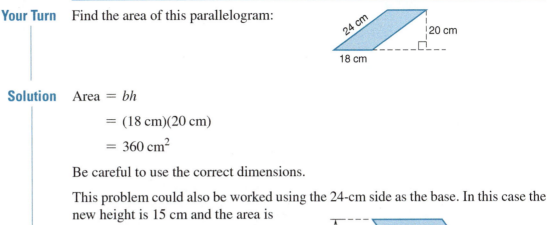

Solution   Area $= bh$

$= (18 \text{ cm})(20 \text{ cm})$

$= 360 \text{ cm}^2$

Be careful to use the correct dimensions.

This problem could also be worked using the 24-cm side as the base. In this case the new height is 15 cm and the area is

$A = (24 \text{ cm})(15 \text{ cm})$

$= 360 \text{ cm}^2$

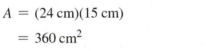

To find the perimeter of a parallelogram, simply add the lengths of the four sides or use the following formula:

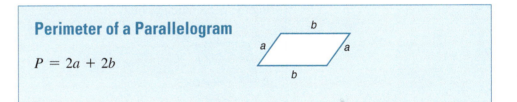

Perimeter of a Parallelogram

$P = 2a + 2b$

EXAMPLE 9   Using the figure in the previous Your Turn,

$P = 2a + 2b$

$= 2(24 \text{ cm}) + 2(18 \text{ cm})$

$= 48 \text{ cm} + 36 \text{ cm} = 84 \text{ cm}$ ●

More Practice   Find the perimeter and area for each parallelogram shown.

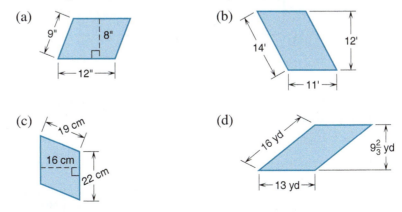

(a)   9"   8"   12"

(b)   14'   12'   11'

(c)   19 cm   16 cm   22 cm

(d)   16 yd $9\frac{2}{3}$ yd   13 yd

(e) A parallelogram-shaped opening has a base of 2 ft 9 in. and a height of 3 ft 3 in. Find its area. Express your answer in square feet rounded to two decimal places.

Answers   (a) $P = 42$ in.; $A = 96$ in.^{2} (b) $P = 50$ ft; $A = 132$ ft^2

(c) $P = 82$ cm; $A = 352$ cm^2 (d) $P = 58$ yd; $A = 125\frac{2}{3}$ yd^2

(e) $A \approx 8.94$ ft^2

2.75 ⊗ **3.25** ⊜ →   *8.9375* ≈ 8.94

Area of a Parallelogram

You may be interested in where the formula for the area of a parallelogram came from. If so, follow this explanation.

Here is a typical parallelogram. As you can see, opposite sides $\overline{AB}$ and $\overline{DC}$ are parallel, and opposite sides $\overline{AD}$ and $\overline{BC}$ are also parallel.

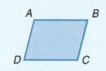

Now let's cut off a small triangle from one side of the parallelogram by drawing the perpendicular line AE.

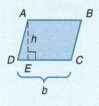

Next, reattach the triangular section to the other side of the figure, forming a rectangle. Notice that $EF = DC = b$.

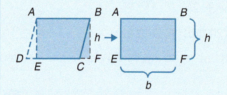

The area of the rectangle $ABFE$ is $A = bh$. This is exactly the same as the area of the original parallelogram $ABCD$.

Area and Perimeter of Trapezoids   A trapezoid is a four-sided figure with only one pair of sides parallel. Here are a few examples of trapezoids:

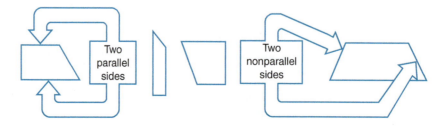

To find the **area** of a trapezoid, use the following formula:

Area of a Trapezoid

$$A = \left(\frac{b_1 + b_2}{2}\right)h \qquad \text{or} \qquad A = \frac{h}{2}(b_1 + b_2)$$

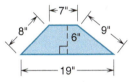

The factor $\left(\dfrac{b_1 + b_2}{2}\right)$ is the average length of the two parallel sides b_1 and b_2. The height h is the perpendicular distance between the two parallel sides.

The **perimeter** of a trapezoid is simply the sum of the lengths of all four sides.

EXAMPLE 10   Find the area of this trapezoid-shaped metal plate.

The parallel sides, b_1 and b_2, have lengths 7 in. and 19 in. The height h is 6 in.

First, write the formula.

$$A = \left(\frac{b_1 + b_2}{2}\right)h$$

Next, substitute the given values.

$$= \left(\frac{7 \text{ in.} + 19 \text{ in.}}{2}\right)(6 \text{ in.})$$

Then, evaluate the quantity $\dfrac{7 + 19}{2}$.

$$= \left(\frac{26}{2} \text{ in.}\right)(6 \text{ in.})$$

$$= (13 \text{ in.})(6 \text{ in.})$$

Finally, multiply 13×6.

$$= 78 \text{ in.}^2$$

 7 ⊞ 1 9 ⊜ ⊟ 2 ⊠ 6 ⊜ → [*78.*]

The perimeter of the trapezoid is $P = 8 \text{ in.} + 7 \text{ in.} + 9 \text{ in.} + 19 \text{ in.} = 43 \text{ in.}$ ●

Your Turn   Find the area and perimeter of each of the following trapezoids.

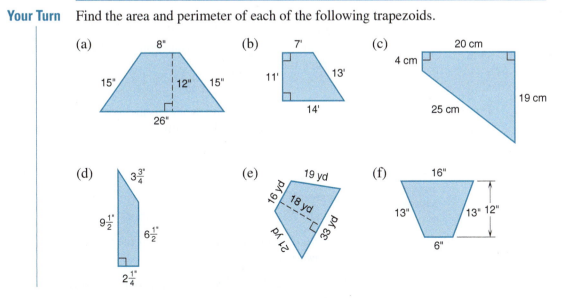

Solutions   (a) $A = \left(\dfrac{8 \text{ in.} + 26 \text{ in.}}{2}\right)(12 \text{ in.})$ $P = 8 \text{ in.} + 15 \text{ in.} + 26 \text{ in.} + 15 \text{ in.}$

  $ = (17 \text{ in.})(12 \text{ in.})$ $ = 64 \text{ in.}$

  $ = 204 \text{ in.}^2$

(b) $A = \left(\dfrac{7 \text{ ft} + 14 \text{ ft}}{2}\right)(11 \text{ ft})$ $P = 11 \text{ ft} + 7 \text{ ft} + 13 \text{ ft} + 14 \text{ ft}$

  $ = (10.5 \text{ ft})(11 \text{ ft})$ $ = 45 \text{ ft}$

  $ = 115.5 \text{ ft}^2$

(c) $A = \left(\dfrac{4 \text{ cm} + 19 \text{ cm}}{2}\right)(20 \text{ cm})$ $P = 4 \text{ cm} + 20 \text{ cm} + 19 \text{ cm} + 25 \text{ cm}$

  $ = (11.5 \text{ cm})(20 \text{ cm})$ $ = 68 \text{ cm}$

  $ = 230 \text{ cm}^2$

(d) $A = \left(\dfrac{9\frac{1}{2} \text{ in.} + 6\frac{1}{2} \text{ in.}}{2}\right)(2\frac{1}{4} \text{ in.})$ $P = 3\frac{3}{4} \text{ in.} + 6\frac{1}{2} \text{ in.} + 2\frac{1}{4} \text{ in.} + 9\frac{1}{2} \text{ in.}$

  $ = (8 \text{ in.})(2\frac{1}{4} \text{ in.})$ $ = 22 \text{ in.}$

  $ = (8 \text{ in.})(2.25 \text{ in.})$

  $ = 18 \text{ in.}^2$

(e) $A = \left(\dfrac{16 \text{ yd} + 33 \text{ yd}}{2}\right)(18 \text{ yd})$ $P = 16 \text{ yd} + 19 \text{ yd} + 33 \text{ yd} + 21 \text{ yd}$

  $ = (24.5 \text{ yd})(18 \text{ yd})$ $ = 89 \text{ yd}$

  $ = 441 \text{ yd}^2$

(f) $A = \left(\dfrac{16 \text{ in.} + 6 \text{ in.}}{2}\right)(12 \text{ in.})$ $P = 13 \text{ in.} + 16 \text{ in.} + 13 \text{ in.} + 6 \text{ in.}$

  $ = (11 \text{ in.})(12 \text{ in.})$ $ = 48 \text{ in.}$

  $ = 132 \text{ in.}^2$

Note   In a real job situation, you will usually need to measure the height of the figure. Remember, the bases of a trapezoid are the parallel sides and h is the distance between them. ●

Practical Applications   When solving application problems involving area, you will sometimes need to convert units. You may find it useful to review these conversion factors in the tables on pages 308, 323, and 335.

EXAMPLE 11   **Construction**   What is the cost of fencing a rectangular playground measuring 45 yd by 58 yd at a cost of $16 per foot?

Fencing is installed along the *sides* of the playground. Therefore, we must **first** find its *perimeter*:

$$P = 2L + 2W = 2(58 \text{ yd}) + 2(45 \text{ yd}) = 116 \text{ yd} + 90 \text{ yd} = 206 \text{ yd}$$

Next, note that the cost is quoted per *foot* and the perimeter is in *yards*. We must convert one or the other so that our units agree. Converting the perimeter to feet, we have:

$$206 \text{ yd} \times \frac{3 \text{ ft}}{1 \text{ yd}} = 618 \text{ ft}$$

Finally, each linear foot costs $16, so we multiply the perimeter by $16 to find the total cost:

$$\frac{\$16}{1 \text{ ft}} \times 618 \text{ ft} = \$9888$$ ●

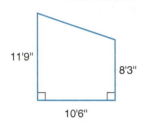

EXAMPLE 12   **Construction**   The cathedral ceiling in a family room creates two opposite facing walls each with the shape and dimensions shown in the figure. The homeowner wishes to have them surfaced with textured drywall at a cost of $0.59 per square foot. What will be the total cost for the two walls?

Two clues indicate that we must find the *area* of the walls: First, the textured drywall covers the entire *surface* of the walls; second, cost is quoted per *square foot*, and square feet are units of area. Notice that the walls are in the shape of trapezoids, with parallel bases of 11'9" and 8'3" and a height of 10'6".

First, convert each dimension to a decimal number:

$$11'9'' = 11\frac{9}{12}' = 11.75'$$

$$8'3'' = 8\frac{3}{12}' = 8.25'$$

$$10'6'' = 10\frac{6}{12}' = 10.5'$$

Next, substitute these into the area formula for trapezoids:

$$A = \left(\frac{b_1 + b_2}{2}\right)h = \left(\frac{11.75 \text{ ft} + 8.25 \text{ ft}}{2}\right)10.5 \text{ ft} = (10 \text{ ft})(10.5 \text{ ft}) = 105 \text{ sq ft}$$

Now multiply by 2 to find the area of both walls:

$$2(105 \text{ sq ft}) = 210 \text{ sq ft}$$

Each square foot costs $0.59, so we multiply the area by $0.59 to find the total cost:

$$\frac{\$0.59}{1 \text{ sq ft}} \times 210 \text{ sq ft} = \$123.90$$

Using a calculator, we have:

 `( 11.75 + 8.25 ) ÷ 2 × 10.5 × 2 × .59 = →` $\boxed{123.9}$ ●

A Closer Look   A clever way to shorten the calculation in this example is to notice that, instead of multiplying the area of one trapezoid by 2, you can simply not divide by 2 to calculate the area of both trapezoids. ●

EXAMPLE 13   **HVAC**   A square hot-air duct must have the same area as two rectangular ones with dimensions 5 in. × 8 in. and 4 in. × 6 in. How long are the sides of the square duct?

To solve this problem, **first** find the sum of the areas of the two rectangular ducts. Using the formula $A = LW$, we have:

Total area $= (5 \text{ in.} \times 8 \text{ in.}) + (4 \text{ in.} \times 6 \text{ in.}) = 40 \text{ in.}^2 + 24 \text{ in.}^2 = 64 \text{ in.}^2$

Then, use the formula $s = \sqrt{A}$ to find the side length of the square duct that has this total area:

$$s = \sqrt{64 \text{ in.}^2} = 8 \text{ in.}$$

●

Your Turn

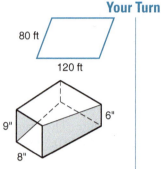

(a) **Construction**   To fence the parallelogram-shaped lot shown in the figure, fence posts must be installed every 8 feet along the border of the lot. How many fence posts are needed? (Make sure there is a post at each corner of the lot.)

(b) **Landscaping**   A homeowner wishes to cover a driveway with decomposed granite at a cost of $65 per square yard. If the driveway is a square measuring 24 feet 6 in. on a side, what is the cost of the job rounded to the nearest dollar?

(c) **Sheet Metal Trades**   A sheet metal worker must build a heating duct in which a rectangular vent 6 in. wide must have the same area as a rectangular opening 8 in. by 9 in. Find the length of the vent. (See the figure.)

(d) **HVAC**   In designing a heating system for a house, an HVAC technician must first calculate the area of each type of interior surface that borders the exterior of the house. This will allow him to calculate the total heat loss of the house. Suppose a certain house is rectangular in shape with outside dimensions of 64 ft 6 in. by 44 ft 6 in. and a 9-ft ceiling. There are 596 sq ft of windows, sliding glass doors, and wooden doors. The rest of the interior wall surface is composed of standard drywall.

   (1) Find the area of the interior drywall surfaces that border the exterior of the house.

   (2) Find the area of the ceiling.

Solutions

(a) **First,** note that each side length, 120 ft and 80 ft, is divisible by 8. We can place a post at each corner, space them every 8 feet, and not need any extra posts. Find the perimeter of the lot:

$$P = 2a + 2b = 2(120 \text{ ft}) + 2(80 \text{ ft}) = 240 \text{ ft} + 160 \text{ ft} = 400 \text{ ft}$$

Then, divide this by 8 to determine how many 8-ft sections are in 400 ft:

$$400 \text{ ft} \div 8 \text{ ft/post} = 50 \text{ fence posts}$$

(b) The decomposed granite covers the *surface* of the driveway, so we must find the area. Converting 24 ft 6 in. to 24.5 ft, we have:

$$A = s^2 = (24.5 \text{ ft})^2 = 600.25 \text{ ft}^2$$

Because the cost of the granite is quoted per square *yard*, we must convert our area to square yards so that our units agree. Recall that $1 \text{ yd}^2 = 9 \text{ ft}^2$, so we have:

$$600.25 \text{ ft}^2 \times \frac{1 \text{ yd}^2}{9 \text{ ft}^2} = 66.694 \ldots \text{ yd}^2 \approx 66.69 \text{ yd}^2$$

Now multiply by $65 per yd^2 to find the total cost of the job:

$$66.69 \text{ yd}^2 \times \frac{\$65}{1 \text{ yd}^2} = \$4334.85 \approx \$4335$$

Using a calculator, we can perform the entire calculation as follows:

$$2\,4\,.\,5 \;\boxed{x^2}\; \boxed{\div}\; 9 \;\boxed{\times}\; 6\,5\; \boxed{=} \;\rightarrow\; \boxed{4335.138889} \text{ or } \$4335, \text{ rounded}$$

(c) The area of the opening is $A = LW$

$$= (8 \text{ in.})(9 \text{ in.}) = 72 \text{ in.}^2$$

The length of the vent is $\quad L = \dfrac{A}{W}$

$$= \frac{72 \text{ in.}^2}{6 \text{ in.}} = \frac{72}{6} \frac{\text{in.} \times \text{in.}}{\text{in.}}$$

$$= 12 \text{ in.}$$

(d) (1) To solve this problem, we must first calculate the area of the outside walls and then subtract the area of the windows and doors.

The height of each rectangular wall is 9 ft—the same as the height of the ceiling. We will call this the width W of the walls. There are two walls with bases or lengths of 64 ft 6 in., or 64.5 ft. Their total area A_1 is

$$A_1 = 2LW = 2(64.5 \text{ ft})(9 \text{ ft}) = 1161 \text{ sq ft}$$

There are two walls with bases of 44 ft 6 in., or 44.5 ft. Their total area A_2 is

$$A_2 = 2LW = 2(44.5 \text{ ft})(9 \text{ ft}) = 801 \text{ sq ft}$$

The area A of all four outside walls is

$$A = A_1 + A_2 = 1161 \text{ sq ft} + 801 \text{ sq ft} = 1962 \text{ sq ft}$$

Finally, subtract the area of the windows and doors to find the area of the interior drywall surfaces:

$$1962 \text{ sq ft} - 596 \text{ sq ft} = 1366 \text{ sq ft of interior drywall surfaces}$$

(2) The dimensions of the ceiling are the same as the outside dimensions of the house. The area A is

$$A = LW = (64.5 \text{ ft})(44.5 \text{ ft}) = 2870.25 \text{ sq ft}$$

Turn to Exercises 8-2 for practice in finding the perimeter of polygons and the area of quadrilaterals.

| Exercises 8-2 | **Perimeter of Polygons and Area of Quadrilaterals** |

A. Find the perimeter of each polygon.

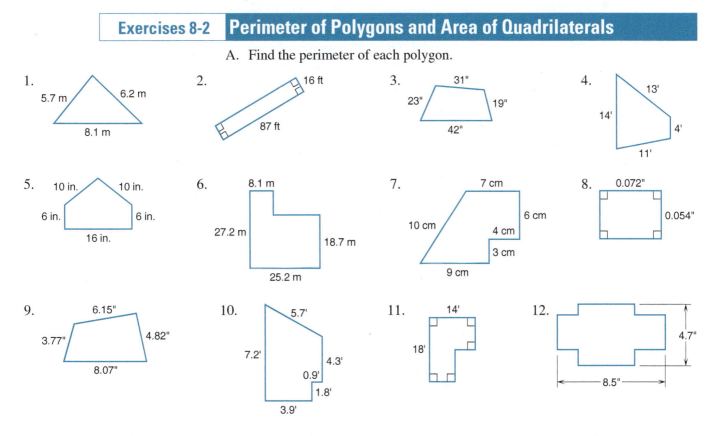

B. Find the perimeter and area of each figure. (Round to the nearest tenth if necessary. Assume right angles and parallel sides except where obviously otherwise.)

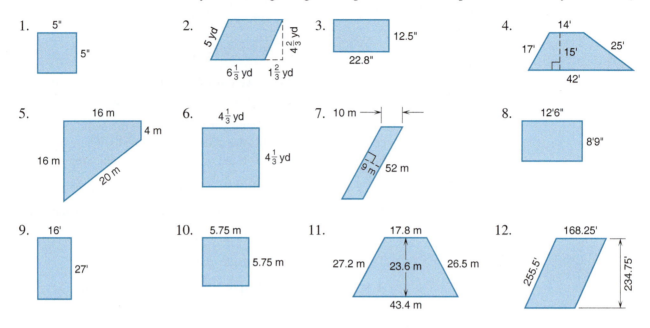

1. 5" 5"

2. 5 yd $4\frac{2}{3}$ yd $6\frac{1}{3}$ yd $1\frac{2}{3}$ yd

3. 12.5"   22.8"

4. 14'   17'   15'   25'   42'

5. 16 m   4 m   16 m   20 m

6. $4\frac{1}{3}$ yd $4\frac{1}{3}$ yd

7. 10 m   9 m   52 m

8. 12'6"   8'9"

9. 16'   27'

10. 5.75 m   5.75 m

11. 17.8 m   27.2 m   23.6 m   26.5 m   43.4 m

12. 168.25'   255.5'   234.75'

C. Practical Applications

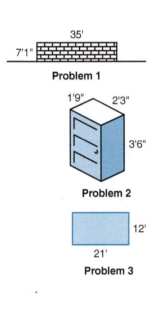

35'
7'1"
Problem 1

1'9"   2'3"   3'6"
Problem 2

12'
21'
Problem 3

1. **Masonry**   How many bricks will it take to build the wall shown if each brick measures $4\frac{1}{4}$ in. × $8\frac{3}{4}$ in. including mortar? (*Hint:* Change the wall dimensions to inches.)

2. **Carpentry**   How many square feet of wood are needed to build the cabinet shown in the figure? (Assume that wood is needed for all six surfaces.)

3. **Flooring and Carpeting**   The floor plan of a room is shown in the figure. If the entire room is to be carpeted, how many square *yards* of carpet are needed? (*Hint:* Change each dimension to yards first.) At $39.95 per square yard, how much will the carpet cost?

4. **Construction**   The amount of horizontal wood siding (in square feet) needed to cover a given wall area can be calculated from the formula

   Area of siding (sq ft) = K × total area to be covered (sq ft)

   where K is a constant that depends on the type and size of the siding used. How much 1-by-6 rustic shiplapped drop siding is needed to cover a rectangular wall 60 ft by 14 ft? Use $K = 2.19$ and round to the nearest ten square feet.

5. **Flooring and Carpeting**   How many bundles of strip flooring are needed to cover a rectangular floor area 16 ft by 12 ft 6 in. in size if a bundle of strip flooring contains 24 sq ft, and if 30% extra must be allowed for side and end matching?

6. **Construction**   A large rectangular window opening measures $72\frac{5}{8}$ in. by $60\frac{3}{8}$ in. Calculate the area of this opening. (Round to the nearest square inch.)

7. **Masonry**   Find the area of one stretcher course of 20 concrete blocks if the blocks are $7\frac{5}{8}$ in. long and 4 in. high and if the mortar joints are $\frac{3}{8}$ in. thick. (*Hint:* Twenty blocks require 19 mortar joints.)

8. **Masonry**   To determine the approximate number of concrete blocks needed to construct a wall, the following formula is often used:

Number of blocks per course $= N \times$ area of one course (sq ft)

where N is a number that depends on the size of the block. How many 8-in. by 8-in. by 16-in. concrete blocks are needed to construct a foundation wall 8 in. thick with a total length of 120 ft if it is laid five courses high? Use $N = \frac{9}{8}$ for this block size. (*Hint:* The blocks are set lengthwise. Translate all dimensions to inches.)

9. **Automotive Trades**   A roll of gasket material is 46.5 cm wide. What length is needed to obtain 1.20 sq m of the material? (*Careful:* The numbers are not expressed in compatible units. Round to the nearest centimeter.)

10. **Welding**   A type of steel sheet weighs 3.125 lb per square foot. What is the weight of a piece measuring 24 in. by 108 in.? (*Hint:* Be careful of units.)

11. **Printing**   A rule of thumb in printing says that the area of the typed page should be half the area of the paper page. If the paper page is 5 in. by 8 in., what should the length of the typed page be if the width is 3 in.?

12. **Printing**   A ream of 17-in. by 22-in. paper weighs 16.0 lb. Find the weight of a ream of 19-in. by 24-in. sheets of the same density of paper.

13. **Printing**   A sheet of 25-in. by 38-in. paper must be cut into 6-in. by 8-in. cards. The cuts can be made in either direction, but they must be consistent. Which plan will result in the least amount of waste, cutting the 6-in. side of the card along the 25-in. side of the sheet or along the 38-in. side of the sheet? State the amount of waste created by each possibility. (*Hint:* Make a cutting diagram for each possibility.)

14. **HVAC**   To determine the size of a heating system needed for a home, an installer needs to calculate the heat loss through surfaces such as walls, windows, and ceilings exposed to the outside temperatures. The formula

$$L = kDA$$

gives the heat loss L (in Btu per hour) if D is the temperature difference between the inside and outside, A is the area of the surface in sq ft, and k is the insulation rating of the surface. Find the heat loss per hour for a 7-ft by 16-ft rectangular single-pane glass window ($k = 1.13$) if the outside temperature is 40° and the desired inside temperature is 68°. (Round to the nearest hundred.)

15. **Construction**   A playground basketball court 94 ft long and 46 ft wide is to be resurfaced at a cost of $3.25 per sq ft. What will the resurfacing cost?

16. **Landscaping**   At $6.75 per square foot, plus a $60 delivery charge, how much will it cost to purchase artificial turf for a square lawn 24 ft 6 in. on a side? (Round to the nearest cent.)

17. **Agriculture**   Fertilizer must be applied to the planted area shown in the figure. If each bag covers 300 sq ft, how many bags are needed? (Be careful of units!)

18. **Agriculture**   In problem 17, the plot of land must also be fenced. At $24 per linear foot, how much will it cost to install fencing?

19. **Agriculture**   The parallelogram-shaped plot of land shown in the figure is put up for sale at $2500 per acre. To the nearest dollar, what is the total price of the land? (*Hint:* 1 acre $= 43,560$ sq ft.)

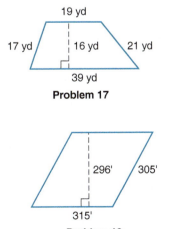

19 yd

17 yd   16 yd   21 yd

39 yd

Problem 17

296'   305'

315'

Problem 19

20. **Construction**   A rectangular structure 56 ft long and 38 ft wide is being built. The walls require studs 16 in. o.c., that is, 16 in. is the center-to-center distance separating the studs. Eight additional studs are needed for starters and corners. Determine the total number of studs required.

21. **Construction**   A contractor needed a small workshop. He found a preengineered steel building advertised for $12,619. If the building is a 36-ft by 36-ft square, what is the cost of the building per square foot? (Round to the nearest cent.)

22. **Flooring and Carpeting**   When expressed in U.S. customary units, imported area rugs seem to come in odd sizes because they are converted from metric measurements to feet and inches. One such area rug measured 5 ft 3 in. by 7 ft 9 in. and cost $399.

    (a) What is the area of this rug in square feet? (*Hint:* Convert the inch measurements to fractions of a foot before calculating the area.)

    (b) What is the cost of the rug per square foot?

23. **Flooring and Carpeting**   Because carpeting typically comes in 12-ft widths, the carpet for a 22-ft by 16-ft room must be laid out as shown in the diagram.

    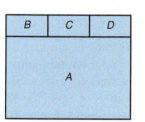

    Problem 23

    The four sub-regions shown have the following dimensions:

    A 12 ft × $22\frac{1}{4}$ ft B 4 ft × $7\frac{1}{2}$ ft

    C 4 ft × $7\frac{1}{2}$ ft D 4 ft × $7\frac{1}{2}$ ft

    (*Note:* An extra $\frac{1}{4}$ ft has been added to each length to give the installer a margin of error.)

    (a) Determine the total area of carpet needed for this room.

    (b) At $5.65 per square foot installed, how much will it cost to carpet the room?

24. **Electrical Trades**   At a solar energy plant being built in the Mojave Desert, power is generated using 173,500 heliostats. Each heliostat consists of two rectangular mirrors, and each mirror measures 7.2 ft by 10.5 ft. How many total square feet of mirrors are there at the plant?

25. **Carpentry**   A certain 3-ft by 3-ft rubber shop-floor mat costs $44. A 3-ft by 4-ft foam mat costs $63. Which of these mats has the lower cost per square foot, and how much lower is it?

26. **Welding**   A tensile specimen had an original rectangular cross section measuring 0.615 in. by 0.188 in. After a pull, the cross section measured 0.506 in. by 0.094 in. What was the percent reduction in area? (Assume a rectangular cross section and round to the nearest tenth.)

27. **Carpentry**   A room is 15'6" wide, 19'9" long, and has an 8-ft ceiling. Doors and window openings total 86 sq ft. Find the amount of wood paneling needed for all four walls. Add 15% for waste, and be sure to deduct the area of the openings.

28. **Construction**   A certain grade of sheet glass weighs 11.9 kilograms per square meter. How many *pounds* would 8 sheets weigh if they measure 50.8 cm by 91.4 cm each? (Be careful of units and round to the nearest tenth.)

29. **Agriculture**   Tomatoes are being planted in a 12-ft by 6-ft raised rectangular bed. They must be planted 24-in. apart with 36 in. between rows.

What is the maximum number of tomato plants that can be planted in this bed? (Assume that you can plant along all four edges of the bed.)

30. **Flooring and Carpeting** How many 9-inch square tiles are needed to cover 64 square *feet* of floor area if we add 10% for waste?

When you have finished these problems, check your answers to the odd-numbered problems in the Appendix, then continue in Section 8-3 with the study of other polygons.

8-3 Triangles, Regular Hexagons, and Irregular Polygons

learning catalytics™

1. How many sides does a hexagon have?
2. Which of these appears to be a right triangle?

(a) (b)

(c)

(d)

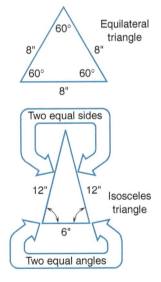

A **triangle** is a polygon having three sides and therefore three angles.

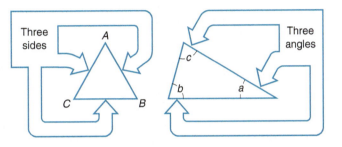

Just as there are several varieties of four-sided figures—squares, rectangles, parallelograms, and trapezoids—there are also several varieties of triangles. Fortunately, one area formula can be used with all triangles. First, you must learn to identify the many kinds of triangles that appear in practical work.

Triangles may be classified according to their sides.

An **equilateral triangle** is one in which all three sides have the same length. An equilateral triangle is also said to be **equiangular** because, if the three sides are equal, the three angles will also be equal. In fact, because the interior angles of any triangle always add up to 180°, each angle of an equilateral triangle must equal 60°.

An **isosceles triangle** is one in which at least two of the three sides are equal. It is always true that the two angles opposite the equal sides are also equal.

A **scalene triangle** is one in which no sides are equal.

Triangles may be classified also according to their angles. An **acute triangle** is one in which all three angles are acute (less than 90°). An **obtuse triangle** contains one obtuse angle (greater than 90°).

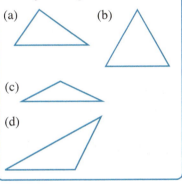

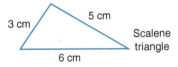

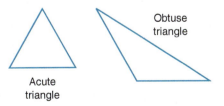

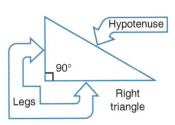

A **right triangle** contains a 90° angle. Two of the sides, called the **legs**, are perpendicular to each other. The longest side of a right triangle is always the side opposite to the right angle. This side is called the **hypotenuse** of the triangle. An isosceles or scalene triangle can be a right triangle, but an equilateral triangle can never be a right triangle.

Your Turn   Identify the following triangles as being equilateral, isosceles, or scalene. Also name the ones that are right triangles.

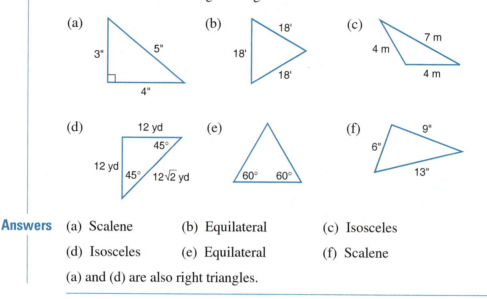

(a)   (b)   (c)

(d)   (e)   (f)

Answers   (a) Scalene   (b) Equilateral   (c) Isosceles

(d) Isosceles   (e) Equilateral   (f) Scalene

(a) and (d) are also right triangles.

A Closer Look   Did you have trouble with (d) or (e)? The answer to each depends on the fact that the interior angles of a triangle add up to 180°. In (d) the two given angles add up to 90°. Therefore, the missing angle must be 90° if the three angles sum to 180°. This makes (d) a right triangle.

In (e) the two angles shown sum to 120°, so the third angle must be 180° − 120° or 60°. Because the three angles are equal, we know the triangle is an equiangular triangle, and this means it is also equilateral. ●

Pythagorean Theorem   The **Pythagorean theorem** is a rule or formula that allows us to calculate the length of one side of a right triangle when we are given the lengths of the other two sides. Although the formula is named after the ancient Greek mathematician Pythagoras, it was known to Babylonian engineers and surveyors more than a thousand years before Pythagoras lived.

> ## Pythagorean Theorem
>
> For any right triangle, the square of the hypotenuse is equal to the sum of the squares of the other two sides.
>
> $c^2 = a^2 + b^2$
>
>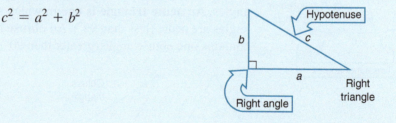

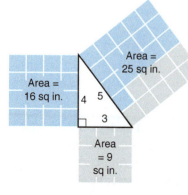

Geometrically, this means that if the squares are built on the sides of the triangle, the area of the larger square is equal to the sum of the areas of the two smaller squares. For the right triangle shown

$$3^2 + 4^2 = 5^2$$

$$9 + 16 = 25$$

Notice that it does not matter whether leg a is 3 and leg b is 4, or vice versa. However, the longest side, opposite the right angle, must be the hypotenuse c.

EXAMPLE 1 Let's use this formula to find the distance d between points A and B for this rectangular plot of land.

The Pythagorean theorem tells us that for the right triangle ABC,

$$d^2 = 36^2 + 20^2$$

$$= 1296 + 400$$

$$= 1696$$

Taking the square root of both sides of the equation, we have

$$d \approx 41 \text{ ft} \quad \text{rounded to the nearest foot}$$

Here are two different ways to find d using a calculator:

$\sqrt{}$ **3 6** $\boxed{x^2}$ $\boxed{+}$ **2 0** $\boxed{x^2}$ $\boxed{=}$ $\rightarrow$ 41.18252056

Or **3 6** $\boxed{x^2}$ $\boxed{+}$ **2 0** $\boxed{x^2}$ $\boxed{=}$ $\sqrt{}$ $\boxed{\text{ANS}}$ $\boxed{=}$ $\rightarrow$ 41.18252056

With the first option, if your calculator does not automatically open parentheses with square root, then you must press $\boxed{(}$ after $\sqrt{}$. ●

The steps we used to calculate the hypotenuse in Example 1 can be summarized as follows:

If $c^2 = a^2 + b^2$, then $c = \sqrt{a^2 + b^2}$.

This rule may also be used to find either one of the legs if the hypotenuse and the other leg are known. For example, to solve algebraically for a,

First, write the original formula: $\qquad c^2 = a^2 + b^2$

Next, subtract b^2 from both sides: $\qquad c^2 - b^2 = a^2$

Finally, take the square root of both sides: $\sqrt{c^2 - b^2} = a \quad \text{or} \quad a = \sqrt{c^2 - b^2}$

Similarly, if we solve for b, we obtain: $\qquad b = \sqrt{c^2 - a^2}$

! Careful The square root of a sum or difference is not equal to the sum or difference of the square roots. For example, $\sqrt{9 + 16} = \sqrt{25} = 5$, and this is not the same as $\sqrt{9} + \sqrt{16}$, which equals 3 + 4, or 7. Similarly, $\sqrt{100 - 64} = \sqrt{36} = 6$, and this is not the same as $\sqrt{100} - \sqrt{64}$, which equals 10 − 8, or 2. The algebraic expressions $\sqrt{a^2 + b^2}$ and $\sqrt{c^2 - b^2}$ cannot be simplified further. ●

These results are summarized in the following box.

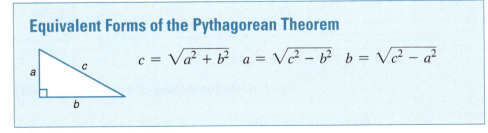

Equivalent Forms of the Pythagorean Theorem

$$c = \sqrt{a^2 + b^2} \quad a = \sqrt{c^2 - b^2} \quad b = \sqrt{c^2 - a^2}$$

! Careful These three formulas are valid only for a *right* triangle—a triangle with a right or 90° angle. ●

EXAMPLE 2   (a) In Figure (a), the hypotenuse is unknown. Substitute into $c = \sqrt{a^2 + b^2}$ to get:

$$x = \sqrt{8^2 + 15^2} = \sqrt{64 + 225} = \sqrt{289} = 17$$

Figure (a)   Figure (b)

(b) In Figure (b), one of the legs is unknown. Substitute into $a = \sqrt{c^2 - b^2}$ to get:

$$y = \sqrt{(22.3)^2 - (13.8)^2} = \sqrt{497.29 - 190.44} = \sqrt{306.85} = 17.517\ldots$$

$$\approx 17.5, \text{ rounded}$$

Using a calculator, we have:

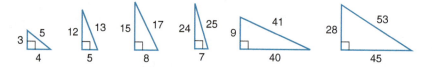

For some right triangles, the three side lengths are all whole numbers rather than fractions or decimals. Such special triangles have been used in technical work since Egyptian surveyors used them 2000 years ago to lay out rectangular fields for farming.

Here are a few "Pythagorean triple" right triangles:

It will be helpful to know that any "multiple" of a Pythagorean triple is also a Pythagorean triple. For example, if we double all sides of a 3-4-5 right triangle, we get a 6-8-10 right triangle, and this too is a Pythagorean triple. Similarly, the side lengths 15-36-39 form a right triangle because each side is three times the length of the 5-12-13 triple.

More Practice Use the Pythagorean theorem to find the missing side in each of the following right triangles. (If your answer does not come out exactly, round to the nearest tenth.)

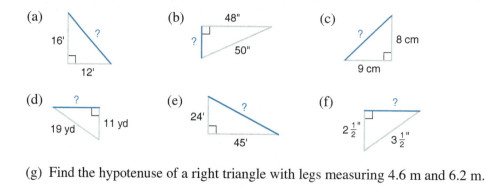

(g) Find the hypotenuse of a right triangle with legs measuring 4.6 m and 6.2 m.

Answers   (a) 20 ft   (b) 14 in.   (c) 12.0 cm

(d) 15.5 yd   (e) 51 ft   (f) 2.4 in.

(g) 7.7 m

🔍 **A Closer Look** The missing side lengths in problems (a), (b), and (e) can be found without using the Pythagorean theorem if you recognize that they are multiples of three of the Pythagorean triples.

In (a), notice that 12 is 4 · 3 and 16 is 4 · 4, so this right triangle must be a multiple of the 3-4-5 triple. The hypotenuse is 4 · 5, or 20.

In (b), 48 is 2 · 24 and 50 is 2 · 25, so this right triangle must be a multiple of the 7-24-25 triple. The missing leg is 2 · 7, or 14.

Finally, in (e), 24 = 3 · 8 and 45 = 3 · 15, so this right triangle must be a multiple of the 8-15-17 triple. The hypotenuse is 3 · 17, or 51. ●

Applications of the Pythagorean Theorem

EXAMPLE 3 **Masonry** A mason wants to use a 12-ft ladder to reach the top of a 10-ft wall. How far must the base of the ladder be from the base of the wall?

The ladder, the ground, and the wall form a right triangle with hypotenuse equal to 12 ft and one leg equal to 10 ft.

Use the formula $a = \sqrt{c^2 - b^2}$ to get

$$x = \sqrt{12^2 - 10^2}$$
$$= \sqrt{144 - 100}$$
$$= \sqrt{44}$$

≈ 6.63 ft or 6 ft 8 in., rounded to the nearest inch

If you need help with square roots, pause here and return to page 406 for a review. ●

Your Turn (a) **Carpentry** A rectangular table measures 4.0 ft by 6.0 ft. Find the length of a diagonal brace placed beneath the table top. (Give your answer in feet and inches rounded to the nearest half-inch.)

(b) **Manufacturing** What is the distance between the centers of two pulleys if one is placed 9 in. to the left and 6 in. above the other? (See the figure. Round to the nearest tenth.)

(c) **Machine Trades** Find the length of the missing dimension x in the part shown in the figure. (Round to the nearest tenth.)

(d) **Carpentry** Find the diagonal length of the stairway shown in the drawing. (Give your final answer in feet and inches rounded to the nearest inch.)

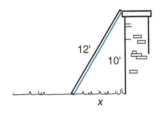

6"
9"
Problem (b)

22 cm
18 cm
x
8 cm
22 cm
Problem (c)

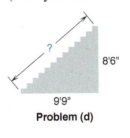

?
8'6"
9'9"
Problem (d)

Answers (a) 7 ft 2½ in.   (b) 10.8 in.   (c) 21.4 cm   (d) 12 ft 11 in.

(d) In feet: ☑ **8.5** $\boxed{x^2}$ ➕ **9.75** $\boxed{x^2}$ 🟰 → 　*12.93493332*　 ≈ 12.9 ft

Converting the decimal part to inches:

➖ **12** 🟰 ✖ **12** 🟰 → 　*11.21919984*　 ≈ 11 in.

🔍 **A Closer Look**   In problem (c), the first difficulty is to locate the correct triangle.

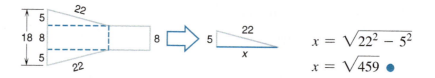

$$x = \sqrt{22^2 - 5^2}$$

$$x = \sqrt{459} \quad \bullet$$

Area of a Triangle   The area of any triangle, no matter what its shape or size, can be found using the same simple formula.

> ## Area of a Triangle
>
> Area, $A = \frac{1}{2}bh$ or $A = \dfrac{bh}{2}$
>
> b = base h = height

Any side of a triangle can be used as the **base**. For a given base, the **height** is the perpendicular distance from the opposite vertex to the base. The following three identical triangles show the three different base–height combinations.

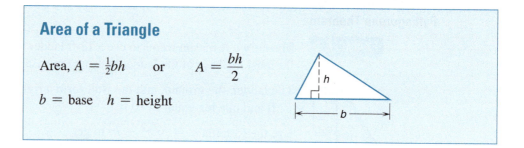

Notice that for the last two cases, the height falls outside the triangle, and the base must be extended to meet the height at a right angle. For any such triangle, the three area calculations are equal:

$$A = \tfrac{1}{2}b_1h_1 = \tfrac{1}{2}b_2h_2 = \tfrac{1}{2}b_3h_3$$

EXAMPLE 4   (a) In triangle PQR, the base b is 13 in., and the height h is 8 in. Applying the formula, we have

Area, $A = \tfrac{1}{2}bh$

$\qquad = \tfrac{1}{2}(13 \text{ in.})(8 \text{ in.})$

$\qquad = 52 \text{ in.}^2$

(b) In triangle STU,

$b = 25$ cm

$h = 16$ cm

Area, $A = \tfrac{1}{2}(25 \text{ cm})(16 \text{ cm})$

$\qquad = 200 \text{ cm}^2$

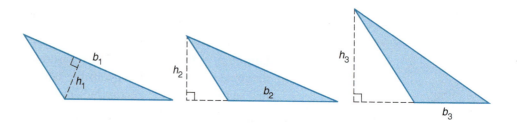

(c) And in triangle *VWX*,

$b = 20$ cm

$h = 12$ cm

Area, $A = \frac{1}{2}(20\text{ cm})(12\text{ cm})$

$= 120 \text{ cm}^2$

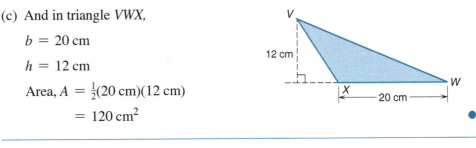

Your Turn Find the area of triangle *ABC*.

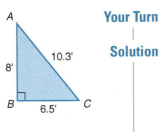

Solution Base, $b = 6.5$ ft

Height, $h = 8$ ft

Area, $A = \frac{1}{2}bh$

$= \frac{1}{2}(6.5\text{ ft})(8\text{ ft})$

$= 26 \text{ ft}^2$

Notice that for a right triangle, the two sides meeting at the right angle can be used as the base and height.

Area of a Triangle

The area of any triangle is equal to one-half of its base times its height, $A = \frac{1}{2}bh$.

To learn where this formula comes from, follow this explanation:

First, for any triangle *ABC*,

draw a second identical triangle *DEF*.

Second, flip the second triangle over and attach it to the first,

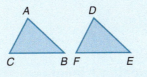

to get a parallelogram.

Finally, if the base and height of the original triangle are *b* and *h*, then these are also the base and height for the parallelogram.

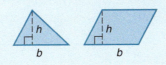

(continued)

The area of the parallelogram is

$A = bh$

so the area of the original triangle must be

$A = \frac{1}{2}bh$

EXAMPLE 5   Suppose we need to find the area of the isosceles triangle shown.

We must first find the height.

If we draw the height perpendicular to the 6-in. base, it will divide the base in half. Then in the triangle shown

$h = \sqrt{9^2 - 3^2}$

$= \sqrt{81 - 9}$

$= \sqrt{72} \approx 8.485$ in.

Now we can find the area:

$A = \frac{1}{2}bh$

$\approx \frac{1}{2}(6 \text{ in.})(8.485 \text{ in.}) \approx 25.455 \text{ in.}^2$

$\approx 25 \text{ in.}^2$ rounded

If you use a calculator, the entire calculation looks like this:

As an alternative to the method shown in Example 5, you may instead use the following formula:

Area of an Isosceles Triangle

Area, $A = \frac{1}{2}b \sqrt{a^2 - \left(\frac{b}{2}\right)^2}$

Your Turn   **Construction**   An attic wall has the shape shown. How many square feet of insulation are needed to cover the wall? (Round to the nearest square foot.)

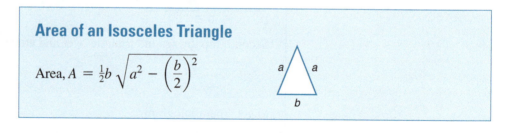

Solution   The triangle is isosceles because two sides are equal. Substituting $a = 7$ ft and $b = 12$ ft into the formula, we have

Area, $A = \frac{1}{2}(12 \text{ ft}) \sqrt{(7 \text{ ft})^2 - \left(\frac{12 \text{ ft}}{2}\right)^2}$

$= 6\sqrt{49 - 36}$

$= 6\sqrt{13} \approx 21.63 \text{ ft}^2$, or 22 ft² rounded

.5 ✕ **1 2** ✕ √ **7** x² − **6** x² = → ⟶ 21.63330765

For an equilateral triangle, an even simpler formula may be used.

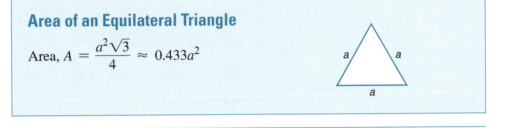

Area of an Equilateral Triangle

Area, $A = \dfrac{a^2\sqrt{3}}{4} \approx 0.433a^2$

Your Turn **Sheet Metal Trades** What area of sheet steel is needed to make a triangular pattern where each side is 3 ft long? Use the given formula and round to one decimal place.

Solution Area, $A \approx 0.433a^2$, where $a = 3$ ft

$\approx 0.433(3 \text{ ft})^2$

$\approx 0.433(9 \text{ ft}^2)$

$\approx 3.9 \text{ ft}^2$ rounded

More Practice In the following problems find the area of each triangle. (Round each answer to the nearest whole number.)

(a) 15" base 18"

(b) 21.0' 24.7' 13.0'

(c) 22.8 cm 32.1 cm 31.5 cm 35.6 cm

(d) $9\frac{1}{3}$ yd $9\frac{1}{3}$ yd $13\frac{1}{5}$ yd

(e) 15 m 17 m 30 m 8 m 18 m

(f) 10' 10' 5'

(g) 16" 16" 22.6"

(h) 22 cm 22 cm 22 cm

Answers

(a) 135 in.²   (b) 137 ft²   (c) 359 cm²   (d) 44 yd²

(e) 135 m²   (f) 24 ft²   (g) 128 in.²   (h) 210 cm²

For (h), use the formula for an equilateral triangle and enter

.4 3 3 ✕ **2 2** x² = → ⟶ 209.572

A Closer Look   In problem (e), you should have used 18 m as the base and *not* 26 m: $b = 18$ m, $h = 15$ m. ●

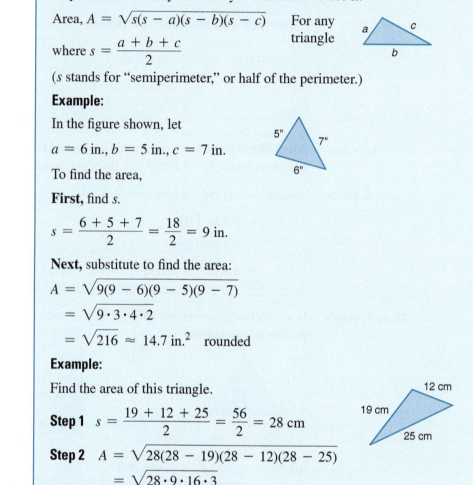

Hero's Formula

It is possible to find the area of any triangle from the lengths of its sides. The formula that enables us to do this was first devised almost 2000 years ago by Hero or Heron, a Greek mathematician. It is a complicated formula, and you may feel like a hero yourself if you learn how to use it.

Area, $A = \sqrt{s(s - a)(s - b)(s - c)}$ For any triangle

where $s = \dfrac{a + b + c}{2}$

(*s* stands for "semiperimeter," or half of the perimeter.)

Example:

In the figure shown, let

$a = 6$ in., $b = 5$ in., $c = 7$ in.

To find the area,

First, find *s*.

$$s = \frac{6 + 5 + 7}{2} = \frac{18}{2} = 9 \text{ in.}$$

Next, substitute to find the area:

$$A = \sqrt{9(9 - 6)(9 - 5)(9 - 7)}$$
$$= \sqrt{9 \cdot 3 \cdot 4 \cdot 2}$$
$$= \sqrt{216} \approx 14.7 \text{ in.}^2 \quad \text{rounded}$$

Example:

Find the area of this triangle.

Step 1 $s = \dfrac{19 + 12 + 25}{2} = \dfrac{56}{2} = 28 \text{ cm}$

Step 2 $A = \sqrt{28(28 - 19)(28 - 12)(28 - 25)}$
$$= \sqrt{28 \cdot 9 \cdot 16 \cdot 3}$$
$$= \sqrt{12{,}096} = 109.98 \ldots$$
$$\approx 110 \text{ cm}^2, \quad \text{rounded}$$

On a calculator, Step 2 looks like this:

$\boxed{\checkmark}\,2\,8\,\boxed{\times}\,\boxed{(}\,2\,8\,\boxed{-}\,1\,9\,\boxed{)}\,\boxed{\times}\,\boxed{(}\,2\,8\,\boxed{-}\,1\,2\,\boxed{)}\,\boxed{\times}\,\boxed{(}\,2\,8\,\boxed{-}\,2\,5\,\boxed{)}\,\boxed{=}$

→   *109.9818167*

Applications of Triangles

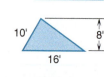

EXAMPLE 6   **Flooring and Carpeting**   At a cost of $3.95 per square foot, what would it cost to tile the floor of a triangular workspace shown?

To solve this problem, **first** calculate the area of the triangle:

$$A = \tfrac{1}{2}bh = \tfrac{1}{2}(16 \text{ ft})(8 \text{ ft}) = 64 \text{ ft}^2$$

Then, because each square foot costs $3.95, multiply this cost by the area to determine the total cost to tile the floor:

$$\text{Cost} = \frac{\$3.95}{1 \text{ ft}^2} \times 64 \text{ ft}^2 = \$252.80$$ ●

Your Turn (a) **Carpentry** A window in the shape of an equilateral triangle is bordered by a wooden frame as shown in the figure. How many square inches of wood are in the frame? (Round to the nearest whole number.)

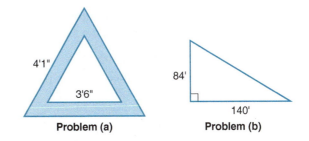

Problem (a)   Problem (b)

(b) **Agriculture** A farmer wishes to fertilize the right-triangular plot of land shown in the figure. If one bag of fertilizer covers 1500 sq ft, how many bags will he need? (Answer must be a whole number.)

Solutions (a) **First,** convert the side lengths to inches:

$3 \text{ ft } 6 \text{ in.} = (3 \cdot 12 + 6) \text{ in.} = 42 \text{ in.} \quad 4 \text{ ft } 1 \text{ in.} = (4 \cdot 12 + 1) \text{ in.} = 49 \text{ in.}$

Then, to find the area of the frame, subtract the inner triangle area A_I (the glass) from the outer triangle area A_O (the glass plus the frame). Using the formula for the area of an equilateral triangle, we have

$$A_O - A_I = 0.433(49 \text{ in.})^2 - 0.433(42 \text{ in.})^2$$

$$= 1039.633 \text{ in.}^2 - 763.812 \text{ in.}^2 = 275.821 \text{ in.}^2$$

Rounded to the nearest whole number, the area of the frame is 276 square inches.

(b) Because this is a right triangle, the base and height are the two perpendicular sides, 140 ft and 84 ft. The area is

$$A = \tfrac{1}{2}bh = \tfrac{1}{2}(140 \text{ ft})(84 \text{ ft}) = 5880 \text{ ft}^2$$

Each bag covers 1500 sq ft, so we must divide this into the area to determine the number of bags needed. Setting up unity fractions helps us see why we divide.

$$5880 \text{ sq ft} \times \frac{1 \text{ bag}}{1500 \text{ sq ft}} = 3.92 \text{ bags}$$

The farmer will need four bags of fertilizer.

Regular Hexagons A polygon is a plane geometric figure with three or more sides. A **regular polygon** is one in which all sides are the same length and all angles are equal. The equilateral triangle and the square are both regular polygons.

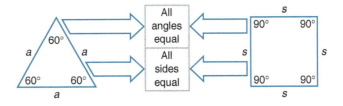

A **pentagon** is a five-sided polygon. A **regular pentagon** is one whose sides are all the same length and whose angles are equal.

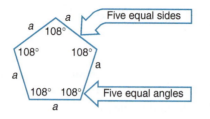

The **regular hexagon** is a six-sided polygon in which each interior angle is 120° and all sides are the same length.

Regular hexagon:
All sides equal.
All angles 120°.

If we draw the diagonals of a regular hexagon, six equilateral triangles are formed:

Because you already know that the area of one equilateral triangle is approximately $A \approx 0.433a^2$, the area of the complete hexagon will be 6 times $0.433a^2$ or $2.598a^2$.

Area of a Regular Hexagon

Area, $A = \dfrac{3a^2\sqrt{3}}{2} \approx 2.598a^2$

Learning Help   There is no need to memorize this formula if you have already memorized the formula for the area of an equilateral triangle. Simply find the area of each equilateral triangle within the regular hexagon and multiply by six. ●

EXAMPLE 7   If each side of a regular hexagon is 6.00 cm, then the area A is: $A \approx 2.598(6.00 \text{ cm})^2$
First, square the number in parentheses. $\approx 2.598(36 \text{ cm}^2)$
Then multiply. $\approx 93.5 \text{ cm}^2$, rounded

 2.598 ✕ **6** x^2 ⊟ →   93.528   ●

Your Turn   **Machine Trades**   Use this formula to find the area of a hexagonal plate 4.0 in. on each side. (Round to the nearest whole number.)

Solution   Area, $A \approx 2.598a^2$ $a = 4.0$ in.

$\approx 2.598(4.0 \text{ in.})^2$

$\approx 2.598(16 \text{ in.}^2)$

$\approx 42 \text{ in.}^2$, rounded

Because the hexagon is used so often in practical and technical work, several other formulas may be helpful to you.

A hexagon is often measured by specifying the distance across the corners or the distance across the flats.

Distance *d* across   Distance *f* across
the corners   the flats

Dimensions of a Regular Hexagon

Distance across the corners, $d = 2a$ or $a = 0.5d$

Distance across the flats, $f \approx 1.732a$ or $a \approx 0.577f$

EXAMPLE 8   **Machine Trades**   For a piece of hexagonal bar stock where $a = \frac{1}{2}$ in., the distance across the corners would be

$$d = 2a \qquad \text{or} \qquad d = 1 \text{ in.}$$

The distance across the flats would be

$$f \approx 1.732a \qquad \text{or} \qquad f \approx 0.866 \text{ in.} \qquad \bullet$$

Your Turn   **Machine Trades**   If the cross section of a hexagonal nut measures $\frac{3}{4}$ in. across the flats, find each of the following:

(a) The side length, a

(b) The distance across the corners, d

(c) The cross-sectional area

(Round to three decimal places.)

Solutions $f = \frac{3}{4}$ in.   or   0.75 in.

(a) $a \approx 0.577f$ (b) $d = 2a$
  $\approx 0.577(0.75 \text{ in.})$ $\approx 2(0.433 \text{ in.})$
  ≈ 0.433 in. ≈ 0.866 in.

(c) Area, $A \approx 2.598a^2$
  $\approx 2.598(0.433 \text{ in.})^2$
  $\approx 2.598(0.1875 \text{ in.}^2)$
  $\approx 0.487 \text{ in.}^2$

More Practice   Fill in the blanks with the missing dimensions of each regular hexagon. (Round to one decimal place if necessary.)

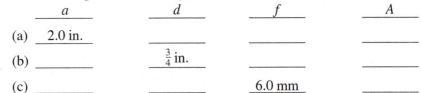

	_____a_____	_____d_____	_____f_____	_____A_____
(a)	2.0 in.			
(b)		$\frac{3}{4}$ in.		
(c)			6.0 mm	

Answers

	a	d	f	A
(a)	2 in.	4 in.	3.5 in.	10.4 in.^{2}
(b)	$\frac{3}{8}$ in.	$\frac{3}{4}$ in.	0.6 in.	0.4 in.^{2}
(c)	3.5 mm	6.9 mm	6 mm	31.1 mm^2

For problem (c),

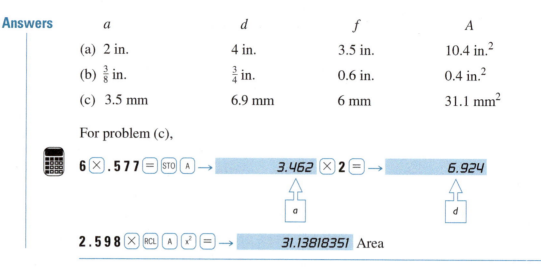

Applications of Hexagons

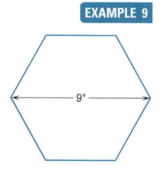

EXAMPLE 9 **Flooring and Carpeting**   A homeowner has selected the hexagonal tile shown in the figure to cover the floor of his 7-ft by 10-ft rectangular entryway. To estimate the number of tiles needed, we can divide the area of the entryway by the area of each tile.

For the tile, we are given the distance d across the corners of the hexagon, but to use the area formula, we need the length of a side. Using the proper dimension formula, we have

$$a = 0.5d = 0.5(9 \text{ in.}) = 4.5 \text{ in.}$$

Now we can find the area A_T of each tile:

$$A_T \approx 2.598a^2 \approx 2.598(4.5 \text{ in.})^2 \approx 52.6095 \text{ sq in.}$$

The area A_E of the entryway is $A_E = LW = (10 \text{ ft})(7 \text{ ft}) = 70 \text{ sq ft}$

Before dividing, we must convert this to square inches so the units agree with the area of the tile. Using the conversion factor 1 sq ft = 144 sq in., we have

$$A_E = 70 \text{ sq ft} \times \frac{144 \text{ sq in.}}{1 \text{ sq ft}} = 10{,}080 \text{ sq in.}$$

Finally, divide this by the area A_T of each tile to get the number of tiles needed:

$$10{,}080 \text{ sq in.} \div 52.6095 \text{ sq in.} = 191.6 \ldots \approx 192 \text{ tiles} \qquad \bullet$$

Your Turn   (a) **Construction**   An artist is building a studio in the shape of a regular hexagon. If the distance across the flats is 22 ft 9 in., how many square feet of floor space will the studio contain? (Round your final answer to the nearest whole number.)

(b) **Carpentry**   A hex nut has a side length of $\frac{1}{4}$ in. From the following list, pick the smallest size wrench that will fit it.

   (1) $\frac{1}{4}$ in.   (2) $\frac{3}{8}$ in.   (3) $\frac{7}{16}$ in.   (4) $\frac{1}{2}$ in.   (5) $\frac{5}{8}$ in.

Solutions   (a) **Step 1**   Find the side length a of the hexagon. Note that 9 in. = 0.75 ft.

$$a \approx 0.577f \approx 0.577(22.75 \text{ ft}) \approx 13.12675 \text{ ft}$$

Step 2   Calculate the area A of the studio.

$$A \approx 2.598a^2 \approx 2.598(13.12675 \text{ ft})^2 \approx 447.665 \ldots \approx 448 \text{ sq ft}$$

Using a calculator, we enter:

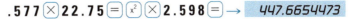

$.577 \boxed{\times} 22.75 \boxed{=} \boxed{x^2} \boxed{\times} 2.598 \boxed{=} \rightarrow$ *447.6654473*

(b) By drawing a picture we can see that we must find f. Because $a = \frac{1}{4}$ in., we have

$$f \approx 1.732a$$

$$\approx 1.732(0.25)$$

$$\approx 0.433 \text{ in.}$$

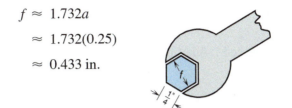

If we then convert each of the fractional sizes to decimal form, we find that $\frac{7}{16}$ in. = 0.4375 in., and so choice (3) is the smallest size wrench that will fit the nut.

CLASSIFICATION OF POLYGONS

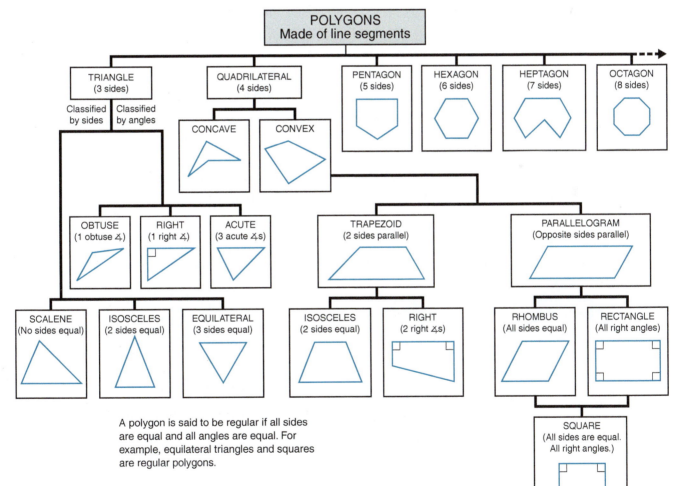

A polygon is said to be regular if all sides are equal and all angles are equal. For example, equilateral triangles and squares are regular polygons.

Irregular Polygons   Often, the shapes of polygons that appear in practical work are not the simple geometric figures that you have seen so far. The easiest way to work with irregular polygon shapes is to divide them into simpler, more familiar figures.

For example, look at this L-shaped figure:

There are two ways to find the area of this shape.

Method 1: Addition

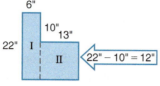

By drawing the dashed line, we have divided this figure into two rectangles, I and II. I is 22 in. by 6 in. for an area of 132 in.^{2}. II is 13 in. by 12 in. for an area of 156 in.^{2}. The total area is

$$132 \text{ in.}^2 + 156 \text{ in.}^2 = 288 \text{ in.}^2$$

Method 2: Subtraction

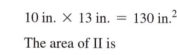

In this method, we subtract the area of the small rectangle (I) from the area of the large rectangle (II). The area of I is

$$10 \text{ in.} \times 13 \text{ in.} = 130 \text{ in.}^2$$

The area of II is

$$22 \text{ in.} \times 19 \text{ in.} = 418 \text{ in.}^2$$

$$418 \text{ in.}^2 - 130 \text{ in.}^2 = 288 \text{ in.}^2$$

Notice that in the addition method, we had to calculate the height of rectangle II as 22 in. − 10 in., or 12 in. In the subtraction method we had to calculate the base of rectangle II as 6 in. + 13 in., or 19 in.

Which method you use to find the area of an irregular polygon depends on your ingenuity and on which dimensions are given.

EXAMPLE 10   Find the area of the shape *ABCD*.

You may at first wish to divide this into two triangles and add their areas.

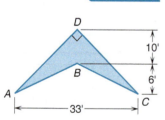

However, the dimensions provided do not allow you to do this.

The only option here is to subtract the area of triangle I from the area of triangle II.

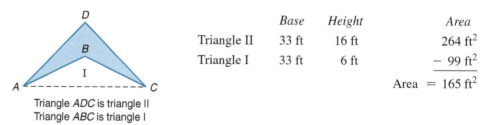

Triangle *ADC* is triangle II
Triangle *ABC* is triangle I

	Base	Height	Area
Triangle II	33 ft	16 ft	264 ft^2
Triangle I	33 ft	6 ft	− 99 ft^2
		Area =	165 ft^2

If both methods will work, pick the one that looks easier and requires the simpler arithmetic.   ●

Your Turn Find the area of this shape. (Round to the nearest square inch.)

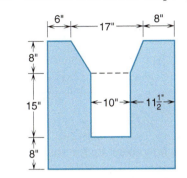

Solution You might have been tempted to split the shape into the five regions indicated in the figure on the left. This will work using the addition method, but it is easier to apply the subtraction method using the figure on the right.

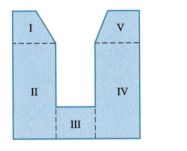 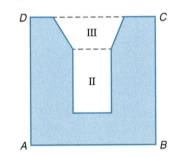

Instead of calculating the sum of five areas, we need only find three areas: Simply subtract the areas of regions II and III from the area of rectangle *ABCD*, which we will call region I.

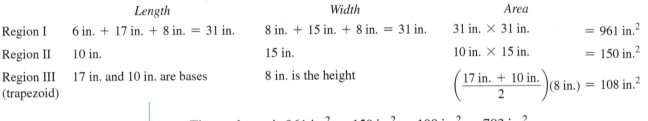

	Length	Width	Area	
Region I	6 in. + 17 in. + 8 in. = 31 in.	8 in. + 15 in. + 8 in. = 31 in.	31 in. × 31 in.	= 961 in.^{2}
Region II	10 in.	15 in.	10 in. × 15 in.	= 150 in.^{2}
Region III (trapezoid)	17 in. and 10 in. are bases	8 in. is the height	$\left(\dfrac{17\text{ in.} + 10\text{ in.}}{2}\right)(8\text{ in.})$	= 108 in.^{2}

The total area is 961 in.^{2} − 150 in.^{2} − 108 in.^{2} = 703 in.^{2}

Learning Help When working with complex figures, organize your work carefully. Neatness will help eliminate careless mistakes. ●

More Practice Now, for practice, find the areas of the following shapes. (Round to the nearest tenth.)

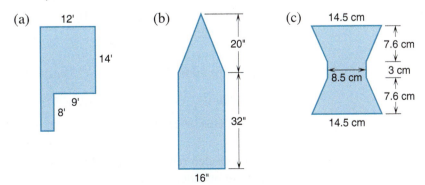

(a)   12'   14'   9'   8'

(b)   20"   32"   16"

(c)   14.5 cm   7.6 cm   3 cm   8.5 cm   7.6 cm   14.5 cm

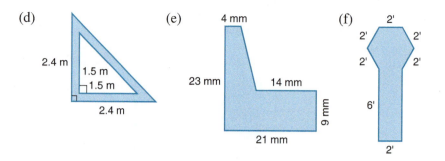

(d)   (e) 4 mm   (f) 2'

2.4 m   1.5 m   1.5 m   2.4 m

23 mm   14 mm   21 mm   9 mm

2'   2'   2'   2'   6'   2'

Solutions

(a) The figure shows how to set up the shape to use the subtraction method.

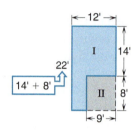

←— 12' —→

I   14'

22'

14' + 8'

II   8'

←— 9' —→

Region I is the larger rectangle: Area $= LW$
$$= (22 \text{ ft})(12 \text{ ft})$$
$$= 264 \text{ ft}^2$$

Region II is the smaller rectangle: Area $= (9 \text{ ft})(8 \text{ ft})$
$$= 72 \text{ ft}^2$$

Total area $= 264 \text{ ft}^2 - 72 \text{ ft}^2 = 192 \text{ ft}^2$

(The addition method works just as easily with this shape.)

(b) Divide this shape into a triangle (I) and a rectangle (II). Find the sum of the two areas.

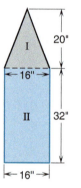

I   20"

←— 16" —→

II   32"

←— 16" —→

Region I: Area $= \frac{1}{2}bh = \frac{1}{2}(16 \text{ in.})(20 \text{ in.}) = 160 \text{ in.}^2$

Region II: Area $= (16 \text{ in.})(32 \text{ in.}) = 512 \text{ in.}^2$

Total area $= 160 \text{ in.}^2 + 512 \text{ in.}^2 = 672 \text{ in.}^2$

(c) Divide this shape into two identical trapezoids (I and III) and a rectangle (II).

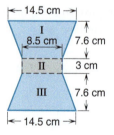

←— 14.5 cm —→

I

8.5 cm   7.6 cm

II   3 cm

III   7.6 cm

←— 14.5 cm —→

Region I: Area $= \frac{1}{2}(b_1 + b_2)h$
$$= \frac{1}{2}(14.5 \text{ cm} + 8.5 \text{ cm})(7.6 \text{ cm})$$
$$= \frac{1}{2}(23 \text{ cm})(7.6 \text{ cm})$$

Now, to obtain the area of *both* trapezoids, double this quantity by just eliminating the $\frac{1}{2}$.

Regions I and III: Area $= (23 \text{ cm})(7.6 \text{ cm}) = 174.8 \text{ cm}^2$

Region II: Area $= (8.5 \text{ cm})(3 \text{ cm}) = 25.5 \text{ cm}^2$

Total area $= 174.8 \text{ cm}^2 + 25.5 \text{ cm}^2 = 200.3 \text{ cm}^2$

(d) The shaded region is created by removing the smaller inside triangle from the larger outside triangle. Therefore, the area A of the shaded region is equal to the area A_L of the larger triangle minus the area A_S of the smaller triangle. Using the formula $A = \frac{1}{2}bh$, we have:

$$A = A_L - A_S = \frac{1}{2}(2.4 \text{ m})(2.4 \text{ m}) - \frac{1}{2}(1.5 \text{ m})(1.5 \text{ m})$$

$$= 2.88 \text{ m}^2 - 1.125 \text{ m}^2 = 1.755 \text{ m}^2, \text{ or } 1.8 \text{ m}^2, \text{ rounded}$$

(e) Divide this shape into a trapezoid (I) and a rectangle (II), as shown. Notice that the bottom base of the trapezoid is 21 mm − 14 mm = 7 mm, and the height of the trapezoid is 23 mm − 9 mm = 14 mm.

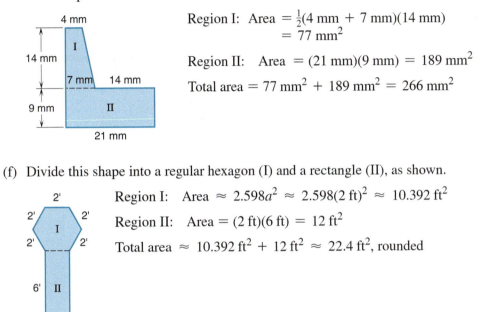

Region I: Area $= \frac{1}{2}(4 \text{ mm} + 7 \text{ mm})(14 \text{ mm})$
$= 77 \text{ mm}^2$

Region II: Area $= (21 \text{ mm})(9 \text{ mm}) = 189 \text{ mm}^2$

Total area $= 77 \text{ mm}^2 + 189 \text{ mm}^2 = 266 \text{ mm}^2$

(f) Divide this shape into a regular hexagon (I) and a rectangle (II), as shown.

Region I: Area $\approx 2.598a^2 \approx 2.598(2 \text{ ft})^2 \approx 10.392 \text{ ft}^2$

Region II: Area $= (2 \text{ ft})(6 \text{ ft}) = 12 \text{ ft}^2$

Total area $\approx 10.392 \text{ ft}^2 + 12 \text{ ft}^2 \approx 22.4 \text{ ft}^2$, rounded

Applications of Irregular Polygons

EXAMPLE 11   **Masonry**   A contemporary fire pit has the shape shown in the figure. A homeowner wishes to cover the edge around the pit (the shaded region of the figure) in marble tile priced at $6.75 per square foot.

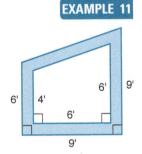

To determine the total cost of the tile, we must first calculate the area of the shaded region. Notice that the fire pit consists of a smaller right trapezoid inside a larger right trapezoid. The subtraction method works best here: Subtract the area of the smaller trapezoid (A_S) from the area of the larger trapezoid (A_L) to find the area A of the region to be covered in tile:

$$A_L = \left(\frac{b_1 + b_2}{2}\right)h = \left(\frac{6 \text{ ft} + 9 \text{ ft}}{2}\right)9 \text{ ft} = (7.5 \text{ ft})9 \text{ ft} = 67.5 \text{ sq ft}$$

$$A_S = \left(\frac{4 \text{ ft} + 6 \text{ ft}}{2}\right)6 \text{ ft} = (5 \text{ ft})6 \text{ ft} = 30 \text{ sq ft}$$

$$A = A_L - A_S = 67.5 \text{ sq ft} - 30 \text{ sq ft} = 37.5 \text{ sq ft}$$

Now calculate the cost of the tile by multiplying the number of square feet by the cost per square foot:

$$\text{Cost} = 37.5 \text{ sq ft} \times \frac{\$6.75}{1 \text{ sq ft}} = \$253.125 \approx \$253.13$$

Your Turn   **Construction**   A homeowner wants to cover the L-shaped driveway shown in the figure with concrete pavers. If each paver covers 72 sq in., how many pavers will be needed?

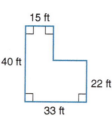

Solution   **First,** use either the addition or the subtraction method to find the area of the driveway:

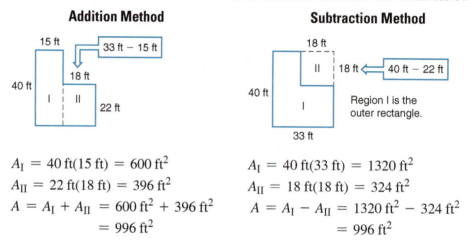

Addition Method

$A_I = 40 \text{ ft}(15 \text{ ft}) = 600 \text{ ft}^2$

$A_{II} = 22 \text{ ft}(18 \text{ ft}) = 396 \text{ ft}^2$

$A = A_I + A_{II} = 600 \text{ ft}^2 + 396 \text{ ft}^2$

$= 996 \text{ ft}^2$

Subtraction Method

$A_I = 40 \text{ ft}(33 \text{ ft}) = 1320 \text{ ft}^2$

$A_{II} = 18 \text{ ft}(18 \text{ ft}) = 324 \text{ ft}^2$

$A = A_I - A_{II} = 1320 \text{ ft}^2 - 324 \text{ ft}^2$

$= 996 \text{ ft}^2$

The area of the pavers is given in square inches, and the area of the driveway is in square feet. We must convert one of the areas so that the units agree. Recall that 1 sq ft = 144 sq in., and notice that the area of each paver, 72 sq in., is exactly half of 144 sq in. In this case, it is easier to convert the area of the pavers to square feet.

$$72 \text{ sq in.} \times \frac{1 \text{ sq ft}}{144 \text{ sq in.}} = 0.5 \text{ sq ft}$$

Finally, divide this into the area of the driveway to determine how many pavers are needed:

$$996 \text{ ft}^2 \div 0.5 \text{ ft}^2/\text{paver} = 996 \text{ ft}^2 \times \frac{1 \text{ paver}}{0.5 \text{ ft}^2} = 1992 \text{ pavers}$$

It should be noted that the homeowner or contractor would normally order a few additional pavers to account for waste.

Now continue with Exercises 8-3 for a set of problems on the work of this section.

Exercises 8-3   **Triangles, Regular Hexagons, and Irregular Polygons**

A. Find the missing dimensions of each figure shown. All hexagons are regular. Round to the nearest tenth unless otherwise directed in parentheses.

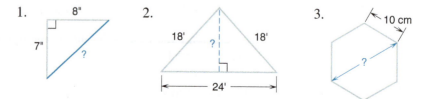

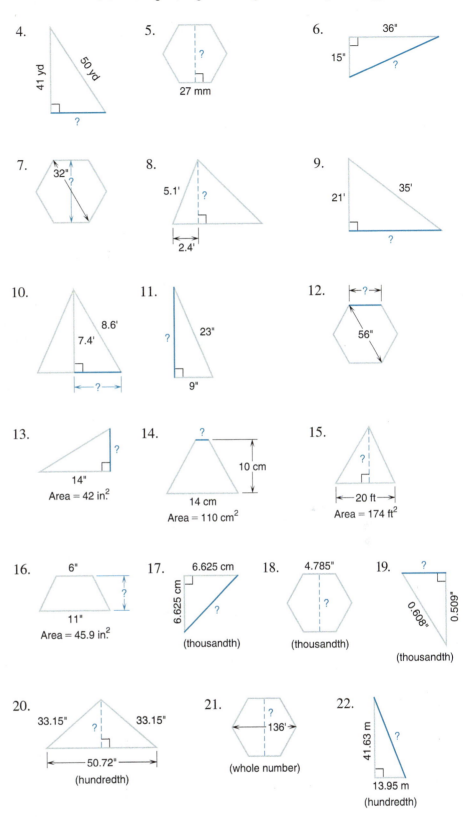

4.

5. 27 mm

6. 36" 15" ?

7. 32" ?

8. 5.1' ? 2.4'

9. 35' 21' ?

10. 8.6' 7.4' ?

11. ? 23" 9"

12. ? 56"

13. 14" ? Area = 42 in.²

14. ? 10 cm 14 cm Area = 110 cm²

15. ? 20 ft Area = 174 ft²

16. 6" ? 11" Area = 45.9 in.²

17. 6.625 cm 6.625 cm ? (thousandth)

18. 4.785" ? (thousandth)

19. ? 0.608" 0.509" (thousandth)

20. 33.15" ? 33.15" 50.72" (hundredth)

21. ? 136' (whole number)

22. 41.63 m ? 13.95 m (hundredth)

B. Find the area of each figure. All hexagons are regular. Round to the nearest
    whole number unless otherwise directed in parentheses.

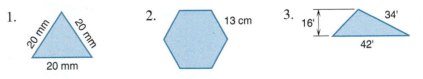

1. 20 mm 20 mm 20 mm

2. 13 cm

3. 16' 34' 42'

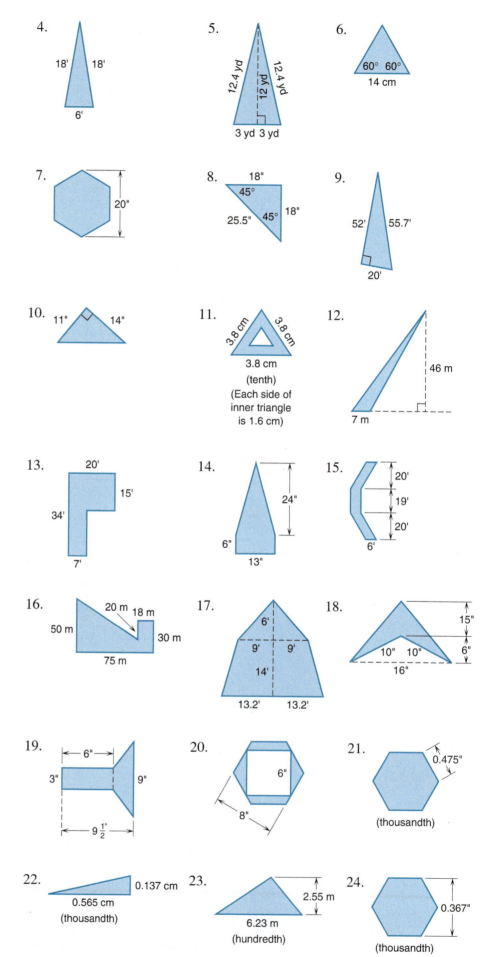

4.

18'   18'

6'

5.

12.4 yd   12.4 yd   12 yd

3 yd 3 yd

6.

60° 60°

14 cm

7.

20"

8.

18"
45°
25.5"   45° 18"

9.

52'   55.7'

20'

10.

11"   14"

11.

3.8 cm   3.8 cm

3.8 cm
(tenth)
(Each side of
inner triangle
is 1.6 cm)

12.

46 m

7 m

13.

20'
15'
34'

7'

14.

24"

6"
13"

15.

20'
19'
20'

6'

16.

20 m 18 m
50 m
30 m
75 m

17.

6'
9'   9'
14'
13.2'   13.2'

18.

15"
10" 10"   6"
16"

19.

6"
3"   9"
9 1/2"

20.

6"
8"

21.

0.475"
(thousandth)

22.

0.137 cm
0.565 cm
(thousandth)

23.

2.55 m
6.23 m
(hundredth)

24.

0.367"
(thousandth)

C. Practical Applications

1. **Painting** At 460 sq ft per gallon, how many gallons of paint are needed to cover the outside walls of a house as pictured below? Do not count windows and doors and round up to the nearest gallon.

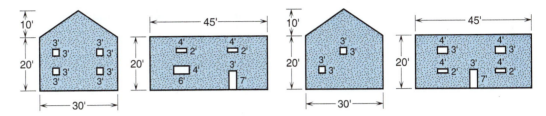

2. **Sheet Metal Trades** A four-sided vent connection must be made out of sheet metal. If each side of the vent is like the one pictured, how many total square inches of sheet metal will be used?

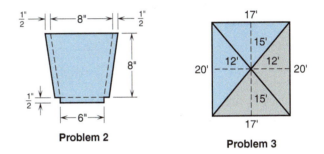

Problem 2   Problem 3

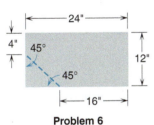

Problem 5

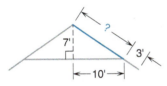

Problem 6

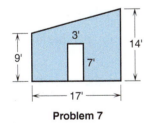

Problem 7

3. **Roofing** The aerial view of a roof is shown in the figure. How many squares (100 sq ft) of shingles are needed for the roof?

4. **Roofing** In problem 3, how many feet of gutters are needed?

5. **Carpentry** Allowing for a 3-ft overhang, how long a rafter is needed for the gable of the house shown in the figure? (Round to the nearest tenth.)

6. **Metalworking** A cut is to be made in a piece of metal as indicated by the dashed line in the figure. Find the length of the cut to the nearest tenth of an inch.

7. **Construction** How many square feet of drywall are needed for the wall shown in the figure?

8. **Construction** Find the area of the gable end of the house shown, not counting windows. Each window opening is 4 ft $3\frac{1}{4}$ in. by 2 ft $9\frac{1}{4}$ in. (Round to the nearest square foot.)

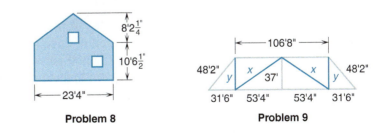

Problem 8   Problem 9

9. **Construction** Find the missing dimensions x and y on the bridge truss shown.

10. **Welding** The steel gusset shown is made in the shape of a right triangle. It will be welded in place along all three sides. Find the length of the weld.

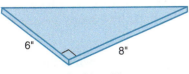

Problem 10

11. **Roofing**   A roof is designed to have an $\frac{8}{12}$ slope, that is, 8 inches of rise per foot of run. The total run of one side of the roof is 16 ft. (See the figure below.) Find the total length of the rafter by doing parts (a) through (d):

   (a) Use the Pythagorean theorem to find the *ULL*, unit line length.

   (b) Multiply the *ULL* by the run to find *R* in inches.

   (c) Add the length of the overhang as shown.

   (d) Give the answer in feet and inches to the nearest inch.

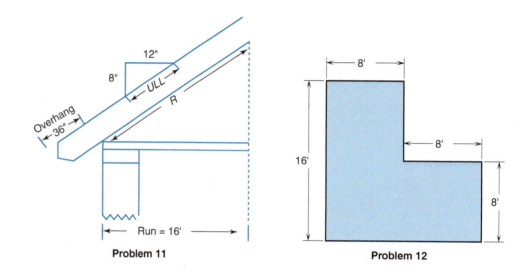

Problem 11   **Problem 12**

12. **Construction**   A builder needs to pour a 4-in.-thick concrete slab that is shaped as shown in the figure.

   According to the manufacturer's guidelines for this thickness, approximately $\frac{3}{4}$ of a bag of concrete mix is required per square foot.

   (a) Find the area of the slab.

   (b) How many bags of the mix are needed for this job?

   (c) At $4.68 per bag, what will be the cost for purchasing the required amount of mix?

13. **Carpentry**   A hex nut measures 15 mm across the corners. From the following list, pick the smallest size metric wrench that will fit:

   (a) 12 mm   (b) 13 mm   (c) 14 mm   (d) 15 mm   (e) 16 mm

14. **Agriculture**   A plot of land in the shape of a right triangle is formed by the intersection of three roads. A farmer wishes to purchase the land, which is priced at $6600 per acre. Calculate the cost of the land, given the dimensions shown in the figure. (*Hint:* One acre = 43,560 square feet. Round to the nearest hundred dollars.)

15. **Masonry**   A driveway is being surfaced with concrete pavers in the shape of regular hexagons. If the driveway is a 24-ft by 28-ft rectangle, and the pavers measure 6 inches across the flats, how many pavers are needed?

16. **Flooring and Carpeting**   The floor plan of a house is shown in the figure. If the entire surface is to be covered in carpet costing $38 per square *yard*,

Problem 14

what will be the total cost of the carpet? (Assume that all corners are right angles. Round to the nearest dollar.)

Problem 16

Problem 17

17. **Construction** The back side of an A-frame house is shown in the figure. All panels except for the two shaded triangular sections are windows. How many square feet of glass are needed for the windows?

18. **Construction** A yoga studio is built in the shape of a regular hexagon. The interior distance across the flats is 32 feet. Six-foot high mirrors are to be installed across the full width of all the inside walls. How many square feet of mirrors are needed? (Round to the nearest square foot.)

When you have finished these problems, check your answers to the odd-numbered problems in the Appendix, then continue in Section 8-4 with the study of circles.

8-4   Circles

1. What is the diameter of this circle?

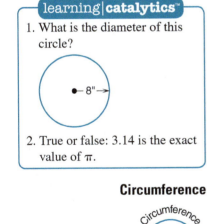

2. True or false: 3.14 is the exact value of π.

A circle is probably the most familiar and the simplest plane figure. Certainly, it is the geometric figure that is most often used in practical and technical work. Mathematically, a **circle** is a closed curve representing the set of points some fixed distance from a given point called the **center**.

In this circle the center point is labeled O. The **radius** r is the distance from the center of the circle to the circle itself. The **diameter** d is the straight-line distance across the circle through the center point. It should be obvious from the drawing that the diameter is twice the radius.

Circumference

The **circumference** of a circle is the distance around it. It is a distance similar to the perimeter measure for a polygon. For all circles, the ratio of the circumference to the diameter, the distance around the circle to the distance across, is the same number.

$$\frac{\text{circumference}}{\text{diameter}} \approx 3.14159\ldots = \pi$$

The value of this ratio has been given the label π, the Greek letter "pi." The value of π is approximately 3.14, but the number has no simple decimal form—the digits continue without ending or repeating. Pressing the $\boxed{\pi}$ button on a scientific calculator will display the rounded ten-digit approximation of π:

$\boxed{\pi}\ \boxed{=}\ \rightarrow$ 3.141592654

The relationships among radius, diameter, and circumference can be summarized as follows:

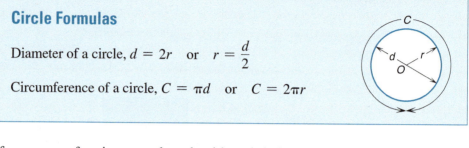

> **Circle Formulas**
>
> Diameter of a circle, $d = 2r$ or $r = \dfrac{d}{2}$
>
> Circumference of a circle, $C = \pi d$ or $C = 2\pi r$

📝 **Note**   If you are performing mental math with π, it is important to remember that π is slightly larger than 3. If you are doing calculations by hand or with a simple calculator, it is important to know that $\pi \approx 3.14$ or 3.1416. To reinforce this knowledge, we will sometimes instruct you to use one of these decimal equivalents for π. Otherwise, you may use the $\boxed{\pi}$ button on your scientific calculator. ●

EXAMPLE 1   We can use these formulas to find the radius and circumference of a hole with a diameter of 4.50 in.

Radius, $r = \dfrac{d}{2}$

$$= \frac{4.50 \text{ in.}}{2} = 2.25 \text{ in.}$$

Circumference, $C = \pi d$

$$= (\pi)(4.50 \text{ in.})$$

$$= 14.137 \ldots \approx 14.1 \text{ in.} \quad \text{rounded}$$

$\boxed{\pi}\,\boxed{\times}\,\mathbf{4.5}\,\boxed{=}\rightarrow$ `14.13716694`   ●

Your Turn   Calculate the missing quantities for each circle:

	r	d	C
(a)	2.7 cm	_____	_____
(b)	_____	_____	22.8 ft

Solutions   (a) $d = 2r = 2(2.7 \text{ cm}) = 5.4 \text{ cm}$

$C = \pi d = \pi(5.4 \text{ cm}) = 16.96 \ldots \approx 17 \text{ cm, rounded}$

(b) **First,** substitute 22.8 ft for C into $C = \pi d$

$22.8 \text{ ft} = \pi d$

Then, divide both sides by π:

$$\frac{22.8}{\pi} = \frac{\pi d}{\pi}$$

$$d = 7.257 \ldots \text{ ft} \approx 7.26 \text{ ft, rounded}$$

Finally, divide this by 2 to find r:

$$r \approx \frac{7.26 \text{ ft}}{2} \approx 3.63 \text{ ft}$$

Applications of Circumference

EXAMPLE 2 **Manufacturing** A circular redwood hot tub has a diameter of 5 ft 3 in. Allowing 4 in. extra for fastening, what length of steel band is needed to encircle the tub and hold the boards in place?

The length of the steel band is the circumference of the hot tub plus 4 in. We know that $d = 5$ ft 3 in., or 5.25 ft; therefore,

$C = \pi d$

$\approx \pi(5.25 \text{ ft})$

$\approx 16.5 \text{ ft} \approx 16 \text{ ft } 6 \text{ in.}$

Adding 4 in., the length needed is 16 ft 10 in.

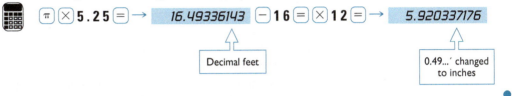

↑		↑
Decimal feet		0.49...′ changed to inches

Your Turn (a) **Sheet Metal Trades** An 8.0-ft strip of sheet metal is bent into a circular tube. What will be the diameter of this tube rounded to the nearest 0.1 ft?

(b) **Machine Trades** If eight bolts are to be equally spaced around a circular steel plate at a radius of 9.0 in., what must be the spacing between the centers of the bolts along the curve? (Round to the nearest tenth of an inch.)

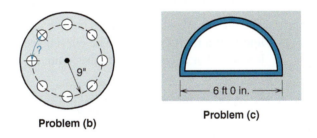

Problem (b)

Problem (c)

6 ft 0 in.

(c) **Construction** At $18.50 per foot, what will be the cost of the molding around the semicircular window shown? (Round the length to the nearest half-foot.)

Solutions (a) When the 8.0-ft strip is bent into a tube, it becomes the circumference of the tube.

Use $C = \pi d$, and solve for d:

If $C = \pi d$ then $d = \dfrac{C}{\pi}$

and $d = \dfrac{8.0 \text{ ft}}{\pi}$

$\approx 2.5 \text{ ft}$ rounded

(b) $C = 2\pi r$

$= 2(\pi)(9.0 \text{ in.})$

$\approx 56.55 \text{ in.}$

Because eight bolts must be spaced evenly around the circumference of the circle, divide by 8. Spacing $= 56.55 \div 8 \approx 7.1$ in.

(c) **First,** find the length of the curved portion:

$$C = \pi d \text{ for a complete circle}$$
$$= (\pi)(6.0 \text{ ft})$$
$$\approx 18.8496 \text{ ft}$$

The curved portion is a semicircle, or half-circle. The length is

$$18.8496 \text{ ft} \div 2 = 9.4248 \text{ ft} \approx 9.5 \text{ ft, rounded to the nearest half-foot}$$

Next, add the length of the straight portion along the bottom of the window:

$$9.5 \text{ ft} + 6 \text{ ft} = 15.5 \text{ ft, rounded}$$

Finally, multiply this total length by \$18.50 per foot to calculate the total cost:

$$(\$18.50)(15.5 \text{ ft}) = \$286.75$$

Parts of Circles   Carpenters, plumbers, sheet metal workers, and other trades people often work with parts of circles. A common problem is to determine the length of a piece of material that contains both curved and straight segments.

EXAMPLE 3   **Metalworking**   Suppose that we need to determine the total length of steel rod needed to form the curved piece shown. The two straight segments are no problem: We will need a total of 8 in. + 6 in. or 14 in. of rod for these. When a rod is bent, the material on the outside of the curve is stretched, and the material on the inside is compressed. We must calculate the circumference for the curved section from a *neutral line* midway between the inside and outside radius.

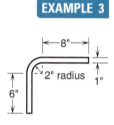

The inside radius is given as 2 in. The stock is 1 in. thick, so the midline or neutral line is at a radius of $2\frac{1}{2}$ in. The curved section is one-quarter of a full circle, so the length of bar needed for the curved arc is

$$C = \frac{2\pi r}{4} \quad \leftarrow \boxed{\text{Divide the circumference by 4 to find the length of the quarter-circle arc.}}$$

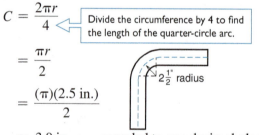

$$= \frac{\pi r}{2}$$

$$= \frac{(\pi)(2.5 \text{ in.})}{2}$$

$$\approx 3.9 \text{ in.} \quad \text{rounded to one decimal place}$$

The total length of bar needed for the piece is

$$L \approx 3.9 \text{ in.} + 14 \text{ in.}$$
$$\approx 17.9 \text{ in.}$$

    ●

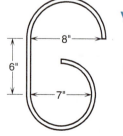

Your Turn   **Metalworking**   Find the total length of ornamental iron $\frac{1}{2}$ in. thick needed to bend into this shape. (Round final answer to one decimal place.)

Solution   Measuring to the midline we find that the upper curve (A) has a radius of $4\frac{1}{4}$ in. and the lower curve (C) has a radius of $3\frac{3}{4}$ in. Calculate the lengths of the two curved pieces this way:

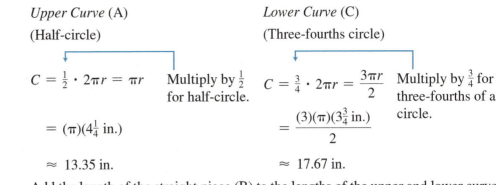

Upper Curve (A)
(Half-circle)

Lower Curve (C)
(Three-fourths circle)

$$C = \tfrac{1}{2} \cdot 2\pi r = \pi r \quad \text{Multiply by } \tfrac{1}{2} \text{ for half-circle.}$$

$$C = \tfrac{3}{4} \cdot 2\pi r = \frac{3\pi r}{2} \quad \text{Multiply by } \tfrac{3}{4} \text{ for three-fourths of a circle.}$$

$$= (\pi)(4\tfrac{1}{4} \text{ in.})$$

$$= \frac{(3)(\pi)(3\tfrac{3}{4} \text{ in.})}{2}$$

$$\approx 13.35 \text{ in.}$$

$$\approx 17.67 \text{ in.}$$

Add the length of the straight piece (B) to the lengths of the upper and lower curves to find the total length of straight stock needed to create this shape:

13.35 in. + 17.67 in. + 6 in. ≈ 37.0 in., rounded to one decimal place

Once you have your work organized, you can perform the entire calculation on a calculator as follows:

$\boxed{\pi} \boxed{\times} \mathbf{4.25} \boxed{+} \mathbf{3} \boxed{\times} \boxed{\pi} \boxed{\times} \mathbf{3.75} \boxed{\div} \mathbf{2} \boxed{+} \mathbf{6} \boxed{=} \rightarrow$ *37.02322745*

More Practice Find the length of stock needed to create each of the following shapes. (Round to one decimal place.)

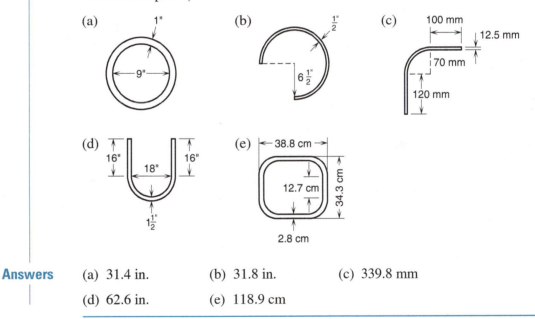

Answers

(a) 31.4 in. (b) 31.8 in. (c) 339.8 mm

(d) 62.6 in. (e) 118.9 cm

A Closer Look For problem (e), a bit of careful reasoning will convince you that the dimensions are as follows:

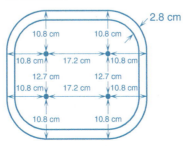

Each of the four corner arcs has radius 9.4 cm to the midline, and each is a quarter-circle.

Radius = 10.8 cm − 1.4 cm
       = 9.4 cm ●

Area of a Circle   To find the *area* of a circle, use one of the following formulas:

> ### Area of a Circle
>
> $$A = \pi r^2 \quad \text{or} \quad A = \frac{\pi d^2}{4} \approx 0.7854d^2$$
>
>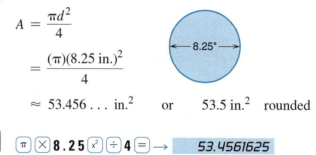

If you are curious about how these formulas are obtained, don't miss the box on page 592.

EXAMPLE 4   Let's use the formula to find the area of a circle with a diameter of 8.25 in.

$$A = \frac{\pi d^2}{4}$$

$$= \frac{(\pi)(8.25 \text{ in.})^2}{4}$$

8.25"

$$\approx 53.456 \dots \text{ in.}^2 \quad \text{or} \quad 53.5 \text{ in.}^2 \quad \text{rounded}$$

$\pi \times 8.25 \; x^2 \div 4 = \rightarrow$ **53.4561625**

Of course, you would find exactly the same area if you used the radius $r = 4.125$ in. in the first formula. Most people who need the area formula in their work memorize the formula $A = \pi r^2$ and calculate the radius r if they are given the diameter d. ●

Your Turn   Find the areas of each of the following figures.

(a)

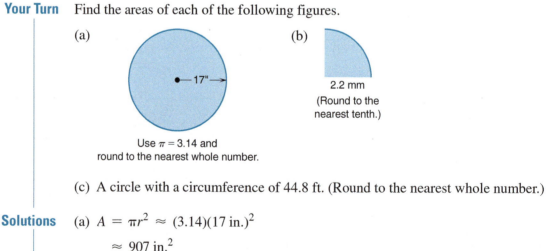

17"

Use $\pi = 3.14$ and
round to the nearest whole number.

(b)

2.2 mm
(Round to the
nearest tenth.)

(c) A circle with a circumference of 44.8 ft. (Round to the nearest whole number.)

Solutions   (a) $A = \pi r^2 \approx (3.14)(17 \text{ in.})^2$

$$\approx 907 \text{ in.}^2$$

(b) This is a quarter-circle.

$$A = \tfrac{1}{4}(\pi r^2) = \tfrac{1}{4}(\pi)(2.2 \text{ mm})^2$$

$$\approx 3.8 \text{ mm}^2 \quad \text{rounded}$$

(c) **First,** substitute into the formula $C = \pi d$ and solve for d:

$$C = \pi d \quad \Rightarrow \quad 44.8 \text{ ft} = \pi d$$

Divide by π. $\dfrac{44.8 \text{ ft}}{\pi} = \dfrac{\pi d}{\pi}$

$$14.26 \text{ ft} \approx d$$

Then, substitute 14.26 ft for d into the area formula:

$$A = \frac{\pi d^2}{4} = \frac{\pi (14.26 \text{ ft})^2}{4} \approx 160 \text{ ft}^2$$

We can enter all of the steps on a calculator as follows:

$$\underbrace{44.8 \div \pi =}_{\text{This gives us } d.} \underbrace{x^2 \times \pi \div 4 =}_{\text{This gives us } A.} \rightarrow \boxed{159.7151685}$$

Applications of Area

EXAMPLE 5   **Forestry**   After wildfires destroyed a portion of Glacier National Park, forest service biologists needed to determine the germination rate of new seedlings. Using a rope as a radius, they marked off a circular area of forest with a 40-ft radius and then counted 450 new seedlings that had germinated in that circle. What was the germination rate in seedlings per acre? (Round to the nearest whole number.)

Step 1   Find the area of the circle:

$$A = \pi r^2 = \pi (40)^2 = 1600\pi \text{ sq ft}$$

Keep the area as 1600π to avoid a rounding error in Step 2.

Step 2   Use a proportion to find the germination rate in seedlings per acre.

$$\frac{450 \text{ seedlings}}{x \text{ seedlings}} = \frac{1600\pi \text{ sq ft}}{43,560 \text{ sq ft}} \qquad \boxed{\text{1 acre} = 43,560 \text{ sq ft}}$$

$$1600\pi x = 19,602,000$$

$$x \approx 3900$$

The germination rate is approximately 3900 new seedlings per acre.   ●

Your Turn   (a) **Painting**   Find the area of wood that must be stained to finish the top and bottom of a circular table 6 ft 6 in. in diameter. (Use $\pi \approx 3.14$ and round to the nearest tenth.)

(b) **Industrial Technology**   A pressure of 860 lb per square foot is exerted on the bottom of a cylindrical water tank. If the bottom has a diameter of 15.0 ft, what is the total force on the bottom? (Round the final answer to the nearest thousand pounds.)

Solutions   (a) To find the area of the top and bottom, we must multiply the area of the table by 2:

$$A = 2\pi r^2 \approx 2(3.14)(3.25 \text{ ft})^2 \qquad \text{(If } d = 6 \text{ ft 6 in., } r = 3 \text{ ft 3 in. or 3.25 ft)}$$

$$\approx 66.3 \text{ ft}^2$$

(b) $A = \pi r^2 = (\pi)(7.5 \text{ ft})^2 = 56.25\pi \text{ ft}^2$

$$\text{Force} = 860 \frac{\text{lb}}{\text{ft}^2} \times 56.25\pi \text{ ft}^2$$

$$\approx 152,000 \text{ lb, rounded}$$

$$\pi \times 7.5 \, x^2 \times 860 = \rightarrow \boxed{151974.5446}$$

More Practice   Now try these problems.

(a) **Construction**   Find the area of stained glass used in a semicircular window with a radius of $9\frac{3}{4}$ in. (Use $\pi \approx 3.14$ and round to the nearest whole number.)

(b) **Sheet Metal Trades**   What is the area of the largest circle that can be cut out of a square piece of sheet metal 4 ft 6 in. on a side? (Round to the nearest tenth.)

Solutions   (a) For a semicircle, multiply the area by $\frac{1}{2}$ (or divide by 2):

$$A = \tfrac{1}{2}(\pi r^2) \approx \tfrac{1}{2}(3.14)(9.75 \text{ in.})^2$$

$$\approx 149 \text{ in.}^2$$

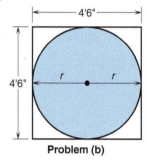

Problem (b)

(b) From the figure, we see that the diameter of the largest circle is equal to the side length of the square. Using 4 ft 6 in. = 4.5 ft, we have

$$A = \frac{\pi d^2}{4} = \frac{\pi (4.5 \text{ ft})^2}{4} \approx 15.9 \text{ ft}^2$$

Area of a Circle

To find the formula for the area of a circle, first divide the circle into many pie-shaped sectors, then rearrange them like this:

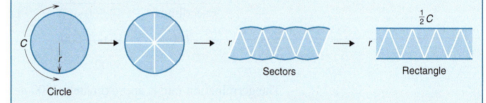

Circle   Sectors   Rectangle

When the sectors are rearranged, they fit into a rectangle whose height is r, the radius of the original circle, and whose base length is roughly half the circumference of the circle. The more sectors the circle is divided into, the better the fit. The area of the sectors is approximately equal to the area of the rectangle.

Area of circle, $A = r \cdot \tfrac{1}{2}C$

$$= r \cdot \tfrac{1}{2}(2\pi r)$$

$$= \pi r^2$$

Rings   A circular **ring**, or annulus, is defined as the area between two concentric circles—two circles having the same center point. A washer is a ring; so is the cross section of a pipe, a collar, or a cylinder.

To find the area of a ring, subtract the area of the inner circle from the area of the outer circle. Use the following formula:

Area of a Ring (Annulus)

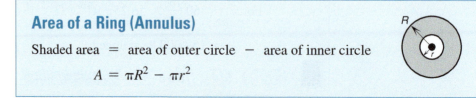

Shaded area = area of outer circle − area of inner circle

$$A = \pi R^2 - \pi r^2$$

Note This area formula can be written as $A = \pi(R^2 - r^2)$, but the version in the box is easier for most students to understand and remember. ●

EXAMPLE 6 To find the cross-sectional area of the wall of a ceramic pipe whose I.D. (inside diameter) is 8 in. and whose O.D. (outside diameter) is 10 in., substitute $r = 4$ in. and $R = 5$ in. into the formula.

$$A = \pi(5 \text{ in.})^2 - \pi(4 \text{ in.})^2$$
$$= (25\pi - 16\pi) \text{ in.}^2$$
$$= 9\pi \text{ in.}^2 \approx 28 \text{ in.}^2, \text{ rounded}$$

●

A Closer Look Notice in Example 6 that we did not substitute for π until the last step. It is usually easier to simplify all π-terms before substituting to get a decimal approximation. ●

Your Turn Find the area of a washer with an outside radius of $\frac{7}{16}$ in. and a thickness of $\frac{1}{8}$ in. (Round to three decimal digits.)

Solution $R = \frac{7}{16}$ in. $r = \frac{7}{16}$ in. $- \frac{1}{8}$ in. or $r = \frac{5}{16}$ in. then

$$A = \pi R^2 - \pi r^2$$
$$= \pi\left(\frac{7}{16} \text{ in.}\right)^2 - \pi\left(\frac{5}{16} \text{ in.}\right)^2$$
$$= \left(\frac{49}{256}\pi - \frac{25}{256}\pi\right) \text{ in.}^2$$
$$= \frac{24}{256}\pi \text{ in.}^2 = \frac{3}{32}\pi \text{ in.}^2 \approx 0.295 \text{ in.}^2$$

$\boxed{\pi} \boxed{\times} \boxed{7} \boxed{\text{A}^{\text{b}}_{\text{c}}} \boxed{1\ 6} \boxed{x^2} \boxed{-} \boxed{\pi} \boxed{\times} \boxed{5} \boxed{\text{A}^{\text{b}}_{\text{c}}} \boxed{1\ 6} \boxed{x^2} \boxed{=} \rightarrow$ `0.294524311`

Applications of Rings

More Practice Solve the following problems involving rings. (Round as directed in parentheses.)

(a) **Plumbing** Find the cross-sectional area of the wall of the pipe shown in the figure. (nearest ten)

(b) Find the area of a washer with an inside diameter (I.D.) of $\frac{1}{2}$ in. and an outside diameter (O.D.) of $\frac{7}{8}$ in. (nearest hundredth)

(c) **Machine Trades** What is the cross-sectional area of a steel collar 0.6 cm thick with an inside diameter of 9.4 cm? (nearest whole number)

Problem (a)

(d) **Landscaping** A landscape company charges $2.10 per square foot to install sod. What would the company charge for a circular ring of lawn, 6 ft wide, surrounding a flower garden with a radius of 20 ft? (nearest $10)

(e) **Plumbing** A pipe has an inside diameter of 8 in. How much larger is the cross-sectional area with a $\frac{1}{2}$-in.-thick wall than with a $\frac{3}{8}$-in.-thick wall? (nearest tenth)

Answers (a) 350 in.^{2} (b) 0.40 in.^{2} (c) 19 cm^2
(d) $1820   (e) 3.5 in.^{2}

A Closer Look   In problem (c),

Radius of inner hole $= \dfrac{9.4 \text{ cm}}{2} = 4.7 \text{ cm}$

Radius of outer edge $= 4.7 \text{ cm} + 0.6 \text{ cm} = 5.3 \text{ cm}$

Then,

Area $= \pi(5.3 \text{ cm})^2 - \pi(4.7 \text{ cm})^2 \approx 19 \text{ cm}^2$

In problem (e), organize your work like this:

	Thicker-Walled Pipe	*Thinner-Walled Pipe*
Inside radius	4 in.	4 in.
Outside radius	4 in. $+ \frac{1}{2}$ in. $= 4.5$ in.	4 in. $+ \frac{3}{8}$ in. $= 4.375$ in.
Cross-sectional area	$\pi(4.5 \text{ in.})^2 - \pi(4 \text{ in.})^2$	$\pi(4.375 \text{ in.})^2 - \pi(4 \text{ in.})^2$
	$= 13.3518$ sq in.	≈ 9.8666 sq in.
Difference	13.3518 sq in. $- 9.8666$ sq in. ≈ 3.5 sq in.	

$\boxed{\pi} \boxed{\times} \textbf{4.375} \boxed{x^2} \boxed{-} \boxed{\pi} \boxed{\times} \textbf{4} \boxed{x^2} \boxed{=} \rightarrow$ 9.866564428

$\boxed{\pi} \boxed{\times} \textbf{4.5} \boxed{x^2} \boxed{-} \boxed{\pi} \boxed{\times} \textbf{4} \boxed{x^2} \boxed{-} \boxed{\text{ANS}} \boxed{=} \rightarrow$ 3.48520435 ●

Now continue with Exercises 8-4 for a set of review problems on circles.

Exercises 8-4   Circles

A. Find the missing quantities for each circle. Use the $\boxed{\pi}$ key for π and round to the nearest tenth.

	Radius	Diameter	Circumference
1.	26.4 cm	_____	_____
2.	_____	7.50 yd	_____
3.	_____	_____	326 in.
4.	19.5 ft	_____	_____
5.	_____	11.4 m	_____
6.	_____	_____	1370 m

B. Find the circumference and area of each circle. Use the $\boxed{\pi}$ key for π and round to the nearest tenth.

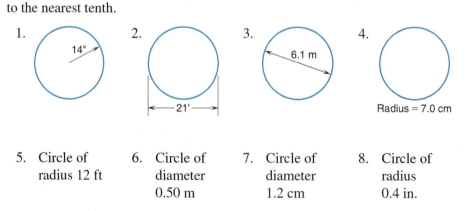

1. 14"
2. 21'
3. 6.1 m
4. Radius = 7.0 cm

5. Circle of radius 12 ft

6. Circle of diameter 0.50 m

7. Circle of diameter 1.2 cm

8. Circle of radius 0.4 in.

9. Find the area of a circle with a circumference of 73.6 in.

10. Find the area of a circle with a circumference of 8.72 m.

C.   Find the area of each figure. Use the $\boxed{\pi}$ key for π and round to the nearest tenth.

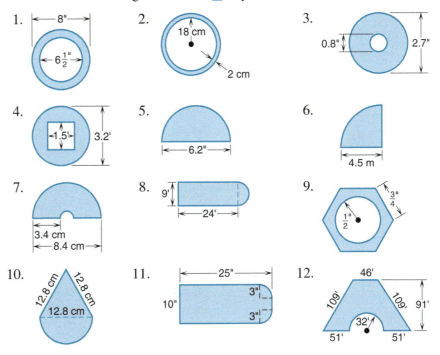

Find the length of stock needed to create each shape. Round to the nearest tenth.

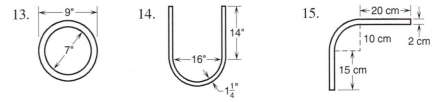

D.   Practical Applications

Problem 1

Problem 3

Problem 6

1. **Metalworking**   What length of 1-in. stock is needed to bend a piece of steel into the shape shown in the figure? (Use $\pi \approx 3.14$ and round to the nearest tenth.)

2. **Landscaping**   (a) Calculate, to the nearest square foot, the cultivable area of a circular garden 38 ft in diameter if the outer 2 ft are to be used for a path. (b) Find the area of the path.

3. **Machine Trades**   What diameter must a circular piece of stock be to mill a hexagonal shape with a side length of 0.3 in.? (See the figure.)

4. **Machine Trades**   In problem 3, how many square inches of stock will be wasted? (Round to the nearest hundredth.)

5. **Plumbing**   What is the cross-sectional area of a cement pipe 2 in. thick with an I.D. of 2 ft? (Use $\pi \approx 3.14$ and round to the nearest whole number.)

6. **Machine Trades**   In the piece of steel shown in the figure, $4\frac{1}{2}$-in.-diameter holes are drilled in a 22-in.-diameter circular plate. Find the area remaining. (Use $\pi \approx 3.14$ and round to the nearest whole number.)

7. **Plumbing**   How much additional cross-sectional area is there in a 30-in.-I.D. pipe 2 in. thick than one 1 in. thick? (Round to one decimal place.)

8. **Machine Trades**   What diameter round stock is needed to mill a hexagonal nut measuring $1\frac{1}{2}$ in. across the flats? (Round up to the nearest quarter of an inch.)

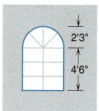

Problem 10

Problem 11

9. **Landscaping** How many plants spaced every 6 in. are needed to surround a circular walkway with a 25-ft radius? (Use $\pi \approx 3.14$.)

10. **Machine Trades** What diameter circular stock is needed to mill a square end 3 cm on a side? (See the figure. Round to the nearest hundredth.)

11. **Carpentry** At $11.25 per foot for the curved portion and $1.75 per foot for the straight portion, how much will it cost to put molding around the window pictured? (Use $\pi \approx 3.14$.)

12. **Sports and Leisure** If the radius of the semicircular ends of the track in the figure is 32 m, how long must each straightaway be to make a 400-m track? (Use $\pi \approx 3.14$.)

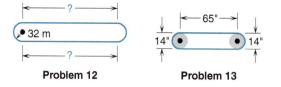

Problem 12   **Problem 13**

13. **Manufacturing** What is the total length of belting needed for the pulley shown in the figure? (Round to the nearest whole number.)

14. **Electrical Trades** An electrician bends a $\frac{1}{2}$-in. conduit as shown in the figure to follow the bend in a wall. Find the total length of conduit needed between A and B. (Use $\pi \approx 3.14$.)

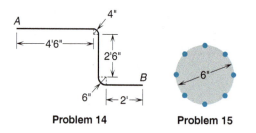

Problem 14   **Problem 15**

15. **Metalworking** How many inches apart will eight bolts be along the circumference of the metal plate shown in the figure? (Round to the nearest tenth of an inch.)

16. **Landscaping** How many square *yards* of concrete surface are there in a walkway 8 ft wide surrounding a tree if the inside diameter is 16 ft? (Round to one decimal place. See the figure.)

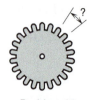

Problem 16

17. **Carpentry** The circular saw shown in the figure has a diameter of 18 cm and 22 teeth. What is the spacing between teeth? (Round to the nearest tenth.)

18. **Automotive Trades** A hose must fit over a cylindrical opening 57.2 mm in diameter. If the hose wall is 3.0 mm thick, what length of clamp strap is needed to go around the hose connection? (Round to the nearest whole number.)

Problem 17

19. **Automotive Trades** What length on the circumference would a 20° angle span on a 16-in. flywheel? (Round to the nearest hundredth.)

20. **Automotive Trades** In making a certain turn on a road, the outside wheels of a car must travel 8.75 ft farther than the inside wheels. If the tires are 23 in. in diameter, how many additional turns do the outside wheels make during this turn? (*Hint:* In one "turn," a tire travels a distance equal to its circumference. Convert 8.75 ft to inches.)

21. **Water/Wastewater Treatment** To find the velocity of flow (in feet per second, or fps) through a pipe, divide the rate of flow (in cubic feet per second,

or cfs) by the cross-sectional area of the pipe (in square feet). If the rate of flow through an 8.0-in.-diameter pipe is 0.75 cfs, find the velocity of flow. (Round to the nearest tenth.)

22. **Plumbing** The pressure exerted by a column of water amounts to 0.434 psi of surface for every foot of height or *head.* What is the total force (pressure times area) exerted on the bottom surface of a cylindrical container 16.0 in. in diameter if the head is 4.50 ft? (Round to the nearest whole number.)

23. **Plumbing** To construct a bracket for a drain pipe, Jerry the plumber needs to know its circumference. Calculate the circumference of a pipe 42.3 cm in diameter. (Round to the nearest centimeter.)

24. **Culinary Arts** A small pizza at Papa's Pizza Place is 9 in. in diameter and a medium pizza is 12 in. in diameter. If a small pizza serves four customers, how many servings does a medium pizza provide?

25. **Sports and Leisure** Two cars are racing around a circular track. The red car is traveling in a lane that is 82 ft from the center. The blue car is in a lane that is 90 ft from the center. If both cars are traveling at 120 mph, how much later, in seconds, will the blue car finish if both cars race one lap around the track? (Round to the nearest 0.1 sec. *Hint:* 60 mph $\approx$ 88 fps.)

26. **Forestry** Forest service biologists roped off a circular area of recently burned forest to determine how well the forest was recovering. The circular sample area had a 32-ft radius and contained 220 new seedlings. Calculate the germination rate in seedlings per acre. (Use $\pi \approx 3.14$ and 1 acre = 43,560 sq ft. Round to the nearest 10 seedlings.) See Example 5 on page 591.

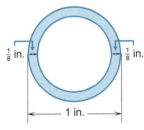

$\frac{1}{8}$ in. $\frac{1}{8}$ in.

1 in.

Problem 27

27. **Welding** To determine the ultimate tensile strength (UTS) of a small-diameter pipe, a welder uses the formula

$$\text{UTS} = \frac{L}{A}$$

where L is the maximum load of the pipe, in pounds, as recorded on a tensile machine, and A is the cross-sectional area of the pipe. If the pipe shown in the figure had a maximum load of 24,280 lb, find the UTS for the pipe in pounds per square inch (psi). (Round to the nearest 10 psi.)

28. **Construction** The size of fiber line required to lift a given load is given by the formula

$$C = \sqrt{15W}$$

where C is the circumference of the line in inches and W is the weight of the load in tons. What *diameter* line is needed to lift a 4.5-ton load?

29. **Electrical Trades** The **circular mil** is a unit used in electrical trades to indicate the cross-sectional area of wires. The cross-sectional area of a wire in circular mils is given by the formula $A = d^2$, where d is the diameter of the wire in thousandths of an inch. For example, the area of wire with diameter 0.105 in. in circular mils is $A = 105^2$. Find the cross-sectional area of a wire in circular mils if the diameter is 0.125 in.

30. **Sheet Metal Trades** What is the area of the largest circle that can be cut out of a square piece of sheet metal 3 ft 3 in. on a side? (Round to the nearest tenth of a square foot.)

Check your answers to the odd-numbered problems in the Appendix, then turn to Problem Set 8 on page 602 for practice on the work of Chapter 8. If you need a quick review of the topics in this chapter, visit the chapter Summary first.

CHAPTER 8
SUMMARY

Practical Plane Geometry

Objective	Review
Measure angles with a protractor. (p. 533)	Place the protractor over the angle so that the zero-degree mark is aligned with one side of the angle and the center mark is on the vertex. Read the measure, in degrees, where the other side of the angle intersects the scale of the protractor.

Example:

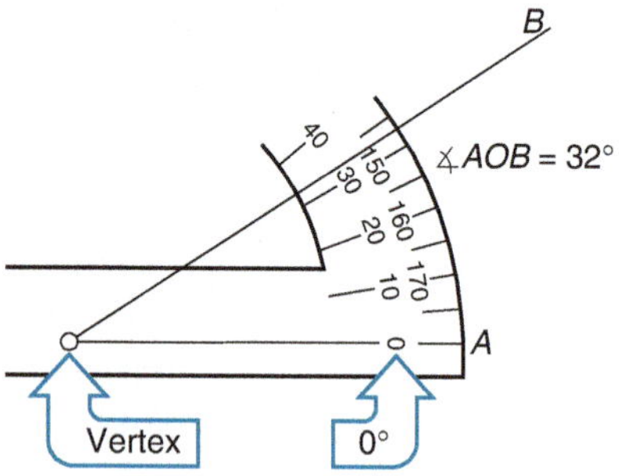

$\angle AOB = 32°$

Vertex   0°

Classify angles. (p. 534)	An acute angle measures less than 90°. A right angle measures exactly 90°. An obtuse angle measures greater than 90° and less than 180°. A straight angle measures exactly 180°.

Example:

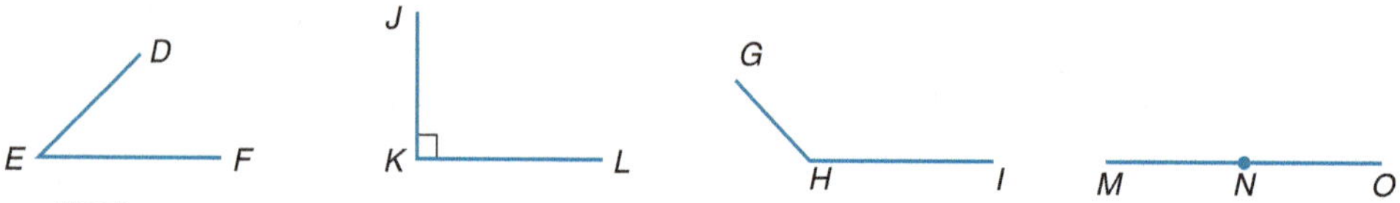

$\angle DEF$ is acute. $\angle JKL$ is right. $\angle GHI$ is obtuse. $\angle MNO$ is a straight angle.

Use simple geometric relationships involving intersecting lines and triangles. (p. 536)	When two lines intersect, the vertical angles are always equal.

Example:

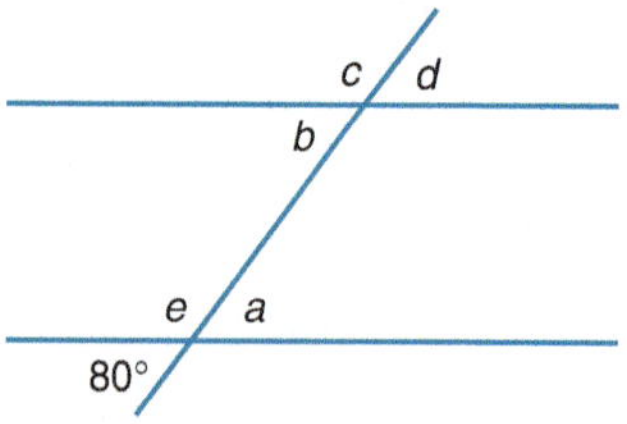

$\angle a = 80°$

80°

When parallel lines are cut by a third line, the corresponding angles are equal, and the alternate interior angles are equal.

Example: If the horizontal lines are parallel in the preceding figure, $\angle a$ and $\angle d$ are equal corresponding angles; therefore, $\angle d = 80°$. Also, $\angle a$ and $\angle b$ are equal alternate interior angles. Therefore $\angle b = 80°$.

When two lines intersect, each pair of adjacent angles sums to 180°.

Example: In the preceding figure, $\angle b$ and $\angle c$ are adjacent angles; therefore:

$$\angle b + \angle c = 180°   \text{ or }   80° + \angle c = 180°$$

$$\angle c = 180° - 80° = 100°$$

Objective	Review

The interior angles of a triangle always sum to 180°.

Example: $x + 70° + 90° = 180°$

$x + 160° = 180°$

$x = 180° - 160° = 20°$

Identify polygons, including triangles, squares, rectangles, parallelograms, trapezoids, and hexagons. (pp. 545, 561, 572)

Triangles may be classified according to their sides or their angles as follows:

Equilateral: all sides equal
Isosceles: at least two sides equal
Scalene: no sides equal

Acute: all three angles acute
Obtuse: contains one obtuse angle
Right: contains a 90° angle

Example:

ΔABC is an obtuse triangle and a scalene triangle.

The following quadrilaterals (four-sided polygons) are useful in the trades:

Parallelogram: opposite sides parallel and equal
Rectangle: parallelogram with four right angles
Square: rectangle with all sides equal
Trapezoid: two parallel sides and two nonparallel sides

Example:

DEFG is a parallelogram.   *HIJK* is a trapezoid.

A hexagon is a six-sided polygon. In a regular hexagon, all six sides are equal, and all six angles are equal.

Use the Pythagorean theorem. (p. 562)

In a right triangle, the square of the hypotenuse is equal to the sum of the squares of the other two sides (the legs).

Example:

$x^2 + (1.2)^2 = (1.9)^2$
$x^2 + 1.44 = 3.61$
$x^2 = 2.17$
$x \approx 1.5"$

Find the area and perimeter of geometric figures. (pp. 543–594)

Use the formulas listed in the following summary.

Summary of Formulas for Plane Figures

Figure		Area	Perimeter
Rectangle		$A = LW$	$P = 2L + 2W$
Square		$A = s^2$	$P = 4s$

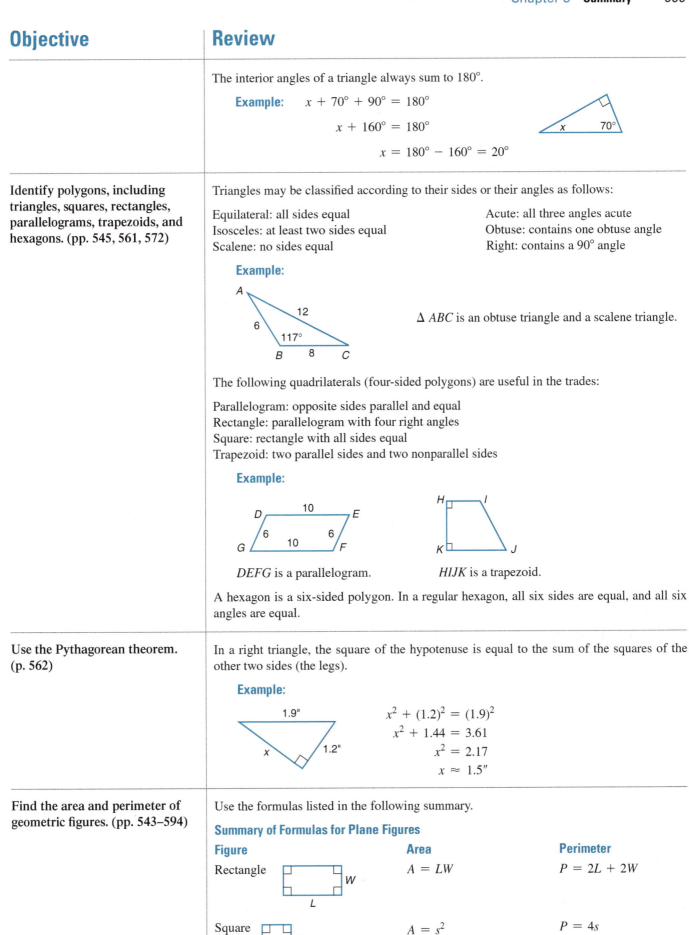

Objective

Review

Figure | Area | Perimeter

Parallelogram

$A = bh$

$P = 2a + 2b$

Triangle

$A = \dfrac{1}{2}bh$ or $\dfrac{bh}{2}$

$P = a + b + c$

$A = \sqrt{s(s-a)(s-b)(s-c)}$

where $s = \dfrac{a + b + c}{2}$

Equilateral triangle

$A = \dfrac{a^2\sqrt{3}}{4} \approx 0.433a^2$

$P = 3a$

Isosceles triangle

$A = \dfrac{b}{2}\sqrt{a^2 - \left(\dfrac{b}{2}\right)^2}$

$P = 2a + b$

Right triangle

$A = \dfrac{1}{2}ab$

$P = a + b + c$

Trapezoid

$A = \left(\dfrac{b_1 + b_2}{2}\right)h$

or $A = \dfrac{h}{2}(b_1 + b_2)$

$P = a + c + b_1 + b_2$

Regular hexagon

$A = \dfrac{3a^2\sqrt{3}}{2} \approx 2.598a^2$

$P = 6a$

Circle

$A = \pi r^2$

$C = \pi d$

$A = \dfrac{\pi d^2}{4} \approx 0.7854\, d^2$

$C = 2\pi r$

Ring

$A = \pi R^2 - \pi r^2$

$C = 2\pi R$, the outside circumference

Example:

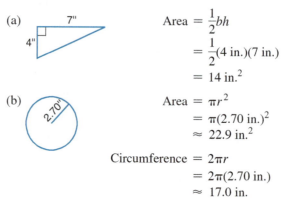

(a) 7" 4"

Area $= \dfrac{1}{2}bh$

$= \dfrac{1}{2}(4 \text{ in.})(7 \text{ in.})$

$= 14 \text{ in.}^2$

(b) 2.70"

Area $= \pi r^2$

$= \pi(2.70 \text{ in.})^2$

$\approx 22.9 \text{ in.}^2$

Circumference $= 2\pi r$

$= 2\pi(2.70 \text{ in.})$

$\approx 17.0 \text{ in.}$

Objective	Review
Solve practical problems involving area and perimeter of plane figures. (pp. 554, 570, 574, 579, 591, 593)	Use the appropriate formula from the preceding summary to find the area or perimeter of the figure in the problem. Then perform any further calculations as needed.

Example: A circular flowerbed with a radius of 8.5 ft is surrounded by a circular ring of lawn 12 ft wide. At $2.10 per square foot, how much will it cost to sod the lawn?

First, find the area of the ring of lawn:

Notice that the radius of the outer circle is the radius of the inner circle plus the width of the lawn, or 8.5 ft + 12 ft = 20.5 ft.

$$A = \pi R^2 - \pi r^2 = \pi(20.5 \text{ ft})^2 - \pi(8.5 \text{ ft})^2$$
$$= 348\pi \text{ sq ft}$$

Then, multiply by $2.10 per sq ft:

$$(348\pi)(\$2.10) \approx \$2295.88$$

PROBLEM SET 8   Practical Plane Geometry

Answers to the odd-numbered problems are given in the Appendix.

A. Solve the following problems involving angles.

Name each angle and tell whether it is acute, obtuse, or right.

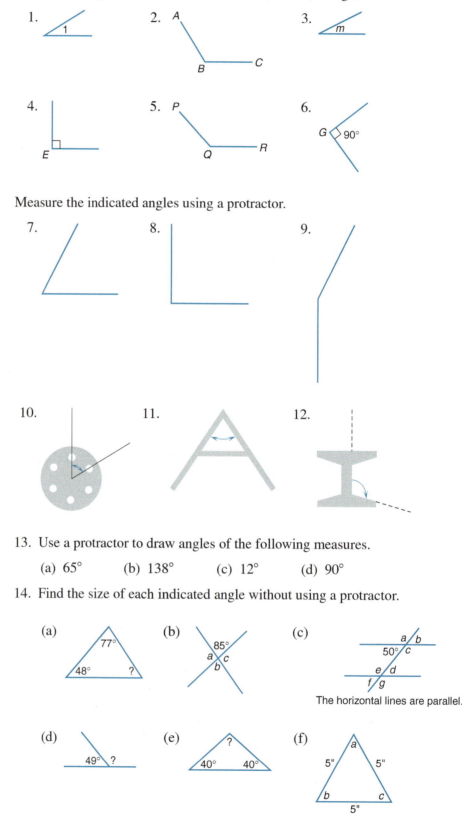

1.

2. A   B   C

3. m

4. E

5. P   Q   R

6. G   90°

Measure the indicated angles using a protractor.

7.

8.

9.

10.

11.

12.

13. Use a protractor to draw angles of the following measures.

(a) 65°   (b) 138°   (c) 12°   (d) 90°

14. Find the size of each indicated angle without using a protractor.

(a)   77°   48°   ?

(b)   85°   a c   b

(c)   a b   50° c   e d   f g
The horizontal lines are parallel.

(d)   49° ?

(e)   ?   40°   40°

(f)   a   5"   5"   b   c   5"

Name

Date

Course/Section

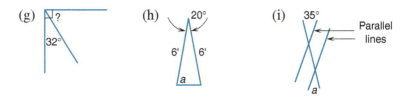

(g) ? 32°

(h) 20° 6' 6' a

(i) 35° Parallel lines a

B. Find the outside perimeter (or circumference) and shaded area of each figure. Round to the nearest tenth if necessary.

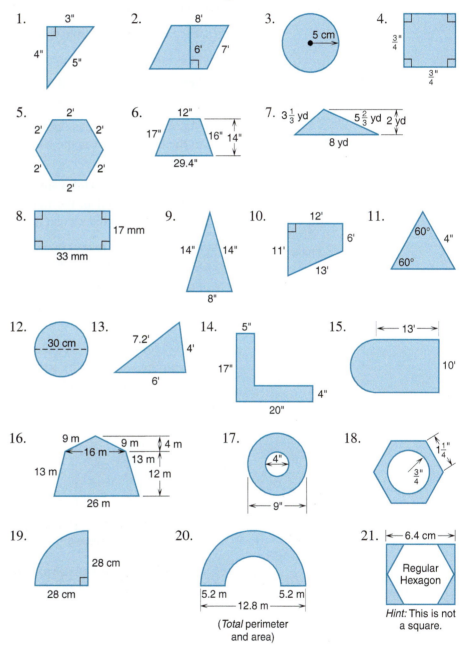

1. 3" 4" 5"

2. 8' 6' 7'

3. 5 cm

4. $\frac{3}{4}$" $\frac{3}{4}$"

5. 2' 2' 2' 2' 2' 2'

6. 12" 17" 16" 14" 29.4"

7. $3\frac{1}{3}$ yd $5\frac{2}{3}$ yd 2 yd 8 yd

8. 17 mm 33 mm

9. 14" 14" 8"

10. 12' 6' 11' 13'

11. 60° 4" 60°

12. 30 cm

13. 7.2' 4' 6'

14. 5" 17" 4" 20"

15. 13' 10'

16. 9 m 9 m 4 m 16 m 13 m 13 m 12 m 26 m

17. 4" 9"

18. $1\frac{1}{4}$" $\frac{3}{4}$"

19. 28 cm 28 cm

20. 5.2 m 5.2 m 12.8 m (*Total* perimeter and area)

21. 6.4 cm Regular Hexagon *Hint:* This is not a square.

C. Find the missing dimensions of the following figures. Round to the nearest hundredth if necessary.

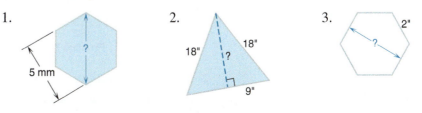

1. ? 5 mm

2. 18" 18" ? 9"

3. 2" ?

4.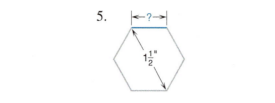

5.

6.

7. Find the radius of a circle with a circumference of 8.75 yd.

8. Find the diameter of a circle with a circumference of 64.8 cm.

9. Find the area of a circle with a circumference of 168.50 in.

D. Practical Applications

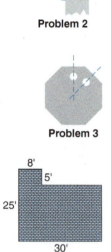

1. **Construction**   What will it cost to pave a rectangular parking lot 220 ft by 85 ft at $5 per square foot?

2. **Metalworking**   A steel brace is used to strengthen the table leg as shown in the figure. What is the length of the brace? (Round to the nearest hundredth of an inch.)

Problem 2

3. **Machine Trades**   Holes are punched in a regular octagonal (8-sided) steel plate as indicated in the figure. Through what angle must the plate be rotated to locate the second hole?

Problem 3

4. **Masonry**   How many square feet of brick are needed to lay a patio in the shape shown in the figure?

5. **Electrical Trades**   A coil of wire has an average diameter of 8.0 in. How many feet of wire are there if it contains 120 turns? (Round to the nearest ten feet.)

Problem 4

6. **Sheet Metal Trades**   A sheet metal worker needs to make a vent connection in the shape of the trapezoidal prism shown in the drawing. Use a protractor to find the indicated angle.

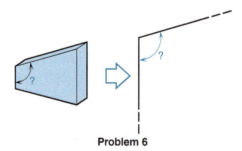

Problem 6

7. **Landscaping**   At $16 per foot, how much does it cost to build a chain-link fence around the yard shown in the figure?

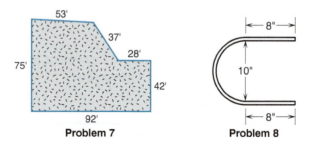

Problem 7   **Problem 8**

8. **Metalworking**   Find the length of straight stock needed to bend $\frac{1}{2}$-in. steel into the shape indicated. (Use $\pi \approx 3.14$ and round to the nearest tenth.)

9. **Machine Trades** A hexagonal piece of steel 9 in. on a side must be milled by a machinist. What diameter round stock does he need? (See the figure. Round to the nearest tenth of an inch.)

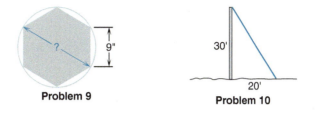

Problem 9

Problem 10

10. **Construction** How much guy wire is needed to anchor a 30-ft pole to a spot 20 ft from its base? Allow 4 in. for fastening and round to the nearest inch. (See the figure.)

11. **Carpentry** How many 4-ft by 8-ft sheets of exterior plywood must be ordered for the 4 walls of a building 20 ft long, 32 ft wide, and 12 ft high? Assume that there are 120 sq ft of window and door space. (*Note:* You can cut the sheets to fit, but you may not order a fraction of a sheet.)

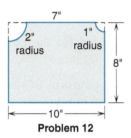

Problem 12

12. **Metalworking** Find the perimeter of the steel plate shown in the figure. (Round to the nearest tenth of an inch.)

13. **Construction** At what angle must Sheetrock be cut to conform to the shape of the wall shown in the figure? Measure with a protractor.

14. **Metalworking** A triangular shape with a base of 6 in. and a height of 9 in. is cut from a steel plate weighing 5.1 lb/sq ft. Find the weight of the shape (1 sq ft = 144 sq in.). (Round to the nearest ounce.)

15. **Construction** An air shaft is drilled from the indicated spot on the hill to the mine tunnel below. (See the figure.) How long is the shaft to the nearest foot?

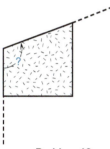

Problem 13

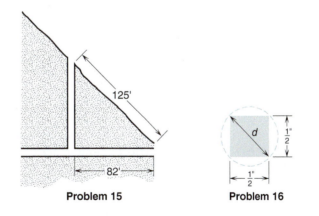

Problem 15

Problem 16

16. **Machine Trades** What diameter circular stock is needed to mill a square bolt $\frac{1}{2}$ in. on a side? (Round to the nearest thousandth. See the figure.)

17. **Life Skills** A 17-ft roll of weather stripping costs $5.98, including tax. Suppose you need to weather-strip the following: 6 windows measuring 4 ft by 6 ft, 8 windows measuring 3 ft by 2 ft, and all sides except the bottom of two doors 3 ft wide and 7 ft high.
    (a) How many rolls will you need?
    (b) What will it cost to purchase enough weather stripping?

18. **Construction**   How many square feet must be plastered to surface the wall shown?

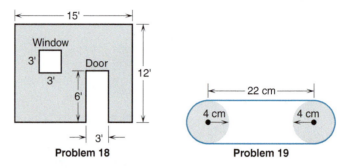

Problem 18   **Problem 19**

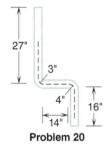

Problem 20

Problem 21

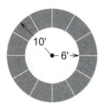

Problem 22

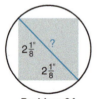

Problem 24

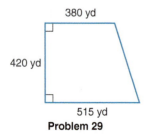

Problem 29

19. **Manufacturing**   What length of belt is needed for the pulley shown? (Round to the nearest tenth.)

20. **Electrical Trades**   Electrical conduit must conform to the shape and dimensions shown. Find the total length of conduit in inches. (Use $\pi \approx 3.14$ and round to the nearest tenth.)

21. **Construction**   At $11 per square foot, how much will it cost to cover the circular pathway shown in the figure with pavers? (Use $\pi \approx 3.14$.)

22. **Carpentry**   Find the size of $\angle 2$ when the carpenter's square is placed over the board as shown in the figure.

23. **Welding**   The top to a container is made of steel weighing 4.375 lb/sq ft. If the top is a 30.0 in. by 30.0 in. square, find its weight to the nearest tenth.

24. **Machine Trades**   A square bolt must be machined from round stock. If the bolt must be $2\frac{1}{8}$ in. on each side, what diameter stock is needed? (See the figure. Round to the nearest thousandth.)

25. **Automotive Trades**   A fan pulley has a diameter of $3\frac{7}{8}$ in. What is its circumference? (Round to the nearest hundredth.)

26. **Fire Protection**   Firefighters report that a wildfire with a front about 10 miles long has consumed approximately 25,000 acres. What is the average width of the destruction in miles? (1 sq mi = 640 acres. Round to the nearest mile.)

27. **Automotive Trades**   If the tires of a car have a diameter of 24 in., how many miles will the car travel when the tires make 16,000 revolutions? (*Remember:* During one revolution of the tires, the car travels a distance equal to the circumference of the tire. Round to the nearest mile.)

28. **Agriculture**   At the bean farm, the automatic seedling planter creates rows 30 in. apart and places seedlings 24 in. apart. If the bean field is a square 680 ft on a side, how many bean plants will the machine plant? (*Hint:* Add one row and one plant per row to account for rows and/or plants on all four sides of the square.)

29. **Agriculture**   The plot of farmland shown in the figure is for sale at $6850 per acre. What is the total price of the land, rounded to the nearest thousand dollars? (*Hint:* One acre = 4840 square yards.)

30. **Construction**   At a price of $14.25 per foot, how much will it cost to purchase molding for the perimeter of the hexagonal window shown in the figure on the next page?

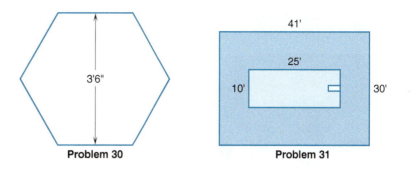

Problem 30   Problem 31

31. **Masonry**   The figure shows a rectangular swimming pool surrounded by a rectangular patio. The surface of the patio will be covered with hexagonal pavers measuring 12 in. across the corners. How many pavers will be needed?

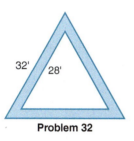

Problem 32

32. **Landscaping**   The figure shows a flower garden in the shape of an equilateral triangle surrounded by a pathway of lawn. The outer triangle has a side length of 32 ft, and the inner triangle has a side length of 28 ft.
    (a) If bulbs are to be planted at a density of 15 per square foot, how many bulbs will be needed for the flower garden?
    (b) At a cost of $9.25 per square foot, how much will it cost to plant artificial turf on the pathway? (Round to the nearest dollar.)

33. **Construction**   The size of a wire rope needed to lift a given load is given by the formula

$$C = \sqrt{2.5W}$$

where C is the circumference of the wire rope in inches, and W is the weight of the load in tons. What *diameter* wire rope is needed to lift a 3.9-ton load?

34. **Flooring and Carpeting**   Strip flooring must be installed on a rectangular floor with dimensions of 14.5 ft by 12 ft. The contractor estimates that this type of flooring generates 30% waste. How much flooring should the contractor order to account for the waste?

35. **Masonry**   The front side of a "slim jumbo modular" brick measures $2\frac{1}{8}$ in. by $7\frac{5}{8}$ in. With a $\frac{1}{4}$-inch mortar joint, how many bricks would be needed per square yard (1296 square inches) on the front side of a wall? (Round to the nearest whole number.)

36. **Agriculture**   Sweet corn is being planted in a 24-ft by 18-ft rectangular plot. Sweet corn must be planted 15-in. apart with 30 in. between rows. How many sweet corn seeds can be planted in this plot?

37. **Flooring and Carpeting**   How many 9-inch square tiles are needed to cover 64 square *feet* of floor area if we add 10% for waste?

38. **Construction**   A rectangular room measures 14 ft by 20 ft and has an 8-ft ceiling. The cost of lath and plaster is $42 per square *yard* with no deductions for openings. What is the total cost to lath and plaster all four walls and the ceiling?

CASE STUDY: The Basketball Arena

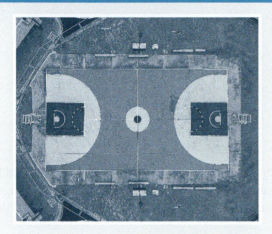

Funds have recently become available to renovate the basketball arena at Shady Acres Community College. The main project will be to repaint the court itself (See the figure at the bottom of the page.). Leftover funds will be used for a couple of additional projects.

1. Donated paint will be used for repainting the blue and red areas of the gym floor. Each gallon of donated paint covers 200 square feet.

    (a) Blue paint will be used in the following areas: (1) the circular ring at center court; (2) the two semicircles at the top of each key; and (3) the four alleys on the outside edges of the key. How many gallons of blue paint will be needed? (Round up to the nearest whole gallon.)

    (b) Red paint will be needed for the following areas: (1) the smaller circle at the center of the court; (2) the key areas under the baskets. An additional gallon will be needed to repaint the 3-point line. How many gallons of red paint will be needed? (Round up to the nearest whole gallon.)

2. The court itself measures 84 ft by 50 ft with a 4-ft-wide walking perimeter around the gym floor. Two coats of clear polyurethane will be used to refinish this entire surface. If the polyurethane covers 350 square feet per gallon, how many gallons will be needed?

3. The estimated time to sand, repaint, and resurface the gym floor is 60 hours. If three employees work on the project for 8 hours a day, how many days will it take to finish the work on the gym floor?

4. If the employees are paid \$15.25 per hour, what is the total cost of the labor for the gym floor resurfacing?

5. A new type of sander will require 8% less sanding time compared to the old sander. If $\frac{1}{4}$ of the labor estimate is dedicated to sanding, how much time will be saved by switching to the new sander?

6. The stairs at the service entrance to the gymnasium are in need of some additional bracing on the underside between the rise and tread of each stair. One end of the bracing bar will be attached to the backside of the rise and the other to the underside of the tread. The stairs have a rise of 18 cm and a tread (run) of 26 cm. The school has a pile of braces left over from a previous job with lengths of 15 cm, 28 cm, 34 cm, and 45 cm. The braces cannot be cut. Find which length of brace would provide the best support for the stairs.

7. The school is hoping to erect a hexagon-shaped concession stand in an area near the entrance of the gym. If the available area allows for the distance across the corners to be 13.2 ft, what would the area of the concession stand be?

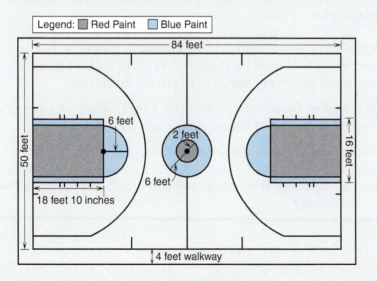

9 Solid Figures

Objective	Sample problems	For help, go to page

When you finish this chapter you will be able to:

1. Identify solid figures, including prisms, cubes, cones, cylinders, pyramids, spheres, and frustums.

(a) _____ 613

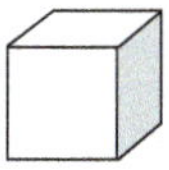

(b) _____

(c) _____ 646

2. Find the surface area and volume of solid objects. (Round to one decimal place.)

(a) 2' 4'

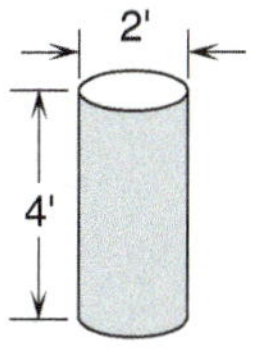

Volume = _____ 634

Total surface area = _____

(b) Sphere
$r = 2.1$ cm

Volume = _____ 638

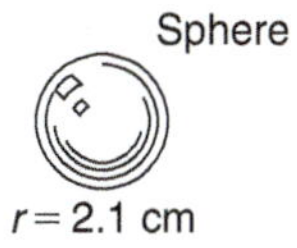

Surface area = _____

(c) 4" 0.6"

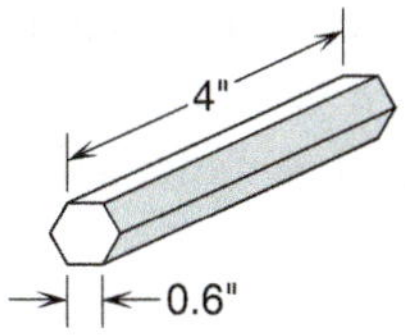

Volume = _____ 614

(d) 8" 6"

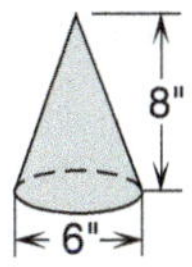

Lateral surface area = _____ 643

Volume = _____

Name _____

Date _____

Course/Section _____

Objective	Sample problems	For help, go to page

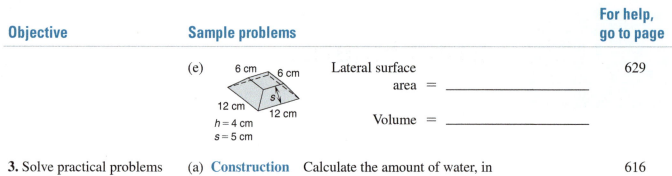

(e)   Lateral surface
                 area = _____

       Volume = _____

629

3. Solve practical problems involving solid figures.

(a) **Construction**   Calculate the amount of water, in gallons, needed to fill the swimming pool shown. (1 cu ft ≈ 7.48 gal.) (Round to the nearest 100 gallons.)

616

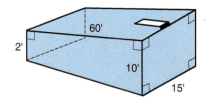

(b) **Metalworking**   Find the weight, in pounds, of the steel pin shown if its density is 0.2963 lb/cu in. (Round to the nearest tenth.)

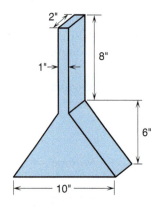

(c) **Roofing**   The roof of a building is in the shape of a pyramid having a square base 84 ft on each side. If the slant height of the roof is 60 ft, what would be the cost of the roofing material needed, at $0.95 per sq ft?

628

(Answers to these preview problems are given in the Appendix. Also, worked solutions to many of these problems appear in the chapter Summary.)

If you are certain that you can work *all* these problems correctly, turn to page 654 for a set of practice problems. If you cannot work one or more of the preview problems, turn to the page indicated to the right of the problem. Those who wish to master this material with the greatest success should turn to Section 9-1 and begin work there.

Solid Figures

In the previous chapter, we studied plane figures—those with only two dimensions. For example, a rectangle has just length and width and can be drawn on a flat plane surface. In this chapter, we will study solid figures—those with three dimensions. For example, a rectangular prism has length, width, and height, and making an exact model of such a figure requires shaping it with paper, clay, wood, metal, or plastic.

Trades workers encounter solid or three-dimensional figures in the form of tanks, pipes, ducts, boxes, birthday cakes, and buildings. They must be able to identify the solid and its component parts and compute its surface area and volume.

CASE STUDY: The Majestic Sign Company

In the Case Study at the end of this chapter, you will experience a specific example illustrating how volume calculations can help The Majestic Sign Company make a critical decision regarding the installation of billboards.

9-1   Prisms

learning | catalytics™

1. What is the perimeter of a regular hexagon with side length 4 cm?
2. Find the area:

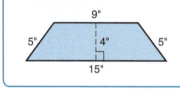

A **prism** is a solid figure having at least one pair of parallel surfaces that create a uniform cross section. The figure below shows a hexagonal prism.

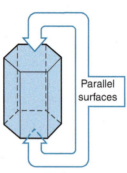

Parallel surfaces

Hexagonal prism

Cutting the prism anywhere parallel to the hexagonal surfaces produces the same hexagonal cross section.

All the polygons that form the prism are called **faces**. The faces that create the uniform cross section are called the **bases**. They give the prism its name. The others are called **lateral faces**.

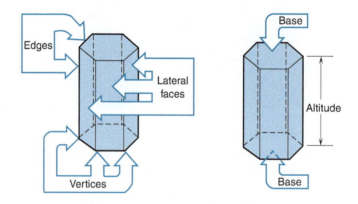

Edges

Lateral faces

Vertices

Base

Altitude

Base

The sides of the polygons are the **edges** of the prism, and the corners are still referred to as **vertices**. The perpendicular distance between the bases is called the **altitude** of the prism.

In a **right prism** the lateral edges are perpendicular to the bases, and the lateral faces are rectangles. In this section you will learn about right prisms only.

Right prisms

Not right prisms

Your Turn The components of a prism are its faces, vertices, and edges. Count the number of faces, vertices, and edges on this prism:

Number of faces = _____

Number of vertices = _____

Number of edges = _____

Answers There are 7 faces (2 bases and 5 lateral faces), 15 edges, and 10 vertices on this prism.

Here are some examples of simple prisms.

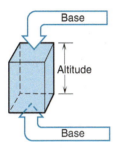

Base

Altitude

Base

Rectangular Prism The most common solid in practical work is the *rectangular prism* (see the figure in the margin). All opposite faces are parallel, so any pair can be chosen as bases. Every face is a rectangle. The angles at all vertices are right angles.

Cube A *cube* is a rectangular prism in which all edges are the same length.

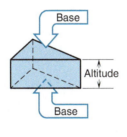

Triangular Prism In a **triangular prism** the bases are identical triangles.

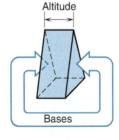

Trapezoidal Prism In a **trapezoidal prism** the bases are identical trapezoids.

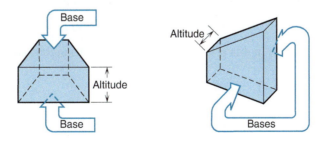

We know that the trapezoids and not the other faces are the bases because cutting the prism anywhere parallel to the trapezoid faces produces another trapezoidal prism.

Three important quantities can be calculated for any prism: the lateral surface area, the total surface area, and the volume. The following formulas apply to all right prisms:

Right Prisms

Lateral surface area	$L = ph$
Total surface area	$T = L + 2A$
Volume	$V = Ah$

h = altitude
p = perimeter of the base
A = area of the base

The **lateral surface area** L is the area of all surfaces *excluding* the two bases.

The **total surface area** T is the lateral surface area *plus* the area of the two bases.

The **volume** or capacity of a prism is the total amount of space inside it. Volume is measured in cubic units. (You may want to review these units in Chapter 5, pages 303 and 329.)

✎ **Note**   As with units of area, units of volume can be written using exponents. For example, cubic inches (cu in.) result from multiplying inches by inches by inches, which in exponent form is in.^{3}. Similarly, cubic yards (cu yd) can be expressed as yd^3, cubic meters (cu m) as m^3, and so on. Workers in the trades use both types of notation, and we shall use both in the remainder of this text. ●

EXAMPLE 1   Let's look at an example. Find the (a) lateral surface area, (b) total surface area, and (c) volume of the triangular cross-section duct shown.

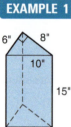

(a) The perimeter of the base is

$$p = 6 \text{ in.} + 8 \text{ in.} + 10 \text{ in.} = 24 \text{ in. and } h = 15 \text{ in.}$$

Therefore, the lateral surface area is

$$L = ph = (24 \text{ in.})(15 \text{ in.}) = 360 \text{ in.}^2$$

(b) In this prism the bases are right triangles, so the area of each base is

$$A = \frac{1}{2}bh$$
$$= \frac{1}{2}(6 \text{ in.})(8 \text{ in.})$$
$$= 24 \text{ in.}^2$$

Therefore, the total surface area is

$$T = L + 2A$$
$$= 360 \text{ in.}^2 + 2(24 \text{ in.}^2)$$
$$= 408 \text{ in.}^2$$

(c) The volume of the prism is

$$V = Ah$$

$$= (24 \text{ in.}^2)(15 \text{ in.})$$

$$= 360 \text{ in.}^3$$

Your Turn Find L, T, and V for the right trapezoidal prism shown in the margin.

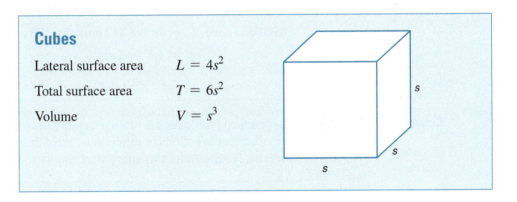

Solution From the figure, $h = 10$ ft and $p = 4$ ft + 15 ft + 12 ft + 17 ft = 48 ft

The area of each trapezoidal base is

$$A = \left(\frac{b_1 + b_2}{2}\right)h$$

$$= \left(\frac{4 \text{ ft} + 12 \text{ ft}}{2}\right)(15 \text{ ft})$$

$$= 120 \text{ ft}^2$$

Therefore,

$L = ph$	$T = L + 2A$	$V = Ah$
$= (48 \text{ ft})(10 \text{ ft})$	$= 480 \text{ ft}^2 + 2(120 \text{ ft}^2)$	$= (120 \text{ ft}^2)(10 \text{ ft})$
$= 480 \text{ ft}^2$	$= 720 \text{ ft}^2$	$= 1200 \text{ ft}^3$

Because all edges of a cube are the same length, we can simplify the formulas for L, T, and V for cubes as follows:

Cubes

Lateral surface area	$L = 4s^2$	
Total surface area	$T = 6s^2$	
Volume	$V = s^3$	

Be sure to use these formulas in part (d) below.

More Practice Find the lateral surface area, total surface area, and volume for each of the following prisms. Round to the nearest whole number.

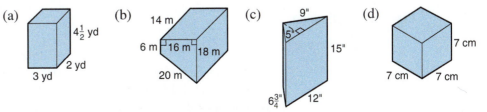

(a)   (b)   (c)   (d)

Solutions

(a) This is a rectangular prism—that is, the base is a rectangle.

First, find the altitude, h, of the prism and the perimeter, p, of the base:

$$h = 4\tfrac{1}{2} \text{ yd} \qquad p = 3 \text{ yd} + 2 \text{ yd} + 3 \text{ yd} + 2 \text{ yd} = 10 \text{ yd}$$

Then, find the area, A, of the base: $A = (3 \text{ yd})(2 \text{ yd}) = 6 \text{ yd}^2$

Now, we can proceed to calculate lateral area, L, total surface area, T, and volume, V:

$L = ph$	$T = L + 2A$	$V = Ah$
$= (10 \text{ yd})(4.5 \text{ yd})$	$= 45 \text{ yd}^2 + 2(6 \text{ yd}^2)$	$= (6 \text{ yd}^2)(4.5 \text{ yd})$
$= 45 \text{ yd}^2$	$= 57 \text{ yd}^2$	$= 27 \text{ yd}^3$

(b) This is a trapezoidal prism. The base of the prism is a right trapezoid with bases measuring 6 m and 18 m and a height of 16 m.

The altitude of the prism is 14 m and the perimeter of the trapezoidal base is 60 m.

The area of the trapezoidal base is 192 m². Therefore,

$$L = 840 \text{ m}^2 \qquad T = 1224 \text{ m}^2 \qquad V = 2688 \text{ m}^3$$

(c) This is a triangular prism. The base of the prism is a triangle with $b = 12$ in. and $h = 5$ in.

The altitude of the prism is 15 in. and the perimeter, p, of the triangular base is

$$p = 9 \text{ in.} + 12 \text{ in.} + 6\tfrac{3}{4} \text{ in.} = 27\tfrac{3}{4} \text{ in.}$$

The area of the triangular base is $A = \tfrac{1}{2} bh = \tfrac{1}{2}(12 \text{ in.})(5 \text{ in.}) = 30 \text{ in.}^2$

Therefore,

$L = 416.25 \text{ in.}^2$	$T = 476.25 \text{ in.}^2$	$V = 450 \text{ in.}^3$
$\approx 416 \text{ in.}^2$, rounded	$\approx 476 \text{ in.}^2$	

(d) For a cube, $L = 4s^2 = 4(7 \text{ cm})^2 = 196 \text{ cm}^2$

$$T = 6s^2 = 6(7 \text{ cm})^2 = 294 \text{ cm}^2$$
$$V = s^3 = (7 \text{ cm})^3 = 343 \text{ cm}^3$$

Converting Volume Units   In practical problems it is often necessary to convert volume units from cubic inches or cubic feet to gallons or other units, including metric units. Use the following conversion factors and set up unity fractions as shown in Chapter 5.

1 cu ft (ft³) $= 1728$ cu in. (in.³)	1 cu yd (yd³) $= 27$ cu ft (ft³)
≈ 7.48 gallons	
≈ 28.32 liters (L)	1 cu in. (in.³) ≈ 16.39 cu cm (cm³)
≈ 0.0283 cu m (m³)	1 gal $= 231$ cu in. (in.³)
≈ 0.806 bushel	1 liter $= 1000$ mL $= 1000$ cu cm (cm³)

Practical Applications

EXAMPLE 2   **Construction**   A rectangular plot of land measuring 40.0 ft by 80.0 ft must be excavated to a benchmark elevation of 130.0 ft. The four corners of the plot have elevations of 136.5 ft, 138.3 ft, 139.4 ft, and 137.3 ft. The contractor needs to estimate the volume of earth to be removed in cubic yards.

First, find the area A of the plot of land. $A = LW = (40 \text{ ft})(80 \text{ ft})$

$$A = 3200 \text{ sq ft}$$

Second, find the average difference between the elevations of the four corners and the benchmark elevation. This will be the average depth in feet of the excavation.

$$\text{Average of the four corners} = \frac{136.5 + 138.3 + 139.4 + 137.3}{4}$$

$$= \frac{551.5}{4}$$

$$= 137.875 \text{ ft}$$

Average difference from the benchmark $= 137.875 \text{ ft} - 130.0 \text{ ft} = 7.875 \text{ ft}$

Third, multiply this average depth by the area of the plot to find the volume V of the excavated earth.

$$V = (3200 \text{ sq ft})(7.875 \text{ ft})$$

$$V = 25{,}200 \text{ cu ft}$$

Finally, convert this to cubic yards using the conversion factor 1 cu yd $= 27$ cu ft.

$$V = 25{,}200 \text{ cu ft} \times \frac{1 \text{ cu yd}}{27 \text{ cu ft}}$$

$$V = 933.33 \ldots \text{ cu yd} = 933\tfrac{1}{3} \text{ cu yd}$$

Therefore, approximately 933 cubic yards of earth must be removed.   ●

Your Turn **Construction**   Calculate the volume of water needed to fill this lap pool in gallons. (Round to the nearest hundred.)

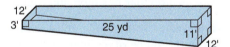

Solution The pool is a prism with trapezoidal bases. The bases of the trapezoid are 3 ft and 11 ft, the height of the trapezoid is 25 yd or 75 ft, and the altitude of the prism is 12 ft.

First, find the area A of the base:

$$A = \left(\frac{b_1 + b_2}{2}\right)h = \left(\frac{3 \text{ ft} + 11 \text{ ft}}{2}\right)75 \text{ ft} = (7 \text{ ft})(75 \text{ ft}) = 525 \text{ ft}^2$$

Next, find the volume in cubic feet:

$$V = Ah = (525 \text{ ft}^2)(12 \text{ ft}) = 6300 \text{ ft}^3$$

Finally, use the conversion factor, 1 ft^3 ≈ 7.48 gal, to form a unity fraction that cancels cubic feet and replaces it with gallons:

$$V \approx 6300 \text{ ft}^3 \times \frac{7.48 \text{ gal}}{1 \text{ ft}^3}$$

$$V \approx 47{,}124 \text{ gal} \qquad \text{or} \qquad 47{,}100 \text{ gal, rounded}$$

3 $+$ 1 1 $=$ $\div$ 2 $\times$ 7 5 $\times$ 1 2 $\times$ 7 . 4 8 $=$ → 47124.

In many practical problems, the prism encountered has an irregular polygon for its base.

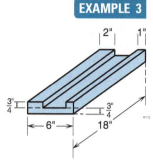

EXAMPLE 3   **Metalworking**   Steel weighs approximately 0.283 lb/in.^{3}. Suppose we need to find the weight of the steel bar shown.

First, we must calculate the volume of this prism. To find this, we first calculate the cross-sectional area by subtracting.

$$A = (6 \text{ in.} \times 1.5 \text{ in.}) - (3 \text{ in.} \times \tfrac{3}{4} \text{ in.})$$

$$= 9 \text{ in.}^2 - 2.25 \text{ in.}^2$$

$$= 6.75 \text{ in.}^2$$

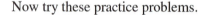

The volume of the bar is

$$V = Ah$$

$$= (6.75 \text{ in.}^2)(18 \text{ in.})$$

$$= 121.5 \text{ in.}^3$$

Using 1 in.$^3 \approx 0.283$ lb, we multiply by the unity fraction that will cancel cubic inches and replace it with pounds. The weight of the bar is

$$W = 121.5 \text{ in.}^3 \times \frac{0.283 \text{ lb}}{1 \text{ in.}^3}$$

$$W = 34.3845 \text{ lb} \qquad \text{or} \qquad 34.4 \text{ lb} \quad \text{rounded}$$

✋ Learning Help   The most common source of errors in most calculations of this kind is carelessness. Organize your work neatly. Work slowly and carefully. ●

More Practice   Now try these practice problems.

(a) What volume of liquid, in liters, is needed to fill the container shown? Assume that the base is a regular hexagon and round your answer to the nearest liter.

(b) **Water/Wastewater Treatment**   What is the capacity, to the nearest gallon, of a septic tank in the shape of a rectangular prism 12 ft by 16 ft by 6 ft?

(c) **Metalworking**   What is the weight of the steel V-block in the figure if the density of steel is 0.0173 lb/cm^3? Round to the nearest pound.

(d) **Metalworking**   Find the weight of the piece of brass pictured if the density of brass is 0.2963 lb/in.^{3}. (Round to the nearest tenth.)

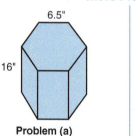

Problem (a)

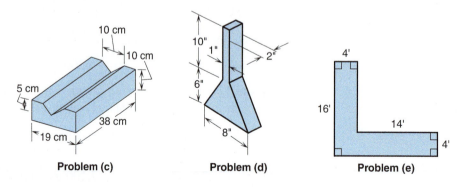

Problem (c)   Problem (d)   Problem (e)

(e) **Landscaping**   The figure shows an overhead view of a raised bed built in a vegetable garden. A landscaper needs to fill the bed to a depth of 2 ft with soil. How many cubic *yards* of soil should she order? (Round to the nearest cubic yard.)

Solutions

(a) As shown in Chapter 8, the area of a hexagon can be calculated as

$$A \approx 2.598a^2 \approx 2.598(6.5 \text{ in.})^2 \approx 109.7655 \text{ in.}^2$$

And the volume of the prism is

$$V = Ah \approx (109.7655 \text{ in.}^2)(16 \text{ in.}) \approx 1756.248 \text{ in.}^3$$

To convert this to liters, first convert to either cubic feet or cubic centimeters, then use a second unity fraction to convert to liters. If we use $1 \text{ ft}^3 = 1728 \text{ in.}^3$ and $1 \text{ ft}^3 \approx 28.32 \text{ L}$, the calculation looks like this:

$$V \approx 1756.248 \text{ in.}^3 \times \frac{1 \text{ ft}^3}{1728 \text{ in.}^3} \times \frac{28.32 \text{ L}}{1 \text{ ft}^3}$$

$$\approx 28.78\ldots \approx 29 \text{ L, rounded.}$$

To convert to cubic centimeters first and then liters, use $1 \text{ in.}^3 \approx 16.39 \text{ cm}^3$ and $1 \text{ L} = 1000 \text{ cu cm}$ as follows:

$$V \approx 1756.248 \text{ in.}^3 \times \frac{16.39 \text{ cm}^3}{1 \text{ in.}^3} \times \frac{1 \text{ L}}{1000 \text{ cm}^3}$$

$$\approx 28.78\ldots \approx 29 \text{ L, rounded}$$

(b) $V = Ah = (12 \text{ ft})(16 \text{ ft})(6 \text{ ft})$

$\qquad = 1152 \text{ ft}^3 \quad$ but $1 \text{ ft}^3 \approx 7.48 \text{ gallons}$

Converting to gallons, we have:

$$V \approx 1152 \text{ ft}^3 \times \frac{7.48 \text{ gal}}{1 \text{ ft}^3}$$

$$\approx 8616.96 \text{ gal} \approx 8617 \text{ gal}$$

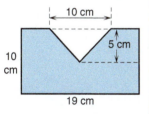

10 cm
5 cm
10 cm
19 cm

(c) **First,** find the area of the base. The cross-sectional view shown indicates that the base of the block is a rectangle with a triangle cut out. The area of the base is

$$A = \underbrace{(10 \text{ cm})(19 \text{ cm})}_{\text{rectangle}} - \underbrace{\tfrac{1}{2}(10 \text{ cm})(5 \text{ cm})}_{\text{triangle}}$$

$$= 190 \text{ cm}^2 - 25 \text{ cm}^2 = 165 \text{ cm}^2$$

Next, find the volume.

$$V = Ah = (165 \text{ cm}^2)(38 \text{ cm})$$

$$= 6270 \text{ cm}^3$$

Finally, form a unity fraction from the density and multiply by the volume to calculate the weight W.

$$W \approx 6270 \text{ cm}^3 \times \frac{0.0173 \text{ lb}}{1 \text{ cm}^3}$$

$$\approx 108.471 \text{ lb} \approx 108 \text{ lb}$$

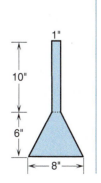

1"
10"
6"
8"

(d) The brass object is in the form of a prism with a base as shown. The area of the base is the area of the narrow rectangle plus the area of the trapezoid.

$$A = \underbrace{(10 \text{ in.})(1 \text{ in.})}_{\text{rectangle}} + \underbrace{\tfrac{1}{2}(1 \text{ in.} + 8 \text{ in.})(6 \text{ in.})}_{\text{trapezoid}}$$

$$= 10 \text{ in.}^2 + 27 \text{ in.}^2 = 37 \text{ in.}^2$$

The altitude of the prism is the thickness of the brass, or 2 in. The volume of the object is

$$V = Ah = (37 \text{ in.}^2)(2 \text{ in.}) = 74 \text{ in.}^3$$

The weight of the object is

$$W \approx 74 \text{ in.}^3 \times \frac{0.2963 \text{ lb}}{1 \text{ in.}^3} \approx 21.9262 \text{ lb} \approx 21.9 \text{ lb}$$

(e) One of several ways to approach this solution is to split the base of this L-shaped prism into rectangles I and II (see the figure). The total area of the base is therefore

$$A = \underbrace{(16 \text{ ft})(4 \text{ ft})}_{\text{Area I}} + \underbrace{(14 \text{ ft})(4 \text{ ft})}_{\text{Area II}} = 64 \text{ ft}^2 + 56 \text{ ft}^2 = 120 \text{ ft}^2$$

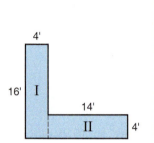

The altitude, h, of the prism is the depth of the soil, so that $h = 2$ ft. Therefore, the volume, V, of soil in cubic feet is

$$V = Ah = (120 \text{ ft}^2)(2 \text{ ft}) = 240 \text{ ft}^3$$

Soil must be ordered by the cubic yard. Using 1 yd^3 = 27 ft^3, we have

$$V = 240 \text{ ft}^3 \times \frac{1 \text{ yd}^3}{27 \text{ ft}^3} = 8.888 \ldots \approx 9 \text{ yd}^3$$

Now continue with Exercises 9-1 for more practice on calculating the surface area and volume of prisms.

| Exercises 9-1 | **Prisms** |

A. Find the volume of each of the following right prisms. Round to the nearest tenth unless indicated otherwise.

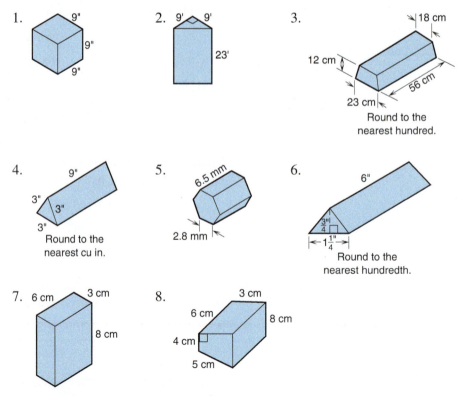

B. Find the lateral surface area and the volume of each of the following right prisms. (Round to the nearest tenth.)

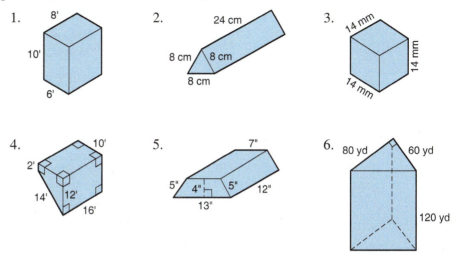

1. 8' 10' 6'
2. 24 cm 8 cm 8 cm 8 cm
3. 14 mm 14 mm 14 mm
4. 10' 2' 14' 12' 16'
5. 7" 5" 4" 5" 12" 13"
6. 80 yd 60 yd 120 yd

C. Find the total outside surface area and the volume of each of the following right prisms. (Round to the nearest tenth.)

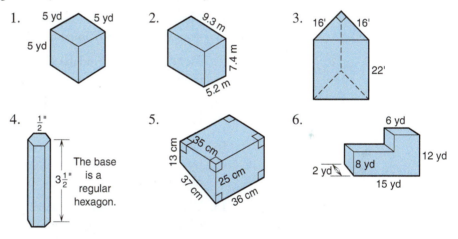

1. 5 yd 5 yd 5 yd
2. 9.3 m 7.4 m 5.2 m
3. 16' 16' 22'
4. $\frac{1}{2}$" $3\frac{1}{2}$" The base is a regular hexagon.
5. 35 cm 13 cm 37 cm 25 cm 36 cm
6. 6 yd 2 yd 8 yd 12 yd 15 yd

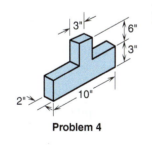

Problem 4

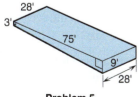

Problem 5

D. Practical Applications. Round to the nearest tenth unless otherwise directed.

1. **Construction** A concrete pour requires 9.83 cubic yards of cement. How many one cubic foot sacks must be ordered?

2. **Masonry** A standard brick has dimensions of $2\frac{1}{2}$ in. by $3\frac{3}{4}$ in. by 8 in. With a $\frac{1}{4}$-in. mortar joint added to each side, determine the number of bricks required per cubic foot. (*Hint:* Add the mortar thickness just once to each dimension.)

3. **Manufacturing** How many cubic *feet* of warehouse space are needed for 450 boxes 16 in. by 8 in. by 10 in.?

4. **Metalworking** Find the weight of the piece of steel pictured. (Steel weighs 0.2833 lb/cu in.)

5. **Construction** How many gallons of water are needed to fill a swimming pool that approximates the shape in the figure? (Round to the nearest gallon.)

6. **Painting** How many gallons of paint are needed for the outside walls of a building 26 ft high by 42 ft by 28 ft if there are 480 sq ft of windows? One gallon covers 400 sq ft. (Assume that you cannot buy a fraction of a gallon.)

7. **Construction**   Dirt must be excavated for the foundation of a building 30 yards by 15 yards to a depth of 3 yards. How many trips will it take to haul the dirt away if a truck with a capacity of 3 cu yd is used?

8. **Masonry**   If brick has a density of 103 lb/cu ft, what will be the weight of 500 bricks, each $3\frac{3}{4}$ in. by $2\frac{1}{4}$ in. by 8 in.? (Round to the nearest pound.)

9. **Landscaping**   Dirt cut 2 ft deep from a section of land 38 ft by 50 ft is used to fill a section 25 ft by 25 ft to a depth of 3 ft. How many cubic *yards* of dirt are left after the fill?

10. **Metalworking**   Find the weight of the steel V-block in the figure if its density is 0.0173 lb/cm^3.

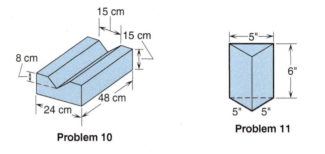

Problem 10

Problem 11

11. **Metalworking**   Find the weight of the cast iron shape in the figure. Cast iron weighs 0.2607 lb/cu in. The base is an equilateral triangle.

12. **Metalworking**   Find the weight of the piece of brass shown. Brass weighs 0.296 lb/cu in.

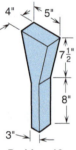

Problem 12

13. **Manufacturing**   How high must a 400-gallon rectangular tank be if the base is a square 3 ft 9 in. on a side? (1 cu ft ≈ 7.48 gallons.)

14. **Landscaping**   How many pounds of rock will you need to fill an area 25 ft by 35 ft to a depth of 2 in. if the rock weighs 1050 lb/cu yd? (Round to the nearest 10 lb.)

15. **Water/Wastewater Treatment**   A rectangular sludge bed measuring 80 ft by 120 ft is filled to a depth of 2 ft. If a truck with a capacity of 3 cu yd is used to haul away the sludge, how many trips must the truck make to empty the sludge bed?

16. **Water/Wastewater Treatment**   A reservoir with a surface area of 2.5 acres loses 0.3 in. per day to evaporation. How many gallons does it lose in a week? (1 acre = 43,560 sq ft, 1 sq ft = 144 sq in., 1 gal = 231 in.^{3}) (Round to the nearest thousand.)

17. **Construction**   A construction worker needs to pour a rectangular slab 28 ft long, 16 ft wide, and 6 in. thick. How many cubic yards of concrete does the worker need? (Round to the nearest tenth.)

18. **Construction**   Concrete weighs 150 lb/cu ft. What is the weight of a concrete wall 18 ft long, 4 ft 6 in. high, and 8 in. thick?

19. **Masonry**   The concrete footings for piers for a raised foundation are cubes measuring 18 in. by 18 in. by 18 in. How many cubic yards of concrete are needed for 28 footings?

20. **Plumbing**   A flush tank 21 in. by 6.5 in. contains water to a depth of 12 in. How many gallons of water will be saved per flush if a conservation device reduces the capacity to two-thirds of this amount? (Round to the nearest tenth of a gallon.)

21. **Roofing** A flat roof 28 ft long and 16 ft wide has a 3-in. depth of water sitting on it. What is the weight of the water on the roof? (Water weighs 62.4 lb/cu ft.) (Round to the nearest hundred pounds.)

22. **Plumbing** In a 4-hour percolation test, an 8-ft depth of water seeped out of a square pit 4 ft on a side. The plumbing code says that the soil must be able to absorb 5000 gallons in 24 hours. At this rate, will the soil be up to code?

23. **Plumbing** The Waste-Away Plumbing Company installs a rectangular septic tank 10 ft long by 5 ft wide and 6 ft deep. Calculate the capacity of the tank in gallons.

24. **Automotive Trades** The fuel tank of a motor vehicle has the shape of a rectangular prism measuring 16 in. wide, 36 in. long, and 10 in. high. How many gallons will the tank hold? (1 gal = 231 cu. in.) (Round to the nearest gallon.)

25. **Plumbing** A pipe trench is roughly rectangular with dimensions 75 ft long by 2 ft wide and with an average depth of $3\frac{1}{2}$ ft. What volume of dirt was removed?

26. **Welding** The parts bin shown in the figure is made of 16-gauge sheet metal. Calculate the volume in cubic feet and the total length of weld needed in feet.

27. **Construction** BiltWell Construction needs to create a concrete foundation 5 ft deep measuring 60 ft by 32 ft, outside dimensions, with walls 7 in. thick. How many cubic yards of concrete will the construction crew need? (See the figure in the margin.)

28. **Metalworking** Find the weight (in ounces) of the steel pin shown in the figure if its density is 0.0173 lb/cm^3. (Round to the nearest hundredth).

29. **Landscaping** A landscaper has built a U-shaped raised bed in a vegetable garden as shown in the figure. How many cubic yards of soil should be ordered to fill the bed to a depth of 18 in.? (Round to the nearest cu yd.)

30. **Construction** A foundation wall 4 ft high and 9 in. thick consists of lengths of 38 ft, 46 ft, 16 ft, 6.5 ft, and 8.5 ft. What is the total number of cubic *yards* of concrete needed for the wall? (Round to the nearest cubic yard.)

31. **Construction** A rectangular plot of land measuring 65.5 ft by 92.0 ft must be excavated to a benchmark elevation of 152.5 ft. The four corners of the plot have elevations of 158.0 ft, 162.3 ft, 155.8 ft, and 159.5 ft. Use the method of Example 2 to estimate the total volume of earth to be removed in cubic yards. (Round to the nearest ten cubic yards.)

32. **Landscaping** A rectangular pool measuring 200 in. by 400 in. has a leak that causes the pool to lose $2\frac{3}{8}$ in. of water depth per day.

   (a) How many cubic inches of water are lost per day?

   (b) How many cubic feet of water are lost per day?
       (*Hint:* 1 cu ft = 1728 cu in.) (Round to the nearest cubic foot.)

   (c) The local water district charges $4.15 per HCF (hundred cubic feet). If the leak is not fixed and the lost water is replaced, how much will this add to the resident's *monthly* water bill? (Assume a 31-day month.)

33. **General Interest** Ocean engineers estimated that the earthquake that hit Japan in 2011 pushed a section of sea floor 250 miles long and 50 miles wide downward by an average of one yard. This movement caused a tsunami. How many cubic yards of water suddenly shifted position? (Express your answer in scientific notation.)

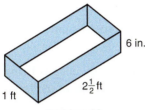

Problem 26

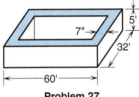

Problem 27

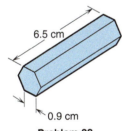

Problem 28

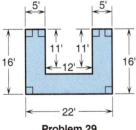

Problem 29

34. **Agriculture**   Consider the following information about barley moving along a grain conveyor:

    - The conveyor moves at a speed of 250 ft/min.
    - The conveyor belt is 18 in. wide, and barley is piled on the full width of the belt at an average depth of 6 in.
    - Barley has a density of approximately 38.4 lb/cu ft.
    - One bushel of barley weighs 48 lb.

    How many bushels of barley does the grain conveyor unload in an hour? (Round to the nearest bushel.)

35. **Construction**   A regular hexagonal foundation 4 in. deep is being poured for a hexagonal building. If the distance across the flats of the foundation is 26 ft, how many cubic *yards* of concrete are needed?

When you have completed these exercises, check your answers to the odd-numbered problems in the Appendix, then turn to Section 9-2 to study pyramids.

9-2   Pyramids and Frustums of Pyramids

Pyramids

A **pyramid** is a solid object with one base and three or more **lateral faces** that taper to a single point opposite the base. This single point is called the **apex**.

learning|**catalytics**™

1. What type of triangle has two equal sides?
2. What is the formula for the area of an equilateral triangle?

A **right pyramid** is one whose base is a regular polygon and whose apex is centered over the base. We will examine only right pyramids here.

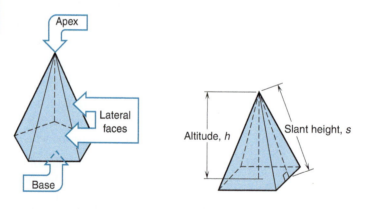

The **altitude** of a pyramid is the perpendicular distance from the apex to the base. The **slant height** of the pyramid is the height of one of the lateral faces.

As with prisms, pyramids are identified according to the shape of the base:

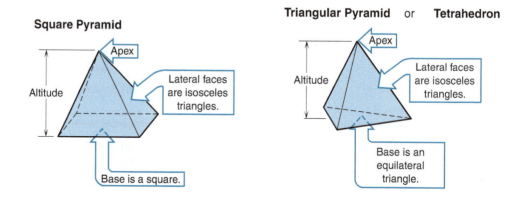

There are several very useful formulas relating to pyramids:

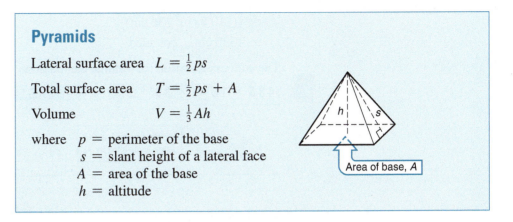

> **Pyramids**
>
> Lateral surface area $L = \frac{1}{2}ps$
>
> Total surface area $T = \frac{1}{2}ps + A$
>
> Volume $V = \frac{1}{3}Ah$
>
> where p = perimeter of the base
> s = slant height of a lateral face
> A = area of the base
> h = altitude
>
> Area of base, A

EXAMPLE 1   The square pyramid shown has a slant height $s \approx 10.77$ ft and an altitude $h = 10$ ft. To find the lateral surface area, first calculate the perimeter p of the base.

$$p = 8 \text{ ft} + 8 \text{ ft} + 8 \text{ ft} + 8 \text{ ft} = 32 \text{ ft}$$

The lateral surface area is

$$L = \frac{1}{2}ps$$
$$\approx \frac{1}{2}(32 \text{ ft})(10.77 \text{ ft})$$
$$\approx 172 \text{ ft}^2 \quad \text{This is the area of the four triangular lateral faces.}$$

To find the volume, first calculate the area A of the square base.

$$A = (8 \text{ ft})(8 \text{ ft}) = 64 \text{ ft}^2$$

The volume is

$$V = \frac{1}{3}Ah$$
$$= \frac{1}{3}(64 \text{ ft}^2)(10 \text{ ft})$$
$$\approx 213 \text{ ft}^3 \quad \text{rounded}$$

Your Turn   Find the lateral surface area and volume of this triangular pyramid. (*Remember:* The base is an equilateral triangle.) (Round to the nearest whole unit.)

Solutions   To find the lateral surface area, first find the perimeter p of the base.

$$p = 16 \text{ cm} + 16 \text{ cm} + 16 \text{ cm}$$
$$= 48 \text{ cm}$$

Second, note that the slant height $s \approx 36.3$ cm.

Therefore, $L = \frac{1}{2}ps$
$$\approx \frac{1}{2}(48 \text{ cm})(36.3 \text{ cm})$$
$$\approx 871.2 \text{ cm}^2 \text{ or } 871 \text{ cm}^2 \quad \text{rounded}$$

To find the volume, first calculate the area A of the base.

$$A \approx 0.433s^2 \text{ for an equilateral triangle}$$
$$\approx 0.433(16 \text{ cm})^2$$
$$\approx 110.8 \text{ cm}^2$$

36.3 cm

16 cm
$h = 36$ cm

The volume is

$$V = \frac{1}{3}Ah$$

$$\approx \frac{1}{3}(110.8 \text{ cm}^2)(36 \text{ cm})$$

$$\approx 1330 \text{ cm}^3 \quad \text{rounded}$$

The entire calculation for volume can be done as follows using a calculator:

$$.433 \boxed{\times} 16 \boxed{x^2} \boxed{\times} 36 \boxed{\div} 3 \boxed{=} \rightarrow \quad \boxed{1330.176}$$

A

Multiplying by $\frac{1}{3}$ is the same as dividing by 3.

To calculate the total surface area, T, of a pyramid, add the area of the base to the lateral surface area.

EXAMPLE 2   Here's how to find the total surface area of the hexagonal pyramid shown in the margin:

First, find the lateral surface area, L.

$$p = 6(3.5 \text{ yd}) = 21 \text{ yd}$$

$$s \approx 12.38 \text{ yd}$$

$$L = \frac{1}{2}ps \approx \frac{1}{2}(21 \text{ yd})(12.38 \text{ yd})$$

$$\approx 129.99 \text{ yd}^2$$

Then, find the area of the hexagonal base.

$$A \approx 2.598 (3.5 \text{ yd})^2 \approx 31.8255 \text{ yd}^2$$

Finally, add these areas.

$$T = L + A \approx 129.99 \text{ yd}^2 + 31.8255 \text{ yd}^2$$

$$\approx 161.8155 \text{ yd}^2 \approx 162 \text{ yd}^2, \text{ rounded}.$$

The entire calculation can be performed on a calculator as follows:

$$6 \boxed{\times} 3.5 \boxed{\times} 12.38 \boxed{\times} .5 \boxed{+} 2.598 \boxed{\times} 3.5 \boxed{x^2} \boxed{=} \rightarrow \quad \boxed{161.8155}$$

Lateral area   Base area

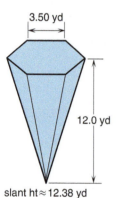

3.50 yd

12.0 yd

slant ht $\approx$ 12.38 yd

More Practice   (a) Find the lateral surface area and volume of each pyramid. (Round to the nearest tenth.)

1.
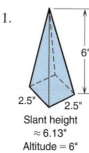

6"

2.5"   2.5"

Slant height
$\approx 6.13"$
Altitude = 6"

2.
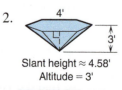

4'

3'

Slant height $\approx 4.58'$
Altitude = 3'

(b) Find the total surface area and volume of this triangular pyramid. (Round to the nearest whole number.)

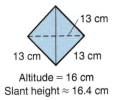

13 cm

13 cm · 13 cm

Altitude = 16 cm
Slant height ≈ 16.4 cm

(c) Find the volume of an octagonal pyramid if the area of the base is 245 in.^{2} and the altitude is 16 in. (Round to the nearest cubic inch.)

Solutions

(a) 1. $p = 4(2.5 \text{ in.}) = 10 \text{ in.}$ $s \approx 6.13 \text{ in.}$

$L = \frac{1}{2}ps \approx \frac{1}{2}(10 \text{ in.})(6.13 \text{ in.})$

$\approx 30.65 \text{ in.}^2 \approx 30.7 \text{ in.}^2$, rounded

$V = \frac{1}{3}Ah = \frac{1}{3}(2.5 \text{ in.})^2(6 \text{ in.})$

$= 12.5 \text{ in.}^3$

2. $p = 6(4 \text{ ft}) = 24 \text{ ft}$ $s \approx 4.58 \text{ ft}$

$L = \frac{1}{2}ps \approx \frac{1}{2}(24 \text{ ft})(4.58 \text{ ft})$

$\approx 54.96 \text{ ft}^2 \approx 55.0 \text{ ft}^2$, rounded

area of hexagonal base

$V = \frac{1}{3}Ah \approx \frac{1}{3}(2.598)(4 \text{ ft})^2(3 \text{ ft})$

$\approx 41.568 \text{ ft}^3 \approx 41.6 \text{ ft}^3$, rounded

(b) **First,** find the lateral surface area, L.

$p = 3(13 \text{ cm}) = 39 \text{ cm}$ $s \approx 16.4 \text{ cm}$

$L = \frac{1}{2}ps \approx \frac{1}{2}(39 \text{ cm})(16.4 \text{ cm})$

$\approx 319.8 \text{ cm}^2$

Next, find A, the area of the base, an equilateral triangle.

$A \approx 0.433a^2$

$\approx 0.433(13 \text{ cm})^2 \approx 73.177 \text{ cm}^2$

Finally, add $L + A$ to obtain the total surface area, T.

$T \approx 319.8 \text{ cm}^2 + 73.177 \text{ cm}^2$

$\approx 392.977 \text{ cm}^2 \approx 393 \text{ cm}^2$, rounded

Having calculated the area of the base, we can now find the volume.

$V = \frac{1}{3}Ah \approx \frac{1}{3}(73.177 \text{ cm}^2)(16 \text{ cm})$

$\approx 390.277 \ldots \approx 390 \text{ cm}^3$, rounded

To use a calculator efficiently, find the area of the base first and store it in memory. You can then recall it when you calculate the volume.

Base area   Lateral area

Surface area: $.433 \times 13 \boxed{x^2} = \boxed{\text{STO}} \boxed{A} + 3 \times 13 \times 16.4 \times .5 =$

→ 392.977

Volume: RCL A × **1 6** ÷ **3** = → **390.2773333**

(c) $V = \frac{1}{3}Ah = \frac{1}{3}(245 \text{ in.}^2)(16 \text{ in.})$

$= 1306.666 \ldots \approx 1307 \text{ in.}^3$

Applications of Pyramids

EXAMPLE 3   **Painting**   One section of the roof of a building is in the shape of a square pyramid with a side length of 28 ft and a slant height of 13.25 ft. A heat-reflecting paint with a spreading rate of 150 square feet per gallon is to be applied to this section. Suppose we wish to know the number of gallons needed in order to apply two coats of this paint.

First, calculate the lateral surface area of the pyramid:

$p = 4(28 \text{ ft}) = 112 \text{ ft}$ and $s = 13.25 \text{ ft}$

$L = \frac{1}{2}ps = \frac{1}{2}(112 \text{ ft})(13.25 \text{ ft}) = 742 \text{ sq ft}$

Next, multiply this by 2 to allow for two coats of paint:

$2(742 \text{ sq ft}) = 1484 \text{ sq ft}$

Then, divide this result by the spreading rate of 150 square feet per gallon:

$1484 \text{ sq ft} \div 150 \text{ sq ft/gal} = 1484 \text{ sq ft} \times \dfrac{1 \text{ gal}}{150 \text{ sq ft}} = 9.893 \ldots \text{ gal}$

Finally, rounding up to the nearest whole gallon, we would need 10 gallons of paint.   ●

Your Turn   **Manufacturing**   A candlemaker is producing pyramid-shaped candles with an 8.0-cm square base and an altitude of 15 cm. If paraffin wax weighs 0.93 grams per cubic centimeter, how many kilograms of wax will be needed to produce 60 of these candles? (Round to the nearest whole number.)

Solution   **Step 1**   Calculate the area A of the square base of each candle:

$A = s^2 = (8.0 \text{ cm})^2 = 64 \text{ cm}^2$

Step 2   Calculate the volume V of each candle:

$V = \frac{1}{3}Ah = \frac{1}{3}(64 \text{ cm}^2)(15 \text{ cm}) = 320 \text{ cm}^3$

Step 3   Multiply by 60 to determine the total volume of wax needed:

$60(320 \text{ cm}^3) = 19,200 \text{ cm}^3$

Step 4   Multiply this by 0.93 g/cm³ to determine the weight W of wax needed:

$W = 19,200 \text{ cm}^3 \times \dfrac{0.93 \text{ g}}{1 \text{ cm}^3} = 17,856 \text{ g}$

Step 5   Use the fact that 1 kg = 1000 g to convert this to kilograms:

$W = 17,856 \text{ g} \times \dfrac{1 \text{ kg}}{1000 \text{ g}} = 17.856 \text{ kg} \approx 18 \text{ kg of paraffin wax}$

Frustum of a Pyramid   A **frustum** of a pyramid is the solid figure remaining after the top of the pyramid is cut off parallel to the base. Frustum shapes appear as containers, as building foundations, and as transition sections in ducts.

Most frustums used in practical work, such as sheet metal construction, are frustums of square pyramids or cones.

Every frustum has two bases, upper and lower, that are parallel and different in size. The altitude is the perpendicular distance between the bases.

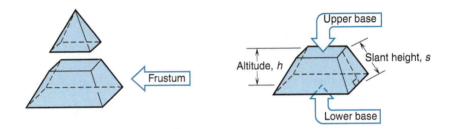

The following formulas enable you to find the lateral area and volume of any pyramid frustum:

Frustum of a Pyramid

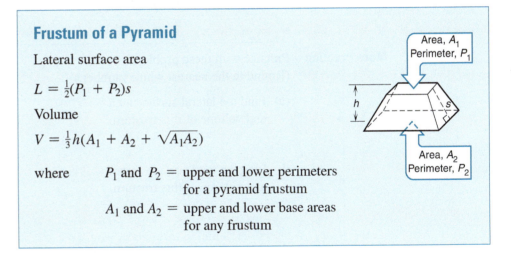

Lateral surface area

$$L = \tfrac{1}{2}(P_1 + P_2)s$$

Volume

$$V = \tfrac{1}{3}h(A_1 + A_2 + \sqrt{A_1 A_2})$$

where P_1 and P_2 = upper and lower perimeters
                    for a pyramid frustum

A_1 and A_2 = upper and lower base areas
                    for any frustum

EXAMPLE 4

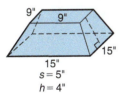

$s = 5"$
$h = 4"$

To find the lateral surface area and volume of the pyramid frustum shown, first calculate the upper and lower perimeters.

Upper perimeter, P_1 = 9 in. + 9 in. + 9 in. + 9 in. = 36 in.

Lower perimeter, P_2 = 15 in. + 15 in. + 15 in. + 15 in. = 60 in.

Now substitute these values into the formula for lateral surface area.

$$L = \tfrac{1}{2}(36 \text{ in.} + 60 \text{ in.})(5 \text{ in.}) = 240 \text{ in.}^2$$

To find the volume of this frustum, first calculate the upper and lower areas.

Upper area, A_1 = (9 in.)(9 in.) = 81 in.^{2}

Lower area, A_2 = (15 in.)(15 in.) = 225 in.^{2}

Now substitute these values into the formula for volume.

$$V = \tfrac{1}{3}(4 \text{ in.})\left(81 \text{ in.}^2 + 225 \text{ in.}^2 + \sqrt{81 \cdot 225} \text{ in}^2\right)$$

$$= \tfrac{1}{3}(4 \text{ in.})(81 + 225 + 135) \text{ in.}^2 = 588 \text{ in.}^3$$

Use the following calculator sequence to find the volume:

 9 x^2 **+ 1 5** x^2 **+** $\sqrt{}$ *** 9** x^2 **× 1 5** x^2 **)** **=** **× 4 ÷ 3** **=** → [___] **588.**

*Open parentheses here if your calculator does not do so automatically.

Notice that we multiply by the altitude 4 near the end to avoid needing a second set of parentheses. Notice also that dividing by 3 is easier than multiplying by $\frac{1}{3}$. ●

To find the **total surface area** of the frustum, add the areas A_1 and A_2 to the lateral surface area.

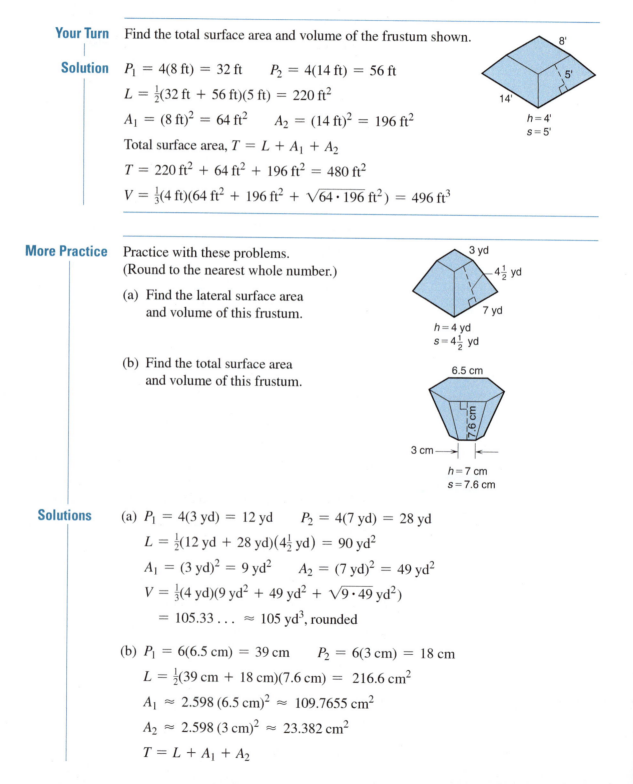

Your Turn   Find the total surface area and volume of the frustum shown.

Solution $P_1 = 4(8\text{ ft}) = 32\text{ ft}$ $P_2 = 4(14\text{ ft}) = 56\text{ ft}$

$L = \frac{1}{2}(32\text{ ft} + 56\text{ ft})(5\text{ ft}) = 220\text{ ft}^2$

$A_1 = (8\text{ ft})^2 = 64\text{ ft}^2$ $A_2 = (14\text{ ft})^2 = 196\text{ ft}^2$

Total surface area, $T = L + A_1 + A_2$

$T = 220\text{ ft}^2 + 64\text{ ft}^2 + 196\text{ ft}^2 = 480\text{ ft}^2$

$V = \frac{1}{3}(4\text{ ft})(64\text{ ft}^2 + 196\text{ ft}^2 + \sqrt{64 \cdot 196}\text{ ft}^2) = 496\text{ ft}^3$

More Practice   Practice with these problems.
(Round to the nearest whole number.)

(a) Find the lateral surface area and volume of this frustum.

(b) Find the total surface area and volume of this frustum.

Solutions   (a) $P_1 = 4(3\text{ yd}) = 12\text{ yd}$ $P_2 = 4(7\text{ yd}) = 28\text{ yd}$

$L = \frac{1}{2}(12\text{ yd} + 28\text{ yd})(4\frac{1}{2}\text{ yd}) = 90\text{ yd}^2$

$A_1 = (3\text{ yd})^2 = 9\text{ yd}^2$ $A_2 = (7\text{ yd})^2 = 49\text{ yd}^2$

$V = \frac{1}{3}(4\text{ yd})(9\text{ yd}^2 + 49\text{ yd}^2 + \sqrt{9 \cdot 49}\text{ yd}^2)$

$= 105.33 \ldots \approx 105\text{ yd}^3$, rounded

(b) $P_1 = 6(6.5\text{ cm}) = 39\text{ cm}$ $P_2 = 6(3\text{ cm}) = 18\text{ cm}$

$L = \frac{1}{2}(39\text{ cm} + 18\text{ cm})(7.6\text{ cm}) = 216.6\text{ cm}^2$

$A_1 \approx 2.598\,(6.5\text{ cm})^2 \approx 109.7655\text{ cm}^2$

$A_2 \approx 2.598\,(3\text{ cm})^2 \approx 23.382\text{ cm}^2$

$T = L + A_1 + A_2$

$$T \approx 216.6 \text{ cm}^2 + 109.7655 \text{ cm}^2 + 23.382 \text{ cm}^2$$

$$\approx 349.7475 \text{ cm}^2 \approx 350 \text{ cm}^2, \text{ rounded}$$

$$V \approx \tfrac{1}{3}(7 \text{ cm})(109.7655 \text{ cm}^2 + 23.382 \text{ cm}^2 + \sqrt{109.7655 \times 23.382} \text{ cm}^2)$$

$$\approx 428.8865 \text{ cm}^3 \approx 429 \text{ cm}^3, \text{ rounded}$$

To make efficient use of a calculator, find A_1 and A_2 first and store them in different memory locations. You can then recall them when calculating both surface area and volume.

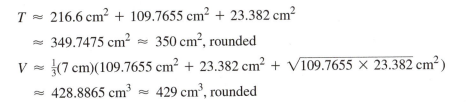

A_1: **2.598** ✕ **6.5** x^2 = STO A →   *109.7655*

A_2: **2.598** ✕ **3** x^2 = STO B →   *23.382*

T: **6** ✕ **6.5** + **6** ✕ **3** = ✕ **7.6** ✕ **.5** + RCL A + RCL B = →   *349.7475*

             ⎵Lateral area⎵ A_1 A_2

V: RCL A + RCL B + √ * RCL A ✕ RCL B) = ✕ **7** ÷ **3** =

                                                     →   *428.8865*

*Open parentheses here if your calculator does not do so automatically.

Applications of Frustums

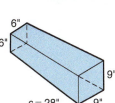

EXAMPLE 5

HVAC A connection must be made between two square vent openings in an air-conditioning system. If the sheet metal used to make the connecting piece weighs 0.906 lb/ft^2, what is the total weight of the piece shown in the figure?

To solve this problem, we must **first** find the amount of sheet metal needed. This would be the lateral area of the frustum. We see that $P_1 = 24$ in. and $P_2 = 36$ in.; therefore:

$$L = \tfrac{1}{2}(24 \text{ in.} + 36 \text{ in.})(28 \text{ in.}) = 840 \text{ in.}^2$$

Next, use the fact that 1 sq ft = 144 sq in. to convert this to square feet.

$$840 \text{ sq in.} \times \frac{1 \text{ sq ft}}{144 \text{ sq in.}} \approx 5.83 \text{ sq ft}$$

Finally, multiply this by 0.906 lb/ft^2 to determine the total weight W of the piece.

$$W \approx 5.83 \text{ sq ft} \times \frac{0.906 \text{ lb}}{1 \text{ sq ft}} \approx 5.285 \text{ lb} \approx 5.3 \text{ lb, rounded}$$

Using a calculator, we can perform the entire calculation as follows:

24 + **36** = ÷ **2** ✕ **28** ÷ **144** ✕ **.906** = →   *5.285*   ●

⚠ **Careful** Before entering a multi-step calculation such as this on your calculator, be sure to organize your work on paper first to avoid mistakes. ●

Your Turn **Construction** How many cubic yards of concrete are needed to pour the foundation shown in the figure? (Watch your units!) (Round to the nearest tenth.)

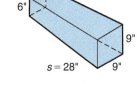

Solution To solve, we must find the volume of the frustum in cubic feet and then convert to cubic yards. Recall that $1 \text{ yd}^3 = 27 \text{ ft}^3$.

$$A_1 = (22 \text{ ft})^2 = 484 \text{ ft}^2 \qquad A_2 = (35 \text{ ft})^2 = 1225 \text{ ft}^2$$

$$V = \tfrac{1}{3}(0.5 \text{ ft})(484 \text{ ft}^2 + 1225 \text{ ft}^2 + \sqrt{484 \cdot 1225} \text{ ft}^2)$$

$$= 413.166\ldots \text{ ft}^3$$

$$\approx 413.166 \text{ ft}^3 \times \frac{1 \text{ yd}^3}{27 \text{ ft}^3} \approx 15.3 \text{ yd}^3$$

Notice that the height, 6 in., was converted to 0.5 ft to agree with the other units. The answer was then converted from cu ft to cu yd.

↩ Learning Help   Always be certain that your units agree before entering them into a calculation. ●

Now continue with Exercises 9-2 for more practice finding the surface area and volume of pyramids and frustums of pyramids.

Exercises 9-2 | **Pyramids and Frustums of Pyramids**

A. Find the lateral surface area and volume of the solid objects shown. If necessary, round to the nearest whole number unless otherwise directed.

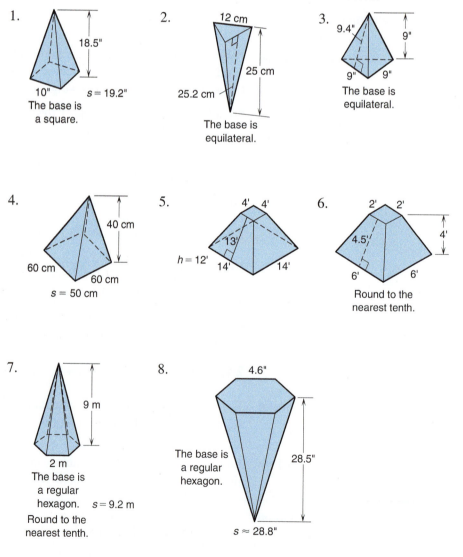

1. 18.5"
    10"   s = 19.2"
    The base is
    a square.

2. 12 cm
    25 cm
    25.2 cm
    The base is
    equilateral.

3. 9.4"   9"
    9"   9"
    The base is
    equilateral.

4. 40 cm
    60 cm   60 cm
    s = 50 cm

5. 4'   4'
    13'
    h = 12'   14'   14'

6. 2'   2'
    4.5'   4'
    6'   6'
    Round to the
    nearest tenth.

7. 9 m
    2 m
    The base is
    a regular
    hexagon.   s = 9.2 m
    Round to the
    nearest tenth.

8. 4.6"
    The base is
    a regular
    hexagon.   28.5"
    s ≈ 28.8"

B. Find the total outside surface area and volume of the following solid objects. If necessary, round to the nearest whole number unless otherwise directed. (Assume that all bases are regular.)

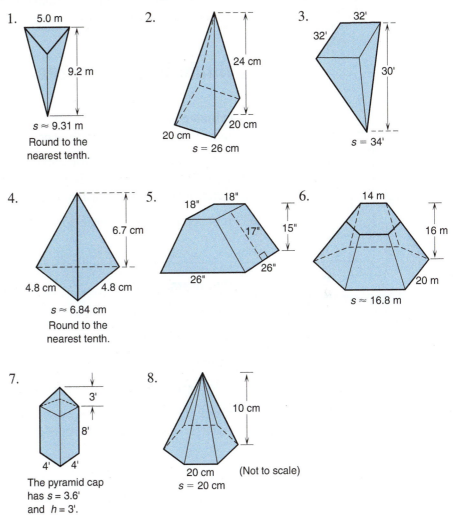

1. 5.0 m
9.2 m
$s \approx 9.31$ m
Round to the nearest tenth.

2. 24 cm
20 cm
20 cm
$s = 26$ cm

3. 32'
32'
30'
$s = 34'$

4. 6.7 cm
4.8 cm   4.8 cm
$s \approx 6.84$ cm
Round to the nearest tenth.

5. 18"   18"
17"
15"
26"   26"
26"

6. 14 m
16 m
20 m
$s \approx 16.8$ m

7. 3'
8'
4'   4'
The pyramid cap has $s = 3.6'$ and $h = 3'$.

8. 10 cm
20 cm   (Not to scale)
$s = 20$ cm

C. Practical Applications

1. **Roofing** The roof of a building is in the shape of a square pyramid 25 m on each side. If the slant height of the pyramid is 18 m, how much will roofing material cost at $10.23 per sq m?

2. **Metalworking** A pyramid-shaped piece of steel has a triangular base measuring 6.8 in. on each side. If the height of the piece is 12.4 in., find the weight of two dozen of these pyramids. Assume that the steel has a density of 0.283 lb/cu in. (Round to the nearest pound.)

3. **HVAC** The Last Chance gambling casino is designed in the shape of a square pyramid with a side length of 75 ft. The height of the building is 32 ft. An air-filtering system can circulate air at 20,000 cu ft/hr. How long will it take the system to filter all the air in the room?

4. **Agriculture** How many bushels will the bin in the figure hold? (1 cu ft = 1.24 bushels.) (Round to the nearest bushel.)

5. **Sheet Metal Trades** How many square inches of sheet metal are used to make the vent transition shown? (The ends are open.)

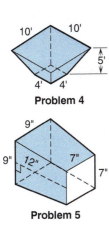

10'   10'
5'
4'   4'
Problem 4

9"
9"   12"   7"
7"
Problem 5

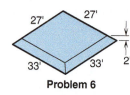

Problem 6

6. **Construction** How many cubic yards of concrete are needed to pour the building foundation shown in the figure?

When you have completed these exercises, check your answers to the odd-numbered problems in the Appendix, then continue with Section 9-3 to study cylinders and spheres.

9-3 Cylinders and Spheres

Cylinders

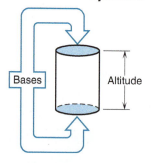

A **cylinder** is a solid object with two identical circular bases. The **altitude** of a cylinder is the perpendicular distance between the bases.

A **right cylinder** is one whose curved side walls are perpendicular to its circular base. Whenever we mention the radius, diameter, or circumference of a cylinder, we are referring to those dimensions of its circular base.

Two important formulas enable us to find the lateral surface area and volume of a cylinder:

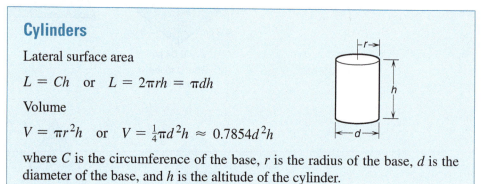

Cylinders

Lateral surface area

$$L = Ch \quad \text{or} \quad L = 2\pi rh = \pi dh$$

Volume

$$V = \pi r^2 h \quad \text{or} \quad V = \tfrac{1}{4}\pi d^2 h \approx 0.7854 d^2 h$$

where C is the circumference of the base, r is the radius of the base, d is the diameter of the base, and h is the altitude of the cylinder.

To find the total surface area of a cylinder, simply add the areas of the two circular bases to the lateral area. The formula is

Total surface area: $T = 2\pi r^2 + 2\pi rh = 2\pi r(r + h)$

learning|catalytics™

1. What is the formula for the circumference of a circle in terms of the radius?
2. Find the area of a circle with diameter 10 cm. Give the answer as a multiple of π.

Learning Help You can visualize the lateral surface area by imagining the cylinder wall unrolled into a rectangle as shown.

Lateral surface area = area of rectangle

$$= L \times W$$

$$= \pi d \times h$$

$$= \pi dh$$

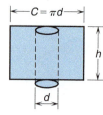

EXAMPLE 1 Use these formulas to find the lateral surface area and volume of the cylindrical container shown in the margin.

First, note that

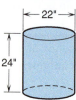

$$d = 22 \text{ in.} \qquad r = \frac{d}{2} = 11 \text{ in.} \qquad h = 24 \text{ in.}$$

Then, the lateral surface area is

$$L = \pi dh$$

$$= (\pi)(22 \text{ in.})(24 \text{ in.})$$

$$\approx 1659 \text{ in.}^2 \quad \text{rounded}$$

and the volume is

$$V = \pi r^2 h$$

$$= (\pi)(11 \text{ in.})^2(24 \text{ in.})$$

$$= (\pi)(121 \text{ in.}^2)(24 \text{ in.})$$

$$\approx 9123 \text{ in.}^3 \quad \text{rounded}$$

●

Your Turn   (a) Find the lateral surface area and volume of each cylinder. (Round to the nearest whole number.)

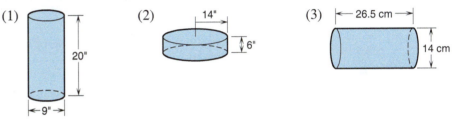

(1)   20"   9"

(2)   14"   6"

(3)   26.5 cm   14 cm

(b) Find the total surface area of a cylinder 9 yd high and 16 yd in diameter. Round to the nearest whole number.

Solutions   (a) (1) 565 in.², 1272 in.³   (2) 528 in.², 3695 in.³   (3) 1166 cm², 4079 cm³

(b) $d = 16$ yd $r = 8$ yd $h = 9$ yd

$$T = 2\pi r^2 + 2\pi rh$$

$$= 2\pi(8 \text{ yd})^2 + 2\pi(8 \text{ yd})(9 \text{ yd})$$

$$= 128\pi \text{ yd}^2 + 144\pi \text{ yd}^2 = 272\pi \text{ yd}^2$$

$$\approx 854.513\ldots \approx 855 \text{ yd}^2, \text{rounded}$$

Applications of Cylinders   Here are some applications involving surface area of cylinders.

EXAMPLE 2   **Industrial Technology**   The bottom and the inside walls of the cylindrical tank shown in the figure must be lined with sheet copper. How many square feet of sheet copper are needed? (Round to the nearest whole number.)

5'
12'

To solve this problem, note that the inside wall is the lateral surface area.

$$L = 2\pi rh = 2\pi(5 \text{ ft})(12 \text{ ft})$$

$$= 120\pi \text{ ft}^2$$

The bottom wall is a circle.

$$A = \pi r^2 = \pi(5 \text{ ft})^2$$

$$= 25\pi \text{ ft}^2$$

The total area of sheet copper is

$$L + A = 120\pi \text{ ft}^2 + 25\pi \text{ ft}^2 = 145\pi \text{ ft}^2$$

$$\approx 455.53\ldots \text{ ft}^2 \approx 456 \text{ ft}^2, \text{rounded}$$

●

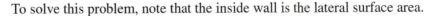

Learning Help   When you are adding two or more circular areas, it is usually simpler to express each in terms of π and substitute for π after finding the sum. ●

Your Turn   **Painting**   How many quarts of paint are needed to cover the outside wall (not including the top and bottom) of a cylindrical water tank 10 ft in diameter and 10 ft in height if one quart covers 100 sq ft? Use 3.14 for π. (Assume that you cannot buy a fraction of a quart.)

Solution   **First,** find the lateral surface area.

$$L = \pi dh \approx 3.14(10 \text{ ft})(10 \text{ ft})$$

$$\approx 314 \text{ ft}^2$$

Then, use a unity fraction to determine the amount of paint used.

$$L \approx 314 \text{ ft}^2 \times \frac{1 \text{ qt}}{100 \text{ ft}^2} \approx 3.14 \text{ qt}$$

Because you cannot buy a fraction of a quart, you will need 4 qt of paint.

Here are some applications involving volume of cylinders.

EXAMPLE 3   **Water/Wastewater Treatment**   A water district discovers a break in a 10-in. diameter pipe that has contaminated the water over a 1.2-mile length. To treat the water flowing through the pipe, the water district needs to know the number of gallons of water flowing through this stretch of pipe. How many gallons is this? (Round to the nearest gallon. One cubic foot of water contains approximately 7.48 gallons.)

To solve this problem:

First, convert the dimensions of the pipe so the units agree. There is no single "correct" choice, but we shall convert both units to feet:

Diameter, $d = 10 \text{ in.} = 10 \text{ in.} \times \frac{1 \text{ ft}}{12 \text{ in.}} = \frac{10}{12} \text{ ft} = \frac{5}{6} \text{ ft}$

Altitude, h (the length of the pipe) $= 1.2 \text{ mi} = 1.2 \text{ mi} \times \frac{5280 \text{ ft}}{1 \text{ mi}} = 6336 \text{ ft}$

Next, calculate the volume of this stretch of pipe:

$$V \approx 0.7854d^2h \approx 0.7854\left(\frac{5}{6} \text{ ft}\right)^2 (6336 \text{ ft}) \approx 3455.76 \text{ ft}^3$$

Finally, convert this volume to gallons:

$$V \approx 3455.76 \text{ ft}^3 \times \frac{7.48 \text{ gal}}{1 \text{ ft}^3} \approx 25{,}849 \text{ gal}$$

Here is how we would key this in a calculator:

.7854 ⊠ 5 [Aᵇ/c] 6 [x²] ⊠ 6336 ⊠ 7.48 ⊜ → 25849.0848   ●

EXAMPLE 4   **Manufacturing**   Cylindrical steel cable can be manufactured by heating a steel block and forming the cable from the heated steel. This process is called "extrusion." Suppose the original block is 16 ft long with an 18-in. by 18-in. cross section. Suppose further that a particular extrusion process is 96% efficient—that is, 96% of the volume of the original block ends up as steel cable. To determine the number of linear feet of 2-in. diameter cable that can be produced, follow these steps:

Step 1   Convert the 16-ft length to inches so that all units agree.

$$16 \text{ ft} \times \frac{12 \text{ in.}}{1 \text{ ft}} = 192 \text{ in.}$$

Step 2   Calculate the volume of the original block, a rectangular prism.

$$V = Ah = (18 \text{ in.} \times 18 \text{ in.})(192 \text{ in.}) = 62{,}208 \text{ in.}^3$$

Step 3   Find 96% of this volume.

$$0.96(62{,}208 \text{ in.}^3) = 59{,}719.68 \text{ in.}^3$$

Step 4   Using $r = 1$ in., write the formula for the volume of the cylindrical cable. Note that h represents the length of cable, which is the unknown in the problem.

$$V = \pi r^2 h = \pi(1 \text{ in.})^2 h = \pi h \text{ in.}^3$$

Step 5   Set this equal to the volume of heated steel from Step 3 and solve for h.

$$\pi h = 59{,}719.68$$

$$\frac{\pi h}{\pi} = \frac{59{,}719.68}{\pi}$$

$$h \approx 19{,}009.36 \text{ in.}$$

Step 6   Convert this to feet.

$$19{,}009.36 \text{ in.} \times \frac{1 \text{ ft}}{12 \text{ in.}} = 1584.11 \ldots \approx 1584 \text{ linear feet of cable}   \quad \bullet$$

Your Turn   (a) **Construction**   How many cubic *yards* of concrete will it take to pour a cylindrical column 14 ft high with a diameter of 2 ft? (Round to the nearest tenth.)

(b) **Industrial Technology**   A pipe 3 in. in diameter and 40 ft high is filled to the top with water. If 1 cu ft of water weighs 62.4 lb, what is the weight at the base of the pipe? (Be careful of your units. Round to the nearest pound.)

(c) **Industrial Technology**   A cylindrical tank must fit into a space that allows for a 3-ft diameter. If the tank must have a capacity of 180 gallons, how high must it be? ($1 \text{ ft}^3 \approx 7.48$ gal. Round to the nearest inch.)

Solutions   (a) **First,** find the volume in cubic feet.

$$V \approx 0.7854 d^2 h$$

$$\approx 0.7854(2 \text{ ft})^2 (14 \text{ ft})$$

$$\approx 43.9824 \text{ ft}^3$$

Then, convert this result to cubic yards.

$$V \approx 43.9824 \text{ ft}^3 \times \frac{1 \text{ yd}^3}{27 \text{ ft}^3}$$

$$\approx 1.6289 \ldots \approx 1.6 \text{ yd}^3, \text{ rounded.}$$

With a calculator, the entire operation may be keyed in as follows:

.7854 ⊠ 2 x^2 ⊠ 1 4 ⊟ 2 7 ⊜ → 1.628977778

(b) **First,** convert the diameter, 3 in., to 0.25 ft to agree with the other units in the problem. **Next,** find the volume in cubic feet.

$$V \approx 0.7854\, d^2 h \approx 0.7854(0.25 \text{ ft})^2(40 \text{ ft})$$

$$\approx 1.9635 \text{ ft}^3$$

Finally, convert the volume to pound units.

$$V \approx 1.9635 \text{ ft}^3 \times \frac{62.4 \text{ lb}}{1 \text{ ft}^3}$$

$$\approx 123 \text{ lb, rounded}$$

 .7854 ☒ .25 $\boxed{x^2}$ ☒ 40 ☒ 62.4 ⊟ → ░░░░ *122.5224*

(c) **Step 1**   Convert 180 gallons to cubic feet:

$$180 \text{ gal} \times \frac{1 \text{ ft}^3}{7.48 \text{ gal}} \approx 24.064 \text{ ft}^3$$

Step 2   Write the formula for the volume of the tank and set it equal to the result in Step 1.

$$V = \frac{\pi d^2 h}{4} = \frac{\pi (3 \text{ ft})^2 h}{4} = 24.064 \text{ ft}^3$$

Step 3   Solve the equation in Step 2 for the height h.

Square the 3.

$$\frac{\pi(9)h}{4} = 24.064$$

Multiply both sides by 4.

$$\pi(9)h = 96.256$$

Divide both sides by 9π.

$$h \approx 3.4 \text{ ft}$$

Convert 0.4 ft to inches.

$$0.4 \text{ ft} \times \frac{12 \text{ in.}}{1 \text{ ft}} = 4.8 \text{ in.}$$

Rounded to the nearest inch, the height of the tank must be 3 ft 5 in.

Spheres

The **sphere** is the simplest of all solid geometric figures. Geometrically, it is defined as the surface whose points are all equidistant from a given point called the **center**. The **radius** is the distance from the center to the surface. The **diameter** is the straight-line distance across the sphere on a line through its center.

The following formulas enable you to find the surface area and volume of any sphere:

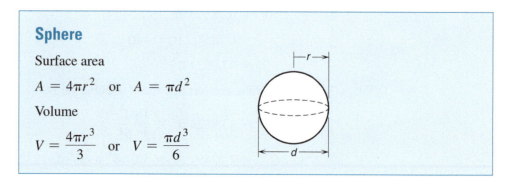

Sphere

Surface area

$$A = 4\pi r^2 \quad \text{or} \quad A = \pi d^2$$

Volume

$$V = \frac{4\pi r^3}{3} \quad \text{or} \quad V = \frac{\pi d^3}{6}$$

The volume formulas can be written approximately as

$$V \approx 4.1888 r^3 \quad \text{and} \quad V \approx 0.5236 d^3$$

EXAMPLE 5 Here are two examples of finding the surface area and volume of a sphere:

(a) For a sphere with a radius of 15 in.:

The surface area is $A = 4\pi r^2 = 4\pi(15 \text{ in.})^2 \approx 2826 \text{ in.}^2$

The volume is $V = \dfrac{4\pi r^3}{3} = \dfrac{4\pi(15 \text{ in.})^3}{3} \approx 14{,}130 \text{ in.}^3$

(b) For a hemisphere (half-sphere) with a diameter of 22 ft:

The area of the curved outside surface is $A = \dfrac{1}{2}\pi d^2 = \dfrac{1}{2}\pi(22 \text{ ft})^2 \approx 760 \text{ ft}^2$

The volume is $V = \dfrac{1}{2} \cdot \dfrac{\pi d^3}{6} = \dfrac{\pi(22 \text{ ft})^3}{12} \approx 2786 \text{ ft}^3$ ●

Your Turn Find the surface area and volume of each of the following. (Round to the nearest whole unit.)

(a) Hemisphere: radius = 6.5 cm   (b) Sphere: diameter = 8 ft 6 in.

Answers (a) $A = 2\pi r^2 \approx 265 \text{ cm}^2$; $V = \dfrac{2\pi r^3}{3} \approx 575 \text{ cm}^3$

(b) $A \approx 227 \text{ ft}^2$; $V \approx 322 \text{ ft}^3$

Applications of Spheres

EXAMPLE 6 **Water/Wastewater Treatment** A spherical water tank is 50 inches in diameter. Water treatment personnel need to know how many gallons of water it will hold. Follow these steps:

First, calculate the volume of the tank in cubic inches:

$$V = \frac{\pi d^3}{6} = \frac{\pi(50 \text{ in.})^3}{6} = 65{,}449.8 \ldots \approx 65{,}450 \text{ in.}^3$$

Then, use the fact that 1 gallon equals 231 in.^{3} to convert the volume to gallons:

$$65{,}450 \text{ in.}^3 \times \frac{1 \text{ gal}}{231 \text{ in.}^3} = 283.33 \ldots \approx 283 \text{ gallons}$$ ●

Your Turn **Industrial Technology** What will be the weight of an open-topped hemispherical steel tank of radius 9 ft when it is full of water? (See the figure.) The steel used weighs approximately 127 lb/ft^2 and water weighs 62.4 lb/ft^3. (Round to the nearest 100 lb.)

9 ft

Solution **Step 1** Calculate the surface area of the tank. For a hemisphere,

$$A = 2\pi r^2 = 2\pi(9 \text{ ft})^2 = 508.938 \ldots \approx 508.9 \text{ ft}^2$$

Step 2 Multiply by a unity fraction to find the weight of the empty tank.

$$508.9 \text{ ft}^2 \times \frac{127 \text{ lb}}{1 \text{ ft}^2} \approx 64{,}635 \text{ lb}$$

Step 3 Calculate the volume of the tank. For a hemisphere,

$$V = \frac{2\pi r^3}{3} = \frac{2\pi(9 \text{ ft})^3}{3} = 1526.814 \ldots \approx 1526.8 \text{ ft}^3$$

Step 4 Multiply by a unity fraction to find the weight W of the water filling the tank.

$$W \approx 1526.8 \text{ ft}^3 \times \frac{62.4 \text{ lb}}{1 \text{ ft}^3} \approx 95{,}272 \text{ lb}$$

Step 5   Add the weight of the tank (Step 2) to the weight of the water (Step 4) and round the final sum to the nearest hundred pounds.

$$\text{Total weight of the full tank} \approx 64{,}635 \text{ lb} + 95{,}272 \text{ lb} \approx 159{,}907 \text{ lb}$$

$$\approx 159{,}900 \text{ lb}$$

We can perform the entire calculation on a calculator as follows.

First, calculate the weight of the tank.

2 $\times$ π $\times$ **9** x^2 $\times$ **1 2 7** $=$ →   `64635.12725`

Then, calculate the weight of the water and add it to the last answer, which is the weight of the tank.

2 $\times$ π $\times$ **9** $\wedge$ **3** $\div$ **3** $\times$ **6 2 . 4** $+$ ANS $=$ →   `159908.3227`

Now continue with Exercises 9-3 for a set of problems on cylinders and spheres.

Exercises 9-3   Cylinders and Spheres

A. Find the lateral surface area and volume of each of the following solid objects. (Round as directed in parentheses and use the $\boxed{\pi}$ key for π.)

1.
46.2 cm
85.4 cm
(hundred)

2.
28 mm
17 mm
(ten)

3.
3"
13"
(whole number)

4.
←12 yd→
8 yd
(whole number)

5.
8.25'
(whole number)

6.
16.2 m
(whole number)

B. Find the total surface area and volume of each of the following solid objects. (Round as directed in parentheses and use the $\boxed{\pi}$ key for π.)

1.
Diameter = 1.5"
(hundredth)

2.
$r = 18'$
(hundred)

3.
8.25'
16.5'
(ten)

4.
$3\frac{1}{4}"$
$1\frac{3}{8}"$
(tenth)

5.
44 m
17 m
(hundred)

6.
18 yd
52 yd
(hundred)

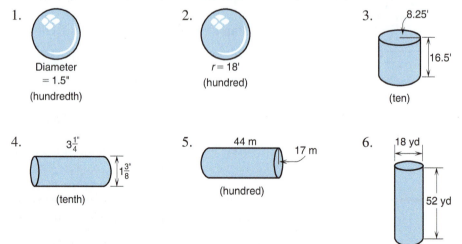

C. Practical Applications. (Round to the nearest tenth unless otherwise directed.)

1. **Painting**   A cylindrical tank has a radius of 6.5 ft and an altitude of 14 ft. If a gallon of paint will cover 120 ft^2 of surface, how much paint is needed to put two coats of paint on the entire surface of the tank? (Round to the nearest gallon.)

2. **Plumbing**   A 4-ft-high cylindrical pipe has a diameter of 3 in. If water weighs 62.4 lb/ft^3, what is the weight of water in the pipe when filled to the top? (*Hint:* Change the diameter to feet.)

3. **Landscaping**   A layer of crushed rock must be spread over a circular area 23 ft in diameter. How deep a layer will be obtained using 100 ft^3 of rock? (Round to the nearest hundredth of a foot.)

4. **Water/Wastewater Treatment**   How many liters of water can be stored in a cylindrical tank 8 m high with a radius of 0.5 m? (1 m^3 contains 1000 liters.)

5. **Water/Wastewater Treatment**   What is the capacity in gallons of a cylindrical tank with a radius of 8 ft and an altitude of 20 ft? (Round to the nearest 100 gallons.)

6. **Industrial Technology**   Find the surface area and the volume in gallons of a spherical tank 8 ft in radius. (Round to the nearest whole unit.)

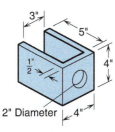

Problem 7

7. **Metalworking**   What is the weight of the piece of aluminum shown in the figure at 0.0975 lb/in.^{3}? (*Hint:* Subtract the volume of the cylindrical hole.)

8. **Industrial Technology**   How high should a 50-gal cylindrical tank be if it must fit into an area allowing a 16-in. diameter? (1 gal = 231 in.^{3})

9. **Plumbing**   A marble-top bathroom sink has the shape of a hemispherical (a half-sphere) porcelain bowl with an inside diameter of 15 in. How much water, in gallons, will the bowl hold? (Round to the nearest gallon.)

10. **Sheet Metal Trades**   How many square inches of sheet copper are needed to cover the inside wall of a cylindrical tank 6 ft high with a 9-in. radius? (*Hint:* Change the altitude to inches.)

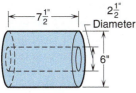

Problem 11

11. **Machine Trades**   What is the weight of the bushing shown in the figure if it is made of steel weighing 0.2833 lb/in.^{3}? (The inner cylinder is hollow.)

12. **Manufacturing**   Find the capacity in gallons of the oil can shown in the figure. (1 gal = 231 in.^{3})

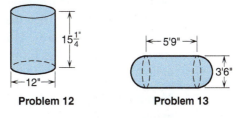

Problem 12   **Problem 13**

13. **Plumbing**   A septic tank has the shape shown in the figure. How many gallons does it hold? (1 ft^3 ≈ 7.48 gallons.) (Round to the nearest gallon.)

14. **Construction**   How many cubic yards of concrete are needed to pour eight cylindrical pillars 12 ft 6 in. high, each with a diameter of 1 ft 9 in.?

15. **Machine Trades** At a density of 0.0925 lb/in.³, calculate the weight of the aluminum piece shown in the figure. (Round to the nearest hundredth.)

Diameter of four outside holes: $\frac{1}{4}$ in.

Diameter of inside hole: $\frac{1}{3}$ in.

Thickness of piece: $\frac{1}{2}$ in.

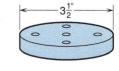

16. **Painting** A spherical tank has a diameter of 16.5 ft. If a gallon of paint will cover 80 ft², how many gallons are needed to cover the tank? (Round to the nearest gallon.)

17. **Industrial Technology** A cylindrical tank 72 cm in diameter must have a capacity of 320 liters. How high must it be? (1 L = 1000 cm³. Round to the nearest centimeter.)

18. **Agriculture** The water tower shown in the figure consists of a 16-ft-high cylinder and a 10-ft-radius hemisphere. How many gallons of water does it contain when it is filled to the top of the cylinder? (Round to the nearest hundred.)

19. **Automotive Trades** In an automobile engine, the bore of a cylinder is its diameter, and the stroke of a cylinder is the height swept by the piston. Find the swept volume of a cylinder with a bore of 3.65 in. and a stroke of 4.00 in. (Round to the nearest tenth.)

20. **Automotive Trades** A cylindrical hose 24 in. long has an inside diameter of 1.4 in. How many quarts of coolant will it hold? (1 qt = 2 pt and 1 pt = 28.875 cu in.) (Round to the nearest hundredth.)

21. **Agriculture** The water trough shown in the figure is constructed with semicircular ends. Calculate its volume in gallons if the diameter of the end is 15 in. and the length of the trough is 8 ft. (Round to the nearest 0.1 gallon.) (*Hint:* Be careful of units.)

22. **Painting** The metal silo shown in the figure has a hemispherical top. It is to be painted with exterior weatherproofing paint. If a gallon of this paint covers 280 ft², how much paint is needed for two coats? (Do not include the bottom of the silo. Round your answer to the nearest gallon.)

23. **Construction** An outdoor cylindrical fire pit has an interior radius of 30 in. Filling the fire pit with fireglass will require 91 lb of glass per cubic foot. How many pounds will be needed to fill the pit to a depth of 4 in.? (Round to the nearest pound.)

24. **Water/Wastewater Treatment** Due to a rupture in a 6-in. diameter pipe, a water district must treat the contaminated flow of water in the pipe over a distance of 720 yards. How many gallons of water flow through this stretch of pipe? (See Example 3 on page 636. Round to the nearest gallon.)

25. **Manufacturing** If a particular extrusion process is 92% efficient, how many linear feet of 1.5-in. diameter steel cable can be produced from a block of steel 10 ft long with a 16-in. by 16-in. cross section? (See Example 4 on page 636. Round to the nearest foot.)

When you have completed these exercises, check your answers to the odd-numbered problems in the Appendix, then turn to Section 9-4 to learn about cones.

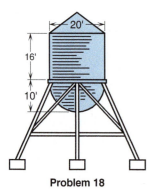

Problem 18

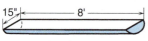

Problem 21

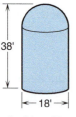

Problem 22

9-4   Cones and Frustums of Cones

Cones

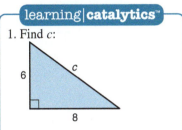

Apex

Slant height, *s*

h

r

Base

A **cone** is a pyramid-like solid figure with a circular base. The radius and diameter of a cone refer to its circular base. The **altitude** is the perpendicular distance from the apex to the base. The **slant height** is the apex-to-base distance along the surface of the cone.

The following formulas enable us to find the lateral surface area and volume of a cone:

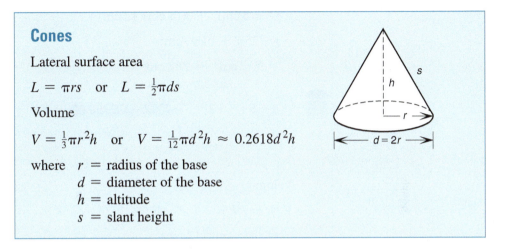

Cones

Lateral surface area

$$L = \pi rs \quad \text{or} \quad L = \tfrac{1}{2}\pi ds$$

Volume

$$V = \tfrac{1}{3}\pi r^2 h \quad \text{or} \quad V = \tfrac{1}{12}\pi d^2 h \approx 0.2618 d^2 h$$

where r = radius of the base
         d = diameter of the base
         h = altitude
         s = slant height

learning|catalytics™

1. Find *c*:

6

c

8

2. Find the volume of a square pyramid with base length 4 in. and altitude 9 in.

To find the total surface area of a cone, simply add the area of the circular base to the lateral area. The formula would be

Total surface area $T = \pi r^2 + \pi rs = \pi r(r + s)$

EXAMPLE 1

15"

18"

10"

To find the lateral surface area and volume of the cone shown, first note that $h = 15$ in., $s = 18$ in., $r = 10$ in.

Then the lateral surface area is

$$L = \pi rs$$

$$= \pi(10 \text{ in.})(18 \text{ in.})$$

$$= 565.486 \ldots \text{ in.}^2 \quad \text{or} \quad 565 \text{ in.}^2 \quad \text{rounded}$$

and the volume is

$$V = \tfrac{1}{3}\pi r^2 h$$

$$= \tfrac{1}{3}\pi(10 \text{ in.})^2(15 \text{ in.})$$

$$= 1570.79 \ldots \text{ in.}^3 \quad \text{or} \quad 1571 \text{ in.}^3 \quad \text{rounded}$$

Using a calculator to find the volume, we obtain

π ÷ 3 × 1 0 x^2 × 1 5 = → *1570.796327*

Your Turn   Find the lateral surface area, total surface area, and volume of this cone. (Round to the nearest whole unit.)

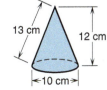

13 cm

12 cm

10 cm

Solution   First, note that $h = 12$ cm $r = 5$ cm $s = 13$ cm

Lateral surface area:

$L = \pi rs$

$= \pi(5 \text{ cm})(13 \text{ cm})$

$\approx 204.203 \ldots \text{cm}^2$ or 204 cm^2 rounded

Total surface area:

$T = \pi r^2 + \pi rs$

$= \pi(5 \text{ cm})^2 + \pi(5 \text{ cm})(13 \text{ cm})$

$= (25\pi + 65\pi) \text{ cm}^2$

$= 90\pi \text{ cm}^2 \approx 283 \text{ cm}^2$

$\boxed{\pi} \boxed{\times} \mathbf{5} \boxed{\times} \mathbf{1\,3} \boxed{=} \rightarrow$ **204.2035225** $\boxed{+} \boxed{\pi} \boxed{\times} \mathbf{5} \boxed{x^2} \boxed{=} \rightarrow$ **282.7433388**

L $\qquad$ T

Volume:

$V = \frac{1}{3}\pi r^2 h$

$= \frac{1}{3}\pi(5 \text{ cm})^2(12 \text{ cm})$

$= 314.159 \ldots \text{cm}^3$ or 314 cm^3 rounded

Learning Help   You should realize that the altitude, radius, and slant height for any cone are always related by the Pythagorean theorem:

$s^2 = h^2 + r^2$

If you are given any two of these quantities, you can use this formula to find the third.

Notice that the volume of a cone is exactly one-third the volume of the cylinder that just encloses it. ●

More Practice   Solve the following practice problems.

(a) Find the lateral surface area and volume of each of the following cones. (Round to the nearest whole number.)

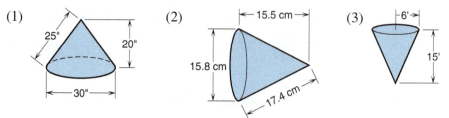

(1) 25", 20", 30"

(2) 15.5 cm, 15.8 cm, 17.4 cm

(3) 6', 15'

(b) Find the total surface area of a cone with radius $\frac{3}{4}$ in., altitude 1 in., and slant height $1\frac{1}{4}$ in. (Round to two decimal places.)

Solutions   (a) (1) 1178 in.^{2}, 4712 in.^{3} (2) 432 cm^2, 1013 cm^3

(3) **First,** we must calculate the slant height using the Pythagorean theorem.

$$s^2 = h^2 + r^2$$

$$= 15^2 + 6^2 = 261$$

$$s = \sqrt{261} \approx 16.155 \text{ ft}$$

The lateral surface area, L, is

$$L = \pi r s \approx \pi(6 \text{ ft})(16.155 \text{ ft})$$

$$\approx 96.93\pi \text{ ft}^2 \approx 305 \text{ ft}^2$$

The volume, V, is

$$V = \tfrac{1}{3}\pi r^2 h = \tfrac{1}{3}\pi(6 \text{ ft})^2(15 \text{ ft})$$

$$= 180\pi \text{ ft}^3 \approx 565 \text{ ft}^3$$

(b) $T = \pi r(r + s) = \pi(\tfrac{3}{4} \text{ in.})(\tfrac{3}{4} \text{ in.} + 1\tfrac{1}{4} \text{ in.})$

$$= \pi(0.75 \text{ in.})(2 \text{ in.})$$

$$\approx 4.71 \text{ in.}^2$$

Applications of Cones

EXAMPLE 2   **Industrial Technology**   A conical drum 4 ft high has a radius of 3 ft. To find the capacity of the drum in gallons:

First, find the volume in cubic feet.

$$V = \tfrac{1}{3}\pi r^2 h = \tfrac{1}{3}(\pi)(3 \text{ ft})^2(4 \text{ ft})$$

$$= 37.699 \ldots \text{ ft}^3$$

Then, convert this to gallons (1 cu ft $\approx$ 7.48 gal).

$$V \approx 37.699 \text{ ft}^3 \times \frac{7.48 \text{ gal}}{1 \text{ ft}^3} \approx 282 \text{ gal}$$

Of course, with a calculator you would not need to round the intermediate result.

 $\boxed{\pi}\,\boxed{\div}\,\boxed{3}\,\boxed{\times}\,\boxed{3}\,\boxed{x^2}\,\boxed{\times}\,\boxed{4}\,\boxed{\times}\,\boxed{7.48}\,\boxed{=} \rightarrow$ `281.9893566`   ●

EXAMPLE 3   **Construction**   A conical pile of sand has a base diameter of 18 ft 3 in. and an altitude of 8 ft 6 in. A pickup truck with a bed capacity of 3.1 cu yd is being used to haul away the sand. Suppose we wish to know the number of truckloads it will take to transport the sand.

First, we convert the dimensions of the pile to decimal feet:

18 ft 3 in. = 18.25 ft   and   8 ft 6 in. = 8.5 ft

Next, we calculate the volume of the sand in cubic feet:

$$V = \tfrac{1}{12}\pi d^2 h = \tfrac{1}{12}\pi(18.25 \text{ ft})^2(8.5 \text{ ft})$$

$$\approx 235.9\pi \text{ ft}^3$$

Then, we use the fact that 1 cu yd = 27 cu ft to convert this to cubic yards:

$$235.9\pi \text{ ft}^3 \times \frac{1 \text{ yd}^3}{27 \text{ ft}^3} \approx 27.45 \text{ yd}^3$$

Finally, we use a unity fraction to determine the number of truckloads needed:

$$27.45 \text{ yd}^3 \times \frac{1 \text{ load}}{3.1 \text{ yd}^3} \approx 8.85 \text{ loads}$$

The answer must be a whole number, so it will take 9 truckloads to transport the sand.   ●

Your Turn

(a) **Industrial Technology**   Find the capacity in gallons (to the nearest tenth) of a conical oil container 15 in. high with a diameter of 16 in. (1 gal = 231 in.^{3})

(b) **Metalworking**   Find the weight of the solid cast iron shape shown in the figure at 0.26 lb/in.^{3} (Round to the nearest pound.)

Problem (b)

6"
13"
10"

Solutions

(a) The capacity, in cubic inches, is

$$V \approx 0.2618 d^2 h \approx 0.2618 (16 \text{ in.})^2 (15 \text{ in.}) \quad \left(\tfrac{\pi}{12} \approx 0.2618\right)$$

$$\approx 1005.312 \text{ in.}^3$$

The capacity in gallons is

$$V \approx 1005.312 \text{ in.}^3 \times \frac{1 \text{ gal}}{231 \text{ in.}^3}$$

$$\approx 4.352 \text{ gal} \approx 4.4 \text{ gal, rounded to the nearest tenth}$$

.2618 ✕ 16 x^2 ✕ 15 ÷ 231 = → **4.352**

(b) **First,** find the volume in cubic inches.

$$d = 10'' \text{ so } r = 5''$$

For the cylinder:

$$V = \pi r^2 h = \pi (5 \text{ in.})^2 (13 \text{ in.})$$

$$= 325\pi \text{ in.}^3$$

For the cone:

$$V = \tfrac{1}{3}\pi r^2 h = \tfrac{1}{3}\pi (5 \text{ in.})^2 (6 \text{ in.})$$

$$= 50\pi \text{ in.}^3$$

Total volume, $V = (325\pi + 50\pi) \text{ in.}^3 = 375\pi \text{ in.}^3$

Then, convert to pounds.

$$\text{Weight} = 375\pi \text{ in.}^3 \times \frac{0.26 \text{ lb}}{1 \text{ in.}^3} \approx 306 \text{ lb}$$

The entire calculation can be done as follows using a calculator:

π ✕ 5 x^2 ✕ 13 + π ÷ 3 ✕ 5 x^2 ✕ 6 = ✕ .26 = → **306.3052837**

Learning Help

The three formulas for the volume of a cone were provided to give you a choice. You might want to memorize only the formula $V = \tfrac{1}{3}\pi r^2 h$, from which the others are derived. ●

Frustum of a Cone

Frustum
of a cone

The **frustum of a cone**, like the frustum of a pyramid, is the figure remaining when the top of the cone is cut off parallel to its base.

Frustum shapes appear as containers, as funnels, and as transition sections in ducts.

Every frustum has two bases, upper and lower, that are parallel and different in size. The altitude is the perpendicular distance between the bases.

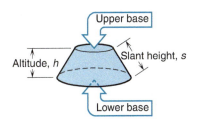

The following formulas enable you to find the lateral area and volume of any frustum of a cone:

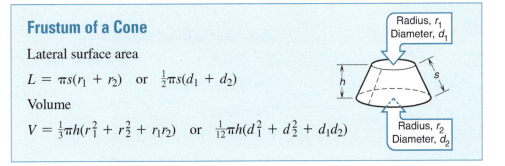

Frustum of a Cone

Lateral surface area

$$L = \pi s(r_1 + r_2) \quad \text{or} \quad \tfrac{1}{2}\pi s(d_1 + d_2)$$

Volume

$$V = \tfrac{1}{3}\pi h(r_1^2 + r_2^2 + r_1 r_2) \quad \text{or} \quad \tfrac{1}{12}\pi h(d_1^2 + d_2^2 + d_1 d_2)$$

To find the total surface area of the frustum, add the areas of the circular ends to the lateral surface area.

EXAMPLE 4

To find the lateral surface area, total surface area, and volume of the frustum shown, use $s = 9.0$ ft, $r_1 = 2$ ft, $r_2 = 5$ ft, and $h = 8.5$ ft.

Lateral surface area, $\qquad L = \pi(9)(2 + 5)$

$$= 63\pi \text{ ft}^2 \approx 198 \text{ ft}^2 \quad \text{rounded}$$

Total surface area, $\qquad T = L + \pi r_1^2 + \pi r_2^2$

$$= 63\pi + \pi(2^2) + \pi(5^2)$$

$$= 92\pi \text{ ft}^2 \approx 289 \text{ ft}^2 \quad \text{rounded}$$

Volume, $\qquad V = \tfrac{1}{3}\pi(8.5)(2^2 + 5^2 + 2\cdot 5)$

$$= 110.5\pi \text{ ft}^3 \approx 347 \text{ ft}^3 \quad \text{rounded} \qquad \bullet$$

Note As long as we know the units of the final answer, we need not write the units in every intermediate step. Eliminating them, as we did in Example 4, helps to simplify complicated formulas like these. ●

More Practice Practice with these problems.

Find the lateral surface area and volume of each of the following frustums. (Round to the nearest whole unit.)

(a)

(b)
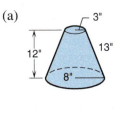

(c)
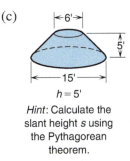

$h = 5'$

Hint: Calculate the slant height *s* using the Pythagorean theorem.

Solutions   (a) 449 in.^{2}; 1219 in.^{3} (b) 271 in.^{2}; 545 in.^{3}

(c) **First,** find the slant height this way:

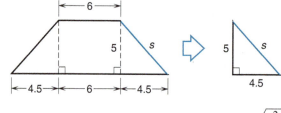

$$s = \sqrt{5^2 + 4.5^2} = \sqrt{45.25} \approx 6.73 \text{ ft}$$

Then, find the lateral surface area.

$$L = \tfrac{1}{2}\pi s(d_1 + d_2) \approx \tfrac{1}{2}\pi(6.73)(6 + 15)$$
$$\approx 70.665\pi \text{ ft}^2 \approx 222 \text{ ft}^2$$

The volume is

$$V = \tfrac{1}{12}\pi h(d_1^2 + d_2^2 + d_1 \cdot d_2)$$
$$= \tfrac{1}{12}\pi(5)(6^2 + 15^2 + 6 \cdot 15)$$
$$= 146.25\pi \text{ ft}^3 \approx 459 \text{ ft}^3$$

Applications of Conical Frustums

EXAMPLE 5

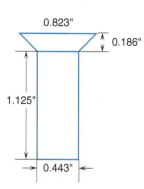

Metalworking   The figure in the margin shows the dimensions of a round counter-sunk head rivet. The rivet is made of solid copper, which weighs 0.324 lb/in.^{3}. Suppose we need to find the total weight of the rivet in ounces:

First, find the volume. The rivet consists of two parts: A head in the shape of the frustum of a cone, and a shaft in the shape of a cylinder.

For the head:

$$V = \tfrac{1}{12}\pi h(d_1^2 + d_2^2 + d_1 \cdot d_2)$$
$$= \tfrac{1}{12}\pi(0.186)[0.823^2 + 0.443^2 + (0.823)(0.443)]$$
$$\approx 0.06029 \text{ in.}^3$$

For the shaft:

$$V = \tfrac{1}{4}\pi d^2 h$$
$$= \tfrac{1}{4}\pi(0.443)^2(1.125)$$
$$\approx 0.17340 \text{ in.}^3$$

The total volume is approximately 0.06029 in.^{3} + 0.17340 in.^{3} $\approx$ 0.23369 in.^{3}

Then, use the given conversion factor to find the weight in pounds:

$$\text{Weight} \approx 0.23369 \text{ in.}^3 \times \frac{0.324 \text{ lb}}{1 \text{ in.}^3} \approx 0.0757 \text{ lb}$$

Finally, use 1 lb = 16 oz to convert the weight to ounces:

$$\text{Weight} \approx 0.0757 \text{ lb} \times \frac{16 \text{ oz}}{1 \text{ lb}} \approx 1.21 \text{ oz, rounded}$$

The entire calculation can be done on a calculator as follows:

$\boxed{\pi}\, \boxed{\div}\, \mathbf{12}\, \boxed{\times}\, \mathbf{.186}\, \boxed{\times}\, \boxed{(}\, \mathbf{.823}\, \boxed{x^2}\, \boxed{+}\, \mathbf{.443}\, \boxed{x^2}\, \boxed{+}\, \mathbf{.823}\, \boxed{\times}\, \mathbf{.443}\, \boxed{)}\, \boxed{+}$

$\boxed{\pi}\, \boxed{\div}\, \mathbf{4}\, \boxed{\times}\, \mathbf{.443}\, \boxed{x^2}\, \boxed{\times}\, \mathbf{1.125}\, \boxed{=} \rightarrow$ `0.2336924581` $\Leftarrow$ Total volume

$\boxed{\times}\, \mathbf{.324}\, \boxed{\times}\, \mathbf{16}\, \boxed{=} \rightarrow$ `1.211461703` $\Leftarrow$ Weight in ounces

Your Turn **Industrial Technology** What is the capacity in gallons of the oil can shown in the figure? (1 gal = 231 in.^{3} Round to the nearest tenth.)

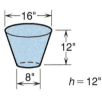

Solution **First,** find the volume in cubic inches.

$$V = \tfrac{1}{12}\pi h(d_1^2 + d_2^2 + d_1 \cdot d_2)$$
$$= \tfrac{1}{12}(\pi)(12)(16^2 + 8^2 + 16 \cdot 8)$$
$$\approx 1407.43 \text{ in.}^3$$

Then, translate this to gallons.

$$V \approx 1407.43 \text{ in.}^3 \times \frac{1 \text{ gal}}{231 \text{ in.}^3}$$

$$\approx 6.1 \text{ gal, rounded}$$

Using a calculator on this problem,

 π ÷ 1 2 × 1 2 × (1 6 x^2 + 8 x^2 + 1 6 × 8) ÷ 2 3 1 =

→ 6.092785752

Now continue with Exercises 9-4 for more practice with cones and frustums of cones.

Exercises 9-4 ## Cones and Frustums of Cones

A. Find the lateral surface area and volume of each solid object. (Round to the nearest whole number.)

1.

16.5 cm

11.6 cm

2.

15'

33.5'   30'

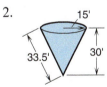

3.

4.20 m

9.79 m   9.70 m

6.80 m

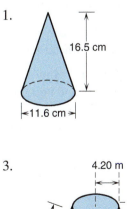

4.

26.0"

25.1"   24.0"

11.0"

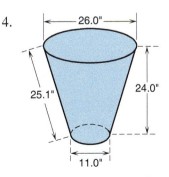

B. Find the total surface area and volume of each solid object. (Round to the nearest whole number.)

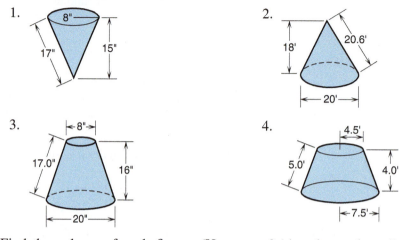

1.   2.

3.   4.

C. Find the volume of each figure. (Use $\pi \approx 3.14$ and round as directed in parentheses.)

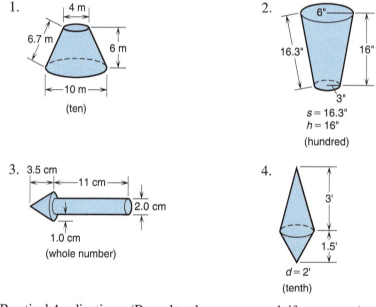

1.

(ten)

2.

$s = 16.3"$
$h = 16"$

(hundred)

3. 3.5 cm

(whole number)

4.

$d = 2'$

(tenth)

D. Practical Applications. (Round to the nearest tenth if necessary.)

1. **Landscaping**   How many cubic meters of dirt are there in a pile, conical in shape, 10 m in diameter and 4 m high?

2. **Sheet Metal Trades**   How many square centimeters of sheet metal are needed for the sides and bottom of the pail shown in the figure?

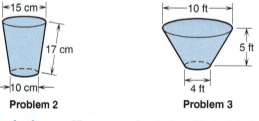

Problem 2   Problem 3

3. **Agriculture**   How many bushels will the bin in the figure hold? (1 cu ft $\approx$ 1.24 bushels.)

4. **Landscaping**   Decorative Moon Rocks cost $92 per cu yd. What will be the cost of a conical pile of this rock $3\frac{1}{3}$ yd in diameter and $2\frac{1}{4}$ yd high? (Round to the nearest dollar.)

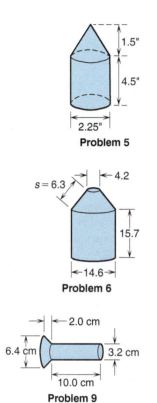

Problem 5

Problem 6

Problem 9

5. **Metalworking**   Calculate the weight of the piece of steel shown in the drawing. (This steel weighs 0.283 lb/cu in.)

6. **Sheet Metal Trades**   How many square centimeters of sheet metal will it take to make the open-top container shown in the figure? All measurements are in centimeters. (*Hint:* No metal is needed for the small opening at the top.)

7. **Manufacturing**   A conical oil cup with a radius of 3.8 cm must be designed to hold 64 cu cm of oil. What should be the altitude of the cup?

8. **Landscaping**   At an excavation site, a conical pile of earth 9 ft high and 14 ft in diameter at the base must be hauled away by a small truck with a capacity of 2.6 cu yd. How many truckloads will it take to transport the soil? (Be careful with units!)

9. **Metalworking**   Find the weight in ounces of the steel rivet shown in the figure. (Steel weighs 0.0173 lb/cu cm.)

10. **Welding**   A cone-shaped hopper is constructed by first cutting a wedge from a circular piece of sheet metal [See figure (a).] The center of the circle is then hoisted up and the two edges of the missing wedge are welded together. [See figure (b).] Note that the radius of the original piece, 3.18 m, is now the slant height of the cone, and the radius of the cone is 2.56 m. Find the volume of the cone.

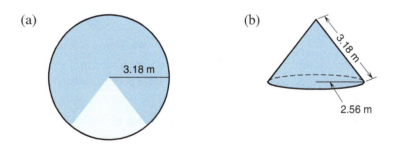

11. **Manufacturing**   A yogurt carton is in the shape of the frustum of a cone. The upper diameter is 2 in., the lower diameter is $2\frac{5}{16}$ in., and the height is 3 in. How many ounces of yogurt does the carton hold? (*Hint:* 1 oz = 1.805 cu in.)

12. **Construction**   The spire of the new Wilshire Grand Center in Los Angeles is in the shape of a conical frustum. It is 78.0 inches in diameter at the base and tapers to 32.0 inches in diameter at the top. If the spire is 175 feet tall, what is its volume in cubic feet? (Round to the nearest ten cubic feet.)

13. **Landscaping**   A landscape contractor must fill 18 pots with potting soil in order to plant flowers inside of them. Each pot is in the shape of a conical frustum with an upper diameter of 19 in., a lower diameter of 13 in., and a height of 16 in. If potting soil comes in bags containing 1.5 cu ft, how many bags does the contractor need to fill all 18 pots?

When you have completed these exercises, go to Problem Set 9 on page 654 for a set of problems covering the work of this chapter. If you need a quick review of the topics in this chapter, visit the chapter Summary first.

CHAPTER 9
SUMMARY
Solid Figures

Objective	Review
Identify solid figures, including prisms, cubes, cones, cylinders, pyramids, spheres, and frustums.	A **prism** is a solid figure having at least one pair of parallel surfaces that create a uniform cross section. A **cube** is a rectangular prism in which all edges are the same length. A **pyramid** is a solid object with one base that is a polygon and three or more lateral faces that taper to a single point, the apex. A **cylinder** is a solid object with two identical circular bases. A **cone** is a pyramid-like solid figure with a circular base. A **frustum** is the solid figure remaining after the top of a pyramid or cone is cut off by a plane parallel to the base. A **sphere** is the round surface created by all points that are equidistant from a given point, the center.

Examples:

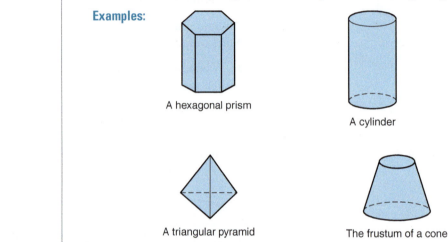

A hexagonal prism

A cylinder

A triangular pyramid

The frustum of a cone

Find the surface area and volume of solid objects.

The following table provides a handy summary of the formulas for solid figures presented in this chapter.

Summary of Formulas for Solid Figures

Figure	Lateral Surface Area		Volume	
Prism		$L = ph$		$V = Ah$
Pyramid		$L = \dfrac{1}{2}ps$		$V = \dfrac{1}{3}Ah$
Cylinder		$L = \pi dh$		$V = \pi r^2 h$
	or	$L = 2\pi rh$	or	$V \approx 0.7854 d^2 h$
Cone		$L = \dfrac{1}{2}\pi ds$		$V = \dfrac{1}{3}\pi r^2 h$
	or	$L = \pi rs$	or	$V \approx 0.2618 d^2 h$
Frustum of pyramid		$L = \dfrac{1}{2}(P_1 + P_2)s$		$V = \dfrac{1}{3}h(A_1 + A_2 + \sqrt{A_1 A_2})$
Frustum of cone		$L = \dfrac{1}{2}\pi s(d_1 + d_2)$		$V = \dfrac{1}{12}\pi h(d_1^2 + d_2^2 + d_1 d_2)$
	or	$L = \pi s(r_1 + r_2)$	or	$V = \dfrac{1}{3}\pi h(r_1^2 + r_2^2 + r_1 r_2)$
Sphere		$A = 4\pi r^2$		$V = \dfrac{4}{3}\pi r^3$
	or	$A = \pi d^2$	or	$V \approx 4.1888 r^3$
				$V = \dfrac{1}{6}\pi d^3$
			or	$V \approx 0.5236 d^3$

Objective	Review

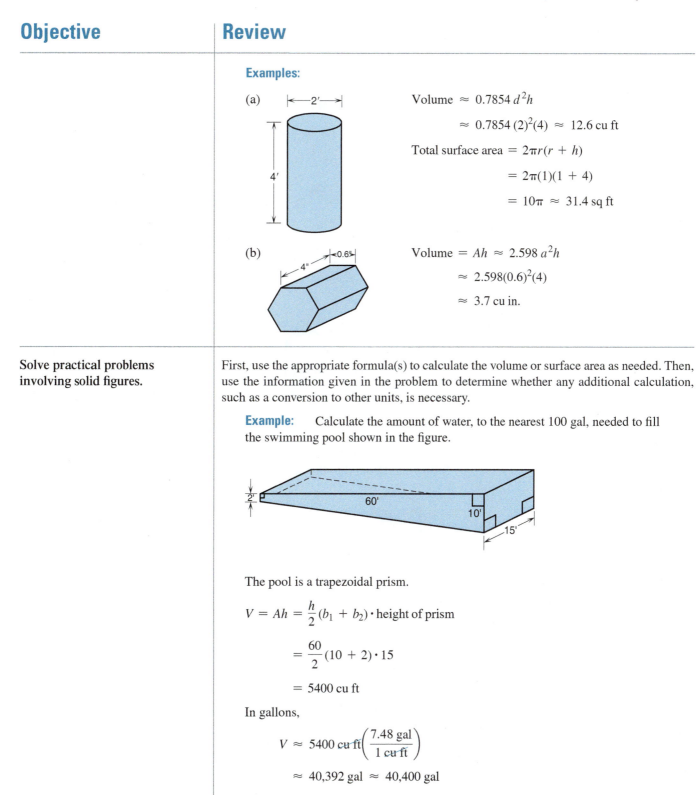

Examples:

(a) |←—2′—→|

Volume $\approx 0.7854\, d^2 h$

$\approx 0.7854\,(2)^2(4) \approx 12.6 \text{ cu ft}$

Total surface area $= 2\pi r(r + h)$

$= 2\pi(1)(1 + 4)$

$= 10\pi \approx 31.4 \text{ sq ft}$

(b)

Volume $= Ah \approx 2.598\, a^2 h$

$\approx 2.598(0.6)^2(4)$

$\approx 3.7 \text{ cu in.}$

Solve practical problems involving solid figures.

First, use the appropriate formula(s) to calculate the volume or surface area as needed. Then, use the information given in the problem to determine whether any additional calculation, such as a conversion to other units, is necessary.

Example: Calculate the amount of water, to the nearest 100 gal, needed to fill the swimming pool shown in the figure.

The pool is a trapezoidal prism.

$$V = Ah = \frac{h}{2}(b_1 + b_2) \cdot \text{height of prism}$$

$$= \frac{60}{2}(10 + 2) \cdot 15$$

$$= 5400 \text{ cu ft}$$

In gallons,

$$V \approx 5400 \text{ cu ft}\left(\frac{7.48 \text{ gal}}{1 \text{ cu ft}}\right)$$

$$\approx 40{,}392 \text{ gal} \approx 40{,}400 \text{ gal}$$

PROBLEM SET 9   Solid Figures

Answers to odd-numbered problems are given in the Appendix.

A. Solve the following problems involving solid figures.

Find (a) the lateral surface area and (b) the volume of each of the following. (Round to the nearest tenth.)

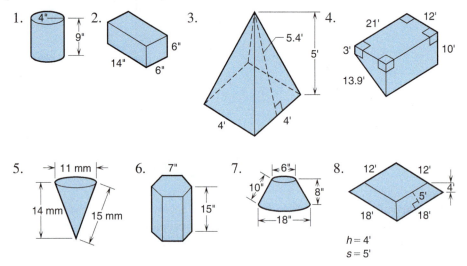

1.
2.
3.
4.

5.
6.
7.
8.

$h = 4'$
$s = 5'$

Find (a) the total outside surface area and (b) the volume of each of the following. (Round to the nearest tenth.)

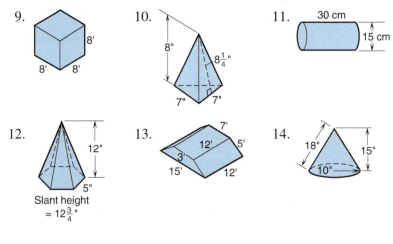

9.
10.
11.   30 cm

12.
13.
14.

    Slant height
    $= 12\frac{3}{4}"$

15. A sphere with a radius of 10 cm.

16.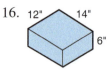

17. A hemisphere with a diameter of 13 in.

Name

_____

Date

_____

Course/Section

_____

18.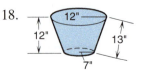

19.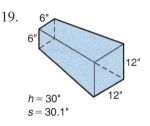

$h = 30"$
$s = 30.1"$

20.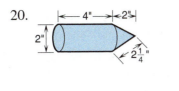

21.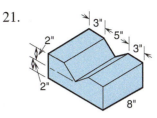

B. Practical Applications

1. **Manufacturing** How many boxes 16 in. by 12 in. by 10 in. will fit in 1250 ft³ of warehouse space? (1 ft³ = 1728 in.³)

2. **Construction** How many cubic *yards* of concrete are needed to pour the highway support shown in the figure? (1 yd³ = 27 ft³.) (Round to nearest cubic yard.)

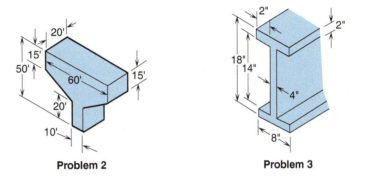

Problem 2   Problem 3

3. **Metalworking** What is the weight of the 10-ft-long steel I-beam shown in the figure if the density of steel is 0.2833 lb/in.³? (Round to the nearest pound.)

4. **Plumbing** What is the capacity in gallons of a cylindrical water tank 3 ft in diameter and 8 ft high? (1 ft³ ≈ 7.48 gal.) (Round to one decimal place.)

5. **Painting** How many quarts of paint will it take to cover a spherical water tank 25 ft in diameter if one quart covers 50 ft²?

6. **Construction** A hole must be excavated for a swimming pool in the shape shown in the figure. How many trips will be needed to haul the dirt away if the truck has a capacity of 9 m³? (*Remember:* A fraction of a load constitutes a trip.)

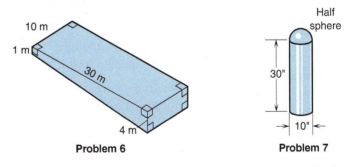

Problem 6   Problem 7

7. **Manufacturing** How many cubic feet of propane will the tank in the figure contain? (1 ft³ = 1728 in.³) (Round to the nearest tenth of a cubic foot.)

8. **Industrial Technology** What must be the height of a cylindrical 750-gal tank if it is 4 ft in diameter? (1 ft³ ≈ 7.48 gal.) (Round to the nearest inch.)

9. **Automotive Trades** At a density of 42 lb/ft³, what is the weight of fuel in the rectangular gas tank pictured?

10. **Masonry** A "slim jumbo modular" brick has dimensions of $3\frac{1}{8}$ in. by $2\frac{3}{4}$ in. by $7\frac{5}{8}$ in. With a $\frac{5}{16}$-in. mortar joint added to each side, determine the number

Problem 9

of bricks required per cubic foot. (*Hint:* Add the mortar thickness just once to each dimension and round your answer up to the next whole number.)

11. **Construction**   A foundation wall 3 ft high and 9 in. thick consists of lengths of 56 ft, 40.5 ft, 26 ft, 8.5 ft, and 25 ft. What is the total number of cubic *yards* of concrete needed for the wall?

12. **Manufacturing**   How many square inches of paper are needed to produce 2000 conical cups like the one shown in the figure? (Round to the nearest 100 in.^{2}.)

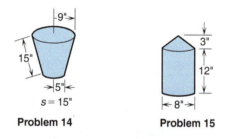

Problem 12   Problem 13

13. **Construction**   How many cubic yards of concrete are needed to pour the monument foundation shown in the figure? (Round to the nearest cubic yard.)

14. **Sheet Metal Trades**   How many square inches of sheet metal are needed to make the bottom and sides of the pail shown in the figure? (Round to the nearest square inch.)

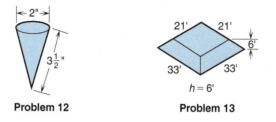

Problem 14   Problem 15

15. **Industrial Technology**   Find the capacity in gallons of the oil can shown in the figure. It is a cylinder with a cone on top. (1 gal = 231 in.^{3}) (Round to the nearest tenth of a gallon.)

16. **Machine Trades**   At a density of 7.7112 g/cm^3, find the weight of 1200 hex nuts, each 1.00 cm on a side and 0.50 cm thick with a hole 1.00 cm in diameter. (See figure in margin. Round to the nearest hundred grams or tenth of a kilogram.)

17. **Industrial Technology**   A cylindrical tank can be no more than 6 ft high. What diameter must it have in order to hold 1400 gal?

18. **Welding**   A rectangular tank is made using steel strips 8 in. wide. The sides are 18 in. by 8 in., the ends are 8 in. by 8 in., and the top and bottom are 18 in. by 8 in. Calculate the volume of the tank and the length of the weld needed. (See the figure.)

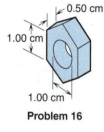

Problem 16

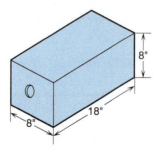

Problem 18

19. **Automotive Trades**   An oil pan has the shape of a rectangular prism measuring 64 cm long by 32 cm wide by 16 cm deep. How many liters of oil will it hold? (1 liter = 1000 cu cm.) (Round to the nearest liter.)

20. **Welding**   Calculate the weight of a solid steel roller bar. It is 14.0 in. long and 3.0 in. in diameter. Use density of steel ≈ 0.2833 lb/cu in. (Round to the nearest pound.)

21. **Agriculture**   A feeding trough is constructed from sheet metal as a half-cylinder, 26.0 ft long and 3.0 ft in diameter. (See the figure.) When it is 80% filled with feed, what volume of feed does it hold? (Round to the nearest cubic foot.)

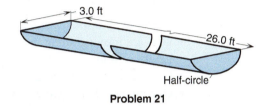

3.0 ft

26.0 ft

Half-circle

Problem 21

22. **Welding**   Calculate the weight of the steel end-plate shown in the figure if it is 20.0 cm on each side and 2.0 cm thick. The bolt holes are 1.0 cm in diameter. The density of steel is 0.7849 g/cu cm. (Round to the nearest ten grams.)

$1\frac{3}{4}$"   8"

$15\frac{5}{8}$"

$15\frac{5}{8}$"

Problem 22   **Problem 23**

23. **Masonry**   A square pillar is built using $15\frac{5}{8}$ in.-square concrete blocks 8 in. high with $1\frac{3}{4}$-in.-thick walls. This creates a hollow square prism inside each block. (See the figure.) If the pillar is 10 blocks high, how many cubic *yards* of concrete will it require to fill the pillar? (Round to the nearest 0.1 cu yd.)

24. **Landscaping**   In the figure shown below, the solid slanted line represents the current grade of soil on a pathway between a building and a fence. The broken horizontal line represents the desired grade line. If the length of the pathway is 24 ft, what volume of soil (in cubic yards) must be brought in to achieve the desired grade? (*Hint:* The new soil forms a trapezoidal prism. Round to the nearest whole number.)

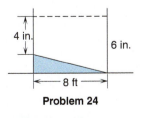

4 in.

6 in.

8 ft

Careful: The figure is not drawn to scale, and the units are not all the same.

Problem 24

25. **Carpentry**   A granite kitchen countertop is $1\frac{5}{16}$ in. thick. If the countertop is a rectangular shape measuring 28.8 in. by 82.5 in., and granite weighs 0.1 lb/cu in., how much does the countertop weigh? (Round to the nearest pound.)

26. **Landscaping**   A landscaper installing a garden takes delivery of 5 cu yd of topsoil. If the topsoil is spread evenly over an area 18 ft by 30 ft, how many inches deep is the topsoil? (Be careful of units!)

27. **Manufacturing**   Steel cable can be manufactured from a heated steel block. The volume of the resulting cable will be equal to the volume of the original block minus a small percentage of waste that occurs during the process. Suppose that a 20-ft long steel block with a 12-in. by 12-in. cross section is heated to produce 3-in. diameter cable. If there is a waste factor of 5%, how many linear feet of cable can be produced? (Round to the nearest foot.)

28. **Agriculture**   Consider the following information about oats moving along a grain conveyor:

    • The conveyor moves at a linear speed of 360 ft/min.
    • Oats are moving along the full 12-in. width of the conveyor at an average height of 4 in.
    • Oats have a density of 25.6 lb/cu ft.
    • One bushel of oats weighs 32 lb.

    How many bushels of oats does the grain conveyor unload in an hour?

29. **Water/Wastewater Treatment**   The amount of chlorine C, in pounds, needed to disinfect a particular well is given by the formula $C = 834V$, where V is the volume of water in the well expressed in *millions* of gallons. Calculate the amount of chlorine needed for a cylindrical well 18 in. in diameter and filled with water to a depth of 200 feet. (*Hints:* One cubic foot of water contains approximately 7.48 gallons. Multiply gallons by 10^{-6} to convert to millions of gallons. Round to the nearest tenth of a pound.)

30. **Construction**   A rectangular plot of land measuring 85.5 ft by 112.0 ft must be excavated to a benchmark elevation of 180.0 ft. The four corners of the plot have elevations of 186.0 ft, 187.5 ft, 189.8 ft, and 182.5 ft. Use the method of Example 2 on page 616 to estimate the total volume of earth to be removed in cubic yards. (Round to the nearest ten cubic yards.)

31. **Landscaping**   A landscape contractor needs to fill 25 pots with potting soil in preparation for planting. Each pot is in the shape of a conical frustum with an upper diameter of 16 in., a lower diameter of 10 in., and a height of 13 in. If potting soil comes in bags containing 1.5 cu ft, how many bags does the contractor need to fill all 25 pots?

CASE STUDY: The Majestic Sign Company

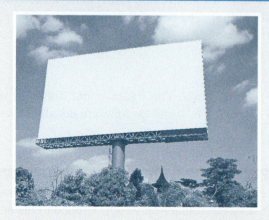

The Majestic Sign Company produces and sets up billboards across the United States. They are trying to decide whether to use one steel cylindrical support or four steel I-beam supports for a new order of billboards. They will choose the option that minimizes their costs, thus maximizing their profit.

The base of each billboard will be 22 ft 6 in. above ground level. The supports must go an additional 11 ft 9 in. underground and an additional 30" above the base of the sign for attachment purposes. Rafael is put in charge of comparing the two options.

continued...

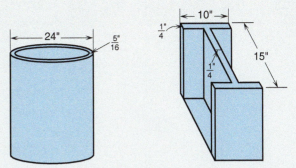

1. Rafael first compares the costs of the steel supports themselves. He checks the blueprints shown in the figure and then looks up the costs for each type of support given the required dimensions.

   a. The hollow cylindrical supports cost $124.48 per linear foot. Determine the cost of one of these supports.

   b. The I-beam supports cost $22.61 per linear foot. Determine the cost of the four I-beam supports.

2. Rafael also must consider the concrete needed to hold each support in the ground. Concrete costs the company $80 per cubic yard.

   a. For the cylindrical support, a 4-ft diameter hole is dug 11 ft 9 in. into the ground. Concrete is poured into the hole around the support but not inside the cylinder. Determine the amount of concrete needed and the cost.

   b. For each I-Beam support, a 2'6" diameter hole is dug 11 ft 9 in. into the ground. Concrete is poured into each hole around the support. Determine the amount of concrete needed and the cost for the four supports.

3. For shipping purposes, Rafael next needs to determine the weight of each option. He uses the dimensions on the blueprint along with the fact that the density of steel is approximately 490 lb/cu ft.

   a. What is the weight of one cylindrical support?

   b. What is the weight of four I-beams?

4. The semi-trailers that will be moving the supports can carry up to 20 tons.

   a. How many supports can the semi-trailer handle if the cylindrical support is chosen?

   b. How many sets of four supports can the semi-trailer handle if the I-beam supports are chosen?

   c. If 75 billboards are going to be set up over the next few months in a particular area, how many truckloads would it take to transport each type of support?

5. The trucking company quotes Rafael a price of $1800 per truckload to transport either type of support to the given destination. What is the total cost of transporting each type of support?

6. For an order of 75 billboards, what is the total cost of materials (supports and concrete) and transportation for each type of support?

7. Based on your answers to question 6, which type of support should the company use for the billboards?

10 Triangle Trigonometry

Objective	Sample problems		For help, go to page

When you finish this chapter, you will be able to:

1. Convert angles between decimal degrees and degrees and minutes.

    (a) Convert 43.4° to degrees and minutes. _____ 666

    (b) Convert 65°15′ to decimal degrees. _____ 667

2. Work with angles in radian measure.

    (a) Express 46° in radians. _____ 668

    (b) Express 2.5 radians in degrees. _____

    (c) Calculate the arc length and area of the circular sector shown. (Round to one decimal place.) _____ 669

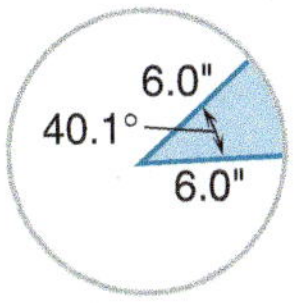

    (d) What is the angular speed in radians per second of a large flywheel that rotates through an angle of 200° in 2 sec? (Round to the nearest tenth.) _____ 670

Name _____

Date _____

Course/Section _____

Objective	Sample problems	For help, go to page
3. Use the special right triangle relationships to find missing parts.	Find the missing parts of these triangles. (Round sides to the nearest tenth.) (a) (b)	673

(a)

(a) $a =$ _____

$b =$ _____

$A =$ _____

(b)

(b) $d =$ _____

$e =$ _____

$F =$ _____

4. Find the values of trig ratios.	(a) $\sin 26°$	_____	684
	(b) $\cos 84°$	_____	686
	(c) $\tan 43°20'$	_____	687
	(Round to three decimal places.)		

5. Find the angle when given the value of a trig ratio.	Find $\sphericalangle x$ to the nearest minute.		
	(a) $\sin x = 0.242$	$\sphericalangle x =$ _____	689
	(b) $\cos x = 0.549$	$\sphericalangle x =$ _____	
	Find $\sphericalangle x$ to the nearest tenth of a degree.		
	(c) $\tan x = 3.821$	$\sphericalangle x =$ _____	
	(d) $\sin x = 0.750$	$\sphericalangle x =$ _____	

6. Solve problems involving right triangles.	Round sides to the nearest tenth and angles to the nearest minute.		

(a)

$X =$ _____   691

$Y =$ _____

(b)

$\sphericalangle m =$ _____

$X =$ _____

(c) **Plumbing**   In a pipe-fitting job, what is the run if the offset is $22\frac{1}{2}°$ and the length of set is $16\frac{1}{2}$ in.?   run = _____   686

Objective	Sample problems	For help, go to page

(d) **Machine Trades** Find the angle of taper on the figure.

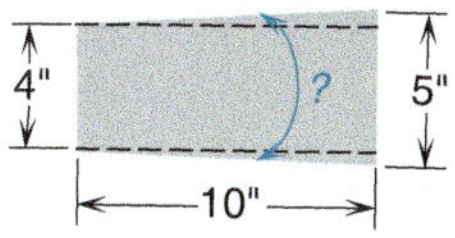

angle = _____

7. Solve oblique triangles.

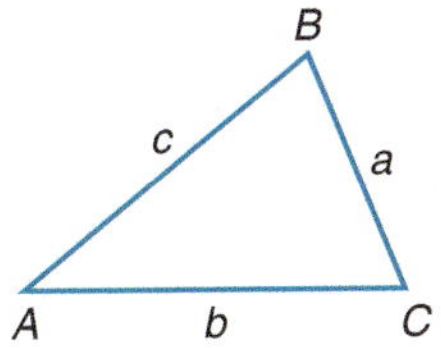

Solve these triangles. (See the figure. Round sides to the nearest tenth and angles to the nearest degree.)

(a) $a = 6.5$ in., $b = 6.0$ in., $c = 3.5$ in.

_____ 705

(b) $A = 68°, B = 79°, b = 8.0$ ft

_____

(Answers to these preview problems are given in the Appendix. Also, worked solutions to many of these problems appear in the chapter Summary.)

If you are certain that you can work all of these problems correctly, turn to page 720 for a set of practice problems. If you cannot work one or more of the preview problems, turn to the page indicated to the right of the problem. Those who wish to master this material with the greatest success should turn to Section 10-1 and begin work there.

The word *trigonometry* means simply "triangle measurement." We know that the ancient Egyptian engineers and architects of 4000 years ago used practical trigonometry in building the pyramids. By 140 B.C., the Greek mathematician Hipparchus had made trigonometry a part of formal mathematics and taught it as a tool for studying astronomy. In this chapter, we look at only the simple practical trigonometry used in a number of various trades applications.

CASE STUDY: Ordering Wood for Roof Trusses

In the Case Study at the end of this chapter, we will see how triangle trigonometry can be very useful in designing and building the infrastructure for a roof.

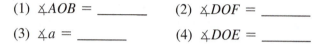

10-1 Angles and Right Triangles

In Chapter 8, you learned some of the vocabulary of angles and triangles. To be certain you remember this information, try these problems.

Your Turn

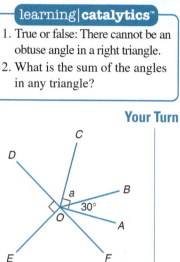

(a) Find each of the angles shown.

  (1) $\angle AOB =$ _____   (2) $\angle DOF =$ _____

  (3) $\angle a =$ _____   (4) $\angle DOE =$ _____

(b) Label each of these angles with the correct name: acute angle, obtuse angle, straight angle, or right angle.

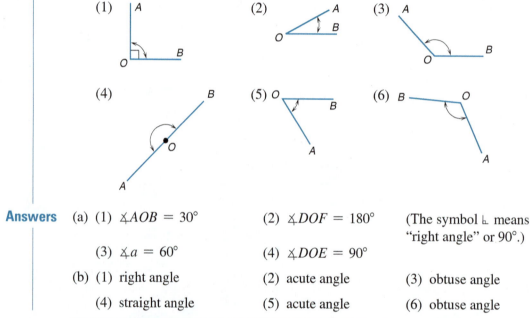

Answers

(a) (1) $\angle AOB = 30°$ (2) $\angle DOF = 180°$ (The symbol ∟ means "right angle" or 90°.)

    (3) $\angle a = 60°$ (4) $\angle DOE = 90°$

(b) (1) right angle   (2) acute angle   (3) obtuse angle

    (4) straight angle   (5) acute angle   (6) obtuse angle

If you missed any of these, you should return to Section 8-1 on page 532 for a quick review; otherwise, continue here.

Converting Angle Units

In most practical work, angles are measured in degrees and fractions of a degree. Smaller units have been defined as follows:

> 60 minutes = 1 degree   abbreviated, $60' = 1°$
>
> 60 seconds = 1 minute   abbreviated, $60'' = 1'$

For most technical purposes, angles can be rounded to the nearest minute. In the trades, angles are usually rounded to the nearest degree.

Sometimes in your work you will need to convert an angle measured in decimal degrees to its equivalent measure in degrees and minutes.

EXAMPLE 1   To convert $17\frac{1}{2}°$ to degrees and minutes, follow these two steps:

Step 1   Write the angle as a sum of a whole number and a fraction.

    $17\frac{1}{2}° = 17° + \frac{1}{2}°$

Step 2　Use a unity fraction to convert $\frac{1}{2}^\circ$ to minutes.

$$= 17^\circ + \left(\frac{1^\circ}{2} \times \frac{60'}{1^\circ} \right)$$

$$= 17^\circ 30'$$

By choosing to multiply by $\frac{60'}{1^\circ}$, the degree units canceled out, leaving us with minutes.　●

Your Turn　Write this angle to the nearest minute:

$36.25^\circ = $ _____

Solution　$36.25^\circ = 36^\circ + 0.25^\circ$

$$= 36^\circ + \left(0.25^\circ \times \frac{60'}{1^\circ} \right) \quad \Longleftarrow \boxed{\text{Multiply the decimal part by a unity fraction.}}$$

$$= 36^\circ 15'$$

The reverse procedure is also useful.

EXAMPLE 2　To convert the angle $72^\circ 6'$ to decimal form:

Step 1　Write the angle as a sum of degrees and minutes.

$$72^\circ 6' = 72^\circ + 6'$$

Step 2　Use a unity fraction to convert minutes to degrees.

$$= 72^\circ + \left(6' \times \frac{1^\circ}{60'} \right)$$

$$= 72^\circ + 0.1^\circ$$

$$= 72.1^\circ$$

By choosing to multiply by $\frac{1^\circ}{60'}$, the minutes units canceled out, leaving us with degrees.　●

Converting Angle Units Using a Calculator

Scientific calculators have special keys designed for entering angles in degrees and minutes and for performing angle conversions. There is normally a key marked $\boxed{\circ\,'\,''}$ that is used for entering and converting angles. You may instead see a key marked $\boxed{\blacktriangleright \text{DMS}}$ or $\boxed{\blacktriangleright \text{DD}}$ to be used for this purpose. Because the procedures for doing these conversions vary greatly from model to model, we leave it to each student to learn the key sequences from his or her calculator's instruction manual. You can always perform the conversions on your calculator using the methods of Examples 1 and 2, and we will show them this way in the text. For example, to convert the angle 67.28° from decimal degrees to degrees and minutes, write down the degree portion, 67°, and then calculate the minutes portion as follows:

$.28 \boxed{\times} 6\,0 \boxed{=} \rightarrow$ ▭ 16.8

Therefore, $67.28^\circ = 67^\circ 16.8'$ or $67^\circ 17'$, rounded to the nearest minute.

To convert $49^\circ\,27'$ to decimal degrees, enter

$49 \boxed{+} 2\,7 \boxed{\div} 6\,0 \boxed{=} \rightarrow$ ▭ 49.45

Therefore, $49^\circ 27' = 49.45^\circ$.

More Practice   For practice in working with angles, rewrite each of the following angles as shown.

(a) Write in degrees and minutes.

(1) $24\frac{1}{3}° = $ _____   (2) $64\frac{3}{4}° = $ _____

(3) $15.3° = $ _____   (4) $124.8° = $ _____

(b) Write in decimal form.

(1) $10°48' = $ _____   (2) $96°9' = $ _____

(3) $33' = $ _____   (4) $1°12' = $ _____

Answers   (a) (1) 24°20'   (2) 64°45'   (3) 15°18'   (4) 124°48'

(b) (1) 10.8°   (2) 96.15°   (3) 0.55°   (4) 1.2°

Radian Measure   In some scientific and technical work, angles are measured or described in an angle unit called the **radian**. By definition,

$$1 \text{ radian} = \frac{180°}{\pi} \text{ or about } 57.296°$$

One radian is the angle at the center of a circle that corresponds to an arc exactly one radius in length.

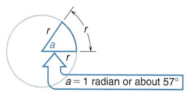

$a = 1$ radian or about 57°

To understand where the definition comes from, set up a proportion:

$$\frac{\text{angle } a}{\text{total angle in the circle}} = \frac{\text{arc length}}{\text{circumference of the circle}}$$

$$\frac{a}{360°} = \frac{r}{2\pi r}$$

$$a = \frac{360° \cancel{r}}{2\pi \cancel{r}}$$

$$a = \frac{180°}{\pi} \approx 57.2957795°$$

Degrees to Radians   To convert angle measurements from degrees to radians (rad), use the following formula:

$$\text{Radians} = \text{degrees} \cdot \frac{\pi}{180°}$$

Use the $\boxed{\pi}$ button on your calculator unless you are told otherwise.

EXAMPLE 3   (a) $43° = 43°\left(\dfrac{\pi}{180°}\right) = \dfrac{43 \cdot \pi}{180}$ Notice that the degree units cancel out.

$\approx 0.75 \text{ rad}$ rounded

(b) $60° = 60°\left(\dfrac{\pi}{180°}\right) = \dfrac{60 \cdot \pi}{180}$ The degree units cancel out.

$= \dfrac{\pi}{3} \text{ rad} \approx 1.0472 \text{ rad}$ ●

Radians to Degrees   To convert angles from radians to degrees, use the following formula:

$$\text{Degrees} = \text{radians} \cdot \dfrac{180°}{\pi}$$

EXAMPLE 4   (a) $1.3 \text{ rad} = 1.3\left(\dfrac{180°}{\pi}\right) = \dfrac{1.3 \cdot 180°}{\pi}$

$= \dfrac{234°}{\pi}$ or about $74.5°$

Notice that multiplying by $\dfrac{180°}{\pi}$ adds degree units to the answer.

(b) $0.5 \text{ rad} = 0.5\left(\dfrac{180°}{\pi}\right) = \dfrac{0.5 \cdot 180°}{\pi}$

$= \dfrac{90°}{\pi}$ or about $28.6°$ ●

Your Turn   Try these problems for practice in using radian measure.

(a) Convert to radians. (Round to two decimal places.)

   (1) $10°$ (2) $35°$ (3) $90°$ (4) $120°$

(a) Convert to degrees. (Round to the nearest tenth.)

   (1) 0.3 rad (2) 1.5 rad (3) 0.8 rad (4) 0.05 rad

Answers   (a) (1) $\dfrac{10° \cdot \pi}{180°} \approx 0.17 \text{ rad}$ (2) 0.61 rad (3) 1.57 rad (4) 2.09 rad

(b) (1) $\dfrac{0.3 \cdot 180°}{\pi} \approx 17.2°$ (2) $85.9°$ (3) $45.8°$ (4) $2.9°$

Sectors   The wedge-shaped portion of a circle is called a **sector**. Both its area and the length of arc S can be expressed most directly using radian measure.

Sectors

Arc length $S = ra$

Area $A = \frac{1}{2}r^2a$ where a is the central angle given in radians.

! Careful   The sector formulas will not work unless the central angle a is in radians.   ●

EXAMPLE 5 **Sheet Metal Trades** A sheet metal worker wants to know the arc length and area of a sector with central angle 150° cut from a circular sheet of metal with radius 16.0 in.

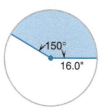

First, calculate angle a in radians: $\quad a = 150°\left(\dfrac{\pi}{180°}\right) \approx 2.618$ rad

Second, calculate the arc length S: $\quad S = ra \approx (16.0 \text{ in.})(2.618)$
$$\approx 41.9 \text{ in.}$$

Finally, calculate the area A: $\quad A = \tfrac{1}{2}r^2a \approx \tfrac{1}{2}(16.0)^2(2.618)$
$$\approx 335 \text{ in.}^2$$

Using a calculator, we have

$S:$ $16 \boxed{\times} 150 \boxed{\times} \boxed{\pi} \boxed{\div} 180 \boxed{=} \rightarrow$ ┃41.88790205┃

$A:$ $.5 \boxed{\times} 16 \boxed{x^2} \boxed{\times} 150 \boxed{\times} \boxed{\pi} \boxed{\div} 180 \boxed{=} \rightarrow$ ┃335.1032164┃ ●

Your Turn Calculate the arc length and area of a sector with central angle 80° and radius 25.0 ft.

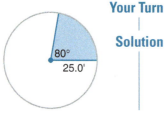

Solution The central angle is $a = 80°\left(\dfrac{\pi}{180°}\right) \approx 1.396$ rad.

The arc length is $S = ra \approx (25.0 \text{ ft})(1.396) \approx 34.9$ ft.

The area is $A = \tfrac{1}{2}r^2a \approx \tfrac{1}{2}(25.0 \text{ ft})^2(1.396) \approx 436 \text{ ft}^2$.

Linear and Angular Speed Radian units are also used in science and technology to describe the rotation of an object. When an object moves along a straight line for a distance d in time t, its **average linear speed** v is defined as

$$v = \frac{d}{t} \quad \text{Linear speed}$$

Notice that this formula matches one of the distance, rate, and time formulas on p. 428. Here we use v for rate.

EXAMPLE 6 If a car travels 1800 ft in 24 seconds, its average linear speed for the trip is

$$v = \frac{d}{t} = \frac{1800 \text{ ft}}{24 \text{ sec}} = 75 \text{ ft/sec}$$ ●

When an object moves along a circular arc for a distance S in time t, it goes through an angle a in radians, and its **average angular speed** w is defined as

$$w = \frac{a}{t} \quad \text{Angular speed}$$

⚠ Careful This formula will not work unless the angle a is in radians. ●

EXAMPLE 7 **Manufacturing** If a 24-in.-diameter flywheel rotates through an angle of 140° in 1.1 seconds, what is the average angular speed of the flywheel?

First, calculate the angle in radians. The angle traveled is $a = 140°\left(\dfrac{\pi}{180°}\right) \approx 2.44$ rad.

Then, calculate the average angular speed, $w = \dfrac{a}{t} \approx \dfrac{2.44 \text{ rad}}{1.1 \text{ sec}} \approx 2.2$ rad/sec.

$140 \boxed{\times} \boxed{\pi} \boxed{\div} 180 \boxed{\div} 1.1 \boxed{=} \rightarrow$ ┃2.221328139┃ ●

Your Turn   For practice in working with linear and angular speeds, try these problems.

(a) **Life Skills**   If a bicycle travels 24 mi in 1.5 hr, find its average linear speed.

(b) **Manufacturing**   The blade on a shop fan rotates at 8.0 rev/sec. Calculate its average angular speed. (Round to the nearest whole number.) (*Hint:* Each revolution is 360°.)

(c) **Manufacturing**   A belt-driven drum of radius 28.4 cm makes one revolution every 0.250 sec. What is the linear speed of the belt driving the drum? (Round to the nearest whole number.) (*Hint:* If the belt doesn't slip, the distance traveled by the belt in one revolution is equal to the circumference of the drum.)

(d) **General Interest**   A Ferris wheel in an amusement park takes an average of 40 sec for one nonstop revolution. Calculate its average angular speed. (Round to the nearest hundredth.)

Solutions   (a) $v = \dfrac{d}{t} = \dfrac{24 \text{ mi}}{1.5 \text{ hr}} = 16 \text{ mi/hr}$

(b) Each revolution is 360°; therefore, the fan blade travels through an angle of $8 \times 360° = 2880°$ in one second. Converting to radians,

$$2880°\left(\frac{\pi}{180°}\right) \approx 50.265 \text{ rad}$$

Using the formula for angular speed,

$$w = \frac{a}{t} \approx \frac{50.265 \text{ rad}}{1 \text{ sec}} \approx 50 \text{ rad/sec}$$

(c) The circumference of the drum is

$$C = 2\pi r = 2\pi(28.4 \text{ cm}) \approx 178.4 \text{ cm}$$

This is the distance traveled by any point on the belt in 0.250 sec. Substitute these values into the linear speed formula.

$$v = \frac{d}{t} \approx \frac{178.4 \text{ cm}}{0.25 \text{ sec}} \approx 714 \text{ cm/sec}$$

(d) One revolution is 360°. In radians, this is

$$\frac{360° \times \pi}{180°} \approx 6.28 \text{ rad}$$

Using the formula for angular speed,

$$w = \frac{a}{t} \approx \frac{6.28 \text{ rad}}{40 \text{ sec}} \approx 0.16 \text{ rad/sec}$$

Right Triangles   Trigonometry is basically triangle measurement, and to keep it as simple as possible, we can begin by studying only *right* triangles. You should remember from Chapter 8 that a **right triangle** is one that contains a right angle, a 90° angle. All of the following are right triangles. In each triangle, the right angle is marked.

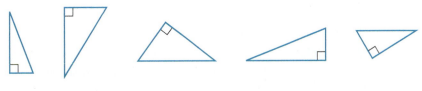

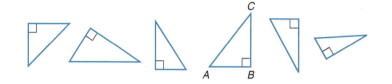

Your Turn   Is triangle *ABC* a right triangle?

Solution   You should remember that the sum of the three angles of any triangle is 180°. For triangle *ABC*, two of the angles total 37° + 53° or 90°; therefore, angle *A* must equal 180° − 90° or 90°. The triangle *is* a right triangle.

Learning Help   To make your study of triangle trigonometry even easier, you may want to place a right triangle in *standard position*. For this collection of right triangles

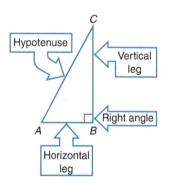

only triangle *ABC* is in standard position.

Place the triangle so that the right angle is on the right side, one leg (*AB*) is horizontal, and the other leg (*BC*) is vertical. The hypotenuse (*AC*) will slope up from left to right.

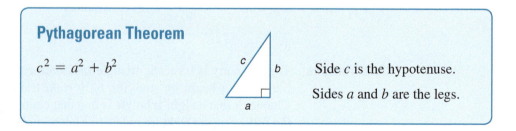

If you do not place a triangle in standard position, remember that the legs are the two sides forming the right angle, and the hypotenuse is the side opposite the right angle. The hypotenuse is always the longest side in a right triangle. ●

Pythagorean Theorem   You should also recall from Chapter 8 that the **Pythagorean theorem** is an equation relating the lengths of the sides of any right triangle.

> **Pythagorean Theorem**
>
> $$c^2 = a^2 + b^2$$
>
> Side *c* is the hypotenuse.
>
> Sides *a* and *b* are the legs.

This formula is true for every right triangle.

Careful   In the formula, the letters *a* and *b*, representing the legs, are interchangeable, but *c* must always represent the hypotenuse. ●

EXAMPLE 8

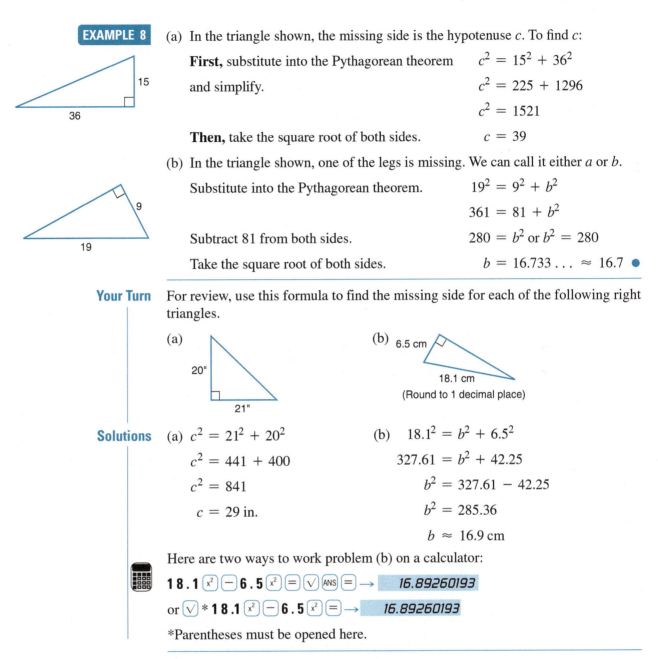

(a) In the triangle shown, the missing side is the hypotenuse c. To find c:

First, substitute into the Pythagorean theorem and simplify.

$$c^2 = 15^2 + 36^2$$
$$c^2 = 225 + 1296$$
$$c^2 = 1521$$

Then, take the square root of both sides.

$$c = 39$$

(b) In the triangle shown, one of the legs is missing. We can call it either a or b.

Substitute into the Pythagorean theorem.

$$19^2 = 9^2 + b^2$$
$$361 = 81 + b^2$$

Subtract 81 from both sides.

$$280 = b^2 \text{ or } b^2 = 280$$

Take the square root of both sides.

$$b = 16.733\ldots \approx 16.7 \ \bullet$$

Your Turn For review, use this formula to find the missing side for each of the following right triangles.

(a)

20"

21"

(b) 6.5 cm

18.1 cm

(Round to 1 decimal place)

Solutions

(a) $c^2 = 21^2 + 20^2$

$c^2 = 441 + 400$

$c^2 = 841$

$c = 29$ in.

(b) $18.1^2 = b^2 + 6.5^2$

$327.61 = b^2 + 42.25$

$b^2 = 327.61 - 42.25$

$b^2 = 285.36$

$b \approx 16.9$ cm

Here are two ways to work problem (b) on a calculator:

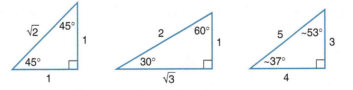

18.1 x^2 $-$ **6.5** x^2 $=$ $\sqrt{}$ ANS $=$ → 16.89260193

or $\sqrt{}$ * **18.1** x^2 $-$ **6.5** x^2 $=$ → 16.89260193

*Parentheses must be opened here.

Special Right Triangles Three very special right triangles are used very often in practical work. Here they are:

$\sqrt{2}$ 45°
1
45°
1

2 60°
1
30°
$\sqrt{3}$

5 ~53°
3
~37°
4

$\sqrt{2}$ 45°
1
45°
1

45°–45°–90° △ The first triangle is a **45°–45° right triangle**. Its angles are 45°, 45°, and 90°, and it is formed by the diagonal and two sides of a square. The two shorter sides, or legs, are the same length, and the hypotenuse is $\sqrt{2}(\approx 1.4)$ times the length of each leg. For the triangle shown, we can use the Pythagorean theorem to show why this is true:

According to the Pythagorean theorem, $c^2 = a^2 + b^2$

Therefore, $c^2 = 1^2 + 1^2$

$$c^2 = 2$$
$$c = \sqrt{2}$$

Our triangle has legs one unit long, but we can draw a 45° right triangle with any length sides. If we increase the length of one side, the others increase in proportion while the angles stay the same size.

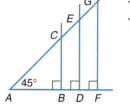

Triangle *ABC* is a 45° right triangle.
Triangle *ADE* is also a 45° right triangle.
Triangle *AFG* is also a 45° right triangle, and so on.

No matter what the size of the triangle, if it is a right triangle and if one angle is 45°, the third angle is also 45°. The two legs are equal, and the hypotenuse is $\sqrt{2} \approx 1.4$ times the length of a leg.

📝 **Note**   We can also refer to a 45°–45° right triangle as an **isosceles right triangle** because the two legs are the same length. ●

EXAMPLE 9   Refer to the previous figure for the following problems.

(a) If *AD* = 2 in., then

$$DE = 2 \text{ in.}$$ The legs are always equal in length. If *AD* is 2 in. long, *DE* is also 2 in. long.

and

$$AE = 2\sqrt{2} \text{ in. or about 2.8 in.}$$ The hypotenuse of a 45° right triangle is $\sqrt{2}$ times the length of each leg.

(b) If *AC* = 5 in., then

$$AC = AB\sqrt{2}, \text{ so } AB = \frac{AC}{\sqrt{2}} = \frac{5}{\sqrt{2}}$$

Using a calculator gives

 5 ÷ √ 2 = → 3.535533906 ≈ 3.5 in.   ●

👆 **Learning Help**   To summarize the results of Example 9, for any 45°–45° right triangle:

- When given the length of a leg, *multiply* by $\sqrt{2}$ to find the length of the hypotenuse.

- When given the length of the hypotenuse, *divide* by $\sqrt{2}$ to find the length of each leg. ●

Your Turn

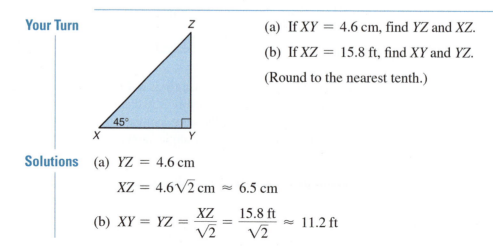

(a) If *XY* = 4.6 cm, find *YZ* and *XZ*.

(b) If *XZ* = 15.8 ft, find *XY* and *YZ*.

(Round to the nearest tenth.)

Solutions   (a) $YZ = 4.6$ cm

$$XZ = 4.6\sqrt{2} \text{ cm} \approx 6.5 \text{ cm}$$

(b) $XY = YZ = \dfrac{XZ}{\sqrt{2}} = \dfrac{15.8 \text{ ft}}{\sqrt{2}} \approx 11.2 \text{ ft}$

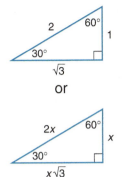

or

30°–60°–90°△ The second of our special triangles is the **30°–60° right triangle**. In a 30°–60° right triangle, the length of the side opposite the 30° angle, the smallest side, is exactly one-half the length of the hypotenuse. The length of the third side is $\sqrt{3}$ times the length of the smallest side.

To see where this relationship comes from, start with an equilateral triangle with sides two units long.

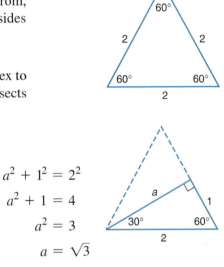

Draw a perpendicular line from one vertex to the opposite side. This segment also bisects the opposite side.

Then in the 30°–60° right triangle, the Pythagorean theorem gives

$$a^2 + 1^2 = 2^2$$
$$a^2 + 1 = 4$$
$$a^2 = 3$$
$$a = \sqrt{3}$$

Again, we may form any number of 30°–60° right triangles, and in each triangle the hypotenuse is twice the length of the smallest side. The third side is $\sqrt{3}$ (≈ 1.7) times the length of the smallest side. When you find a right triangle with one angle equal to 30° or 60°, you automatically know a lot about it.

📝 **Note** For a 30°–60° right triangle, we will often refer to the side opposite the 60° angle as the *longer* leg, and the side opposite the 30° angle as the *shorter* leg. ●

EXAMPLE 10 (a) In this right triangle, if $A = 60°$ and $b = 8$ ft, then

$$c = 2b = 2 \cdot 8 \text{ ft} = 16 \text{ ft}$$

and $a = b\sqrt{3} = 8\sqrt{3} \approx 13.9$ ft

(b) If $A = 60°$ and $c = 24$ in., then

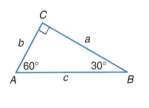

$$b = \frac{1}{2}c = \frac{1}{2}(24 \text{ in.}) = 12 \text{ in.}$$

and $a = b\sqrt{3} = 12\sqrt{3} \approx 20.8$ in.   ●

📝 **Note** The triangle shown in Example 10 is labeled according to a standard convention. Angles are labeled with uppercase letters, and sides are labeled with lowercase letters. Furthermore, side a is opposite angle A; side b is opposite angle B; and side c, the hypotenuse, is opposite angle C, the right angle. ●

Suppose you are given the longer leg a, opposite the 60° angle, and you need to find the shorter leg b, opposite the 30° angle. (See the figure for Example 10.)

We know that $a = b\sqrt{3}$

Now divide both sides by $\sqrt{3}$ to get $b = \dfrac{a}{\sqrt{3}}$

EXAMPLE 11 If $A = 60°$ and $a = 32$ yd,

then $b = \dfrac{32}{\sqrt{3}} \approx 18.5$ yd

and $c = 2b \approx 37.0$ yd.   ●

Learning Help　To summarize all three cases for a 30°–60° right triangle:

- When given the length of the shorter leg (opposite the 30° angle), multiply it by 2 to get the length of the hypotenuse and multiply it by $\sqrt{3}$ to get the length of the longer leg (opposite the 60° angle).

- When given the length of the hypotenuse, divide it by 2 to get the length of the shorter leg, then multiply this result by $\sqrt{3}$ to get the length of the longer leg.

- When given the length of the longer leg, divide it by $\sqrt{3}$ to get the length of the shorter leg, then multiply this result by 2 to get the length of the hypotenuse. ●

Your Turn　(a) In this right triangle, if $A = 30°$ and $a = 2$ in., calculate c, b, and B.

(b) If $A = 30°$ and $c = 3$ in., calculate b and a.

(c) If $A = 30°$ and $b = 18$ cm, calculate a and c.

Solutions　(a) $c = 4$ in., $b = 2\sqrt{3}$ in. ≈ 3.5 in., and $B = 60°$

(b) In a 30°−60° right triangle, $a = \dfrac{1}{2}c = \dfrac{1}{2}(3) = 1.5$ in.

$$b = a\sqrt{3} = 1.5(\sqrt{3}) \approx 2.6 \text{ in.}$$

(c) $a = \dfrac{18}{\sqrt{3}} \approx 10.4$ cm

$$c = 2a \approx 20.8 \text{ cm}$$

3–4–5 △　The third of our special right triangles is the **3–4–5 triangle**, which was discussed in Chapter 8. If a triangle is found to have sides of length 3, 4, and 5 units (any units—inches, feet, centimeters), it must be a right triangle. The 3–4–5 triangle is the smallest right triangle with sides whose lengths are whole numbers.

Construction workers or surveyors can use this triangle to set up right angles.

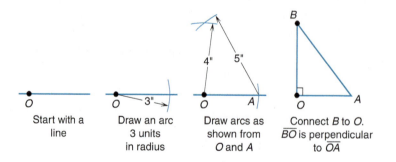

| Start with a line | Draw an arc 3 units in radius | Draw arcs as shown from O and A | Connect B to O. $\overline{BO}$ is perpendicular to $\overline{OA}$ |

Notice that in any 3–4–5 triangle, the acute angles are approximately 37° and 53°. The smallest angle is always opposite the shortest side. If the side lengths of a triangle are proportional to the numbers 3, 4, and 5, they will also create a right triangle with these same angles.

EXAMPLE 12 For the right triangle shown in the figure, notice that the length of the given leg is equal to $3 \cdot 2$, and the hypotenuse is equal to $5 \cdot 2$. This right triangle is therefore similar to a 3–4–5 right triangle, and the missing leg a must be $4 \cdot 2$, or 8 units in length.

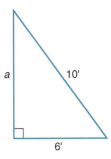

Applications of Right Triangles

EXAMPLE 13

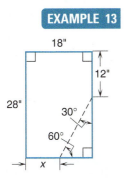

(a) **Sheet Metal Trades** The dashed line in the figure shows where a cut is to be made in a rectangular piece of sheet metal. Suppose we need to find the length x.

First, notice that the vertical leg of the triangular cut must be 28 in. $-$ 12 in., or 16 in., because opposite sides of a rectangle are equal in length.

Second, because the cut forms a $30°$–$60°$ right triangle, the horizontal leg must be equal to $\dfrac{16}{\sqrt{3}}$, or approximately 9.2 in.

Finally, because the entire width of the rectangle was 18 in. before the cut, the remaining part x must be approximately 18 in. $-$ 9.2 in., or 8.8 in.

(b) **Construction** A cathedral ceiling creates the interior wall shown in Figure (1). Suppose we need to find the length y of a beam needed to go along the top of the wall.

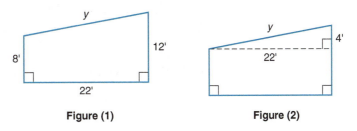

Figure (1)　　　　　　　　**Figure (2)**

First, draw the dashed line shown in Figure (2) to divide the wall into a rectangle and a right triangle.

Next, notice that the vertical leg of the triangle is 12 ft – 8 ft, or 4 ft, in length, and the horizontal leg is 22 ft in length. These lengths do not correspond to one of the three special right triangles, so we must use the Pythagorean theorem to find the hypotenuse y:

$$y^2 = 4^2 + 22^2$$
$$y^2 = 16 + 484$$
$$y^2 = 500$$
$$y = \sqrt{500} \approx 22.4 \text{ ft}$$

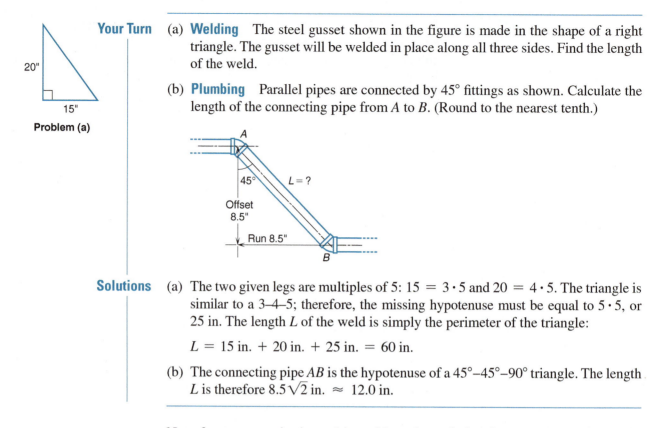

Problem (a)

Your Turn

(a) **Welding**   The steel gusset shown in the figure is made in the shape of a right triangle. The gusset will be welded in place along all three sides. Find the length of the weld.

(b) **Plumbing**   Parallel pipes are connected by 45° fittings as shown. Calculate the length of the connecting pipe from A to B. (Round to the nearest tenth.)

Solutions

(a) The two given legs are multiples of 5: $15 = 3 \cdot 5$ and $20 = 4 \cdot 5$. The triangle is similar to a 3–4–5; therefore, the missing hypotenuse must be equal to $5 \cdot 5$, or 25 in. The length L of the weld is simply the perimeter of the triangle:

$$L = 15 \text{ in.} + 20 \text{ in.} + 25 \text{ in.} = 60 \text{ in.}$$

(b) The connecting pipe AB is the hypotenuse of a 45°–45°–90° triangle. The length L is therefore $8.5\sqrt{2}$ in. ≈ 12.0 in.

Now, for some practice in working with angles and triangles, turn to Exercises 10-1.

Exercises 10-1	**Angles and Right Triangles**

A. Label the shaded angles as acute, obtuse, right, or straight.

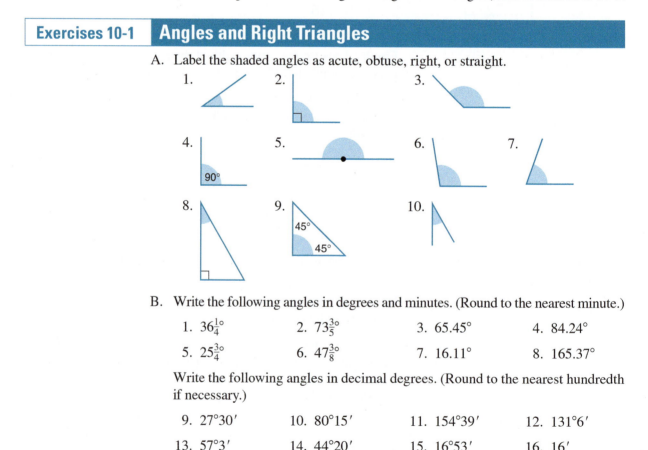

B. Write the following angles in degrees and minutes. (Round to the nearest minute.)

1. $36\frac{1}{4}°$ 2. $73\frac{3}{5}°$ 3. $65.45°$ 4. $84.24°$

5. $25\frac{3}{4}°$ 6. $47\frac{3}{8}°$ 7. $16.11°$ 8. $165.37°$

Write the following angles in decimal degrees. (Round to the nearest hundredth if necessary.)

9. $27°30'$ 10. $80°15'$ 11. $154°39'$ 12. $131°6'$

13. $57°3'$ 14. $44°20'$ 15. $16°53'$ 16. $16'$

Write the following angles in radians. (Round to the nearest hundredth.)

17. $24.75°$ 18. $185.8°$ 19. $67\frac{3}{8}°$ 20. $9.65°$

Write the following angles in degrees. (Round to the nearest hundredth.)

21. 0.10 rad   22. 0.84 rad   23. 2.1 rad   24. 3.5 rad

C. Solve the following problems involving triangles.

Find the lengths of the missing sides on the following triangles using the Pythagorean theorem. (Round to one decimal place.) (*Hint*: Four of these are Pythagorean triples. Try to find them before using your calculator.)

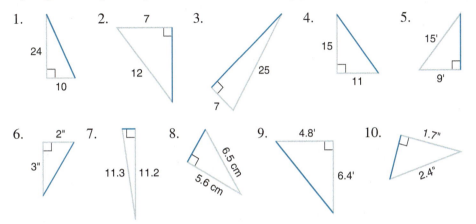

1. 24, 10

2. 7, 12

3. 7

4. 15, 25, 11

5. 15', 9'

6. 2", 3"

7. 11.3, 11.2

8. 6.5 cm, 5.6 cm

9. 4.8', 6.4'

10. 1.7", 2.4"

Each of the following right triangles is an example of one of the three "special triangles" described in this chapter. Find the quantities indicated without using the Pythagorean theorem. (Round the sides to the nearest tenth if necessary.)

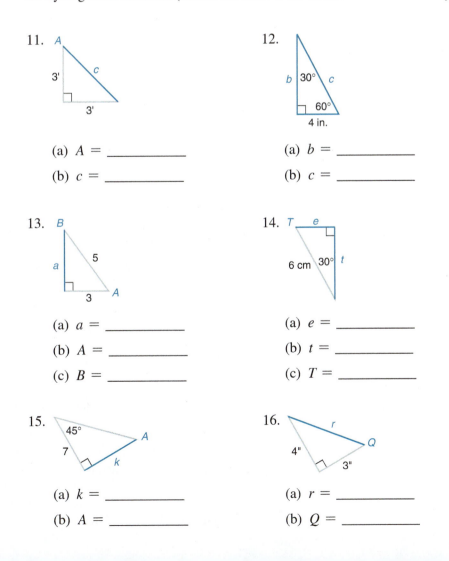

11. A, 3', 3', c

(a) $A =$ _____

(b) $c =$ _____

12. b, 30°, c, 60°, 4 in.

(a) $b =$ _____

(b) $c =$ _____

13. B, a, 5, 3, A

(a) $a =$ _____

(b) $A =$ _____

(c) $B =$ _____

14. T, e, 6 cm, 30°, t

(a) $e =$ _____

(b) $t =$ _____

(c) $T =$ _____

15. 45°, 7, A, k

(a) $k =$ _____

(b) $A =$ _____

16. r, Q, 4", 3"

(a) $r =$ _____

(b) $Q =$ _____

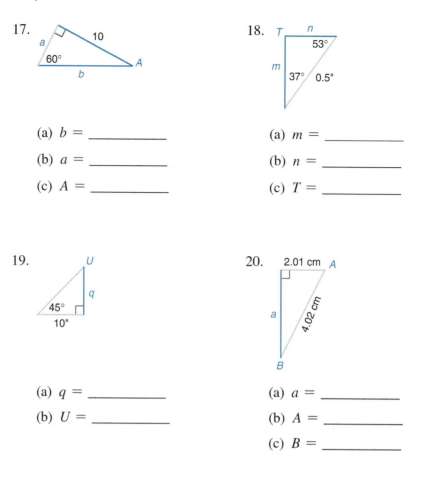

17.   10   60°   a   b   A

(a) $b =$ _____

(b) $a =$ _____

(c) $A =$ _____

18.   T   n   53°   m   37°   0.5"

(a) $m =$ _____

(b) $n =$ _____

(c) $T =$ _____

19.   U   q   45°   10"

(a) $q =$ _____

(b) $U =$ _____

20.   2.01 cm   A   a   4.02 cm   B

(a) $a =$ _____

(b) $A =$ _____

(c) $B =$ _____

Calculate the arc length S and area A for each of the following sectors. (Round to the nearest whole number if necessary.)

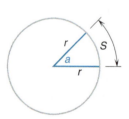

21. $a = 40°$ 22. $a = 120°$ 23. $a = 0.8$ rad   24. $a = 2$ rad
    $r = 30$ cm $r = 24$ ft $r = 110$ ft $r = 18$ m

25. $a = 72°$ 26. $a = 3$ rad   27. $a = 0.5$ rad   28. $a = 180°$
    $r = 5$ in. $r = 80$ cm $r = 10$ m $r = 30$ in.

D. Practical Applications

1. **Sheet Metal Trades**   A transition duct is constructed from a circular sector of radius 18 in. and central angle 115°. (a) What is the area of the metal used? (b) Calculate the arc length of the sector. (Round to the nearest tenth.)

2. **Landscaping**   Calculate the area (in square feet) of a flower garden shaped like a circular sector with radius 60 ft and central angle 40°.

3. **Landscaping**   In problem 2, if shrubs are planted every 2 ft along the outer border of the garden, how many shrubs are needed?

4. **Construction**   The outfield fencing for a Little League field forms a circular sector with home plate as the center. (See the figure at the top of page 681.) The fence is placed at a uniform distance of 215 ft from home plate. The boundaries of the fence, which extends partway into foul territory, create an angle of 105° with home plate. At $28 per foot, how much will the fence cost? (Round to the nearest $10.)

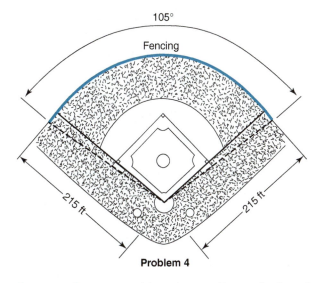

105°

Fencing

215 ft

215 ft

Problem 4

5. **Construction** In problem 4, topsoil must be bought to cover the sector to a depth of 4 in. How many cubic yards of topsoil are needed? (Be careful with the units.)

6. **Life Skills** If a car travels 725 miles in 12.5 hours, find its average linear speed.

7. **Manufacturing** The blades of a fan have a radius of 16.5 in. and they turn one revolution every 0.125 sec. Calculate the linear speed of the tip of a blade. (Round to the nearest whole number.)

8. **Manufacturing** A pulley with a radius of 36.2 cm rotates at 6.5 rev/sec. Find the linear speed of the belt driving the pulley. (Round to the nearest hundred.)

9. **Industrial Technology** If the indicator on a flow rate meter moves 75° in 16.4 sec, what is its average angular speed during this time? (Round to two decimal places.)

10. **Machine Trades** If a lathe makes 25 rev/sec, calculate its average angular speed. (Round to the nearest whole number.)

11. **General Interest** A Ferris wheel in an amusement park takes an average of 16 sec for each revolution. What is its average angular speed? (Round to two decimal places.)

12. **Plumbing** For the connection shown here, calculate the rise, the distance H, and the length AB. The elbows are 45° fittings. (Round to the nearest tenth.)

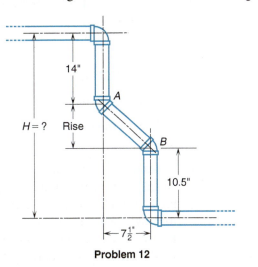

14"

$H = ?$ Rise

A

B

10.5"

$7\frac{1}{2}"$

Problem 12

13. **Automotive Trades**   During a wheel alignment, the camber on a certain car is measured to be $1°48'$. The maximum acceptable camber for this car is $1.750°$. Is the camber within this limit?

14. **Automotive Trades**   The caster for a certain vehicle is measured to be $2.150°$. The maximum allowable caster is specified to be $2°10'$. Is the caster within the maximum limit?

15. **Agriculture**   A grain conveyor 70 feet long is designed to reach the top of a grain bin 32 feet high and with a radius of 8 feet. As shown in the figure, the conveyor extends to the center of the top of the bin. How far from the front edge of the bin should the base of the conveyor be placed? Express your answer in feet and inches, rounded to the nearest inch.

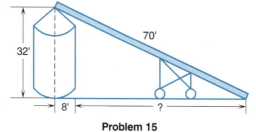

Problem 15

16. **Machine Trades**   Find the dimension a on the metal plate shown in the figure.

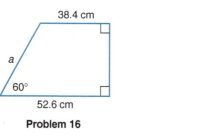

Problem 16

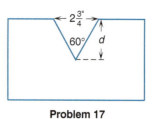

Problem 17

17. **Metalworking**   Find the depth d of the V-slot shown in the figure. (Round to the nearest hundredth of an inch.)

18. **Metalworking**   Find the length w in the casting shown in the figure.

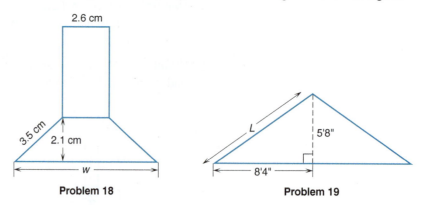

Problem 18   **Problem 19**

19. **Roofing**   Find the length L of the rafter in the figure. (Round to the nearest inch.)

20. **Carpentry**   A carpenter is building a shelf for a corner cupboard that is in the shape of a $270°$-circular sector (a three-quarter circle). If the radius of the shelf is $11\frac{1}{4}$ inches, answer the following questions:
    (a) How many square inches of space will the shelf contain?
    (b) How many inches of bender board will it take to form a border around the entire shelf, including both radii?

Check your answers to the odd-numbered problems in the Appendix, then continue in Section 10-2.

Check your answers to the odd-numbered problems in the Appendix, then continue in Section 10-2.

10-2 Trigonometric Ratios

In Section 10-1 we showed that in *all* 45° right triangles, the adjacent side is equal in length to the opposite side, and the hypotenuse is roughly 1.4 times this length. Similarly, in *all* 30°–60° right triangles, the hypotenuse is twice the length of the side opposite the 30° angle, and the adjacent side is about 1.7 times this length. In this section, we will see that there are fixed ratios among the sides in *all* right triangles.

Labeling Right Triangles

The first skill needed for trigonometry is the ability to label the three sides of a right triangle according to convention: the hypotenuse, the side *adjacent* to a given acute angle, and the side *opposite* that angle.

learning|**catalytics**™

1. If one acute angle of a right triangle measures 40°, what is the measure of the other acute angle?

2. Find *x*:

6 in. 60° 30° *x*

In triangle *ABC*, the side labeled *c* is the hypotenuse. It is the longest side and it is opposite the right angle.

For acute angle *A*:

b is the adjacent side.

a is the opposite side.

For acute angle *B*:

a is the adjacent side.

b is the opposite side.

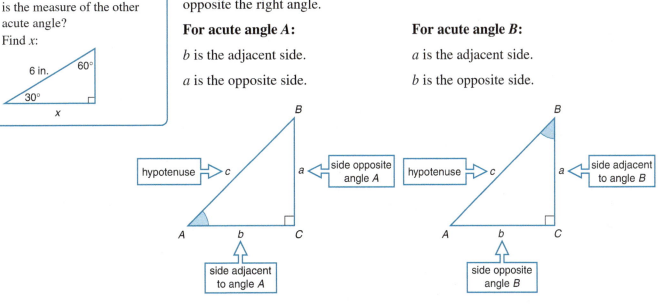

⚠ **Careful**   The hypotenuse is never considered to be either the adjacent side or the opposite side. It is not a leg of the triangle. ●

Your Turn   Check your understanding of this vocabulary by putting triangle *GHI* in standard position, labeling the sides, and completing the following statements.

(a) The hypotenuse is _____.

(b) The opposite side for angle *H* is _____.

(c) The adjacent side for angle *H* is _____.

(d) The opposite side for angle *I* is _____.

(e) The adjacent side for angle *I* is _____.

Answers   (a) *g*   (b) *h*   (c) *i*

(d) *i*   (e) *h*

Note   The terms *adjacent side*, *opposite side*, and *hypotenuse* are used only with right triangles. ●

More Practice   For each of the following right triangles, mark the hypotenuse with the letter *c*, mark the side adjacent to the shaded angle with the letter *b*, and mark the side opposite the shaded angle with the letter *a*.

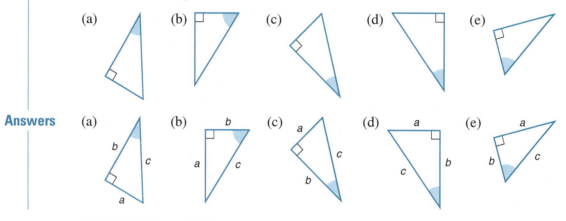

Answers   (a)   (b)   (c)   (d)   (e)

Learning Help   If you find it difficult to identify the adjacent side and the opposite side of a given angle in a right triangle, try placing the triangle in standard position. Position the triangle so that the given angle is at the lower left and the right angle is at the lower right. (This may require you to "flip" the triangle.) When the triangle is in this position, the adjacent side is always the horizontal leg, *b*, and the opposite side will always be the vertical leg, *a*. ●

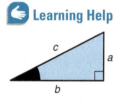

Sine Ratio   The key to understanding trigonometry is to realize that in *every* right triangle there is a fixed relationship connecting the size of either of the acute angles to the lengths of its adjacent side, its opposite side, and the hypotenuse.

For example, in all right triangles that contain the angle 20°,

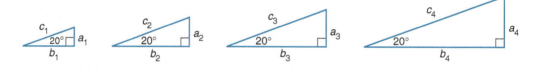

the ratio $\dfrac{\text{opposite side}}{\text{hypotenuse}}$ will always have the same value. Therefore,

$$\frac{a}{c} = \frac{a_1}{c_1} = \frac{a_2}{c_2} = \frac{a_3}{c_3} = \dots \text{ and so on}$$

This ratio will be approximately 0.342.

This kind of ratio of the side lengths of a right triangle is called a **trigonometric ratio**. It is possible to write down six of these ratios, but we will use only three here. Each ratio is given a special name.

For the triangle

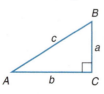

the **sine ratio**, abbreviated **sin**, is defined as follows:

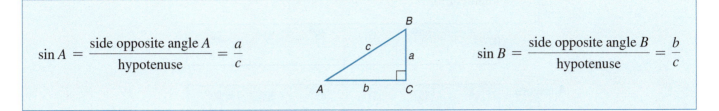

$$\sin A = \frac{\text{side opposite angle } A}{\text{hypotenuse}} = \frac{a}{c} \qquad\qquad \sin B = \frac{\text{side opposite angle } B}{\text{hypotenuse}} = \frac{b}{c}$$

(Pronounce "sin" as "sign" to rhyme with "dine," not with "sin." *Sin* is an abbreviation for *sine*. Sin *A* is read "sine of angle *A*.")

EXAMPLE 1　Suppose we wish to find the approximate value of sin 36° using the triangle shown. Notice that the side opposite the 36° angle has length 52.9, and the hypotenuse has length 90.0.

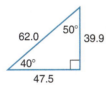

$$\sin 36° = \frac{\text{opposite side}}{\text{hypotenuse}} = \frac{52.9}{90.0} \approx 0.588$$

●

Your Turn　Use this triangle to find the approximate value of sin 50° to three decimal places.

Solution　The side opposite the 50° angle has length 47.5, and the hypotenuse has length 62.0.

$$\sin 50° = \frac{\text{opposite side}}{\text{hypotenuse}} = \frac{47.5}{62.0} \approx 0.766$$

Every possible angle *x* will have some number sin *x* associated with it. If *x* is an acute angle, we can calculate the value of sin *x* from any right triangle that contains *x*.

Now, let's look again at those three special right triangles from Section 10-1.

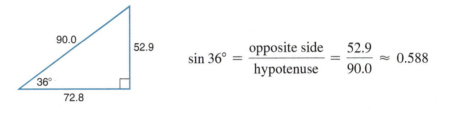

We can use these labeled figures to calculate the approximate values of sine for angles of 30°, 37°, 45°, 53°, and 60°.

EXAMPLE 2　To find sin 60°, use the 30°–60° right triangle. Notice that the side opposite the 60° angle has length $\sqrt{3}$, and the hypotenuse has length 2. Therefore,

$$\sin 60° = \frac{\text{opposite side}}{\text{hypotenuse}} = \frac{\sqrt{3}}{2} = 0.866\ldots \approx 0.87$$

●

Your Turn   Use the three special right triangles to complete the table. (Round to the nearest hundredth.)

A	30°	37°	45°	53°	60°
sin A					0.87

Answers

A	30°	37°	45°	53°	60°
sin A	0.50	0.60	0.71	0.80	0.87

Cosine Ratio   A second useful trigonometric ratio is the **cosine ratio**, abbreviated **cos**. This ratio is defined as follows:

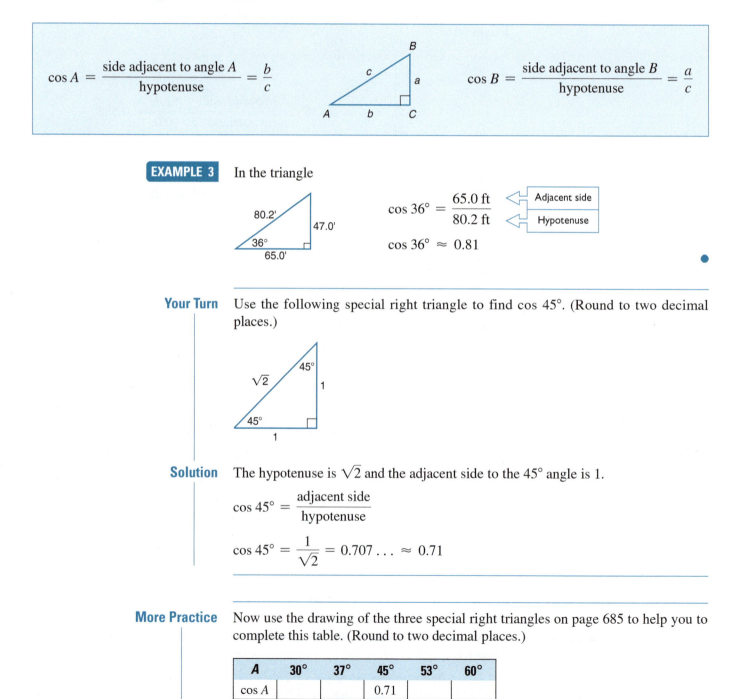

$$\cos A = \frac{\text{side adjacent to angle } A}{\text{hypotenuse}} = \frac{b}{c}$$

$$\cos B = \frac{\text{side adjacent to angle } B}{\text{hypotenuse}} = \frac{a}{c}$$

EXAMPLE 3   In the triangle

$$\cos 36° = \frac{65.0 \text{ ft}}{80.2 \text{ ft}} \quad \begin{array}{l} \leftarrow \text{Adjacent side} \\ \leftarrow \text{Hypotenuse} \end{array}$$

$$\cos 36° \approx 0.81$$

Your Turn   Use the following special right triangle to find cos 45°. (Round to two decimal places.)

Solution   The hypotenuse is $\sqrt{2}$ and the adjacent side to the 45° angle is 1.

$$\cos 45° = \frac{\text{adjacent side}}{\text{hypotenuse}}$$

$$\cos 45° = \frac{1}{\sqrt{2}} = 0.707 \ldots \approx 0.71$$

More Practice   Now use the drawing of the three special right triangles on page 685 to help you to complete this table. (Round to two decimal places.)

A	30°	37°	45°	53°	60°
cos A			0.71		

Answers

A	30°	37°	45°	53°	60°
cos A	0.87	0.80	0.71	0.60	0.50

 Note   The value of sin A or cos A can never be greater than 1 because the legs of any right triangle can never be longer than the hypotenuse. ●

Tangent Ratio   A third trigonometric ratio may also be defined. The **tangent ratio**, abbreviated **tan**, is defined as follows:

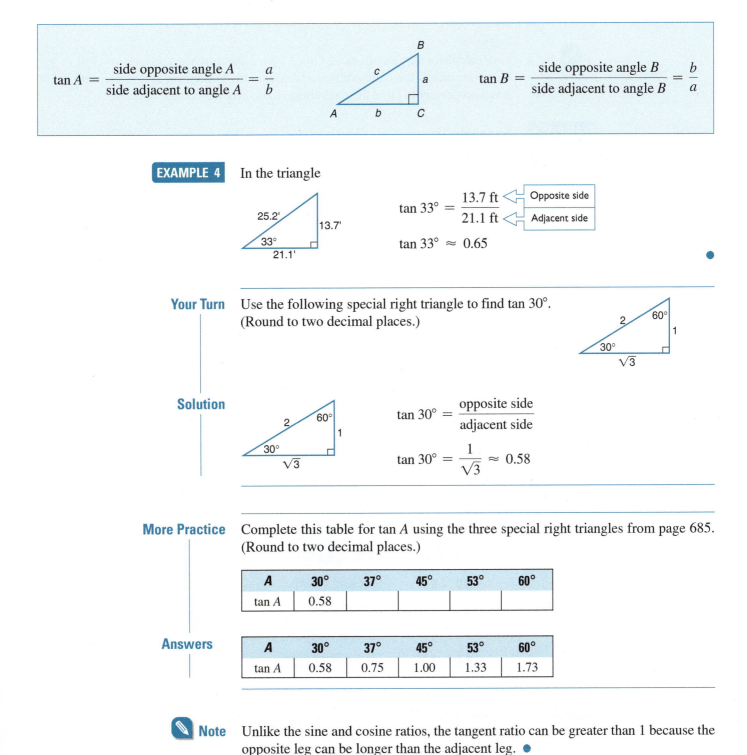

$$\tan A = \frac{\text{side opposite angle } A}{\text{side adjacent to angle } A} = \frac{a}{b}$$

$$\tan B = \frac{\text{side opposite angle } B}{\text{side adjacent to angle } B} = \frac{b}{a}$$

EXAMPLE 4   In the triangle

$$\tan 33° = \frac{13.7 \text{ ft}}{21.1 \text{ ft}} \quad \begin{array}{l} \longleftarrow \text{Opposite side} \\ \longleftarrow \text{Adjacent side} \end{array}$$

$$\tan 33° \approx 0.65$$

●

Your Turn   Use the following special right triangle to find tan 30°. (Round to two decimal places.)

Solution

$$\tan 30° = \frac{\text{opposite side}}{\text{adjacent side}}$$

$$\tan 30° = \frac{1}{\sqrt{3}} \approx 0.58$$

More Practice   Complete this table for tan A using the three special right triangles from page 685. (Round to two decimal places.)

A	30°	37°	45°	53°	60°
tan A	0.58				

Answers

A	30°	37°	45°	53°	60°
tan A	0.58	0.75	1.00	1.33	1.73

 Note   Unlike the sine and cosine ratios, the tangent ratio can be greater than 1 because the opposite leg can be longer than the adjacent leg. ●

Learning Help Many students find it helpful to remember the ratios for sine, cosine, and tangent using the acronym *soh-cah-toa*: "soh" stands for "*s*ine: *o*pposite over *h*ypotenuse"; "cah" stands for "*c*osine: *a*djacent over *h*ypotenuse"; "toa" stands for "*t*angent: *o*pposite over *a*djacent." ●

Finding Values of Trigonometric Functions You have been asked to calculate these "trig" ratios to help you get a feel for how the ratios are defined and to encourage you to memorize a few values. In actual practical applications of trigonometry, the values of the ratios are found by using a calculator with built-in trigonometric function keys.

To find the value of a trigonometric ratio using a scientific calculator, first press the key of the given function, then enter the value of the angle, and finally press $\boxed{=}$.

Note Your calculator may automatically open a parentheses when you press $\boxed{\text{sin}}$, $\boxed{\text{cos}}$, or $\boxed{\text{tan}}$. However, unless there are additional calculations to perform before pressing $\boxed{=}$, it is not neccessary to close the parentheses to display the correct value. ●

EXAMPLE 5 To find sin 38°, enter this sequence:

$\boxed{\text{sin}}$ **3 8** $\boxed{=}$ → $\boxed{0.615661475}$ sin 38° ≈ 0.616   ●

Careful Angles can be measured in units of radians and gradients as well as degrees, and most calculators are capable of accepting all three units of measure. Before entering an angle in degrees, make certain the calculator is in "degree mode" by either using the $\boxed{\text{DRG}}$ key or the $\boxed{\text{MODE}}$ key. Look for the "DEG" indication on the display. On some calculators, the absence of any mode indication means that it is in degree mode. ●

Your Turn Now you try it. Find the value of each of the following to two decimal places.

(a) cos 26° (b) tan 67.8° (c) sin 8°

(d) tan 14.25° (e) cos $47\frac{1}{2}°$ (f) sin 39.46°

(g) sin 138° (h) cos 151.6° (i) tan 141.12°

Solutions (a) $\boxed{\text{cos}}$ **2 6** $\boxed{=}$ → $\boxed{0.898794046}$ cos 26° ≈ 0.90

(b) $\boxed{\text{tan}}$ **6 7.8** $\boxed{=}$ → $\boxed{2.450425198}$ tan 67.8° ≈ 2.45

(c) $\boxed{\text{sin}}$ **8** $\boxed{=}$ → $\boxed{0.139173101}$ sin 8° ≈ 0.14

(d) $\boxed{\text{tan}}$ **1 4.2 5** $\boxed{=}$ → $\boxed{0.253967646}$ tan 14.25° ≈ 0.25

(e) $\boxed{\text{cos}}$ **4 7.5** $\boxed{=}$ → $\boxed{0.675590208}$ cos $47\frac{1}{2}°$ ≈ 0.68

(f) $\boxed{\text{sin}}$ **3 9.4 6** $\boxed{=}$ → $\boxed{0.63553937}$ sin 39.46° ≈ 0.64

(g) $\boxed{\text{sin}}$ **1 3 8** $\boxed{=}$ → $\boxed{0.669130606}$ sin 138° ≈ 0.67

(h) $\boxed{\text{cos}}$ **1 5 1.6** $\boxed{=}$ → $\boxed{-0.879648573}$ cos 151.6° ≈ −0.88

(i) $\boxed{\text{tan}}$ **1 4 1.1 2** $\boxed{=}$ → $\boxed{-0.806322105}$ tan 141.12° ≈ −0.81

A Closer Look • Notice in problems (g), (h), and (i) that the trigonometric functions do exist for angles greater than 90°. The values of the cosine and tangent functions are negative for angles between 90° and 180°.

• Notice in problem (g) that

$$\sin 138° = 0.669130606\ldots$$

but $\sin 42° = 0.669130606\ldots$ Also note that $42° = 180° - 138°$

This suggests that the following formula is true:

$$\sin A = \sin (180° - A)$$

This fact will be useful in Section 10-4. ●

Angles in Degrees and Minutes   When finding the sine, cosine, or tangent of an angle given in degrees and minutes, either enter the angle as given using the ⌈°'"⌉ key, or first convert the angle to decimal form as shown in the following example. See your calculator's instruction manual for the key sequence that will allow you to enter degrees and minutes directly.

EXAMPLE 6   To find cos 54° 28′, we recommend the following sequence:

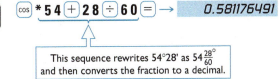

 $\boxed{\text{cos}}$ *$\mathbf{54}\boxed{+}\mathbf{28}\boxed{÷}\mathbf{60}\boxed{=}$ → $\boxed{0.581176491}$

> This sequence rewrites 54°28' as $54\frac{28}{60}°$ and then converts the fraction to a decimal.

*If your calculator does not automatically open parentheses with $\boxed{\text{cos}}$, then you must press $\boxed{(}$ here.   ●

Your Turn   Try these problems for practice. Find each value to three decimal places.

(a) sin 71°26′   (b) tan 18°51′   (c) cos 42°16′   (d) cos 117°22′

Answers   (a) 0.948   (b) 0.341   (c) 0.740   (d) −0.460

Finding the Angle   In some applications of trigonometry, we may know the value of the trigonometric ratio number and we may need to find the angle associated with it. This is called finding the **inverse of the trigonometric function**.

To find the angle associated with any given value of a trigonometric function, first press the $\boxed{2^{nd}}$ key (also labeled $\boxed{2^{nd}\text{F}}$ or $\boxed{\text{SHIFT}}$), second press the key of the given function ($\boxed{\text{sin}}$, $\boxed{\text{cos}}$, or $\boxed{\text{tan}}$), third enter the trig value, and finally press $\boxed{=}$. These inverse functions are labeled $\boxed{\text{sin}^{-1}}$, $\boxed{\text{cos}^{-1}}$, and $\boxed{\text{tan}^{-1}}$ on your keyboard. Because these are second functions on all calculators, we will always include the $\boxed{2^{nd}\text{F}}$ key in our key sequences as a reminder to press it first.

✎ **Note**   If your calculator automatically opens parentheses with the inverse trigonometric functions, it is not necessary to close them unless there are additional calculations to perform before pressing $\boxed{=}$. ●

EXAMPLE 7   To find A if $\sin A = 0.728$, enter the following:

 $\boxed{2^{nd}}$ $\boxed{\text{sin}^{-1}}$ $\mathbf{.728}\boxed{=}$ → $\boxed{46.718988}$

The angle is approximately 46.7°; therefore, $\sin 46.7° \approx 0.728.$* We can also express this result as $\sin^{-1} 0.728 \approx 46.7°$. (Here, the superscript -1 is not an exponent. It is simply a notation indicating the inverse sine function.)   ●

---

*We have found the angle between 0° and 90° that corresponds to the given trigonometric function value, but there are many angles greater than 90° whose sine ratio has this same value. Because such values of the inverse are generally not very important in the trades, we will not discuss them in detail here.

! Careful   Be certain your calculator is in the degree mode before you enter this kind of calculation. ●

Your Turn   Use your calculator to find the angle A for the following trigonometric function values. (Round to the nearest tenth of a degree.)

(a) $\cos A = 0.589$ (b) $\sin A = 0.248$ (c) $\tan A = 1.75$

Answers   (a) $A = \cos^{-1} 0.589 \approx 53.9°$ (b) $A = \sin^{-1} 0.248 \approx 14.4°$

(c) $A = \tan^{-1} 1.75 \approx 60.3°$

To find the angle in degrees and minutes, first follow the procedure described to find the angle in decimal degrees. Then convert to degrees and minutes using either the method of the following example or the method described in your calculator's instruction manual.

EXAMPLE 8   To find A in degrees and minutes for $\cos A = 0.296$, we recommend the following procedure:

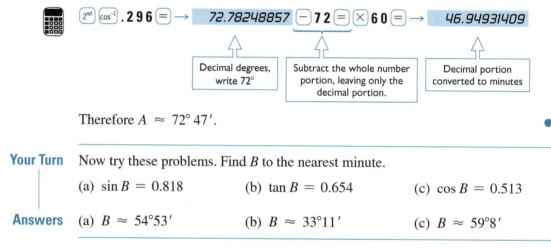

Therefore $A \approx 72° 47'$.   ●

Your Turn   Now try these problems. Find B to the nearest minute.

(a) $\sin B = 0.818$ (b) $\tan B = 0.654$ (c) $\cos B = 0.513$

Answers   (a) $B \approx 54°53'$ (b) $B \approx 33°11'$ (c) $B \approx 59°8'$

Now continue with Exercises 10-2 for a set of practice problems on trigonometric ratios.

Exercises 10-2   Trigonometric Ratios

A. For each of the following triangles, calculate the indicated trigonometric ratios of the given angle. Do not use the trigonometric function keys on your calculator. (Round to the nearest hundredth.)

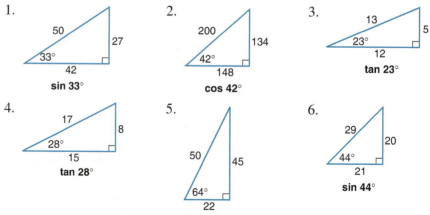

B. Find each of the following trig values. (Round to three decimal places.)

1. sin 27°	2. cos 38°	3. tan 12°	4. sin 86°
5. cos 79°	6. tan 6°	7. cos 87°	8. cos 6°30′
9. tan 50°20′	10. sin 75°40′	11. cos 41.25°	12. cos 81°25′
13. sin 50.4°	14. tan 74.15°	15. tan 81.06°	16. sin 12.6°
17. cos 98°	18. sin 106°	19. sin 144.2°	20. cos 134.5°

C. Find the acute angle A. (Round to the nearest minute.)

1. sin A = 0.974	2. cos A = 0.719
3. tan A = 2.05	4. sin A = 0.077
5. cos A = 0.262	6. sin A = 0.404
7. sin A = 0.168	8. cos A = 0.346

Find the acute angle A. (Round to the nearest tenth of a degree.)

9. tan A = 1.165	10. cos A = 0.662
11. cos A = 0.437	12. tan A = 0.225
13. tan A = 0.872	14. sin A = 0.472
15. sin A = 0.605	16. cos A = 0.154

Check your answers to the odd-numbered problems in the Appendix, then turn to Section 10-3 to learn how to use trigonometric ratios to solve problems involving right triangles.

10-3 Solving Right Triangles

In the first two sections of this chapter, you learned about angles, triangles, and the trigonometric ratios that relate the angles to the side lengths in right triangles. Now it's time to look at a few of the possible applications of these trig ratios.

A right triangle has the following six parts:

- Three angles: A right angle and two acute angles.
- Three sides: The hypotenuse and two legs.

We can use trig ratios to find all unknown sides and angles of a right triangle if we know either of the following combinations:

1. The length of one side and the measure of one of the acute angles.
2. The lengths of any two sides.

Here is an example of the first combination.

EXAMPLE 1

Suppose we need to find side *a* in the triangle shown. Use the following three-step approach:

Step 1 Decide which trig ratio is the appropriate one to use. In this case, we are given the hypotenuse (21.0 cm) and acute angle *B* (50.0°), and we need to find the side adjacent to the given angle *B*. The trig ratio that relates the adjacent side and the hypotenuse is the cosine. (Remember: The "cah" in soh-cah-toa means "cosine: adjacent over hypotenuse.")

$$\cos B = \frac{\text{adjacent side}}{\text{hypotenuse}}$$

Step 2   Write the cosine of the given angle in terms of the sides of the given right triangle.

$$\cos 50.0° = \frac{a}{21.0 \text{ cm}}$$

Step 3   Solve for the unknown quantity. In this case multiply both sides of the equation by 21.

$$a = 21 \, (\cos 50°)$$
$$\approx 21(0.64278\ldots)$$
$$\approx 13.4985 \text{ cm} \quad \text{or} \quad 13.5 \text{ cm}, \quad \text{rounded}$$

2 1 ✕ cos **5 0** = → | *13.4985398* | ●

Rounding When Solving Right Triangles

For practical applications involving triangle trigonometry, rounding will be dictated by specific situations or by job standards. To simplify rounding instructions in the remainder of this chapter, we shall adopt the following convention:

Sides rounded to:	correspond to	angles rounded to the nearest:
2 significant digits		degree
3 significant digits		0.1° or 10′
4 significant digits		0.01° or 1′

If side lengths are given as fractions, assume they are accurate to three significant digits.

If you need a review of significant digits, refer to Section 5-1, p. 280.

Your Turn   Find the length of the hypotenuse c in the triangle shown. Work it out using the three steps shown in Example 1.

Solution   **Step 1**   We are given the side opposite the angle 71.0°, and we need to find the hypotenuse. The trig ratio relating the opposite side and the hypotenuse is the sine. (Recall that the "soh" in "soh-cah-toa" stands for "sine: opposite over hypotenuse.")

$$\sin A = \frac{\text{opposite side}}{\text{hypotenuse}}$$

Step 2 $\sin 71.0° = \dfrac{10.5 \text{ in.}}{c}$

Step 3   Solve for c. $c \cdot \sin 71° = 10.5 \text{ in.}$

$$\text{or} \quad c = \frac{10.5 \text{ in.}}{\sin 71°} \approx \frac{10.5}{0.9455\ldots}$$

$$\approx 11.1 \text{ in.}$$

1 0 . 5 ÷ sin **7 1** = → | *11.10501715* |

Here is an example of the second combination, where we are given two sides.

EXAMPLE 2   To find the measure of angle A in the triangle shown, follow these steps:

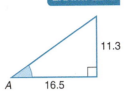

Step 1   We are given the opposite and adjacent sides to angle A. The tangent ratio involves these three parts. (The final part of "soh-cah-toa" stands for "tangent: opposite over adjacent.")

$$\tan A = \frac{\text{opposite side}}{\text{adjacent side}}$$

Step 2   Substitute the given information.

$$\tan A = \frac{11.3}{16.5}$$

Step 3   To find the missing *angle*, use the *inverse* tangent.

$$A = \tan^{-1}\left(\frac{11.3}{16.5}\right) = 34.405\ldots \approx 34.4°$$

Here are two calculator options for this problem.

1.   Calculate the trig ratio first, then find tangent inverse of "last answer":

 $11.3 \boxed{\div} 16.5 \boxed{=} \boxed{2^{nd}} \boxed{\tan^{-1}} \boxed{ANS} \boxed{=} \rightarrow$ **34.40523429**

2.   Perform the calculation exactly as it is written:

$\boxed{2^{nd}} \boxed{\tan^{-1}} *11.3 \boxed{\div} 16.5 \boxed{=} \rightarrow$ **34.40523429**

*If your calculator does not automatically open parentheses here, you must press $\boxed{(}$.

There is no need to close parentheses in this calculation.   ●

In the previous two examples, we were asked to find only one missing part of each right triangle. If we are asked to "solve the right triangle," then we must find *all* unknown sides and angles.

EXAMPLE 3   To *solve* the right triangle in the figure, we must find the lengths of the two unknown legs, a and b, and the unknown acute angle B.

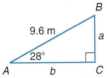

First, find angle B. This is the easy part. Recall that the three angles of any triangle sum to 180°. Because angle C is a right angle, acute angles A and B will always sum to 90°.

Therefore,

$$B = 90° - A = 90° - 28° = 62°$$

Next, use a trig ratio to find one of the unknown legs. We are given the hypotenuse (9.6 m). If we use the given angle A to find its adjacent side b, we must use cosine.

$$\cos A = \frac{\text{adjacent side}}{\text{hypotenuse}}$$

$$\cos 28° = \frac{b}{9.6 \text{ m}}$$

$$b = 9.6\,(\cos 28°)$$

$$b = 8.476\ldots \approx 8.5 \text{ m}   \text{rounded}$$

Finally, find the length of the third side a. Because we now have two sides of the right triangle, we could use either the Pythagorean theorem or a trig equation. To avoid rounding errors, it is best to choose whichever method uses the original given information. The Pythagorean theorem would require the use of the calculated value from the previous step, which is rounded, so in this case we will use a trig equation. The sine ratio relates the given parts, angle A and the hypotenuse, to the unknown side a opposite angle A.

$$\sin A = \frac{\text{opposite side}}{\text{hypotenuse}}$$

$$\sin 28° = \frac{a}{9.6 \text{ m}}$$

$$a = 9.6 \, (\sin 28°)$$

$$a = 4.5069 \ldots \approx 4.5 \text{ m} \quad \text{rounded}$$

Therefore, the missing parts of the given triangle are $B = 62°$, $b \approx 8.5$ m, $a \approx 4.5$ m.

●

Your Turn   Solve the right triangle shown. Express all angles in degrees and minutes.

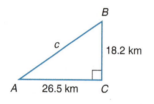

Solution   **First,** find one of the acute angles. To find angle A, notice that we are given the opposite and adjacent sides to angle A, so we must use the tangent ratio.

$$\tan A = \frac{\text{opposite side}}{\text{adjacent side}}$$

$$\tan A = \frac{18.2 \text{ km}}{26.5 \text{ km}}$$

$$A = \tan^{-1}\left(\frac{18.2}{26.5}\right)$$

$$A = 34.48 \ldots° \approx 34° \, 30' \quad \text{rounded to the nearest } 10'$$

Next, find the other acute angle by subtraction.

$$B = 90° - A \approx 90° - 34° \, 30' \approx 55° \, 30'$$

Finally, determine the length of the third side. In this case, to use the given information, we choose the Pythagorean theorem.

$$c = \sqrt{a^2 + b^2}$$

$$c = \sqrt{(18.2)^2 + (26.5)^2}$$

$$c \approx 32.1 \text{ km}$$

Therefore, the missing parts of the given triangle are $A \approx 34°30'$, $B \approx 55°30'$, $c \approx 32.1$ km

More Practice   (a) Find c.   (b) Find angle B in decimal degrees.

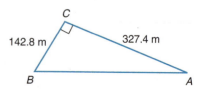

(c) Solve the triangle. Express angles in decimal degrees and round a to the nearest eighth of an inch.

(d) Solve the triangle. Express angles in degrees and minutes.

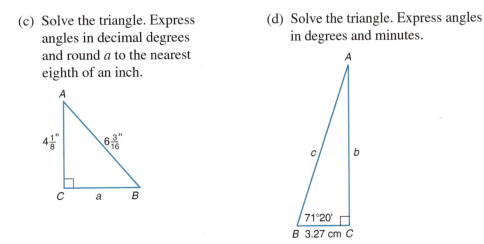

Solutions

(a) For the given angle, 12 ft is the length of the opposite side and c is the hypotenuse. Use the sine ratio.

$$\sin 42° = \frac{12 \text{ ft}}{c}$$

$$c = \frac{12 \text{ ft}}{\sin 42°} = 17.93\ldots \approx 18 \text{ ft rounded}$$

(b) For the unknown angle B, 142.8 m is the length of the adjacent side and 327.4 m is the length of the opposite side. Use the tangent ratio.

$$\tan B = \frac{327.4 \text{ m}}{142.8 \text{ m}}$$

$$B = \tan^{-1}\left(\frac{327.4}{142.8}\right) = 66.434\ldots° \approx 66.43° \text{ rounded}$$

(c) **First,** for angle A, we have the adjacent side and the hypotenuse. Use cosine:

$$\cos A = \frac{4\frac{1}{8} \text{ in.}}{6\frac{3}{16} \text{ in.}} = \frac{2}{3}$$

$$A = \cos^{-1}\left(\frac{2}{3}\right) \approx 48.2°$$

With a calculator, we recommend dividing the fractions first and then finding the inverse cosine of "last answer."

4 $\boxed{A\frac{b}{c}}$ 1 $\boxed{A\frac{b}{c}}$ 8 $\boxed{\div}$ 6 $\boxed{A\frac{b}{c}}$ 3 $\boxed{A\frac{b}{c}}$ 16 $\boxed{=}$ → ☐ *2/3*

$\boxed{2^{nd}}$ $\boxed{\cos^{-1}}$ $\boxed{\text{ANS}}$ $\boxed{=}$ → ☐ *48.1896851*

Next, to find angle B, subtract angle A from $90°$.

$$B = 90° - A \approx 90 - 48.2° \approx 41.8°$$

Finally, to find side a, use the Pythagorean theorem.

$$a = \sqrt{c^2 - b^2}$$

$$= \sqrt{\left(6\frac{3}{16}\right)^2 - \left(4\frac{1}{8}\right)^2}$$

$$\approx 4.61 \text{ in.}$$

To the nearest eighth of an inch, $a \approx 4\frac{5}{8}$ in.

$$6 \boxed{A\frac{b}{c}} 3 \boxed{A\frac{b}{c}} 1 6 \boxed{x^2} \boxed{-} 4 \boxed{A\frac{b}{c}} 1 \boxed{A\frac{b}{c}} 8 \boxed{x^2} \boxed{=} \boxed{\sqrt{}} \boxed{ANS} \boxed{=} \rightarrow \boxed{4.611890204}$$

Therefore, the missing parts of the given triangle are $A \approx 48.2°$, $B \approx 41.8°$, $a \approx 4\frac{5}{8}$ in.

(d) **First,** find the unknown acute angle A.

$$A = 90° - B = 90° - 71°20' = 18°40'$$

Next, find one of the missing sides. We have the acute angle B (71° 20′) and its adjacent side (3.27 cm). We could use tangent to find the opposite side b or cosine to find the hypotenuse c. We will use tangent:

$$\tan 71°20' = \frac{b}{3.27 \text{ cm}}$$

$$b = 3.27 \, (\tan 71° \, 20')$$

$$b \approx 9.68 \text{ cm}$$

$$3.27 \boxed{\times} \boxed{\tan} *71 \boxed{+} 20 \boxed{\div} 60 \boxed{=} \rightarrow \boxed{9.679337948}$$

*If your calculator does not automatically open parentheses with $\boxed{\tan}$, you must press $\boxed{(}$ here. You may also use your $\boxed{°\,'\,''}$ key to enter degrees and minutes.

Finally, find the hypotenuse. Using the given angle B and its adjacent side, we must use the cosine ratio.

$$\cos 71° \, 20' = \frac{3.27 \text{ cm}}{c}$$

$$c = \frac{3.27 \text{ cm}}{\cos 71° \, 20'}$$

$$c \approx 10.2 \text{ cm}$$

$$3.27 \boxed{\div} \boxed{(} \boxed{\cos} 71 \boxed{+} 20 \boxed{\div} 60 \boxed{=} \rightarrow \boxed{10.21677459}$$

Therefore, the missing parts of the given triangle are $A = 18° \, 40'$, $b \approx 9.68$ cm, $c \approx 10.2$ cm.

Practical Applications   Practical problems are usually a bit more difficult to set up, but they are solved using the same approach. Sometimes you must redraw a given figure in order to find a right triangle.

EXAMPLE 4   **Surveying**   Suppose you need to measure the distance between two points A and B on opposite sides of a river. You have a tape measure and a protractor but no way to stretch the tape measure across the river. How can you measure the distance AB?

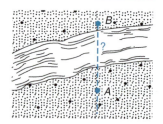

Try it this way:

1. Choose a point C in line with points A and B. If you stand at C, A and B will appear to "line up."

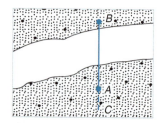

2. Construct a perpendicular to line *BC* at *A* and pick any point *D* on this perpendicular.

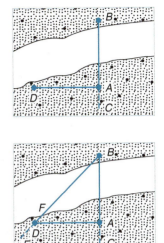

3. Find a point *E* where *B* and *D* appear to "line up." Draw *DE* and extend *DE* to *F* to form angle *D*.

4. Triangle *ABD* is a right triangle. You can measure *AD* and angle *D*, and, using the trigonometric ratios, you can calculate *AB*.

Suppose *AD* = 105 ft and angle *D* = 49.0°. To find *AB*, first draw the following right triangle:

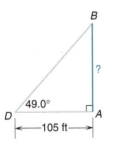

AD is the adjacent side to angle *D*.

AB is the side opposite angle *D*.

The trig ratio that relates the opposite to the adjacent side is the tangent.

$$\tan D = \frac{\text{opposite side}}{\text{adjacent side}}$$

$$\tan D = \frac{AB}{AD}$$

$$\tan 49° = \frac{AB}{105 \text{ ft}}$$

Multiply both sides of this equation by 105 to get

$$105(\tan 49°) = AB$$

Using a calculator, we have

1 0 5 ⊗ tan **4 9** = → *120.7886828*

Therefore, *AB* ≈ 121 ft

Using trig ratios, you have managed to calculate the distance from *A* to *B* without ever coming near point *B*. ●

Your Turn **Metalworking** Find the angle of taper, *x*. (Round to the nearest degree.)

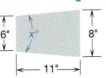

Solution   We draw a new figure and draw dashed line segment BC to create triangle ABC with angle m as indicated. Notice that angle x, the *taper angle,* is twice as large as angle m. We can find angle m from triangle ABC.

$$\tan m = \frac{\text{opposite side}}{\text{adjacent side}} = \frac{AC}{BC}$$

$$\tan m = \frac{1 \text{ in.}}{11 \text{ in.}} \approx 0.090909 \ldots$$

$$m \approx \tan^{-1} 0.090909 \ldots \approx 5.2°$$

$$x = 2m \approx 10.4° \approx 10°$$

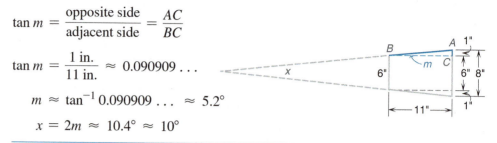

Here are a few more applications for you to try.

More Practice   (a) **Metalworking**   Find the angle a in the casting shown. (Round to the nearest $0.1°$.)

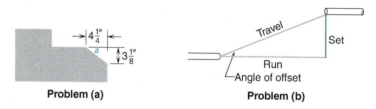

Problem (a)   Problem (b)

(b) **Plumbing**   In a pipe-fitting job, what is the length of set if the offset angle is $22\frac{1}{2}°$ and the travel is $32\frac{1}{8}$ in.? (See the figure above. Round to the nearest eighth of an inch.)

(c) **Metalworking**   Find the width, w, of the V-slot shown.

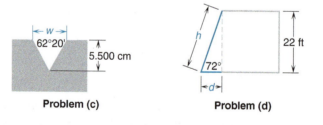

Problem (c)   Problem (d)

(d) **Machine Trades**   Find the dimensions d and h on the metal plate shown above.

Solutions   (a) In this problem, the angle a is unknown. The information given allows us to calculate $\tan a$:

$$\tan a = \frac{\text{opposite side}}{\text{adjacent side}} = \frac{3\frac{1}{8} \text{ in.}}{4\frac{1}{4} \text{ in.}}$$

$$\tan a = \frac{25}{34}$$

$$a = \tan^{-1}\left(\frac{25}{34}\right) \approx 36.3°$$

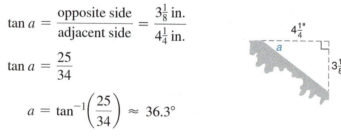

(b) In this triangle, the opposite side is unknown and the hypotenuse is given. Use the formula for sine:

$$\sin 22\frac{1}{2}° = \frac{\text{opposite side}}{\text{hypotenuse}}$$

Then substitute, $\sin 22.5° = \dfrac{s}{32\frac{1}{8} \text{ in.}}$

And solve, $s = (32\frac{1}{8} \text{ in.})(\sin 22.5°) = 12.29\ldots \approx 12\frac{2}{8}$ or $12\frac{1}{4}$ in.

(c) The V-slot creates an isosceles triangle whose height is 5.500 cm. The height divides the triangle into two identical right triangles, as shown in the figure.

To find y:

$\tan 31°10' = \dfrac{y}{5.500 \text{ cm}}$

or

$y = (5.500 \text{ cm})(\tan 31°10')$

$y \approx 3.3265 \text{ cm}$

$w = 2y \approx 6.653 \text{ cm}$

(d) $\sin 72° = \dfrac{22 \text{ ft}}{h}$

$h = \dfrac{22 \text{ ft}}{\sin 72°} \approx 23 \text{ ft}$

$\tan 72° = \dfrac{22 \text{ ft}}{d}$

$d = \dfrac{22 \text{ ft}}{\tan 72°} \approx 7.1 \text{ ft}$

Now continue with Exercises 10-3 for a set of problems on solving right triangles using trigonometric ratios.

Exercises 10-3 | Solving Right Triangles

Unless otherwise directed, round according to the rules in the box on page 692.

A. For each right triangle, find the missing quantity indicated below the figure. Express angles in decimal degrees.

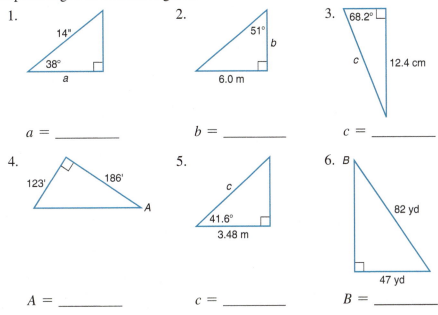

1.
14"
38°
a

a = _____

2.
51°
b
6.0 m

b = _____

3.
68.2°
c
12.4 cm

c = _____

4.
123'
186'
A

A = _____

5.
c
41.6°
3.48 m

c = _____

6. B
82 yd
47 yd

B = _____

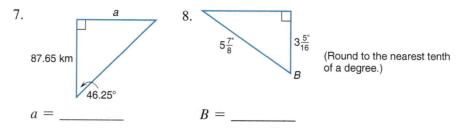

7.

$a =$ _____

8. (Round to the nearest tenth of a degree.)

$B =$ _____

B. For each right triangle, find the missing quantity indicated below the figure. Express angles in degrees and minutes.

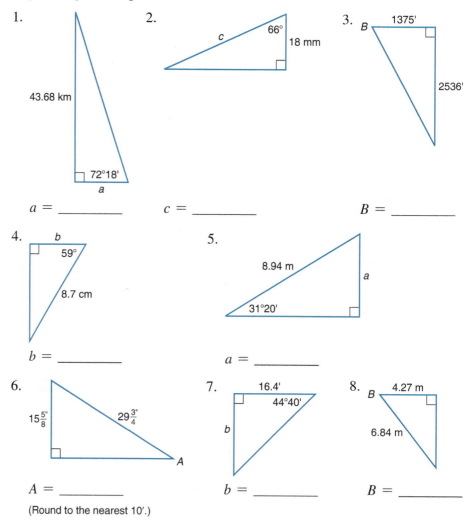

1.

43.68 km

72°18'

a

$a =$ _____

2.

66°

18 mm

c

$c =$ _____

3.

1375'

B

2536'

$B =$ _____

4.

b

59°

8.7 cm

$b =$ _____

5.

8.94 m

a

31°20'

$a =$ _____

6.

$15\frac{5}{8}''$ $29\frac{3}{4}''$

A

$A =$ _____

(Round to the nearest 10'.)

7.

16.4'

44°40'

b

$b =$ _____

8.

B 4.27 m

6.84 m

$B =$ _____

C. For each problem, use the given information to solve right triangle ABC for all missing parts. Express angles in decimal degrees.

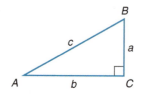

B

c a

A b C

1. $a = 36$ ft, $B = 68°$

2. $b = 52.5$ m, $c = 81.6$ m

3. $A = 31.5°$, $a = 6\frac{3}{4}$ in.

4. $B = 58°$, $c = 5.4$ cm

5. $a = 306.5$ km, $c = 591.3$ km

6. $A = 47.4°$, $b = 158$ yd

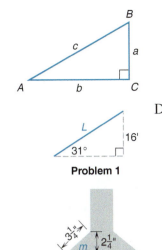

Problem 1

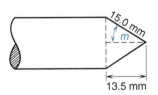

Problem 2

7. $B = 74°, b = 16$ ft

9. $A = 21° 15', c = 24.75$ m

8. $a = 12\frac{3}{8}$ in., $b = 16\frac{5}{16}$ in.

(Round c to the nearest eighth of an inch and angles to the nearest tenth of a degree.)

D. Practical Applications

1. **Manufacturing** The most efficient operating angle for a certain conveyor belt is 31°. If the parts must be moved a vertical distance of 16 ft, what length of conveyor is needed? (See the figure in the margin.)

2. **Metalworking** Find the angle m in the casting shown. (Round to the nearest tenth of a degree.)

3. **Plumbing** A pipe fitter must connect a pipeline to a tank as shown in the figure. The run from the pipeline to the tank is 62 ft 6 in., while the set (rise) is 38 ft 9 in.

   (a) How long is the connection? (Round to the nearest inch.)

   (b) Will the pipe fitter be able to use standard pipe fittings (i.e., $22\frac{1}{2}°$, 30°, 45°, 60°, or 90°)?

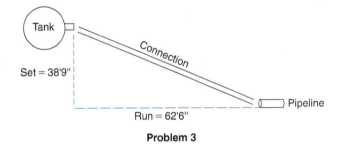

Problem 3

4. **Aviation** A helicopter, flying directly over a fishing boat at an altitude of 1200 ft, spots an ocean liner at a 15° angle of depression. How far from the boat is the liner?

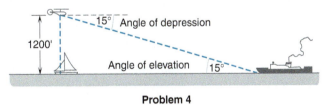

Problem 4

5. **Metalworking** Find the angle of taper on the steel bar shown in the figure if it is equal to twice m. (Round to the nearest 10'.)

6. **Construction** A road has a rise of 6.0 ft in 82 ft. (See the figure.) What is the gradient angle of the road?

Problem 5

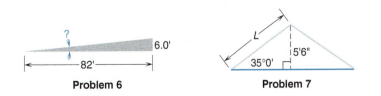

Problem 6

Problem 7

7. **Roofing** Find the length of the rafter shown. (Round to the nearest inch.)

8. **Metalworking** Three holes are drilled into a steel plate. Find the distances A and B as shown in the figure. (Round to the nearest sixteenth of an inch.)

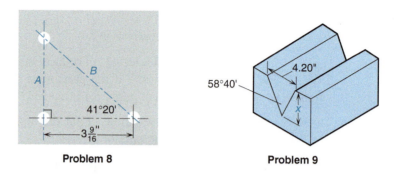

Problem 8   Problem 9

9. **Machine Trades** Find the depth of cut x needed for the V-slot shown. (Assume that the V-slot is symmetric.)

10. **General Trades** A 20-ft ladder leans against a building at a 65° angle with the ground. (See the figure.) Will it reach a window 17 ft above the ground?

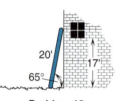

Problem 10

11. **Metalworking** Find the distance B in the special countersink shown in the figure. (Round to the nearest sixteenth of an inch.)

Problem 11   Problem 12

12. **Drafting** Find angle a in the figure. (Express your answer in decimal degrees.)

13. **Construction** A road has a slope of 2°25'. Find the rise in 6500 ft of horizontal run.

14. **Machine Trades** Find the missing dimension d in the $\frac{5}{16}$-in. flathead screw shown.

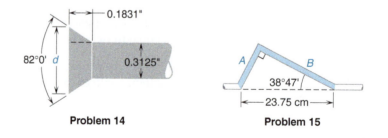

Problem 14   Problem 15

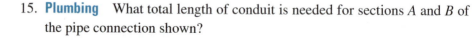

Problem 16

15. **Plumbing** What total length of conduit is needed for sections A and B of the pipe connection shown?

16. **Machine Trades** Find the included angle m of the taper shown. (Express your answer in degrees and minutes.)

17. **Machine Trades**   Ten holes are spaced equally around a flange on a 5.0-in.-diameter circle. Find the center-to-center distance x in the figure.

Problem 17
Problem 18

18. **Police Science**   Detectives investigating a crime find a bullet hole in a wall at a height of 7 ft 6 in. from the floor. The bullet passed through the wall at an angle of 34°. If they assume that the gun was fired from a height of 4 ft above the floor, how far away from the wall was the gun when it was fired? (Round to the nearest foot. See the figure.)

19. **Construction**   The Santa Teresa municipal building code specifies that public buildings must have a handicapped access ramp with an incline of no more than 10°. What length of ramp is needed if the ramp climbs 3 ft 8 in.? Express the answer in feet and inches to the nearest inch. (*Hint:* Work the problem in inches only, then convert the answer to feet and inches.)

20. **Construction**   A gable roof is to be constructed with a slope of 28.0° and a run of 27 ft 6 in. Calculate the rise of the roof. (Express the answer in feet and inches to the nearest inch.)

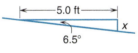

Problem 21

21. **Construction**   A crane is being set up on a slope of 6.5°. If the base of the crane is 5.0 ft wide, how many *inches* should the downhill side of the base be raised in order to level the crane? (See the figure. Round to the nearest tenth.)

22. **General Interest**   The dispersion angle of a speaker indicates the maximum range of undistorted sound for listeners directly in front of the speaker. A dispersion angle of 90° means that the listener can be within 45 degrees on either side of the speaker and still receive good sound quality. The two speakers being used for a particular concert have a dispersion angle of 120°, are perpendicular to the stage, and are 10 ft from the audience. (See the figure.) What is the maximum distance between the speakers so that no one in the first row of the audience is outside the dispersion angle?

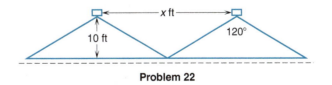

Problem 22

23. **Electrical Trades**   An electrician is installing 1-m-square solar panels on the roof of an apartment building. The roof is flat and measures 20 m wide in the north–south direction. As shown in the figure, the panels are installed at an angle of 55° and must be separated by 0.2 m in the north–south direction to avoid shadowing. With these specifications, how

many panels will fit along the 20-m side? (Round down to the next whole number.)

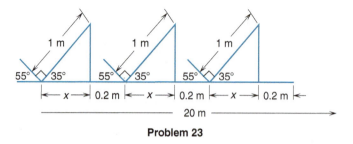

Problem 23

24. **Surveying**   A surveyor wants to estimate the width *EF* of the ravine shown in the figure without crossing over to the other side. He walks 55 ft perpendicular to *EF* to point *G* and sights point *E*. He then measures angle *EGF* to be 36°. Calculate the width *EF*.

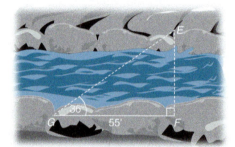

Problem 24

25. **Forestry**   A forest ranger needs to estimate the height of the tree shown in the figure. She sights the top of the tree through a clinometer, a device that gives her the angle of elevation as 68°. If she is standing 32 ft from the tree, and it is 5.5 ft from the ground to her eye level, how tall is the tree?

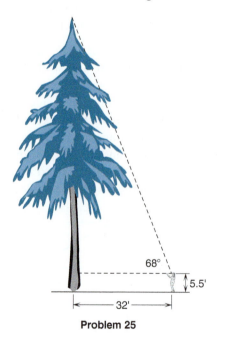

Problem 25

26. **Agriculture**   A grain conveyor is set at an angle of 30° with respect to the ground. The conveyor must reach the top of a storage bin at height of 28 ft. How long should the conveyor be?

Turn to the Appendix to check your answers to the odd-numbered problems. Then go to Section 10-4 to study oblique triangles.

|10-4| **Oblique Triangles**

learning|**catalytics**™

1. True or false: A triangle can have more than one obtuse angle.

2. Solve for x: $\dfrac{6}{x} = \dfrac{3}{8}$

We have already seen how the trigonometric functions can be used to solve right triangles. To solve for the missing parts of a triangle that does not contain a right angle, we need two new formulas, the *law of sines* and the *law of cosines*. In this section, we will derive these formulas and show how they may be used to solve triangles.

Oblique Triangles Any triangle that is not a right triangle is called an **oblique triangle**. There are two types of oblique triangles: acute and obtuse. In an **acute triangle** all three angles are acute—each is less than 90°. In an **obtuse triangle** one angle is obtuse—that is, greater than 90°. Notice in these figures that the angles have been labeled A, B, and C, and the sides opposite these angles have been labeled a, b, and c, respectively.

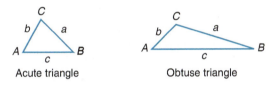

Acute triangle   Obtuse triangle

Law of Sines The law of sines states that the lengths of the sides of a triangle are proportional to the sines of the corresponding angles. To show this for an acute triangle, first construct a perpendicular line from C to side AB. Then in triangle ACD,

$$\sin A = \frac{h}{b} \quad \text{or} \quad h = b \sin A$$

and in triangle BCD,

$$\sin B = \frac{h}{a} \quad \text{or} \quad h = a \sin B$$

Then $b \sin A = a \sin B$ or $\dfrac{\sin A}{a} = \dfrac{\sin B}{b}$

By repeating this process with a perpendicular line from B to side AC, we can obtain a similar equation involving angles A and C and sides a and c. Combining these results gives the **law of sines**.

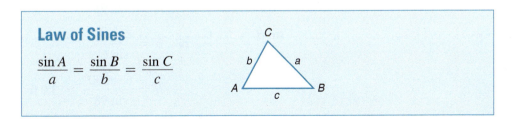

Law of Sines

$$\frac{\sin A}{a} = \frac{\sin B}{b} = \frac{\sin C}{c}$$

To *solve* a triangle means to calculate the values of all the unknown sides and angles from the information given. The law of sines enables us to solve any oblique triangle for two situations that we shall refer to as Case 1 and Case 2.

Case 1   Two angles and a side are known.

Case 2   Two sides and the angle opposite one of them are known.

Case 1 is simple and straightforward. First, find the third angle by subtracting the given angles from 180°. Then use the law of sines twice to find each of the two missing sides.

EXAMPLE 1   For the triangle shown, $A = 47°$, $B = 38°$, and $a = 8.0$ in. Find all its unknown parts, in this case angle C and sides b and c.

First, sketch the triangle as shown. Then, because the angles of a triangle sum to 180°, find angle C by subtracting.

$$C = 180° - A - B = 180° - 47° - 38° = 95°$$

Next, apply the law of sines. Substitute A, B, and a to find b.

$$\frac{\sin A}{a} = \frac{\sin B}{b} \quad \text{gives} \quad b = \frac{a \sin B}{\sin A}$$

$$b = \frac{8.0 \sin 38°}{\sin 47°} \approx 6.7 \text{ in.}$$

 Using a calculator, we have **8** ✕ sin **3 8**) ÷ sin **4 7** = →   **6.734486736**

Finally, apply the law of sines again, using A, C, and a to find c.

$$\frac{\sin A}{a} = \frac{\sin C}{c} \quad \text{gives} \quad c = \frac{a \sin C}{\sin A}$$

$$c = \frac{8.0 \sin 95°}{\sin 47°} \approx 11 \text{ in.}$$

In any triangle, the largest angle is opposite the longest side, and the smallest angle is opposite the shortest side. Always check to make certain that this is true for your solution.   ●

! Careful   Never use the Pythagorean theorem with an oblique triangle. It is valid only for the sides of a right triangle. ●

Your Turn   Solve the triangle for which $c = 6.0$ ft, $A = 52°$, and $C = 98°$.

Solution $B = 180° - 52° - 98° = 30°$

$$\frac{\sin A}{a} = \frac{\sin C}{c} \quad \text{or} \quad a = \frac{c \sin A}{\sin C}$$

$$a = \frac{6 \sin 52°}{\sin 98°} \approx 4.8 \text{ ft}$$

$$\frac{\sin B}{b} = \frac{\sin C}{c} \quad \text{or} \quad b = \frac{c \sin B}{\sin C}$$

$$b = \frac{6 \sin 30°}{\sin 98°} \approx 3.0 \text{ ft}$$

In Case 2, when two sides and the angle opposite one of these sides are known, there are three possible outcomes: There may be no solution, one solution, or two solutions to the triangle. Because of this, Case 2 is often called the "ambiguous case."

If the given angle is obtuse, there can only be one solution or no solution. Here is what the two possibilities look like:

- If $A > 90°$ and $a > b$, then the triangle looks like this ⟶

- If $A > 90°$ and $a \leq b$, then there will be no solution, like this

As the next example will illustrate, there is no need to memorize the inequalities associated with each possibility. Your calculator will automatically give you the correct solution.

EXAMPLE 2 (a) Suppose that $A = 109.0°$, $a = 14.8$ in., and $b = 19.5$ in. If we substitute these values into the law of sines,

$$\frac{\sin A}{a} = \frac{\sin B}{b} \quad \text{gives us} \quad \frac{\sin 109°}{14.8} = \frac{\sin B}{19.5}$$

Solving for $\sin B$, we have $\quad \sin B = \dfrac{19.5 \sin 109°}{14.8}$

or $\quad \sin B \approx 1.25$

But the value of the sine function can never be greater than 1, so there is no triangle corresponding to the given values. If you forget this fact about sine and press $\boxed{\text{2nd}}\ \boxed{\sin^{-1}}$ to solve for angle B, your calculator will give you an error message, verifying that there is no solution.

A Closer Look In the previous problem, you could have avoided using the law of sines entirely if you had noticed that the longer of the two given sides (side b) was not opposite the obtuse angle (angle A). As we learned in Chapter 8, the largest angle of a triangle must be opposite the longest side, and there can only be one obtuse angle in a triangle. ●

(b) Suppose that $A = 121.4°$, $a = 28.4$ cm, and $b = 11.2$ cm. Here the longer of the two given sides is opposite the obtuse angle, so there will be a solution. Substituting into the law of sines,

$$\frac{\sin A}{a} = \frac{\sin B}{b} \quad \text{gives us} \quad \frac{\sin 121.4°}{28.4} = \frac{\sin B}{11.2}$$

Solving for $\sin B$, we have $\quad \sin B = \dfrac{11.2 \sin 121.4°}{28.4}$

or $\quad \sin B = 0.3366\ldots$

Taking the inverse sine of this, $\quad B \approx 19.67° \approx 19.7°$

Next, find angle C: $\quad C = 180° - 121.4° - 19.7° \approx 38.9°$

Finally, use the law of sines again to find side c:

$$\frac{\sin A}{a} = \frac{\sin C}{c} \quad \text{gives us} \quad \frac{\sin 121.4°}{28.4} = \frac{\sin 38.9°}{c}$$

or $\quad c = \dfrac{28.4 \sin 38.9°}{\sin 121.4°} \approx 20.9$ cm

Here are the calculator steps for this problem:

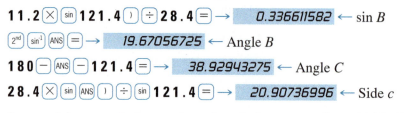

Summarizing our solution, we have the missing parts as $B \approx 19.7°$, $C \approx 38.9°$, and $c \approx 20.9$ cm. ●

If the given angle in a Case 2 problem is acute, then all three outcomes are possible: We could have no solution, one solution, or two solutions.

Let's consider the easiest situation first.

- If $A < 90°$ and $a = b \sin A$, then the triangle is a right triangle. You can solve it using the methods of Section 10-3. ⟶

Again, you do not need to memorize these requirements in order to solve the triangle. If you simply substitute the given information into the law of sines, your calculator will come up with the correct answer.

EXAMPLE 3   Given $A = 30.0°$, $a = 4.62$ ft, and $b = 9.24$ ft, then the law of sines gives us

$$\frac{\sin 30°}{4.62} = \frac{\sin B}{9.24}$$

Solving for $\sin B$, $\sin B = \dfrac{9.24 \sin 30°}{4.62} = 1$

The inverse sine of 1 is 90°, so this is a right triangle. The missing angle C is 60°, and the longer leg c can be found using our knowledge of 30°–60° right triangles. The longer leg is equal to the square root of 3 times the shorter leg, or in this case $4.62\sqrt{3}$. If you did not remember this fact, then you could alternatively use the law of sines again to find side c, and the calculation will give you $c \approx 8.00$, which is the decimal equivalent of $4.62\sqrt{3}$. ●

🔍 A Closer Look

If you had made a quick sketch of Example 3 (see the figure), you might have noticed that the given information made this a 30°–60° right triangle. The side adjacent to the 30° angle (the hypotenuse) is exactly twice as long as the side opposite this angle (the short leg). Making this observation will save you the trouble of using the law of sines. ●

The other straightforward situation when the given angle A is acute occurs when $a < b \sin A$, where a and b are the two given sides. This will result in no solution, although, as shown by the next example, your calculator will automatically tell you this.

EXAMPLE 4   Given $A = 30.0°$, $a = 4.62$ ft, and $b = 9.54$ ft, the law of sines gives us

$$\frac{\sin 30°}{4.62} = \frac{\sin B}{9.54}$$

Solving for sin B, $\sin B = \dfrac{9.54 \sin 30°}{4.62} = 1.032 \ldots$

But the value of sine cannot be greater than 1, so our calculator is telling us that there is no solution. If you fail to recognize that this is an impossible value of sine, and you take the inverse sine of this last answer, an error message will tell you that there is no solution.

Here is what this looks like:

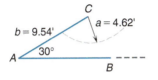

No matter how we draw side a, it will not intersect line AB. There is no triangle with these measurements.

●

The most difficult situation in Case 2 occurs when the given angle A is acute and $a > b \sin A$. Here we will have just one solution if $a \geq b$, but we will have two valid solutions if $a < b$. Here is what these two scenarios look like:

$$a > b \sin A \text{ and } a \geq b \qquad\qquad a > b \sin A \text{ and } a < b$$

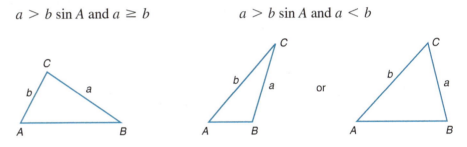

The following example illustrates how to distinguish between these two scenarios without memorizing the formulas.

EXAMPLE 5   Solve triangle ABC if $A = 30.0°$, $a = 4.62$ ft, and (a) $b = 5.48$ ft (b) $b = 3.75$ ft.

Again, we are given two sides and the angle opposite one of these sides. We must first recognize that this is Case 2, and we must keep in mind that the possibility of two solutions exists.

For part (a), we first substitute into the law of sines to get

$$\frac{\sin 30°}{4.62} = \frac{\sin B}{5.48}$$

Solving for sin B, we have $\sin B = \dfrac{5.48 \sin 30}{4.62} = 0.593 \ldots$

Taking the inverse sine of this, we get $B \approx 36.38° \approx 36.4°$

Now here is the key step in finding the second solution if it exists: Recall from page 689 that for every acute angle A there is an obtuse angle $(180° - A)$ that has the same sine value. You must check this obtuse angle to see if it will give you a valid triangle. In this case,

If $B_1 \approx 36.38°$, then $B_2 = 180 - B_1 \approx 180° - 36.38° \approx 143.62°$

To determine if this creates a valid second triangle, we add $B_2 \approx 143.62°$ to the original given angle $A = 30°$ to get $173.62°$. Because this is still less than $180°$, there does exist an angle C that creates a valid triangle. Notice that this is confirmed by the fact that $a > b \sin A$ and $a < b$. Now we must complete both solutions.

For $B_1 \approx 36.38°$, $C_1 \approx 180° - 36.38° - 30° \approx 113.62°$.

Now use the law of sines again to find side c_1:

$$\frac{\sin 30°}{4.62} = \frac{\sin 113.62°}{c_1}$$

Solving for c_1, $$c_1 = \frac{4.62 \sin 113.62°}{\sin 30°} \approx 8.47 \text{ ft}$$

For $B_2 \approx 143.62°$, $C_2 \approx 180° - 143.62° - 30° \approx 6.38°$.

Using the law of sines to find side c_2,

$$\frac{\sin 30°}{4.62} = \frac{\sin 6.38°}{c_2}$$

$$c_2 = \frac{4.62 \sin 6.38°}{\sin 30°} \approx 1.03 \text{ ft}$$

In summary, the two solutions, properly rounded, are:

$B_1 \approx 36.4°$, $C_1 \approx 113.6°$, $c_1 \approx 8.47$ ft, and $B_2 \approx 143.6°$, $C_2 \approx 6.4°$, $c_2 \approx 1.03$ ft

For part (b), we first substitute into the law of sines to get

$$\frac{\sin 30°}{4.62} = \frac{\sin B}{3.75}$$

Solving for $\sin B$, $$\sin B = \frac{3.75 \sin 30°}{4.62} = 0.4058\ldots$$

Taking the inverse sine of this, $B \approx 23.94° \approx 23.9°$

As in part (a), we must now test for the possibility of a second solution. The obtuse angle with the same sine value as $23.94°$ is $180° - 23.94°$ or $156.06°$. But if we add this to the original given angle A of $30°$, we get $186.06°$. Because this is already greater than $180°$, there is no valid second solution. Notice that this is confirmed by the fact that $a > b \sin A$ and $a \geq b$. Now we must complete the one valid solution.

First find angle C: $C = 180° - B - A \approx 180° - 23.94° - 30° \approx 126.06°$

Then use the law of sines to calculate side c:

$$\frac{\sin 30°}{4.62} = \frac{\sin 126.06°}{c}$$

$$c = \frac{4.62 \sin 126.06°}{\sin 30°} \approx 7.47 \text{ ft}$$

The one solution, properly rounded, is: $B \approx 23.9°$, $C \approx 126.1°$, $c \approx 7.47$ ft   ●

Your Turn   Use the law of sines to solve the following triangles:

	B	c	b
(a)	36.0°	16.4 cm	20.5 cm
(b)	36.0°	16.4 cm	15.1 cm
(c)	36.0°	16.4 cm	9.40 cm
(d)	30°	24.6 ft	12.3 ft
(e)	118.0°	8.73 in.	6.42 in.
(f)	125.0°	108 in.	129 in.

Solutions   First, note that these are all Case 2 problems: The given angle is opposite one of the given sides.

(a) One solution: $C \approx 28.0°, A \approx 116.0°, a \approx 31.4$ cm.

When you find angle C in Step 1, be sure to test for a second solution. You will discover that $C_2 \approx 180° - 28.0° \approx 152.0°$. But $B + C_2 \approx 36° + 152° \approx 188°$, so the second solution does not exist.

(b) Two solutions:

$C_1 \approx 39.7°, A_1 \approx 104.3°, a_1 \approx 24.9$ cm

$C_2 \approx 140.3°, A_2 \approx 3.7°, a_2 \approx 1.64$ cm

When you find angle C_1 in the first step, test for a possible second solution. You should find that $C_2 \approx 180° - 39.67° \approx 140.33°$. This does create a valid second triangle because $C_2 + B \approx 140.33° + 36° \approx 176.33°$, which is less than $180°$.

(c) No solution. Substituting into the law of sines and solving for angle C, you should get $\sin C = 1.025\ldots$, which is impossible. Pressing $\boxed{2^{nd}}$ $\boxed{\sin^{-1}}$ verifies this by displaying an error message.

(d) One solution—a right triangle: $C = 90°, A = 60°, a \approx 21.3$ ft. If you had recognized this as a $30°$–$60°$ right triangle to begin with, you could have avoided using the law of sines. The key clue is that the longer given side is twice as long as the side opposite the $30°$ angle.

(e) No solution. After substituting into the law of sines, your calculator should give you an error message when you solve for angle C. However, you need not use the law of sines if you begin with a quick sketch. You should see that the side opposite the obtuse angle is shorter than the other given side. As we noted earlier, this is impossible because the longest side of a triangle must be opposite the largest angle.

(f) One solution: $C \approx 43.3°, A \approx 11.7°, a \approx 31.9$ in. Because this is an obtuse triangle, there is no need to consider the possibility of a second solution.

Law of Cosines   In addition to the two cases already mentioned, it is also possible to solve any oblique triangle if the following information is given:

Case 3   Two sides and the angle included by them.

Case 4   All three sides.

But the law of sines is not sufficient for solving triangles given this information. We need an additional formula called the **law of cosines**.

To obtain this law, construct the perpendicular from C to side AB in the triangle shown. We can use the Pythagorean theorem. In triangle BCD,

$a^2 = (c - x)^2 + h^2$

$= c^2 - 2cx + x^2 + h^2$

and in triangle ACD,

$b^2 = x^2 + h^2$

Therefore,

$a^2 = c^2 - 2cx + b^2$

$= b^2 + c^2 - 2cx$

In triangle ACD, $\cos A = \dfrac{x}{b}$ or $x = b \cos A$; therefore,

$$a^2 = b^2 + c^2 - 2bc \cos A$$

By rotating the triangle ABC, we can derive similar equations for b^2 and c^2.

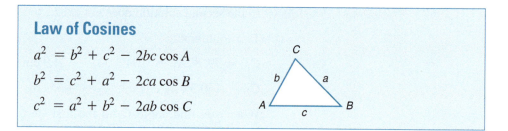

Law of Cosines

$$a^2 = b^2 + c^2 - 2bc \cos A$$
$$b^2 = c^2 + a^2 - 2ca \cos B$$
$$c^2 = a^2 + b^2 - 2ab \cos C$$

 **Learning Hint** Notice the similarity in these formulas. Rather than memorize all three formulas, you may find it easier to remember the following version of the law of cosines expressed in words:

The square of the length of any side of a triangle equals the sum of the squares of the lengths of the other two sides minus twice the product of the lengths of these two sides and the cosine of the angle between them. ●

The law of cosines may be used to solve any oblique triangle for which the information in Case 3 and Case 4 is given.

First, let's consider an example of Case 3, where two sides and the included angle are given.

EXAMPLE 6   Solve the triangle with $a = 22.8$ cm, $b = 12.3$ cm, and $C = 42.0°$.

First, make a sketch and notice that the given angle C is the angle formed by the given sides a and b. Use the form of the law of cosines that involves the given angle to find the third side c.

$$c^2 = a^2 + b^2 - 2ab \cos C$$
$$c^2 = 22.8^2 + 12.3^2 - 2(22.8)(12.3) \cos 42°$$

Using a calculator, we get

Therefore, $c \approx 15.9$ cm

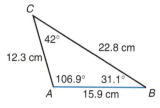

Next, because we now know all three sides and one angle, we can use the law of cosines to find one of the remaining angles. (Use $c \approx 15.95$ to calculate the first angle.)

$$a^2 = b^2 + c^2 - 2bc \cos A$$

or $\cos A = \dfrac{b^2 + c^2 - a^2}{2bc} \approx \dfrac{12.3^2 + 15.95^2 - 22.8^2}{2(12.3)(15.95)}$

Using a calculator gives us

Therefore, $A \approx 106.9°$

Finally, find the remaining angle by subtracting.

$$B \approx 180° - 106.9° - 42.0° \approx 31.1°$$

●

Your Turn Practice using the law of cosines by solving the triangle with $B = 34.4°$, $a = 145$ ft, and $c = 112$ ft.

Solution Use the law of cosines to find the third side b.

$$b^2 = a^2 + c^2 - 2ac \cos B$$

$$b^2 = 145^2 + 112^2 - 2(145)(112) \cos 34.4°$$

$$b \approx 82.28 \text{ ft} \approx 82.3 \text{ ft}$$

Now find angle A. $\cos A = \dfrac{b^2 + c^2 - a^2}{2bc} \approx \dfrac{82.28^2 + 112^2 - 145^2}{2(82.28)(112)}$

$$A \approx 95.3°$$

$$C \approx 180° - 34.4° - 95.3° \approx 50.3°$$

Now let's consider an example of Case 4, where all three sides are given.

EXAMPLE 7 Solve the triangle with sides $a = 106$ m, $b = 135$ m, and $c = 165$ m.

First, use the law of cosines to find the largest angle, the angle opposite the longest side. Side c is the longest side, so find C.

$$c^2 = a^2 + b^2 - 2ab \cos C$$

Solve for $\cos C$.

$$c^2 - a^2 - b^2 = -2ab \cos C$$

or

$$2ab \cos C = a^2 + b^2 - c^2$$

$$\cos C = \frac{a^2 + b^2 - c^2}{2ab} = \frac{106^2 + 135^2 - 165^2}{2(106)(135)}$$

$$\cos C = 0.078127\ldots$$

$$C \approx 85.52° \approx 85.5°$$

Next, use the law of sines to find one of the remaining angles. To find angle A,

$$\frac{\sin A}{a} = \frac{\sin C}{c} \quad \text{or} \quad \sin A = \frac{a \sin C}{c}$$

These are both known.

$$\sin A \approx \frac{106 \sin 85.52°}{165} \approx 0.6404\ldots$$

$$A \approx 39.8°$$

Finally, we can calculate the remaining angle by subtracting because the angles of a triangle must sum to 180°.

$$B \approx 180° - 85.5° - 39.8° \approx 54.7°$$

! Careful To avoid difficulties that can arise in the second step of the last example, you must find the largest angle first. Remember, the largest angle is always opposite the longest side. ●

Your Turn Solve the triangle with sides $a = 9.5$ in., $b = 4.2$ in., and $c = 6.4$ in.

Solution **First,** find the largest angle A using the law of cosines.

$$a^2 = b^2 + c^2 - 2bc \cos A$$

$$\cos A = \frac{b^2 + c^2 - a^2}{2bc} = \frac{4.2^2 + 6.4^2 - 9.5^2}{2(4.2)(6.4)}$$

$$A \approx 126.07° \approx 126°$$

Next, use the law of sines to find angle B.

$$\frac{\sin B}{b} = \frac{\sin A}{a} \qquad \text{gives} \qquad \sin B = \frac{b \sin A}{a} \approx \frac{4.2 \sin 126.07°}{9.5}$$

$$B \approx 21°$$

Finally, subtract to find angle C.

$$C \approx 180° - 126° - 21° \approx 33°$$

The following table provides a summary showing when to use each law in solving oblique triangles. In each situation, remember that when two of the angles are known, the third may be found by subtracting from 180°.

When Given ...	Use ...	To Find ...
Any two angles and a side	Law of sines	A second side opposite one of the given angles. You may need to find the third angle first.
	Law of sines	The third side, opposite the third angle.
Two sides and the angle opposite one of them	Law of sines	A second angle, opposite one of the given sides. This is the ambiguous case.
	Law of sines	The third side, after finding the third angle by subtraction.
Two sides and the angle included by them	Law of cosines	The third side.
	Law of cosines	Any one of the remaining angles. (Then find the third angle by subtraction.)
All three sides	Law of cosines	The largest angle (opposite the longest side).
	Law of sines	One of the remaining angles. (Then find the third angle by subtraction.)

Now continue with Exercises 10-4 for a set of problems on solving oblique triangles.

Exercises 10-4   Oblique Triangles

A. Solve each triangle.

    1. $a = 6.5$ ft, $A = 43°$, $B = 62°$

    2. $b = 17.2$ in., $C = 44.0°$, $B = 71.0°$

    3. $b = 165$ m, $B = 31.0°$, $C = 110.0°$

    4. $c = 2300$ yd, $C = 120°$, $B = 35°$

5. $a = 96.0$ in., $b = 58.0$ in., $B = 30.0°$

6. $b = 265$ ft, $c = 172$ ft, $C = 27.0°$

7. $a = 8.5$ m, $b = 6.2$ m, $C = 41°$

8. $b = 19.3$ m, $c = 28.7$ m, $A = 57.0°$

9. $a = 625$ ft, $c = 189$ ft, $B = 102.0°$

10. $b = 1150$ yd, $c = 3110$ yd, $A = 125.0°$

11. $a = 27.2$ in., $b = 33.4$ in., $c = 44.6$ in.

12. $a = 4.8$ cm, $b = 1.6$ cm, $c = 4.2$ cm

13. $a = 7.42$ m, $c = 5.96$ m, $B = 99.7°$

14. $b = 0.385$ in., $c = 0.612$ in., $A = 118.5°$

15. $a = 1.25$ cm, $b = 5.08$ cm, $c = 3.96$ cm

16. $a = 6.95$ ft, $b = 9.33$ ft, $c = 7.24$ ft

B. **Practical Problems**

1. **Landscaping**   To measure the height of a tree, Steve measures the angle of elevation of the treetop as 46°. He then moves 15 ft closer to the tree and from this new point measures the angle of elevation to be 59°. How tall is the tree? (See the figure.)

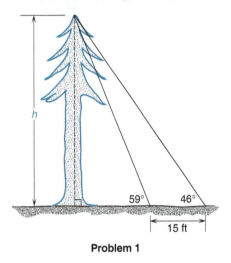

h

59°   46°

15 ft

Problem 1

Problem 2

2. **Construction**   In the channel shown in the figure, angle $B = 122.0°$, angle $A = 27.0°$, and side $BC = 32.0$ ft. Find the length of the slope AB.

3. **Construction**   A triangular traffic island has sides 21.5, 46.2, and 37.1 ft. What are the angles at the corners?

4. **Construction**   The lot shown in the figure is split along a diagonal as indicated. What length of fencing is needed for the boundary line?

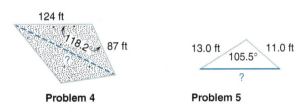

124 ft

118.2°   87 ft

?

13.0 ft   11.0 ft

105.5°

?

Problem 4   **Problem 5**

5. **Carpentry**   Two sides of the sloped ceiling shown in the figure meet at an angle of 105.5°. If the distances along the sides to the opposite walls are 11.0 and 13.0 ft, what length of beam is needed to join the walls?

6. **Machine Trades** For the crankshaft shown in the figure, $AB = 4.2$ in. and the connecting rod $AC = 12.5$ in. Calculate the size of angle A when the angle at C is $12°$.

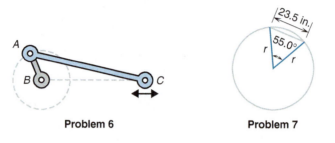

Problem 6   Problem 7

7. **Industrial Technology** If the chord of the circle shown in the figure is 23.5 in. long and subtends a central angle of 55.0°, what is the radius of the circle?

Turn to the Appendix to check your answers to the odd-numbered problems. Then go to Problem Set 10 on page 720 for a set of problems covering the work of this chapter. If you need a quick review of the topics in this chapter, visit the chapter Summary first.

CHAPTER 10
SUMMARY

Triangle Trigonometry

Objective	Review
Convert angles between decimal degrees and degrees and minutes. (p. 666)	Use the fact that $1° = 60'$ and set up unity fractions to convert. **Examples:** (a) $43.4° = 43° + 0.4°$ $= 43° + \left(0.4° \times \dfrac{60'}{1°}\right)$ $= 43°24'$ (b) $65°15' = 65° + 15'$ $= 65° + \left(15' \times \dfrac{1°}{60'}\right)$ $= 65.25°$
Convert angles between degree measure and radian measure. (p. 668)	To convert, Degrees to radians → multiply by $\dfrac{\pi}{180°}$ Radians to degrees → multiply by $\dfrac{180°}{\pi}$ **Examples:** (a) $46° = 46\left(\dfrac{\pi}{180°}\right) \approx 0.80$ radian (b) $2.5 \, rad = 2.5\left(\dfrac{180°}{\pi}\right) \approx 143.2°$
Calculate the arc length and area of a sector. (p. 669)	Use the following formulas: Arc length $S = ra$ Area of a sector $A = \dfrac{1}{2}r^2a$ where r is the radius and a is the measure of the central angle in radians.

Objective	Review

Example: Calculate the arc length and area of the circular sector shown.

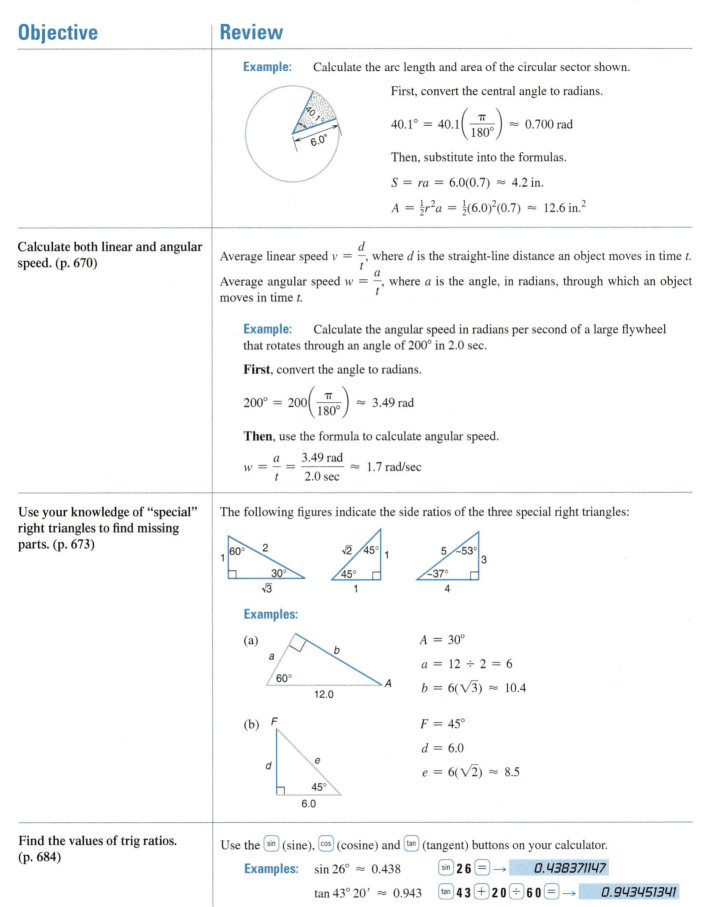

First, convert the central angle to radians.

$$40.1° = 40.1\left(\frac{\pi}{180°}\right) \approx 0.700 \text{ rad}$$

Then, substitute into the formulas.

$$S = ra = 6.0(0.7) \approx 4.2 \text{ in.}$$

$$A = \tfrac{1}{2}r^2a = \tfrac{1}{2}(6.0)^2(0.7) \approx 12.6 \text{ in.}^2$$

Calculate both linear and angular speed. (p. 670)

Average linear speed $v = \dfrac{d}{t}$, where d is the straight-line distance an object moves in time t.

Average angular speed $w = \dfrac{a}{t}$, where a is the angle, in radians, through which an object moves in time t.

Example: Calculate the angular speed in radians per second of a large flywheel that rotates through an angle of 200° in 2.0 sec.

First, convert the angle to radians.

$$200° = 200\left(\frac{\pi}{180°}\right) \approx 3.49 \text{ rad}$$

Then, use the formula to calculate angular speed.

$$w = \frac{a}{t} = \frac{3.49 \text{ rad}}{2.0 \text{ sec}} \approx 1.7 \text{ rad/sec}$$

Use your knowledge of "special" right triangles to find missing parts. (p. 673)

The following figures indicate the side ratios of the three special right triangles:

Examples:

(a)

$A = 30°$

$a = 12 \div 2 = 6$

$b = 6(\sqrt{3}) \approx 10.4$

(b)

$F = 45°$

$d = 6.0$

$e = 6(\sqrt{2}) \approx 8.5$

Find the values of trig ratios. (p. 684)

Use the ⟨sin⟩ (sine), ⟨cos⟩ (cosine) and ⟨tan⟩ (tangent) buttons on your calculator.

Examples: $\sin 26° \approx 0.438$ ⟨sin⟩ 2 6 ⟨=⟩ → 0.438371147

$\tan 43° 20' \approx 0.943$ ⟨tan⟩ 4 3 ⟨+⟩ 2 0 ⟨÷⟩ 6 0 ⟨=⟩ → 0.943451341

Objective	Review

Find the angle when given the value of a trig ratio. (p. 689)

Use the $\boxed{2^{nd}}$ key on your calculator in combination with the $\boxed{\sin}$, $\boxed{\cos}$, and $\boxed{\tan}$ buttons to find $\sin^{-1}$, $\cos^{-1}$, and $\tan^{-1}$.

Example: If $\cos x = 0.549$

then $x = \cos^{-1} 0.549 \approx 56.7°$ or $56° \, 42'$

$\boxed{2^{nd}} \boxed{\cos^{-1}} . 5 \, 4 \, 9 \boxed{=} \rightarrow \boxed{56.70156419}$

Solve problems involving right triangles. (p. 686)

Use the following definitions to set up equations used to determine the missing quantities. For a right triangle,

$$\sin A = \frac{\text{side opposite angle } A}{\text{hypotenuse}} = \frac{a}{c} \qquad \sin B = \frac{\text{side opposite angle } B}{\text{hypotenuse}} = \frac{b}{c}$$

$$\cos A = \frac{\text{side adjacent to angle } A}{\text{hypotenuse}} = \frac{b}{c} \qquad \cos B = \frac{\text{side adjacent to angle } B}{\text{hypotenuse}} = \frac{a}{c}$$

$$\tan A = \frac{\text{side opposite angle } A}{\text{side adjacent to angle } A} = \frac{a}{b} \qquad \tan B = \frac{\text{side opposite angle } B}{\text{side adjacent to angle } B} = \frac{b}{a}$$

Examples:

(a) To find x, $\sin 38° = \dfrac{x}{17}$ and $x = 17 \sin 38° \approx 10$

To find y, $\cos 38° = \dfrac{y}{17}$ and $y = 17 \cos 38° \approx 13$

(b) In a pipe-fitting job, what is the run if the offset is $22\frac{1}{2}°$ and the length of set is $16\frac{1}{2}$ in.?

$$\tan 22\frac{1}{2}° = \frac{\text{set}}{\text{run}} = \frac{16\frac{1}{2}}{\text{run}}$$

$$\text{run} = \frac{16\frac{1}{2}}{\tan 22\frac{1}{2}°}$$

$$\approx 39.8 \text{ in.}$$

Solve oblique triangles. (p. 705)

When given two angles and a side (Case 1), or two sides and the angle opposite one of them (Case 2), use the law of sines to solve the triangle. Remember that Case 2 is the ambiguous case, and you must check for the possibility of two solutions.

$$\frac{\sin A}{a} = \frac{\sin B}{b} = \frac{\sin C}{c}$$

Example: Solve the triangle where

$A = 68°$, $B = 79°$, and $b = 8.0$ ft.

Subtract from $180°$ to find the missing angle.

$C = 180° - 68° - 79° = 33°$

Objective

Review

Use the law of sines to find both side a and side c.

$$\frac{\sin 68°}{a} = \frac{\sin 79°}{8.0}$$

$$a = \frac{8.0\,(\sin 68°)}{\sin 79°} \approx 7.6\ \text{ft}$$

$$\frac{\sin 33°}{c} = \frac{\sin 79°}{8.0}$$

$$c = \frac{8.0\,(\sin 33°)}{\sin 79°} \approx 4.4\ \text{ft}$$

When given three sides, or two sides and the included angle, use the law of cosines first.

$$c^2 = a^2 + b^2 - 2ab\cos C \quad \text{or} \quad \cos C = \frac{a^2 + b^2 - c^2}{2ab}$$

Example: (a) Solve the triangle where

$a = 6.5$ in., $b = 6.0$ in., and $c = 3.5$ in.

Use the second formula to first find the largest angle.

$$\cos A = \frac{6.0^2 + 3.5^2 - 6.5^2}{2(6.0)(3.5)} \approx 0.143$$

$$A = \cos^{-1} 0.143 \approx 81.78° \approx 82°$$

Then use the law of sines to find one of the remaining angles:

$$\frac{\sin C}{3.5} = \frac{\sin 81.78°}{6.5}$$

$$\sin C = \frac{3.5\sin 81.78°}{6.5}$$

$$C \approx 32.2° \approx 32°$$

Find the third angle by subtracting the other two from $180°$.

$$B \approx 180° - 32° - 82° \approx 66°$$

(b) Solve the triangle where

$a = 12.4$ cm, $b = 15.6$ cm, and $C = 58.2°$

Use the law of cosines to find the third side.

$$c^2 = 12.4^2 + 15.6^2 - 2(12.4)(15.6)\cos 58.2°$$

$$c^2 = 193.251\ldots$$

$$c \approx 13.90\ \text{cm} \approx 13.9\ \text{cm}$$

Then use the law of cosines again to find one of the remaining angles.

$$\cos B = \frac{a^2 + c^2 - b^2}{2ac} = \frac{12.4^2 + 13.9^2 - 15.6^2}{2(12.4)(13.9)} = 0.30056\ldots$$

$$B \approx 72.5°$$

Find the third angle by subtracting.

$$A = 180° - 58.2° - 72.5° \approx 49.3°$$

PROBLEM SET 10 Triangle Trigonometry

Answers to odd-numbered problems are given in the Appendix.

A. Convert the following angles as indicated.

Express in degrees and minutes. (Round to the nearest minute.)

    1. 87.8°   2. 39.3°   3. 51.78°   4. 16.23°

Express in decimal degrees. (Round to the nearest tenth.)

    5. 41°12′   6. 76°24′   7. 65°51′   8. 32°7′

Express in radians. (Round to three decimal places.)

    9. 35°   10. 21.4°   11. 74°30′   12. 112.2°

Express in decimal degrees. (Round to the nearest tenth.)

    13. 0.45 rad   14. 1.7 rad   15. 2.3 rad   16. 0.84 rad

B. Calculate the arc length *S* and area *A* for each of the following sectors.

(Round to nearest whole number if necessary.)

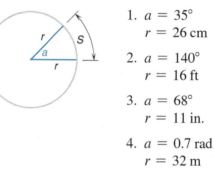

1. $a = 35°$
   $r = 26$ cm

2. $a = 140°$
   $r = 16$ ft

3. $a = 68°$
   $r = 11$ in.

4. $a = 0.7$ rad
   $r = 32$ m

C. Use a calculator to find the following trig values and angles.

Find each trig value to three decimal places.

    1. cos 67°   2. tan 81°   3. sin 4°   4. cos 63°10′

    5. tan 35.75°   6. sin 29.2°   7. sin 107°   8. cos 123°

Find the acute value of *A* to the nearest minute.

    9. $\sin A = 0.242$ 10. $\tan A = 1.54$ 11. $\sin A = 0.927$

    12. $\cos A = 0.309$ 13. $\tan A = 0.194$ 14. $\cos A = 0.549$

Find the acute value of *A* to the nearest tenth of a degree.

    15. $\tan A = 0.506$ 16. $\cos A = 0.723$ 17. $\sin A = 0.488$

    18. $\sin A = 0.154$ 19. $\cos A = 0.273$ 20. $\tan A = 2.041$

Name

_____

Date

_____

Course/Section

_____

D. Find the missing dimensions in the right triangles as indicated.

Use the special right triangle relationships to solve problems 1–4. (Round all lengths to the nearest tenth and all angles to the nearest degree.)

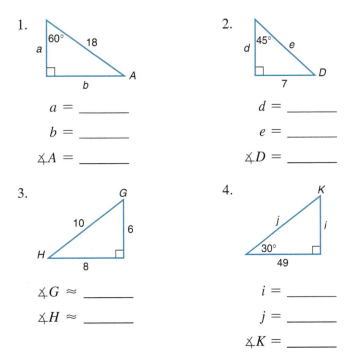

1.

a = _____

b = _____

∡A = _____

2.

d = _____

e = _____

∡D = _____

3.

∡G ≈ _____

∡H ≈ _____

4.

i = _____

j = _____

∡K = _____

For each problem, use the given information to solve right triangle ABC for all missing parts. Express angles in decimal degrees.

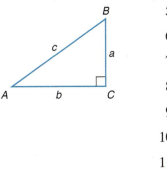

5. $a = 18$ mm, $A = 29°$

6. $B = 51°$, $c = 26$ ft

7. $b = 5.25$ m, $c = 7.35$ m

8. $a = 652$ yd, $B = 39.5°$

9. $a = 3\frac{1}{4}$ in., $b = 4\frac{1}{2}$ in.

10. $B = 65.5°$, $a = 32.4$ cm

11. $a = 12.0$ in., $c = 15.5$ in.

12. $A = 47°$, $c = 16$ ft

E. Solve the following oblique triangles.

1. $a = 17.9$ in., $A = 65.0°$, $B = 39.0°$

2. $a = 721$ ft, $b = 444$ ft, $c = 293$ ft

3. $a = 260$ yd, $c = 340$ yd, $A = 37°$

4. $b = 87.5$ in., $c = 23.4$ in., $A = 118.5°$

5. $a = 51.4$ m, $b = 43.1$ m, $A = 64.3°$

6. $a = 166$ ft, $c = 259$ ft, $B = 47.0°$

7. $a = 1160$ m, $c = 2470$ m, $C = 116.2°$

8. $a = 7.6$ in., $b = 4.8$ in., $B = 30°$

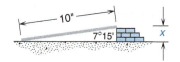

Problem 1

F.   Practical Applications

1. **Machine Trades**   What height of gauge blocks is required to set an angle of 7°15′ for 10-in. plots? (Round to the nearest tenth. See the figure.)

2. **Aviation**   The destination of an airplane is due north, and the plane flies at 220 mph. There is a crosswind of 20.0 mph from the west. (See the figure.) What heading angle *a* should the plane take? What is its relative ground speed *v*? (Round *a* to the nearest minute and *v* to the nearest whole number.)

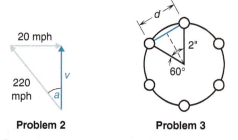

Problem 2   **Problem 3**

3. **Carpentry**   Six holes are spaced evenly around a circular piece of wood with a 4-in. diameter. (See the figure.) Find the distance, *d*, between the holes. (*Hint:* The dashed line creates two identical right triangles.)

4. **Machine Trades**   Find the angle of taper of the shaft shown in the figure. (Round to the nearest minute.)

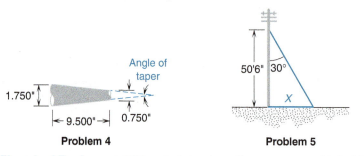

Problem 4   **Problem 5**

5. **Electrical Trades**   A TV technician installs an antenna 50 ft 6 in. tall on a flat roof. Safety regulations require a minimum angle of 30° between mast and guy wires. (See the figure.)
   (a) Find the minimum value of *X* to the nearest inch.
   (b) Find the length of the guy wires to the nearest inch.

6. **Construction**   A bridge approach shown in the figure is 12 ft high. The maximum slope allowed is 15%—that is, $\frac{15}{100}$.
   (a) What is the length *L* of the approach?
   (b) What is the angle *a* of the approach?

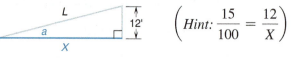

$$\left(Hint: \frac{15}{100} = \frac{12}{X} \right)$$

Problem 6

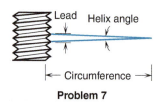

Problem 7

7. **Machine Trades**   The *helix angle* of a screw is the angle at which the thread is cut. The *lead* is the distance advanced by one turn of the screw. The circumference refers to the circumference of the screw as given by $C = \pi d$. (See the figure.) The following formula applies:

$$\text{Tangent of helix angle} = \frac{\text{lead}}{\text{circumference}}$$

   (a) Find the helix angle for a 2-in.-diameter screw if the lead is $\frac{1}{8}$ in. (Round to the nearest minute.)
   (b) Find the lead of a 3-in.-diameter screw if the helix angle is 2°0′.

8. **Construction** Find the lengths x and y of the beams shown in the bridge truss. (Round to the nearest tenth.)

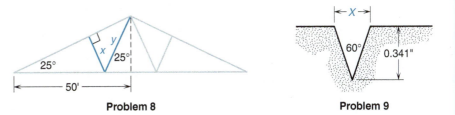

Problem 8   Problem 9

9. **Machine Trades** Find the width X of the V-thread shown. (Round to the nearest thousandth.)

10. **Machine Trades** Find the head angle a of the screw shown.

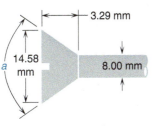

Problem 10

11. **Machine Trades** A machinist makes a cut 13.8 cm long in a piece of metal. Then another cut is made that is 18.6 cm long and at an angle of $62°0'$ with the first cut. How long a cut must be made to join the two endpoints?

12. **Drafting** Determine the center-to-center measurement x in the adjustment bracket shown in the figure.

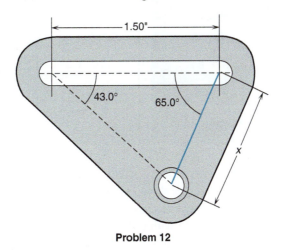

Problem 12

13. **Machine Trades** If a lathe makes 286 revolutions in 10 sec, calculate its average angular speed. (Round to the nearest whole number.)

14. **Automotive Trades** The wheels on a car have a radius of 12 in. and make 1 revolution every 0.12 sec. What is the linear speed of the tire in inches per second? (Round to the nearest ten.)

15. **Construction** An amphitheater is in the shape of a circular sector with a smaller sector removed. (See the figure.) Find the area of the amphitheater shown as the shaded portion of the figure. (Round to the nearest thousand square feet.)

Problem 15

16. **Construction** The frontage of a lot at the end of a cul-de-sac is defined by an arc with central angle of $42°30'$ and radius of 55.0 ft. Find the frontage distance in linear feet.

17. **Agriculture** Barley has an angle of repose R of 28 degrees. The following formula gives the height H of a pile of barley:

$$H = \tfrac{1}{2}d \times \tan R$$

where d is the diameter of the pile.

What is the height of a pile of barley with a diameter of 28 inches? (Round to the nearest tenth.)

CASE STUDY: Ordering Wood for Roof Trusses

Pitch-Perfect Roofers have been subcontracted to install the roof on a bungalow-style home. David has been put in charge of ordering the wood for the trusses. Trusses are the triangular structures that support the roof. The blueprint shows certain horizontal distances on the home and the pitch that is to be used on each part of the roof. Recall that pitch is defined as rise:run.

1. To determine the length of the rooflines and the angles of his cuts, David first needs to calculate the angles for each of the pitches he will use. Calculate the angles, to the nearest hundredth of a degree, for the pitches of 8:12, 6:12, and 2:12, which will be used for various parts of the roof.

2. For the house roof and dormer roof, Pitch-Perfect decides to use a king-post truss configuration as shown in the figure. Because the porch roof slopes in only one direction, its truss will be in the shape of a right triangle with only one rafter, one strut, and one-half of a tie beam. (Think of it as the right half of the figure shown.). Using the pitch angles from Step 1, David must now calculate (a) the rise of the roof (the king-post length), (b) the rafter length, and (c) the strut length for each of the three sections of the roof. Calculate these three lengths for each section and write your answers in feet and inches to the nearest $\frac{1}{8}$ inch.

3. David next must determine the number of trusses needed for each section of the roof. The trusses will be 24 in. apart. As shown in the blueprint, the porch roof and house roof have lengths of 24 ft 6 in., and the dormer has a length of 5 ft 9 in. Determine the number of trusses that will be needed for (a) the porch roof, (b) the house roof, and (c) the dormer. (*Hint:* There must be trusses at both ends of each section.)

4. Next, David needs to determine the amount of wood needed for each truss based on the lengths of the king post, the rafters, the struts, and the tie beams (see question 2). Determine the amount of wood needed per truss for (a) the porch roof, (b) the house roof, and (c) the dormer. (Give your answers in feet rounded to four decimal places.)

5. Now calculate the total amount of wood David needs for all parts of the trusses for (a) the porch roof, (b) the house roof, and (c) the dormer. (d) Finally, calculate the grand total for all three sections rounded up to the next foot.

6. To account for waste, David decides to order 10% more than the calculated total. How much wood should he order?

PREVIEW

11 Intermediate Algebra

Objective	Sample problems		For help, go to page
When you finish this chapter, you will be able to:			
1. Solve a system of two linear equations in two variables.	(a) $5x + 2y = 20$ $3x - 2y = 12$	$x =$ ________ $y =$ ________	728
	(b) $x = 3y$ $2y - 4x = 30$	$x =$ ________ $y =$ ________	
	(c) $3x - 5y = 7$ $6x - 10y = 14$	$x =$ ________ $y =$ ________	
	(d) $2y = 5x - 7$ $10x - 4y = 9$	$x =$ ________ $y =$ ________	
2. Solve word problems involving systems of equations in two variables.	(a) The sum of two numbers is 35. Twice the smaller decreased by the larger is 10. Find them.	________	743
	(b) **HVAC** The perimeter of a rectangular vent is 28 in. The length of the vent is 2 in. greater than its width. Find the dimensions.	________	

Name ________

Date ________

Course/Section ________

Objective	**Sample problems**	**For help, go to page**

(c) **Carpentry**   A mixture of 800 specialty nails costs $42. If the mixture consists of one type costing 4 cents each and another type costing 6 cents each, how many of each kind are there?   _____

3. Solve quadratic equations.

(a) $x^2 = 16$ $x =$ _____   751

(b) $x^2 - 7x = 0$ $x =$ _____

(c) $x^2 - 5x = 14$ $x =$ _____

(d) $3x^2 + 2x - 16 = 0$ $x =$ _____

(e) $2x^2 + 3x + 11 = 0$ $x =$ _____

4. Solve word problems involving quadratic equations.

(a) **General Trades**   Find the side length of a square opening whose area is 36 cm^2.   _____   756

(b) **Manufacturing**   Find the radius of a circular pipe whose cross-sectional area is 220 sq in.   _____

(c) **Electrical Trades**   In the formula $P = RI^2$, find I (in amperes) if $P = 2500$ watts and $R = 15$ ohms.   _____

(d) **HVAC**   The length of a rectangular vent is 3 in. longer than its width. Find the dimensions if the cross-sectional area is 61.75 sq in.   _____

(Answers to these preview problems are given in the Appendix. Also, worked solutions to many of these problems appear in the chapter Summary.)

If you are certain that you can work *all* these problems correctly, turn to page 766 for a set of practice problems. If you cannot work one or more of the preview problems, turn to the page indicated to the right of the problem. Those who wish to master this material with the greatest success should turn to Section 11-1 and begin work there.

In this chapter, we will cover two topics of intermediate algebra: solving systems of equations and solving quadratic equations. Although these concepts are used mostly in technical occupations, they can be powerful tools in many trades applications as well.

CASE STUDY: Fighting Fires in Yellowstone National Park

In the Case Study at the end of this chapter, you will learn how fires are fought in wildland areas such as national parks. As you work through the Case Study, you will see how the algebraic skills covered in this chapter could be used to solve a variety of problems dealing with fire hoses, pumps, aerial drops, and forest recovery.

11-1 Systems of Equations

Before you can solve a system of two equations in two unknowns, you must be able to solve a single linear equation in one unknown. Let's review what you learned in Chapter 7 about solving linear equations.

Solution A **solution** to an equation, such as $3x - 4 = 11$, is a number that we may use to replace x in order to make the equation a true statement. To find such a number, we *solve* the equation by changing it to an *equivalent* equation with only x on one side of the equation. The process is explained in Example 1.

EXAMPLE 1 **Solve:** $3x - 4 = 11$

Step 1 Add 4 to both sides of the equation

$$3x - \underbrace{4 + 4}_{0} = \underbrace{11 + 4}_{15}$$

$$3x = 15$$

Step 2 Divide both sides of the equation by 3.

$$\frac{\cancel{3}x}{\cancel{3}} = \frac{15}{3}$$

$$x = 5$$ This is the solution, and you can check it by substituting 5 for x in the original equation.

✔ $3(5) - 4 = 11$

$15 - 4 = 11$ which is true.   ●

Your Turn Solve each of the following equations and check your answer.

(a) $\dfrac{3x}{2} = 9$ (b) $17 - x = 12$

(c) $2x + 7 = 3$ (d) $3(2x + 5) = 4x + 17$

Answers (a) $x = 6$ (b) $x = 5$ (c) $x = -2$ (d) $x = 1$

A **system of equations** is a set of equations in two or more variables that may have a common solution.

EXAMPLE 2 The system of equations

$2x + y = 11$

$4y - x = 8$

has the common solution $x = 4$, $y = 3$. This pair of numbers will make each equation a true statement. If we substitute 4 for x and 3 for y, the left side of the first equation becomes

$2(4) + 3 = 8 + 3$ or 11, which equals the right side,

and the left side of the second equation becomes

$4(3) - 4 = 12 - 4$ or 8, which equals the right side.

The pair of values $x = 4$ and $y = 3$ satisfies *both* equations. This solution is often written $(4, 3)$, where the x-value is listed first and the y-value is listed second.   ●

Note   Each of these equations by itself has an infinite number of solutions. For example, the pairs $(0, 11)$, $(1, 9)$, $(2, 7)$, and so on, all satisfy $2x + y = 11$, and the pairs $(0, 2)$, $(1, \frac{9}{4})$, $(2, \frac{5}{2})$, and so on, all satisfy $4y - x = 8$. However, $(4, 3)$ is the only pair that satisfies *both* equations. ●

Your Turn   By substituting, show that the numbers $x = 2$, $y = -5$ give the solution to the system of equations

$$5x - y = 15$$

$$x + 2y = -8$$

Solution   The first equation is

$$5(2) - (-5) = 15$$

$$10 + 5 = 15, \text{ which is correct.}$$

The second equation is

$$(2) + 2(-5) = -8$$

$$2 - 10 = -8, \text{ which is also correct.}$$

Therefore, $(2, -5)$ is a solution to the original system of two equations.

Solution by Substitution   In this chapter, we will show you two different methods of solving a system of two linear equations in two unknowns. The first method is called **solution by substitution**.

EXAMPLE 3   To solve the system of equations

$$y = 3 - x$$

$$3x + y = 11$$

follow these steps:

Step 1   *Solve* the first equation for x or y.

The first equation is already solved for y:

$$y = 3 - x$$

Step 2   *Substitute* this expression for y in the second equation.

$$3x + y = 11 \quad \text{becomes}$$

$$3x + \boxed{(3 - x)} = 11$$

Substitute this for y.

Step 3   *Solve* the resulting equation.

$$3x + (3 - x) = 11$$

becomes $2x + 3 = 11$

Subtract 3 from each side. $2x = 8$

Divide each side by 2. $x = 4$

Step 4   *Substitute* this value of x into the first equation and find a value for y.

$$y = 3 - x$$

$$y = 3 - (4)$$

$$y = -1$$

The solution is $x = 4, y = -1$ or $(4, -1)$.

x-value

y-value

☑ **Step 5**   *Check* your solution by substituting it back into the second equation. (We use the second equation for the check because we used the first equation in the solution.)

$$3x + y = 11 \quad \text{becomes}$$

$$3(4) + (-1) = 11$$

$$12 - 1 = 11 \quad \text{which is correct.}$$

Your Turn   Use this substitution procedure to solve the system of equations

$$x - 2y = 3$$
$$2x - 3y = 7$$

Solution   **Step 1**   *Solve* the first equation for x by adding $2y$ to both sides of the equation.

$$x - 2y + 2y = 3 + 2y$$
$$x = 3 + 2y$$

Step 2   *Substitute* this expression for x in the second equation.

$$2x - 3y = 7 \quad \text{becomes}$$

$$2(3 + 2y) - 3y = 7$$

Step 3   *Solve:*

Multiply to remove parentheses. $6 + 4y - 3y = 7$

Simplify by combining the y-terms. $6 + y = 7$

Subtract 6 from each side. $y = 1$

Step 4   *Substitute* this value of y in the first equation to find x.

$$x - 2y = 3$$

becomes $x - 2(1) = 3$

or $x - 2 = 3$

Add 2 to each side. $x = 5$

The solution is $x = 5, y = 1$ or $(5, 1)$.

☑ **Step 5**   *Check* the solution in the second equation.

$$2x - 3y = 7 \quad \text{becomes}$$

$$2(5) - 3(1) = 7$$

$$10 - 3 = 7 \quad \text{which is correct.}$$

Of course, it does not matter which variable, x or y, we solve for in Step 1, or which equation we use in Step 4.

EXAMPLE 4   In the system of equations

$$2x + 3y = 22$$

$$x - y = 1$$

the simplest procedure is to solve the *second* equation for x to get $x = 1 + y$ and substitute this expression for x into the first equation.

The entire solution looks like this:

Step 1   In the second equation, add y to both sides to get $x = 1 + y$.

Step 2   When we substitute into the first equation,

$$2x + 3y = 22 \quad \text{becomes}$$

$$2(1 + y) + 3y = 22$$

Step 3   *Solve:*

Remove parentheses.	$2 + 2y + 3y = 22$
Combine terms.	$2 + 5y = 22$
Subtract 2 from each side.	$5y = 20$
Divide each side by 5.	$y = 4$

Step 4   *Substitute* 4 for y in the second equation.

$$x - y = 1 \quad \text{becomes}$$

$$x - (4) = 1 \quad \text{or}$$

$$x = 5 \quad \text{The solution is } x = 5, y = 4 \text{ or } (5, 4).$$

✓ **Step 5**   *Check* the solution by substituting these values into the first equation.

$$2x + 3y = 22 \quad \text{becomes}$$

$$2(5) + 3(4) = 22$$

$$10 + 12 = 22 \quad \text{which is correct.} \qquad \bullet$$

Your Turn   Solve the following systems of equations by using the substitution method.

(a) $x = 1 + y$ $\qquad$ (b) $3x + y = 1$ $\qquad$ (c) $x - 3y = 4$
$\quad\ \ 2y + x = 7$ $\qquad\qquad\ \ y + 5x = 9$ $\qquad\qquad\ \ 3y + 2x = -1$

(d) $y + 2x = 1$ $\qquad$ (e) $x - y = 2$ $\qquad$ (f) $y = 4x$
$\quad\ \ 3y + 5x = 1$ $\qquad\qquad\ \ y + x = 1$ $\qquad\qquad\ \ 2y - 6x = 0$

Solutions   (a) **Step 1**   The first equation is already solved for x.

Step 2   Substitute this expression for x into the second equation.

$$2y + (1 + y) = 7$$

Step 3   Solve for y.

$$3y + 1 = 7$$

$$3y = 6$$

$$y = 2$$

Step 4   Substitute 2 for y in the first equation.

$$x = 1 + (2)$$

$$\text{or} \quad x = 3$$

Step 5 The solution is $x = 3$, $y = 2$ or $(3, 2)$.
Be sure to check it.

(b) **Step 1** You can easily solve either equation for y. Using the first equation, subtract $3x$ from both sides to obtain

$$y = 1 - 3x$$

Step 2 Substitute this expression for y into the second equation.

$$(1 - 3x) + 5x = 9$$

Step 3 Solve for x.

$$1 + 2x = 9$$
$$2x = 8$$
$$x = 4$$

Step 4 Substitute 4 for x in the equation from Step 1.

$$y = 1 - 3(4)$$
$$y = 1 - 12 = -11$$

Step 5 The solution is $(4, -11)$. Check your answer.

(c) **Step 1** Use the first equation to solve for x.

$$x = 3y + 4$$

Step 2 Substitute this expression for x into the second equation.

$$3y + 2(3y + 4) = -1$$

Step 3 Solve for y.

$$3y + 6y + 8 = -1$$
$$9y + 8 = -1$$
$$9y = -9$$
$$y = -1$$

Step 4 Substitute -1 for y in the equation from Step 1.

$$x = 3(-1) + 4$$
$$x = 1$$

Step 5 The solution is $x = 1$, $y = -1$.

(d) $x = 2$, $y = -3$ (e) $x = 1\frac{1}{2}$, $y = -\frac{1}{2}$ (f) $x = 0$, $y = 0$

Dependent and Inconsistent Systems   So far we have looked only at systems of equations with a single solution—one pair of numbers. Such a system of equations is called a **consistent** and **independent** system. However, it is possible for a system of equations to have no solution at all or to have an infinite number of solutions.

EXAMPLE 5   The system of equations

$$y + 3x = 5$$
$$2y + 6x = 10$$

has *no unique* solution. If we solve for y in the first equation

$$y = 5 - 3x$$

and substitute this expression into the second equation

$$2y + 6x = 10$$

or $\quad 2(5 - 3x) + 6x = 10$

This resulting equation simplifies to

$$10 = 10$$

There is no way of solving to get a unique value of x or y because both variables have dropped out of the equation. ●

A system of equations that does not have a single unique number-pair solution but has an unlimited number of solutions is said to be **dependent**. The two equations are essentially the same. For the system shown in Example 5, the second equation is simply twice the first equation. There are infinitely many pairs of numbers that will satisfy this system of equations. For example,

$x = 0, y = 5$

$x = 1, y = 2$

$x = 2, y = -1$

$x = 3, y = -4$

and so on.

If our efforts to solve a system produce a *true* statement containing *no* variables, then it is a dependent system with an infinite number of solutions. As we shall see in the next example, if a system of equations is such that our efforts to solve it produce a false statement, the equations are said to be **inconsistent**. The system of equations has no solution.

EXAMPLE 6 The system of equations

$y - 1 = 2x$

$2y - 4x = 7 \quad$ is said to be inconsistent.

If we solve the first equation for y

$y = 2x + 1$

and substitute this expression into the second equation,

$$2y - 4x = 7 \quad \text{becomes}$$

$2(2x + 1) - 4x = 7$

$\quad 4x + 2 - 4x = 7$

or $\quad\quad\quad\quad 2 = 7 \quad$ which is false.

All the variables have dropped out of the equation, and we are left with an incorrect statement. The original system of equations is said to be inconsistent, and it has no solution. ●

Your Turn Try solving the following systems of equations.

(a) $2x - y = 5 \quad\quad$ (b) $3x - y = 5$

$\quad 2y - 4x = 3 \quad\quad\quad\quad 6x - 10 = 2y$

Solutions (a) Solve the first equation for y.

$\quad y = 2x - 5$

Substitute this into the second equation.

$$2y - 4x = 3 \quad \text{becomes}$$

$$2\,(2x - 5) - 4x = 3$$

$$4x - 10 - 4x = 3$$

$$-10 = 3 \qquad \text{This is impossible. There is no solution for this system of equations. The system of equations is inconsistent.}$$

(b) Solve the first equation for y.

$$y = 3x - 5$$

Substitute this into the second equation.

$$6x - 10 = 2y \quad \text{becomes}$$

$$6x - 10 = 2\,(3x - 5)$$

$$6x - 10 = 6x - 10$$

$$0 = 0 \qquad \text{This is true, but all of the variables have dropped out and we cannot get a single unique solution. The system of equations is dependent. There are an infinite number of solutions.}$$

Solution by Elimination   The second method for solving a system of equations is called the **elimination method** or the **addition method**. When it is difficult or "messy" to solve one of the two equations for either x or y, the method of elimination may be the simplest way to solve the system of equations.

EXAMPLE 7   In the system of equations

$$2x + 3y = 7$$

$$4x - 3y = 5$$

neither equation can be solved for x or y without introducing fractions that are difficult to work with. But we can simply add these equations, and the y-terms will be eliminated.

$$2x + 3y = 7$$
$$\underline{4x - 3y = 5}$$
$$6x + 0 = 12 \qquad \Longleftarrow \boxed{\text{Adding like terms eliminates } y.}$$
$$6x = 12$$
$$x = 2$$

Now substitute this value of x back into either one of the original equations to obtain a value for y.

The first equation	$2x + 3y = 7$
becomes	$2(2) + 3y = 7$
	$4 + 3y = 7$
Subtract 4 from each side.	$3y = 3$
Divide each side by 3.	$y = 1$

The solution is $x = 2$, $y = 1$, or $(2, 1)$.

☑ Check the solution by substituting it back into the second equation.

$$4x - 3y = 5$$

$$4(2) - 3(1) = 5$$

$$8 - 3 = 5 \qquad \text{which is correct.} \qquad \bullet$$

Your Turn Solve the following system of equations by adding.

$$2x - y = 3$$

$$y + x = 9$$

Solution **First,** rearrange the terms in the second equation $2x - y = \quad 3$

so that like terms are in the same column. ——————→ $x + y = \quad 9$

Then, add the columns of like terms to $\overline{3x + 0 = 12}$

eliminate y and solve for x. $x = \quad 4$

Finally, substitute 4 for x in the simpler $y + (4) = \quad 9$

second equation to solve for y. $y = \quad 5$

The solution is $x = 4$, $y = 5$, or $(4, 5)$. Check it.

⚠ **Careful** It is important to rearrange the terms in the equations so that the x- and y-terms appear in the same order in both equations:

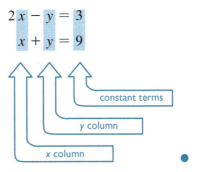

constant terms

y column

x column

More Practice Solve the following systems of equations by this process of elimination.

(a) $\quad x + 5y = 17$ (b) $x + y = 16$

$\quad\quad -x + 3y = \quad 7$ $\quad\quad x - y = \quad 4$

(c) $3x - y = -5$ (d) $\frac{1}{2}x + 2y = 10$

$\quad\quad y - 5x = 9$ $\quad\quad y + 1 = \frac{1}{2}x$

Solutions (a) **First,** add like terms to eliminate x and solve for y.

$$x + 5y = 17$$

$$-x + 3y = \quad 7$$

$$\overline{0 + 8y = 24}$$

$$8y = 24$$

$$y = \quad 3$$

Then, substitute 3 for y in the first equation, and solve for x.

$x + 5(3) = 17$

$x + \quad 15 = 17$

$\qquad x = \quad 2$ The solution is $x = 2$, $y = 3$, or $(2, 3)$.

✓

$\qquad -x + 3y = 7$

$-(2) + 3(3) = 7$

$\qquad -2 + 9 = 7$ which is correct.

(b) **First,** add like terms to eliminate y and solve for x.

$\qquad x + y = 16$

$\qquad \underline{x - y = \quad 4}$

$\qquad 2x + 0 = 20$

$\qquad\qquad 2x = 20$

$\qquad\qquad x = 10$

Then, substitute 10 for x in the first equation, and solve for y.

$\qquad (10) + y = 16$

$\qquad\qquad\qquad y = 6$ The solution is $x = 10$, $y = 6$, or $(10, 6)$.

✓

$\qquad x - y = 4$

$(10) - (6) = 4$

$\qquad 10 - 6 = 4$ which is correct.

(c) **First,** note that x- and y-terms are not in the same order.

$\boxed{\text{x-term is first}} \Rightarrow \qquad 3x - y = -5$

$\boxed{\text{y-term is first}} \Rightarrow \qquad y - 5x = 9$

Rearrange terms in the second equation. ⟶ $\qquad 3x - y = -5$

$\qquad\qquad\qquad\qquad\qquad\qquad \underline{-5x + y = \quad 9}$

Then, add like terms to eliminate y. $\qquad -2x + 0 = \quad 4$

$\qquad\qquad\qquad\qquad\qquad\qquad\qquad -2x = \quad 4$

Next, solve for x. $\qquad\qquad\qquad\qquad\qquad x = -2$

Finally, substitute -2 for x in the $\qquad y - 5(-2) = \quad 9$

second equation and solve for y. $\qquad\qquad y + 10 = \quad 9$

$\qquad\qquad\qquad\qquad\qquad\qquad\qquad\qquad y = -1$

The solution is $x = -2$, $y = -1$, or $(-2, -1)$.

Be sure to check your answer.

(d) **First,** note that like terms are not in the same order in the two equations.

| x-term is first | $\Rightarrow$ | $\frac{1}{2}x + 2y = 10$ |

| y-term is first | $\Rightarrow$ | $y + 1 = \frac{1}{2}x$ |

Therefore, we must rearrange terms in one of the equations to match the order in the other. Let's rearrange the second to match the first.

Subtract 1 from both sides, and $\frac{1}{2}x + 2y = 10$

subtract $\frac{1}{2}x$ from both sides. $\longrightarrow$ $\underline{-\frac{1}{2}x + y = -1}$

Add like terms to eliminate x. $0 + 3y = 9$

$$3y = 9$$

Solve for y. $y = 3$

Finally, substitute 3 for y in the first equation, and solve for x.

$$\frac{1}{2}x + 2(3) = 10$$

$$\frac{1}{2}x + 6 = 10$$

$$\frac{1}{2}x = 4$$

$$x = 8 \qquad \text{The solution is } x = 8, y = 3, \text{ or } (8, 3).$$

The check is left to you.

Multiplication with the Elimination Method   With some systems of equations, neither x nor y can be eliminated by simply adding like terms.

EXAMPLE 8   In the system

$$3x + y = 17$$

$$x + y = 7$$

adding like terms will not eliminate either variable. Recall from Chapter 7 that we can multiply both sides of an equation by any non-zero number. To solve this system of equations, multiply all terms of the second equation by -1 so that

| $x + y = 7$ | becomes $\Rightarrow$ | $-x - y = -7$ |

and the system of equations becomes

$$3x + y = 17$$

$$-x - y = -7$$

The system of equations may now be solved by adding the like terms as before.

$$3x + y = 17$$

$$\underline{-x - y = -7}$$

$$2x + 0 = 10 \qquad \Longleftarrow \boxed{\text{Adding like terms eliminates } y.}$$

$$2x = 10$$

$$x = 5$$

Substitute 5 for x in the first equation, and solve for y.

$3(5) + y = 17$

$15 + y = 17$

$y = 2$ The solution is $x = 5, y = 2$, or $(5, 2)$.

✓ $x + y = 7$

$(5) + (2) = 7$

$5 + 2 = 7$ which is correct.   ●

! Careful   When you multiply an equation by some number, be careful to multiply *all* terms on *both* sides by that number. It is very easy to forget to multiply on the right side. ●

Your Turn   Use this "multiply and add" procedure to solve the following system of equations:

$2x + 7y = 29$

$2x + y = 11$

Solution   **First,** multiply all terms in the second equation by -1. ⟶ $2x + 7y = 29$

$\underline{-2x - y = -11}$

Then, add like terms to eliminate x. $0 + 6y = 18$

Next, solve for y. $6y = 18$

$y = 3$

Finally, substitute 3 for y in the first equation, and solve for x.

$2x + 7(3) = 29$

$2x + 21 = 29$

$2x = 8$

$x = 4$ The solution is $x = 4, y = 3$, or $(4, 3)$.

✓   Check the solution by substituting in the second equation.

$2x + y = 11$

$2(4) + (3) = 11$

$8 + 3 = 11$

Solving by the multiply and add procedure may involve multiplying by constants other than -1, of course.

EXAMPLE 9   Let's use this method to solve the following system of equations.

$7x + 4y = 25$

$3x - 2y = 7$

First, look at these equations carefully. Notice that the y-terms can be eliminated easily if we multiply all terms in the second equation by 2.

The y column $\boxed{\begin{array}{c}+4y\\-2y\end{array}}$ becomes $\boxed{\begin{array}{c}+4y\\-4y\end{array}}$ when we multiply by 2.

$$\text{sum} = 0$$

The second equation becomes $2\,(3x) - 2\,(2y) = 2\,(7)$

or $\qquad\qquad\qquad\qquad 6x - 4y = 14$

and the system of equations is converted to the equivalent system

$$7x + 4y = 25$$
$$\underline{6x - 4y = 14}$$
$$13x + 0 = 39 \quad \longleftarrow \boxed{\text{Adding like terms now eliminates } y.}$$
$$13x = 39$$
$$x = 3$$

Substitute 3 for x in the first equation, and solve for y.

$$7(3) + 4y = 25$$
$$21 + 4y = 25$$
$$4y = 4$$
$$y = 1 \qquad \text{The solution is } x = 3, y = 1, \text{ or } (3, 1).$$

Check the solution by substituting it back into the original equations. ●

More Practice Try these problems to make certain you understand this procedure.

(a) $3x - 2y = 14$ (b) $6x + 5y = 14$
   $5x - 2y = 22$ $-2x + 3y = -14$

(c) $5y - x = 1$ (d) $-x - 2y = 1$
   $2x + 3y = 11$ $19 - 2x = -3y$

Solutions (a) Multiply each term in the first equation by -1.

$$(-1)(3x) - (-1)(2y) = (-1)(14)$$
$$-3x + 2y = -14$$

The system of equations is therefore

$$-3x + 2y = -14$$
$$\underline{5x - 2y = 22}$$
$$2x + 0 = 8 \quad \longleftarrow \boxed{\text{Adding like terms now eliminates } y.}$$
$$x = 4$$

Substitute 4 for x in the first equation, and solve for y.

$$3(4) - 2y = 14$$
$$12 - 2y = 14$$
$$-2y = 2$$
$$y = -1 \qquad \text{The solution is } x = 4, y = -1, \text{ or } (4, -1).$$

☑

$$3x - 2y = 14 \qquad 5x - 2y = 22$$
$$3(4) - 2(-1) = 14 \qquad 5(4) - 2(-1) = 22$$
$$12 + 2 = 14 \qquad 20 + 2 = 22$$
$$14 = 14 \qquad 22 = 22$$

(b) Multiply each term in the second equation by 3.

$$(3)(-2x) + (3)(3y) = (3)(-14)$$
$$-6x + 9y = -42$$

The system of equations is now

$$6x + 5y = 14$$
$$\underline{-6x + 9y = -42}$$
$$0 + 14y = -28 \qquad \text{←} \boxed{\text{Adding like terms eliminates } x.}$$
$$y = -2$$

Substitute -2 for y in the first equation, and solve for x.

$$6x + 5(-2) = 14$$
$$6x - 10 = 14$$
$$6x = 24$$
$$x = 4 \qquad \text{The solution is } x = 4, y = -2, \text{ or } (4, -2).$$

Be certain to check your solution.

(c) Rearrange to put the terms in the first equation in the same order as they are in the second equation.

$$-x + 5y = 1$$
$$2x + 3y = 11$$

Multiply each term in the first equation by 2.

$$-2x + 10y = 2$$
$$\underline{2x + 3y = 11}$$
$$0 + 13y = 13 \qquad \text{←} \boxed{\text{Adding like terms eliminates } x.}$$
$$13y = 13$$
$$y = 1$$

Substitute 1 for y in the first equation, and solve for x.

$$5(1) - x = 1$$
$$5 - x = 1$$
$$-x = -4$$
$$x = 4 \qquad \text{The solution is } x = 4, y = 1, \text{ or } (4, 1).$$

Check it.

(d) Rearrange the terms in the second equation in the same order as they are in the first equation. (Subtract 19 from both sides and add $3y$ to both sides.)

$$-x - 2y = 1$$
$$-2x + 3y = -19$$

Multiply each term in the first equation by -2.

$$(-2)(-x) - (-2)(2y) = (-2)(1)$$
$$2x + 4y = -2$$

The system of equations is now

$$2x + 4y = -2$$
$$\underline{-2x + 3y = -19}$$
$$0 + 7y = -21 \qquad \boxed{\text{Adding like terms eliminates } x.}$$
$$7y = -21$$
$$y = -3$$

Substitute -3 for y in the first equation of the original problem, and solve for x.

$$-x - 2(-3) = 1$$
$$-x + 6 = 1$$
$$-x = -5$$
$$x = 5 \qquad \text{The solution is } x = 5, y = -3, \text{ or } (5, -3).$$

Check your solution.

Multiplying Both Equations   If you examine the system of equations

$$3x + 2y = 7$$
$$4x - 3y = -2$$

you will find that there is no single integer we can use as a multiplier that will allow us to eliminate one of the variables when the equations are added. Instead we must convert each equation separately to an equivalent equation so that when the new equations are added, one of the variables is eliminated.

EXAMPLE 10   With the system of equations given, if we wish to eliminate the y variable, note that the least common multiple of the y coefficients is 6. To make them both 6, multiply the first equation by 3 and the second equation by 2.

First equation: $\boxed{3x + 2y = 7}$ $\boxed{\text{Multiply by 3}}$ $\Rightarrow$ $\boxed{9x + 6y = 21}$

Second equation: $\boxed{4x - 3y = -2}$ $\boxed{\text{Multiply by 2}}$ $\Rightarrow$ $\boxed{8x - 6y = -4}$

The new system of equations is

$$9x + 6y = 21$$
$$\underline{8x - 6y = -4}$$
$$17x + 0 = 17 \qquad \boxed{\text{Adding like terms now eliminates } y.}$$
$$x = 1$$

Substitute 1 for x in the original first equation to solve for y.

$$3x + 2y = 7$$
$$3(1) + 2y = 7$$
$$3 + 2y = 7$$
$$2y = 4$$
$$y = 2 \qquad \text{The solution is } x = 1, y = 2, \text{ or } (1, 2).$$

✔
Check this solution by substituting it back into the original pair of equations.

$$3x + 2y = 7 \qquad 4x - 3y = -2$$
$$3(1) + 2(2) = 7 \qquad 4(1) - 3(2) = -2$$
$$3 + 4 = 7 \qquad 4 - 6 = -2$$
$$7 = 7 \qquad -2 = -2$$

●

Your Turn   Use this same procedure to solve the following system of equations:

$$2x - 5y = 9$$
$$3x + 4y = 2$$

Solution   Because their coefficients already have opposite signs, it is easier to eliminate y from these two equations. Note first that the least common multiple of 5 and 4 is 20, so we will multiply the first equation by 4 and the second equation by 5.

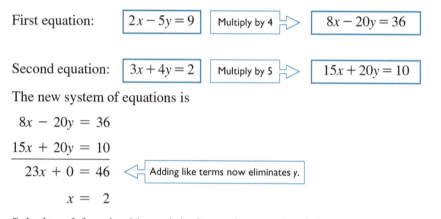

First equation: $2x - 5y = 9$ Multiply by 4 $8x - 20y = 36$

Second equation: $3x + 4y = 2$ Multiply by 5 $15x + 20y = 10$

The new system of equations is

$$8x - 20y = 36$$
$$\underline{15x + 20y = 10}$$
$$23x + 0 = 46 \qquad \text{◁⌐ Adding like terms now eliminates } y.$$
$$x = 2$$

Substitute 2 for x in either original equation to solve for y. Using the second equation,

$$3(2) + 4y = 2$$
$$6 + 4y = 2$$
$$4y = -4$$
$$y = -1$$

The solution is $x = 2$, $y = -1$, or $(2, -1)$.

🔍 **A Closer Look**   Of course, we could have chosen to eliminate x instead of y in the first step. This would have required us to multiply one of the equations by a negative to get opposite signs so that we could eliminate x by adding the two resulting equations. ●

More Practice   When you are ready to continue, practice your new skills by solving the following systems of equations.

(a) $2x + 2y = 4$
     $5x + 7y = 18$

(b) $5x + 2y = 11$
     $6x - 3y = 24$

(c) $3x + 2y = 10$
     $2x = 5y - 25$

(d) $-7x - 13 = 2y$
     $3y + 4x = 0$

Answers   (a) $x = -2$, $y = 4$
(c) $x = 0$, $y = 5$

(b) $x = 3$, $y = -2$
(d) $x = -3$, $y = 4$

In (a), multiply the first equation by -5 and the second equation by 2.

In (b), multiply the first equation by 3 and the second equation by 2.

In (c), rearrange the terms of the second equation, then multiply the first equation by -2 and the second equation by 3.

In (d), rearrange the terms of the first equation to agree with the second equation, then multiply the first equation by 3 and the second equation by 2.

Word Problems In many practical situations, not only must you be able to solve a system of equations, you must also be able to write the equations in the first place. You must be able to set up and solve word problems. In Chapter 7 we listed some "signal words" and showed how to translate English sentences and phrases to mathematical equations and expressions. If you need to review the material on word problems in Chapter 7, turn to page 485 now; otherwise, continue here.

EXAMPLE 11 Translate the following sentence into *two* equations.

The sum of two numbers is 26 and their difference is 2. (Let x and y represent the two numbers.)

The sum of two numbers is 26   . . . their difference is 2.

$$x + y = 26 \qquad\qquad x - y = 2$$

The two equations are

$$x + y = 26$$

$$x - y = 2$$

We can solve this system of equations using the elimination method.

The solution is $x = 14$, $y = 12$. Check the solution by seeing if it fits the original problem. The sum of these numbers is 26 ($14 + 12 = 26$) and their difference is 2 ($14 - 12 = 2$). ●

Your Turn Ready for another word problem? Translate and solve this one:

The difference of two numbers is 14, and the larger number is three more than twice the smaller number.

Solution The first phrase in the sentence should be translated as

"The difference of two numbers is 14 . . ."   The larger number is L; the smaller number is S; the

$$L - S = 14$$ difference must be $L - S$.

and the second phrase should be translated as

". . . the larger number is three more than twice the smaller . . ."

$$L = 3 + 2S$$

The system of equations is

$$L - S = 14$$

$$L = 3 + 2S$$

To solve this system of equations, substitute the value of L from the second equation into the first equation. Then the first equation becomes

$(3 + 2S) - S = 14$

or $3 + S = 14$

$S = 11$

Now substitute this value of S into the first equation to find L.

$L - (11) = 14$

$L = 25$ The solution is $L = 25, S = 11$.

Check the solution by substituting it back into both of the original equations. Never neglect to check your answer.

More Practice   Translating word problems into systems of equations is a very valuable and very practical algebra skill. Translate each of the following problems into a system of equations, then solve.

(a) **Machine Trades**   The total value of an order of two sizes of carriage bolts is $2.80. The smaller bolts cost 15 cents each, and the larger bolts cost 25 cents each. If the number of smaller bolts is four more than twice the number of larger ones, how many of each are there? (*Hint:* Keep all money values in cents to avoid decimals.)

(b) **Landscaping**   The perimeter of a rectangular lot is 350 ft. The length of the lot is 10 ft more than twice the width. Find the dimensions of the lot.

(c) **Construction**   A materials yard wishes to make a 500-cu-ft mixture of two different types of decorative rock. One type of rock costs $4 per cubic foot and the other costs $6 per cubic foot. If the cost of the mixture is to be $2700, how many cubic feet of each should go into the mixture?

(d) **Allied Health**   A lab technician wishes to mix a 5% salt solution and a 15% salt solution to obtain 4 liters of a 12% salt solution. How many liters of each solution must be added?

Solutions   (a) In problems of this kind, it is often helpful to first set up a table:

Item	Number of Items	Cost per Item	Total Cost
Smaller bolts	S	15	$15S$
Larger bolts	L	25	$25L$

We can write the first equation as

"The total value of an order . . . is 280 cents."

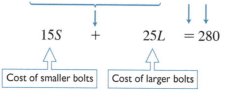

$15S + 25L = 280$

Cost of smaller bolts   Cost of larger bolts

The second equation would be

"... the number of smaller bolts is four more than twice the number of larger ones ..."

$$S = 4 + 2L$$

The system of equations to be solved is

$$15S + 25L = 280$$
$$S = 4 + 2L$$

Use substitution. Because S is equal to $4 + 2L$, replace S in the first equation with $4 + 2L$. Then solve for L.

$$15(4 + 2L) + 25L = 280$$

Remove parentheses. $\qquad 60 + 30L + 25L = 280$

Add like terms. $\qquad 60 + 55L = 280$

Subtract 60. $\qquad 55L = 220$

Divide by 55. $\qquad L = 4$

When we replace L with 4 in the second equation,

$$S = 4 + 2(4)$$
$$= 4 + 8$$
$$= 12$$

There are 4 of the larger bolts and 12 of the smaller bolts.

$15S + 25L = 280$	$S = 4 + 2L$
$15(12) + 25(4) = 280$	$12 = 4 + 2(4)$
$180 + 100 = 280$	$12 = 4 + 8$
$280 = 280$	$12 = 12$

(b) Recalling the formula for the perimeter of a rectangle, we have

"The perimeter of a rectangular lot is 350 ft."

$$2L + 2W = 350$$

The second sentence gives us

"The length of the lot is 10 ft more than twice the width."

$$L = 10 + 2W$$

The system of equations to be solved is

$$2L + 2W = 350$$
$$L = 10 + 2W$$

Using substitution, we replace L with $10 + 2W$ in the first equation. Then solve for W.

$$2\,(10 + 2W) + 2W = 350$$
$$20 + 4W + 2W = 350$$
$$20 + 6W = 350$$
$$6W = 330$$
$$W = 55 \text{ ft}$$

Now substitute 55 for W in the second equation:

$$L = 10 + 2(55)$$
$$= 10 + 110$$
$$= 120 \text{ ft}$$

The lot is 120 ft long and 55 ft wide. The check is left to you.

(c) First set up the following table:

Item	Amount (cu ft)	Cost per cu ft	Total Cost
Cheaper rock	x	$4	$4x$
More expensive rock	y	$6	$6y$

The first equation comes from the statement

"... a 500-cu ft mixture of two different types of rock."

$$500 \qquad = \qquad x + y$$

Consulting the table, we write the second equation as follows:

"... the cost of the mixture is to be $2700."

$$4x + 6y \qquad = \qquad 2700$$

Our system of equations is:

$$x + y = 500$$
$$4x + 6y = 2700$$

To use the elimination method, multiply the first equation by -4 and add it to the second equation:

$$\begin{array}{r} -4x - 4y = -2000 \\ 4x + 6y = 2700 \\ \hline 2y = 700 \\ y = 350 \text{ cu ft} \end{array}$$

Replacing y with 350 in the first equation, we have:

$$x + (350) = 500$$
$$x = 150 \text{ cu ft}$$

There should be 150 cu ft of the $4 per cu ft rock, and 350 cu ft of the $6 per cu ft rock. Be sure to check your answer.

(d)

Solution	Amount (liters)	Salt Fraction	Total Salt
5%	A	0.05	0.05A
15%	B	0.15	0.15B
12%	4	0.12	0.12(4)

The final solution is to contain 4 liters, so we have

$A + B = 4$

The second equation represents the total amount of salt:

$0.05A + 0.15B = (0.12)4$

Multiplying this last equation by 100 to eliminate the decimals, we have the system

$A + B = 4$

$5A + 15B = 48$

To solve this system, multiply each term in the first equation by -5 and add to get

$$-5A - 5B = -20$$
$$\underline{5A + 15B = \quad 48}$$
$$10B = \quad 28$$
$$B = 2.8 \text{ liters}$$

Substituting back into the first equation, we have

$A + 2.8 = 4$

$\quad A = 1.2 \text{ liters}$

The technician must mix 1.2 liters of the 5% solution with 2.8 liters of the 15% solution.

$$
\begin{array}{ll}
A + B = 4 & 5A + 15B = 48 \\
1.2 + 2.8 = 4 & 5(1.2) + 15(2.8) = 48 \\
4 = 4 & 6 + 42 = 48 \\
& 48 = 48
\end{array}
$$

Now continue with Exercises 11-1 for a set of problems on systems of equations.

Exercises 11-1   Systems of Equations

A. Solve each of the following systems of equations using the method of substitution. If the system is inconsistent or dependent, say so.

1. $y = 10 - x$
   $2x - y = -4$

2. $3x - y = 5$
   $2x + y = 15$

3. $2x - y = 3$
   $x - 2y = -6$

4. $2y - 4x = -3$
   $y = 2x + 4$

5. $3x + 5y = 26$
   $x + 2y = 10$

6. $x = 10y + 1$
   $y = 10x + 1$

7. $3x + 4y = 5$
   $x - 2y = -5$

8. $2x - y = 4$
   $4x - 3y = 11$

B. Solve each of the following systems of equations. If the system is inconsistent or dependent, say so.

1. $x + y = 5$
   $x - y = 13$

2. $2x + 2y = 10$
   $3x - 2y = 10$

3. $2y = 3x + 5$
   $2y = 3x - 7$

4. $2y = 2x + 2$
   $4y = 5 + 4x$

5. $x = 3y + 7$
   $x + y = -5$

6. $3x - 2y = -11$
   $x + y = -2$

7. $5x - 4y = 1$
   $3x - 6y = 6$

8. $y = 3x - 5$
   $6x - 3y = 3$

9. $y - 2x = -8$
   $x - \frac{1}{2}y = 4$

10. $x + y = a$
    $x - y = b$

11. $3x + 2y = 13$
    $5x + 3y = 20$

12. $3x + 4y = 11$
    $2x + 2y = 7$

C. Word Problems

Translate each problem statement into a system of equations and solve.

1. The sum of two numbers is 39 and their difference is 7. What are the numbers?

2. The sum of two numbers is 14. The larger is two more than three times the smaller. What are the numbers?

3. Separate a collection of 20 objects into two parts so that twice the larger amount equals three times the smaller amount.

4. The average of two numbers is 25 and their difference is 8. What are the numbers?

5. **Automotive Trades** Four bottles of leather cleaner and three bottles of protectant cost $64. Three bottles of leather cleaner and four bottles of protectant cost $62. How much does a single bottle of each cost?

6. **Carpentry** The perimeter of a rectangular window is 14 ft, and its length is 2 ft less than twice the width. What are the dimensions of the window?

7. Harold exchanged a $1 bill for change and received his change in nickels and dimes, with seven more dimes than nickels. How many of each coin did he receive?

8. If four times the larger of two numbers is added to three times the smaller, the result is 26. If three times the larger number is decreased by twice the smaller, the result is 11. Find the numbers.

9. **Office Services** An office manager bought three cases of paper and five ink cartridges, but he forgot what each cost. He knows that the total cost was $295, and he recalls that a case of paper cost $13 more than an ink cartridge. What was the cost of a single unit of each?

10. **Sheet Metal Trades** The length of a piece of sheet metal is twice the width. The difference in length and width is 20 in. What are the dimensions?

11. **Electrical Trades** A 30-in. piece of wire is to be cut into two parts, one part being four times the length of the other. Find the length of each.

12. **Painting** A painter wishes to mix paint worth $30 per gallon with paint worth $36 per gallon to make a 12-gal mixture worth $33.50 per gallon. How many gallons of each should he mix?

13. **Allied Health** A lab technician wishes to mix a 10% salt solution with a 2% salt solution to obtain 6 liters of a 4% salt solution. How many liters of each should be added?

14. **Landscaping** The perimeter of a rectangular field is 520 ft. The length of the field is 20 ft more than three times the width. Find the dimensions.

15. **Landscaping** A 24-ton mixture of crushed rock is needed in a construction job; it will cost $1200. If the mixture is composed of rock costing $45 per ton and $60 per ton, how many tons of each should be added?

When you have completed these problems check your answers to the odd-numbered problems in the Appendix. Then continue with Section 11-2 to learn about quadratic equations.

11-2 Quadratic Equations

Thus far in your study of algebra you have worked only with linear equations. In a linear equation, the variable appears only to the first power. For example, $3x + 5 = 2$ is a linear equation. The variable appears as x or x^1. No powers of x, such as x^2, x^3, or x^4, appear in the equations.

An equation in which the variable appears to the second power, but to no higher power, is called a **quadratic equation**.

Your Turn Which of the following are quadratic equations?

(a) $x^2 = 49$ (b) $2x - 1 = 4$ (c) $3x - 2y = 19$

(d) $5x^2 - 8x + 3 = 0$ (e) $x^3 + 3x^2 + 3x + 1 = 0$

Answers Equations (a) and (d) are quadratic equations. Equation (b) is a linear equation in one variable. Equation (c) is a linear equation in two variables, x and y. Equation (e) is a cubic or third-order equation. Because (e) contains an x^3-term, it is not a quadratic.

Standard Form Every quadratic equation can be put into a **standard quadratic form**.

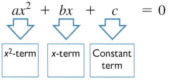

$$ax^2 + bx + c = 0 \qquad \text{where } a \text{ cannot equal zero}$$

x^2-term x-term   Constant term

Every quadratic equation must have an x^2-term, but the x-term or the constant term may be missing.

EXAMPLE 1 $2x^2 + x - 5 = 0$ is a quadratic equation in standard form: the x^2-term is first, the x-term second, and the constant term last on the left side of the equation.

$x^2 + 4 = 0$ is also a quadratic equation in standard form. The x-term is missing, but the other terms are in the proper order. We could rewrite this equation as

$x^2 + 0x + 4 = 0$ ●

Your Turn Which of the following quadratic equations are written in standard form?

(a) $7x^2 - 3x + 6 = 0$ (b) $8x - 3x^2 - 2 = 0$ (c) $2x^2 - 5x = 0$

(d) $x^2 = 25$ (e) $x^2 - 5 = 0$ (f) $4x^2 - 5x = 6$

Answers Equations (a), (c), and (e) are in standard form.

To solve a quadratic equation, it may be necessary to rewrite it in standard form.

EXAMPLE 2   The equation

$x^2 = 25x$	becomes	$x^2 - 25x = 0$	in standard form
$8x - 3x^2 - 2 = 0$	becomes	$-3x^2 + 8x - 2 = 0$	in standard form
	or	$3x^2 - 8x + 2 = 0$	if we multiply all terms by -1
$4x^2 - 5x = 6$	becomes	$4x^2 - 5x - 6 = 0$	in standard form

In each case we add or subtract a term on both sides of the equation until all terms are on the left, then rearrange terms until the x^2 is first on the left, the x-term next, and the constant term third.   ●

Your Turn   Rearrange the following quadratic equations in standard form.

(a) $5x - 19 + 3x^2 = 0$ (b) $7x^2 = 12 - 6x$

(c) $9 = 3x - x^2$ (d) $2x - x^2 = 0$

(e) $5x = 7x^2 - 12$ (f) $x^2 - 6x + 9 = 49$

(g) $3x + 1 = x^2 - 5$ (h) $1 - x^2 + x = 3x + 4$

Answers   (a) $3x^2 + 5x - 19 = 0$ (b) $7x^2 + 6x - 12 = 0$

(c) $x^2 - 3x + 9 = 0$ (d) $-x^2 + 2x = 0$ or $x^2 - 2x = 0$

(e) $7x^2 - 5x - 12 = 0$ (f) $x^2 - 6x - 40 = 0$

(g) $x^2 - 3x - 6 = 0$ (h) $x^2 + 2x + 3 = 0$

Solutions to Quadratic Equations   The solution to a linear equation is a single number. The solution to a quadratic equation is usually a *pair* of numbers, each of which satisfies the equation.

EXAMPLE 3   The quadratic equation

$$x^2 - 5x + 6 = 0$$

has the solutions

$$x = 2 \quad \text{or} \quad x = 3$$

To see that either 2 or 3 is a solution, substitute each into the equation.

For $x = 2$ For $x = 3$

$(2)^2 - 5(2) + 6 = 0$ $(3)^2 - 5(3) + 6 = 0$

$4 - 10 + 6 = 0$ $9 - 15 + 6 = 0$

$-6 + 6 = 0$ $-6 + 6 = 0$ ●

Your Turn   Show that $x = 5$ or $x = 3$ gives a solution to the quadratic equation

$$x^2 - 8x + 15 = 0$$

Solution $x^2 - 8x + 15 = 0$

For $x = 5$ For $x = 3$

$(5)^2 - 8(5) + 15 = 0$ $(3)^2 - 8(3) + 15 = 0$

$25 - 40 + 15 = 0$ $9 - 24 + 15 = 0$

$-15 + 15 = 0$ $-15 + 15 = 0$

Solving $x^2 - c = 0$ The easiest kind of quadratic equation to solve is one in which the x-term is missing.

Rather than write these in standard form, put the x^2-term on one side of the equation and the constant term c on the other side.

EXAMPLE 4   To solve the quadratic equation

$$x^2 - 25 = 0$$

add 25 to both sides to rewrite it as

$$x^2 = 25$$

and take the square root of both sides of the equation.

$$\sqrt{x^2} = \sqrt{25}$$

so $x = +\sqrt{25} = 5$ or $x = -\sqrt{25} = -5$

Both 5 and -5 satisfy the original equation.

For $x = 5$ For $x = -5$

$\qquad x^2 - 25 = 0$ and $\qquad x^2 - 25 = 0$

$\qquad (5)^2 - 25 = 0$ $\qquad (-5)^2 - 25 = 0$

$\qquad 25 - 25 = 0$ $\qquad 25 - 25 = 0$ ●

⚠ **Careful**   As we mentioned in Chapter 6, every positive number has two square roots, one positive and the other negative. Therefore, when you take the square root of both sides of a quadratic equation, you must include both the positive and the negative square roots in your answer. ●

Your Turn   Solve each of the following quadratic equations and check *both* solutions. (Round to two decimal places if necessary.)

(a) $x^2 - 36 = 0$ (b) $x^2 = 8$ (c) $x^2 - 1 = 2$

(d) $4x^2 = 81$ (e) $3x^2 = 27$ (f) $\dfrac{3x^2}{5} = 33.3$

Solutions   (a) $x^2 = 36$ (b) $x^2 = 8$

$\qquad\quad x = \pm\sqrt{36}$ $\qquad x = \pm\sqrt{8}$

$\qquad\quad x = 6$ or $x = -6$ $\qquad x \approx 2.83$ or $x \approx -2.83$ rounded

(c) $x^2 - 1 = 2$ (d) $4x^2 = 81$

$\qquad\quad x^2 = 3$ $\qquad x^2 = \dfrac{81}{4}$

$\qquad\quad x = \pm\sqrt{3}$

$\qquad\quad x \approx 1.73$ or $x \approx -1.73$ $\qquad x = \pm\sqrt{\dfrac{81}{4}}$

$\qquad\qquad\qquad\qquad\qquad\qquad\qquad\qquad x = \dfrac{9}{2}$ or $x = -\dfrac{9}{2}$

(e) $3x^2 = 27$

$x^2 = 9$

$x = \pm\sqrt{9}$

$x = 3$ or $x = -3$

(f) $\dfrac{3x^2}{5} = 33.3$

$x^2 = \dfrac{(5)(33.3)}{3} = 55.5$

$x = \pm\sqrt{55.5}$

$x \approx 7.45$ or $x \approx -7.45$

You can do problem (f) in one of the following two ways on a calculator:

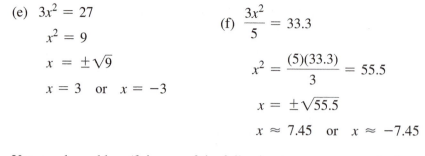

*Parentheses must be opened here.

Notice in each case that first we rewrite the equation so that x^2 appears alone on the left and a number appears alone on the right. Second, take the square root of both sides. The equation will have two solutions.

You should also notice that the equation

$x^2 = -4$

has no solution. There is no number x whose square is a negative number.

The Quadratic Formula   The method just explained will work only for quadratic equations in which the x-term is missing. A more general method that will work for *all* quadratic equations is to use the **quadratic formula**. The solution of any quadratic equation

$ax^2 + bx + c = 0$ (where $a \neq 0$)

is

$$x = \frac{-b + \sqrt{b^2 - 4ac}}{2a} \quad \text{or} \quad x = \frac{-b - \sqrt{b^2 - 4ac}}{2a}$$

or

The Quadratic Formula

$$x = \frac{-b \pm \sqrt{b^2 - 4ac}}{2a}$$

! **Careful**   Note that the entire expression in the numerator is divided by $2a$. ●

EXAMPLE 5   To solve the quadratic equation

$2x^2 + 5x - 3 = 0$

follow these steps.

Step 1   *Identify* the coefficients a, b, and c for the quadratic equation.

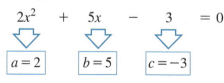

Step 2 *Substitute* these values of a, b, and c into the quadratic formula.

$$x = \frac{-(5) \pm \sqrt{(5)^2 - 4(2)(-3)}}{2(2)}$$

The $\pm$ sign means that there are two solutions, one to be calculated using the $+$ sign and the other solution calculated using the $-$ sign.

Step 3 *Simplify* these equations for x.

$$x = \frac{-5 \pm \sqrt{25 + 24}}{4}$$

$$= \frac{-5 \pm \sqrt{49}}{4} = \frac{-5 \pm 7}{4}$$

$$x = \frac{-5 + 7}{4} \quad \text{or} \quad x = \frac{-5 - 7}{4}$$

$$x = \frac{2}{4} = \frac{1}{2} \quad \text{or} \quad x = \frac{-12}{4} = -3$$

The solution is $x = \frac{1}{2}$ or $x = -3$.

Step 4 *Check* the solution numbers by substituting them into the original equation.

$$2x^2 + 5x - 3 = 0$$

For $x = \frac{1}{2}$

$$2\left(\frac{1}{2}\right)^2 + 5\left(\frac{1}{2}\right) - 3 = 0$$

$$2\left(\frac{1}{4}\right) + 5\left(\frac{1}{2}\right) - 3 = 0$$

$$\frac{1}{2} + 2\frac{1}{2} - 3 = 0$$

$$3 - 3 = 0$$

For $x = -3$

$$2(-3)^2 + 5(-3) - 3 = 0$$

$$2(9) + 5(-3) - 3 = 0$$

$$18 - 15 - 3 = 0$$

$$3 - 3 = 0$$

Your Turn Use the quadratic formula to solve $x^2 + 4x - 5 = 0$.

Solution **Step 1** $\quad x^2 + 4x - 5 = 0.$

$a = 1 \quad b = 4 \quad c = -5$

Step 2 $\quad x = \dfrac{-(4) \pm \sqrt{(4)^2 - 4(1)(-5)}}{2(1)}$

Step 3 Simplify

$$x = \frac{-4 \pm \sqrt{36}}{2}$$

$$= \frac{-4 \pm 6}{2}$$

$$x = \frac{-4 + 6}{2} \quad \text{or} \quad x = \frac{-4 - 6}{2}$$

$$x = 1 \quad \text{or} \quad x = -5 \quad \text{The solution is } x = 1 \text{ or } x = -5.$$

✓ **Step 4** $x^2 + 4x - 5 = 0$

For $x = 1$ For $x = -5$

$(1)^2 + 4(1) - 5 = 0$ $(-5)^2 + 4(-5) - 5 = 0$

$1 + 4 - 5 = 0$ $25 - 20 - 5 = 0$

$5 - 5 = 0$ $5 - 5 = 0$

EXAMPLE 6   Here is one that is a bit tougher. To solve $3x^2 - 7x = 5$, follow these steps.

Step 1   Rewrite the equation in standard form.

$$3x^2 - 7x - 5 = 0$$

$$a = 3 \qquad b = -7 \qquad c = -5$$

$$-(-7) = +7$$

Step 2 $x = \dfrac{-(-7) \pm \sqrt{(-7)^2 - 4(3)(-5)}}{2(3)}$

$$(-7)^2 = (-7) \cdot (-7) = 49$$

Step 3   Simplify

$$x = \frac{7 \pm \sqrt{109}}{6}$$

$$x = \frac{7 + \sqrt{109}}{6} \qquad \text{or} \qquad x = \frac{7 - \sqrt{109}}{6}$$

$$\approx \frac{7 + 10.44}{6} \qquad \text{or} \qquad \approx \frac{7 - 10.44}{6} \quad \begin{array}{l}\text{rounded to two decimal}\\\text{places}\end{array}$$

$$\approx \frac{17.44}{6} \qquad \text{or} \qquad \approx \frac{-3.44}{6}$$

$$\approx 2.91 \qquad \text{or} \qquad \approx -0.57 \qquad \begin{array}{l}\text{The solution is}\\ x \approx 2.91 \quad \text{or} \quad x \approx -0.57.\end{array}$$

Step 4   Check both answers by substituting them back into the original quadratic equation.

To find both solutions efficiently using a calculator, compute $\sqrt{b^2 - 4ac}$ first and store it in memory. That way, you can recall it to find the second answer more easily. Also note that $(-7)^2$ is the same as $(+7)^2$, so enter the positive version to avoid having to square a negative. Here is the entire recommended sequence for finding both solutions:

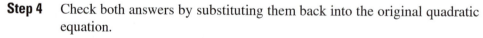

$$\boxed{\sqrt{}} \; * \; \mathbf{7} \; \boxed{x^2} \; \boxed{-} \; \mathbf{4} \; \boxed{\times} \; \mathbf{3} \; \boxed{\times} \; \boxed{(-)} \; \mathbf{5} \; \boxed{=} \; \boxed{\text{STO}} \; \boxed{A} \; \boxed{+} \; \mathbf{7} \; \boxed{=} \; \boxed{\div} \; \mathbf{6} \; \boxed{=}$$

$$\rightarrow \quad \fbox{2.906717751} \; \approx 2.91 \; \text{(First answer)}$$

$$\mathbf{7} \; \boxed{-} \; \underbrace{\boxed{\text{RCL}} \; \boxed{A}}_{\sqrt{b^2 - 4ac}} \; \boxed{=} \; \boxed{\div} \; \mathbf{6} \; \boxed{=} \; \rightarrow \quad \fbox{-0.573384418} \; \approx -0.57 \; \text{(Second answer)}$$

*Parentheses must be opened here.

More Practice Use the quadratic formula to solve each of the following equations. (Round to two decimal places if necessary.)

(a) $6x^2 - 13x + 2 = 0$ (b) $3x^2 - 13x = 0$

(c) $2x^2 - 5x + 17 = 0$ (d) $8x^2 = 19 - 5x$

Solutions (a) $6x^2 \quad - \quad 13x \quad + \quad 2 \quad = 0$

$$\boxed{a = 6} \qquad \boxed{b = -13} \qquad \boxed{c = 2}$$

$$x = \frac{-(-13) \pm \sqrt{(-13)^2 - 4(6)(2)}}{2(6)}$$

$$x = \frac{13 \pm \sqrt{121}}{12} = \frac{13 \pm 11}{12}$$

The solution is $x = \dfrac{13 + 11}{12}$ or $x = \dfrac{13 - 11}{12}$

$$x = 2 \qquad \text{or} \qquad x = \frac{1}{6}$$

The solution is $x = 2$ or $x = \dfrac{1}{6}$.

Check it.

(b) $3x^2 - 13x = 0$

$$3x^2 \quad - \quad 13x \quad + \quad 0 \quad = 0$$

$$\boxed{a = 3} \qquad \boxed{b = -13} \qquad \boxed{c = 0}$$

$$x = \frac{-(-13) \pm \sqrt{(-13)^2 - 4(3)(0)}}{2(3)}$$

$$= \frac{13 \pm \sqrt{169}}{6}$$

The solution is $x = \dfrac{13 + 13}{6}$ or $x = \dfrac{13 - 13}{6}$

$$x = \frac{13}{3} \qquad \text{or} \qquad x = 0$$

The solution is $x = \dfrac{13}{3}$ or $x = 0$.

(c) $2x^2 \quad - \quad 5x \quad + \quad 17 \quad = 0$

$$\boxed{a = 2} \qquad \boxed{b = -5} \qquad \boxed{c = 17}$$

$$x = \frac{-(-5) \pm \sqrt{(-5)^2 - 4(2)(17)}}{2(2)}$$

$$= \frac{5 \pm \sqrt{-111}}{4}$$

But the square root of a negative number is not acceptable if our answer must be a real number. This quadratic equation has no real number solution.

(d) $8x^2 = 19 - 5x$ or, in standard form,

$$8x^2 \quad + \quad 5x \quad - \quad 19 \quad = 0$$

$$a = 8 \qquad b = 5 \qquad c = -19$$

$$x = \frac{-(5) \pm \sqrt{(5)^2 - 4(8)(-19)}}{2(8)}$$

$$= \frac{-5 \pm \sqrt{633}}{16}$$

$$\approx \frac{-5 \pm 25.159}{16} \qquad \qquad \text{Find the square root to three decimal places to ensure an accurate answer.}$$

The solution is $x \approx 1.26$ or $x \approx -1.88$ rounded

 Note When you check a solution that includes a rounded value, the check may not give an exact fit. The differences should be very small if you have the correct solution. ●

Applications of Quadratic Equations

EXAMPLE 7 **Electrical Trades** In a dc circuit, the power P dissipated in the circuit is given by the equation

$$P = RI^2$$

where R is the circuit resistance in ohms and I is the current in amperes. What current will produce 1440 watts of power in a 10-ohm resistor?

Substituting into the equation yields

$$1440 = 10I^2$$

or $\quad I^2 = 144$

$$I = \pm\sqrt{144}$$

The solution is $I = 12$ amperes or $I = -12$ amperes. (Both positive and negative amperages are used to show the direction in which current is flowing.)   ●

EXAMPLE 8 In many practical situations, it is necessary to determine what size circle has the same area as two or more smaller circles.

Plumbing Suppose we must determine the diameter of a water main with the same total area as two mains with diameters of 3 in. and 4 in. Because the problem involves diameters, we will use the formula

Area of a circle $\approx 0.7854d^2$

We then write the following equation:

Area of the large main $\qquad$ Sum of the areas of the two smaller mains

$$0.7854d^2 \quad = \quad 0.7854(3)^2 + 0.7854(4)^2$$

Solving this quadratic equation for d,

$$0.7854d^2 = 7.0686 + 12.5664$$

$$0.7854d^2 = 19.635$$

$$d^2 = 25$$

$$d = 5 \text{ in.}$$

A diameter cannot be negative, so only the positive square root is a valid solution.   ●

EXAMPLE 9   **HVAC**   One side of a rectangular opening for a heating pipe is 3 in. longer than the other side. The total cross-sectional area is 70 sq in. We wish to find the dimensions of the cross section.

Let L = length, W = width.

Then $L = 3 + W$

and the area is

   area $= LW$

or $70 = LW$

Substituting $L = 3 + W$ into the area equation, we have

$$70 = (3 + W)\,W$$

or $70 = 3W + W^2$

or $W^2 + 3W - 70 = 0$ in standard quadratic form.

$a = 1$ $b = 3$ $c = -70$

To find the solution, substitute a, b, and c into the quadratic formula.

$$W = \frac{-(3) \pm \sqrt{(3)^2 - 4(1)(-70)}}{2(1)}$$

or $$W = \frac{-3 \pm \sqrt{289}}{2}$$

$$= \frac{-3 \pm 17}{2}$$

The solution is

$W = 7$ or $W = -10$

Only the positive value makes sense. The answer is $W = 7$ in.

Substituting back into the equation $L = 3 + W$, we find that $L = 10$ in.

Check to see that the area is indeed 70 sq in.   ●

Your Turn   In Chapter 9, the total surface area of a cylinder was found using the formula

$$S = 2\pi r^2 + 2\pi rh$$

Find the radius of a 15-in.-high cylinder with a total surface area of 339 sq in. (Use $\pi \approx 3.14$ and round to the nearest inch.)

Solution   First, substitute the given information into the formula:

$$S = 2\pi r^2 + 2\pi rh$$

$$339 = 2(3.14)r^2 + 2(3.14)r(15)$$

$$339 = 6.28r^2 + 94.2r$$

Then, subtract 339 from both sides and solve using the quadratic formula:

$$0 = 6.28r^2 + 94.2r - 339$$

Here $a = 6.28$, $b = 94.2$, and $c = -339$

$$r = \frac{-(94.2) \pm \sqrt{(94.2)^2 - 4(6.28)(-339)}}{2(6.28)}$$

$$r = 2.999 \ldots \approx 3 \text{ in.} \quad \text{or} \quad r = -17.999 \ldots \approx -18 \text{ in.}$$

A negative radius cannot exist, so the radius of the cylinder is approximately 3 in.

More Practice   Solve the following problems. (Round each answer to two decimal places if necessary.)

(a) Find the side length of a square whose area is 200 sq m.

(b) **Sheet Metal Trades**   The cross-sectional area of a rectangular duct must be 144 sq in. If one side must be twice as long as the other, find the length of each side.

(c) **Machine Trades**   The power P (in watts) required by a metal punch machine to punch six holes at a time is given by the formula

$$P = 3.17t^2$$

where t is the thickness of the metal being punched. What thickness can be punched using 0.750 watt of power?

(d) **Plumbing**   Find the diameter of a circular pipe whose cross-sectional area is 3.00 sq in.

(e) **Automotive Trades**   The SAE rating of an engine is given by $R = \dfrac{D^2N}{2.5}$, where D is the bore of the cylinder in inches, and N is the number of cylinders. What must the bore be for an eight-cylinder engine to have an SAE rating of 33.8?

(f) **Metalworking**   One side of a rectangular plate is 6 in. longer than the other. The total area is 216 sq in. How long is each side?

(g) **Hydrology**   A field worker has a long strip of sheet steel 12 ft wide. She wishes to make an open-topped water channel with a rectangular cross section. If the cross-sectional area must be 16 ft², what should the dimensions of the channel be? (See the figure. *Hint:* $H + H + W = 12$ ft.)

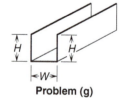

Problem (g)

Solutions   (a) $A = s^2$

$$200 = s^2$$

$$s = \pm\sqrt{200}$$

$$\approx 14.14 \text{ m, rounded} \qquad \text{The negative solution is not possible.}$$

(b) $L \cdot W = A$

and $L = 2W$

Therefore, $(2W) W = A$

or $\qquad 2W^2 = A$

$$2W^2 = 144$$

$$W^2 = 72$$

$$W = +\sqrt{72}$$

$$\approx 8.485 \approx 8.5 \text{ in.} \quad \text{rounded} \quad \text{for the positive root}$$

Then $L = 2W$

or $\quad L \approx 17.0 \text{ in.} \quad$ rounded

(c) First, substitute 0.750 for P:

$$0.750 = 3.17t^2$$

Then, solve the equation

$$t^2 = \frac{0.750}{3.17}$$

$$t^2 \approx 0.2366$$

$$t \approx 0.486 \text{ in.}$$

Because thickness cannot be negative, we need only consider the positive square root. The given power will allow the machine to punch holes in metal 0.486 in. thick.

(d) $\qquad A = \dfrac{\pi d^2}{4}$

$$3.00 = \frac{\pi d^2}{4}$$

or $\quad d^2 = \dfrac{4(3.00)}{\pi}$

$$d^2 = 3.8197 \ldots$$

$$d \approx +\sqrt{3.8197}$$

$$\approx 1.95 \text{ in.} \quad \text{rounded}$$

$\boxed{\sqrt{}}\ \mathbf{4}\ \boxed{\times}\ \mathbf{3}\ \boxed{\div}\ \boxed{\pi}\ \boxed{=} \rightarrow$ *1.954410048*

(e) $\qquad R = \dfrac{D^2 N}{2.5}$

$$33.8 = \frac{D^2 \cdot 8}{2.5}$$

or $\quad D^2 = \dfrac{(33.8)(2.5)}{8}$

$$D^2 = 10.5625$$

$$D = +\sqrt{10.5625}$$

$$D = 3.25 \text{ in.}$$

(f) "One side . . . is 6 in. longer than the other."

$$L = 6 + W$$

For a rectangle, area $= LW$.

Then,

$$216 = LW$$

or $216 = (6 + W)W$

$$216 = 6W + W^2$$

$$W^2 + 6W - 216 = 0 \qquad \text{in standard quadratic form}$$

$\boxed{a = 1} \quad \boxed{b = 6} \quad \boxed{c = -216}$

When we substitute into the quadratic formula,

$$W = \frac{-(6) \pm \sqrt{(6)^2 - 4(1)(-216)}}{2(1)}$$

$$= \frac{-6 \pm \sqrt{900}}{2}$$

$$= \frac{-6 \pm 30}{2}$$

The solution is

$W = 12$ or $W = -18$

Only the positive value is a reasonable solution. The answer is $W = 12$ in.

Substituting into the first equation yields

$L = 6 + W$

$L = 6 + (12) = 18$ in.

(g) 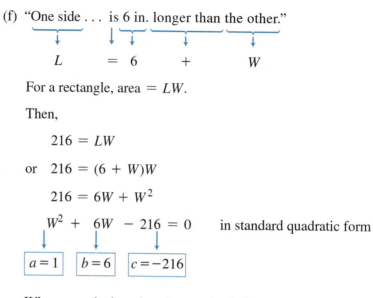 $H + H + W = 12$

$2H + W = 12$

or $W = 12 - 2H$

Area $= HW$

or $16 = HW$ Substitute $W = 12 - 2H$ into the area equation.

$16 = H (12 - 2H)$

$16 = 12H - 2H^2$ or $8 = 6H - H^2$ Divide each term by 2 to get a simpler equation.

$$H^2 - 6H + 8 = 0 \qquad \text{in standard quadratic form}$$

$\boxed{a = 1} \quad \boxed{b = -6} \quad \boxed{c = 8}$

Substituting a, b, and c into the quadratic formula yields

$$H = \frac{-(-6) \pm \sqrt{(-6)^2 - 4(1)(8)}}{2(1)}$$

$$H = \frac{6 \pm \sqrt{4}}{2}$$

$$H = \frac{6 \pm 2}{2}$$

The solution is $H = 4$ ft   or $H = 2$ ft

Here, both solutions are positive and both produce valid answers for W.

For $H = 4$ ft, $W = 4$ ft, because $16 = HW$

For $H = 2$ ft, $W = 8$ ft

The channel can be either 4 ft by 4 ft or 2 ft by 8 ft. Both dimensions give a cross-sectional area of 16 sq ft.

Now continue with Exercises 11-2 for more practice solving quadratic equations.

Exercises 11-2   **Quadratic Equations**

A. Which of the following are quadratic equations?

1. $5x - 13 = 23$
2. $2x + 5 = 3x^2$
3. $2x^3 - 6x^2 - 5x + 3 = 0$
4. $x^2 = 0$
5. $8x^2 - 9x = 0$

Which of the following quadratic equations are in standard form? For those that are not, rearrange them into standard form.

6. $7x^2 - 5 + 3x = 0$
7. $14 = 7x - 3x^2$
8. $13x^2 - 3x + 5 = 0$
9. $23x - x^2 = 5x$
10. $4x^2 - 7x + 3 = 0$

B. Solve each of these quadratic equations. (Round to two decimal places if necessary.)

1. $x^2 = 25$
2. $3x^2 - 27 = 0$
3. $5x^2 = 22x$
4. $2x^2 - 7x + 3 = 0$
5. $4x^2 = 81$
6. $6x^2 - 13x - 63 = 0$
7. $15x = 12 - x^2$
8. $4x^2 - 39x = -27$
9. $0.4x^2 + 0.6x - 0.8 = 0$
10. $0.001 = x^2 + 0.03x$
11. $x^2 - x - 13 = 0$
12. $2x - x^2 + 11 = 0$
13. $7x^2 - 2x = 1$
14. $3 + 4x - 5x^2 = 0$
15. $2x^2 + 6x + 3 = 0$
16. $5x^2 - 7x + 1 = 0$
17. $3x^2 = 8x - 2$
18. $x = 1 - 7x^2$
19. $0.2x^2 - 0.9x + 0.6 = 0$
20. $1.2x^2 + 2.5x = 1.8$

C. Practical Applications. (Round to the nearest hundredth.)

1. The area of a square is 625 mm^2. Find its side length.

2. **Manufacturing**   The capacity in gallons of a cylindrical tank can be found using the formula

$$C = \frac{\pi D^2 L}{924}$$

where C = capacity in gallons
D = diameter of tank in inches
L = length of tank in inches

(a) Find the diameter of a 42-in.-long tank that has a capacity of 30 gal.

(b) Find the diameter of a 60-in.-long tank that has a capacity of 50 gal.

3. **Plumbing**   The length of a rectangular pipe is three times longer than its width. Find the dimensions that give a cross-sectional area of 75 sq in.

4. **HVAC**   Find the radius of a circular vent that has a cross-sectional area of 250 cm^2.

5. **Sheet Metal Trades**   The length of a rectangular piece of sheet metal must be 5 in. longer than the width. Find the exact dimensions that will provide an area of 374 sq in.

6. **Construction**   The building code in a certain county requires that public buildings be able to withstand a wind pressure of 25 lb/ft^2. The pressure P is related to the wind speed v in miles per hour by the formula $P = v^2/390$. What wind speed will produce the maximum pressure allowed? (Round to the nearest whole number.)

7. **Plumbing**   Find the diameter of a circular pipe whose cross-sectional area is 40 sq in.

8. **Electrical Trades**   If $P = RI^2$ for a direct current circuit, find I (current in amperes) if

(a) The power (P) is 405 watts and the resistance (R) is 5 ohms.

(b) The power is 800 watts and the resistance is 15 ohms.

9. **Automotive Trades**   Total piston displacement is given by the following formula:

$$\text{P.D.} = \frac{\pi}{4} D^2 LN$$

where   P.D. = piston displacement, in cubic inches
D = diameter of bore of cylinder
L = length of stroke, in inches
N = number of cylinders

Find the diameter if

(a) P.D. = 400 cu in., L = 4.5 in., N = 8

(b) P.D. = 392.7 cu in., L = 4 in., N = 6

10. **Hydrology**   An open-topped channel must be made out of a 20-ft-wide piece of sheet steel. What dimensions will result in a cross-sectional area of 48 sq ft?

11. **Sheet Metal Trades**   A cylindrical water heater has a total surface area of 4170 in.^{2}. If the height of the water heater is 75 in., find the radius to the nearest inch.

12. **Construction** The area of an octagon can be approximated using the formula

$$A = 4.8275S^2$$

where S is the length of a side. How long should the length of a side be to construct an octagonal gazebo with an area of 400 ft^2? (Round to the nearest 0.01 ft.)

13. **Construction** The section modulus S of a beam in cubic inches is given by the formula

$$S = \frac{bd^2}{6}$$

where b and d are the breadth and depth of the beam in inches.

Find the depth of a beam if its breadth is 3.0 in. and the section modulus is 7.5 in.^{3}. (Round to the nearest eighth of an inch.)

14. **Police Science** The braking distance d, in meters, of a vehicle traveling at a velocity v, in meters per second, is given by the formula

$$d = \frac{v^2}{2\mu g}$$

where μ is the coefficient of friction and g is the acceleration due to gravity. On Earth, $g \approx 9.8 \text{ m/s}^2$.

(a) If an accident report showed skid marks that were 49 meters in length on dry pavement ($\mu = 0.80$), how fast was the vehicle traveling in kilometers per hour?

(b) If an accident report showed skid marks that were 86 feet in length on wet pavement ($\mu = 0.45$), how fast was the vehicle traveling in miles per hour?

15. **Forestry** After wildfires destroyed a portion of Glacier National Park, forest service biologists calculated the germination rate of new seedlings. Using a rope as a radius, they marked off a circular area of forest equivalent to one-tenth of an acre. They then counted the number of new seedlings and multiplied by 10 to obtain an estimate of the germination rate in seedlings per acre. What length of rope is needed to create the circle? (Round to the nearest 0.1 ft.)

16. **Water/Wastewater Treatment** Two small drain pipes merge into one larger pipe by means of a Y-connector. (See the figure.) The two smaller pipes have inside diameters of $1\frac{1}{4}$ in. each. The cross-sectional area of the large pipe must be at least as large as the sum of the areas of the two smaller pipes. What is the minimum diameter of the larger pipe? (Round to the nearest tenth of an inch.)

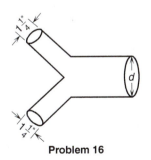

Problem 16

17. **Fire Protection** The pressure loss F (in pounds per square inch) of a hose is given by

$$F = 2Q^2 + Q$$

where Q is the flow rate (in hundreds of gallons per minute) of water through the hose. If the pressure loss is 42.0 lb/sq in., what is the flow rate to the nearest gallon per minute?

18. **Automotive Trades** The fuel economy E (in miles per gallon) of a certain vehicle is given by the formula,

$$E = -0.0088v^2 + 0.75v + 14$$

where v is the average speed of the vehicle (in miles per hour). What average speed must be maintained in order to have a fuel economy of 26 mi/gal? (Round to the nearest mile per hour.)

19. **Water/Wastewater Treatment**   According to **Torricelli's Law**, the volume V of water remaining in a certain tank with an open drain at the bottom is given by

$$V = 40\left(1 - \frac{t}{15}\right)^2$$

where t represents the time in minutes that the drain has been open. If the volume of water remaining in the tank is 10 gallons, how long has the tank been draining?

Check your answers to the odd-numbered problems in the Appendix. Then turn to Problem Set 11 on page 766 for a problem set covering both systems of equations and quadratic equations. If you need a quick review of the topics in this chapter, visit the chapter Summary first.

CHAPTER 11
SUMMARY

Intermediate Algebra

Objective	Review
Solve a system of two linear equations in two variables. (p. 728)	Use either the substitution method or the elimination method to solve a system of equations. If both variables in the system drop out, the system has no unique solution.

Examples:

(a) To solve the system of equations

$$\begin{aligned} 5x + 2y &= 20 \\ 3x - 2y &= 12 \end{aligned}$$

Add the equations $\overline{8x + 0 \ \ = 32}$

Solve for x $x = 4$

Substitute this back into the first equation: $5(4) + 2y = 20$

Solve for y: $20 + 2y = 20$

$$2y = 0$$

$$y = 0$$

The solution to the system of equations is $x = 4$, $y = 0$ or $(4,0)$

(b) Solve the system

$$\begin{aligned} x &= 3y \\ 2y - 4x &= 30 \end{aligned}$$

Substitute $3y$ for x in the second equation.

$$2y - 4(3y) = 30$$

Solve for y. $2y - 12y = 30$

$$-10y = 30$$

$$y = -3$$

Substitute this back into the first equation and solve for x.

$$x = 3(-3) = -9$$

The solution to the system of equations is $x = -9$, $y = -3$.

| **Solve word problems involving systems of equations in two variables.** (p. 743) | When there are two unknowns in a word problem, assign two different variables to the unknowns. Then look for two facts that can be translated into two equations. Solve the resulting system. |

Example: A mixture of 800 drywall screws costs $42. If the mixture consists of one type costing 4 cents each and another type costing 6 cents each, how many of each kind are there?

Objective	Review
	Let $x =$ the number of 4-cent screws and $y =$ the number of 6-cent screws. The total number of screws is $$x + y = 800$$ The total cost in cents is $$4x + 6y = 4200$$ Solve this system of equations by substitution or elimination. The solution is $x = 300$, $y = 500$.
Solve quadratic equations. (p. 751)	If the x-term is missing from a quadratic equation to be solved, isolate the x^2-term on one side, and take the square root of both sides. Remember that every positive number has two square roots. **Example:**　Solve $x^2 = 16$ Take the square root of both sides: $x = 4$ or $x = -4$
Solve quadratic equations using the quadratic formula. (p. 752)	If the x-term is not missing, use the quadratic formula $$x = \frac{-b \pm \sqrt{b^2 - 4ac}}{2a}$$ where $ax^2 + bx + c = 0$ is the equation being solved. **Example:**　Solve $3x^2 + 2x - 16 = 0$ $$a = 3, b = 2, c = -16$$ Substitute into the quadratic formula $$x = \frac{-(2) \pm \sqrt{(2)^2 - 4(3)(-16)}}{2(3)}$$ $$= \frac{-2 \pm \sqrt{196}}{6}$$ $$x = \frac{-2 + 14}{6} \quad \text{or} \quad x = \frac{-2 - 14}{6}$$ $$x = 2 \quad \text{or} \quad x = -2\tfrac{2}{3}$$
Solve word problems involving quadratic equations. (p. 756)	To solve word problems involving quadratic equations, first assign a variable to the unknown quantity. If there is a second unknown, represent it with an expression involving the same variable. Finally, set up an equation based on the facts given in the problem statement. **Example:**　　The length of a rectangular vent is 3 in. longer than its width. Find the dimensions of the vent if its cross-sectional area is 61.75 sq in. Let　　　　$W =$ the width then　　$W + 3 =$ the length and　$W(W + 3) =$ the cross-sectional area Therefore, the quadratic equation is $W(W + 3) = 61.75$ or $W^2 + 3W - 61.75 = 0$ and the quadratic formula becomes $$W = \frac{-3 \pm \sqrt{256}}{2} = \frac{-3 \pm 16}{2}$$ $W = 6.5$ or $W = -9.5$ But the width cannot be a negative number, so the width is 6.5 in., and the length is $6.5 + 3 = 9.5$ in.

PROBLEM SET 11 **Intermediate Algebra**

Answers to odd-numbered problems are given in the Appendix.

A. **Solve each of the following systems of equations. If the system is inconsistent or dependent, say so.**

1. $x + 4y = 27$
   $x + 2y = 21$

2. $3x + 2y = 17$
   $x = 5 - 2y$

3. $5x + 2y = 20$
   $3x - 2y = 4$

4. $x = 10 - y$
   $2x + 3y = 23$

5. $3x + 4y = 45$
   $x - \frac{1}{3}y = 5$

6. $2x - 3y = 11$
   $4x - 6y = 22$

7. $2x + 3y = 5$
   $3x + 2y = 5$

8. $5x = 1 - 3y$
   $4x + 2y = -8$

9. $3x - 2y = 10$
   $4x + 5y = 12$

B. **Solve each of the following quadratic equations.**

1. $x^2 = 9$

2. $x^2 - 3x - 28 = 0$

3. $3x^2 + 5x + 1 = 0$

4. $4x^2 = 3x + 2$

5. $x^2 = 6x$

6. $2x^2 - 7x - 15 = 0$

7. $x^2 - 4x + 4 = 9$

8. $x^2 - x - 30 = 0$

9. $\frac{5x^2}{3} = 60$

10. $7x + 8 = 5x^2$

C. **Practical Applications**

For each of the following, set up either a system of equations in two variables or a quadratic equation and solve. (Round to two decimal places if necessary.)

1. The sum of two numbers is 38. Their difference is 14. Find them.

2. The area of a square is 196 sq in. Find its side length.

3. The difference of two numbers is 21. If twice the larger is subtracted from five times the smaller, the result is 33. Find the numbers.

4. **Carpentry**   The perimeter of a rectangular door is 22 ft. Its length is 2 ft more than twice its width. Find the dimensions of the door.

5. **HVAC**   One side of a rectangular heating pipe is four times as long as the other. The cross-sectional area is 125 cm^2. Find the dimensions of the pipe.

6. **Carpentry**   A mixture of 650 nails costs $43.50. If some of the nails cost 5 cents apiece, and the rest cost 7 cents apiece, how many of each are there?

7. **Electrical Trades**   In the formula $P = RI^2$ find the current I in amperes if
   (a) the power P is 1352 watts and the resistance R is 8 ohms.
   (b) the power P is 1500 watts and the resistance is 10 ohms.

8. **Sheet Metal Trades**   The length of a rectangular piece of sheet steel is 2 in. longer than the width. Find the exact dimensions if the area of the sheet is 168 sq in.

9. **Electrical Trades**   A 42-in. piece of wire is to be cut into two parts. If one part is 2 in. less than three times the length of the other, find the length of each piece.

Name

Date

Course/Section

10. **Landscaping** The perimeter of a rectangular field is 750 ft. If the length is four times the width, find the dimensions of the field.

11. **Sheet Metal Trades** Find the diameter of a circular vent with a cross-sectional area of 200 in.^{2}.

12. **Painting** A painter mixes paint worth $28 per gallon with paint worth $34 per gallon. He wishes to make 15 gal of a mixture worth $32 per gallon. How many gallons of each kind of paint must be included in the mixture?

13. **Construction** According to Doyle's log rule, the volume V (in board feet) produced from a log of length L (in feet) and diameter D (in inches) can be estimated by the formula

$$V = \frac{L(D^2 - 8D + 16)}{16}$$

What diameter log of length 20 ft will produce 90 board feet of lumber?

14. **Plumbing** The length of a rectangular pipe is 3 in. less than twice the width. Find the dimensions if the cross-sectional area is 20 sq in.

15. **Landscaping** Fifty yards of a mixture of decorative rock cost $6600. If the mixture consists of California Gold costing $120 per yard and Palm Springs Gold costing $180 per yard, how many yards of each are used to make the mixture?

16. **Aviation** The lift L (in pounds) on a certain type of airplane wing is given by the formula

$$L = 0.083v^2$$

where v is the speed of the airflow over the wing (in feet per second). What air speed is required to provide a lift of 2500 lb? (Round to the nearest ten feet per second.)

17. **Plumbing** Four small drain pipes flow into one large pipe as shown in the figure. Two of the pipes have 1-in. diameters and the other two have $1\frac{1}{2}$-in. diameters. The cross-sectional area of the large main pipe must be at least as large as the sum of the cross-sectional areas of the four smaller ones. What should be the minimum diameter of the large pipe? (Round to the nearest 0.01 in.)

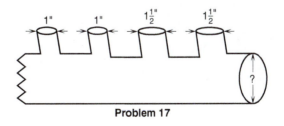

Problem 17

18. **Electronics** For a certain temperature-sensitive electronic device, the voltage output V (in millivolts) is given by the formula

$$V = 3.1T - 0.014T^2$$

where T is the temperature (in degrees Celsius). What temperature is needed to produce a voltage of 150 mV?

19. **Transportation**   The stopping distance d (in feet) of a certain vehicle is approximated by the formula

$$d = 0.0611v^2 + 0.796v + 5.72$$

where v is the speed of the vehicle (in miles per hour) before the brakes are applied. If the vehicle must come to a stop within 140 ft, what is the maximum speed at which it can be traveling before the brakes are applied? (Round to the nearest mile per hour.)

20. **Automotive Trades**   The fuel economy F of a certain hybrid electric car at speeds between 50 mi/hr and 100 mi/hr is given by the formula

$$F = 0.005v^2 - 1.25v + 96$$

where F is in miles per gallon and v is the speed in miles per hour. What speed will result in a fuel economy of 39 mi/gal?

21. **Construction**   An historic home in the Ojai Valley of Southern California is in the shape of an equilateral triangle. If the home contains 4366 square feet, what is the length of each of the three sides of the home? (*Hint:* The area A of an equilateral triangle is given by $A \approx 0.433a^2$, where $a =$ side length.) Give the answer in feet and inches, rounded to the nearest inch.

22. **Construction**   The section modulus S of a beam in cubic inches is given by the formula

$$S = \frac{bd^2}{6}$$

where b and d are the breadth and depth of the beam in inches.

Find the depth of a beam if its breadth is $3\frac{7}{8}$ in. and the section modulus is 7.25 in.^{3}. (Round to the nearest eighth of an inch.)

CASE STUDY: Fighting Fires in Yellowstone National Park

In the summer of 1988, the approximately 80,000 km^2 Yellowstone National Park (YNP) had 7 major forest fires as well as other fire centers for a total of 248 fires ignited within the park boundaries (nps.gov).

More than 25,000 firefighters fought the fires. On the ground, firefighters built fire lines and sprayed water from portable pumps placed in park waterways. From the air, water and fire retardant were dropped from planes and helicopters to boost the work on the ground. The total cost of fighting the fire was more than $120 million (nps.gov).

1.   Suppose that firefighters running portable pumps from the rivers and lakes mixed two primary lengths of hose connected together. Some portable pumps are able to maintain pressure for a maximum hose length of 175 feet if there are no more than five sections of hose: three long hose sections and two short hose sections. If a long section of hose was 10 feet longer than a short section, and the maximum total hose length was used, what were the lengths of each section of hose?

2.   Two pumps, pumping at different rates, were able to discharge a combined total of 3450 gallons of water per hour when they were run at partial capacity. One pump was running at 50% capacity while the

continued...

other pump was operating at 35% capacity. At full capacity, the pumps together would discharge 7245 gallons per hour. What was the full capacity of each of the two pumps?

3. Aerial water and retardant dumps were crucial to slowing fire advances toward historic structures. Two different types of aircraft were used. Each larger aircraft carried 2.5 times more water than a smaller aircraft. How many gallons of water can each aircraft carry if one drop from each released a total of 161,238 gallons?

4. Firefighters need hoses that provide strong flow rates to the nozzles when they hook up to a water source. Some materials used to manufacture fire hoses cause the flow rate of the water to slow down more than other materials. Calculate which hose manufacturer provides the highest flow rate. Below are the formulas to calculate the flow rates (x) in gallons per minute of two hose manufacturers:

    Hoses R Us $375 = 0.08x^2 + 0.05x$

    HoseMart $0.02x^2 + 2.5x = 425$

5. Lodgepole pines produce pinecones annually yet require fire to force the germination of the seeds. Scientists studied the germination rate of the seeds from the cones based on the age of a collection of cones, called a cone bin. They determined that the germination percent G was related to the age A of the bin by a linear equation in the form $G = mA + b$. Suppose that a cone bin approximately 10 years old had a 69% germination rate, and that a cone bin approximately 25 years old had a 36% germination rate. Using this data, we can set up the following system of equations to solve for the constants m and b:

    $69 = 10m + b$

    $36 = 25m + b$

    (a) Solve this system of equations for m and b.
    (b) Substitute your answers from part (a) into $G = mA + b$, leaving G and A as variables.
    (c) Use your equation from part (b) to determine the age of a cone bin that would produce a germination rate of 45%.

6. After wildfires destroyed a portion of the forest in Yellowstone National Park, foresters studied the recovery rate for native grasses and lodgepole pines in heavily burned areas. If the foresters roped off half-acre rectangular areas with lengths 22 feet longer than their widths, what were the dimensions, in feet, of the rectangular study areas?

7. Foresters predict fire danger using many different variables, including the time of year and the type of available fuel. One type of calculation for the fire danger for branch type fuel relates the fire danger F to the number of months Y since the beginning of the calendar year. Use the equation $F = 0.14Y^2 - Y + 16.66$ to find the time of year that a lodgepole pine typically has a fire danger value of 19.3.

References

Ekey, R., Mayer, L., & Bellinghausen, P. (1995). "Yellowstone on fire!" Billings, MT: Billings Gazette.

National Park Service. (2014). Yellowstone National Park: Frequently asked questions: 1988 fires. Retrieved from https://www.nps.gov/yell/learn/nature/1988firefaqs.htm

12 Statistics

Objective	Sample problems	For help, go to page

When you finish this chapter, you will be able to:

1. Read bar graphs, line graphs, and circle graphs.

Trades Management   From the bar graph below,

776

(a) Determine the number of frames assembled by the Tuesday day shift.   _____

(b) Calculate the percent decrease in output from the Monday day shift to the Monday night shift.   _____

Weekly Frame Assembly

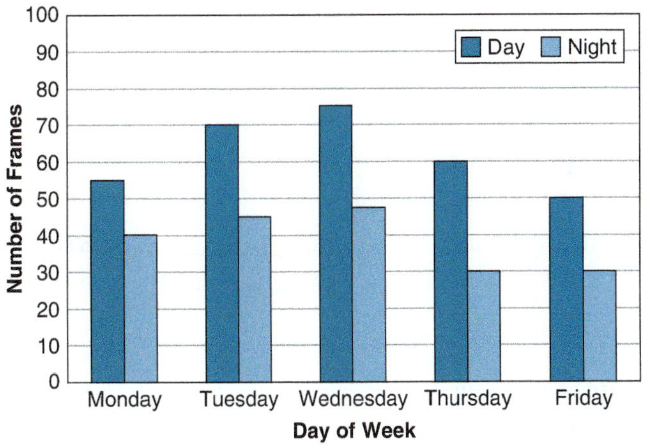

Monthly Paint Jobs at Autobrite

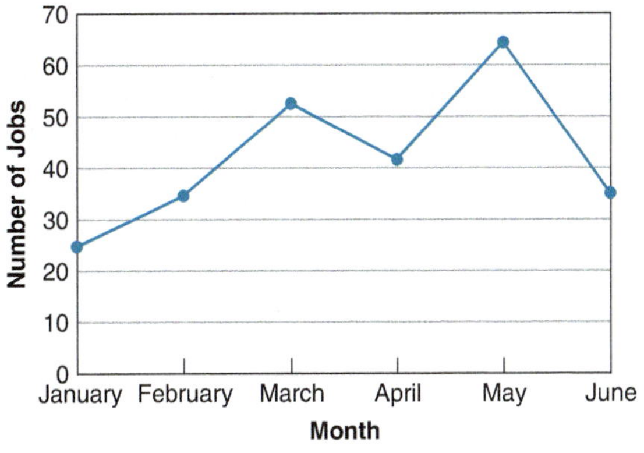

Problems (c) and (d) refer to the line graph above.   786

(c) Determine the maximum number of paint jobs and the month during which they occurred.   _____

(d) Calculate the percent increase in the number of paint jobs from January to February.   _____

(e) The average job for ABC Plumbing generates $227.50. Use the circle graph on the next page to calculate what portion of this amount is spent on advertising.   _____   790

Name   _____

Date   _____

Course/Section   _____

Objective	Sample problems	For help, go to page

Percent of Business Expenditures

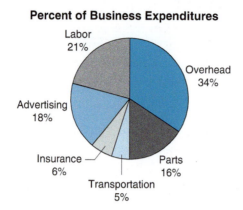

2. Draw bar graphs, line graphs, and circle graphs from tables of data.

Work Output	Day
150	Mon
360	Tues
435	Wed
375	Thurs
180	Fri

Trades Management For the data in the table, draw

(a) A bar graph.   780
(b) A line graph.   789
(c) A circle graph.   791

3. Calculate measures of central tendency: mean, median, and mode.

Find (a) the mean, (b) the median, and (c) the mode for the following set of lengths. All measurements are in meters.

12.7   16.2   15.5   13.9   13.2   17.1   15.5

(a) _____   806
(b) _____   807
(c) _____   808

4. Calculate measures of dispersion: range and standard deviation.

Find (a) the range and (b) the standard deviation for the set of lengths in objective 3.

(a) _____   818
(b) _____   819

5. Calculate mean and standard deviation for data grouped in a frequency distribution.

Welding The regulator pressure settings on an oxygen cylinder used in oxyacetylene welding (OAW) must fall between a range of 110 and 160 pounds per square inch (psi) when welding material is 1 in. thick.

The following frequency distribution shows the pressures for 40 such welding jobs. Use these data to find (a) the mean and (b) the standard deviation.

(a) _____   810
(b) _____   822

Class	Frequency, F
110–119 psi	4
120–129 psi	8
130–139 psi	12
140–149 psi	10
150–159 psi	6

Objective	**Sample problems**	**For help, go to page**
6. Use the concept of normal distribution to predict the spread of data values.	A set of data contains 80 measurements with a mean value of 6.24 cm and a standard deviation of 0.28 cm. If the measurements are distributed normally, how many of them would you expect to find in the interval between 5.96 cm and 6.52 cm?	_____ 825

(Answers to these preview problems are given in the Appendix. Also, worked solutions to many of these problems appear in the chapter Summary.)

If you are certain you can work *all* these problems correctly, turn to page 837 for a set of practice problems. If you cannot work one or more of the preview problems, turn to the page indicated after the problem. Those who wish to master this material with the greatest success should turn to Section 12-1 and begin work there.

Statistics

Working with the flood of information available to us today is a major problem for scientists, technicians, business and finance personnel, and those in management positions. We need ways to organize and analyze numerical data to make it meaningful and useful. **Statistics** is a branch of mathematics that provides us with the tools we need to do this. In this chapter, you will learn the basics of how to prepare and read statistical graphs, calculate some statistical measures, and use these to help analyze data.

According to the recruiting company Glassdoor, the best job in America is that of a **data scientist**. This number-one ranking is based on potential for career and earnings growth as well as work–life balance. Data scientists help companies analyze mountains of data, so they must have a solid background in math, statistics, and computer science. In this chapter we will study the basic statistics that will give any aspiring data scientist a foundation on which to build further knowledge.

CASE STUDY: Las Vegas or Honolulu?

In the Case Study for this chapter, we will examine weather data for two cities: Las Vegas and Honolulu. We will see that if we rely on just one statistical measure, we may reach a misleading conclusion from a given set of data—in this case, monthly and daily temperature records. However, if we use all of the statistical tools covered in this chapter, we can clarify the picture and therefore not be misled so easily.

12-1 Reading and Constructing Graphs

Note In discussing statistics, we often use the word "data." The word "data" refers to a collection of measurement numbers that describe some specific characteristic of an object or person or a group of objects or people. ●

Reading Bar Graphs

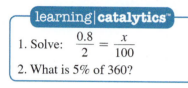

1. Solve: $\dfrac{0.8}{2} = \dfrac{x}{100}$

2. What is 5% of 360?

Graphs allow us to transform a collection of measurement numbers into a visual form that is useful, simplified, and brief. Every graph tells a story that you need to be able to read.

A **bar graph** is used to display and compare the sizes of different but related quantities. The lengths of the bars are meant to convey a general sense of magnitude, but not necessarily precise numerical quantities. The bars are usually arranged in either ascending or descending order of magnitude. On a **vertical bar graph**, the bars are oriented vertically, and their labels appear underneath them along the **horizontal axis**. The numerical scale appears along the left **vertical axis**. By aligning the top of a bar horizontally with this scale, we can estimate the numerical value of the quantity represented by the bar.

EXAMPLE 1 **Allied Health** The following vertical bar graph shows the average annual health care cost per person in six different countries in 2015.

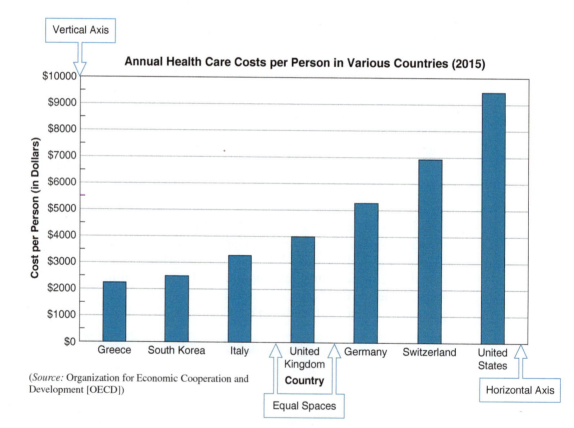

(*Source:* Organization for Economic Cooperation and Development [OECD])

The countries being compared are listed, equally spaced, along the horizontal axis. The cost, in dollars, is shown along the vertical axis. The numbers are listed in thousand-dollar increments, and tick marks halfway between the numbers represent $500, $1500, $2500, and so on. Horizontal lines are drawn across the graph at the thousand-dollar intervals. This scale allows the reader to estimate the annual cost to about the nearest $250. For example,

- The average cost of health care during 2015 in Switzerland was about $7000 per person because the top of the Switzerland bar is just below the $7000 line.

- The top of the bar representing Greece is about halfway between $2000 and $2500, so we can estimate that the annual cost of health care per person in Greece was about $2250 in 2015.

We can also use the graph to make some rough calculations. For example, suppose we wish to know the *total* cost of health care in the United States during 2015. From the graph, we can estimate the cost per person to be about $9500. The average population of the United States during 2015 was about 321,000,000. Multiplying these two numbers, we can conclude that the total cost of health care in the United States in 2015 was approximately $3,050,000,000,000, or a little over $3 trillion. ●

Your Turn   **Allied Health**   Use the graph in Example 1 to answer the following questions:

(a) What was the average cost of health care per person in South Korea in 2015, to the nearest $250?

(b) By what percent did the cost per person in the United Kingdom exceed that of Italy?

Answers   (a) $2500   (b) 23%

On a horizontal bar graph, the bars are oriented horizontally, the category labels are listed along the left vertical axis, and the numerical scale is on the horizontal axis.

EXAMPLE 2   **Automotive Trades**   The following bar graph compares the approximate costs of driving seven different fuel-efficient cars over a distance of 25 miles. These are based on gas costing $2.25 per gallon and electricity costing $0.12 per kilowatt-hour.

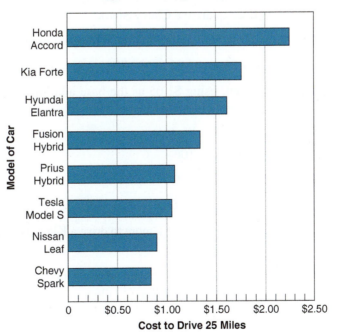

Driving Costs of Eight Fuel-Efficient Cars

(*Source:* EPA)

This is a horizontal bar graph. The different car models are listed on the vertical axis, and the numerical scale is on the horizontal axis. Labels are placed on the numerical scale at fifty-cent increments, but each tick mark represents a $0.10 increment. Therefore, we can estimate that a Prius hybrid costs about $1.10 to drive

25 miles, while a Ford Fusion hybrid costs $1.35. We can calculate the approximate cost to drive a Hyundai Elantra for 5000 miles using a proportion, as follows:

$$\frac{5000}{25} = \frac{x}{\$1.60}$$

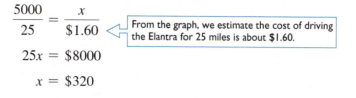
From the graph, we estimate the cost of driving the Elantra for 25 miles is about $1.60.

$$25x = \$8000$$

$$x = \$320$$

Your Turn   **Automotive Trades**   Use the graph in Example 2 to answer the following questions:

(a) How much does it cost to drive a Nissan Leaf a distance of 25 miles?

(b) How much more would it cost to drive a Honda Accord 1000 miles compared to a Chevy Spark?

(c) By what percent does the cost of driving a Kia Forte exceed that of a Tesla Model S?

Solutions   (a) $0.90

(b) **First,** subtract to find the cost difference for driving 25 miles:

$$\$2.25 - \$0.85 = \$1.40$$

Then, use a proportion to calculate the cost difference for 1000 miles:

$$\frac{1000}{25} = \frac{x}{\$1.40}$$

$$25x = \$1400$$

$$x = \$56$$

It costs approximately $56 more to drive a Honda Accord 1000 miles than it does to drive a Chevy Spark.

(c) **First,** subtract to find the difference in the 25-mile costs of the two models:

$$\$1.75 - \$1.05 = \$0.70$$

Then, using $1.05 as the base and $0.70 as the amount of increase, set up a percent change proportion and solve:

$$\frac{\$0.70}{\$1.05} = \frac{x}{100}$$

$$\$1.05x = \$70$$

$$x = 66.66\ldots\% \approx 67\%$$

It costs about 67% more to drive a Kia Forte than it does to drive a Tesla Model S.

If the quantity being displayed consists of two or more parts, we can split each bar into sections using different colors or shading to show the magnitude of each part. This type of graph is often called a "stacked" bar graph.

EXAMPLE 3   **Automotive Trades**   The following stacked bar graph compares the average cost per gallon of gas in nine different countries during the spring of 2016. The lighter, lower section of each bar shows the portion of each price represented by taxes. The scale of the graph allows us to estimate the costs to the nearest 25 cents. For example, a gallon of gas in New Zealand had an average price of about $5.25, of which about $2.00 represented taxes.

Cost of Gas and Gas Taxes Worldwide (Summer of 2016)

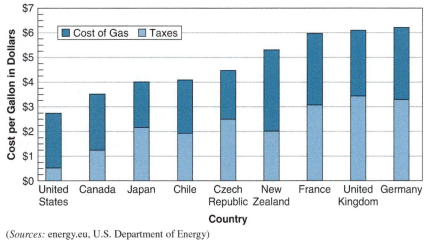

(*Sources:* energy.eu, U.S. Department of Energy)

Your Turn   **Automotive Trades**   Use the graph in Example 3 to answer the following questions. Estimate the costs per gallon to the nearest $0.25.

(a) What was the total cost for a gallon of gas in Canada?

(b) In the United Kingdom, what was the dollar amount per gallon represented by taxes?

(c) In Germany, what percent of the total cost per gallon was represented by taxes?

(d) How much more did it cost to fill up a 16-gallon tank in France than it did in the United States?

(e) By what percent did the total cost of a gallon of gas in the Czech Republic exceed the cost in Japan?

Answers   (a) $3.50   (b) $3.50   (c) 52%   (d) $52   (e) 12.5%

A **multiple bar graph** allows us to show side-by-side comparisons of related quantities on the same graph.

EXAMPLE 4   **Allied Health**   The following horizontal double bar graph shows the amounts of fat and protein per ounce for four different types of nuts. Both are measured in grams.

Amounts of Fat and Protein in Four Varieties of Nuts

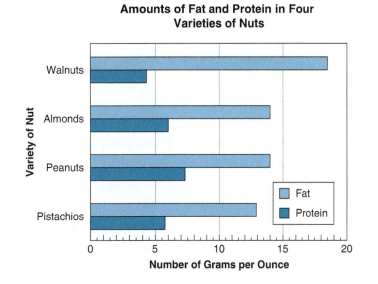

The main purpose of the graph is to compare the relative amounts of fat and protein contained in each type of nut. For example, we can see that walnuts contain the largest amount of fat per ounce and the smallest amount of protein per ounce. The scale also allows us to estimate the actual amounts of fat and protein to about the nearest 0.25 grams. For example, we can estimate that one ounce of almonds contains about 14 grams of fat, while one ounce of pistachios contains about 5.75 grams of protein. Finally, we can use these approximate numbers to perform various calculations. For example, if we ate four ounces of almonds, we would be consuming approximately (4 oz) $\times$ (14 g/oz), or 56 grams, of fat. If we wanted to eat enough pistachios to supply us with 20 grams of protein, we would divide 20 grams by 5.75 grams per ounce and find that we would need to eat about 3.5 ounces of pistachios.   ●

Your Turn   **Allied Health**   Use the graph in Example 4 to answer the following questions:

(a) Which nut contains the least amount of fat per ounce?

(b) Which nut contains the most amount of protein per ounce?

(c) How many grams of fat are contained in six ounces of peanuts?

(d) How many ounces of almonds do we need to eat in order to consume 60 grams of protein?

Solutions   (a) Pistachios

(b) Peanuts

(c) (6 oz)(14 g/oz) = 84 g

(d) (60 g) ÷ (6 g/oz) = 10 oz

Drawing Bar Graphs   Usually, a bar graph is created not from mathematical theory or abstractions, but from a set of measurement numbers.

EXAMPLE 5   **Retail Merchandising**   Suppose you want to display the following data in a bar graph showing sales of DVD players and televisions in five stores:

Quarterly Sales of DVD Players and TVs in Five Stores

Store	DVD Players	TVs
Ace	140	65
Wilson's	172	130
Martin's	185	200
Cost Less	195	285
Shop-Rite	190	375

To draw a bar graph, follow these steps:

Step 1   Decide what type of bar graph to use. Because we are comparing two different items, we should use a double bar graph. The bars can be placed either horizontally or vertically. In this case, let's make the bars horizontal.

Step 2   Choose a suitable spacing for the vertical (side) axis and a suitable scale for the horizontal (bottom) axis. Label each axis. When you label the vertical or side axis it should read in the normal way.

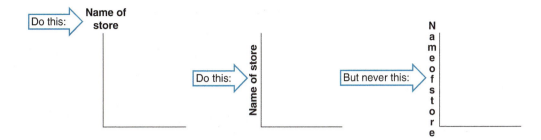

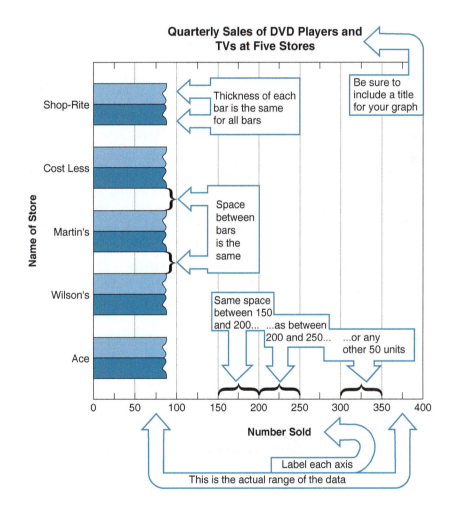

Notice that the numbers on the horizontal axis are evenly spaced and end at 400, just past the highest value in the set of data. Because each bar provides a visual indication of magnitude, it is preferable to begin the numerical scale at zero.

Also notice that the stores were arranged in order of amount of sales. The biggest seller, Shop-Rite, is at the top, and the smallest seller, Ace, is at the bottom. This is not necessary, but it makes the graph easier to read.

Step 3 Use a straight edge to mark the length of each bar according to the data given. Round the numbers if necessary. Drawing your bar graph on graph paper will make the process easier.

The final bar graph will look like this:

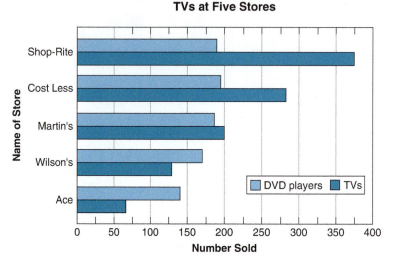

Quarterly Sales of DVD Players and TVs at Five Stores

Note In order to identify the color that represents each item, a small box has been added inside the graph. This box is called a **legend**. ●

Sometimes we use a bar graph to illustrate data containing both positive and negative numbers. The next example illustrates how we may do this.

EXAMPLE 6 **Water/Wastewater Treatment**   The following table shows the change in water usage between 2005 and 2010 for eight different groups of users in the United States:

User Group	Change in Usage from 2005 to 2010
Mining	+32.3%
Aquaculture	+7.3%
Public Supply	−5.0%
Self-Supplied Domestic	−6.0%
Livestock	−6.5%
Irrigation	−10.2%
Self-Supplied Industrial	−12.1%
Thermoelectric Power	−20.0%

(*Source:* U.S. Geological Survey.)

To make a horizontal bar graph of this information, draw the vertical scale down the center of the graph. Then show the positive numbers as bars drawn to the right of this scale, and the negative numbers as bars drawn to the left. The category names can be written on the side opposite the bars. The completed graph is shown below:

Change in Water Usage by Various Groups: 2005–2010

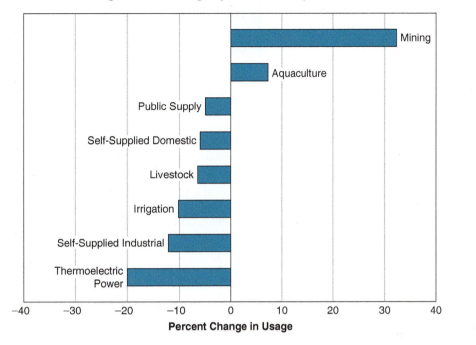

Percent Change in Usage

✎ **Note**   Graphs displaying both positive and negative numbers can also be oriented vertically, with positive bars extending up and negative bars extending down. •

Your Turn   (a) **Construction**   Use the following data to make a vertical bar graph.

Average Ultimate Compression Strength of Common Materials

Material	Compression Strength (psi)
Hard bricks	12,000
Light red bricks	1,000
Portland cement	3,000
Portland concrete	1,000
Granite	19,000
Limestone/Sandstone	9,000
Trap rock	20,000
Slate	14,000

(b) **General Interest**   The following table shows the total enrollments and the total female enrollments at colleges and universities from 1965 to 2015, in five-year increments. Construct a bar graph of these data similar to the one in Example 3, where each bar consists of two sections, one representing female enrollment and the other representing male enrollment.

Year	Total Enrollment (in millions)	Total Female Enrollment (in millions)
1965	5.9	2.3
1970	8.6	3.5
1975	11.2	5.0
1980	12.1	6.2
1985	12.3	6.4
1990	13.8	7.5
1995	14.3	7.9

(continued)

Year	Total Enrollment (in millions)	Total Female Enrollment (in millions)
2000	15.3	8.6
2005	17.5	10.0
2010	21.0	12.0
2015	20.3	11.5

(*Source:* National Center for Educational Statistics, U.S. Department of Education)

(c) **Transportation** The following table shows what percent of teenagers and young adults had driver's licenses in 1983 and in 2014. Construct a double bar graph illustrating these data. The categories along the horizontal axis will be the age groups, and for each age group there will be two bars: one for 1983 and one for 2014.

Age	1983	2014
16	46	24
17	69	45
18	80	60
19	87	69
20–24	92	77
25–29	96	85

(*Source:* Michael Sivak and Brandon Schoettle, University of Michigan)

(d) **General Interest** The following table shows the percent change from 2000 to 2013 in the number of jobs that existed in eight selected occupations in the United States. Make a horizontal bar graph illustrating this information.

Occupation	Percent Change in Number of Jobs from 2000–2013
Physical Therapists	+57%
Computer Software Engineers	+49%
Registered Nurses	+32%
Financial Managers	+28%
Carpenters	−22%
Travel Agents	−46%
Telemarketers	−58%
Computer Operators	−70%

(*Source:* Bureau of Labor Statistics)

Solutions (a)

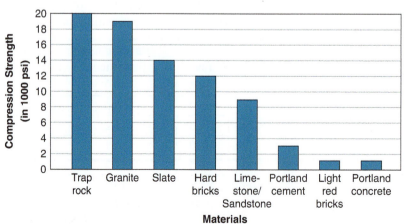

Average Ultimate Compression Strength of Common Materials

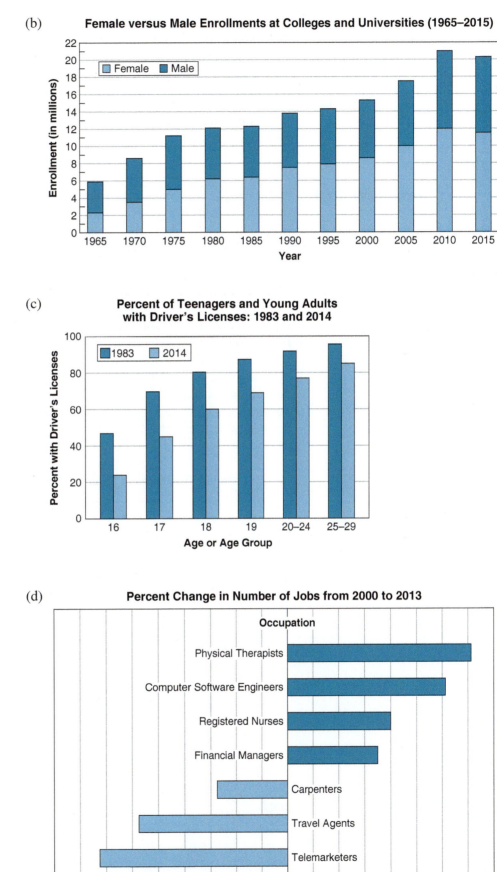

(b) **Female versus Male Enrollments at Colleges and Universities (1965–2015)**

(c) **Percent of Teenagers and Young Adults with Driver's Licenses: 1983 and 2014**

(d) **Percent Change in Number of Jobs from 2000 to 2013**

Don't worry if your graphs do not look exactly like these. The only requirement is that all of the data be displayed clearly and accurately.

A Closer Look   Notice that the vertical scales in graphs (a) and (b) have units of "*thousands* of pounds per square inch" and "millions of students." They were drawn this way to save space. ●

Reading Line Graphs   A **line graph**, or **broken-line graph**, is a display that shows the change in a quantity, usually as it changes over a period of time. Time is always displayed along the horizontal axis, and the numerical scale is always shown on the vertical axis. Because line graphs emphasize changes and patterns over time more than they do magnitude, it is not necessary to begin the vertical axis at zero.

The following broken-line graph shows monthly sales of tires at Treadwell Tire Company. The months of the year are indicated along the horizontal axis, while the numbers of tires sold are shown along the vertical axis. The actual data ranges from 550 to 950. To avoid having a large empty space below the broken line, we begin the vertical axis at 500. Each interval represents 50 tires, and the vertical axis ends at 1000 tires, just slightly above the largest data value. Each dot represents the number of tires sold in the month that is directly below the dot, and straight line segments connect the dots to show the monthly fluctuations in sales.

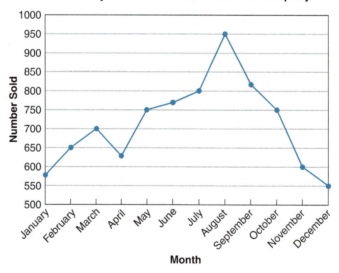

The following example illustrates how to read information from a broken-line graph.

EXAMPLE 7   **Trades Management**   To find the number of tires sold in May, find May along the horizontal axis and follow the perpendicular line up to the graph. From this point, look directly across to the vertical axis and read or estimate the number sold. There were approximately 750 tires sold in May.

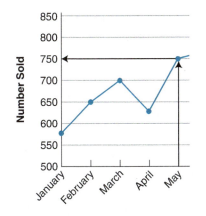

Other useful information can be gathered from the graph:

The highest monthly total was 950 in August.
The largest jump in sales occurred from July to August, with sales rising from 800 to 950 tires. To calculate the percent increase in August, set up the following proportion:

$$\frac{950 - 800}{800} = \frac{R}{100} \quad \text{or} \quad \frac{150}{800} = \frac{R}{100}$$

$$R = \frac{150 \cdot 100}{800}$$

$$= 18.75\%$$

The increase in sales from July to August was approximately 19%.   ●

Your Turn   **Trades Management**   Answer the following questions about the graph of monthly tire sales.

(a) What was the lowest monthly total, and when did this occur?

(b) How many tires were sold during the first three months of the year combined?

(c) During which two consecutive months did the largest drop in sales occur? By what percent did sales decrease?

Solutions   (a) Only 550 tires were sold in December.

(b) We estimate that 580 tires were sold in January, 650 in February, and 700 in March. Adding these, we get a total of 1930 for the three months.

(c) The largest drop in sales occurred from October to November, when sales decreased from 750 to 600. Calculating the percent decrease, we have

$$\frac{750 - 600}{750} = \frac{R}{100} \quad \text{or} \quad \frac{150}{750} = \frac{R}{100}$$

$$R = \frac{150 \cdot 100}{750}$$

$$R = 20\%$$

Sales decreased by 20% from October to November.

If we wish to compare changes over time of two or more related quantities, we can use a multiple-line graph.

EXAMPLE 8   **Agriculture**   The following double-line graph shows the U.S. production of oats and barley from 2003 to 2014:

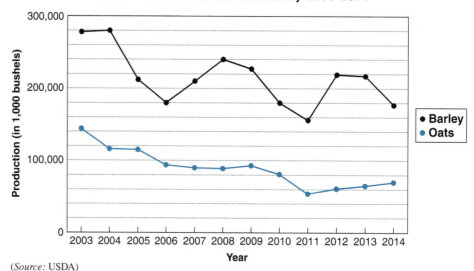

U.S. Production of Oats and Barley: 2003–2014

(*Source:* USDA)

Here are some observations that we can make from the graph:

- The production of barley was always greater than the production of oats.

- The amounts on the vertical axis are in thousands of bushels, so we must add three zeros to the numbers that we read from the graph. For example, the highest production of barley was approximately 280,000,000 bushels in 2004.

- The lowest production of oats was approximately 55,000,000 bushels in 2011.

- The largest gap in production between the two crops was approximately 165,000,000 bushels in 2004.

- The steepest decline in barley production occurred between 2004 and 2005. We can calculate the percent decrease in production as follows:

$$\frac{280 - 210}{280} = \frac{R}{100} \quad \text{or} \quad \frac{70}{280} = \frac{R}{100}$$

$$R = \frac{70 \cdot 100}{280}$$

$$R = 25\%$$

Note   Because all of the numbers end in at least six zeros, we can eliminate these zeros when performing our percent calculation. ●

───────────────

Your Turn   **Agriculture**   Use the graph in Example 8 to answer the following questions:

(a) Which crop showed the most fluctuation in production over the time period shown?

(b) What was the lowest production of barley and when did it occur?

(c) What was the highest production of oats and when did it occur?

(d) What was the smallest gap in production between the two crops and when did this occur?

(e) What was the only year in which production of oats increased?

(f) By what percent did production of barley increase over the two-year period from 2006 to 2008?

Answers   (a) Barley

(b) 155,000,000 bushels in 2011

(c) 145,000,000 bushels in 2003

(d) 85,000,000 bushels in 2006

(e) 2009

(f) Approximately 33%

Drawing Line Graphs   Constructing a broken-line graph is similar to drawing a bar graph. If possible, begin with data arranged in a table. Then follow the steps shown in the next example.

EXAMPLE 9   **Manufacturing**   The following table shows a company's average unit production cost for an electronic component during the years 2012–2017.

Year	Production Cost (per unit)
2012	$5.16
2013	$5.33
2014	$5.04
2015	$5.57
2016	$6.55
2017	$6.94

Step 1   Draw and label the axes. According to convention, time is plotted on the horizontal axis. In this case, production cost is placed on the vertical axis. Space the years equally on the horizontal axis, placing them directly on a graph line and not between lines. Choose a suitable scale for the vertical axis. In this case, each graph line represents an interval of $0.20. Notice that we begin the vertical axis at $5.00 to avoid a large gap at the bottom of the graph. Be sure to title the graph.

Step 2   For each pair of numbers, locate the year on the horizontal scale and the cost for that year on the vertical scale. Imagine a line extended up from the year and another line extended horizontally from the cost. Place a dot where these two lines intersect.

From the table, the cost in 2013 is $5.33.

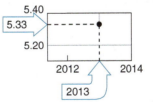

Step 3   After all the number pairs have been placed on the graph, connect adjacent points with straight line segments.

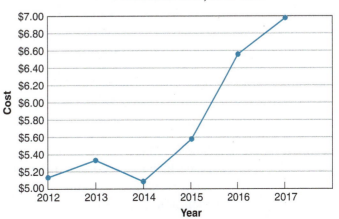

Production Cost, 2012–2017

Your Turn   **Manufacturing**   Plot the following data on the graph in the previous example to create a double-line graph.

Year	Shipping Cost (per unit)
2012	$5.30
2013	$5.61
2014	$6.05
2015	$6.20
2016	$6.40
2017	$6.50

Solution

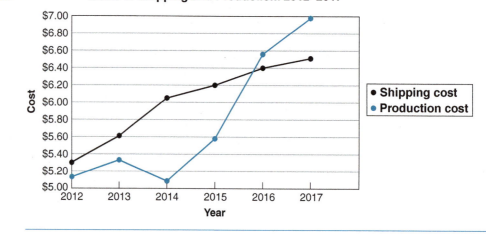

Costs of Shipping and Production: 2012–2017

Reading Circle Graphs   A **circle graph**, or **pie chart**, is used to show what percent of the whole of some quantity is represented by its separate parts.

EXAMPLE 10   **Welding**   The following circle graph gives the distribution of questions contained on a comprehensive welding exam. The area of the circle represents the entire exam, and the wedge-shaped sectors represent the percentage of questions in each part of the exam. The percents on the sectors add up to 100%.

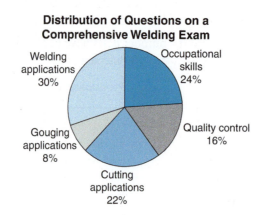

Distribution of Questions on a Comprehensive Welding Exam

The size of each sector is proportional to the percent it represents. If we wanted to know which topic was asked about the most, we could visually pick out the largest sector, welding applications, or compare the percents shown to reach the same conclusion.

We can also use the percents on the graph to calculate additional information. For example, suppose we know that there are 150 questions on the exam, and we want to know exactly how many questions deal with occupational skills. From the graph, we see that 24% of the questions relate to occupational skills, so we find 24% of 150 as follows:

$$\frac{S}{150} = \frac{24}{100}$$

$$S = \frac{150 \cdot 24}{100}$$

$$S = 36$$

There are 36 questions relating to occupational skills.   ●

Your Turn   **Welding**   Use the graph from Example 10 to answer the following questions:

(a) What percent of the exam dealt with cutting applications?

(b) Which question topic was asked about least?

(c) If there were 125 questions on the exam, how many questions would deal with quality control?

Solutions   (a) 22%   (b) Gouging Applications

(c) From the graph, we see that 16% of the questions deal with quality control. Using a proportion, we can calculate 16% of 125 as follows:

$$\frac{Q}{125} = \frac{16}{100}$$

$$Q = \frac{125 \cdot 16}{100}$$

$$Q = 20$$

There would be 20 questions on quality control.

Constructing Circle Graphs   If the data to be used for a circle graph is given in percent form, determine the number of degrees for each sector using the quick calculation method explained in Chapter 4. Convert each percent to a decimal and multiply by 360°.

EXAMPLE 11   **General Interest**   The following table shows the share of total electricity generated by major U.S. energy sources in 2014.

Energy Source	Percent of Total Electricity Generated
Coal	39%
Natural Gas	27%
Nuclear	19%
Hydropower	6%
Wind and Solar	4.8%
Other	4.2%

(*Source:* U.S. Energy Information Administration.)

To help us organize our sector calculations, we first extend the table as shown:

Energy Source	Percent	Calculation of Angle Measure for Each Sector
Coal	39%	39% of 360° = 0.39 × 360° ≈ 140°
Natural Gas	27%	27% of 360° = 0.27 × 360° ≈ 97°
Nuclear	19%	19% of 360° = 0.19 × 360° ≈ 68°
Hydropower	6%	6% of 360° = 0.06 × 360° ≈ 22°
Wind and Solar	4.8%	4.8% of 360° = 0.048 × 360° ≈ 17°
Other	4.2%	4.2% of 360° = 0.042 × 360° ≈ 15°

Sum = 100%

Round angles to the nearest degree

 Note The sum of the angles should be 360° most of the time, but it may be a degree off due to rounding. ●

Finally, using a protractor, mark off the sectors and complete the circle graph. Notice that each sector is labeled with a category name and its percent.

Share of Total Electricity from Major U.S. Energy Sources

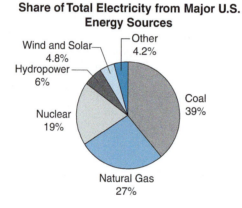

If the data for a circle graph are not given in percent form, first convert the data to percents and then convert the percents to degrees.

EXAMPLE 12   **General Interest**   In recent years there has been a push in the United States to raise the minimum wage to $15 an hour. The following table shows the breakdown by age group of the approximate number of hourly workers making less than $15 per hour in 2014. Draw a circle graph illustrating these numbers.

Age Group	Workers Making Less than $15/Hour (in millions)
16–19	4.04
20–24	9.41
25–34	10.57
35–44	6.75
45–54	6.70
55–64	4.97
65+	1.95

(*Source:* Bureau of Labor Statistics.)

Unlike the data in Example 11, the numbers are not given in percent form. We must convert the numbers to percents before we can determine the proper angles for a circle graph.

To do this, **first** find the total number of workers:

$$4.04 + 9.41 + 10.57 + 6.75 + 6.70 + 4.97 + 1.95 = 44.39 \text{ million}$$

Second, use this sum as the base to calculate each percent. For example, we would convert the number of workers in the first age group (16–19) to a percent using a proportion as follows:

$$\frac{4.04}{44.39} = \frac{x}{100}$$

Solving for x, $x = \dfrac{4.04(100)}{44.39} \approx 9.1\%$

Continuing in this manner, we calculate the percents for the remaining age groups as follows:

20–24: 21.2%   25–34: 23.8%   35–44: 15.2%
45–54: 15.1%   55–64: 11.2%   65+: 4.4%

Third, use the percent values to calculate the angles for the sectors of the circle graph:

$$(0.091)(360°) \approx 33° \quad (0.212)(360°) \approx 76° \quad (0.238)(360°) \approx 86°\text{...and so on.}$$

We now extend the original table to summarize our calculations:

Age Group	Number of Workers (in millions)	Percent of Total	Angle
16–19	4.04	9.1%	33°
20–24	9.41	21.2%	76°
25–34	10.57	23.8%	86°
35–44	6.75	15.2%	55°
45–54	6.70	15.1%	54°
55–64	4.97	11.2%	40°
65+	1.95	4.4%	16°

Finally, use the angles in the last column to draw the circle graph. Label each sector as shown and include a title with the graph.

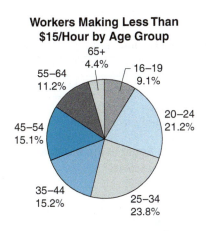

Workers Making Less Than $15/Hour by Age Group

Your Turn   (a) **Allied Health**   Recent U.S. dietary guidelines have recommended that Americans limit their intake of added sugars to no more than 10% of their total calories. The following table shows the major sources of added sugars and what percent of total added sugars each source represents. Make a circle graph illustrating these data.

Source of Added Sugar	Percent of Total Added Sugar
Beverages	47%
Snacks and Sweets	31%
Grains	8%
Mixed Dishes	6%
Dairy	4%
Condiments, Spreads, Dressings	2%
Fruits and Vegetables	2%

(*Source:* U.S. Department of Health and Human Services.)

(b) **Plumbing**   Jane the plumber keeps a record of the number of trips she makes to answer emergency calls. A summary of her records looks like this:

Trip Length	Number of Trips
Less than 5 miles	152
5–9 miles	25
10–19 miles	49
20–49 miles	18
50 miles or more	10

Calculate the percents and the angles for each category, and plot Jane's data in a circle graph.

Solutions   (a) The information is already given in percent form. We need only calculate the angles to draw the graph. Here is the extended table:

Source of Added Sugar	Percent of Total Added Sugar	Angle
Beverages	47%	$(0.47)(360°) \approx 169°$
Snacks and Sweets	31%	$(0.31)(360°) \approx 112°$
Grains	8%	$(0.08)(360°) \approx 29°$
Mixed Dishes	6%	$(0.06)(360°) \approx 22°$
Dairy	4%	$(0.04)(360°) \approx 14°$
Condiments, Spreads, Dressings	2%	$(0.02)(360°) \approx 7°$
Fruits and Vegetables	2%	7°
	100%	360°

Here is the completed circle graph:

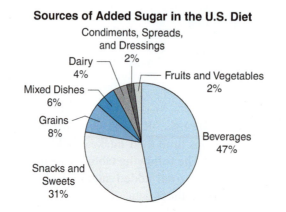

Sources of Added Sugar in the U.S. Diet

(b) The completed table shows the percents and angles for each category:

Trip Length	Number of Trips	Percent	Angle
Less than 5 miles	152	60%	216°
5–9 miles	25	10%	36°
10–19 miles	49	19%	68°
20–49 miles	18	7%	25°
50 miles or more	10	4%	14°
	254	100%	359°

Because of rounding, the percents will not always sum to exactly 100% and the angles will not always sum to exactly 360°. In this case, the angles added up to 359°. This will not noticeably affect the appearance of the graph.

The circle graph is shown here.

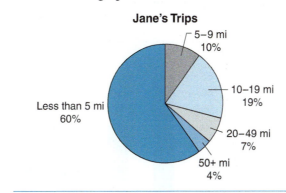

Jane's Trips

Now continue with Exercises 12-1 for more practice in reading and constructing bar graphs, line graphs, and circle graphs.

Exercises 12-1   Reading and Constructing Graphs

A. Answer the questions following each graph.

1. **Automotive Trades**   The following bar graph shows the annual U.S. sales of hybrid vehicles from 2006–2014. Study the graph and answer the questions below.

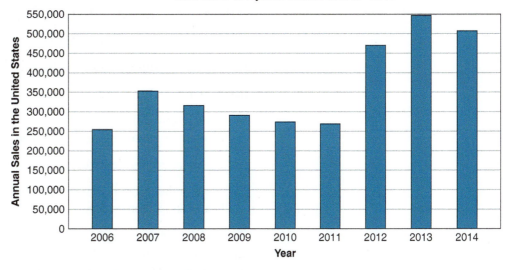

U.S. Sales of Hybrid Vehicles: 2006–2014

(*Source:* Bureau of Transportation Statistics)

(a) During which year did the largest increase in hybrid sales occur?

(b) During which year did the largest decrease in hybrid sales occur?

(c) Approximately how many hybrid vehicles were sold in 2008?

(d) How many more hybrid vehicles were sold in 2007 than in 2006? (Estimate your answer to the nearest hundred thousand.)

(e) By what percent did sales increase from 2011 to 2013? (Round to the nearest hundred percent.)

(f) By what percent did sales decrease from 2007 to 2011? (Round to the nearest percent.)

2. **Trades Management**   The following graph shows the hourly energy consumption of eight different appliances. Study the graph and answer the questions that follow.

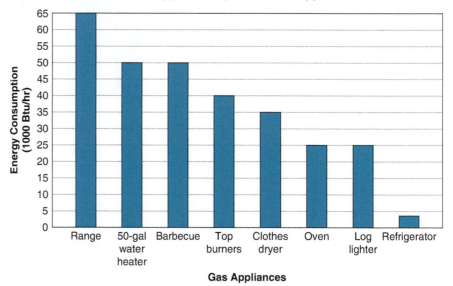

(a) Which appliance uses the most energy in an hour?

(b) Which appliance uses the least energy in an hour?

(c) How many Btu/hr does a gas barbecue use?

(d) How many Btu/*day* (24 hr) would a 50-gal water heater use?

(e) Is there any difference between the energy consumption of the range and that of the top burners plus oven?

(f) How many Btus are used by a log lighter in 15 min?

(g) What is the difference in energy consumption between a 50-gal water heater and a clothes dryer?

3. **General Interest**   The following double bar graph shows the median annual earnings (in 2011–2012) of both recent college graduates and those with graduate degrees in six different occupational areas. Study the graph and answer the questions that follow.

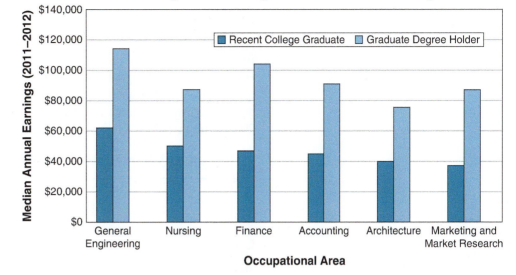

Annual Earnings: Recent College Graduates vs. Graduate Degree Holders

(a) How much more did the average electrical engineer make compared to the average accountant if they were both recent college graduates?

(b) In marketing and market research, how much more did the average graduate degree holder make compared to the average recent college graduate?

(c) For architects, by what percent did the median earnings increase for those who held a graduate degree compared to those who were recent college graduates?

(d) Answer the question in part (c) for nurses.

(e) For which of the occupations shown did the median earnings increase the most for holders of graduate degrees?

4. **General Interest** The following stacked bar graph shows total movie ticket sales worldwide from 2010–2015. The lower section of each bar indicates box-office receipts for the United States and Canada, while the upper section shows the international receipts. The vertical scale is in billions of dollars. Study the graph and answer the questions that follow.

Worldwide Movie Ticket Sales (2010–2015)

(*Source:* MPAA and Rentrak.)

(a) What does the graph tell you in general about the growth of international ticket sales versus the growth of U.S. and Canadian sales?

(b) What was the U.S. and Canadian total in 2011?

(c) What was the international total for 2014?

(d) By what percent did international ticket sales increase from 2010 to 2015?

(e) What percent of total ticket sales came from the U.S. and Canada in 2010? In 2015?

5. **Allied Health**   When Dr. Friedrich began working at the Zizyx County Hospital in 2008, his goal was to improve the quality and quantity of bone marrow transplants performed at the hospital. The number of successful bone marrow transplants performed at the Zizyx County Hospital from 2008 through 2017 is illustrated in the following broken-line graph.

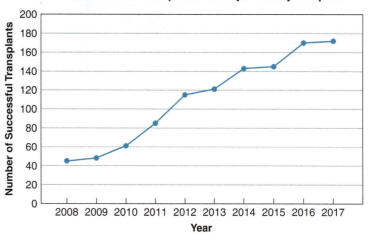

Bone Marrow Transplants at Zizyx County Hospital

(a) Approximately how many successful bone marrow transplants were performed in 2008?

(b) Approximately how many successful bone marrow transplants were performed in 2017?

(c) By about what percent did the number of successful bone marrow transplants increase in the five-year period from 2012 to 2017?

(d) If the bone marrow transplant program increases at the same rate as it did from 2012 to 2017, how many successful bone marrow transplants can the Zizyx County Hospital expect to perform in 2022?

6. **Automotive Trades**   The following line graph shows how the average fuel economy of a selected group of automobiles varies according to the speed of the vehicle.

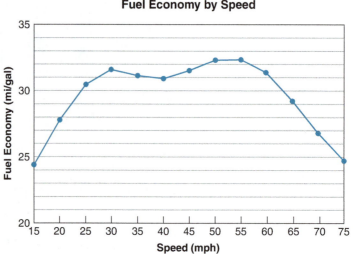

Fuel Economy by Speed

(a) At what speed is fuel economy the best? The worst?

(b) What is the fuel economy at 65 mph? At 40 mph?

(c) At what speed is the fuel economy 29 mi/gal?

(d) After which two speeds does fuel economy begin to decrease?

7. **Construction** The following double-line graph compares the number of existing home sales to the number of homes for sale (inventory) at the end of the year in the United States from 1996 through 2016. Note that the vertical scale is in thousands—meaning that 5000, for example, actually means 5,000,000. Study the graph and answer the questions that follow.

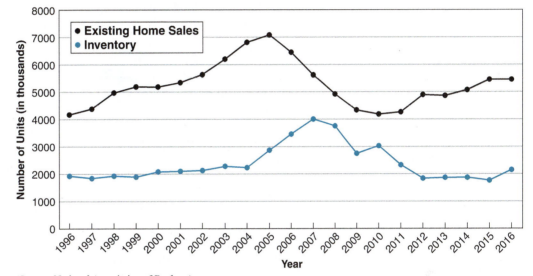

U.S. Existing Home Sales Versus Inventory: 1996–2016

(*Source:* National Association of Realtors)

(a) In what year did the biggest drop in home sales occur? By approximately how many units did sales drop?

(b) Identify at least two years when the number of sales and the inventory both increased.

(c) In what year were home sales at their lowest level? In what year were home sales at their highest level?

(d) Based on your answer to part (c), how many more units were sold in the highest-selling year than in the lowest-selling year?

(e) By approximately what percent did the inventory increase between 2004 and 2007?

(f) By approximately what percent did home sales drop between 2005 and 2010?

8. **General Interest** Study the circle graph at the top of page 800 and answer the questions that follow.

(a) Which are the largest two continents on the earth?

(b) Which two are the smallest?

(c) Which continent covers 12% of the earth?

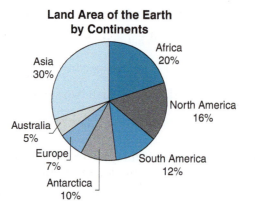

Land Area of the Earth by Continents

9. **Life Skills**   Study this circle graph and answer the questions that follow.

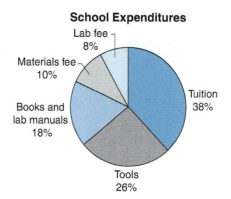

School Expenditures

(a) Which category represents the largest expenditure?

(b) Which category represents the smallest expenditure?

(c) If a student spent a total of $2500, how much of this went toward tools and lab fees combined?

(d) If a student spent $300 on materials, how much would she spend on books and lab manuals?

10. **Allied Health**   An assistant at a pharmaceutical company summarized the use of anti-obesity drugs in Zizyx County. Based on a survey of local pharmacies, the assistant estimated the percent of patients using each of the most common weight-loss medications and presented the results in a circle graph.

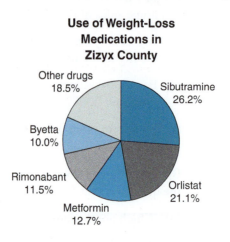

Use of Weight-Loss Medications in Zizyx County

(a) What is the most commonly used weight-loss medication in Zizyx County?

(b) If a total of 2780 people in Zizyx County take weight-loss medication, how many of these are taking Sibutramine? How many more people take Orlistat than Metformin?

(c) After a fitness center was opened in Zizyx County, the number of people taking weight-loss medications [see part (b)] decreased by 43%. How many people in Zizyx County were taking weight-loss medications after the opening of the fitness center?

(d) If the percents in the circle graph remained the same after the opening of the fitness center, how many patients were still taking Byetta?

B.   From the following data, construct the type of graph indicated.

1. **Fire Protection**   Plot the following data as a bar graph.

Causes of Fires in District 12

Cause of Fire	Number of Fires
Appliance	6
Arson	8
Electrical	18
Flammable materials	7
Gas	2
Lightning	2
Motor vehicle	10
Unknown	9

2. **Transportation**   The following table lists the total carbon emissions per person for various forms of transportation for a round trip from Los Angeles to San Francisco. Construct a bar graph of these data.

Form of Transportation	Carbon Emissions (pounds)
Flying: smaller, newer plane	300
Flying: larger, older plane	406
Driving alone (car with 47 mi/gal)	273
Driving alone (car with 25 mi/gal)	527
Driving alone (car with 16 mi/gal)	793
Carpooling (2 people, car with 25 mi/gal)	263
Train	284

(*Source:* Terra Pass)

3. **Metalworking**   Draw a bar graph from the following information. (*Hint:* Use multiples of 5 on the vertical axis.)

Linear Thermal Expansion Coefficients for Materials

Material	Coefficient (in parts per million per °C)
Aluminum	23
Copper	17
Gold	14
Silicon	3
Concrete	12
Brass	19
Lead	29

4. **Trades Management**   Plot the following data as a bar graph.

Radish Tool and Die Worker Experience

Length of Service	Number of Workers
20 years or more	9
15–19	7
10–14	14
5–9	20
1–4	33
Less than 1 year	5

5. **Construction**   Plot the following data as a double bar graph.

Sales of Construction Material: CASH IS US

Year	Wood (× $1000)	Masonry (× $1000)
2012	$289	$131
2013	$325	$33
2014	$296	$106
2015	$288	$92
2016	$307	$94
2017	$412	$89

6. **General Interest**   The following table shows the total annual costs of tuition, room, and board for full-time undergraduate students at both public and private colleges and universities. The costs are shown in ten-year increments beginning in 1980–1981, and they are given in terms of 2011 dollars. Construct a double bar graph comparing these costs. (*Hint:* Round each amount to the nearest $500 before drawing the graph.)

Year	Public Institutions	Private Institutions
1980–1981	$2373	$5470
1990–1991	$4757	$12,910
2000–2001	$7586	$21,373
2010–2011	$13,564	$32,026

7. **Automotive Trades**   The following table shows the average U.S. prices per gallon of both regular gasoline and diesel fuel from 2007 through 2016. (The prices are quoted for November of each year.) Construct a double bar graph illustrating these prices.

Year	Cost of Regular Gasoline	Cost of Diesel Fuel
2007	$2.80	$2.89
2008	$3.25	$3.80
2009	$2.35	$2.47
2010	$2.78	$2.99
2011	$3.52	$3.84
2012	$3.62	$3.97
2013	$3.51	$3.92
2014	$3.36	$3.83
2015	$2.43	$2.71
2016	$2.18	$2.44

8. **Trades Management**   The following table shows the monthly breakdown of sales of new cars, new trucks, and used vehicles at a particular auto dealership over the first six months of the year. Construct a bar graph like the one in Example 3. (*Hint:* In this case, each bar should be divided into *three* sections.)

Month	New Cars	New Trucks	Used Vehicles
January	38	18	22
February	32	12	20
March	30	16	28
April	24	10	19
May	28	11	21
June	33	15	18

9. **General Interest**   The following table shows, by their major, the percent change in the starting salaries of college graduates between 2007 and 2014. Construct a horizontal bar graph displaying this information. (See Example 6 on pp. 782–3.)

College Major	Percent Change in Starting Salary from 2007–2014
Petroleum Engineering	+29.9%
Mining Engineering	+15.8%
Computer Science	+12.6%
Electrical Engineering	+11.2%
Aerospace Engineering	+5.8%
Chemical Engineering	+2.8%
Nursing	+1.0%
Nuclear Engineering	−3.8%
English	−5.9%
Psychology	−6.8%
Elementary Education	−9.6%
Architecture	−11.2%

(*Source:* National Association of Colleges and Employers.)

10. **Agriculture**   The following table shows the price per pound of cotton on the New York Board of Trade in mid-June from 2006–2016. Construct a broken-line graph illustrating the fluctuation in these prices over the time period indicated.

Year	Price of Cotton per Pound	Year	Price of Cotton per Pound
2006	$0.49	2012	$0.74
2007	$0.53	2013	$0.84
2008	$0.68	2014	$0.80
2009	$0.49	2015	$0.63
2010	$0.77	2016	$0.65
2011	$1.46		

11. **Allied Health**   The following table shows the percent of 12th graders who smoked cigarettes daily at various times from 1995–2014. Plot a broken-line graph illustrating these numbers.

Year	Percent of 12th Graders Who Smoked Daily
1995	21.6
2000	20.6
2005	13.6
2010	10.7
2014	6.7

(*Source:* World Almanac 2016 from University of Michigan Institute for Social Research; National Institute on Drug Abuse)

12. **General Interest**   The following data show the U.S. production of renewable energy (in quadrillion—10^{15}—Btu) from 1960–2014. Use these data to construct a broken-line graph.

Year	U.S. Renewable Energy Production (in quadrillion Btu)
1960	2.93
1970	4.08
1980	5.49
1985	6.18
1990	6.04
1995	6.56
2000	6.11
2005	6.24
2010	8.07
2014	9.62

(*Source:* Energy Information Administration, U.S. Department of Energy, World Almanac 2016)

13. **Fire Protection**   The following data show the number of total acres burned in wildfires in the United States from 2002 through 2015. The numbers have been rounded to the nearest hundred thousand. Plot a broken-line graph of these data.

Year	Acres Burned	Year	Acres Burned
2002	7,200,000	2009	5,900,000
2003	4,000,000	2010	3,400,000
2004	8,100,000	2011	8,700,000
2005	8,700,000	2012	9,300,000
2006	9,900,000	2013	4,000,000
2007	9,300,000	2014	3,600,000
2008	5,300,000	2015	10,100,000

(*Source:* National Interagency Fire Center)

14. **General Interest**   The following table shows the percent of 18–34 year olds with a bachelor's degree or higher during four selected time periods. Construct a broken-line graph illustrating this information.

Year	Percent with Bachelor's Degree or Higher
1980	15.7%
1990	17.0%
2000	19.5%
2009–2013	22.3%

15. **Hydrology**   The following table shows the daily evaporation and usage totals, in acre-feet, during a summer week at Bradbury Dam. Use these data to construct a double broken-line graph.

Day	Daily Evaporation (in acre-ft)	Daily Usage (in acre-ft)
Monday	62.3	108.7
Tuesday	58.2	130.2
Wednesday	51.5	117.9
Thursday	58.6	126.5
Friday	67.3	130.5
Saturday	67.3	130.2
Sunday	62.8	128.0

16. **Business and Finance**   Plot a double broken-line graph for these data.

Actual and Projected Sales, IMD Corp. (× $1000)

Month	Actual	Projected
January	$10	$45
February	$18	$50
March	$12	$47
April	$22	$50
May	$40	$65
June	$39	$71
July	$50	$76
August	$42	$75
September	$35	$55
October	$37	$60
November	$41	$50
December	$44	$87

17. **Electrical Engineering**   Plot the following data as a circle graph.

California's In-State Sources of Electricity

Source	Percent	Source	Percent
Renewable	22.5%	Natural gas	61.3%
Nuclear	8.6%	Large hydroelectric	7.1%
Coal	0.5%		

18. **Water/Wastewater Treatment**   The following table shows how much water per year is distributed to seven different states from the Colorado River. Convert each amount into a percent of the total, and then construct a circle graph to illustrate this distribution.

State	Water Allotment from the Colorado River (in millions of acre-ft per year)
California	4.40
Colorado	3.90
Arizona	2.85
Utah	1.70
Wyoming	1.00
New Mexico	0.85
Nevada	0.30

(*Source:* USGS, Southern Nevada Water Authority.)

19. **Aviation**   An aircraft mechanic spends 12.5% of a 40-hr week working on aircraft airframes, 37.5% of the week on landing gear, 43.75% of the week on power plants, and 6.25% of the week on avionics. Plot a circle graph using this information.

20. **General Interest**   Recent surveys have shown that college graduates are still paying off loans years and even decades after graduation. The following table shows the percent of graduates still in debt by age group. Make a circle graph of these data.

Age Group	Percent of Total Graduates in Debt
Under 30	39%
30 to 39	27%
40 to 49	15%
50 to 59	12%
60+ or unknown	7%

(*Source:* Federal Reserve Bank of New York)

Check your answers to the odd-numbered problems in the Appendix, then turn to Section 12-2 to learn about measures of central tendency.

12-2   Measures of Central Tendency

1. Find the average of these numbers: 6, 8, 12, 14
2. What number is halfway between 12 and 15?

A **measure of central tendency** is a single number that summarizes an entire set of data. The phrase *central tendency* implies that these measures represent a central or middle value of the set. Those who work in the trades and other fields, as well as consumers, use these measures to make other important calculations, comparisons, and projections of future data. In this chapter, we shall study three measures of central tendency: the mean, the median, and the mode.

Mean   As we learned in Chapter 3, the **arithmetic mean,** also referred to as simply the **mean** or the **average,** of a set of data values is given by the following formula:

$$\text{Mean} = \frac{\text{sum of the data values}}{\text{the number of data values}}$$

 Note   In this chapter, if the mean is not exact, round it to one decimal digit more than the least precise data value in the set unless otherwise directed. ●

EXAMPLE 1 **Aviation** An airplane mechanic was asked to prepare an estimate of the cost for an annual inspection of a small Bonanza aircraft. The mechanic's records showed that eight prior inspections on the same type of plane had required 9.6, 10.8, 10.0, 8.5, 11.0, 10.8, 9.2, and 11.5 hr. To help prepare his estimate, the mechanic first found the mean inspection time of the past jobs as follows:

$$\text{Mean inspection time} = \frac{\text{sum of prior inspection times}}{\text{number of prior inspections}}$$

$$= \frac{9.6 + 10.8 + 10.0 + 8.5 + 11.0 + 10.8 + 9.2 + 11.5}{8}$$

$$= \frac{81.4}{8} = 10.175 \approx 10.18 \text{ hr}$$

To complete his cost estimate, the mechanic would then multiply this mean inspection time by his hourly labor charge. ●

Your Turn **Machine Trades** A machine shop produces seven copies of a steel disk. The thicknesses of the disks, as measured by a vernier caliper are, in inches, 1.738, 1.741, 1.738, 1.740, 1.739, 1.737, and 1.740. Find the mean thickness of the disks.

Solution $$\text{Mean thickness} = \frac{1.738 + 1.741 + 1.738 + 1.740 + 1.739 + 1.737 + 1.740}{7}$$

$$= \frac{12.173}{7} = 1.739 \text{ in.}$$

1.738 ⊕ 1.741 ⊕ 1.738 ⊕ 1.74 ⊕ 1.739 ⊕ 1.737 ⊕ 1.74 ⊜ ÷ 7 ⊜ →

> *1.739*

Note Most scientific calculators have special "STAT" and "DATA" functions that allow you to enter data and calculate statistical measures. Before using any of these functions, however, you should learn to set up the calculations and compute these measures the "long way." ●

Median Another commonly used measure of central tendency is the median. The **median** of a set of data values is the middle value when all values are arranged in order from smallest to largest.

EXAMPLE 2 **Machine Trades** To determine the median thickness of the disks in the previous Your Turn, first we arrange the seven measurements in order of magnitude:

1.737   1.738   1.738   1.739   1.740   1.740   1.741

The middle value is 1.739, because there are three values less than this measurement and three values greater than it. Therefore, the median thickness is 1.739 in. Notice that for these data, the median is equal to the mean. This will not ordinarily happen. ●

In the previous example, the median was easy to find because there was an odd number of data values. If a set contains an even number of data values, there is no single middle value, so the median is defined to be the mean of the two middle values.

Your Turn **Aviation** Find the median inspection time for the data in Example 1.

Solution First, arrange the eight prior times in order from smallest to largest as follows:

8.5   9.2   9.6   10.0   10.8   10.8   11.0   11.5

Next, because there is an even number of inspection times, there is no single middle value. We must locate the two middle times. To help us find these, we begin crossing out left-end and right-end values alternately until only the two middle values remain.

8̶.̶5̶   9̶.̶2̶   9̶.̶6̶   10.0   10.8   1̶0̶.̶8̶   1̶1̶.̶0̶   1̶1̶.̶5̶

$\underbrace{\qquad\qquad}$

The two middle times

Finally, calculate the mean of the two middle values. This is the median of the set.

$$\text{Median inspection time} = \frac{10.0 + 10.8}{2} = 10.4 \text{ hours}$$

Notice that the median inspection time of 10.4 differs slightly from the mean time of 10.18.

Note   The median and the mean are the two most commonly used "averages" in statistical analysis. The median tends to be a more representative measure when the set of data contains relatively few values that are either much larger or much smaller than the rest of the numbers. The mean of such a set will be skewed toward these extreme values and will therefore be a misleading representation of the data. ●

Mode   The final, and perhaps least used, measure of central tendency that we will consider is the mode. The **mode** of a set of data is the value that occurs most often in the set. If there is no value that occurs most often, there is no mode for the set of data. If there is more than one value that occurs with the greatest frequency, each of these is considered to be a mode.

EXAMPLE 3   **Aviation**   The mode inspection time for the Bonanza aircraft (see Example 1) is 10.8 hr. It is the only value that occurs more than once.   ●

Your Turn   If possible, find the mode(s) of each set of numbers.

(a) 3, 7, 6, 4, 8, 6, 5, 6, 3   (b) 22, 26, 21, 30, 25, 28   (c) 9, 12, 13, 9, 11, 13, 8

Solutions   (a) The value 6 appears three times, the value 3 appears twice, and all others appear once. The mode is 6.

(b) No value appears more than once in the set. There is no mode.

(c) Both 9 and 13 appear twice, and no other value repeats. There are two modes, 9 and 13. A set of data containing two modes is referred to as a **bimodal** set.

More Practice   Find the mean, median, and mode of each set of numbers.

1. 22, 25, 23, 26, 23, 21, 24, 23

2. 6.5, 8.2, 7.7, 6.7, 8.9, 7.3, 6.9

3. 157, 153, 155, 157, 160, 153, 159, 158, 166, 154, 168

4. **Automotive Trades**   A daily survey revealed the following prices for a gallon of regular unleaded gas at various stations in a certain metropolitan area:

$2.39, $2.49, $2.39, $2.59, $2.45, $2.49, $2.35, $2.55, $2.39, $2.65

Find the mean, median, and mode of these prices.

5. **Construction** For quality control purposes, 8-ft-long 2-by-4s are randomly selected to be carefully measured. Find the mean, median, and mode for the following measurements:

$$8'0000''\quad 8'\frac{1''}{8}\quad 8'\frac{1''}{16}\quad 7'\frac{15''}{16}\quad 7'\frac{7''}{8}\quad 8'0000''\quad 8'\frac{1''}{8}\quad 8'\frac{3''}{8}\quad 7'\frac{5''}{8}\quad 8'0000''$$

Solutions

1. Mean: 23.375 or 23.4, rounded; median: 23; mode: 23

2. Mean: 7.457 . . . or 7.46, rounded; median: 7.3; mode: none

3. Mean: 158.18 . . . or 158.2, rounded; median: 157; modes: 153 and 157

4. Mean: $2.474; median: $2.47; mode: $2.39

5. Mean: $8\frac{1'}{80}$ or 8.0125'; median: 8'0000''; mode: 8'0000''

Grouped Frequency Distributions

When a set of data contains a large number of values, it can be very cumbersome to deal with. In such cases, we often condense the data into a form known as a **frequency distribution**. In a frequency distribution, data values are often placed into groups of the same size. These groups are known as **classes**, and the number of values within each class is tallied. The result of each tally is called the **frequency** of that class.

We will often need to refer to the size or width of the classes in a grouped frequency distribution. We define **class width** as the difference between the lower limits of two consecutive classes. For example, if the first two classes of a grouped frequency distribution are 12–16 and 17–21, the class width is $17 - 12$, or 5. Notice that the class width also corresponds to the number of whole number values within a class. In the 12–16 class, five whole number values are possible—namely, 12, 13, 14, 15, and 16.

EXAMPLE 4 **Drafting** Consider the following data values, representing the number of hours during one week that a group of drafting students spent working on the conception, design, and modeling of a crankshaft.

19	27	56	37	61	42	39	53	73	46	30	48	59	26	45	39	21
33	30	37	16	25	15	24	13	23	41	34	27	56	32	45	17	31
46	16	30	32	63	24											

To construct a grouped frequency distribution, we must first decide on an appropriate width for the classes. We shall use widths of 10 hr beginning at 10 for this data.* We then construct a table with columns representing the "Class" and the "Frequency, F," or number of students whose hours on the project fell within each interval. To help determine the frequency within each class, use tally marks to count data points. Cross out each data point as it is counted. The final table looks like this:

Class	Tally	Frequency, F
10–19 hr	⩕I	6
20–29 hr	⩕III	8
30–39 hr	⩕ ⩕ II	12
40–49 hr	⩕II	7
50–59 hr	IIII	4
60–69 hr	II	2
70–79 hr	I	1

---

*Statisticians consider many factors when deciding on the width of the class intervals. This skill is beyond the scope of this text, and all problems will include an appropriate instruction.

Your Turn   **Welding**   Construct a frequency distribution for the following percent scores on a gas tungsten arc welding project. Use classes with a width of 15 percentage points beginning at 41%.

81   75   93   74   56   56   87   93   75   70   46   67   91   72   73   81   76

65   44   83   74   83   93   65   49   62   78   80   42   58   63   54   79   86

77   92   79   66   55   75

Solution

Class	Tally	Frequency, F
41–55%	𝍖|	6
56–70%	𝍖 𝍖	10
71–85%	𝍖 𝍖 𝍖||	17
86–100%	𝍖||	7

Mean of Grouped Data   To calculate the mean of data that are grouped in a frequency distribution, follow these steps:

Step 1   Find the midpoint, M, of each class by calculating the mean of the end-points of the class. List these in a separate column.

Step 2   Multiply each frequency, F, by its corresponding midpoint, M, and list these products in a separate column headed "Product, FM."

Step 3   Find the sum of the frequencies and enter it at the bottom of the F column. We will refer to this sum as n, the total number of data values.

Step 4   Find the sum of the products from Step 2 and enter it at the bottom of the FM column.

Step 5   Divide the sum of the products from Step 4 by n. This is defined as the mean of the grouped data.

EXAMPLE 5   **Drafting**   To find the mean number of hours worked on the crankshaft design (see the previous example), extend the table as shown. The tally column has been omitted to save space. The results of each step in the process are indicated by arrow diagrams.

Step 1   Step 2

Class	Frequency, F	Midpoint, M	Product, FM
10–19 hr	6	14.5	87
20–29 hr	8	24.5	196
30–39 hr	12	34.5	414
40–49 hr	7	44.5	311.5
50–59 hr	4	54.5	218
60–69 hr	2	64.5	129
70–79 hr	1	74.5	74.5
	$n = 40$		Sum = 1430

Step 3   Step 4

Step 5 $\text{Mean} = \dfrac{\text{sum of products, } FM}{n} = \dfrac{1430}{40} = 35.75 \text{ hr}$

$= 35.8 \text{ hr, rounded}$

Using a calculator for Steps 2–5, we enter:

6 ✕ 1 4 . 5 ⊞ 8 ✕ 2 4 . 5 ⊞ 1 2 ✕ 3 4 . 5 ⊞ 7 ✕ 4 4 . 5 ⊞ 4 ✕ 5 4 . 5 ⊞ 2 ✕ 6 4 . 5 ⊞ 7 4 . 5 ⊟ ⊡ 4 0 ⊟ → **_35.75_** ●

✎ **Note** The mean of grouped data is not equal to the mean of the ungrouped data unless the midpoint of every interval is the mean of all the data in that interval. Nevertheless, this method for finding the mean from a frequency distribution provides us with a useful approximation when there is a large number of data values. In the previous example, the mean of the ungrouped data is 35.775 hr, which rounds to the same result obtained from the grouped data. ●

Your Turn **Welding** Find the mean score from the gas tungsten arc welding project (see the previous Your Turn).

Solution

Class	Frequency, F	Midpoint, M	Product, FM
41–55%	6	48	288
56–70%	10	63	630
71–85%	17	78	1326
86–100%	7	93	651
	n = 40		Sum = 2895

Step I → (Class heading)   Step 2 → (Product heading)
Step 3 → (Frequency)   Step 4 → (Sum)

Step 5 $\text{Mean score} = \dfrac{\text{sum of products, } FM}{n} = \dfrac{2895}{40} = 72.375\%$

$= 72.4\%, \text{ rounded}$

🔍 **A Closer Look** The mean of the ungrouped scores is 71.7%. There is a slight discrepancy with the mean of the grouped scores because the classes were relatively large in this case. ●

Now continue with Exercises 12-2 for more practice on measures of central tendency.

Exercises 12-2 **Measures of Central Tendency**

A. Find the mean, median, and mode for each set of numbers.

1. 9, 4, 4, 8, 7, 6, 2, 9, 4, 7

2. 37, 32, 31, 34, 36, 33, 35

3. 98, 79, 99, 79, 54, 52, 98, 58, 73, 62, 54, 54

4. 3.488, 3.358, 3.346, 3.203, 3.307

5. 123, 163, 149, 132, 183, 167, 105, 192

6. 7.9, 8.1, 9.5, 8.1, 9.6, 8.7, 7.7, 9.5, 8.1

7. 1188, 1176, 1128, 1126, 1356, 1151, 1313, 1344, 1367, 1396

8. 0.39, 0.39, 0.84, 0.11, 0.78, 0.18, 0.78

B. Construct an extended frequency distribution for each set of numbers and calculate the mean of the grouped data. Follow the instructions for class width.

1. Use a class width of 0.10 beginning at 1.00. (*Hint:* The first class is 1.00–1.09.)

    1.52   1.68   1.46   1.13   1.89   1.44   1.56   1.09   1.81   1.79   1.23

    1.46   1.91   1.40   1.30   1.35   1.79   1.61   1.80   1.26   1.29   1.40

    1.57   1.53   1.28   1.45   1.72   1.35   1.93   1.86   1.95   1.70   1.88

    1.76   1.05

2. Use a class width of 50 beginning at 100. (*Hint:* The first class is 100–149.)

    245   106   112   321   235   209   263   215   168   350   340   401   179

    433   286   141   358   468   166   498   341   171   119   362   264   325

    225   391   133   127

C. Applications

1. **Aviation**   BF Goodrich produces brake pads for commercial airliners. Two production teams are working to produce 15-lb brake pads over a six-month period. Production team A works on one furnace deck, while production team B works on a second furnace deck. A 15-lb carbon brake pad costs $1000 per pad to produce. Use the information in the table to answer the questions that follow.

Month	Team A	Team B
January	37,750 brake pads	40,000 brake pads
February	34,500	41,500
March	35,250	39,000
April	38,750	35,750
May	39,250	32,500
June	36,500	33,250

(a) Calculate the mean monthly production for team A.

(b) Calculate the mean monthly production for team B.

(c) Find the median monthly production for team A.

(d) Find the median monthly production for team B.

(e) Calculate the average cost per month for the two teams combined.

2. **General Trades**   The U.S. Department of Labor reports the following average (mean) hourly earnings of workers in 28 different occupations in 2015:

Occupation	Mean Hourly Earnings	Occupation	Mean Hourly Earnings
Automotive Tech	$19.85	Hydrologist	$40.11
Brick Mason	$24.13	Machinist	$20.25
Carpenter	$22.49	Painter	$19.47

continued . . .

Occupation	Mean Hourly Earnings	Occupation	Mean Hourly Earnings
Carpet Installer	$20.77	Plaster/Stucco Worker	$20.22
Chef/Head Cook	$22.07	Plumber	$26.49
Drywall Worker	$22.48	Registered Nurse	$34.14
Electrician	$26.23	Roofer	$19.54
Equipment Operator	$23.26	Sheet Metal Worker	$23.95
Firefighter	$23.72	Solar Installer	$19.26
Floor Sander/ Finisher	$18.35	Steel/Iron Worker	$26.32
Food Service Manager	$25.79	Stone Mason	$20.66
Forester	$29.16	Surveyor	$29.75
Glazier	$21.84	Tile/Marble Setter	$21.16
HVAC Installer	$22.78	Water/Wastewater Treatment Operator	$22.49

(a) Find the mean of these average hourly earnings.

(b) Find the median of these average hourly earnings.

(c) In a 40-hr work week, how much more would the average surveyor earn compared to the mean of all 28 occupations?

(d) In a 40-hr work week, how much less would the average automotive tech make compared to the median of all 28 occupations?

(e) By what percent does the average registered nurse's earnings exceed the mean?

(f) By what percent does the average roofer's earnings lag the median?

3. **Forestry** The following table shows both the total number of wildfires and the total number of acres burned by wildfires annually in the United States from 2005 through 2015:

Year	Number of Wildfires	Total Acres Burned
2005	66,753	8,689,389
2006	96,385	9,873,745
2007	85,705	9,328,045
2008	78,979	5,292,468
2009	78,792	5,921,786
2010	71,971	3,422,724
2011	74,126	8,711,367
2012	67,774	9,326,238
2013	47,579	4,319,546
2014	63,312	3,595,613
2015	68,151	10,125,149

(*Source:* National Interagency Fire Center)

For this time period, calculate each of the following to the nearest whole number:

(a) The mean number of annual wildfires

(b) The median number of annual wildfires

(c) The mean number of acres burned

(d) Use your answers to parts (a) and (c) to find the average number of acres burned per wildfire. (Round to the nearest whole number.)

(e) By what percent did the number of wildfires in 2006 exceed the median number for the time period shown?

(f) By what percent was the number of acres burned in 2014 below the mean for the time period shown?

4. **Water/Wastewater Treatment**   The seven-day mean of settleable solids cannot exceed 0.15 milliliters per liter (mL/L). During the past seven days, the measurements of concentration have been 0.13, 0.18, 0.21, 0.14, 0.12, 0.11, and 0.15 mL/L. What was the mean concentration? Was it within the limit?

5. **Automotive Trades**   A mechanic has logged the following numbers of hours in maintaining and repairing a certain Detroit Series 6 diesel engine for ten different servicings.

8.5   15.9   20.0   6.5   11.2   19.1   18.0   7.4   9.8   22.5

(a) Find the mean for the hours logged.

(b) Find the median for the hours logged.

(c) If the mechanic earns $20 per hour, how much does he earn doing the median servicing on this engine?

6. **Forestry**   A forest ranger wishes to determine the velocity of a stream. She throws an object into the stream and clocks the time it takes to travel a premeasured distance of 200 ft. She then divides the time into 200 to calculate the velocity in feet per second. For greater accuracy, she repeats this process four times and finds the mean of the four velocities. If the four trials result in times of 21, 23, 20, and 23 sec, calculate the mean velocity to the nearest 0.1 ft/sec.

7. **Hydrology**   The following table shows the monthly flow through the Darnville Dam from October through April:

Month	Volume (in thousand acre-feet, kAF)
October	92
November	133
December	239
January	727
February	348
March	499
April	1031

(a) Calculate the mean monthly water flow.

(b) Find the median monthly water flow.

(c) By what percent did the flow in March exceed the mean flow?

8. **Meteorology** The National Weather Service provides data on the number of tornados in each state that occur throughout the year. Use the following statistics to answer questions (a) through (e). (Round to the nearest whole number.)

Year	Kansas	Oklahoma
2005	136	28
2006	91	27
2007	141	50
2008	187	77
2009	103	33
2010	88	102
2011	68	119
2012	95	63
2013	55	82
2014	41	16
2015	126	111

(a) Find the mean of the number of tornados in Kansas over the 11-year period.

(b) Find the mean of the number of tornados in Oklahoma over the 11-year period.

(c) Find the median of the number of tornados in Kansas over the 11-year period.

(d) Find the median of the number of tornados in Oklahoma over the 11-year period.

(e) Suppose a tornado causes an average $500,000 worth of damage. Use the mean in part (a) to calculate the average cost per year to repair tornado damage in Kansas.

9. **Automotive Trades** The following table shows both the average fuel economy (in miles per gallon) and the tailpipe emissions of carbon dioxide (in grams per mile) for 12 hybrid cars. Study the table and answer the questions that follow.

Car Model	Average Fuel Economy (mi/gal)	Tailpipe Emissions (g/mi)
Prius c	49.5	178
Fusion (2nd gen.)	47	190
Lincoln MKZ	45	198
Honda Civic (3rd gen.)	44	202
Honda Insight	42.5	212
Prius v	42	212
Lexus CT	41.5	212
Camry LE	41	217
Acura ILX	38.5	234
Hyundai Sonata	37.5	240
Nissan Altima	33	269
Ford Escape FWD	32.5	278

(a) What is the mean fuel economy of these models?

(b) What is the median fuel economy of these models?

(c) What is the mean amount of carbon dioxide emissions of these models?

(d) What is the median amount of carbon dioxide emissions of these models?

(e) For a trip of 500 miles, how many fewer gallons of fuel will a Prius c use than the median of all the models? (Round to the nearest tenth.)

(f) By what percent do the tailpipe emissions of the Nissan Altima exceed the mean emissions of the group?

10. **Hydrology**   The following table shows the daily evaporation and usage totals, in acre-feet, during a summer week at Bradbury Dam. Use these data to answer the questions.

Day	Daily Evaporation (in acre-ft)	Daily Usage (in acre-ft)
Monday	62.3	108.7
Tuesday	58.2	130.2
Wednesday	51.5	117.9
Thursday	58.6	126.5
Friday	67.3	130.5
Saturday	67.3	130.2
Sunday	62.8	128.0

(a) Find the mean daily evaporation.

(b) Find the median daily evaporation.

(c) Find the mean daily usage.

(d) Find the median daily usage.

11. **Allied Health**   The Apgar score is widely used to assess the general health of infants shortly after birth. Scores range from 0 (very poor) to 10 (perfect). To evaluate the success of a new maternal health program, a hospital compared the Apgar scores of 20 infants whose mothers participated in the health program (P = participant) with the scores of 20 infants whose mothers did not participate (NP = nonparticipant). The infants' scores for the two groups are shown in the following table.

P:	7	8	7	10	7	9	8	4	8	10	9	7	8	8	5	10	9	7	8	9
NP:	8	10	5	7	6	8	7	6	9	7	4	6	7	9	2	9	6	10	8	7

(a) Calculate the mean and median of the Apgar scores for the infants whose mothers participated in the program.

(b) Calculate the mean and median of the Apgar scores for the infants whose mothers did not participate in the program.

12. **Allied Health**   A pharmacist keeps careful track of the amount of medication sold each month in order to ensure that supplies are always available for her customers. However, she must avoid storing too much medication so that they do not expire before they are sold. For the antidepressant medication Alljoy, the following amounts were sold last year:

Month	Packages Sold
January	242
February	273
March	187
April	163

continued . . .

Month	Packages Sold
May	155
June	135
July	93
August	148
September	159
October	174
November	238
December	256

(a) Calculate the mean monthly number of packages of Alljoy sold during the year.

(b) Find the median monthly number of packages of Alljoy sold during the year.

(c) Calculate the monthly mean for the winter months of January, February, and March.

(d) Calculate the monthly mean for the summer months of June, July, and August.

(e) What was the percent of decrease from winter [answer (c)] to summer [answer (d)]?

For Problems 13 and 14, construct an extended frequency distribution for each set of numbers and calculate the mean of the grouped data. Follow the instruction for class width.

13. **Automotive Trades** To best determine the fuel economy of a particular car model, an automotive technician calculates the mean gas mileage for 20 full tanks of gas.

(a) Use classes of 1.0 mi/gal, beginning with 19.0–19.9 mi/gal, to compute the mean fuel economy (in miles per gallon) from the following data:

21.7 19.5 20.6 25.6 24.3 20.1 20.7 22.5 20.2 24.5

23.7 20.1 23.1 19.9 23.5 24.4 19.7 25.2 21.8 22.3

(b) At $2.70 per gallon, use the mean fuel economy to calculate the average cost per mile to operate this vehicle.

14. **Allied Health** A registered nurse has been carefully monitoring the blood work of a patient each day over a 24-day period. Each day a blood analysis was performed, and the patient's medication, in milligrams, was titrated (adjusted) accordingly. Use the values of the medication administered to determine the mean dose for the last 24 days. Use classes of 0.50 mg beginning with 78.50–78.99 mg.

Doses in mg

80.63 79.78 78.78 80.79 81.13 79.17 78.66 79.13

79.60 79.29 81.29 81.33 78.71 81.47 79.89 80.41

80.01 79.46 78.84 79.26 79.58 79.47 80.60 81.08

Check your answers to the odd-numbered problems in the Appendix, then turn to Section 12-3 to learn about measures of dispersion.

12-3   Measures of Dispersion

learning|catalytics™
1. Calculate: $65 - 72.5$
2. Calculate: $(-2.5)^2$

Measures of central tendency, such as the mean, give us valuable information about a set of data, but they do not indicate how much variation exists within the set. To analyze whether or not the mean, for example, is typical of the numbers in the set, we use a measure of dispersion. A **measure of dispersion** is a statistical measure that indicates the spread or variability of data values in a set. A large value for a measure of dispersion indicates that the data are spread out, while a small value indicates that the data values are tightly clustered and vary little from the mean value of the set. In this section, we shall study two measures of dispersion: range and standard deviation.

Range   The range is the simplest measure of dispersion. In any set of data, the **range** is the difference between the largest and the smallest value.

> Range = largest data value − smallest data value

EXAMPLE 1   **Electronics**   The numbers below are the resistor values from the E12 Series most recently used by an electronics technician. Determine the range for the following resistors in ohms:

2.2 Ω   3.9 Ω   8.2 Ω   4.7 Ω   3.3 Ω   1.2 Ω   5.6 Ω   2.7 Ω   6.8 Ω

Range = largest resistor value − smallest resistor value

= 8.2 Ω − 1.2 Ω = 7.0 Ω   ●

Your Turn   **Automotive Trades**   The following measurements represent the distance between the ground electrode and the center electrode on a spark plug. This is commonly known as the spark plug gap. Determine the range for these spark plug gaps.

0.030 in.   0.045 in.   0.028 in.   0.032 in.   0.020 in.
0.025 in.   0.035 in.   0.040 in.   0.015 in.

Solution   Range = largest gap − smallest gap

= 0.045 in. − 0.015 in. = 0.030 in.

Deviation from the Mean   The range gives a quick indication of the overall spread of the data, but, if there are just one or two extreme values in a large set of data, this could be a misleading indication of dispersion. The **standard deviation** is a more commonly used measure of dispersion because it gives statisticians an indication of how widely scattered all the values of the data are, not just the two extremes.

A key component in the formula for standard deviation is a value called the deviation from the mean. For any data value, x, its **deviation from the mean**, is given by the difference $x - \bar{x}$, where $\bar{x}$ represents the mean of the set of data.

EXAMPLE 2   **Landscaping**   The following set of data represents the number of cubic yards of a compost mix used in the last eight projects completed by a landscaping company:

9.4   12.8   4.5   15.6   2.2   23.9   8.3   6.1

To calculate the deviations from the mean, first we must find the mean.

$$\text{Mean }(\overline{x}) = \frac{9.4 + 12.8 + 4.5 + 15.6 + 2.2 + 23.9 + 8.3 + 6.1}{8}$$

$$= 10.35 \text{ yd}^3$$

Finally, we construct a table showing the data (referred to as x-values) in one column and the corresponding deviations from the mean $(x - \overline{x})$, for each value of x, in a second column.

Data, x	$(x - \overline{x})$
9.4	$9.4 - 10.35 = -0.95$
12.8	$12.8 - 10.35 = \;\;2.45$
4.5	$4.5 - 10.35 = -5.85$
15.6	$15.6 - 10.35 = \;\;5.25$
2.2	$2.2 - 10.35 = -8.15$
23.9	$23.9 - 10.35 = 13.55$
8.3	$8.3 - 10.35 = -2.05$
6.1	$6.1 - 10.35 = -4.25$

Difference between a data point and the mean

Your Turn **Allied Health** A certified nursing assistant (CNA) recorded the following heart rates in beats per minute (bpm) of a patient over a 12-hr period:

62 69 73 76 67 58 65 73 81 76 71 63

Calculate the mean heart rate, and then prepare a table showing the heart rates (x) in one column and the deviations from the mean $(x - \overline{x})$ in a second column.

Solution $\text{Mean }(\overline{x}) = \dfrac{62 + 69 + 73 + 76 + 67 + 58 + 65 + 73 + 81 + 76 + 71 + 63}{12}$

$$= 69.5 \text{ bpm}$$

Heart Rates	$(x - \overline{x})$
62	$62 - 69.5 = \;\;-7.5$
69	$69 - 69.5 = \;\;-0.5$
73	$73 - 69.5 = \;\;\;\;3.5$
76	$76 - 69.5 = \;\;\;\;6.5$
67	$67 - 69.5 = \;\;-2.5$
58	$58 - 69.5 = -11.5$
65	$65 - 69.5 = \;\;-4.5$
73	$73 - 69.5 = \;\;\;\;3.5$
81	$81 - 69.5 = \;\;11.5$
76	$76 - 69.5 = \;\;\;\;6.5$
71	$71 - 69.5 = \;\;\;\;1.5$
63	$63 - 69.5 = \;\;-6.5$

Standard Deviation If we find the sum of the deviations from the exact (unrounded) value of the mean in each of the last two examples, we obtain zero. This will always happen because the mean is the "center" of the data and the negative and positive deviations from this center will cancel each other out. In calculating standard deviation, we use the squares of the deviations from the mean to eliminate the difficulty with negative deviations. To calculate the **standard deviation** (*s*) for a set of data, follow these steps:

1. Find the mean ($\bar{x}$).

2. Calculate each deviation from the mean: ($x - \bar{x}$).

3. Square each of these deviations: ($x - \bar{x}$)^2.

4. Find the sum of these squares.

5. Divide this sum by $n - 1$, where *n* is the number of values in the set.

6. Take the square root of this quotient.

We can summarize this six-step procedure with the following formula:

$$\text{Standard deviation } (s) = \sqrt{\frac{\text{sum of } (x - \bar{x})^2}{n - 1}}$$

Note   This formula is often called the "sample standard deviation," and it is used primarily when the data represent a sample of the entire population being studied. If the data represents the entire population being studied, an alternate formula called the "population standard deviation" is used. The only difference in this formula is that the denominator under the square root symbol is *n* instead of $n - 1$. For our purposes, we shall assume that all data in this chapter qualify for the formula given in the box. ●

EXAMPLE 3   **Allied Health**   In the previous Your Turn, we found the mean heart rate of a patient (Step 1) and the deviations from the mean (Step 2). To complete the calculation for the standard deviation of this data, we continue as follows:

Step 3   Extend the table to show an $(x - \bar{x})^2$ column, and square each of the deviations from the mean in column 2 to obtain these numbers.

Step 4   Find the sum of these squares.

The results of these two steps are shown in the completed table:

Heart Rates	$(x - \bar{x})$	$(x - \bar{x})^2$
62	−7.5	56.25
69	−0.5	0.25
73	3.5	12.25
76	6.5	42.25
67	−2.5	6.25
58	−11.5	132.25
65	−4.5	20.25
73	3.5	12.25
81	11.5	132.25
76	6.5	42.25
71	1.5	2.25
63	−6.5	42.25
		Sum = 501

Step 5   Divide the sum of the $(x - \bar{x})^2$ values by $n - 1$, which in this case is 11.

$$\frac{\text{sum of } (x - \bar{x})^2}{n - 1} = \frac{501}{11} \approx 45.55$$

Step 6   Take the square root of this quotient.

$$s = \sqrt{45.55} = 6.749\ldots$$

We round the standard deviation to the same number of decimal digits as the mean.

$$s \approx 6.7 \text{ bpm}$$

Notice that standard deviation has the same units as the mean and the original data.

Once you have calculated the mean, use the following sequence to fill in the $(x - \bar{x})^2$ column directly:

First, store the mean in memory: **6 9 . 5** (STO)(A)

Then, use this sequence: **6 2** (−)(RCL)(A)(=)(x²)(=) → | 56.25 | ← **Row 1**

6 9 (−)(RCL)(A)(=)(x²)(=) → | 0.25 | ← **Row 2**

. . . and so on.

After completing the $(x - \bar{x})^2$ column, use your calculator to find the sum of this column. In this case the sum is 501. **Finally,** substitute the sum into the standard deviation formula:

5 0 1 (÷)**1 1**(=) → | 45.54545455 | (√)(ANS)(=) → | 6.748737256 | ●

Learning Help   Remember: The first step in finding the standard deviation is always to calculate the mean. ●

Your Turn   **Automotive Trades**   An automotive technician recorded the number of hours of labor required for each of the last 10 engine rebuilds.

20.3   22.1   18.9   21.7   19.3   20.8   21.6   19.9   20.5   21.3

Calculate the standard deviation of these data. Show all intermediate results in a table as illustrated in the previous example.

Solution   The following is the completed table:

Hours of Labor	(x − x̄) *Step 2: Find the deviations from the mean.*	(x − x̄)² *Step 3: Find the squares of the deviations.*
20.3	$20.3 - 20.64 = -0.34$	0.1156
22.1	$22.1 - 20.64 = 1.46$	2.1316
18.9	$18.9 - 20.64 = -1.74$	3.0276
21.7	$21.7 - 20.64 = 1.06$	1.1236
19.3	$19.3 - 20.64 = -1.34$	1.7956
20.8	$20.8 - 20.64 = 0.16$	0.0256
21.6	$21.6 - 20.64 = 0.96$	0.9216

(continued)

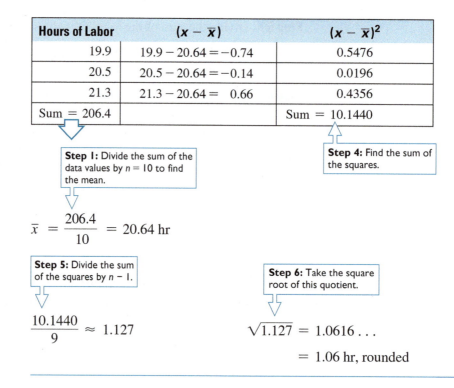

Hours of Labor	$(x - \bar{x})$	$(x - \bar{x})^2$
19.9	$19.9 - 20.64 = -0.74$	0.5476
20.5	$20.5 - 20.64 = -0.14$	0.0196
21.3	$21.3 - 20.64 = 0.66$	0.4356
Sum = 206.4		Sum = 10.1440

Step 1: Divide the sum of the data values by $n = 10$ to find the mean.

Step 4: Find the sum of the squares.

$$\bar{x} = \frac{206.4}{10} = 20.64 \text{ hr}$$

Step 5: Divide the sum of the squares by $n - 1$.

Step 6: Take the square root of this quotient.

$$\frac{10.1440}{9} \approx 1.127$$

$$\sqrt{1.127} = 1.0616\ldots$$
$$= 1.06 \text{ hr, rounded}$$

Visualizing Dispersion

To visualize dispersion, imagine two archers shooting at a target. Think of the bull's-eye as the mean. The first archer is very accurate and consistent. Her shots are clustered around the center of the target, close to the bullseye. They have a small dispersion. The second archer is a beginner; his shots are widely spread out. One way to describe his inconsistency is to say that his shots have a large dispersion.

The standard deviation measures the scatter of the data points from the mean value of the data. The set of measurement numbers (24, 24, 24, 24, 24) has mean equal to 24 and $s = 0$. There is no dispersion. The set of measurements (22, 23, 24, 25, 26) also has a mean of 24 and a standard deviation of approximately 1.6. The dispersion is small. The set of measurements (2, 8, 21, 36, 53) still has a mean of 24, but the standard deviation is approximately 20.8, indicating a large dispersion.

Grouped Data   Recall from Section 12-2 that we often use a frequency distribution to group data when a set is large. The original data are displayed in two columns, one showing the limits of each class and one showing frequencies (F) within each class. To help us calculate the mean of this grouped data, we added two more columns, one showing the midpoints (M) of the classes and the other showing the products of the frequencies and the midpoints (FM). Now, to help us calculate the standard deviation for grouped data, we create the following three additional columns:

1. $(M - \bar{x})$, to list the deviation between each midpoint and the mean.

2. $(M - \bar{x})^2$, to list the square of each difference.

3. $F(M - \bar{x})^2$, to list the product of each square times the frequency of the corresponding interval.

Notice that we must calculate the mean before we can fill in these last three columns. The standard deviation for grouped data is defined by the following formula:

> ## Standard Deviation for Grouped Data
>
> $$s = \sqrt{\frac{\text{sum of } F(M - \bar{x})^2}{n - 1}}$$

For grouped data, the number of values, n, is the same as the sum of the frequencies.

EXAMPLE 4 **Machine Trades** The following frequency distribution represents the diameters of 56 finely machined holes measured to 0.001 in. The diameters are grouped in classes with a width of 0.005 in. beginning with 0.360–0.364 in. The number of measurements in each class is shown in the frequency column.

Class	Frequency, F
0.360–0.364 in.	4
0.365–0.369 in.	10
0.370–0.374 in.	14
0.375–0.379 in.	15
0.380–0.384 in.	8
0.385–0.389 in.	5

To calculate the mean and standard deviation for this grouped data, follow these steps:

Step 1 Extend the table as shown with five additional columns.

Step 2 Fill in the midpoint (M) and product (FM) columns.

Class	Frequency, F	Midpoint, M	Product, FM	$(M - \bar{x})$	$(M - \bar{x})^2$	$F(M - \bar{x})^2$
0.360–0.364 in.	4	0.362	1.448			
0.365–0.369 in.	10	0.367	3.670			
0.370–0.374 in.	14	0.372	5.208			
0.375–0.379 in.	15	0.377	5.655			
0.380–0.384 in.	8	0.382	3.056			
0.385–0.389 in.	5	0.387	1.935			
	$n = 56$		Sum $= 20.972$			

Step 3 Find the sum of the F and FM columns and use these sums to calculate the mean.

$$\text{Mean } (\bar{x}) = \frac{\text{sum of } FM}{n} = \frac{20.972}{56} = 0.3745 \text{ in.}$$

Step 4 Fill in the last three columns.

Class	Frequency, F	Midpoint, M	Product, FM	$(M - \bar{x})$	$(M - \bar{x})^2$	$F(M - \bar{x})^2$
0.360–0.364 in.	4	0.362	1.448	−0.0125	0.00015625	0.000625
0.365–0.369 in.	10	0.367	3.670	−0.0075	0.00005625	0.0005625
0.370–0.374 in.	14	0.372	5.208	−0.0025	0.00000625	0.0000875
0.375–0.379 in.	15	0.377	5.655	0.0025	0.00000625	0.00009375
0.380–0.384 in.	8	0.382	3.056	0.0075	0.00005625	0.00045
0.385–0.389 in.	5	0.387	1.935	0.0125	0.00015625	0.00078125
	$n = 56$		Sum = 20.972			Sum = 0.0026

Step 5 Use the sum of the $F(M - \bar{x})^2$ column and $n - 1$ to calculate the standard deviation.

$$s = \sqrt{\frac{\text{sum of } F(M - \bar{x})^2}{n - 1}} = \sqrt{\frac{0.0026}{55}} \approx 0.00687 \text{ in.} = 0.0069 \text{ in. rounded}$$

 Using a calculator, we can fill in the last column directly as follows:

First, store the mean in memory: **.3745** STO A

Then, calculate the $F(M - \bar{x})^2$ values like this:

First interval → **.362** − RCL A = x^2 × **4** = →   *0.000625*

Second interval → **.367** − RCL A = x^2 × **10** = →   *0.0005625*
... and so on.

Once the final column is filled in, calculate the sum of the entries in this column:

0.000625 + **.0005625** + **.0000875** + **.00009375** + **.00045** +

.00078125 = →   *0.0026*

Finally, substitute into the standard deviation formula:

÷ **55** = √ = →   *0.006875517*   ●

Your Turn **Construction** The following is a grouped frequency distribution showing the number and the cost of projects in the last year that have been overseen by a construction company. Use the five-step method shown in the previous example to calculate the mean and the standard deviation for this data.

Class	Frequency, F
$5,000–$14,999	12
$15,000–$24,999	8
$25,000–$34,999	4
$35,000–$44,999	2
$45,000–$54,999	8
$55,000–$64,999	6
$65,000–$74,999	10

Solution

Step 1: Attach the five additional columns.

Step 2: Fill in these two columns.

Step 4: Fill in these three columns.

Class	Freq., F	Midpt., M	Prod., FM	$(M - \bar{x})$	$(M - \bar{x})^2$	$F(M - \bar{x})^2$
\$5,000–\$14,999	12	9,999.5	119,994	−28,800	829,440,000	9,953,280,000
\$15,000–\$24,999	8	19,999.5	159,996	−18,800	353,440,000	2,827,520,000
\$25,000–\$34,999	4	29,999.5	119,998	−8,800	77,440,000	309,760,000
\$35,000–\$44,999	2	39,999.5	79,999	1,200	1,440,000	2,880,000
\$45,000–\$54,999	8	49,999.5	399,996	11,200	125,440,000	1,003,520,000
\$55,000–\$64,999	6	59,999.5	359,997	21,200	449,440,000	2,696,640,000
\$65,000–\$74,999	10	69,999.5	699,995	31,200	973,440,000	9,734,400,000
	$n = 50$	Sum = 1,939,975				Sum = 26,528,000,000

Step 3: Find the mean using these two sums.

Step 5: Use this sum and $n - 1$ to calculate standard deviation.

$$\bar{x} = \frac{1,939,975}{50} = \$38,799.50$$

$$s = \sqrt{\frac{\text{sum of } F(M - \bar{x})^2}{n - 1}}$$

$$= \sqrt{\frac{26,528,000,000}{49}}$$

$$\approx \$23,267.74 \approx \$23,268$$

Note The standard deviation of \$23,268 for the construction projects is relatively large compared with the mean. This implies that the cost data are widely scattered about the mean and that the mean would not be very useful in predicting costs of future projects.

On the other hand, the standard deviation of 0.0069 in. for the finely machined holes in Example 4 is relatively small. This implies that the data are closely bunched about the mean, and the mean measurement of 0.3745 in. is a reliable representation of the data. ●

Normal Distribution Given a large enough sample, the distribution of measurements of many natural objects or processes tend to approximate what is known as the **normal distribution**. The graph in the figure below illustrates this distribution.

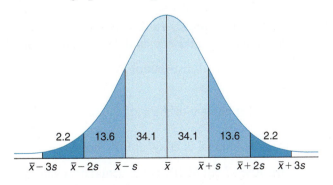

The vertical line at the center of the graph, drawn to the peak of the curve, represents the mean $\bar{x}$ of the data. The first section to the right of the mean, which ends one standard deviation above the mean ($\bar{x} + s$), should contain about 34% of the data values. The same is true for the first section to the left of the mean, ending at $\bar{x} - s$. This means that, in a normal distribution, approximately 68% of the data values are *within one standard deviation* of the mean ($\bar{x} \pm s$). The sections of the graph that are to the right and left of these end at $\bar{x} + 2s$ and $\bar{x} - 2s$, and these should each contain an additional 13.6% of the data values. Adding up all four percentages, we can conclude that in a normal distribution approximately 95% of the data values should fall within two standard deviations of the mean—that is, in the interval between $\bar{x} - 2s$ and $\bar{x} + 2s$. The final two sections shown end at $\bar{x} + 3s$ and $\bar{x} - 3s$. In a normal distribution, approximately 99.7% of all data values should fall within three standard deviations of the mean.

EXAMPLE 5 Suppose a set of data contains 100 values with a mean of 45 and a standard deviation of 4.8. If the data were distributed normally, we would expect that 68 of the 100 values would fall between $45 - 4.8$ and $45 + 4.8$—in other words, between 40.2 and 49.8. All values in this range are within one standard deviation of the mean. ●

Your Turn   Assume that the set of data described in Example 5 were distributed normally. Within what range would you expect to find 95 of the 100 data values?

Solution   With normal distribution, we expect to find 95% of the data values within two standard deviations of the mean. If one standard deviation is 4.8, then two standard deviations is $2(4.8) = 9.6$. Therefore, 95 of the 100 data values should fall between $45 - 9.6$ and $45 + 9.6$, or between 35.4 and 54.6.

Note   You should not expect any given set of real-world data to perfectly reflect normal distribution. However, many large sets of data do tend to approximate normal distribution, and we can therefore use this concept to make predictions and otherwise analyze sets of data. ●

EXAMPLE 6   **Allied Health**   In Example 3 we calculated the standard deviation for the heart rate data to be $s \approx 6.7$ bpm. We had previously found the mean of these data to be 69.5 bpm (see Your Turn on page 819). If this were a much larger set of measurements, we would expect 68% of them to be in the interval 69.5 ± 6.7 bpm, or between 62.8 bpm and 76.2 bpm. Checking the actual numbers, we see that 9 of the 12 heart rate measurements, or 75%, fall within one standard deviation of the mean. Considering the fact that this is a relatively small set of data, this is very close to the expected 68% for a normal distribution. ●

Your Turn   **Automotive Trades**   In the Your Turn on pages 821–2, we found the mean to be 20.64 hours and the standard deviation to be approximately 1.06 hours.

(a) In a normal distribution, what is the interval in which 68% of the data should fall?

(b) What percent of the actual data fell within this interval?

Solutions   (a) The interval is 20.64 ± 1.06, or between 19.58 and 21.70.

(b) 70% (7 out of 10) of the actual data values fell within this interval.

Now continue with Exercises 12-3 for more practice in calculating measures of dispersion.

Exercises 12-3 | **Measures of Dispersion**

A. Calculate the range and the standard deviation for each set of numbers.

1. 5, 4, 2, 5, 7, 5, 2, 4, 3, 8

2. 19.9, 2.5, 19.8, 5.6, 12.4, 17.1, 7.6, 9.3, 20.1

3. 75, 72, 83, 85, 77, 89, 79

4. 364, 350, 355, 320, 349, 310, 353, 347

5. 3.58, 3.45, 3.63, 3.89, 3.13, 3.45, 3.51

6. 48, 35, 25, 33, 38, 31, 17, 25, 44, 32, 36, 14

7. 2643, 1511, 3133, 2507, 3251, 3456, 2469, 2800, 3264, 2295

8. 2.852, 2.893, 2.681, 2.538, 2.507

B. Calculate the standard deviation of the grouped data in each problem. Create an extended table using the five additional columns shown in the text.

1.

Class	Frequency, F
500–999 sq. ft	13
1,000–1,499 sq. ft	9
1,500–1,999 sq. ft	11
2,000–2,499 sq. ft	7
2,500–2,999 sq. ft	5

2.

Class	Frequency, F
0–9 cm	12
10–19 cm	16
20–29 cm	9
30–39 cm	8
40–49 cm	15
50–59 cm	10

3.

Class	Frequency, F
0–1.9 sec	4
2.0–3.9 sec	6
4.0–5.9 sec	9
6.0–7.9 sec	10
8.0–9.9 sec	8
10.0–11.9 sec	7
12.0–13.9 sec	6

4.

Class	Frequency, F
50–149 psi	5
150–249 psi	9
250–349 psi	14
350–449 psi	4
450–549 psi	6
550–649 psi	12
650–749 psi	8
750–849 psi	2

C. For each set of data, (a) calculate the standard deviation, (b) state the interval that, if normally distributed, should contain approximately 68% of the data, and (c) calculate the percent of values that actually fall within this interval.

1. 46, 55, 48, 47, 39, 52, 60, 44, 50, 49, 33, 42, 54, 66, 48, 53, 45, 55, 31, 40

2. 7.48, 7.52, 7.49, 7.51, 7.47, 7.50, 7.49, 7.48, 7.46, 7.49, 7.48, 7.51, 7.47, 7.49

D. For each set of data, (a) calculate the standard deviation, (b) state the interval that, if normally distributed, should contain approximately 95% of the data, and (c) calculate the percent of values that actually fall within this interval.

1. 286, 278, 288, 290, 276, 285, 289, 293, 276, 285, 287, 290, 291, 274, 278

2. 0.788, 0.787, 0.788, 0.790, 0.786, 0.788, 0.785, 0.789, 0.792, 0.787, 0.785, 0.794

E. Applications

1. **Landscaping**   The following set of data represents the number of yards (cubic yards) of a bark mulch used in the last ten projects completed by a landscaping company. Determine the range for these amounts.

   6   8   2   10   5   13   7   3   15   5

2. **Automotive Trades**   An automotive technician has recorded the number of hours of labor required for each of the last ten engine rebuilds. Find the range of these times.

   20.3   22.1   18.9   21.7   19.3   20.8   21.6   19.9   20.5   21.3

3. **Police Science**   The following table shows the number of citations issued by the Pine Valley police department for certain traffic violations from 2012–2017:

	Speeding	Parking
2012	332	612
2013	495	510
2014	380	495
2015	472	550
2016	412	483
2017	366	568

   (a) Which category had the largest range, and what was this value?

   (b) Find the standard deviation for the speeding data.

   (c) Find the standard deviation for the parking data.

4. **Business and Finance**   The following table shows the revenue for each quarter of Megacashron in billions of dollars. Use this information to answer the following questions.

Quarter	2016	2017
First	$6.0	$5.5
Second	$4.0	$7.0
Third	$5.5	$6.5
Fourth	$7.0	$3.0

   (a) During which year was the range of quarterly revenue the largest?

   (b) Find the standard deviation of the revenue for all eight quarters.

5. **Aviation** BF Goodrich produces brake pads for commercial airliners. Two production teams are working to produce 15-lb brake pads over a 6-month period. Team A works on one furnace deck, while production team B works on a second furnace deck. A 15-lb carbon brake pad costs $1000 per pad to produce. The following table shows the total monthly output of each team:

Month	Team A	Team B
January	37,750 brake pads	40,000 brake pads
February	34,500	41,500
March	35,250	39,000
April	38,750	35,750
May	39,250	32,500
June	36,500	33,250

(a) Calculate the mean and standard deviation for the data for production team A.

(b) Calculate the mean and standard deviation for the data for production team B.

(c) As the lead of the two production teams, which team would you prefer and why? (Use the mean and the standard deviation of each to help answer this question.)

6. **Automotive Trades** The following chart shows the number of tires sold per month last year by the Treadwell Tire Company:

Month	Number of Tires Sold
January	580
February	650
March	700
April	630
May	750
June	770
July	800
August	950
September	820
October	750
November	600
December	550

(a) Find the range of the number of tires sold per month.

(b) Find the standard deviation of the monthly sales totals. Round to the nearest whole number.

7. **Machine Trades** A tool and die company machines a precise pin that is supposed to measure 10.25 mm in diameter. For quality control purposes, ten pins are selected at random and measured to determine how closely they match the specified diameter. The results are as follows:

10.28 10.31 10.24 10.23 10.25 10.22 10.23 10.32 10.25 10.22

(a) Find the range of the measurements.

(b) Find the standard deviation of measurements.

8. **Welding**   The pressure in acetylene cylinders used in oxyacetylene welding (OAW) must be monitored carefully since the pure form of acetylene is extremely unstable at a pressure greater than 15 pounds per square inch (psi). Find the standard deviation of the pressure measurements for the eight cylinders shown in the table.

Cylinder	Acetylene Cylinder Pressure (in psi)
A	12
B	13
C	15
D	10
E	14
F	15
G	11
H	12

9. **Meteorology**   The following table shows the number of tornados that have occurred in four different states over a six-year period.

Year	Kansas	Oklahoma	Arkansas	Missouri
2010	88	102	33	64
2011	68	119	75	78
2012	95	63	18	29
2013	55	82	34	49
2014	41	16	20	47
2015	126	111	20	48

(a) In which state was the range of these yearly totals the largest, and what was this range?

(b) Find the standard deviation for the numbers of tornados in Kansas.

(c) Find the standard deviation for the numbers of tornados in Missouri.

(d) Comparing the answers to parts (b) and (c), what does this tell us about the relative reliability of the mean for the two states?

10. **General Interest**   The Energy Information Administration of the U.S. Department of Energy records the annual crude oil field production in the United States. Use the information in the table to find the standard deviation for a recent seven-year period.

Year	Annual U.S. Crude Oil Field Production (billion barrels)
2009	1.95
2010	2.00
2011	2.06
2012	2.37
2013	2.73
2014	3.20
2015	3.44

For Problems 11 and 12, construct an extended frequency distribution for each set of numbers. Calculate the standard deviation of the grouped data in each problem. Create an extended table using the five additional columns shown in the text. Follow the instruction for class width.

11. **Electrical Trades**   The following values are amperages measured for a collection of wires varying in size from 0000 AWG (American Wire Gauge) gauge to 6 AWG gauge. Use classes of 60 AWG beginning with 100–159 AWG.

Amps

383	168	389	179	217	309	186	340	221	156	398	254
106	282	290	111	143	367	264	275	345	128	173	161
326	190	379	267	209	124						

12. **Automotive Trades**   The following data are the average fuel efficiency numbers (in miles per gallon) for passenger cars in the United States from 1991 through 2014. Use classes of 1.0 mi/gal beginning with 27.0–27.9 mi/gal. (*Source:* Bureau of Transportation)

28.4	27.9	28.4	28.3	28.6	28.5	28.7	28.8
28.3	28.5	28.8	29.0	29.5	29.5	30.3	30.1
31.2	31.5	32.9	33.9	33.1	35.4	36.2	36.4

13. **Electrical Trades**   From a series of measurements, the mean voltage in a certain line is found to be 22,680 V with a standard deviation of 128 V. A breaker must be installed to protect against surges in voltage. If the breaker is set so that it is activated only about 2.5% of the time, at what level is it set?

14. **Automotive Trades**   In testing the acceleration from 0 to 60 mph for a certain automobile, the following times (in seconds) resulted from a series of trials:

6.8, 6.9, 6.8, 6.7, 6.8, 6.6, 6.9, 7.0, 6.7, 6.8, 6.6, 6.7, 7.1, 6.9, 6.5

(a) Calculate the mean and standard deviation of these times.

(b) What range of times should the technician report so the interval would normally contain approximately 68% of the actual times?

Check your answers to the odd-numbered problems in the Appendix, then turn to Problem Set 12 on page 837 for more practice on graphs and statistics. If you need a quick review of the topics in this chapter, visit the chapter Summary first.

CHAPTER 12
SUMMARY

Statistics

Objective	Review
Read bar graphs, line graphs, and circle graphs. (pp. 776, 786, 790)	For vertical bar graphs and all line graphs, locate numerical information along the vertical axis and categories or time periods along the horizontal axis. The numerical value associated with a bar or a point is the number on the vertical scale horizontally across from it. For horizontal bar graphs, the labeling of the axes is switched.
	Example: (a) Consider the double bar graph at the top of page 832. To find the number of frames assembled by the Tuesday day shift, locate "Tuesday" on the horizontal axis. Then find the top of the day shift bar, and look

Objective	Review

horizontally across to the vertical axis. The day shift on Tuesday assembled approximately 70 frames.

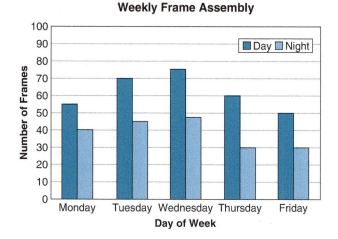

Weekly Frame Assembly

(b) Study the broken-line graph below. Suppose we wish to calculate the percent increase in the number of paint jobs from January to February. From the graph, we read the January job total to be 25, and the February total to be 35. The percent increase is

$$\frac{35 - 25}{25} = \frac{R}{100}$$

$$\frac{10}{25} = \frac{R}{100}$$

$$R = \frac{10 \times 100}{25} = 40\%$$

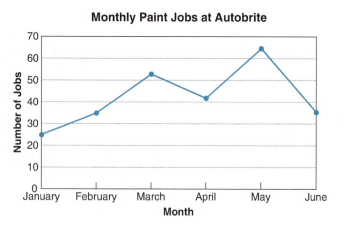

Monthly Paint Jobs at Autobrite

For circle graphs, or pie charts, each sector is usually labeled with a percent.

Example: Consider the circle graph at the top of page 833. Suppose the average job in this particular business generates $227.50. To calculate the portion of this revenue spent on advertising, notice that the advertising sector is labeled 18%. Therefore, the portion of the average job spent on advertising would be

$$0.18 \times \$227.50 = \$40.95$$

Objective	Review

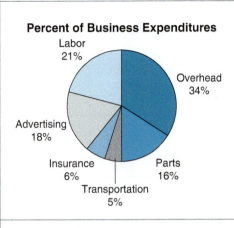

Draw bar graphs, line graphs, and circle graphs. (pp. 780, 789, 791)

For bar graphs and line graphs, construct a horizontal and vertical axis. Label the horizontal axis with categories or time periods and space these equally along the axis. Use the vertical axis to represent the numerical data. (The axes may be switched to make a horizontal bar graph.) Choose a convenient scale that will cover the range of numbers and provide reasonably precise readings for the graph. Be sure to title the graph.

Example: The data in the table were used to create the bar graph that follows.

Work Output	Day
150	Monday
360	Tuesday
435	Wednesday
375	Thursday
180	Friday

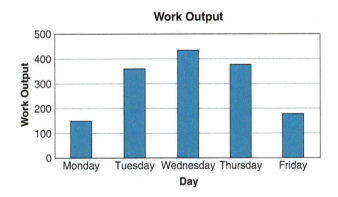

To create a circle graph, we must first convert each numerical value to its percent of the total. Then convert each percent to its corresponding portion of 360°. Finally, use a protractor to measure the sectors of the circle graph. Title the graph and label each sector with its category name and its percent of the total.

Example: For the work output data from the previous example, first we calculate that the total output for the week is 1500. The percent represented by Monday's output can then be determined as follows:

$$\text{Monday output} \rightarrow \frac{150}{1500} = \frac{R}{100} \qquad R = \frac{150 \times 100}{1500} = 10\%$$
$$\text{Weekly output} \rightarrow$$

Converting this percent to degrees, we have:

$$0.10 \times 360° = 36°$$

Objective	Review

Continuing this process for the remaining days of the week, we obtain the results shown in the table. The completed circle graph is shown to the right of the table.

Day	Percent	Degrees
Monday	10%	36°
Tuesday	24%	86°
Wednesday	29%	104°
Thursday	25%	90°
Friday	12%	43°

Daily Work Output

Calculate measures of central tendency: mean, median, and mode. (pp. 806–808)

The following formula is used to find the mean:

$$\text{Mean} = \frac{\text{sum of the data values}}{\text{number of values}}$$

The median is the middle value when a set of data is arranged in order from smallest to largest. If there is an even number of data values, the median is the mean of the middle two values.

The mode is the data value that occurs most often in a set of data. There may be more than one mode or no mode at all.

Example: Find (a) the mean, (b) the median, and (c) the mode for the following set of lengths. All measurements are in meters.

12.7   16.2   15.5   13.9   13.2   17.1   15.5

(a)   The sum of the lengths is 104.1, and there are 7 measurements. The mean is

$$\frac{104.1}{7} = 14.871 \ldots \approx 14.87 \text{ m}$$

(b)   Rearranging the lengths in numerical order, we have

12.7   13.2   13.9   15.5   15.5   16.2   17.1

The median, or middle value, is 15.5 m.

(c)   There are two measurements of 15.5 m—this is the mode.

Calculate measures of dispersion: range and standard deviation. (pp. 818–819)

To find the range, subtract the smallest data value from the largest.

To find the standard deviation, use the following formula, where $(x - \bar{x})^2$ represents the square of the deviation from the mean:

$$\text{Standard deviation } (s) = \sqrt{\frac{\text{sum of } (x - \bar{x})^2}{n - 1}}$$

Example: Find (a) the range and (b) the standard deviation for the set of lengths in the previous example.

(a) The range is

$$17.1 - 12.7 = 4.4 \text{ m}.$$

(b)   Using the mean value of 14.87 from the previous example, we organize our work into the table below:

Objective	**Review**

Data, x	$(x - \bar{x})$	$(x - \bar{x})^2$
12.7	−2.17	4.7089
16.2	1.33	1.7689
15.5	0.63	0.3969
13.9	−0.97	0.9409
13.2	−1.67	2.7889
17.1	2.23	4.9729
15.5	0.63	0.3969

Sum = 15.9743

Using $n = 7$ and the sum of the $(x - \bar{x})^2$ column, we substitute into the standard deviation formula:

$$s = \sqrt{\frac{15.9743}{6}} \approx 1.63 \text{ m}$$

Calculate mean and standard deviation for data grouped in a frequency distribution.
(pp. 809, 822)

For the mean, $\bar{x}$, prepare a table with columns for class, frequency (F), midpoint (M), and product (FM). Find the sum of the FM column and divide this by the sum of the frequencies, n.

For standard deviation, prepare three additional columns: $(M - \bar{x})$, $(M - \bar{x})^2$, and $F(M - \bar{x})^2$. Use the following formula:

$$s = \sqrt{\frac{\text{sum of } F(M - \bar{x})^2}{n - 1}}$$

Example: Welding The regulator pressure settings on an oxygen cylinder used in oxy-acetylene welding (OAW) must fall between a range of 110 and 160 psi when welding material is 1 in. thick. The following frequency distribution shows the pressures for 40 such welding jobs. Use these data to find the mean and the standard deviation.

Class	Frequency, F
110–119 psi	4
120–129 psi	8
130–139 psi	12
140–149 psi	10
150–159 psi	6

The completed table is shown below:

Class	Frequency, F	Midpoint, M	Product, FM	$(M - \bar{x})$	$(M - \bar{x})^2$	$F(M - \bar{x})^2$
110–119 psi	4	114.5	458	−21.5	462.25	1849.0
120–129 psi	8	124.5	996	−11.5	132.25	1058.0
130–139 psi	12	134.5	1614	−1.5	2.25	27.0
140–149 psi	10	144.5	1445	8.5	72.25	722.5
150–159 psi	6	154.5	927	18.5	342.25	2053.5
	$n = 40$		Sum = 5440			Sum = 5710

The mean is:

$$\bar{x} = \frac{5440}{40} = 136 \text{ psi}$$

The standard deviation is:

$$s = \sqrt{\frac{5710}{39}}$$

$$s \approx 12.1 \text{ psi, rounded}$$

Objective	Review
Use the concept of normal distribution to predict the spread of data values. (p. 825)	Given a large enough sample, measurements involving many natural objects or processes tend to approximate a normal distribution. If a set of data values are distributed normally, approximately 68% of them will fall within one standard deviation of the mean, about 95% will fall within two standard deviations of the mean, and approximately 99.7% will fall within three standard deviations of the mean.

Example: Suppose a set of data contained 50 measurements with a mean value of 3.46 in. and a standard deviation of 0.07 in.

(a) If these measurements are distributed normally, we would expect to find 34 of them (68% of 50) in the interval 3.46 ± 0.07 in., or between 3.39 in. and 3.53 in.

(b) Again, if we assume a normal distribution, we would expect to find 47 or 48 or them (95% of 50 is 47.5) in the interval 3.46 ± 0.14 in., or between 3.32 in. and 3.60 in.

PROBLEM SET 12   **Statistics**

Answers to odd-numbered problems are given in the Appendix.

A. Answer the questions based on the graphs pictured.

Questions 1 to 5 refer to Graph I.

Graph I   Electrical Trades

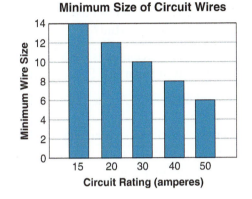

Minimum Size of Circuit Wires

1. In general, as amps increase, the necessary wire size _____.
    (increases, decreases)

2. What is the minimum size of wire needed for a circuit rating of 30 amps?

3. What is the minimum wire size needed for a circuit rating of 15 amps?

4. What circuit ratings could a wire size of 8 be used for?

5. What circuit ratings could a size 14 wire be used for?

Questions 6 to 15 refer to Graph II.

Graph II   Culinary Arts

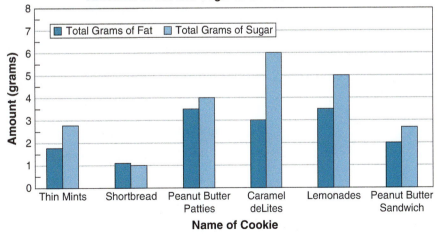

Amounts of Fat and Sugar in Girl Scout Cookies

Name

_____

Date

_____

Course/Section

_____

6. How many grams of fat are contained in a Peanut Butter Sandwich?

7. How many grams of sugar are contained in a Shortbread cookie?

8. How many grams of fat would you consume if you ate three Thin Mints?

9. How many grams of sugar would you consume if you ate four Lemonades?

10. How many more grams of fat are in two Caramel deLites than in two Short-bread cookies?

11. How many more grams of sugar are in three Peanut Butter Patties than in three Peanut Butter Sandwich cookies?

12. Which cookie appears to be the "healthiest" (as measured by the fewest combined amounts of fat and sugar)?

13. Which cookie appears to be the least "healthy" (as measured by the greatest combined amounts of fat and sugar)?

14. By what percent does the amount of fat in a Lemonade exceed the amount of fat in a Thin Mint?

15. By what percent does the amount of sugar in a Caramel deLite exceed the amount of sugar in a Peanut Butter Sandwich cookie?

Retail Merchandising   A small computer store is making purchasing decisions for the upcoming year. To help decide which products to purchase, the owners looked at sales trends over the past year. The stacked bar graph (Graph III) shows the unit sales of desktop computers, laptop computers, and computer tablets over a 12-month period. Study the graph and answer questions 16 through 23 that follow.

Graph III

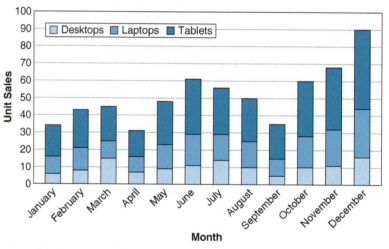

16. During which month were total sales the highest?

17. During which month were total sales the lowest?

18. Which item consistently outsold the other two?

19. In October, what percent of total sales was represented by laptops?

20. In September, what was the ratio of computer tablet sales to desktop sales?

21. In August, what percent of total sales was represented by desktops?

22. During which month did desktops outsell laptops?

23. What was the percent increase in total sales from October to December?

Problems 24 to 30 refer to Graph IV.

Graph IV   Automotive Trades

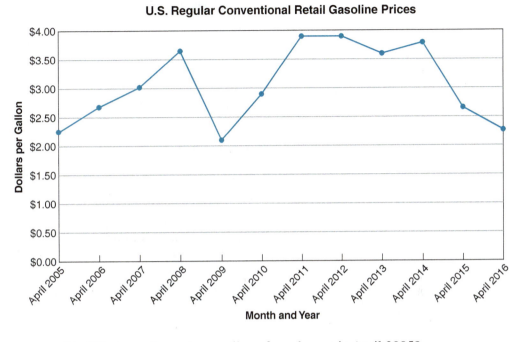

U.S. Regular Conventional Retail Gasoline Prices

24. What was the cost per gallon of regular gas in April 2005?

25. When was the next time that the cost per gallon was lower than it was in April 2005?

26. During which time periods did the cost of gas decrease?

27. During which time period did the largest price increase occur?

28. What was the difference between the highest and lowest cost per gallon over the time frame shown?

29. By what percent did the cost decrease from April 2012 to April 2016?

30. How much more did it cost to fill up a 20-gal tank in 2008 than in 2009?

Problems 31 to 38 refer to Graph V.

Graph V   Business and Finance

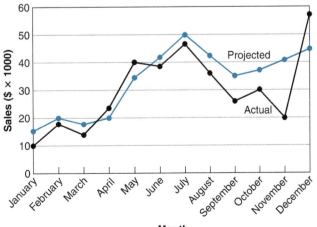

Projected versus Actual Sales

31. What was the actual sales total in October?

32. What was the projected sales total in May?

33. During which month were actual sales highest?

34. During which month were projected sales highest?

35. During which month were actual sales lowest?

36. During which month were projected sales lowest?

37. During which months were actual sales lower than projected sales?

38. During which month was the gap between actual and projected sales the largest? What was the difference?

Problems 39 to 43 refer to Graph VI.

Graph VI   Metalworking

Composition of Marine Bronze

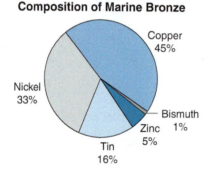

39. What percent of marine bronze is composed of tin?

40. What percent of this alloy is made up of quantities other than copper?

41. Without measuring, calculate how many degrees of the circle should have gone to nickel.

42. How many ounces of zinc are there in 50 lb of marine bronze?

43. How many grams of bismuth are there in 73 grams of this alloy?

B.   Construct graphs from the data given.

1. **Construction**   Construct a horizontal bar graph based on these data:

Fire Resistance Ratings of Various Plywoods

Type of Wood	Fire Resistance Rating
A: $\frac{1}{4}$ in. plywood	10
B: $\frac{1}{2}$ in. plywood	25
C: $\frac{3}{4}$ in. tongue and groove	20
D: $\frac{1}{2}$ in. gypsum wallboard	40
E: $\frac{1}{2}$ in. gypsum—two layers	90
F: $\frac{5}{8}$ in. gypsum wallboard	60

2. **General Interest**   The following table shows the percent changes in various greenhouse gas emissions from 1990 to 2013 in the United States. Construct a horizontal bar graph illustrating these changes.

Gas and Source	Percent Change: 1990–2013
Total Carbon Dioxide (CO_2)	+7.4
CO_2 from Fossil Fuel Consumption	+8.8
Total Methane (CH_4)	−14.6
CH_4 from Animals	+0.2
CH_4 from Natural Gas Systems	−12.1
CH_4 from Landfills	−38.5
CH_4 from Coal Mining	−33.1
Nitrous Oxide (N_2O)	+7.7
N_2O from Agricultural Soil Management	+17.7
Hydrofluorocarbons	+72.8

(*Source:* World Almanac 2016 from the EPA.)

3. **General Interest** The following table shows what percent of jobs in five different careers were held by women in both 1980 and 2014. Construct a horizontal double bar graph illustrating this information.

Career	Percent of Jobs Held by Women in 1980	Percent of Jobs Held by Women in 2014
Civil Engineer	3%	16%
Dentist	4	27
Lawyer	12	36
Physician/Surgeon	14	37
Pharmacist	32	48

(*Source:* U.S. Census Bureau.)

4. **Masonry** Construct a vertical double bar graph based on these data:

Output of Bricklayers for the ABC Construction Co.

	Bricks Laid	
Initials of Employee	March	April
R.M.	1725	1550
H.H.	1350	1485
A.C.	890	1620
C.T.	1830	1950
W.F.	1175	1150
S.D.	2125	1875

5. **Meteorology** Construct a broken-line graph based on these monthly average high temperatures:

Month	Average High Temperature	Month	Average High Temperature
January	42°	July	81°
February	44°	August	84°
March	50°	September	77°
April	56°	October	69°
May	65°	November	57°
June	75°	December	46°

6. **Allied Health** Construct a double broken-line graph based on these recorded blood pressures of a patient over a 12-hr period:

Time	Systolic	Diastolic	Time	Systolic	Diastolic
7 A.M.	124	85	1 P.M.	118	78
8	128	88	2	124	78
9	120	80	3	115	75
10	116	75	4	112	78
11	120	78	5	116	74
12 P.M.	114	80	6	120	80

7. **Metalworking** Construct a circle graph based on the alloy composition shown in the table:

Metal	Alloy Composition
Copper	43%
Aluminum	29%
Zinc	3%
Tin	2%
Lead	23%

8. **Fire Protection** Construct a circle graph based on the portion of calls represented by each category during a particular week at a fire station:

Type of Call	Number of Calls	Type of Call	Number of Calls
Fire	4	False alarm	10
Medical	9	Hazardous leak/spill	2
Traffic accident	12	Other	3

9. **General Interest** Students in a community college statistics class surveyed a large sampling of people in their city about their use of cell phones. The data in the table show the percent of those sampled in six different age groups who use either a smartphone, a basic cell phone, or no cell phone at all. Use these results to create a stacked bar graph like the one in Example 3 on page 779. Let each bar represent an age group and divide each bar into three sections.

Age Group	Smartphone	Basic Cell Phone	No Cell Phone
18–24	84%	10%	6%
25–34	92%	6%	2%
35–44	87%	8%	5%
45–54	68%	22%	10%
55–64	56%	30%	14%
65+	36%	42%	22%

C. Find the mean, median, and mode for each set of numbers.

1. 89, 88, 84, 87, 89, 84, 84, 83, 85, 89, 84

2. 5.8, 3.2, 4.1, 3.5, 4.3, 5.2

3. 20.220, 20.434, 20.395, 20.324, 20.345, 20.356, 20.324, 20.258

4. 4287, 4384, 4036, 6699, 6491, 5460, 3182, 6321, 6642

5. 696, 658, 756, 727, 606, 607, 727, 635, 733, 637

6. 14.66, 17.70, 13.83, 15.53, 17.35, 12.18, 14.80

D. **Calculate (a) the range and (b) the standard deviation for each set of numbers.**

1. 9, 4, 3, 2, 9, 8, 2, 1, 3, 5, 4, 8, 5, 7

2. 105.4, 108.2, 101.9, 100.9, 102.2, 114.1, 117.7

3. 32,500; 36,000; 38,500; 35,500; 37,000; 36,000; 33,000

4. 0.472, 0.798, 0.419, 0.207, 0.639, 0.837, 0.735, 0.849, 0.711, 0.753

E. **Calculate the mean and the standard deviation of the grouped data in each problem.**

1.

Class	Frequency, F
0–4.9 cu cm	12
5.0–9.9 cu cm	9
10.0–14.9 cu cm	3
15.0–19.9 cu cm	2
20.0–24.9 cu cm	6
25.0–29.9 cu cm	8
30.0–34.9 cu cm	10

2.

Class	Frequency, F
22–26 ft	4
27–31 ft	7
32–36 ft	11
37–41 ft	9
42–46 ft	5
47–51 ft	2

F. **Applications**

1. **Printing**   Technica Replica uses high-speed printers for large-volume copying. The following values represent the number of copies requested, followed by the number of impressions that were wasted during each run.
    (a) For each case, calculate the percent of waste and fill in the chart. (Round to the nearest 0.1%.)

Copies Requested	Wasted Impressions	Percent of Waste
15,500	409	
17,500	345	
18,000	432	
16,750	546	
14,000	316	
16,250	845	
18,500	538	

    (b) Find the median of the percent of waste.
    (c) Find the mean of the percent of waste.
    (d) Find the range of the percent of waste.
    (e) Find the standard deviation of the percent of waste.
    (f) There is an upcoming job that requires 17,000 usable copies. The supervisor decides to print 17,500. Based on the median percent of waste, should this provide enough of a cushion to complete the job?

2. **Meteorology** The following table shows the average monthly precipitation in Seattle, Washington, from March through August:

Month	Inches
March	3.51
April	2.77
May	2.16
June	1.63
July	0.79
August	0.97

(a) Find the median monthly total for the six-month period.
(b) Find the mean monthly total.
(c) Find the range of the monthly averages.

3. **Agriculture** The following data from the United States Department of Agriculture represent the U.S. average price for corn in dollars per bushel at the begining of each year:

Year	Price ($/bushel)	Year	Price ($/bushel)
2002	$1.97	2010	$3.66
2003	$2.33	2011	$4.94
2004	$2.39	2012	$6.07
2005	$2.12	2013	$6.96
2006	$2.00	2014	$4.42
2007	$3.05	2015	$3.82
2008	$3.97	2016	$3.66
2009	$4.36		

(a) Determine the median yearly price from 2002 through 2016.
(b) Calculate the mean yearly price over this time period. (Round to the nearest cent.)
(c) Determine the range of these prices.
(d) Calculate the standard deviation of the prices.
(e) If a corn grower annually produces an average 80,000 bushels, use the mean price per bushel to calculate his average yearly revenue.

4. **Industrial Technology** The Energy Information Administration reports the following monthly prices for industrial customer natural gas in the United States (in dollars per *thousand* cubic feet) for the year 2015:

Month	Price	Month	Price
January	$4.76	July	$3.67
February	$4.60	August	$3.73
March	$4.35	September	$3.58
April	$3.86	October	$3.45
May	$3.49	November	$3.18
June	$3.69	December	$3.38

(a) What was the mean monthly price for the year?
(b) What was the median monthly price for the year?
(c) What was the range of the monthly prices?
(d) Calculate the standard deviation of the monthly prices.
(e) Use the median monthly price to compute the cost of 46,500 cu ft of natural gas.

5. **Police Science** To determine the speed that a vehicle was traveling at the time of an accident, police officers carefully measure the lengths of the skid marks left at the scene. There are many factors that determine the skid length, and a quick and simple formula for determining the speed is $S = \sqrt{30 \cdot L \cdot d \cdot e}$, where S is the speed in miles per hour, L is the skid length in feet, d is the drag factor, and e is the braking efficiency. The following table shows the skid lengths for ten accidents that occurred on a particular stretch of an interstate during icy conditions this past winter:

Skid Length, L (ft)	Speed, S (mph)	Skid Length, L (ft)	Speed, S (mph)
514		600	
294		476	
726		384	
350		486	
216		546	

   (a) Use the given formula, with $d = 0.2$ (icy road) and $e = 1.0$, to calculate the speeds for the given skid lengths and complete the table. (Round to the nearest whole number.)
   (b) Find the median speed.
   (c) Find the mean speed.
   (d) Find the range of the speeds.
   (e) Find the standard deviation of the speeds.

6. **Agriculture** The following list of numbers represents the total annual U.S. cotton harvest, in acres, for the years 1982–2015.

9734	7348	10,379	10,229	8468	10,030	11,948	9538
11,732	12,960	11,123	12,783	13,322	16,007	12,888	13,406
10,684	13,425	13,053	13,828	12,417	12,003	13,057	13,803
12,732	10,489	7569	7533	10,699	9460	9322	7544
9347	8058						

   (a) Construct a grouped frequency distribution for these yields using class widths of 1000 and beginning with 7000–7999.
   (b) Calculate the mean and the standard deviation of the grouped data rounded to the nearest whole number.
   (c) In 2015, the average yield per acre was about 1150 pounds of cotton. Based on this figure, what was the mean total yield in pounds?
   (d) Based on an average price of $0.67 per pound, what was the value of the mean total yield? (Round to the nearest hundred thousand dollars.)

7. **Sheet Metal Trades** The following data represent the gauges of the sheet metal used by a sheet metal worker during his last eight jobs:

   28 GA   12 GA   16 GA   24 GA   22 GA   18 GA   30 GA   8 GA

   (a) Use the chart at the top of page 846 to find the median thickness for these jobs in inches.
   (b) Use the chart to find the mean thickness of the gauges used in millimeters.
   (c) Use the chart to find the range of thicknesses of the gauges used in millimeters.

Gauge	Thickness (in.)	Thickness (mm)
30	0.0157	0.3988
28	0.0187	0.4750
26	0.0217	0.5512
24	0.0276	0.7010
22	0.0336	0.8534
20	0.0396	1.0058
18	0.0516	1.3106
16	0.0635	1.6129
14	0.0785	1.9939
12	0.1084	2.7534
10	0.1382	3.5103
8	0.1681	4.2697

8. **Sheet Metal Trades**   The following data represent the number of 48-in. by 96-in. (4-ft by 8-ft) sheets of aluminum used by the sheet metal worker for the 8 jobs in problem 7. Note that the number of sheets correspond to the gauges of sheet metal, respectively, as given in problem 7 (e.g., the first project used five 48-in. by 96-in. sheets of 28-gauge sheet metal).

5 7 10 16 9 12 3 4

(a) Calculate the total surface area, in square inches, of sheet metal used for each job. Then calculate the mean amount of surface area used per job.
(b) Find the total volume of aluminum, in cubic inches, used in each project by multiplying each surface area by the corresponding sheet thickness.
(c) Find the mean of the volumes from part (b).
(d) If aluminum costs approximately $0.093 per cubic inch, what is the cost of the average project?

9. **Machine Trades**   For quality control purposes, 120 supposedly identical steel shafts were weighed, and the mean weight was 2.643 kg with a standard deviation of 0.021 kg. Only those shafts whose weights were within two standard deviations of the mean were accepted.
(a) What was the range of acceptable weights?
(b) Assuming that the weights were normally distributed, about how many shafts would have been rejected?

CASE STUDY: Las Vegas or Honolulu?

In his book *How to Lie with Statistics*, Darrell Huff (Huff & Geis, 1954, Norton) coined the term "well-chosen average." He was referring to the three measures of central tendencies: mean, median, and mode. When all three measures are very close or equal in value, the mean, median, or mode may each fairly represent the average. Unfortunately, sometimes reporters of statistical data selectively choose which measure of central tendency to report as an average, potentially distorting the basic understanding of the conclusion generated by the data. Measures of dispersion, such as range and standard deviation, can allow us to analyze these "averages" even better.

1. The following tables show the average monthly temperatures for Las Vegas and Honolulu (°F) in 2016. Calculate the three measures of central tendency and the two measures of dispersion for these data for each city.

Las Vegas, 2016

January	February	March	April	May	June
47	57	63	67	74	91

July	August	September	October	November	December
94	96	81	72	60	47

Mean _____ Median _____ Mode _____ Range_____ Standard Deviation _____

Honolulu, 2016

January	February	March	April	May	June
74	75	75	78	79	79

July	August	September	October	November	December
81	81	81	80	78	75

Mean _____ Median _____ Mode _____ Range_____ Standard Deviation _____

(a) Which measure of central tendency is the most misleading for both cities? Why?
(b) Suppose the tourist bureau of each city used the mean or the median annual temperature in its advertisements to attract tourists. Which city would be misrepresenting its true climate? Explain.
(c) How do the range and standard deviation help clarify the true nature of both climates?
(d) If your employer allowed you to take your two-week vacation anytime during the year, why might Las Vegas be the more appealing of the two cities?
(e) If your employer randomly selects the time of year for you to take your vacation, why might Honolulu be the more appealing destination?
(f) Suppose you need to take your vacation in September of next year. Would the average monthly temperature data provide you with a clear-cut choice? Explain.

2. The following tables show the daily high temperatures for both cities in September 2016:
    (The top row represents the first six days of the month, the second row represents days 7–12, and so on.)

Las Vegas

Daily Temperatures °F September 1–30, 2016					
92	92	86	84	80	82
82	86	92	92	87	88
86	79	74	78	78	80
83	83	79	75	77	70
76	77	77	76	78	76

Honolulu

Daily Temperatures °F September 1–30, 2016					
85	83	83	83	83	81
80	82	81	80	81	81
81	78	80	83	82	82
79	79	78	79	81	82
80	79	80	80	79	80

(a) As you scan the temperatures from the beginning to the end of the month, how does the trend for Las Vegas differ from that of Honolulu? How might this affect your vacation decision if you needed to go sometime in September of next year?
(b) Construct a grouped frequency distribution for each city. Use class widths of three degrees for Las Vegas (beginning with 70–72) and two degrees for Honolulu (beginning with 78–79).
(c) Calculate the two means from the grouped frequency distributions. (Round to the nearest tenth.) How do these results compare with the means for September that are given in problem 1?

continued...

    (d) Calculate the standard deviation for each city from the grouped frequency distributions. (Round to the nearest tenth.) If you needed to take your vacation in September of next year, how might this result help you decide where to go?

    (e) For each city, calculate the percent of the daily temperatures that are within one standard deviation of the mean from part (b).

    (f) For each city, calculate the percent of the daily temperatures that are within two standard deviations of the mean.

    (g) Based on your answers to (e) and (f), which city's temperatures for September most reflect a normal distribution?

3. Temperatures alone do not tell the whole story about the weather. For example, humidity is a measure of the amount of moisture in the air, which affects one's level of comfort. A measurement known as the Heat Index combines temperature and humidity to indicate what a given temperature actually feels like. Historical data show that the average humidity levels in September are 67% for Honolulu and 23% for Las Vegas. Do an Internet search for "Heat Index Calculator" and input these numbers along with the mean temperatures for September. What is the Heat Index for each city? How might this affect your vacation decision?

4. Other than temperature and humidity, what other aspects of weather can affect you during your vacation?

Answers to Previews

Answers to Preview 1

1. (a) two hundred fifty thousand three hundred seventy-four   (b) 1,065,008

2. (a) 210,000   (b) 214,700

3. (a) 125   (b) 8607

4. (a) 37   (b) 2068   (c) 77

5. (a) 2368   (b) 74,115   (c) 640,140

6. (a) 334 remainder 2   (b) 203

7. (a) 1, 2, 3, 4, 6, 12   (b) $2 \cdot 2 \cdot 3$

8. (a) 33   (b) 33   (c) 33   (d) 48

9. 445 lb

Answers to Preview 2

1. (a) $7\frac{3}{4}$ (b) $\frac{31}{8}$ (c) $\frac{20}{64}$ (d) $\frac{56}{32}$ (e) $\frac{5}{32}$ (f) $1\frac{7}{8}$

2. (a) $\frac{35}{256}$ (b) 3   (c) $\frac{9}{10}$ (d) $1\frac{1}{2}$ (e) $2\frac{3}{10}$ (f) 8

3. (a) $\frac{5}{8}$ (b) $1\frac{15}{16}$ (c) $\frac{11}{20}$ (d) $2\frac{11}{16}$

4. $1\frac{3}{8}$ in.

Answers to Preview 3

1. (a) Twenty-six and thirty-five thousandths   (b) 106.0027

2. (a) 5.916   (b) 2.791   (c) 22.97   (d) 2.18225

   (e) 3256.25   (f) 225   (g) 3.045

3. 4.3

4. (a) 0.1875   (b) 3.452   (c) 9.225   (d) 0.305

5. (a) $583.25   (b) 0.34 lb

Answers to Preview 4

1. (a) 5:2 or 2:5   (b) $8\frac{2}{3}$ to 1

2. (a) $x = 10$ (b) $y = 14.3$

3. (a) 64 ounces   (b) 18 ft   (c) 12 machines

4. (a) 25%   (b) 46%   (c) 500%   (d) 7.5%

5. (a) 0.35   (b) 0.0025   (c) $1\frac{3}{25}$

6. (a) 225   (b) 54   (c) 6.25% or $6\frac{1}{4}$%   (d) 75

   (e) 0.90%   (f) 10.4 lb   (g) $94.09   (h) 80%

   (i) 11.4%

849

Answers to Preview 5

1. (a) nearest hundredth   (b) nearest whole number   (c) nearest ten
   Measurement (a) is most precise.

   (d) 2 significant digits   (e) 3 significant digits   (f) 1 significant digit

   (g) 25 psi   (h) 0.05 kg   (i) $\frac{1}{8}$ in.

2. (a) 44.2 sec   (b) 5.94 lb

3. (a) 94 sq ft   (b) 15 hr   (c) 13,000 lb

4. (a) 1040 oz   (b) 4.7 mi   (c) 504 sq in.

5. (a) 100 lb   (b) 85 qt

6. (a) 6.5 g   (b) 450 m   (c) 2850 sq m

7. (a) 12.7 gal   (b) 72 km/h   (c) 108°F

Answers to Preview 6

1. (a) -14 (b)
   (c) -250

2. (a) -8 (b) $11\frac{3}{4}$ (c) -7.5 (d) -3

3. (a) -3 (b) -8 (c) 10.9   (d) $-1\frac{7}{8}$

4. (a) -72 (b) 6   (c) 30.745   (d) $-\frac{3}{20}$ (e) 1500

5. (a) 64   (b) 4.2025

6. (a) 11   (b) 500   (c) 22

7. (a) 13   (b) 3.81

Answers to Preview 7

1. (a) 7   (b) 5   (c) $A = 12$ (d) $T = 4$ (e) $P = 120$

2. (a) $6ax^2$ (b) $-3x - y$ (c) $4x - 2$ (d) $3x^2 + 7x + 12$

3. (a) $x = 5$ (b) $x = 9$ (c) $x = 12$ (d) $N = 2S - 52$

   (e) $A = \dfrac{8M + BL}{L}$

4. (a) $4A$ (b) i^2R (c) $E = \dfrac{100u}{I}$ (d) $R = \dfrac{12L}{D^2}$

5. (a) \$24.07   (b) 10.37 hr, or 11 hr, rounded up

6. (a) $6y^2$ (b) $-12x^5y^4$ (c) $3xy - 6x^2$ (d) $-5x^5$ (e) $\dfrac{b^2}{3a^3}$

   (f) $4m^2 - 3m + 5$

7. (a) 1.84×10^{-4} (b) 2.13×10^5 (c) 1.44×10^{-3}

   (d) 6.5×10^{-8}

Answers to Preview 8

1. about 32°

2. ∠ *DEF* is acute, ∠ *GHI* is obtuse, ∠ *JKL* is a right angle

3. (a) 80°, 80°, 100°, 80°, 100°   (b) 20°

4. (a) obtuse and scalene triangle   (b) parallelogram   (c) trapezoid

   (d) square   (e) hexagon

5. 1.5 in.

6. (a) 14 sq in.   (b) 24 cm^2 (c) 0.42 sq in., 2.4 in.

   (d) 22.9 sq in., 17.0 in.

7. (a) $2337   (b) $10,386.10

Answers to Preview 9

1. (a) cube   (b) rectangular prism   (c) frustum of a cone

2. (a) 12.6 cu ft, 31.4 sq ft   (b) 38.8 cm^3, 55.4 cm^2 (c) 3.7 cu in.

   (d) 80.5 sq in., 75.4 cu in.   (e) 180 cm^2, 336 cm^3

3. (a) 40,400 gal   (b) 24.3 lb   (c) $9576

Answers to Preview 10

1. (a) 43°24′   (b) 65.25°

2. (a) 0.80   (b) 143.2°   (c) 4.2 in., 12.6 sq in.   (d) 1.7 rad/sec

3. (a) $a = 6.0, b = 10.4, A = 30°$ (b) $d = 16.0, e = 22.6, F = 45°$

4. (a) 0.438   (b) 0.105   (c) 0.943

5. (a) 14°0′   (b) 56°42′   (c) 75.3°   (d) 48.6°

6. (a) $X \approx 10.5, Y \approx 13.4$ (b) $\angle m = 44°25′, X \approx 3.7$ in.

   (c) $\approx$ 39.8 in.   (d) 5°43′

7. (a) $A = 82°, B = 66°, C = 32°$ (b) $C = 33°, a = 7.6$ ft, $c = 4.4$ ft

Answers to Preview 11

1. (a) (4, 0)   (b) (−9, −3)

   (c) dependent   (d) inconsistent, no solution

2. (a) 20 and 15   (b) 8 in. and 6 in.

   (c) 300 of the 4¢ screws, and 500 of the 6¢ screws

3. (a) $x = 4$ or $x = -4$ (b) $x = 0$ or $x = 7$

   (c) $x = 7$ or $x = -2$ (d) $x = -2\frac{2}{3}$ or $x = 2$ (e) no solution

4. (a) 6 cm   (b) $\approx$ 8.4 in.   (c) 12.9 amp   (d) 6.5 in. by 9.5 in.

Answers to Preview 12

1. (a) 70   (b) 27%   (c) 65 in May   (d) 40%   (e) $40.95

2. (a)

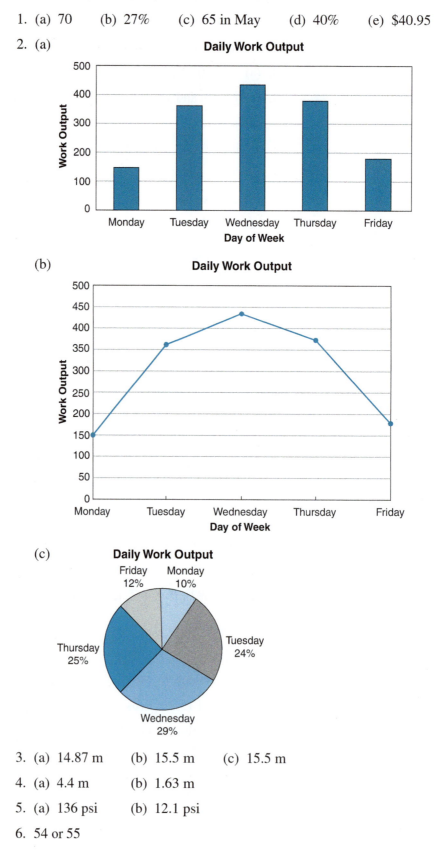

3. (a) 14.87 m   (b) 15.5 m   (c) 15.5 m

4. (a) 4.4 m   (b) 1.63 m

5. (a) 136 psi   (b) 12.1 psi

6. 54 or 55

Answers to Odd-Numbered Problems

A.
10	11	11	12	16	7	9	15	12	13
10	13	15	8	14	9	13	18	9	11
16	8	10	17	9	14	12	7	10	11
13	13	12	16	9	11	10	10	14	17
11	14	11	10	11	14	13	12	13	15

B.
11	12	14	17	18	21	11	20	18	15
15	14	15	18	22	17	14	14	18	16
12	19	8	10	22	12	19	15	14	16

Exercises 1-1, page 15

A. 1. three hundred fifty-seven
3. seventeen thousand ninety-two
5. two million thirty-four
7. seven hundred forty thousand one hundred six
9. one hundred eighteen million one hundred eighty thousand eighteen
11. 3006 13. 11,100 15. 4,040,006
17. 360 19. 4000 21. 230,000

B. 1. 69 3. 80 5. 123 7. 132 9. 806
11. 1390 13. 1009 15. 861 17. 5461 19. 11,428
21. 11,071 23. 25,717 25. 175,728 27. 6095 29. 663,264

C. 1. 1042 3. 6352 5. 6514 7. 2442 9. 7083
11. 4114 13. 64 15. 55 17. 357 19. 1,166,040

D. 1. 4861 ft 3. 1636 screws 5. 1129 min
7. (a) 3607 watts (b) 1997 watts (c) 850 watts
9. 3114 11. $1557 13. 2929 ohms
15. 756 g 17. 3900 W
19. (a) 4264 (b) 3027 (c) 7291

E. 1. 97,001 kHz
3. (a) $307,225 (b) $732,813 (c) $2,298,502 (d) $7156
5. (a) #12 BHD: 11,453 (b) A3: 3530
    #Tx: 258 A4: 8412
    410 AAC: 12,715 B1: 4294
    110 ACSR: 8792 B5: 5482
    6B: 7425 B6: 5073
    C4: 6073
    C5: 7779

Problems, page 21

2	8	7	5	1	3	2	5	4	7	9
9	7	7	7	2	8	2	5	4	9	3
8	5	6	8	6	3	3	7	9	3	3
6	2	1	8	3	5	6	2	7	4	1
1	8	4	6	4	9	4	9	6	5	9
6	7	6	9	1	8	5	5	5	3	7
9	2	8	3	8	2	4	1	1	4	4

Exercises 1-2, page 27

A. 1. 35 3. 25 5. 33 7. 62 9. 13
11. 12 13. 15 15. 38 17. 46 19. 25
21. 189 23. 281 25. 408 27. 273 29. 574
31. 2809 33. 12,518 35. 4741 37. 47,593

B. 1. $459 3. 1758 ft 5. $330,535 7. 3 drums, by 44 liters
9. 174 gal 11. 7750 13. 13,500 Ω 15. 500,000 hertz
17. 122 HCF 19. Nonhybrid costs $1958 less over 5 years

C. 1.

Truck No.	1	2	3	4	5	6	7	8	9	10
Mileage	1675	1167	1737	1316	1360	299	1099	135	1461	2081

Total mileage 12,330

3. $24,431
5. (a) $2065
(b)

Deposits	Withdrawals	Balance
		$6375
	$ 379	5996
$1683		7679
474		8153
487		8640
	2373	6267
	1990	4277
	308	3969
	1090	2879
	814	2065

Problems, page 31

A. 12 32 63 36 12 18 56 24 14
48 16 45 30 10 9 72 35 18
28 15 36 49 8 40 42 54 64
20 6 25 27 81 4 21 24 16

B. 16 30 9 35 18 28 48 12 63
32 20 18 24 25 24 45 10 72
15 49 40 54 36 8 42 64 4
25 27 6 56 36 12 81 14 21

Exercises 1-3, page 38

A. 1. 42 3. 48 5. 63 7. 54 9. 45
11. 296 13. 576 15. 320 17. 290 19. 282
21. 416 23. 792 25. 1404 27. 720 29. 5040
31. 1938 33. 4484 35. 3822

B. 1. 7281 3. 297,591 5. 25,000 7. 10,112 9. 30,780
11. 37,515 13. 89,577 15. 3,532,536 17. 378,012 19. 397,584

C. 1. $3000 3. 1300 ft 5. 8000 envelopes
7. 2430 9. 1200 11. 7650
13. 23,040 15. 115 in. 17. No
or 9 ft 7 in.
19. 37,400 bu 21. 480 A 23. 88,000 mL
25. $33,280 27. 30,000 gal 29. 136,000 lb

D. 1. $681,355 left
    3. (a) 111,111,111; 222,222,222; 333,333,333
       (b) 111,111; 222,222; 333,333
       (c) 1; 121; 12,321; 1,234,321; 123,454,321
       (d) 42; 4422; 444,222; 44,442,222; 4,444,422,222

    5. | Alpha | Beta | Gamma | Delta | Tau |
       |-------|------|-------|-------|-----|
       | $28,080 | $45,560 | $37,120 | $72,520 | $56,160 |

Problems, page 42

6	6	4	6	6	3	4	7
2	7	3	6	5	9	4	8
8	2	2	4	9	8	5	5
4	5	6	4	2	2	3	9
6	7	6	5	7	4	3	8
2	4	9	9	9	3	9	5
9	7	3	2	3	5	7	3
5	2	7	8	7	8	8	8

Exercises 1-4, page 50

A. 1. 9   3. 8   5. 6
    7. 14 *r* 2   9. 10 *r* 1   11. 23 *r* 6
    13. 51 *r* 4   15. 210 *r* 6   17. 37
    19. 222 *r* 2   21. 501   23. 604

B. 1. 23   3. 39   5. 9 *r* 1
    7. 22   9. 8 *r* 35   11. 120
    13. 56 *r* 8   15. 305 *r* 5   17. 119
    19. 96   21. 200   23. 108 *r* 4
    25. 600   27. 102 *r* 98   29. 17 *r* 123

C. 1. (a) 1, 2, 3, 6   (b) 2 × 3   3. (a) 1, 19   (b) 19
    5. (a) 1, 2, 4, 5, 8, 10, 20, 40   (b) 2 × 2 × 2 × 5

D. 1. 27 in.   3. 13 hr   5. 28   7. 7 in.
    9. $72   11. 48   13. 27 reams   15. 6
    17. 230 months or 19 years, 2 months   19. 4 hr 10 min   21. $462

E. 1. (a) 93   (b) 449
    3. 19.148255 or 20 rivets to be sure
    5. 51 hours 40 minutes, or 52 hr, rounded
    7. 42 hours

Exercises 1-5, page 56

A. 1. 50   3. 36   5. 57   7. 4   9. 42   11. 6
    13. 61   15. 112   17. 31   19. 17   21. 84   23. 4
    25. 6   27. 36   29. 2   31. 8   33. 13   35. 7
    37. 6   39. 5

B. 1. 3 × $34 + 5 × $39 = $297   3. 12 × $30 − 3 × $6 = $342
    5. 2 × $12 × 40 + 3 × $20 × 40 + $3240 + $500 = $7100
    7. (33 × $80) + (12 × $40) + (45 × $18) = $3930
    9. China: 346, United States: 330; China "won"
    11. 14 gal

C.   1. 8347   3. 7386   5. 5   7. 1359
      9. 1691   11. 1458   13. 13,920   15. 63

Problem Set 1, page 62

A.   1. five hundred ninety-three
      3. forty-five thousand two hundred six
      5. two million four hundred three thousand five hundred sixty
      7. ten thousand twenty
      9. twelve billion six hundred four million seven hundred thousand two
         hundred fifty
      11. 230,056   13. 64,700   15. 6,047,920,000   17. 5500
      19. 94,700   21. 710,000

B.   1. 96   3. 55   5. 93   7. 528
      9. 934   11. 15   13. 649   15. 195
      17. 1504   19. 1407   21. 13,041   23. 230,384
      25. 37   27. 57   29. 9   31. 18
      33. 6   35. 115   37. 7   39. 1245
      41. 1352

C.   1. (a) 1, 2, 4, 8   (b) $2 \times 2 \times 2$ 3. (a) 1, 31   (b) 31
      5. (a) 1, 2, 3, 4, 6, 9, 12, 18, 36   (b) $2 \times 2 \times 3 \times 3$

D.   1. 43 ft   3. (a) $x = 14$ ft, $y = 27$ ft   (b) $x = 18$ ft, $y = 50$ ft
      5. 6   7. 207 lb   9. $1720   11. Yes
      13. 650 gpm   15. 2839 lb   17. 24 hr   19. 87,780 cu in.
      21. 193 rpm   23. 6 ft   25. $110   27. $365
      29. (a) 20 gal   (b) $40   31. 2860 cal
      33. ES 350: $48,513; ES 300h: $50,088; ES 350 lower by $1575
      35. 24 bushels   37. 36,832 Btu/hr

Chapter 2   *Exercises 2-1, page 84*

A.   1. $4\frac{1}{4}$ 3. $1\frac{3}{8}$ 5. $1\frac{1}{2}$ 7. $16\frac{4}{6}$ or $16\frac{2}{3}$ 9. $2\frac{16}{32}$ or $2\frac{1}{2}$

B.   1. $\frac{7}{3}$ 3. $\frac{67}{8}$ 5. $\frac{23}{8}$ 7. $\frac{8}{3}$ 9. $\frac{29}{6}$

C.   1. 14   3. 8   5. 20   7. 36   9. 5   11. 42

D.   1. $\frac{1}{4}$ 3. $\frac{3}{8}$ 5. $\frac{2}{5}$ 7. $\frac{4}{5}$

      9. $4\frac{1}{4}$ 11. $\frac{21}{32}$ 13. $\frac{5}{12}$ 15. $\frac{19}{12}$

E.   1. $\frac{3}{5}$ 3. $1\frac{1}{2}$ 5. $\frac{7}{8}$ 7. $\frac{6}{4}$ 9. $\frac{13}{5}$ 11. $\frac{5}{12}$

F.   1. $15\frac{3}{4}$ in.   3. $\frac{19}{6}, \frac{25}{8}$ 5. No.   7. $\frac{3}{5}$ 9. $\frac{1}{5}$ 11. $\frac{23}{32}$ in.

      13. (a) $\frac{4}{16}$ or $\frac{1}{4}$ (b) $\frac{10}{16}$ or $\frac{5}{8}$ (c) $\frac{15}{16}$ (d) $1\frac{6}{16}$ or $1\frac{3}{8}$

          (e) $1\frac{8}{16}$ or $1\frac{1}{2}$ (f) $1\frac{12}{16}$ or $1\frac{3}{4}$ (g) $2\frac{5}{16}$ (h) $2\frac{14}{16}$ or $2\frac{7}{8}$

Exercises 2-2, page 90

A.   1. $\frac{1}{8}$ 3. $\frac{2}{15}$ 5. $2\frac{2}{3}$ 7. $1\frac{1}{9}$ 9. $2\frac{1}{2}$ 11. 3

      13. $3\frac{1}{4}$ 15. 69   17. 74   19. $10\frac{3}{8}$ 21. $\frac{1}{8}$ 23. $\frac{1}{15}$

      25. 2

B.   1. $\frac{1}{6}$ 3. $\frac{1}{2}$ 5. $\frac{3}{4}$ 7. $1\frac{5}{16}$ 9. 1

      11. $1\frac{1}{20}$ 13. $2\frac{5}{8}$ 15. 1

C.   1. $137\frac{3}{4}$ in.   3. $11\frac{2}{3}$ ft   5. 14 ft $1\frac{1}{2}$ in.   7. $318\frac{1}{2}$ mi

       9. $356\frac{1}{2}$ lb   11. $\frac{9}{10}$ in.   13. $9\frac{3}{4}$ in.   15. $348\frac{3}{4}$ min

       17. $10\frac{2}{3}$ hr   19. $5\frac{2}{5}$ in.   21. $22\frac{1}{2}$ picas   23. $31\frac{1}{8}$ in.

       25. $3\frac{1}{4}$ in.   27. $\frac{15}{32}$ in.   29. $9\frac{1}{4}$ lb   31. $3\frac{7}{8}$ in.

       33. 74

Exercises 2-3, page 97

A.   1. $1\frac{2}{3}$ 3. $\frac{5}{16}$ 5. $\frac{1}{2}$ 7. $\frac{1}{4}$ 9. 9   11. $1\frac{1}{3}$

       13. $1\frac{1}{5}$ 15. 16   17. 18   19. $7\frac{1}{2}$ 21. $\frac{5}{6}$ 23. $\frac{1}{8}$

B.   1. 8 ft   3. 48   5. 84   7. 18

       9. 210   11. 7 sheets   13. 45 threads   15. 284 sq ft

       17. $\frac{1}{8}$ in. per yr   19. 6   21. (a) 6   (b) $10\frac{1}{2}$

Exercises 2-4, page 113

A.   1. $\frac{1}{4}$ 3. $\frac{3}{4}$ 5. $\frac{1}{2}$ 7. $\frac{2}{5}$ 9. $\frac{15}{16}$

       11. $1\frac{1}{2}$ 13. $\frac{3}{4}$ 15. $\frac{17}{24}$ 17. $\frac{1}{8}$ 19. $\frac{7}{16}$

       21. $\frac{29}{40}$ 23. $\frac{19}{40}$ 25. $\frac{5}{8}$ 27. $1\frac{3}{4}$ 29. $4\frac{1}{8}$

       31. $3\frac{8}{15}$ 33. $2\frac{3}{8}$ 35. $1\frac{7}{60}$

B.   1. $5\frac{1}{8}$ 3. $2\frac{13}{16}$ 5. $\frac{7}{8}$ 7. $2\frac{1}{8}$ 9. $2\frac{11}{16}$ 11. $3\frac{9}{10}$

C.   1. 9 in.   3. $11\frac{5}{8}$ in.   5. $1\frac{9}{16}$ in.   7. $25\frac{1}{4}$ c.i.

       9. $1\frac{1}{8}$ in.   11. $23\frac{5}{16}$ in.   13. $2\frac{13}{16}$ in.   15. $\frac{3}{8}$ in.

       17. $7\frac{3}{4}$ in. by $6\frac{1}{2}$ in.   19. $\frac{9}{16}$ in.   21. $4\frac{33}{64}$ in.   23. $\frac{3}{4}$ qt

       25. $4\frac{2}{3}$ cu yd

Problem Set 2, page 118

A.   1. $\frac{9}{8}$ 3. $\frac{5}{3}$ 5. $\frac{99}{32}$ 7. $\frac{13}{8}$ 9. $2\frac{1}{2}$ 11. $8\frac{1}{3}$

       13. $1\frac{9}{16}$ 15. $8\frac{3}{4}$ 17. $\frac{3}{16}$ 19. $\frac{3}{8}$ 21. $\frac{1}{6}$ 23. $1\frac{4}{5}$

       25. 9   27. 44   29. 68   31. 15   33. $\frac{7}{16}$ 35. $\frac{7}{8}$

       37. $\frac{3}{5}$ 39. $\frac{7}{4}$

B.   1. $\frac{3}{32}$ 3. $\frac{7}{12}$ 5. $1\frac{1}{4}$ 7. $\frac{5}{64}$ 9. $7\frac{1}{2}$ 11. 27

       13. $11\frac{2}{3}$ 15. 2   17. 32   19. $\frac{1}{6}$ 21. $\frac{7}{10}$ 23. $2\frac{4}{9}$

C.   1. $1\frac{1}{4}$ 3. $\frac{7}{32}$ 5. $1\frac{13}{30}$ 7. $\frac{3}{8}$ 9. $\frac{7}{16}$ 11. $\frac{23}{40}$

       13. $3\frac{3}{8}$ 15. $4\frac{1}{2}$ 17. $1\frac{19}{24}$ 19. $1\frac{1}{30}$ 21. $1\frac{1}{6}$ 23. $\frac{2}{5}$

D.   1. $37\frac{1}{8}$ in.   3. $22\frac{5}{16}$ in.; $22\frac{7}{16}$ in.   5. $\frac{25}{32}$ in.   7. $4\frac{3}{5}$ cu ft

       9. $92\frac{13}{16}$ in.   11. $\frac{15}{16}$ in.   13. $4\frac{3}{32}$ in.   15. $244\frac{1}{2}$ in.

       17. $1\frac{1}{16}$ in. and $\frac{11}{16}$ in.   19. $5\frac{3}{16}$ in.   21. $6\frac{2}{3}$ min   23. 15 yd

       25. $\frac{9}{32}$ in.   27. 225   29. 1 in.

       31. (a) $72\frac{1}{12}$ ft   (b) 865 sq ft   (c) $5190   33. 105

Chapter 3   *Exercises 3-1, page 138*

A. 1. seventy-two hundredths
    3. twelve and thirty-six hundredths
    5. three and seventy-two thousandths
    7. three and twenty-four ten-thousandths

9. 0.004	11. 6.7	13. 12.8
15. 10.32	17. 0.116	19. 2.0374

B.

1. 21.01	3. $15.02	5. 1.617	7. 828.6
9. 63.7305	11. 6.97	13. $15.36	15. 42.33
17. $22.02	19. 113.96	21. 45.195	23. $27.51
25. 95.888	27. 15.16	29. 8.618	31. 31.23
33. 17.608	35. 0.0776	37. 24.22	39. 1.748

C. 1. 0.473 in.   3. (a) 0.013 in.   (b) smaller; 0.021 in.   (c) #14
    5. A: 2.20 in. B: 0.45 in.   C: 4.22 in.
    7. 2.267 in.   9. No   11. 3.4 hr   13. 3.37 in.
    15. 0.843 in.   17. 70.15 in.   19. 0.009 in.   21. 7.2 hr   23. 5.58 mm

D. 1. $308.24
    3. (a) 0.7399   (b) 4240.775   (c) 510.436   (d) 7.4262

Exercises 3-2, page 156

A.

1. 0.00001	3. 4	5. 0.84	7. 0.00003
9. 209.1	11. 2.16	13. 6.03	15. 20
17. 0.045	19. 400	21. 6.6	23. 605
25. 0.00378	27. 0.048	29. 45,000	31. 0.00008
33. 28.7375	35. 0.000096	37. 1.705	

B. 1. 36.12   3. 5.52   5. 6.8

C. 1. 42.88   3. 6.5   5. 79.14   7. 3.6494   9. 0.216

D. 1. 3.33   3. 10.53   5. 0.12   7. 33.86   9. 33.3
    11. 0.3   13. 0.2   15. 0.143   17. 65   19. 2.999

E. 1. 82.6 ... ≈ 83   3. 45.975 ≈ 46.0   5. 5.02 ... ≈ 5.0

F. 1. $55   3. 18.8 lb
    5.

	W	C
A	16.91 lb	$16.57
B	23.18	20.63
C	8.64	9.07
D	3.04	6.54

$T = \$52.81$

7. 48 lb	9. 153.37 ft	11. 0.075 volts	13. 336
15. 27 gal	17. $576	19. 2625 cycles	21. $899.57
23. 2660 lb	25. 8400 cu ft	27. $648	29. 5.7 in.

31. (a) $152.70   (b) $420.40   (c) $608.80
33. 27,306 kWh   35. 2.736%   37. $62.88   39. $85.80
41. A: $76.25 B: $65.56; Brand A   43. $1.24
45. (a) ZG&E: $4.20   (b) ZG&E: $3.00
        Ready Edison: $12.80   Ready Edison: $2.00

G. 1. 8.00000007   3. 4.2435 in.   5. $580   7. 26.2 psi
    9. 11.67 mi   11. 23.1 mi/gal

13. (a) 0.255 mi/gal (b) 8.92 people · mi/gal

15. 130.052 therms   17. $143.54   19. 0.240

21. Change bit after time period 7

23. (a) Italy: $0.25; Japan: $0.37; Spain: $0.40; France: $0.32; U.S.: $0.50
    (b) Boston to Washington   (c) Milan to Salerno
    (d) It would be the least expensive.

25. 255,378,000

Exercises 3-3, page 173

A. 1. 0.25   3. 0.75   5. 0.6   7. $0.\overline{285714}$ 9. $0.\overline{857142}$
   11. 0.75   13. 0.3   15. $0.41\overline{6}$ 17. 0.38   19. 0.81
   21. 0.22   23. 0.46   25. 0.03   27. 0.019

B. 1. 4.385   3. 1.5   5. 7.88   7. 1.43   9. 3.64   11. 7.65

C. 1. 1.375 g; 5.2 tablets   3. 2.3 squares
    5. (a) $64.94   (b) $385.94   (c) $316.74
       (d) $295.83; Total: $1063.45
    7. 0.3958 in., rounded
    9. (a) 120.20 ft   (b) 12 ft $6\frac{5}{8}$ in.   11. 0.025 kHz
   13. (a) 0.4688   (b) 0.2656   (c) 0.1875   (d) 0.125
       (e) 0.0375   (f) 0.0102
   15. 54.9 cu yd
   17. (a) $\frac{3}{8}$ in.   (b) $3\frac{5}{8}$ in.
       (c) $\frac{13}{16}$ in.   (d) $\frac{5}{16}$ in.
       (e) $6\frac{9}{16}$ in.   (f) $\frac{1}{4}$ in.
       (g) $\frac{25}{32}$ in.   (h) $7\frac{3}{16}$ in.

Problem Set 3, page 178

A. 1. nine tenths
    3. twenty-three and one hundred sixty-four thousandths
    5. nine and three tenths
    7. ten and six hundred twenty-five ten-thousandths
    9. 0.07   11. 200.8   13. 63.063   15. 5.0063

B. 1. 23.19   3. $19.29   5. 1.94   7. 88.26   9. 277.104
   11. 239.01   13. 83.88   15. 33.672   17. 4.28   19. 1.92

C. 1. 0.00008   3. 0.84   5. 0.108   7. 40   9. 61.7
   11. 3.8556   13. 0.006   15. 18   17. 4.34   19. 0.04
   21. 0.23   23. 2.78   25. 27.007   27. 0.526   29. 214.634
   31. 47.2   33. 1.5   35. 4.6

D. 1. 0.0625   3. 0.15625   5. 1.375   7. 2.55
    9. $2.\overline{6}$ 11. 2.3125   13. 30.67   15. 78.57
   17. 0.13

E. 1. 86   3. 6.48

F. 1. 291   3. 9.772 in.   5. 0.0021 in.   7. 489.4 lb   9. 18
   11. (a) 0.0089 in.   (b) 0.0027 in.   (c) 0.55215 in.   (d) 0.306175 in.
   13. 17.5 ft   15. Max: 2.525 in.; min: 2.475 in.   17. $229.90
   19. 56.7 cu ft   21. 18   23. 11.4 hr   25. 10 tablets
   27. $29.94   29. Last year was greater by $0.83   31. 3.375 V
   33. 25.8 mi/gal   35. 0.018 in.   37. $860.16   39. 100.3
   41. (a) $23.99   (b) $58.77   43. 34,529 tons   45. 34.7 ft

47. $0.17   49. (a) 108.29   (b) $5.41   51. 117.31 gal
53. 7.623 in.

Chapter 4   *Exercises 4-1, page 203*

A. (In each case only the missing answer is given.)
    1. (a) 7:1   (b) 12:7   (c) 6   (d) 6
       (e) 45   (f) 9   (g) 16   (h) 50
       (i) 3:2   (j) 2:5
    3. (a) 16:12   (b) 8 ft   (c) 28 ft   (d) 6.88:12
       (e) 7.2:12   (f) 4 ft   (g) 20 ft   (h) 5 ft 1 in.

B. 1. $x = 12$ 3. $y = 100$ 5. $P = \frac{7}{6}$ 7. $A = 3.25$
    9. $T = 1.8$ 11. $x = 3$ 13. $x = 5.04$ 15. $L = 5$ ft
    17. $x = 4$ cm

C. 1. 5 cu in.   3. 21 pins   5. 5.25 cu yd   7. 288.75 gal
    9. 124.8 lb   11. $856   13. 375 ft   15. 37.9 lb
    17. 12.9 mL   19. $21.63   21. 132 lb   23. 24 lb
    25. (a) 0.5 mL   (b) 0.33 mL   (c) 6.67 mL
    27. (a) 2   (b) $1\frac{1}{2}$ (c) $4\frac{1}{2}$
    29. 7500 bu   31. $15,636   33. $3856   35. 6
    37. 4.5 (min)–6.5 (max) hr

Exercises 4-2, page 217

A. 1. (a) $\frac{5}{8}$ in.   (b) 31.2 cm   (c) 8 ft   (d) 3 m

    3. (a) 100   (b) 9   (c) 25   (d) 1500

B. 1. 24 in.   3. 139 hp   5. 109.1 lb   7. 980 rpm
    9. (a) 5310 sq ft   (b) 8800 sq ft   11. 500 rpm   13. 6 hr
    15. 1470 ft   17. 0.96 sq in.   19. 4 in.   21. 22.4 psi
    23. $377.54   25. 14.4 hr, rounded
    27. (a) 6 in.   (b) 8 to 3   29. 458 lb   31. 16

Exercises 4-3, page 227

A. 1. 32%   3. 50%   5. 25%   7. 4000%
    9. 200%   11. 50%   13. 33.5%   15. 0.5%
    17. 150%   19. 330%   21. 40%   23. 95%
    25. 30%   27. 60%   29. 120%   31. 604%
    33. 125%   35. 35%   37. $83\frac{1}{3}$%   39. 370%

B. 1. 0.06   3. 0.01   5. 0.71   7. 0.0025
    9. 0.0625   11. 0.3   13. 8   15. 0.0025
    17. 0.07   19. 0.56   21. 10   23. 0.90
    25. 1.5   27. 0.0675   29. 0.1225   31. 0.012

C. 1. $\frac{1}{20}$ 3. $2\frac{1}{2}$ 5. $\frac{53}{100}$ 7. $\frac{23}{25}$ 9. $\frac{9}{20}$

    11. $\frac{1}{12}$ 13. $\frac{6}{25}$ 15. $\frac{1}{2000}$ 17. $4\frac{4}{5}$ 19. $\frac{5}{8}$

Exercises 4-4, page 235

A. 1. 80%   3. 15   5. 54   7. 150
    9. $66\frac{2}{3}$%   11. 100   13. $21.25   15. 1.5
    17. 43.75%   19. 2.4   21. 40   23. 5000

B.   1. 225   3. 20¢   5. 160   7. 150%
     9. 17.5   11. 427   13. 2%   15. 460
    17. 50

C.   1. 88%   3. 5.5%   5. 2084 lb
     7. $2.14   9. 9.6 in. by 12 in.   11. 1447 ft
   13. 76%   15. 32%   17. 52,400 linear ft
   19. 148–166 bpm   21. 304.3 cfm   23. 618
   25. $750   27. 0.05 gal   29. 19,200 psi
   31. 697.5 cal/day   33. $299.20   35. (a) 11.2%   (b) $5376
   37. $81,950   39. 30.0 gal

Exercises 4-5, page 260

    1. 64%   3. 63.6%   5. 67.5%   7. 1380 bf
    9. 12.5%   11. 0.9%   13. 81.25 hp   15. 0.13%
   17. (a) 0.03%   (b) 0.44%   (c) ±0.007 in.   19. 37.5%
   21. $9.75   23. $93.75   25. $149   27. B, by $1.74
   29. 25%   31. 15.6 MGD   33. 150 lb   35. 28%
   37. $56.52   39. Approx. 2.8%   41. $36,315.14   43. 179 hp
   45. (a) 8800 Btu/hr   (b) 10,080 Btu/hr
   47. (a) 112 bpm   (b) 50%   49. 8.1%   51. 33.3%
   53. (a) 27 lb   (b) 13%   55. $17,262   57. $16.55
   59. 12.7%   61. (a) Eng. tech.: 16%   (b) Construction managers
                         Comp. sci.: 36%   (c) Engineering technicians
                         Firefighters: 29%
                         Constr. man.: 55%
                         Food serv.: 50%

Problem Set 4, page 268

A.   1. (a) 30 in.   (b) 5 to 2   (c) 25 in.

B.   1. $x = 35$ 3. $x = 20$ 5. $x = 14.3$

C.   1. 72%   3. 60%   5. 130%   7. 400%   9. $16\frac{2}{3}\%$

D.   1. 0.04   3. 0.11   5. 0.0125   7. 0.002   9. 0.03875
   11. 1.15

E.   1. $\frac{7}{25}$ 3. $\frac{81}{100}$ 5. $\frac{1}{200}$ 7. $\frac{7}{50}$

F.   1. 60%   3. 5.6   5. 42   7. $3.50   9. 8000

G.   1. 29.92 in.   3. (a) $6977.25   (b) $598.50   (c) $1699
    5. (a) $20.13   (b) $18.54   (c) 73¢   (d) $125.72   (e) $317.99
    7. 1748 ft   9. (a) 0.06%   (b) 2.82%   (c) ±0.009 in.
                     (d) 0.03%   (e) ±0.62 mm
   11. $7216.25   13. ±135 ohms; 4365 to 4635 ohms
   15. 0.9%   17. 72%   19. 9.12 hp
   21. 3.6%   23. (a) 66 fps   (b) 15 mph
   25. $5\frac{1}{4}$ in.   27. 150 seconds or $2\frac{1}{2}$ minutes
   29. 450 parts   31. $3\frac{3}{4}$ hr   33. 6 sacks
   35. 960 lb   37. 1.47 amperes
   39. 6 bags of cement, 15 cu ft of sand, 30 cu ft of gravel
   41. 35 employees   43. 83.96 psi   45. 158 hp
   47. Petroleum: 15%; Computer: 30%; Geological: 62%
   49. Yes (13%)   51. 74
   53. (a) 12 mL   (b) 6 mL   (c) 0.8 mL
   55. (a) 17.65%   (b) 460%

57. (a) 51.2%　　(b) 9.4%　　(c) Camry (+172%)

59. 55.1%

61. Small: $5875; Medium: $14,125; Large: $12,000; Extra Large: $8000

63. 700,000 ppm　　65. 64 mL　　　　67. 345 lb

Chapter 5　*Exercises 5-1, page 292*

A.

	Precision to nearest	Accuracy in significant digits	Greatest possible error
1.	tenth	2	0.05 gal
3.	whole number	3	0.5 psi
5.	thousandth	4	0.0005 lb
7.	hundred	3	50 tons
9.	tenth	3	0.05 in.

B.　1. 21 in.　　　3. 9.6 in.　　　5. 1.91 gal　　　7. 1.19 oz

　　9. 50 psi

C.　1. 40 sq ft　　3. 7.2 sq ft　　5. 1.9 sq ft　　7. 4.6 cu ft

　　9. 63 mph　　11. 1.6 hr

D.　1. $\frac{15}{16}$ in.　　3. $3\frac{11}{16}$ in.　　5. $1\frac{29}{32}$ in.　　7. $2\frac{3}{32}$ in.

　　9. $\frac{15}{64}$ in.　　11. $\frac{41}{64}$ in.

E.　1. (a) $1\frac{3}{16}$ in.　　(b) 0.0125 in.　　3. (a) $\frac{21}{32}$ in.　　(b) 0.0098 in.

　　5. (a) $2\frac{27}{64}$ in.　　(b) 0.0019 in.

F.　1. Yes　　　3. $\frac{30}{64}$ in.　　5. 4.28 in.　　7. $3\frac{10}{32}$ in. or $3\frac{5}{16}$ in.

　　9. 354 mi　　11. 0.013 sec　　13. 24.7 lb　　15. 1.18 in.

Exercises 5-2, page 308

A.　1. 51　　　　3. 18,000　　　5. 504　　　　7. 1270

　　9. 2.3　　　11. 0.75　　　13. 42　　　　15. 15

　　17. 0.8　　　19. 83　　　　21. 18　　　　23. 20 lb 6 oz

　　25. 12　　　27. 6　　　　29. 44　　　　31. 64

　　33. 160

B.　1. 936　　　3. 468　　　　5. 0.5625　　　7. 0.38

　　9. $0.\overline{3}$ or $\frac{1}{3}$　　11. 12,800　　13. 2700　　　15. 0.30

　　17. 43.6　　19. 62.5　　　21. 3.6　　　　23. 216

　　25. 17,200　　27. 622

C.　1. 2.0 ft/sec　　3. 28 in.　　　5. 32.6 ± 0.6 in.　　7. 1265 sq ft

　　9. 10.2 rps　　11. 1.93 cfs　　13. $466\frac{1}{2}$ in.

　　15. (a) $1\frac{1}{4}$ cups　　(b) $2\frac{1}{2}$ pt

　　17. 19 sec　　　19. 8.5 gal　　21. 18,380 sq ft　　23. 17 mi

　　25. The nonhybrid (by $1312.50)

　　27. (a) 43,560　　(b) 325,829　　(c) 1613

　　29. (a) Prius: 7.6 gal　　(b) Prius: 7.9 gal

　　　　　Yaris: 12.8 gal　　　　Yaris: 11.7 gal

　　31. (a) 25 hands　　(b) 14 rods　　(c) 20 bones　　(d) $6\frac{2}{3}$ hands

　　　　(e) 64 ft　　　33. $102

　　35. 7920　　37. No.　　　39. 142 lb

Exercises 5-3, page 323

A.　1. (c)　　　3. (a)　　　5. (b)　　　7. (a)　　　9. (a)

　　11. (c)　　13. (a)　　15. (a)

B. 1. (b) 3. (a) 5. (b) 7. (b) 9. (b) 11. (a)

C. 1. 56 3. 2500 5. 45 7. 1250 9. 420
11. 16 13. 9620 15. 56.5 17. 9.5 19. 0.58
21. 1.4 23. 0.65 25. 4100 27. 75 29. 0.55
31. 2750 33. 1650

D. 1. (a) 18 kV (b) 0.435 V 3. 150 ppb 5. 6
7. 93.3 g 9. 0.5 mm 11. 167 13. 16 hr 40 min

Exercises 5-4, page 335

A. 1. 15.7 3. 3.2 5. 23.5 7. 68.9 9. 18.7
11. 11.9 13. 6.1 15. 5.0 17. 50 19. 19.2
21. 18.4 23. 3.7 25. 427 27. 229.3 29. 6.3
31. 7.4 33. 37.2 35. 3.1 37. 6.7 39. 67.4
41. 75.1

B. 1.

in.	mm
0.030	0.762
0.035	0.889
0.040	1.016
0.045	1.143

in.	mm
$\frac{1}{16}$	1.588
$\frac{5}{64}$	1.984
$\frac{3}{32}$	2.381
$\frac{1}{8}$	3.175

in.	mm
$\frac{5}{32}$	3.969
$\frac{3}{16}$	4.763
$\frac{3}{8}$	9.525
$\frac{11}{64}$	4.366

3. 12.2 mi/hr; 19.6 km/hr
5. (a) 0.6093 (b) 564 (c) 0.89 (d) 0.057 (e) 1.02
7. 12 km/liter 9. 9.3 sq m
11. (a) 1.6093 (b) 21.59 cm × 27.94 cm (c) 2.54 (d) 804.65
(e) 0.0648 (f) 28.35; 453.6 (g) 7.57 (h) 37.85
13. about 137 m 15. 1.7 oz/qt 17. 338°F 19. 560 mph
21. 62 lb 23. $26\frac{7}{8}$ in.
25. U.S. yield more by 1.9 bu/acre or 0.126 metric tons
27. 488 mi/hr 29. 830 in./min

Exercises 5-5, page 361

A. 1. (a) $\frac{5}{8}$ in. (b) $1\frac{7}{8}$ in. (c) $2\frac{1}{2}$ in. (d) 3 in.
(e) $\frac{1}{4}$ in. (f) $1\frac{1}{16}$ in. (g) $2\frac{5}{8}$ in. (h) $3\frac{1}{2}$ in.
3. (a) $\frac{2}{10}$ in. = 0.2 in. (b) $\frac{5}{10}$ in. = 0.5 in. (c) $1\frac{3}{10}$ in. = 1.3 in.
(d) $1\frac{6}{10}$ in. = 1.6 in. (e) $\frac{25}{100}$ in. = 0.25 in. (f) $\frac{72}{100}$ in. = 0.72 in.
(g) $1\frac{49}{100}$ in. = 1.49 in. (h) $1\frac{75}{100}$ in. = 1.75 in.

B. 1. 0.650 in. 3. 0.287 in. 5. 0.850 in. 7. 0.4068 in.
9. 0.2581 in. 11. 0.0888 in. 13. 11.57 mm 15. 22.58 mm
17. 18.94 mm

C. 1. 3.256 in. 3. 2.078 in. 5. 2.908 in. 7. 2.040 in. 9. 0.826 in.

D. 1. 62°21′ 3. 34°56′ 5. 35°34′ 7. 20°26′

E. 1. (a) 110 psi (b) 167 psi (c) 290 psi (d) 395 psi
3. (a) 4.4 V (b) 8.6 V (c) 13.6 V (d) 19.0 V
5. (a) 140 Ω (b) 45 Ω (c) 16 Ω (d) 3.6 Ω

F. 1. 226,849 cu ft 3. 2523 kWh 5. +2.72 mm 7. −0.330 in.

Problem Set 5, page 368

A.
Precision to nearest	Accuracy in significant digits	Greatest possible error
1. tenth	2	0.05 lb
3. hundredth	3	0.005 in.
5. hundredth	1	0.005 kg
7. thousand	1	500 sq yd

B. 1. 3.5 sec   3. 0.75 in.   5. $16\frac{7}{8}$ in.   7. 21.1 mi/gal
    9. 17.65 psi   11. 3.87 in.   13. 21 ft 3 in.

C. 1. $\frac{10}{16}$ in.; 0.003 in.   3. $2\frac{14}{32}$ in.; 0.0025 in.   5. $\frac{38}{64}$ in.; 0.0058 in.

D. 1. 15.9   3. $\approx 72°F$ 5. 820   7. 458
    9. 31.5   11. 667   13. 23.3   15. 387
    17. 1.4   19. 11.5   21. 990   23. 95
    25. 104   27. 1850   29. 40.4   31. 48
    33. 40.7   35. 16.4   37. 5.6   39. 850
    41. 1.2   43. 4134   45. 1850   47. 1.6
    49. 60   51. 4.7

E. 1. (b)   3. (a)   5. (b)   7. (c)

F. 1. (a) $\frac{5}{10}$ in. = 0.5 in.   (b) $1\frac{3}{10}$ in. = 1.3 in.   (c) $\frac{17}{100}$ in. = 0.17 in.
        (d) $1\frac{35}{100}$ in. = 1.35 in.   (e) $\frac{3}{4}$ in.   (f) $3\frac{3}{4}$ in.   (g) $\frac{5}{16}$ in.
        (h) $1\frac{19}{32}$ in.   (i) $\frac{3}{8}$ in.   (j) $1\frac{27}{64}$ in.

    3. (a) 2.156 in.   (b) 3.030 in.   (c) 0.612 in.   (d) 1.925 in.
        (e) 1.706 in.   (f) 1.038 in.
    5. (a) +3.28 mm   (b) −5.63 mm   (c) +0.187 in.   (d) −0.055 in.

G. 1. $22\frac{9}{16}$ in.   3. 6.1 mph   5. $\frac{15}{32}$ in.   7. 936 sq in.
    9. 293°C   11. 0.073 ft/sec   13. 0.28 lb/cu in.
    15. 3.8 cm × 8.9 cm or about 4 cm by 9 cm   17. 3.4 sq mi
    19. 13.5 mi   21. 8 min   23. 864 bf   25. 36 mi/gal
    27. (a) 6.60 m/sec   (b) 6.34 m/sec   (c) 1 hr 50 min 55 sec
    29. 57,600   31. 1950 mg   33. 558 lb/cu ft   35. 144 mi/hr
    37. 0.005 cm   39. less by about 8 bu/acre

Chapter 6   *Exercises 6-1, page 387*

A. 1. <   3. >   5. >   7. <
    9. <   11. >   13. >   15. <

B. 1.

    3.

C. 1. −6   3. +12,000   5. −9   7. −80
    9. −5   11. +$6000   13. −$17.50

D.   1. 5   3. -13 5. -15 7. 17   9. 46

　　11. -16 13. 9   15. 29   17. -30 19. -0.9

　　21. -28.09 23. -0.14 25. $-1\frac{1}{4}$ 27. -14 29. $-\frac{3}{4}$

　　31. $1\frac{9}{16}$ 33. $-4\frac{5}{12}$ 35. -10 37. -16 39. 2

　　41. 18   43. -420 45. 7   47. 3599   49. $-26{,}620$

E.   1. $+\$22$ ($\$22$ above quota)   3. $-9°$ (9 degrees below zero)

　　5. $-\$136{,}000$ (a loss of $\$136{,}000$)   7. 0 mA

Exercises 6-2, page 394

A.   1. -2 3. -16 5. 9   7. -6

　　9. 8   11. 11   13. -18 15. -31

　　17. 8   19. 26   21. 9   23. -19

　　25. -17 27. 8   29. -87 31. -67

　　33. -10 35. -42 37. 6   39. -1

　　41. $-\frac{5}{9}$ 43. $-\frac{5}{8}$ 45. $-6\frac{1}{2}$ 47. $-8\frac{7}{12}$

　　49. 9.2   51. -63.5 53. 41.02   55. -0.85

B.   1. 14,698 ft   3. 35°F   5. $\$1817$ 7. 83°

　　9. lost $\$5211$ 11. 420 V   13. 573°   15. (a) $+\$2.53$ (b) $-\$7.12$

Exercises 6-3, page 400

A.   1. -63 3. 77   5. -4 7. 12

　　9. 5.1   11. -8 13. $\frac{1}{5}$ or 0.2   15. -4

　　17. $\frac{1}{6}$ 19. $-2\frac{11}{20}$ 21. -90 23. 84

　　25. -28 27. 5   29. $12\frac{3}{8}$ 31. $-19\frac{1}{2}$

　　33. -4.48 35. 0.4   37. -90 39. -20

　　41. -0.4

B.   1. -19.3 3. 0.0019   5. -3.16 7. 0.23

　　9. 46   11. 70,200,000

C.   1. $-8\frac{1}{3}°C$ 3. (a) -120 (b) 40   5. -2000 ft/min

Exercises 6-4, page 410

A.   1. 16   3. 64   5. 1000   7. 256

　　9. 512   11. 625   13. -8 15. 729

　　17. 5   19. 32   21. 196   23. 3375

　　25. 108   27. 1125   29. 2700   31. 18

　　33. 216   35. 19   37. 116   39. 2

　　41. 71

B.   1. 9   3. 6   5. 5   7. 16

　　9. 15   11. 18   13. 2.12   15. 3.52

17. 14.49	19. 28.46	21. 31.62	23. 158.11
25. 12.25	27. 54.77	29. 1.12	31. 13.42
33. 51.26	35. 4.80	37. 375	

C. 1. 13.6 ft   3. 89,000 psi   5. 96 fps   7. 26.6

     9. 73 ft   11. 2.00 m^2 13. $8493.42

Problem Set 6, page 414

A.   1. $-13, -8, -2, 4, 7$

     3. $-180, -160, -150, -140, -120, 0$

B.   1. -10 3. -6 5. -15 7. 25

     9. 2.3   11. $-\$9.15$ 13. -16

C.   1. -60 3. 12   5. $7\frac{1}{32}$ 7. -0.4 9. -120 11. -15.8304

D.   1. -43.75 3. -4 5. 64   7. 289

     9. 0.25   11. 0.0004   13. 625   15. 16.1604

     17. 16   19. 28   21. 0.4   23. 16.29

     25. 121.68   27. 432   29. 10   31. 0.5

     33. 8.94   35. 17.61   37. 2.05   39. 1.04

     41. 2.76   43. 0.78   45. 7.76

E.   1. $+7$ amp   3. $-22°$F   5. 163 A   7. 700

     9. 3200 nanosec 11. (a) $-20°$C   (b) $-4°$C   13. 33 ohms

     15. 0.024%   17. (a) $+\$8.38$ (b) $-\$5.06$ 19. $7214.90

Chapter 7   *Exercises 7-1, page 432*

A.   1. $6yx$ 3. $(a + c)(a - c)$ 5. $(n + 5)/4n$

B.   1. factor   3. factor   5. factor   7. factor

C.   1. 2120 mm^2 3. $I = \$15$ 5. $T = 23$ m^2 7. $P = 40$

     9. $V = 121$ m^3 11. $P = 48$ 13. $V = 8177$ m^3 15. $f = 637$

     17. $A = 55$ cm^2 19. $A = 76$ in.^{2}

D.   1. 1160 sq ft   3. $P = 96$ watts   5. $F = 104°$ 7. $S_T = 1.4$ mm

     9. $L = 3500$ lb   11. ACH $= 1.44$ 13. $S = 26$ squares

     15. $T = 1500$ lb   17. $X = 1000$ ft   19. $P = 2.64$ kW

     21. $L = 920$ cm   23. $T = 36.3$ hr   25. $L_v = 0.525$ in.

     27. (a) $d = 15$ m   29. (a) $R = 0.733$ ohms

         (b) $d = 310$ ft   (b) $R = 0.190$ ohms

     31. $T = 208.9$ cfm   33. $V_{rms} = 1.77$ V   35. 4 weeks

     37. $C = 4500$ metric tons/hr   39. $R = 1321$ 41. 715 mi

     43. 80 ft   45. $L = 18\frac{1}{4}$ in.   47. $L = 5700$ in.

49. (a) $-8.5°C$ (b) $-18.1°C$ or $-0.6°F$ 51. 2900 cc

53. (a) 46 mi/gal   (b) 28 mi/gal   55. Duffy: 0.336; Bryant: 0.378

Exercises 7-2, page 448

A.   1. $9y$ 3. $6E$ 5. $7B$

    7. $-2x^2$ 9. $2R^2 + 5R$ 11. $\frac{1}{8}x$

    13. $1\frac{1}{2} - 3.1W$ 15. $3x + 6xy$ 17. $5x^2 + x^2y + 3x$

B.   1. $3x^2 + 2x - 5$ 3. $4m^2 + 10m$ 5. $2 - x - 5y$

    7. $3a - 8 + 6b$ 9. $-10x^2 - 3x$ 11. $23 - 3x$

    13. $x + 8y$ 15. $-14 - 7w + 3z$ 17. $9x - 12y$

    19. $-56m - 48$ 21. $-9x - 15$ 23. $23m - 42$

    25. $3 - 8x - 12y$ 27. $28 - 6w$ 29. $6x + 8y - 24x^2 + 20y^2$

    31. $4x + 10y$ 33. $3x + 9y$ 35. $-2x^2 - 42x + 62$

Exercises 7-3, page 457

A.   1. $x = 9$ 3. $x = 58$ 5. $a = 8\frac{1}{2}$ 7. $y = 24.66$

    9. $x = -13$ 11. $a = 0.006$ 13. $z = 0.65$ 15. $m = -12$

    17. $y = 6.5$ 19. $Q = 0.7$ 21. $x = 25$ 23. $K = 4.24$

B.   1. 360 volts   3. 28.8 parts per million   5. 1.715 m   7. 21 windings

    9. 32 ft   11. \$131,800   13. 1800 ft · lb

Exercises 7-4, page 468

A.   1. $x = 10$ 3. $x = 17.5$ 5. $m = 2.5$ 7. $n = 10$

    9. $z = -6$ 11. $x = 48$ 13. $n = -9$ 15. $x = \frac{1}{4}$

    17. $a = 7$ 19. $z = -5\frac{1}{2}$ 21. $x = \frac{2}{5}$ 23. $P = 3$

    25. $x = 10$ 27. $x = 5\frac{1}{3}$ 29. $x = -3$ 31. $x = -6$

B.   1. 5.5 hr   3. 32.5 years   5. 0.125 in.

    7. 28.75 hr   9. 7.5 yr   11. 551 lb · ft

    13. 19.8 ft   15. 533 cfm   17. \$2012

Exercises 7-5, page 480

A.   1. $x = 9$ 3. $n = 4\frac{5}{6}$ 5. $x = -7$ 7. $y = 7\frac{1}{2}$

    9. $x = 1$ 11. $c = 5$ 13. $x = -7$ 15. $t = \frac{1}{5}$

    17. $x = -2$ 19. $y = 2$ 21. $x = 2$ 23. $t = -3\frac{3}{4}$

    25. $x = 4\frac{2}{7}$ 27. $x = 3$

B.   1. $L = \dfrac{S}{W}$ 3. $I = \dfrac{V}{R}$ 5. $T = \dfrac{2S - WA}{W}$

    7. $B = \dfrac{P - 2A}{2}$ 9. $P = \dfrac{T(R + 2)}{R}$

C.   1. (a) $a = \dfrac{360L}{2\pi R}$ (b) $R = \dfrac{360L}{2\pi a}$ (c) $L = 5.23$ in.

3. $r = 2700$ rpm

5. (a) $V_1 = \dfrac{V_2 P_2}{P_1}$ (b) $V_2 = \dfrac{V_1 P_1}{P_2}$ (c) $P_1 = \dfrac{V_2 P_2}{V_1}$

(d) $P_2 = \dfrac{V_1 P_1}{V_2}$ (e) $P_1 = 300$ psi

7. (a) $D = \dfrac{CA + 12C}{A}$ (b) 0.14 gram

9. (a) $i_L = \dfrac{i_s T_p}{T_s}$ (b) $i_s = \dfrac{i_L T_s}{T_p}$ (c) $i_L = 22.5$ amp

11. 2.5 ft   13. (a) $V = \dfrac{TF}{24}$ (b) $F = \dfrac{24V}{T}$

(c) 60,000 gal/day   (d) 6100 gal

15. 110.7°

Exercises 7-6, page 495

A.   1. $H = 1.4W$ 3. $V = \frac{1}{4} h\pi d^2$ 5. $V = AL$

7. $D = \dfrac{N}{P}$ 9. $W = 0.785\, hDd^2$

B.   1. $133\frac{1}{3}$ gal and $266\frac{2}{3}$ gal

3. $13\frac{5}{7}$ ft, $6\frac{6}{7}$ ft, $3\frac{3}{7}$ ft

5. 22, 22, 22, 20, 18, 16, 14, 12, 10 blocks

7. Jo: $48,800; Ellen: $36,600

9. 21 (with some left over)

11. $6\frac{2}{3}$ hr   13. 5000 parts   15. $50.02

17. $95 + 80(t - 1) = 655; t = 8$ 19. 36 hr

21. 29,000 mi   23. (a) 103,700 mi   (b) fewer

25. 17 yr   27. 80 fpm

Exercises 7-7, page 503

A.   1. $20x$ 3. $6R^2$ 5. 5^{10} 7. 10^{-3}

9. p^6 11. $5x^7$ 13. $-8x^3y^4$ 15. $0.6a^2$

17. $10y^7$ 19. $\frac{1}{16}Q^4$ 21. $24M^6$ 23. $6x^2 - 2x$

25. $-2 + 4y$ 27. $3p^3$ 29. $-6x^4 + 15x^5$

31. $6a^3b^2 - 10a^2b^3 + 14ab^4$

B.   1. 4^2 3. x^3 5. $\dfrac{1}{10^2}$ 7. $2a^4$

9. $-\dfrac{2}{3y}$ 11. $3ab^3$ 13. $-\dfrac{3}{m^4 n}$ 15. $-\dfrac{3}{5b^3}$

17. $3x^2 - 4$ 19. $-2a^5 + a^3 - 3a$

Exercises 7-8, page 510

A.   1. 5×10^3 3. 9×10^1 5. 3×10^{-3}

     7. 4×10^{-4} 9. 6.77×10^6 11. 2.92×10^{-2}

     13. 1.001×10^3 15. 1.07×10^{-4} 17. 3.14×10^1

     19. 1.25×10^2 21. 2.9×10^7 lb/sq in.   23. 9.55×10^7 watts

B.   1. 200,000   3. 0.00009   5. 0.0017   7. 51,000

     9. 40,500   11. 0.003205   13. 2,450,000   15. 647,000

C.   1. 8×10^7 3. 7.8×10^2 5. 2.8×10^{-9}

     7. 5×10^6 9. 1.4×10^2 11. 6.3×10^{-2}

     13. 2.3×10^{10} 15. 8×10^{-11} 17. 2.8×10^{-2}

     19. 5.9×10^{-12} 21. 6.6×10^{13} 23. 4.0×10^2

D.   1. (a) 5,300,000,000,000/L   (b) 9,100,000,000/L

       (c) 3.6×10^{12}/L   (d) 6.7×10^9/L

     3. 13.3 cal/sec   5. 1.53×10^{-4} μF 7. 0.064 mm

     9. 9.935×10^{15} tons

Problem Set 7, page 516

A.   1. $L = 13$ 3. $I = 144$ 5. $V = 42\frac{21}{64}$ 7. $L = 115.2$

     9. $t = 4$

B.   1. $10x$ 3. $12xy$ 5. $1\frac{11}{24}x$ 7. $0.4G$ 9. $21x^2$

     11. $-10x^2y^2z$ 13. 5^{10} 15. $12x - 21$ 17. $9x + 11$ 19. $9x - 3y$

     21. 6^4 23. $-3m^8$ 25. $\dfrac{4a}{b}$

C.   1. $x = 12$ 3. $e = 44$ 5. $x = -5$ 7. $x = 4$ 9. $y = 32$

     11. $x = -3.5$ 13. $x = -6$ 15. $x = -4$ 17. $g = 10$ 19. $x = 16$

     21. $x = 3$ 23. $y = 3.75$ 25. $n = 9$ 27. $y = 2$ 29. $b = \dfrac{A}{H}$

     31. $L = \dfrac{P - 2W}{2}$ 33. $g = \dfrac{2S + 8}{t}$

D.   1. 7.5×10^3 3. 4.1×10^{-2} 5. 5.72×10^{-3}

     7. 4.47×10^5 9. 8.02×10^7 11. 7.05×10^{-6}

     13. 930,000   15. 0.00029   17. 0.0000005146

     19. 90,710   21. 5.7×10^{10} 23. 3.0×10^{-6}

     25. 4.2×10^{-1} 27. 1.2×10^2 29. 2.8×10^{-2}

     31. 2.4×10^{-8}

E.   1. 3900 fpm   3. 71,400 psi   5. 2687 metric tons/hour

     7. $51\frac{9}{16}$ in.   9. 0.617   11. 330 mi

     13. 62 mi/hr   15. (a) 34 mi/gal   (b) 27 mi/gal

17. 108 in. 19. 5.8 μF

21. (a) 219 in.^{4} (b) Yes. ($I_J = 256$ in.^{4})

23. (a) 31.25% (b) $L = \dfrac{100(H - X)}{CG}$; $L = 166.7$ in.

(c) $H = \dfrac{L \cdot CG + 100\,x}{100}$; $H = 140$ in.

(d) $x = \dfrac{100H - L \cdot CG}{100}$; $x = 156$ in.

25. (a) $L = \dfrac{AR}{K}$ (b) $R_m = \dfrac{R_S \cdot I_S}{I_m}$ (c) $C = \dfrac{1}{2\pi\, FX_C}$

(d) $L_m = \dfrac{L_1 + L_2}{2L_T}$

27. (a) 70% (b) 4.1 dS/m 29. 508 kWh 31. 224 cfm

33. 200 kWh 35. 93.3 lb 37. 24 mi/gal 39. $12,308

41. $8W + 2W = 80$; $W = 8$ cm; $L = 32$ cm

43. $2x + x = 0.048$; $x = 0.016$ in.; $2x = 0.032$ in.

45. 133.3 kWh 47. 0.6875

Chapter 8 *Exercises 8-1, page 540*

A. 1. (a) $\angle B, \angle x, \angle ABC$ (b) $\angle O, \angle POQ$ (c) $\angle a$
    (d) $\angle T$ (e) $\angle 2$ (f) $\angle s, \angle RST$

3. (a) (b) (c)

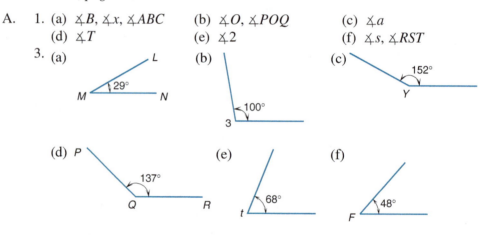

(d) P (e) (f)

B. 1. (a) $a = 55°, b = 125°, c = 55°$ (b) $d = 164°, e = 16°, f = 164°$
    (c) $p = 110°$ (d) $x = 75°, t = 105°, k = 115°$
    (e) $s = 60°, t = 120°$ (f) $p = 30°, q = 30°, w = 90°$
    3. (a) $w = 80°, x = 100°, y = 100°, z = 100°$
    (b) $a = 60°, b = 60°, c = 120°, d = 120°, e = 120°, f = 120°, g = 60°$
    (c) $n = 100°, q = 80°, m = 100°, p = 100°, h = 80°, k = 100°$
    (d) $z = 140°, s = 40°, y = 140°, t = 140°, u = 40°, w = 140°, x = 40°$

C. 1. $\angle ACB \approx 63°$ 3. $B \approx 122°, A \approx 28°, C \approx 30°$ 5. 2″
    7. 29°5′36″ 9. $a = 112°, b = 68°$ 11. $\angle 2 = 106°$
    13. 60°

Exercises 8-2, page 557

A. 1. 20.0 m 3. 115 in. 5. 48 in. 7. 39 cm
    9. 22.81 in. 11. 64 ft

B.	*Perimeter*	*Area*	*Perimeter*	*Area*
	1. 20 in.	25 sq in.	3. 70.6 in.	285 sq in.
	5. 56 m	160 m^2	7. 124 m	468 m^2
	9. 86 ft	432 sq ft	11. 114.9 m	722.2 m^2

C. 1. 960   3. 28 sq yd; $1118.60   5. 11 bundles
7. 638.5 sq in.   9. 258 cm   11. $6\frac{2}{3}$ in.
13. Cut the 6-in. card along the 38-in. side; 18 cards, 86 sq in. of waste
15. $14,053   17. 14 bags   19. $5351
21. $9.74   23. (a) 357 sq ft   (b) $2017.05
25. Rubber is lower by $0.36/sq ft   27. Approx. 550 sq ft
29. 21

Exercises 8-3, page 580

A. 1. 10.6 in.   3. 20 cm   5. 46.8 mm   7. 27.7 in.
9. 28 ft   11. 21.2 in.   13. 6 in.   15. 17.4 ft
17. 9.369 cm   19. 0.333 in.   21. 118 ft

B. 1. 173 mm^2 3. 336 ft^2 5. 36 yd^2 7. 260 in.^{2}
9. 520 ft^2 11. 5.1 cm^2 13. 433 ft^2 15. 354 ft^2
17. 365 ft^2 19. 39 in.^{2} 21. 0.586 in.^{2} 23. 7.94 m^2

C. 1. 7 gal   3. 5 squares   5. 15.2 ft   7. 174.5 sq ft
9. $y = 36$ ft 5 in., $x = 64$ ft 11 in.   11. 22 ft 3 in.   13. (b)
15. 3108   17. 432 sq ft

Exercises 8-4, page 594

A. 1. $d = 52.8$ cm $C \approx 165.9$ cm   3. $r \approx 51.9$ in. $d \approx 103.8$ in.
5. $r = 5.7$ m $C \approx 35.8$ m

B.	*Circumference*	*Area*
	1. 88.0 in.	615.8 sq in.
	3. 19.2 m	29.2 m^2
	5. 75.4 ft	452.4 ft^2
	7. 3.8 cm	1.1 cm^2
	9. Area ≈ 431.1 in.^{2}	

C. 1. 17.1 sq in.   3. 5.2 in.^{2} 5. 15.1 in.^{2} 7. 26.7 cm^2
9. 0.7 sq in.   11. 246.1 in.^{2} 13. 25.1 in.   15. 52.3 cm

D. 1. 30.1 in.   3. 0.6 in.   5. 163 in.^{2} 7. 103.7 sq in.
9. 314   11. $103.11   13. 174 in.   15. 2.4 in.
17. 2.6 cm   19. 2.79 in.   21. 2.1 fps   23. 133 cm
25. 0.3 sec   27. 70,660 psi   29. 15,625 mils

Problem Set 8, page 602

A. 1. $\angle 1$, acute   3. $\angle m$, acute   5. $\angle PQR$ or $\angle Q$, obtuse
7. 63° approx.   9. 155° approx.   11. 70° approx.

B.	*Perimeter*	*Area*	*Perimeter*	*Area*
	1. 12 in.	6 in.^{2}	3. 31.4 cm	78.5 cm^2
	5. 12 ft	10.4 ft^2	7. 17 yd	8 yd^2
	9. 36 in.	53.7 sq in.	11. 12 in.	6.9 sq in.
	13. 17.2 ft	12.0 sq ft	15. 51.7 ft	169.3 ft^2
	17. 28.3 in.	51.1 sq in.	19. 100.0 cm	615.8 cm^2
	21. 23.9 cm	8.9 sq cm		

C. 1. 5.77 mm   3. 3.46 in.   5. $\frac{3}{4}$ in.   7. 1.39 yd   9. 2259.38 in.^{2}

D. 1. $93,500   3. 45°   5. 250 ft   7. $5232
    9. 18 in.   11. 36   13. 110°   15. 94 ft
    17. (a) 14 rolls   (b) $83.72   19. 69.1 cm   21. $2210.56
    23. 27.3 lb   25. 12.17 in.   27. 19 mi   29. $266,000
    31. 1509   33. 1 in.   35. 69   37. 211

Chapter 9   *Exercises 9-1, page 620*

A. 1. 729 cu in.   3. 13,800 cm^3 5. 132.4 mm^3 7. 144 cm^3

B.   *Lateral Surface Area*   *Volume*
    1. 280 sq ft   480 cu ft
    3. 784 mm^2 2744 mm^3
    5. 360 in.^{2} 480 in.^{3}

C.   *Total Surface Area*   *Volume*
    1. 150 sq yd   125 cu yd
    3. 1457.8 ft^2 2816 ft^3
    5. 5290 cm^2 23,940 cm^3

D. 1. 266   3. 333.3 cu ft   5. 94,248 gal   7. 450 trips
    9. 71.3 cu yd   11. 16.9 lb   13. 3.8 ft   15. 238 trips
    17. 8.3 cu yd   19. 3.5 yd^3 21. 7000 lb   23. 2244 gal
    25. 525 ft^3 27. 19.6 yd^3 29. 12 cu yd   31. 1430 cu yd
    33. 3.872×10^{10} cu yd   35. 7.2 cu yd

Exercises 9-2, page 632

A.   *Lateral Surface Area*   *Volume*
    1. 384 sq in.   617 cu in.
    3. 127 sq in.   105 cu in.
    5. 468 sq in.   1072 cu in.
    7. 55.2 sq m   31.2 cu m

B.   *Total Area*   *Volume*   *Total Area*   *Volume*
    1. 80.7 m^2 33.2 m^3 3. 3200 ft^2 10,240 ft^3
    5. 2496 in.^{2} 7340 in.^{3} 7. 173 ft^2 144 ft^3

C. 1. $9207   3. 3 hr   5. 384 sq in.

Exercises 9-3, page 640

A.   *Lateral Surface Area*   *Volume*
    1. 24,800 cm^2 572,700 cm^3
    3. 245 sq in.   368 cu in.
    5. 428 ft^2 1176 ft^3

B.   *Total Area*   *Volume*
    1. 7.07 in.^{2} 1.77 in.^{3}
    3. 1280 ft^2 3530 ft^3
    5. 6500 m^2 39,900 m^3

C. 1. 14 gal   3. 0.24 ft   5. 30,100 gal   7. 2.6 lb
    9. 4 gal   11. 49.6 lb   13. 582 gal   15. 0.43 lb
    17. 79 cm   19. 41.9 in.^{3} 21. 36.7 gal   23. 596 lb
    25. 1333 ft

Exercises 9-4, page 649

A.   *Lateral Surface Area*   *Volume*
    1. 319 cm^2 581 cm^3
    3. 338 m^2 939 m^3

B.   *Total Area*   *Volume*
    1. 628 in.^{2} 1010 in.^{3}
    3. 1112 in.^{2} 2614 in.^{3}

C.   1. 240 m^3 3. 49 cm^3

D.   1. 104.7 m^3 3. 253.2 bu   5. 5.6 lb   7. 4.2 cm
    9. 32.7 oz   11. 6.1 oz

Problem Set 9, page 654

A. (a) *Lateral Surface*   (b) *Volume*   (a) *Lateral Surface*   (b) *Volume*
      Area   *Area*
    1. 226 sq in.   452 cu in.   3. 43 sq ft   27 cu ft
    5. 259 mm^2 443 mm^3 7. 377 ft^2 980 in.^{3}

    (a) *Total Surface*   (b) *Volume*   (a) *Total Surface*   (b) *Volume*
      Area   *Area*
    9. 384 ft^2 512 ft^3 11. 1767 cm^2 5301 cm^3
   13. 450 sq ft   396 cu ft   15. 1257 cm^2 4189 cm^3
   17. 265 sq in.   575 cu in.   19. 1264 in.^{2} 2520 in.^{3}
   21. 329 sq in.   312 cu in.

B.   1. 1125 boxes   3. 2992 lb   5. 40 qt   7. 1.5 ft^3
    9. 294 lb   11. 13 cu yd   13. 165 yd^3 15. 2.8 gal
   17. 6.3 ft   19. 33 L   21. 74 ft^3 23. 0.3 yd^3
   25. 312 lb   27. 387 ft   29. 2.2 lb   31. 17 bags

Chapter 10   *Exercises 10-1, page 678*

A.   1. acute   3. obtuse   5. straight   7. acute
    9. right

B.   1. 36°15′   3. 65°27′   5. 25°45′   7. 16°7′
    9. 27.5°   11. 154.65°   13. 57.05°   15. 16.88°
   17. 0.43   19. 1.18   21. 5.73°   23. 120.32°

C.   1. 26   3. 24   5. 12 ft   7. 1.5
    9. 8.0 ft   11. (a) 45°   (b) 4.2 ft
   13. (a) 4   (b) 53°   (c) 37°
   15. (a) 7   (b) 45°
   17. (a) 11.5 ft   (b) 5.8 ft   (c) 30°
   19. (a) 10 in.   (b) 45°   21. $S = 21$ cm, $A = 314$ cm^2
   23. $S = 88$ ft, $A = 4840$ sq ft   25. $S = 6$ in., $A = 16$ sq in.
   27. $S = 5$ m, $A = 25$ m^2

D.   1. (a) 325.2 in.^{2} (b) 36.1 in.   3. 21
    5. 523 yd^3 7. 829 in./sec   9. 0.08 rad/sec
   11. 0.39 rad/sec   13. No (1.8°)   15. 54 ft 3 in.
   17. 2.38 in.   19. 121 in. or 10 ft 1 in.

Exercises 10-2, page 690

A.   1. 0.54   3. 0.42   5. 0.44

B.   1. 0.454   3. 0.213   5. 0.191   7. 0.052
    9. 1.206   11. 0.752   13. 0.771   15. 6.357
   17. −0.139   19. 0.585

C. 1. 76°54'   3. 64°0'   5. 74°49'   7. 9°40'
    9. 49.4°   11. 64.1°   13. 41.1°   15. 37.2°

Exercises 10-3, page 699

A. 1. 11 in.   3. 13.4 cm   5. 4.65 m   7. 91.56 km

B. 1. 13.94 km   3. 61°32'   5. 4.65 m   7. 16.2 ft

C. 1. $A = 22°, b = 89$ ft, $c = 96$ ft
    3. $B = 58.5°, b = 11.0$ in., $c = 12.9$ in.
    5. $b = 505.7$ km, $B = 58.78°, A = 31.22°$
    7. $A = 16°, a = 4.6$ ft, $c = 17$ ft
    9. $B = 68.75°$ (or 68°45'), $b = 23.07$ m, $a = 8.970$ m

D. 1. 31 ft   3. (a) 73 ft 6 in.   (b) No.   5. 51°40'

    7. 115 in.   9. 3.74 in.   11. $1\frac{1}{8}$ in.   13. 274 ft

    15. 33.39 cm   17. 1.5 in.   19. 21 ft 1 in.   21. 6.8 in.

    23. 19   25. 85 ft

Exercises 10-4, page 714

A. 1. $C = 75°, c = 9.2$ ft, $b = 8.4$ ft   3. $A = 39.0°, a = 202$ m, $c = 301$ m
    5. $A = 55.9°, C = 94.1°, c = 116$ in. or $A = 124.1°, C = 25.9°, c = 50.6$ in.
    7. $c = 5.6$ m, $A = 92°, B = 47°$
    9. $b = 690$ ft, $A = 62.4°, C = 15.6°$
    11. $C = 94.2°, A = 37.5°, B = 48.3°$
    13. $b = 10.3$ m, $A = 45.4°, C = 34.9°$
    15. $B = 149.9°, A = 7.1°, C = 23.0°$

B. 1. 41 ft   3. 27.2°, 52.1°, 100.7°   5. 19.1 ft
    7. 25.4 in.

Problem Set 10, page 720

A. 1. 87°48'   3. 51°47'   5. 41.2°   7. 65.9°
    9. 0.611   11. 1.300   13. 25.8°   15. 131.8°

B. 1. $S = 16$ cm, $A = 206$ cm^2 3. $S = 13$ in., $A = 72$ in.^{2}

C. 1. 0.391   3. 0.070   5. 0.720   7. 0.956
    9. 14°   11. 67°58'   13. 10°59'   15. 26.8°
    17. 29.2°   19. 74.2°

D. 1. $A = 9, B = 15.6, c = 30°$ 3. $g = 53°, h = 37°$
    5. $B = 61°, c = 37$ mm, $b = 32$ mm   7. $a = 5.14$ m, $A = 44.4°, B = 45.6°$
    9. $c = 5.55$ in., $A = 35.8°, B = 54.2°$
    11. $b = 9.81$ in., $A = 50.7°, B = 39.3°$

E. 1. $C = 76.0°, c = 19.2$ in., $b = 12.4$ in.
    3. $C = 52°, B = 91°, b = 430$ yd or $C = 128°, B = 15°, b = 110$ yd
    5. $B = 49.1°, C = 66.6°, c = 52.4$ m
    7. $A = 24.9°, B = 38.9°, b = 1730$ m

F. 1. 1.3 in.   3. 2 in.   5. (a) $X = 29$ ft 2 in.   (b) 58 ft 4 in.
    7. (a) 1°8'   (b) 0.3 in.   9. 0.394 in.   11. 17.2 cm
    13. 180 rad/sec   15. 104,000 sq ft   17. 7.4 in.

Chapter 11   *Exercises 11-1, page 747*

A.   1. $x = 2, y = 8$ 3. $x = 4, y = 5$ 5. $x = 2, y = 4$
  7. $x = -1, y = 2$

B.   1. $x = 9, y = -4$ 3. Inconsistent   5. $x = -2, y = -3$
  7. $x = -1, y = -1\frac{1}{2}$ 9. Dependent   11. $x = 1, y = 5$

C.   1. $x + y = 39$ Solution: $x = 23$
   $x - y = 7$ $y = 16$
  3. $x + y = 20$ Solution: $x = 12$
   $2x = 3y$ $y = 8$
  5. $4L + 3P = 64$ Solution: $L = \$10$ (leather cleaner)
   $3L + 4P = 62$ $P = \$8$ (protectant)
  7. $10d + 5n = 100$ Solution: $n = 2$
   $d = n + 7$ $d = 9$
  9. $3P + 5C = 295$ Solution: $C = \$32$ (ink cartridge)
   $P = C + 13$ $P = \$45$ (case of paper)
  11. $x + y = 30$ Solution: $x = 24$ in.
   $x = 4y$ $y = 6$ in.
  13. $x + y = 6$ Solution: $x = 1.5$ liter (10%)
   $0.1x + 0.02y = 0.24$ $y = 4.5$ liter (2%)
  15. $x + y = 24$ Solution: $x = 16$ tons ($45 rock)
   $45x + 60y = 1200$ $y = 8$ tons ($60 rock)

Exercises 11-2, page 761

A.   1. No   3. No   5. Yes

  7. $3x^2 - 7x + 14 = 0$ 9. $x^2 - 18x = 0$

B.   1. $x = 5$ or $x = -5$ 3. $x = 0$ or $x = 4\frac{2}{5}$
  5. $x = 4\frac{1}{2}$ or $x = -4\frac{1}{2}$ 7. $x \approx 0.76$ or $x \approx -15.76$
  9. $x \approx -2.35$ or $x \approx 0.85$ 11. $x \approx 4.14$ or $x \approx -3.14$
  13. $x \approx 0.55$ or $x \approx -0.26$ 15. $x \approx -2.37$ or $x \approx -0.63$
  17. $x \approx 2.39$ or $x \approx 0.28$ 19. $x \approx 3.69$ or $x \approx 0.81$

C.   1. 25 mm   3. 5 in. by 15 in.   5. 17 in. by 22 in.

  7. 7.14 in.   9. (a) 3.76 in.   (b) 4.56 in.

  11. 8 in.   13. $3\frac{7}{8}$ in.   15. 37.2 ft

  17. 434 gal/min   19. 7.5 min

Problem Set 11, page 766

A.   1. $x = 15, y = 3$ 3. $x = 3, y = 2\frac{1}{2}$ 5. $x = 7, y = 6$

  7. $x = 1, y = 1$ 9. $x = \frac{74}{23}, y = -\frac{4}{23}$

B.   1. $x = 3$ or $x = -3$ 3. $x \approx -1.43$ or $x \approx -0.23$
  5. $x = 0$ or $x = 6$ 7. $x = 5$ or $x = -1$
  9. $x = 6$ or $x = -6$

C.   1. $x + y = 38$ 3. $x - y = 21$ 5. $L = 4W$
   $x - y = 14$ $5y - 2x = 33$ $4W^2 = 125$
   $x = 26$ $x = 46$ $W \approx 5.59$ cm
   $y = 12$ $y = 25$ $L \approx 22.36$ cm
  7. (a) $I = 13$ amp   (b) $I = 12.25$ amp

9. $x + y = 42$ 11. $d = 15.96$ in.   13. 12.5 in.
   $x = 3y - 2$
   $y = 11$ in.
   $x = 31$ in.
15. $x + y = 50$ 17. 2.55 in.   19. 41 mph
   $120x + 180y = 6600$
   $x = 40$ yards (of California Gold)
   $y = 10$ yards (of Palm Springs Gold)
21. 100 ft 5 in.

Chapter 12   *Exercises 12-1, page 795*

A. 1. (a) 2012   (b) 2009   (c) approximately 330,000
       (d) 100,000   (e) 100%   (f) 22%
    3. (a) $17,000   (b) $50,000   (c) 90%
       (d) 74%   (e) Finance
    5. (a) 45   (b) 172   (c) 50%   (d) 258
    7. (a) 2008; 800,000 (b) any two of these: 1996, 1998, 2001, 2002, 2003, 2005
       (c) lowest: 1996, highest: 2005   (d) approximately 4,000,000
       (e) approximately 82%   (f) 41%
    9. (a) tuition   (b) lab fee   (c) $850   (d) $540

B. 1.

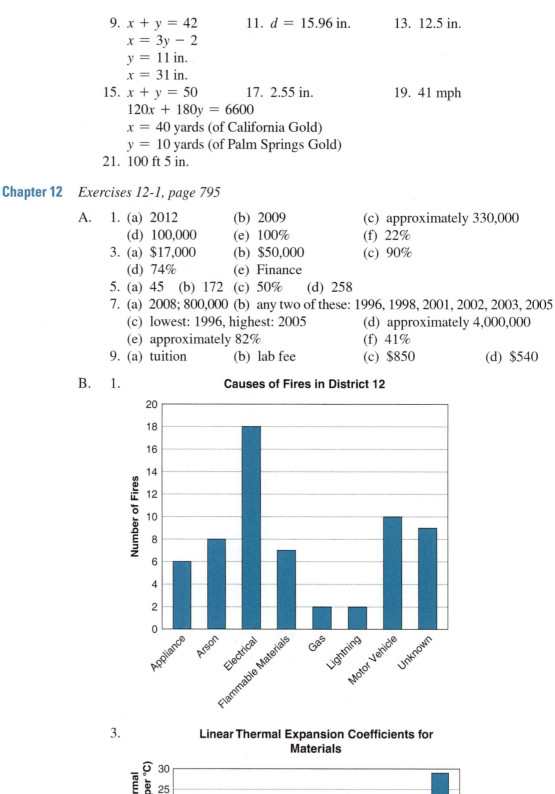

Causes of Fires in District 12

3.

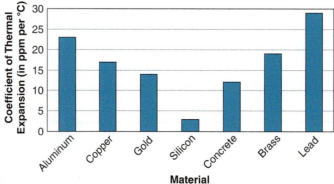

Linear Thermal Expansion Coefficients for Materials

5.

**Yearly Sales of Construction Material for
CASH IS US**

7.

Average U.S. Prices of Regular and Diesel Fuel: 2007–2016

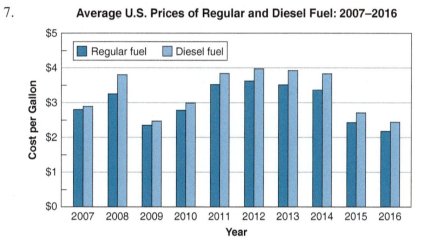

9.   **Changes in Starting Salaries of College Graduates from 2007 to 2014**

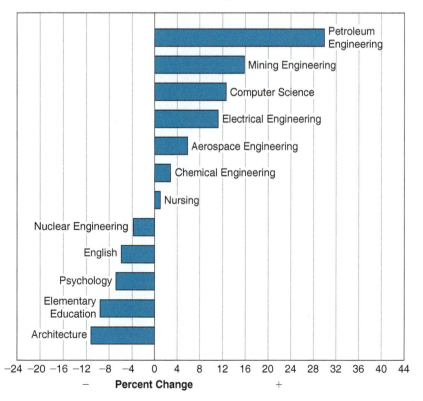

11.

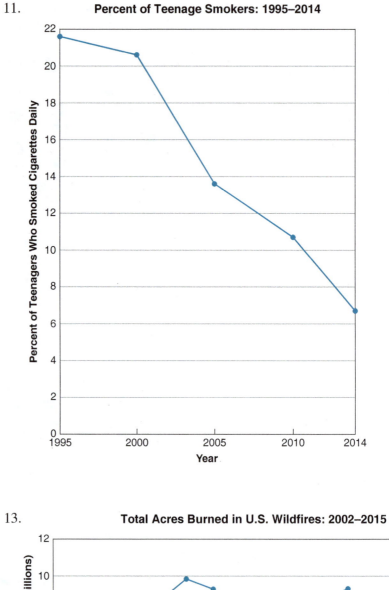

Percent of Teenage Smokers: 1995–2014

13.

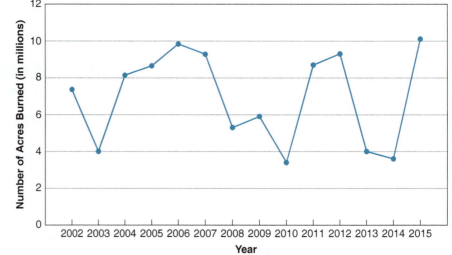

Total Acres Burned in U.S. Wildfires: 2002–2015

15.

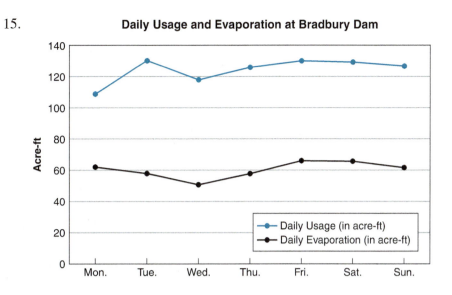

Daily Usage and Evaporation at Bradbury Dam

17. **California's In-State Sources of Electricity**

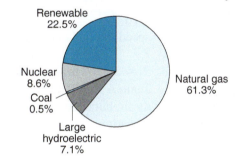

19. **Aviation Mechanics Percent Breakdown of a 40-Hour Work Week**

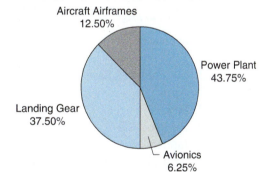

Exercises 12-2, page 811

A.   *Mean*   *Median*   *Mode*

    1. 6   6.5   4

    3. 71.7   67.5   54

    5. 151.8   156   none

    7. 1254.5   1250.5   none

B.   1. 1.548

C.   1. (a) 37,000   (b) 37,000   (c) 37,125

        (d) 37,375   (e) \$74,000,000

    3. (a) 72,684   (b) 71,971   (c) 7,146,006

        (d) 98   (e) 34%   (f) 50%

5. (a) 13.89 hr   (b) 13.55 hr   (c) $271
7. (a) 438.4 kAF   (b) 348 kAF   (c) 13.8%
9. (a) 41.2 mi/gal   (b) 41.75 mi/gal   (c) 220.2 g/mi
   (d) 212 g/mi   (e) 1.9 gal   (f) 22%
11. (a) Mean: 7.9, median: 8   (b) Mean: 7.1, median: 7
13. (a) 22.15 mi/gal   (b) $0.122 (or 12.2¢ per mile)

Exercises 12-3, page 827

A.   *Range*   *Standard Deviation*
      1. 6   2.0
      3. 17   6.0
      5. 0.76   0.229
      7. 1945   582.3

B.   1. 677.5 sq ft   3. 3.59 sec

C.   1. (a) 8.5   (b) 39.35 − 56.35   (c) 75%

D.   1. (a) 6.3   (b) 271.8 − 297.0   (c) 100%

E.   1. 13 cu yd
      3. (a) speeding, 163 citations   (b) 63.2   (c) 49.3
      5. (a) $\bar{x} = 37,000$ brake pads, $s = 1910$ brake pads
         (b) $\bar{x} = 37,000$ brake pads, $s = 3718$ brake pads
         (c) Team A would be preferable to team B. Both teams have the same
             mean, but team A is more consistent since they have a smaller standard
             deviation.
      7. (a) 0.10 mm   (b) 0.054 mm
      9. (a) Oklahoma, 103   (b) 30.6
         (c) 16.7   (d) The mean is much more reliable for Missouri
      11. 89.0 amp   13. 22,936 V

Problem Set 12, page 837

A.   1. decreases   3. 14   5. all circuits in the graph
      7. 1   9. 20   11. 4
      13. Caramel de Lites   15. approximately 122%
      17. April   19. 30%   21. 20%
      23. 50%   25. 2009   27. from 2010 to 2011
      29. approx. 42%   31. $30,000   33. December
      35. January
      37. January, February, March, June, July, August, September, October,
          November
      39. 16%   41. 119   43. 0.73 g

B.   1.   **Fire Resistance Ratings of Various Plywoods**

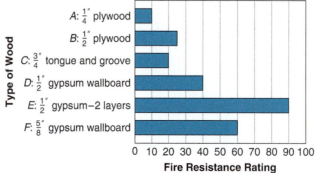

3.

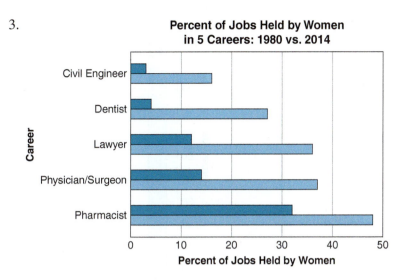

5.

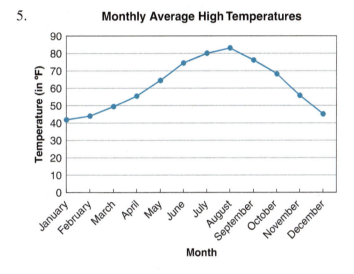

7.

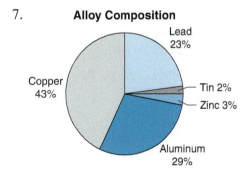

9.

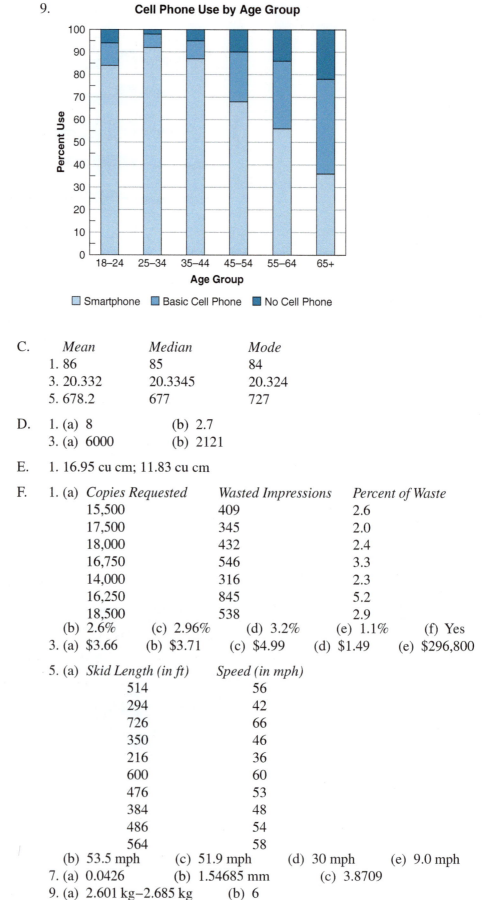

Cell Phone Use by Age Group

C.

	Mean	Median	Mode
1.	86	85	84
3.	20.332	20.3345	20.324
5.	678.2	677	727

D.   1. (a) 8   (b) 2.7
    3. (a) 6000   (b) 2121

E.   1. 16.95 cu cm; 11.83 cu cm

F.   1. (a)

Copies Requested	Wasted Impressions	Percent of Waste
15,500	409	2.6
17,500	345	2.0
18,000	432	2.4
16,750	546	3.3
14,000	316	2.3
16,250	845	5.2
18,500	538	2.9

  (b) 2.6%   (c) 2.96%   (d) 3.2%   (e) 1.1%   (f) Yes

3. (a) $3.66   (b) $3.71   (c) $4.99   (d) $1.49   (e) $296,800

5. (a)

Skid Length (in ft)	Speed (in mph)
514	56
294	42
726	66
350	46
216	36
600	60
476	53
384	48
486	54
564	58

  (b) 53.5 mph   (c) 51.9 mph   (d) 30 mph   (e) 9.0 mph

7. (a) 0.0426   (b) 1.54685 mm   (c) 3.8709

9. (a) 2.601 kg–2.685 kg   (b) 6

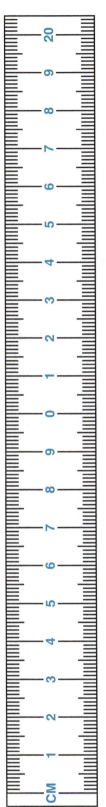

Metric Ruler

Protractor

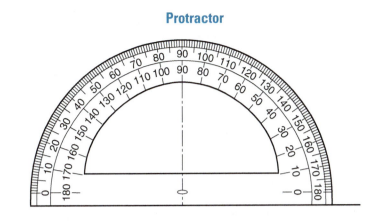

Multiplication Table

×	0	1	2	3	4	5	6	7	8	9	10
0	0	0	0	0	0	0	0	0	0	0	0
1	0	1	2	3	4	5	6	7	8	9	10
2	0	2	4	6	8	10	12	14	16	18	20
3	0	3	6	9	12	15	18	21	24	27	30
4	0	4	8	12	16	20	24	28	32	36	40
5	0	5	10	15	20	25	30	35	40	45	50
6	0	6	12	18	24	30	36	42	48	54	60
7	0	7	14	21	28	35	42	49	56	63	70
8	0	8	16	24	32	40	48	56	64	72	80
9	0	9	18	27	36	45	54	63	72	81	90
10	0	10	20	30	40	50	60	70	80	90	100

Index

Applications Index

Credits